All living organisms develop from germs,
that is to say, they owe their origins to other living beings.
But how did the first living things arise?

A. I. Oparin (1924)

CONTENTS

Section III
SELF-ASSEMBLY OF SUPRAMOLECULAR SYSTEMS

Section IV
ENERGETICS OF LIFE'S ORIGINS

Section V
BIOINFORMATIONAL MOLECULES

REFERENCES ON THE ORIGINS OF LIFE

INDEX OF AUTHORS CITED

FOREWORD

The scientific study of the origins of life is enjoying a second renaissance. The field was born in the 1920's when Oparin and Haldane put forth the idea that the origin of life could be understood in terms of plausible chemical and physical processes occurring on the primitive Earth. The first renaissance occurred in 1953 when Stanley Miller, then a student in the laboratory of Harold Urey, conducted the landmark spark-discharge experiments that placed origins-of-life studies on an experimental footing. Following Miller's results, it seemed only a matter of time before the problems of life's origins would be solved. Considerable progress was made over the subsequent two decades, but by the late 1970's it became clear that the problem was far from being solved—and might not even be solvable if we expected a detailed history of events that led to the emergence of life on Earth.

The second renaissance began in 1982 with the discovery by Thomas Cech that RNA can have catalytic functions. Once again it seemed, to some at least, that the problem of life's origins would soon be solved. Those who had witnessed the first cycle of unbridled optimism, and saw that it had been followed by more sobering reality, were more circumspect. Fashions come and fashions go in the origins-of-life field, as in any discipline. But this second renaissance may be different. It is based not only on the work involving RNA enzymes but also on the advances being made in a number of disparate areas that fall collectively under the origins-of-life banner. There are now so many irons in the fire that continued, substantial progress seems assured.

This volume is intended to provide an overview of the multidisciplinary approach that is now being applied to the origins-of-life problem. Here you will find a smorgasbord of ideas: planetary accretion, impacting comets and asteroids, deep-sea hydrothermal vents, self-replicating polymers, the RNA world, and more. As you feast on these ideas in their remarkable breadth and diversity, keep in mind that what sustains progress in the field—what initiated the serious scientific study of the origins of life in the first place—is a rigorous experimentally-based research program. If origins-of-life research has suffered from a reputation of not being a "hard"

science, it is because too often experimental work that does not meet the usual standards of scientific rigor has been conducted and published, because theory that is not grounded in experimental facts has been proposed. The editors of this volume have made a careful effort to focus on solid experimental work and on theoretical work that has a sound experimental foundation.

The central challenge of origins-of-life research is to describe the transition from inanimate matter to living systems, as occurred on the primitive Earth and as might have occurred elsewhere in the universe. We would do well, therefore, to have an agreed-upon definition of the term *life*.

However, there is no single definition that satisfies the entire scientific community, let alone the public at large. Perhaps the most general working definition is that adopted by the Exobiology Program within the National Aeronautics and Space Agency: *life* is a self-sustained chemical system capable of undergoing Darwinian evolution. The notion of Darwinian evolution subsumes the processes of self-reproduction, material continuity over an historical lineage, genetic variation, and natural selection. The requirement that the system be self-sustained refers to the fact that living systems contain all the genetic information necessary for their own constant production (i.e., metabolism).

By this definition, there is only one life form that we know of: the DNA- and protein-based life-form that is common to all known terrestrial biology. As you will learn from reading this volume, it is reasonable to conclude that a simpler life form, one based on RNA genomes and RNA enzymes, existed during the early history of life on Earth. However, it must be said that we do not know the chemical nature of the first living system on our planet.

It is reasonable to speculate about the possibility of life elsewhere in the universe. While it is highly unlikely that life exists on any of the other planets in our solar system, it is not inconceivable that life exists on Europa or perhaps Titan. It may also be possible that life existed on Mars at one time, but that it has long been extinct. In 1992, one year before this volume's production, the SETI project (the Search for Extra-Terrestrial Intelligence) began a ten-year search of the near galaxy for radiofrequency signals emitted by extraterrestrial civilizations. Those, too, lie within the realm of possibility. Despite what you may have read in the supermarket tabloids, however, there is no credible evidence for the existence of extraterrestrial life.

Finally, it should be noted that rapid progress is being made toward developing synthetic life-forms in the laboratory. Self-replicating chemical systems have been generated, although these systems are not yet capable of undergoing Darwinian evolution. It is amusing to wonder whether or not exobiologists will learn to make life in the laboratory before they find it elsewhere in the universe. Most likely, both of these events will come to pass before we understand the details of how life arose on Earth.

Gerald Joyce
Scripps Research Institute
La Jolla, CA

PREFACE

The excitement of origins-of-life research has its primary source in the obvious significance of its central question: how did life begin on Earth? Yet excitement stems as well from its relative youth as a scientific field, from the freshness of its questions, and from its integration of many diverse areas of scientific inquiry. Chemistry, biophysics, cell and molecular biology, genetics, evolution, geology and geochemistry, atmospheric sciences, and astronomy have all contributed richly. Because of the diversity of origins-of life research efforts, reports and insights have inevitably been lost in the multitude of publications serving those "separate" fields.

That is why we felt it was time for this book. Here we have brought together more than forty papers to lay out the concepts currently shaping origins-of-life research—papers and concepts whose focus is the physico-chemical processes that led to the first cellular life-forms on our planet Earth.

Several related collections have been published, most notably *Geochemistry and the Origin of Life* edited by K. A. Kvenvolden (1974), *Molecular Evolution* edited by E. A. Terzaghi, A. S. Wilkins, and D. Penny (1984), and *The Origin of Eukaryotic Cells* edited by B. D. Dyer and R. Obar (1985). Other related works include all of volume 52 of *Cold Spring Harbor Symposia on Quantitative Biology* (1987), concerning the origin of the genetic code; the several books and articles by M. Eigen and P. Schuster, concerning computer speculations on the origins of the genetic code; and *Artificial Life, Volume VI: Santa Fe Institute Studies in the Sciences of Complexity* edited by C. G. Langton (1989), concerning nanotechnology, computer simulations, and automata (e.g., replicative function of polymers and evolution of morphology). These books, however, do not address the central issue of origins-of-life research, namely, the origins of cellular life on Earth.

In order to sharpen the focus of our book, we have had to exclude papers concerning molecular evolution, the origin of the genetic code, and the antiquity of life (either evidence from the fossil record or inferential models based on metabolic pathways and/or molecular sequence comparisons). The research literature for any one of these topics is voluminous and would have overwhelmed any single collection.

Further, with one exception, all papers collected here are reprinted from English-language publications. (That single exception, the 1938 paper by Groth and Suess in *Naturwissenschaften*, appears here in English translation for the first time). Finally, we have limited our collation to papers that deal with empirical investigation, including as well several theoretical papers and integrative reviews of empirical research in which the authors present a concept or an overview that advances the field or serves as a teaching tool.

As editors, we have found this book enterprise both enlivening and enlightening. We hope that the reader will enjoy using this collection as much as we have enjoyed compiling it.

We want to thank Ron Mistretta, of Boston University Graphics, and the many authors of papers reprinted in this volume for their assistance in obtaining the best possible graphic reproductions of their work. We wish specifically to thank Sherwood Chang, James Ferris, Gerald Joyce, Antonio Lazcano, Stanley Miller, Leslie Orgel, John Oró, and Rod Swenson for their critical reading of various portions of the editorial commentary over the two years and more of this volume's preparation. Special thanks are due to Antonio Lazcano for his informed efforts and unflagging enthusiasm for this book from its inception, to Włodzimierz Ługowski for his early comments regarding the volume's focus and scope, and to Paul Strother for an early conversation with GRF. We also thank Stephanie Henry, our book's designer; Camille Pecoul and Judy Songdahl, our production editors at Jones and Bartlett, who kept their poise and ours throughout the production process.

D. W. Deamer
G. R. Fleischaker

AUTHORS AND ARTICLES REPRINTED

[page number in this volume]

Introduction

ORIGINS OF LIFE
THE CENTRAL CONCEPTS

ORIGINS OF LIFE: THE CENTRAL CONCEPTS

There is a question... about which perhaps I can dare speak to you, because it is accessible to experiment and because I have made it the object of serious and conscientious studies.

It is the question of generation, so-called spontaneous.

Can matter organize itself? In other words, can beings come into the world without parents, without ancestors? Here is the question to resolve.

Louis Pasteur, 7 April 1864

Pasteur, scientific demonstration, and spontaneous generation

Louis Pasteur was probably the first to perform a scientific investigation related to life's origin. In a paper delivered at the Sorbonne in 1864, Pasteur posed the fundamental question quoted above. He also noted that "a belief in spontaneous generation has been the prevailing belief throughout the ages, universally accepted since antiquity.... It is this belief that I have come here to challenge." To illustrate what he meant by the prevailing belief, Pasteur quoted van Helmont, a medical alchemist of the 17th century: "If one stuffs a soiled shirt into the mouth of a jar containing grains of wheat, the ferment released from the soiled shirt, combining with the odor of the grain, transmutes the mixture into mice in about 21 days."

Pasteur went on to describe his now-classical experiments with open and closed flasks of nutrient broths, clearly showing that microbial life can arise only if the broths are exposed to room air containing dust, which carries germs into the broth. Pasteur concluded that "there is no known circumstance today in which one can assert that microscopic beings have originated without germs."

Oparin, Vernadsky, and the biogeochemical system

Pasteur's question of 1864 refers to whether a complete living organism might spontaneously assemble in a few days, and the answer is: certainly not. But sixty years later, in 1924, the thirty-year-old Alexander Ivanovich Oparin suggested that living microorganisms might have been generated from simple organic precursors. After discussing the Earth's geocosmic origin, and how simple organics and colloidal gel-like substances could have arisen on the primitive Earth, Oparin proposed the following scenario:

We can even consider that first piece of organic slime which came into being on the Earth as being the first organism... it was composed of organic substances, it had a definite and complicated structure.... It had a considerable store of chemical energy enabling it to undergo further transformations. Finally, even if it could not metabolize in the full sense of the word, it must certainly have had the ability to nourish itself, to absorb and assimilate substances from its environment, for this is present in every organic gel.

Throughout this early work, Oparin makes clear that, even as metabolism "in the full sense of the word" may distinguish the living from the nonliving, it is the colossal "store of hidden (potential) energy" that gives organic gels or "any physical substance" "the possibility of developing further and increasing their complexity"—that is, of evolving. Here and in his later work (e.g., 1957, 1968, 1972), Oparin points to the "biogeochemical principles" articulated by Vernadsky (1986, orig. 1929), namely that evolution is a single geocosmic energetic phenomenon—that the origin of life comes about in a much larger process, the progressive production of an increasingly higher-ordered global state (see Swenson, 1989).

Haldane, Urey, Miller, and molecular synthesis

In 1929, the British biochemist and geneticist J. B. S. Haldane suggested that

the first living... things were probably large molecules synthesized under the influence of the Sun's radiation, and only capable of reproduction in the particularly favourable medium in which they originated. Each presumably required a variety of highly specialized molecules before it could reproduce itself, and it depended on chance for a supply of them.

More than twenty more years passed before the American scientists Harold Urey and Stanley Miller put speculation and experiment together in the first convincing simulation of chemical events in a prebiotic environment. In his 1952 paper, Urey wrote:

It seems to me that experimentation on the production of organic compounds from water and methane in the presence of ultra-violet light of approximately the spectral distribution estimated for sunlight would be most profitable. The investigation of possible effects of electric discharges on the reactions should also be tried, since electric storms in the reducing atmosphere can be postulated reasonably.

In 1953, Stanley Miller, then a young graduate student in the laboratory of Harold Urey, carried out just such experiments, demonstrating that amino acids could be produced under simulated prebiotic conditions. Urey's suggestion was profitable indeed.

Origins of life as a scientific research field

As a scientific research field, origins of life is quite young. It can be said to date from the 1924 publication of Oparin's small tract, *Proiskhozhdenie zhizny* [*The Origin of Life*]. In 1994, we will celebrate both the seventieth anniversary of that seminal publication and the centenary of its author's birth.

Even in the work of the 1920s, 1930s, and 1940s, origins-of-life questions were of concern to investigators in several established scientific fields— H. J. Muller and J. B. S. Haldane in genetics, for example, J. D. Bernal in crystallography, and Oparin in biochemistry. From the same work, it is also evident that the early origins-of-life field was divided between two conceptual points of view: metabolists (following Oparin) and molecularists (following Muller). Because those two viewpoints have not changed substantially over the intervening years, the field today still retains tension (if not a division) along the same lines—tension that surfaces regularly in arguments over the role of membranes and the genetic molecules and in debates concerning autotrophy or heterotrophy in the first living beings.

Yet, despite such conceptual differences, the origins-of-life field is united in several broad areas of investigation. It is from those scientific research areas that we have drawn the questions presented here as central to the field.

Research themes in biogenesis

Given that we can never know with certainty how life began on the Earth, we can still make real progress—first by understanding the general principles underlying contemporary life processes and then by using those principles to provide increasingly satisfactory explanations of life's origins.

We have sorted the papers reprinted here into five sections, according to five guiding research themes. These themes overlap at their boundaries and ultimately must be integrated into a larger picture. Yet they serve to organize our knowledge in such a way that the immense breadth of the field is less confusing.

The themes are:

1. *The early-Earth environment.* The origin of life depended on the physical environment of our planetary surface. We must therefore make every effort to understand what the primitive Earth, its oceans, and its atmosphere were like approximately 4 billion years ago.

2. *Prebiotic chemistry.* Life processes are physicochemical in nature. We must attempt to find chemical reactions that produce molecules relevant to life as we understand it today and to link those reactions to the prebiotic environment.

3. *Self-assembly processes.* Certain classes of molecules have the ability to self-assemble into larger, more complex structures with properties that are not possessed by the individual components. Biogenesis must have involved self-assembly processes, and we can use model systems of contemporary molecules to learn how mixtures of primitive organic compounds could have formed the first supramolecular systems whose energetic relationship with their environment had the properties of the living state.

4. *Energetics.* All living organisms capture and transduce energy from sources available in the environment. Following their self-assembly, the earliest living systems on Earth must also have been able to capture energy. From investigations of bioenergetic mechanisms under simulated prebiotic conditions, we can construct plausible models of the first living system.

5. *Bioinformational molecules.* At some point, a self-assembled, energy-transducing, supramolecular system must have incorporated mechanisms to store and use information. Contemporary cells use the sequence of bases in nucleic acids for this purpose, and the early supramolecular systems must have included some similar mechanism. With information storage and retrieval, genetics would have emerged in the self-assembled systems, and biological evolution as we know it today could have begun.

Forty-six papers on life's origins

As numerous and diverse ideas now compete for attention, the study of life's origin is in the most exciting stage of any developing science. The field has significant new themes and directions, yet its literature is not readily available, for it is scattered in many journals over nearly a century of historical development.

The central question of research on the origins of life concerns the processes by which systems of molecules led to the earliest cells, capable of maintained metabolism, growth, reproduction, and evolution. This book is about such processes and contains forty-six papers. One might imagine that these are, in some sense, the "top forty" classic papers that all would agree are the most significant publications in this field. If we tried to guess how many papers should ideally be included, however, it would probably be closer to several hundred, each describing important results and concepts. We necessarily had to make some difficult decisions, guided by practical limitations rather than scientific value. We have been able to compensate, in part, by including references to other important papers in the discussion that introduces each section and in the bibliographic section at the end of the book.

References

Bernal, J. D. (1949) The physical basis of life. *Proc. Phys. Soc. A* 62:537-558.

Haldane, J. B. S. (1929) The origin of life. Reprinted in: J. D. Bernal, *The Origin of Life* (1967). London: Weidenfeld and Nicolson, pp. 242-249.

Muller, H. J. (1929) The gene as the basis of life. *Proc. Int. Congr. of Plant Sci.* 1:897-921.

Oparin, A. I. (1924) The Origin of Life [*Proiskhozhdenie zhizny*. Moscow: Izd. Moskovshii Rabochii]. English translation in: J. D. Bernal, *The Origin of Life* (1967). London: Weidenfeld and Nicolson, pp. 199-234.

Oparin, A. I. (1957) *The Origin of Life on the Earth*, 3rd ed. A. Synge, trans., New York: Academic Press.

Oparin, A. I. (1968) *Genesis and Evolutionary Development of Life.* New York: Academic Press.

Oparin, A. I. (1972) The appearance of life in the universe. In: *Exobiology*, C. Ponnamperuma, ed., Amsterdam: North-Holland, pp. 1-15.

Pasteur, L. (1864) (lecture to the Sorbonne on spontaneous generation). In: *Oeuvres de Pasteur*, Volume 2 (1922), S. Pasteur Valley-Padot, ed., Paris: Masson & Cie.

Swenson, R. (1989) The Earth as an incommensurate field at the geo-cosmic interface: Fundamentals to a theory of general evolution. In: *Geo-Cosmic Relations: The Earth and Its Macro-Environment*, G. J. M. Tomassen, W. de Graaf, A. A. Knoop, and R. Hengeveld, eds., Wageningen, The Netherlands: Pudoc Science Publishers, pp. 299-306.

Urey, H. C. (1952) On the early chemical history of the earth and the origin of life. *Proc. Natl. Acad. Sci. USA* 38:351-363.

Vernadsky, V. I. (1986) *The Biosphere* [abridged, based on the original (1929) French edition]. London: Synergetic Press.

Section I

THE EARLY-EARTH ENVIRONMENT

THE EARLY-EARTH ENVIRONMENT

During the critical stages of growth, the infalling of planetesimals is supposed to have continued to be active. The planetesimals are assumed to have contained carbon, sulphur, phosphorus and all other elements found in organic matter; and as they impinged more or less violently upon the surface formed of previous accessions of similar matter, there should have been generated various compounds of these elements.

T. C. Chamberlin and R. T. Chamberlin, 1908

Life on the Earth probably did not begin at a precise moment when inorganic matter somehow acquired the properties of the living state. It is more plausible to think that an imperceptible physico-chemical evolution took place following the accretion of the Earth and that life arose from countless natural experiments in which mixtures of organic molecules were subjected to energy sources such as light, heat, and electrical discharge. Various combinations of molecules were shuffled and recombined into complex molecular systems, some of which happened to have the capacity to capture energy and nutrients for a primitive growth process that would ultimately lead to the first living cells.

To understand this process, it is essential to know about the physical environment. This section includes papers that attempt to define a probable set of prebiotic conditions relevant to the origin of cellular life on the early Earth. The paper quoted above, by Chamberlin and Chamberlin, appeared in *Science* in 1908 and represents probably the first informed opinion about the prebiotic environment. It also proposed that extraterrestrial infall may have been a significant source of the elements required for the later synthesis of organic compounds.

The paper built on an earlier proposal by T.C. Chamberlin—that the Earth was formed through accretion of extraterrestrial objects called planetesimals rather than from a vapor phase in the original solar nebula. (The latter view was, in fact, accepted by Oparin in his original 1924 essay.)

We would say that Chamberlin showed remarkable foresight in developing this concept just after the turn of the century, since accretion of planetesimals is the prevailing view today.

Plausibility arguments and the origin of life

Beginnings hold an endless fascination for the inquiring human mind. It is somehow deeply satisfying to think that the universe apparently exploded into existence 15 or so billion years ago, that our planet and sun are about 4.6 billion years old, and that the origin of life took place at least 3.5 billion years ago. Although we have a modest level of confidence in these statements, the hard fact is that we can arrive at such understanding only by using deductive logic about events long past. The foundations of fact and experiment, so important to all the sciences, are uncomfortably thin in origins-of-life research.

Investigators of biogenesis must therefore often rely on plausible arguments, very much as lawyers argue on circumstantial evidence in a court of law. Plausibility is based on three considerations. The first is the principle of *continuity*. Those models that most clearly demonstrate a continuous evolutionary pathway leading from proposed origins to extant life forms are considered to be more plausible than models requiring a discontinuity between origins and evolved form.

The second principle is *ubiquity*. In general, ubiquitous conditions are considered more plausible than isolated special cases, unless the special case can be shown to be necessary in some fundamental way. A marine site, for example, seems more plausible for biogenic processes than a freshwater site, simply because oceans were ubiquitous on the early Earth and fresh water would have been relatively rare.

The third principle is *robustness*. A robust model is relatively independent of precisely defined environmental parameters. Plausible models for prebiotic chemistry and biogenesis should, therefore, incorporate relatively common ingredients—carbon dioxide, water, nitrogen, silica minerals, heat and light energy—which almost surely would have been abundant in the prebiotic environment.

Finally, plausible models should also include not only synthetic processes but concentration mechanisms and degradation rates as well. If we propose that life began when primitive lipids and catalysts condensed from the prebiotic mix to form protocells, we must be able to demonstrate conditions under which such molecules can be continuously produced and protocellular structure and function can be maintained against degradative reactions.

When did life begin?

Life must have originated some time after the primitive Earth cooled sufficiently for liquid water to exist but before the appearance of stromatolites, the earliest fossil indicators of prokaryotic cellular activity, dated at about 3.5 billion years ago. Our understanding of this interval, the Archaean, must be speculative, arising largely from deductive reasoning rather than physical evidence: oceans appeared, volcanic activity produced the first land masses, and late accretion added the final fraction of the Earth's present mass.

Accretion was largely in the form of meteoritic and cometary infall, which may have contributed a significant fraction of the water and organic carbon available on the Earth's surface. Several investigators have noted that the impact of large extraterrestrial objects—cometary nuclei and asteroids—would also affect early evolutionary processes leading to biogenesis. Maher and Stevenson (1988) were among the first to recognize that high-velocity impacts could release sufficient energy to sterilize the Earth's surface—the so-called impact frustration of the origin of life. Sleep et al.'s later discussion (1989) explores how such impacts might have affected early ecosystems.

What was the source of the Earth's primitive atmosphere?

In his 1929 essay, Haldane was inspired to paint a picture of the early atmosphere that is not too far from what contemporary geophysicists would suggest as a plausible scenario:

> Within a few thousand years from its origin, [the Earth] probably cooled down so far as to develop a fairly permanent solid crust. For a long time, however, this crust must have been above the boiling-point of water, which condensed only gradually. The primitive atmosphere probably contained little or no oxygen, for our present supply of that gas is only enough to burn all the coal and other organic remains found below and on the Earth's surface. On the other hand, almost all the carbon of these organic substances, and much of the carbon now combined in chalk, limestone, and dolomite, were in the atmosphere as carbon dioxide. Probably a good deal of the nitrogen now in the air was combined with metals as nitride in the Earth's crust, so that ammonia was constantly being formed by the action of water. The Sun was perhaps slightly brighter than it is now, and as there was no oxygen in the atmosphere, the chemically active ultra-violet rays from the Sun were not, as they now are, mainly stopped by ozone (a modified form of oxygen) in the upper atmosphere, and oxygen itself lower down. They penetrated to the surface of the land and sea, or at least to the clouds.

In considering the primitive atmosphere, scientists later reasoned that, because most of the matter in the solar system is hydrogen, the early atmosphere must have been reducing rather than oxidizing. That is, the most common biogenic elements—carbon, oxygen, and nitrogen—would have been in their hydrogenated or reduced forms: CH_4 (methane), H_2O (water), and NH_3 (ammonia). More recent evidence argues for a moderately reducing atmosphere and against a strongly reducing atmosphere, chiefly because it seems unlikely that any significant amount of original, volatile, atmospheric components would survive the extreme energies of the accretion process.

What, then, was the source and composition of the primitive atmosphere? The most probable primary source is volcanic activity, in which carbon dioxide and water vapor escaped from the Earth's interior. This means that the atmosphere at the time of life's origin was predominantly carbon dioxide. Probably the most convincing argument in favor of a carbon dioxide atmosphere is that carbonate, now present at the Earth's surface as calcium carbonate sediments, must once have been present as carbon dioxide gas. If all sedimentary carbonate minerals could somehow be turned back into carbon dioxide, they would yield about 60 total atmospheres. This argument is given additional weight by the fact that Venus has an atmosphere today of about 100 atmospheres of carbon dioxide. Venus and the Earth are alike in many ways and probably went through similar planetary accretion processes. Because Venus is closer to the sun, however, it is considerably hotter than the Earth. Thus, water on Venus remained in the vapor phase rather than condensing as oceans and so could not remove carbon dioxide from the atmosphere.

Nitrogen on the early Earth would have been approximately at the partial pressure of Earth's present atmosphere. Oxygen would have been rare and short-lived, produced by high-energy reactions and quickly consumed by various reduced components of the atmosphere and hydrosphere. Trace amounts of hydrogen, methane, hydrogen sulfide, hydrogen cyanide, and formaldehyde would have been continuously added by volcanic outgassing and cometary infall.

Extraterrestrial infall provided carbon compounds

Meteorites, derived largely from the asteroid belt, represent preserved samples of primitive components of the early solar system. Metallic nickel-iron meteorites are easily recognized and are common in museum collections. However, most meteorites are composed of stony minerals and are called *chondrites*. The name refers to chondrules, millimeter-sized structures that are found throughout the matrix. The origin of chondrules is still uncertain, but the most generally accepted view is that they are aggregates of silica minerals that were formed by melting and fusion in the hot solar nebula of the early solar system. Chondrules were then accreted with other minerals into the parent bodies in the asteroid belt.

Most extraterrestrial material that reaches the Earth is very small (in the micron size range) and is referred to as interplanetary dust particles (IDP). Larger particles, the size of small pebbles, are quickly destroyed by the frictional heat generated as they enter the atmosphere. At the other end of the scale, objects a meter or more in diameter are not substantially slowed by atmospheric entry and vaporize when they strike the Earth's surface at velocities of 10–20 kilometers per second. An example is the Arizona meteorite crater, a kilometer across, which was produced by an iron meteorite about 30 meters in diameter. Meteorites of intermediate size, ranging from larger pebbles a few grams in mass up to boulder-sized objects weighing several hundred kilograms, lose their interplanetary velocity in the atmosphere and enter free fall, striking the ground at the same velocity as a stone dropped from an airplane. Although their surfaces briefly reach white heat from atmospheric friction during entry, the interior of such meteorites remains cold, thereby preserving their chemical and physical composition in essentially a pristine state.

Carbonaceous chondrites represent only a small fraction of stony meteorites (perhaps 5 percent). They are characterized by their content of organic carbon, ranging up to several percent by weight. The unusual character of such meteorites has been known since the earliest days of chemistry, and Berzelius, Bertholot, and Wohler all reported that organic material was present in meteorites. Chamberlin and Chamberlin took note of this in their 1908 paper:

> Meteorites contain a very suggestive series of compounds of the essential elements, carbon, oxygen, nitrogen, hydrogen, sulphur and phosphorus.... The hydrocarbons in meteorites embrace, according to Cohen, compounds of carbon and sulphur; of carbon, hydrogen and sulphur; and of carbon, hydrogen, and oxygen, respectively. The hydrocarbons sometimes have a resinous or waxy aspect, and on decomposition may give an oil of bituminous or fatty odor. The compounds of carbon, hydrogen and oxygen afford extracts resembling peat, humus, or lignite.

After Oparin and Haldane provided a paradigm for the origin of life, meteoritic organics increasingly became a focus of attention. There were even serious proposals that the organic material was biogenic in origin, but consensus was soon reached that abiotic chemical synthesis was the most plausible explanation (Hayatsu and Anders, 1981). In supplying organics to the surface of the already-formed planet, meteorites thus have a role in what may be seen as the astronomer's version of the chemist's (Urey-Miller) experiment. As we will see in the next section on prebiotic chemistry, the composition of meteoritic organics continues to guide our thinking about the prebiotic environment of the early Earth.

Prebiotic temperature ranges

Any discussion of temperature must address global mean temperatures, local ambient temperatures, and extremes. The origin of life's processes was most plausible in liquid water—that is, only above water's freezing point and below its boiling point. A reasonable estimate for global mean temperature at the time of life's origin, therefore, would be between 40° C and 80° C, a range of temperatures found during the cooling trend that followed condensation of the oceans. Most forms of life could not exist at temperatures at the upper end of that range, yet contemporary thermophilic bacteria thrive in hot springs associated with volcanic regions.

Daily temperature cycles resulting from solar heating of mineral surfaces would have had an amplitude similar to such cycles today but would have ranged around a higher global mean temperature. The high end of such a cycle would be near 100° C and the low end near 40° C. Because of thermal buffering of warm oceans and higher mean global temperatures, freezing temperatures would have been an uncommon occurrence, with the possible exceptions of polar regions and volcanic mountain peaks.

Plausible sites of origin were not at equilibrium

A central aspect of all life is that its physical processes consist of fluxes: that is, they occur under nonequilibrium conditions. A cell may be seen as a bounded system of fluxes, holding itself at gradients (e.g., electrochemical, ionic, osmotic) relative to the environment. It follows that sites plausible for the origin of cellular life would not be at equilibrium.

That is, they would be subject to cycles of darkness and light, cold and heat, flooding and evaporation—and to fluctuations in pH, ultraviolet and sun light, salinity, and chemical concentrations. Cycles of temperature and dehydration, for instance, would serve to concentrate potential reactants and would provide a ubiquitous source of free energy to drive polymerization reactions necessary for the production of larger molecules.

A marine environment represents a highly plausible site for life's origin—particularly lagoons or intertidal zones that are wetted by daily tide cycles and dried by solar heating. Origins-of-life scenarios featuring intertidal marine sites are supported by evidence that the earliest fossil stromatolites (mineralized bacterial mats) apparently formed in such zones over 3 billion years ago.

What were plausible substrates for early synthesis and concentration of organic compounds?

This question was first addressed by Chamberlin and Chamberlin in 1908. They noted several possibilities:

> The primitive organic synthesis may have taken place (1) in the ocean, (2) in some body of fresh water, or (3) on the land, or, more specifically, (4) in the soil. By soil in this connection is meant merely the earthy mantle of comminuted and weathered material.... May we not take it for granted that the higher presumption will lie in favor of that localization which brings into closest interaction the requisite material in unstable states, attended by the maximum range of concentrations, condensations, catalytic, electrical, nascent and other favoring conditions?

Given that the first robust organic synthesis probably did not occur in the open ocean or on dry land, we can consider several possible mineral substrates. The most abundant surface would be igneous rocks and weathered sediments resembling the lava and volcanic sands of contemporary volcanic islands such as Hawaii and the Galápagos. Such mineral surfaces can approach 100° C by solar heating, such that films of organic material would be subjected to temperatures sufficient to form covalent chemical bonds through condensation reactions. Because, in sands and lava minerals, a less energetic microenvironment exists only a few centimeters below, polymers formed at the surface could migrate downward and be protected there against heat and light-driven degradation reactions.

Reprinted papers on the early-Earth environment

The papers reprinted in this section concern the prebiotic early-Earth environment. We present these papers chronologically so that the little-known 1908 paper by Chamberlin and Chamberlin is given its rightful place. The 1924 essay by Oparin, an English translation of his original 1924 Russian publication, first appeared in Bernal's 1967 book, *The Origin of Life*. (Oparin later expanded his work greatly. It was published in Russian in 1936 and in English translation in 1938.) The next paper here is Haldane's 1929 essay, again decades ahead of its time: recognition of its remarkable scientific insight came only in retrospect.

The 1952 paper by Urey is a classic work that first demonstrated the power of physical and geological concepts for establishing constraints on possible prebiotic environments. The paper by Levine et al. (1982) is the first to consider the stability and lifetime of methane and ammonia in the early atmosphere. Short papers by Walker (1983) and by Kasting and Ackerman (1986) follow, discussing contemporary views of the prebiotic ocean and atmosphere that were first developed by Holland (1978). The last papers, by Sleep et al. (1989) and by Chyba and Sagan (1992), vividly describe the late accretion process in terms of massive impacts of extraterrestrial objects.

References

Chamberlin, T. C. and Chamberlin, R. T. (1908) Early terrestrial conditions that may have favored organic synthesis. *Science* **28**:897-911.

Chyba, C. F. and Sagan, C. (1992) Endogenous production, exogenous delivery, and impact-shock synthesis of organic molecules: An inventory for the origins of life. *Nature* **355**:125-132.

Haldane, J. B. S. (1929) The origin of life. Reprinted in: J. D. Bernal, *The Origin of Life* (1967). London: Weidenfeld and Nicolson, pp. 242-249.

Hayatsu, R. and Anders, E. (1981) Organic compounds in meteorites and their origins. *Top. Curr. Chem.* **99**:1-37.

Holland, H. D. (1978) *The Chemistry of the Atmosphere and Oceans.* New York: Wiley.

Kasting, J. F. and Ackerman, T. F. (1986) Climatic consequences of very high carbon dioxide levels in the Earth's early atmosphere. *Science* 234:1383-1385.

Levine, J. S., Augustsson, T. R. and Natarajan, M. (1982) The prebiological paleoatmosphere: Stability and composition. *Origins of Life* 12:245-259.

Maher, K. A. and Stevenson, D. J. (1988) Impact frustration of the origin of life. *Nature* 331:612-614.

Oparin, A. I. (1924) The Origin of Life [*Proiskhozhdenie zhizny.* Moscow: Izd. Moskovshii Rabochii]. English translation in: J. D. Bernal, *The Origin of Life* (1967). London: Weidenfeld and Nicolson, pp. 199-234.

Oparin, A. I. (1938) *Origin of Life,* S. Morgulis, trans., New York: Macmillan.

Sleep, N. H., Zahnle, K., Kasting, J. F. and Morowitz, H. J. (1989) Annihilation of ecosystems by large asteroid impacts on the early Earth. *Nature* 342:139-142.

Urey, H. C. (1952) On the early chemical history of the Earth and the origin of life. *Proc. Natl. Acad. Sci. USA* 38:351-363.

SCIENCE

A WEEKLY JOURNAL DEVOTED TO THE ADVANCEMENT OF SCIENCE, PUBLISHING THE OFFICIAL NOTICES AND PROCEEDINGS OF THE AMERICAN ASSOCIATION FOR THE ADVANCEMENT OF SCIENCE.

FRIDAY, DECEMBER 25, 1908

CONTENTS

MSS. intended for publication and books, etc., intended for review should be sent to the Editor of SCIENCE, Garrison-on-Hudson, N. Y.

EARLY TERRESTRIAL CONDITIONS THAT MAY HAVE FAVORED ORGANIC SYNTHESIS

THERE is a wide gap between the inorganic carbon compounds, as we now know them in nature, and the much more highly complex carbon compounds which are the material basis of living beings. It is a prevalent view that this gap can not be bridged by natural processes under existing conditions. On the face of things this view seems to be supported by the testimony of experience. This experience, however, when critically examined, is not altogether conclusive. Even if it could be shown beyond question that the chain of carbon compounds necessary to bind the inorganic to the organic never is built up under present conditions, there would still remain a legitimate ground of doubt in the possibility that this may be due to predaceous plants and animals, especially bacteria, which attack the carbon compounds at the first stages at which they become available for food and thus cut off the evolving series before it is complete. In this it is assumed that the formation of the more complex carbon compounds can come about only as the result of a long series of synthetic steps, and that at some of these stages, probably at many of them, the products would be suitable food for existing beings, especially for the almost omnivorous and ubiquitous bacteria. This prolonged evolution may thus be regarded as an extremely precarious process in the presence of predatory organisms; may indeed be regarded as practically prohibitive after the earth was once well planted with the rapacious varieties. This adverse con tingency obtains quite irrespective of the theoretical possibility of such an evolution under favorable conditions, if left undisturbed.

A further element should not be overlooked. Before an organic series can be permanently assured, the power of self-propagation must be acquired. What may be the contingencies attending the addition of the self-propagating function to the acquisition of the requisite chemical constitution is at present quite beyond determination. There may be in this supplementary process, if indeed it be a supplementary process, as great a liability to predatory arrest, as in the chemical synthesis itself. Probably the safest answer which the extreme evolutionist can give to the suggestive question why wholly new orders of living beings have not appeared at frequent intervals in the geologic record, lies in an appeal to the rapaciousness and universality of the attack of organisms already present.

It is merely as a matter of precaution that these considerations are cited here. It does not seem to us safe to make the unqualified affirmation that present terrestrial conditions are wholly incompatible with the natural synthesis of a series of carbon compounds linking the inorganic to the organic if the whole predaceous kingdom were removed so as to leave the organizing agencies unlimited opportunities for interaction for an indefinite period. The purpose of this paper is not to meet a geologic necessity, but merely to consider those conditions in the early history of the globe which may be thought to have been specially favorable to organic synthesis, irrespective of the question whether the natural evolution of life was wholly dependent upon them or was merely facilitated by them. The paper is not the result

of an inquiry into the problem of organic synthesis, as such, but is rather a preliminary statement of supposed geologic conditions intended to suggest inquiry in certain lines which do not seem to have been critically pursued. At the same time, the treatment may serve to connect with this chiefest of synthetic problems the postulates of the planetesimal hypothesis which have been little discussed in this regard.[1] Only a part of the agencies herein suggested as bearing on the synthetic process are however peculiar to this hypothesis. In some notable part they are assignable to any mode of origin which the earth may reasonably be supposed to have had.

The growth of the earth by the planetesimal method leaves a rather wide speculative range relative to the conditions which prevailed at the stage when life was introduced. At the same time, the hypothesis recognizes limitations which shut out certain conditions that have sometimes been thought to be important to the evolution of the higher carbon compounds, notably high atmospheric pressure. Under the older cosmogonies very high atmospheric pressures associated with very high temperatures were postulated. Under the planetesimal hypothesis, it is indeed possible and logical to assign to the atmosphere as great an extent and as high a pressure as the gravity of the earth can control; but the hypothesis, as we hold it, is built on the assumption that molecular activities limit the atmosphere, a principle which, if true, limits it under any hypothesis. If the doctrine of Stoney, modified to meet the new advances in physics, is sound, it can not, of course, be violated by any cosmogonic hypothesis without self-destruction. If this doctrine is true, and the earth-mass lacks power to control an

[1] Chamberlin and Salisbury's "Geology," Vol. II., 1906, pp. 111–116.

atmosphere radically greater than the present one, it is obvious that the notion of an atmospheric pressure appreciably greater than the present should not be entertained. It is because this doctrine of Stoney is believed to be fundamentally sound, subject to some qualifications, that the planetesimal hypothesis has developed its postulates in consonance with it and has not consciously entertained assumptions regarding sequential states that are at variance with it. If the doctrine of Stoney is sound, the gaseous and gaseo-meteoritic hypotheses have no advantage as to the pressures of the early atmospheres; on the contrary, the higher temperatures they assume imply greater molecular activity and hence greater molecular losses and less pressures than prevail around the present relatively cool earth. In this paper, therefore, no appeal will be made to *general* atmospheric pressures appreciably greater than the present; indeed, we shall try to be true to the assumption that the *general* atmospheric pressures in the formative stages of the earth were lower rather than higher than they are at present. What volume of atmosphere and what degree of atmospheric pressure may have been compatible with the gravitative power of the earth at such early stages as were first suitable for the evolution and preservation of organic compounds is not yet definitely postulated by the planetesimal hypothesis. It is merely assumed that a suitable atmosphere, hydrosphere, temperature and other conditions were found at some one of the progressive stages leading up to the present state and that there was a gradual transition from that stage to the existing state of things. So far as gravity alone is concerned, a sufficient atmosphere and hydrosphere might apparently be held when the earth had about one tenth its present mass, *i. e.*, about the present mass of Mars, which ap-

parently holds an atmosphere and perhaps a hydrosphere.

If the earth were formed by a gathering of planetesimals into a nucleus which at an earlier stage was one of the nebulous knots of a spiral nebula, the value of the mass at the time it first became a solid body should have hung upon the ratio of the original knot to the scattered nebulous matter destined to be gathered into it, concerning which there is little to guide opinion. In this uncertainty, let it be assumed that at the stage when the growing earth was ready for the critical synthesis, its mass lay somewhere between one tenth and three fourths of the present mass, and that the mass of the early atmosphere held a proportionate ratio to the mass of the existing atmosphere. Let it also be assumed that at the time at which the hydrosphere began to gather about the globe, the earth's surface was formed of a heterogeneous, uneven, talus-like mantle, such as the infalling planetesimals would naturally produce, and that this constituted a deep porous zone. Until the hydrosphere was added there was a lack of efficient solutions and inwash to cement the fragmental material. The hydrosphere, in its first stages, must have occupied the lower part of this porous zone, and must have crept upwards as its volume increased. It should have appeared at the surface first in the bottoms of the deformation sags, the innumerable craters and the other depressions, and as it gradually extended itself, it must have linked more and more of the isolated water-bodies together. Throughout the earlier stages of growth, the prevailing aspect should have been that of innumerable small water-bodies scattered over the face of the land, rather than that of the great confluent ocean of to-day.

This then is the general physiographic setting which the planetesimal hypothesis, at least in one of its phases, presents for the initiation of the chemical synthesis to which is assigned the task of bridging the gap between the inorganic and the organic compounds.

So far as chemical reactions of the familiar inorganic class are concerned, such a stage of earth-growth must obviously have presented conditions favorable to unusual activity, since the contact surfaces between the atmosphere and the hydrosphere, between the hydrosphere and the rock surfaces, and between the atmosphere and the rock surfaces, were all relatively large and varied, while the temperatures produced by the radiation received from the nebulous sun and by the heat of impact of the planetesimal matter may be presumed to have given a varied range of thermal conditions. As a result, reactions of exceptional variety should have been developed. In view of the vast number of bodies of water and their varied contacts with all sorts of rocks, together with the possibilities thus presented for the isolation of special combinations in their early stages and the correlative possibilities of conjunctions and minglings of these in their subsequent stages, there may well have resulted numerous diverse chains of sequences.

At that stage the atmosphere should have held each of its present constituents but probably not in the same proportions as now. The elements of lower specific gravity should have been smaller in amount relatively than at present, while those of higher specific gravity should have been more abundant relatively, because the controlling power of the smaller earth limited the range of molecular retention. Relative to the lighter constituents, it need only be added that, instead of escaping completely from the control of the earth, as postulated in the earlier phases of the doctrine of Stoney, the lighter molecules should in the main have

escaped merely into the supplementary orbital atmosphere, where they might have persisted for long periods during which they were always subject to being driven back to the earth by occasional encounters, so that, although these constituents may have been rare in the denser collisional atmosphere near the earth's surface, they were doubtless present in sufficient amounts to have participated in any appropriate chemical synthesis. It is to be added further in qualification of this that, with the exception of hydrogen, which might have been derived from its compounds, the lighter atmospheric constituents, such as helium and neon, are chemically inert and there is no specific reason for supposing that they entered into any stage of the organic synthesis. The heavier constituents of the early atmosphere would include the compound gases of carbon, nitrogen, sulphur and phosphorus.

During the critical stages of growth, the infalling of planetesimals is supposed to have continued to be active. The planetesimals are assumed to have contained carbon, sulphur, phosphorus and all other elements found in organic matter; and as they impinged more or less violently upon the surface formed of previous accessions of similar matter, there should have been generated various compounds of these elements. A portion of these compounds should have been gaseous or vaporous and should have been variously disseminated into the atmosphere, or else absorbed into the waters or into the porous earthy mantle. The volcanic action of the period should have contributed its characteristic constituents in similar variety of combination. It may therefore be assumed with plausibility that a larger percentage and a greater variety of volatile compounds of the critical class prevailed in the air, subject to absorption into the earth-mantle and the waters, during the stages of active terrestrial growth than at the present time, when terrestrial activities have more largely settled down into a state of less disturbed equilibrium. The same may doubtless be presumed of the unstable non-volatile compounds of the critical elements, but just here we are considering only the factor which worked through the atmosphere.

We have already denied ourselves any special appeal to *general* atmospheric pressure as an agency of the particular synthesis in question, but in doing this we have limited our abnegation to the sum total of pressure. It is still pertinent to inquire what may have been the effects of *partial* pressure. Starting with the familiar postulate that each constituent of a mixed gas acts in accordance with its own individual partial pressure, it appears that high partial pressures of the *critical* constituents should have been more effective in causing their entrance into combination than high general pressure without such high partial pressures. We have just seen that the chemical and physical conditions attending the stages of active earth-growth might well have given rise to large proportions of the critical compounds. By extending the argument it is not difficult to see that the actual amount of the carbon, sulphur, phosphorus and nitrogen gases might have been sufficiently greater than now to have promoted their union, for they all have high specific weights and could be held in relatively large proportions by the gravity of even a small earth. The gaseous compounds of nitrogen have a place in this class, though the free nitrogen was probably less abundant than now, but since free nitrogen is organically inert, it is negligible except as a source of its compounds.

If we turn to the non-volatile or slightly volatile compounds of the critical elements, we find, as already remarked, reason for thinking that there may have been more of the unstable compounds of this class than now, for secular progress would be in the direction of chemical stability. There are certain probable sources of unstable compounds of carbon, sulphur, phosphorus and nitrogen which call for special notice. Planetesimals are only one form of meteoroids and it is possible that the meteorites which reach the earth in appreciable masses may not often be true planetesimals of the class supposed to have formed the earth, but if the theory of common meteorites advanced by the senior author be the true one,[2] the compositions of the two classes are probably much alike, for according to this theory meteorites are supposed to be the fragments of small planetoidal bodies which have been disrupted by close approach to large bodies. These planetoidal bodies are supposed to have been too small to hold atmospheres and hydrospheres, and hence the planetesimals which formed them were never subjected to atmospheric and hydrospheric agencies, and hence they retained their planetesimal constitution with little modification.[3] This furnishes ground for assuming that the compounds found in meteorites were also present in the planetesimals. This presumption finds support in the nature of the case, for the gases shot forth from the sun to form the planetesimals probably were much dissociated by high temperature at the instant of emerg-

[2] *Astrophys. Jour.*, Vol. XIV., 1901, pp. 17–40; *Jour. Geol.*, Vol. IX., 1901, p. 369, and Chamberlin and Salisbury's "Geology," Vol. II., pp. 22–38.

[3] Dewar states that at a temperature of — 130° C. liquid oxygen has no chemical action on hydrogen, potassium, sodium, phosphorus, etc. (*Proc. Roy. Inst.*, Vol. 11, 1884–6, p. 550), and this is perhaps an additional reason why planetesimals and meteorites enveloped in the low temperatures of space prevalent at considerable distances from the sun have been subjected to little oxidation.

ence and later combined in all available ways as they met appropriate elements at the various lower temperatures through which they passed in the course of their cooling.

Meteorites contain a very suggestive series of compounds of the essential elements, carbon, oxygen, nitrogen, hydrogen, sulphur and phosphorus. Among these compounds it is especially interesting to note that several are unstable under the usual conditions that prevail at the surface of the earth. Among these are the phosphide, schreibersite, the sulphides, troilite, daubréelite and oldhamite, the carbide, cohenite and the chloride, lawrencite. These unstable compounds are sometimes intergrown or otherwise closely associated with one another, so that the products of any reactions that may grow out of their instability are favorably located for reaction upon one another, and such reaction is liable to arise when the celestial equilibrium is disturbed by terrestrial conditions. There is also present a form of graphite which oxidizes more rapidly than terrestrial graphite. Such graphite, as well as amorphous carbon, is intergrown with the unstable sulphide, troilite, and the unstable phosphide, schreibersite. The readily oxidizable graphite sometimes contains within itself sulphur and hydrocarbons. Roscoe extracted from a meteorite a hydrocarbon which contained sulphur.[4] Clöez and Pisani found in the Orgueil meteorite matter of the organic type which had a composition closely similar to terrestrial humus, viz., C, 63.45; H, 5.98; O, 30.57.[5] Lawrence Smith also extracted hydrocarbons from meteorites, among which was a compound of carbon, hydrogen and sulphur having approximately the composition of $C_4H_6S_4$; also one which he

formulated as $C_4H_6S_5$.[6] Ansdell and Dewar favor the view that the hydrocarbons had their origin in carbides, a view supported by the observed presence of carbide in meteorites.

It may be assumed that the graphite resulted from the action of water, gases and other agents, on the carbides of the metals, and that during the chemical interactions which took place, a portion of the carbon became transformed into organic compounds.[7]

The hydrocarbons in meteorites embrace, according to Cohen, compounds of carbon and sulphur; of carbon, hydrogen and sulphur; and of carbon, hydrogen and oxygen, respectively. The hydrocarbons sometimes have a resinous or waxy aspect, and on decomposition may give an oil of bituminous or fatty odor. The compounds of carbon, hydrogen and oxygen afford extracts resembling peat, humus or lignite. All the alkalis, and all the earthy and metallic constituents which are essential to plant life are present in meteorites in variety and abundance. Besides the solid minerals, the gases carbon dioxide, carbon monoxide, methane, hydrogen, nitrogen and sulphuretted hydrogen are either present or are susceptible of development by moderate heat such as the infalls would inevitably produce.[8] It will be noted that these embrace the more common gases associated with ultimate organic reactions.

The chief point of interest here is not simply the presence, but the close intergrowth, of compounds of the organic elements in states of combination which, while stable in the cosmic regions surrounding the earth from which the planetesimal material was gathered, are pronouncedly

[4] *Proc. Phil. Soc. Mon.*, 1862.

[5] *Compt. rend.*, Vol. 59, 1864, pp. 37, 132.

[6] *Am. Jour. Sci.*, 1876, pp. 388–95, also 433–42. Many suggestive details.

[7] *Proc. Roy. Inst.*, Vol. XI., 1884–6, p. 549.

[8] For details see "The Gases in Rocks," R. T. Chamberlin, Carnegie publication No. 106, Washington, 1908.

unstable in the presence of the air and water, even at the ordinary temperatures at which organic synthesis takes place. On reaching the surface of the earth, these compounds would therefore have been no longer stable, but must have sought new relations. The transfer from the one state to the other may well have been accompanied by combinations not likely, or at least less likely, to arise under modern conditions of greater stability.

As already noted, a carbide of iron, nickel and cobalt is found in meteorites. The conditions attending the formation and infall of planetesimals encourage the belief that a wide range of carbides, and probably also of nitrides, may have existed among the accessions to the growing earth. But whether the amount of such material was large or small, it is clear that, as it was borne to the earth in the divided state of planetesimals, its measure of contact with the moisture of the surface was exceptionally large and hence the suggestive reactions which follow under these conditions should have been large in proportion to the amount of carbides and nitrides present. Even with cold water, the carbides of barium, strontium and calcium give rise to acetylene, while the carbides of aluminium and beryllium generate methane. Under the same conditions, uranium carbide gives rise to a gaseous mixture of methane, hydrogen and ethylene, and in addition to this generates considerable quantities of both liquid and solid hydrocarbons.[9] While nitrides are not known to furnish directly such a varied assortment of gases, they may readily give rise to ammonia gas under various conditions. Iron nitride, heated to 200° C. in the presence of hydrogen sulphide, yields iron sulphide, a sulphide of

ammonium, and free hydrogen ($2Fe_2N + 6H_2S = 4FeS + 2NH_4HS + H_2$).[10]

It is impossible to follow the remoter reactions which might spring from these various first products of the carbides and nitrides by further interaction, but it is to be noted that all the new products, *if closely associated*, as the conditions of the case imply, were liable to meet one another while in the state of nascent activity, or in such stimulated activity as may attend previous activities, however this may be interpreted. Reaction might also have been aided by the catalytic relations into which these products came in their intimate association with the débris of the surface, with one another and with the new accessions added constantly to the earth's surface.

As already noted, the superficial portion of the growing earth is supposed to have been a loose, incoherent aggregate of highly fragmental planetesimal matter, and this was subject to weathering and various forms of comminution. The resulting porosity may have led to the condensation of much of the gaseous matter within its capillary and subcapillary pores, as soils are known to do at the present time. The graphite and amorphous carbon may have acted somewhat after the analogy of charcoal which is well known to absorb certain gases to a phenomenal extent. Various porous substances, including earths, possess this property of gas-condensation in appreciable degrees. Not a few metals also possess, irrespective of porosity, certain occlusive affinities or selective powers of absorption for certain gases, as palladium and platinum for hydrogen, etc. While the efficiency of charcoal is clearly due in large part to its extreme porosity, it is doubtless due in part also to the substance itself, for coal manifests a somewhat similar property. Fresh coal rapidly absorbs oxygen from the

[9] Moissan, *Proc. Roy. Soc.*, Vol. 60 (1897), pp. 156–60.

[10] Fowler, *Jour. Chem. Soc.*, Vol. 79, p. 297.

air, condensing it within the pores and crevices; but because of its activity in this condensed form, the oxygen soon unites with the substance of the coal and passes on to chemical union, making room for the absorption of more oxygen from the air.

Pursuing the phenomena of charcoal further by way of illustration, the well-known fact may be recalled that a layer of charcoal spread over decaying organic matter permits it to waste away gradually without giving off offensive odors, owing to the fact that the gases and the volatile products of the decomposition coming in contact with the charcoal are drawn into its pores and there oxidized by the condensed oxygen which the charcoal has absorbed from the air. Owing perhaps to a sort of catalytic action, the oxygen condensed in the pores of the charcoal becomes more active than the ordinary oxygen of the air. While charcoal furnishes the most declared case, soils and comminuted earth matter are known to possess similar properties in a notable degree.

In summation, we conclude that the porous mantle of the earth thus supplied by planetesimal infall with unstable carbides, nitrides, phosphides and sulphides undergoing transformation into more stable compounds, and generating during this process hydrocarbons, ammonia, hydrogen phosphide and hydrogen sulphide gases mingled with the ordinary gases carried by the planetesimals, furnished rather remarkable conditions for interactions and combinations, among which unusual syntheses would not be improbable.

If, with these special possibilities in mind, we turn to the question, what physiographic situation on the surface of the early earth presented the most favorable conditions for the organic synthesis, three general views offer themselves as alternatives, and under one of these there is a localization so specific as to have the force of a fourth view. The primitive organic synthesis may have taken place (1) in the ocean, (2) in some body of fresh water, or (3) on the land, or, more specifically, (4) in the soil. By soil in this connection is meant merely the earthy mantle of comminuted and weathered material; the absence of organic matter at the outset is of course assumed.

May we not take it for granted that the higher presumption will lie in favor of that localization which brings into closest interaction the requisite material in unstable states, attended by the maximum range of concentrations, condensations, catalytic, electrical, nascent and other favoring conditions?

If planetesimals carrying the essential constituents in unstable forms fell into large and deep bodies of water, the soluble and gaseous products of such reactions as followed would have been likely to be widely diffused and diluted, would have received little aid from the catalytic action of rock or earth material, would have been unassisted by soil porosity, and would have been but little favored by concentrations except such as involved an increase of density of all the constituents held in the water body whether favorable, hostile or indifferent in nature. Organic synthesis at present clearly involves a series of very special selections of material from among a miscellaneous association, the larger part of which is either neutrally obstructive or hostile. In the case of a completed organism, provided with the proper selective appliances, the requisite material may be gathered from such a dilute intermixture of the essential and non-essential materials as the water-bodies present, but until these selective appliances are provided, the water-bodies seem to us to be deficient in some of the most propitious conditions.

In the absence of such organic selective structures at the outset, it may be worth while to look for inorganic agencies that may perhaps have performed in a crude way the functions which the organic structures came later to perform in a much superior way. It is not clear that a selective concentrative agency can be found in the large water-bodies, independent of life, for their tendency is diffusive and equalizing rather than concentrative. Can it be found in the soil?

To form a definite picture of the conditions which may have been presented by certain soils, let a foreland lying between extensive uplands and a permanent water-body of appropriate salinity be chosen as a concrete example. In such a situation a constant water-level should have prevailed not only at moderate but at graded depths beneath the surface. The underground water should have received accessions percolating basinwards from the uplands bearing whatever soluble materials the uplands could furnish, while, on the other side, there should have been waters percolating landward from the water-body bearing such salinity as it possessed, while more or less spray from the water-body was scattered by landward winds over the surface and fed the soil from above. The measure of water-movement in the one direction or the other must obviously have depended on the balance between the precipitation and the evaporation on the adjacent land which varied with seasons and localities. The underground waters thus supplied by the slightly fluctuating water-table should have been carried up by the capillary passages of the soil to horizons of evaporation and concentration at or near the surface. The graded depth of soil above the water-table should have furnished unlimited adaptations to the porosity of the soil-mantle and to the meteorologic conditions of the seasons and the special situations.

The horizons of concentration by capillarity and by evaporation were affected by all degrees of insolation from direct sunlight, at the immediate surface, to all measures of shadowing below, furnished by the soil itself. The substances deposited within the soil and the substances leached from it, alike modified its porosity in their own special ways, and such deposits as were formed in the soils should have developed porosities of their own. Isolated cells and tubes may thus have been formed by travertine and similar deposits in the pores of the soil. As already implied, the capillary feeding from the water-table, when the balance set that way, and the partial evaporation of the solutions within the pores of the soil near or at the surface, must have produced concentrations of the non-volatile substances carried by the capillary waters into these upper horizons. Between these upper concentrated solutions and those below there should naturally have followed inter-diffusion and osmotic action, and thus there should have sprung up in the soil a circulation somewhat analogous to that of the plant; indeed, it may not be going too far to suggest that this circulation within the soil is more than a simple analogue of the systematic circulation which is definitely organized in the plant structure; it may possibly be its genetic forerunner. The larger cavities within the soil were inevitably connected by constricted passages which were liable to be partially or wholly filled with porous precipitates or with inorganic colloids, and these chains of pores might thus come to serve a circulatory function which was an actual precursor to that circulation from cell to cell through membranous walls which distinguishes compound plants. It is even more probable that the circulation within an earth-pore, partially isolated from adjacent earth-pores containing solutions of different density, might develop a circulation closely like that of unicellular plants. A constant evaporation at the surface might perpetuate the differences of density sufficiently to continue the circulation for the requisite period. So, on the other hand, if the ground waters were sufficiently saline, the periodic accession of fresh water at the surface might give, in reverse order, density-differences sufficient to perpetuate circulation.

Whether analogy may be pushed so far as this on substantial grounds or not, it is obvious that the soil presents a suggestive assemblage of conditions favorable to chemical synthesis, and that these deserve a critical attention which the limits of this paper forbid us to try to pursue into fuller detail. It is clear that under these conditions evaporation was permitted to develop the greatest variety of concentrative effects which can well be assigned to it, and that at least some facilities were offered for the perpetuation and modification of these concentrations when once established. It is obvious that in the soil there might have been any degree of exposure of the reacting substances to light within very small differences of depth, and that there was easy intercommunication and commingling of the different products of such photosynthesis by means of capillary circulation. It is obvious that in the soils there might have been found the full range of capillary dimensions from coarse tubes and pores down to the limits of sub-capillarity with the full gamut of condensation-effects assignable to these. The same may be said of all catalytic effects referable to the relations of the solids and solutions concentrated in the soils.

An important class of chemical reactions observed in the matured plant of to-day takes place within the organism without much obvious addition or loss of material, apparently an internal readjustment of

matter under its own stimulus. This is familiarly seen in the maturing of seeds after they have been separated from all connection with the plant circulation, in which little external change beyond some possible loss of moisture is observable. In many cases, germination will not take place at once; a period of internal organization or ripening must intervene. This thus represents a last stage of preparation for self-propagation, or, in other words, the last stage of an advanced phase of organic synthesis. It may not accord with conventional usage to call this self-catalytic action, but in some way it seems to be due to stimulus which the partially organized materials exert on one another and by which they thereby push farther forward the synthetic process. The prevalence of this internal action in the last stage of the organic cycle suggests that analogous action may occur far back in the synthetic series, and that much of the more complex part of the cycle may perhaps be accomplished by such action, conditioned like it on the confinement of the partially organized material in a cell-like cavity where it was subject to accessions of the sustaining solutions and to concentrations by the removal of water and volatile matter. If this speculative conception be warranted at all, the quasi-cellular structure of the porous soil might not inappropriately be conceived as serving the function of a rude inorganic husk, or shell, abetting by confinement and protection the internal synthesis of its contents.

If we may appeal to still other agencies whose functions are as yet imperfectly understood, the electrolytic action of earth-currents may be worthy of mention. It is well known that saline waters are better conductors of electricity than fresh waters or dry earth, and hence it is probable that the restoration of electrical equilibrium on the earth's surface after it has suffered disturbance may be attended by electrical passages between the saline water-bodies and the land tracts. There are reasons for believing that electrical interchanges were more active in the growing stages of the earth than at the present time when the solar system may be presumed to have settled down into a condition nearer equilibrium. The planetesimal matter circulating about the planetary nucleus and the sun, and between them, probably served both to disturb the electrical status of these bodies and to afford a means of electrical communication between them and the planetesimals. Under electric laws, planetesimals charged with electricity and circulating at high velocities about the young planet should have had the effect of electrical currents, and should have generated magnetic fields about the magnetizable matter of the earth; and these magnetic fields, in turn, should have modified and perhaps controlled the paths of electrons and ions traversing these fields. It is beyond the scope of this paper to try to follow these into detail, and it may suffice to merely indicate that electrolytic action stimulated by electrical currents traversing the surface may fairly be presumed to have played a more active part during the nebular stages than they do to-day, whatever that part may be.

On similar grounds, it may be suggested that the agitations of radioactivity, and the states of ionization to which it gives rise, may have played a slightly more active part in the chemical processes of early times than they do to-day, whatever that may be, because of secular decay.

Some suggestions respecting the original localization of terrestrial life may be derived from a study of the localizations of to-day. The more primordial types of the vegetal life of the mid-ocean are limited in variety and in susceptibility to variation.

In the judgment of some of our most trusted botanists, pelagic vegetal life does not present pronouncedly the qualities which imply germinal or evolutionary power. A similar statement may be made, with qualifications, relative to the life of the larger fresh-water bodies. In the shore tracts, indeed, there is greater variety of life and greater indication of germinal competency, but even here the forms which lie nearest the hypothetical primitive types do not give signs of conspicuous evolutional potency. On the other hand, the plant life of the land presents much greater variety of form and of organization, and greater signs of germinal virility.

If we may judge of the fitness of soils to serve as a nidus of organic genesis, by the life it fosters to-day, a favorable verdict seems warranted by the remarkable assemblage of low forms which make the soil their habitat and which manifest peculiar adaptations to their earthy conditions and to one another. There are not only a host of simple forms in which photosynthesis plays its usual part, but forms that flourish quite irrespective of sunlight; there are not only species that use carbon dioxide and require oxygen, but species that live quite without free oxygen, and even find it a hostile element; there are forms that oxidize sulphuretted hydrogen and use the energy thus derived for their activities, and there are forms that oxidize ferrous to ferric iron to like ends.[11] Some of the sulphur-bacteria seem to combine the generation of energy from sulphuretted hydrogen with the more common mode of oxidizing carbon compounds, thus uniting in a suggestive way independent chemical processes. There are also bacteria that oxidize free nitrogen into nitrites and others that promote the formation of nitrates. The rich

[11] Jost's "Plant Physiology," Gibson's translation, 1907, pp. 220-31.

realm of soil life seems yet far from being exhausted by research either in respect to the range of forms or the scope of their chemical activities, but it is at least clear that the soil is the foster-ground of remarkable biochemical activities. Some large part of such activities is dependent on preexistent organic matter and is in no sense initial, but in a significant part of it, organic compounds do not seem to be prerequisite. This variety of action and these peculiarities of the life of the soil-mantle are at least suggestive of genetic conditions.

If we seek for such uncertain light as the geological record may throw upon the habitat of the first life, we are confronted by the fact that the record makes only a very distant approach to the real genesis of life. If we permit ourselves to reason from the nature of the Proterozoic formations, we find grounds for a belief in a very early mantling of the land surface with vegetation.[12] A study of the early habitat of some of the leading forms of life seems also to favor the land or the land-waters.[13] While these geological considerations have their obvious limitations and may seem to be too far removed from the specific question of organic synthesis to have much value, they may at least be permitted to serve as an offset to the prepossessions which seem heretofore to have obtained widely in favor of the origin of life in the ocean. They may also help to bring into equitable competition the view that primitive organic synthesis may have found its genetic conditions in some of the lowland soils on the borders of the permanent water-bodies.

[12] Chamberlin and Salisbury, " Geology," Vol. II., pp. 139, 199, 302.

[13] Chamberlin, " The Habitat of the Early Vertebrates," *Jour. Geol.*, Vol. VIII., 1900; Sardeson, " The Phylogenic State of the Cambrian Gasteropods," *Jour. Geol.*, Vol. XI., 1903; Chamberlin and Salisbury, " Geology," Vol. II., p. 480.

The original organic process was undoubtedly vegetative rather than animal in type, *i. e.*, the primitive organisms increased the sum total of organic matter and stored energy in it. The store of energy attained by any given plant consists (1) of that which it inherits with its spore, seed or germinal part and (2) of that which it adds thereto from terrestrial and cosmic sources of energy. If the plant feeds wholly on carbon dioxide, water, nitrates, sulphates, phosphates and similar *wholly oxidized* compounds, its sources of free energy are essentially limited to two classes: (1) radiant energy, derived chiefly from the sun and allied sources, and (2) chemical energy, derived from the oxidation of a portion of the germinal matter. The peculiar cooperation of these two sources of energy forms a distinctive combination. A part of the energy set free by germinal oxidation cooperates with the remainder and with solar energy to build up additional complex compounds. This new complex matter is built up from fully oxidized material by deoxidation combined with a synthetic process which gives it higher complexity. This process obviously requires additional energy. The synthetic function resides primarily in the germinal matter, for the formation of new compounds may proceed to certain lengths without sunlight; but the main source of the energy required for deoxidation resides in the sun. The germinal matter thus seems to have two phases of action: (1) simple oxidation, which gives free energy for its own activities, and (2) synthetic stimulus, by which complex carbon compounds are organized. This last takes place at the outset by the transformation of its own substance, but later, with the cooperation of sunlight, by the synthesis of simpler substances previously in a fully oxidized state.

Now, if any inorganic matter is to take

the place of this germinal matter in synthesis, it must apparently subserve this double function of suffering oxidation of a part to yield the energy required to organize its remaining part into other complex compounds. This peculiar double function seems to furnish a criterion which must probably be met, if we are to find in early terrestrial history any inorganic agency which may have subserved crudely the synthetic function which has since been developed so extraordinarily by self-perpetuating organic agencies. Are there any agencies among those previously under review which act in any such way, and which might be supposed, even by the license of hypothesis, to serve as a crude substitute or forerunner for the germinal matter?

Before trying to answer this crucial question, let another characteristic of organic matter as a source of energy be noted. The energy which can be derived from living matter by oxidation is distributed among many atoms, or, if it be interpreted as concentrated in certain of the atoms among the complex assemblage, it is so diluted or distributed by the neutral atoms present that its effect is distributive. *The end secured by this is a slow—perhaps one may say a controlled—use of energy in oxidation distributed over a prolonged period.* The organism is indeed a thermal engine, but its temperature is phenomenally conservative, uniform and sustained, and its consumption of fuel in any unit of time is small. At the same time its constructive work, both of material and of energy, is so much more conspicuous than its consumptive work as to almost completely mask the latter. This distributive conservative action is perhaps the key to the successful absorption and utilization of the mild but pervasive energies of cosmic origin which furnish the increment of energy necessary

to increase the sum total, notwithstanding the portion expended.

Now if we gather together the essentials of the energy processes previously cited as probably present in the primitive days of the earth, shall we find in them any analogues of this cooperative, controlled and constructive process?

Summarizing these essentials in terms of energy, it appears (1) that several of the elements that take part in organic activities were probably present at the outset in a free state and capable of yielding their maximum of energy by oxidation, such as graphite, amorphous carbon, and perhaps sulphur, together with the free gases, hydrogen, nitrogen and oxygen; (2) that there were probably present simple binary compounds such as the carbides and nitrides, which were capable of reacting with water and of setting energy free, and at the same time generating somewhat more complex compounds; (3) that there were hydrogen compounds susceptible of oxidation with liberation of energy, such as the hydrocarbons, ammonia, sulphuretted and phosphoretted hydrogen; (4) that there were partially oxidized compounds susceptible of further oxidation with a liberation of energy, such as the nitrites and ferrous salts; (5) that there were fully oxidized compounds, end products, such as carbon dioxide and water, from which, *by the expenditure of energy,* the carbohydrates might hypothetically be derived by deoxidation and combination; as well as such end products as nitric acid, nitrates, sulphates and phosphates from which, in combination with the preceding, proteids and the nitrogenous group generally might, hypothetically, be compounded; (6) that there were abundant supplies of potassium, calcium, magnesium and silicon in combination from which these elements could be derived by similar means; (7) that there was abundant diffuse energy in sunlight and other forms of solar radiance, and (8) there were pervasive free energies in states of ionization, in earth currents and in local electric potentials. In short, it appears that there was a group of oxidizable substances which might give up energy by combination, but not in indefinite amounts; that there was an ample supply of substances already oxidized to the full, but which could be partially deoxidized and combined hypothetically to give the complex organic compounds, but only at the expense of energy. A continuous source of pervasive, diffuse energy was afforded by sunlight and other forms of solar radiance, perhaps aided by molecular agitation through ionized states, diffuse electric earth-currents, and local electric potentials; and these sources were ample to meet the requirements of the case.

Recalling the combination of functions required, the most plausible suggestion is offered by the cooperation of the unstable carbides, nitrides, sulphides and phosphides with moisture, sunlight and allied agencies.

Somewhat after the analogy of germinal matter, certain metallic carbides react at ordinary terrestrial temperatures on the accession of water, freeing energy and raising the temperature, but retaining a portion of their power of oxidation and at the same time forming carbon compounds of slightly more complex nature. The phenomena of polymerization may intervene in some cases and favor complex compounds. Some of these compounds are unsaturated, and by additions and substitutions may lead on to other complexities.[14] If the reactions of these carbides took place in the presence of the associated unstable nitrides, sulphides, phosphides and chlorides, there is reason to believe

[14] Moissan, " The Electric Furnace," translation by Moulpied, pp. 244 and 256.

that more complex results would follow, involving other organic elements. Experimental results do not greatly help on this point, because high temperatures, strong reagents and artificial conditions have usually been employed in forming the more complex compounds of the organic elements. Violent agencies and extreme conditions are excluded from organic action, and hence presumably from any antecedent action closely precedent to it. In the case in hand, combination could be abetted by evaporation, by porous condensation, by selective concentration, by catalytic action, by confinement in pores, cells and ducts, by capillary and osmotic action, by genial temperatures, by sunlight and by allied agencies. How far these could go in giving rise to the higher compounds of the organic type experimentation has not yet satisfactorily determined. The analogy of the carbide action does not go very far, but it is something that it takes even a short step in the direction so characteristic of organic synthesis.

While the foregoing combinations and activities seem to us suggestive of the most favorable primitive conditions, they obviously do not warrant anything approaching an affirmation that organic synthesis really took place in this way in the soil, or that it is the offspring of inorganic antecedents solely. It has been our endeavor to trace the early terrestrial conditions and activities into as close an analogy to those that dominate the organic kingdom as present imperfect data permit. The conservative considerations that make it unsafe to assign organic genesis to an early terrestrial age without reserve, simply because it is not now in evidence, find their complement in withholding opinion respecting the possibility of such genesis by any combination of inorganic influences whatsoever, until it shall be experimentally settled. The great problem must lie open yet awhile at least, but every line of approach, however hypothetical, may well be pursued if controlled by due reservations.

As remarked at the outset, many of these considerations are applicable to states of the earth which might have arisen at an early day under any of the cosmogonic hypotheses. We have given precedence to one of these hypotheses partly because it seems to merit a fuller exposition in this particular than it has received, and partly because it seems to us to present a physiographic setting more favorable for synthesis than would probably have arisen under the alternative cosmogonic hypotheses.

T. C. CHAMBERLIN,
R. T. CHAMBERLIN

UNIVERSITY OF CHICAGO

The Origin of Life*

A. I. Oparin

Grau, teurer Freund, ist alle Theorie,
Und grün des Lebens goldener Baum.

GOETHE

The Theory of Spontaneous Generation

EVER since he took the first steps towards a conscious life, Man has tried to solve the problems of cosmogony. The most complicated and also the most interesting of these is that of the origin of life. At different times and at different stages of culture different answers have been given. The religious teachings of all ages and peoples have usually attributed the appearance of life to some creative act by a deity. The first students of nature were very naïve in their answers to this question. Even to a man of such outstanding intelligence as Aristotle in ancient times, the idea that animals, including worms, insects and even fish, could develop from mud presented no special difficulty. On the contrary, this philosopher maintained that any dry body becoming moist or, on the other hand, any wet body becoming dry, would give rise to animals.

The authority of Aristotle had an exceptionally strong influence on the outlook of men of learning in the Middle Ages. In their minds the ideas of this philosopher became interwoven with the doctrines of the fathers of the Church, often giving rise to suppositions which, to our eyes, appear stupid or even ridiculous. In the Middle Ages it was held that although the preparation of a living person, or of something like one in the form of a 'homunculus', in a retort by the mixing and distillation of various chemical substances was extremely difficult and impious, nevertheless it was undoubtedly something which could be done. The production of animals from non-living

*This is a translation by Ann Synge of A. I. Oparin, (1924) *Proiskhozhdenie zhizny*. Moscow. Izd. Moskovshii Rabochii.

Reprinted from J. D. Bernal, *Origin of Life.* London: Weidenfeld & Nicolson, 1967, pp. 199-234.

materials seemed to the scientists of those times to be so simple and ordinary that the well-known alchemist and doctor, van Helmont, actually gave a receipt according to which it was possible to prepare mice artificially by placing damp grain and dirty rags in a covered vessel.

There are a number of writings from the sixteenth and seventeenth centuries describing the transformation of water, stones and other inanimate objects into reptiles, birds and beasts. Grindel von Ach even gives a picture of frogs formed from May dew, while Aldrovandi gives drawings which show how birds and insects arise from the twigs and fruit of trees.

The idea that the maggots in rotting meat, fleas in dung and intestinal worms are generated spontaneously from decaying materials was generally accepted as an unalterable truth which was in full accord with Holy Writ. Also, in the writings of those times, we often meet with numerous texts by means of which the authors hope to convince their readers that the theory of spontaneous generation has the full support of the Bible.

However, the further science developed and the more the study of nature came to involve the use of accurate observations and experiments rather than just argument and philosophizing alone, the narrower became the region in which the learned men believed that spontaneous generation could occur.

As early as the middle of the seventeenth century Redi demonstrated by simple experiments that there was no basis for the opinion that spontaneous generation of maggots takes place in rotting meat. He covered the meat with a thin gauze and thus made it inaccessible to those flies from whose eggs the maggots would develop. Under these conditions the meat putrefied but no maggots whatsoever appeared. It was as simple as that to refute the idea of spontaneous generation of insects.

Thus, so far as living things visible to the naked eye were concerned, the theory of spontaneous generation had no support. However, at the end of the seventeenth century Kircher and van Leeuwenhoek discovered a world of tiny creatures, invisible to the naked eye, and only discernible with the microscope. These 'tiny living beasties' (as van Leeuwenhoek called the bacteria and infusoria which he had discovered) were to be found wherever decay was taking place; in decoctions and infusions of plants which had been allowed to stand for a long time, in decaying meat and broth, in sour milk, in dung and in the fur on teeth. 'There are more of them (microbes)', wrote van Leeuwenhoek, 'than there are people in the United Provinces.' It was only necessary for a substance which sours quickly and decays easily to stand for some time in a warm place for microscopic living creatures which had not been there before to develop in it immediately. Where did these creatures come from ? Had they really arisen from germs which had happened to fall on the decaying material ? How many of these germs there would then have to be everywhere. The idea inevitably arose that it was indeed in the rotting decoctions and infusions that the spontaneous generation of living microbes from non-living material took place. This view received strong support in the middle of the eighteenth century from the work of the Scottish clergyman Needham. He used meat broth or decoctions of vegetable material which he placed in completely closed vessels and boiled for a short while. By doing this Needham reckoned that he must have destroyed all the germs which were present and new ones could not get in from outside because the vessels were completely covered. Nevertheless, after a short while microbes appeared in the liquids. From this demonstration Needham drew the conclusion that he was witnessing the phenomenon of spontaneous generation.

However, there was another learned man who opposed this view. He was the Italian, Spallanzani. He repeated Needham's experiments and became convinced that more prolonged heating of the vessels containing the organic liquids sterilized them completely. A bitter quarrel raged between the proponents of the two opposite viewpoints. Spallanzani showed that, in Needham's experiments, the liquids had not been heated enough and that the germs of living things were still present in them. Needham retorted that it was not he who heated his liquids too little, on the contrary, Spallanzani heated his too much and by such rough treatment destroyed the 'generative power' of the organic infusions which was very tricky and unstable.

Thus, each of the contestants stuck to his opinion and the question of the spontaneous generation of microbes in putrefying liquids was not solved one way or the other for a whole century. During this time quite a large number of attempts were made to prove or disprove by experiment the occurrence of spontaneous generation, but none of them gave a definite result. The question became more and more embroiled and it was not until the middle of the nineteenth century that it was finally solved by the studies of a

French scientist of genius – Pasteur.

Pasteur first showed the extremely wide distribution of microbes. In a series of experiments he showed that everywhere, but especially near human habitations, the air contains tiny germs. They are so light that they float freely in the air, only falling to the ground very slowly. On the slightest breeze they fly up again and are carried around us invisibly. The air of large towns is positively swarming with these crumbs of life. A single cubic metre of the air in Paris in summer contains up to 10,000 viable germs. If they encounter favourable conditions they grow, develop and begin to multiply at an extraordinary speed causing the decomposition of liquids which putrefy easily. Thus, it is not the putrescent liquids which give rise to the microbes but the microbes falling from the air which cause the putrefaction of the liquids.

Pasteur explained the mysterious appearance of micro-organisms in the experiments of earlier authors as being due to incomplete sterilization of the medium or insufficient protection of the liquids against the access of the germs. If the containing retort is carefully boiled and then shielded from germs which might enter it with the air, then, in a hundred per cent of cases, there will be no putrefaction of the liquid and no formation of microbes.

Pasteur used a great variety of devices for sterilizing the air entering his retorts. He sometimes heated it in red-hot tubes of metal or glass; sometimes the neck of the flask was plugged with cotton wool in which all the minute particles carried in the air were trapped; or finally, sometimes the air was passed through a fine glass tube shaped like the letter S, in which case all the germs were trapped mechanically on the damp surfaces of the curves of the tube. Whenever the precautions were sufficiently reliable the appearance of microbes in the liquid was not observed. Maybe, however, the prolonged heating had changed the medium chemically and made it unsuitable for supporting life. Pasteur easily refuted this suggestion. He dropped into the sterilized liquid the cotton-wool plug through which air had passed into the retort and which therefore contained germs. The liquid quickly putrefied. The boiled liquid was, thus, a perfectly suitable soil for the development of bacteria. The only reason why this development did not occur was the absence of seed in the form of germs. As soon as the germs fell in the liquid they began to grow at once and gave a good harvest.

Finally, Pasteur succeeded in showing that it is possible to keep such easily putrefied liquids as blood and wine for long periods even without any heating. It was only necessary to remove them from the animal (where they do not contain bacteria) aseptically, that is, taking precautions against bacteria falling into them from outside.

Thus, Pasteur's experiments showed beyond doubt that the spontaneous generation of microbes in organic infusions does not occur. All living organisms develop from germs, that is to say, they owe their origins to other living things. But how did the first living things arise? How did life originate on the Earth? In what follows theories will be examined which attempt to solve this problem.

The Theory of Panspermia

Pasteur is rightly considered as the father of the science of the simplest organisms, i.e. microbiology. His work gave the impetus to extensive studies of a world of minute creatures, invisible to the naked eye but inhabiting the earth, water and air. The investigations undertaken were not now, as formerly, directed merely towards describing the form of the micro-organisms; bacteria, yeasts, infusoria, amoebae, etc., were studied from the point of view of the conditions necessary for their life, their nutrition, respiration and multiplication; from the point of view of the changes which they bring about in their environment and finally from the point of view of their internal structure in its finest details. The further these studies proceeded the clearer it became that the simplest organisms were by no means so simply constructed as had once been thought.

The body of any organism, whether it be plant, mollusc, worm, fish, bird, beast or man is made up of very small droplets which are only visible under the microscope. It is built up from these droplets or cells as a house is built of bricks. The various organs of different animals and plants are composed of cells of different sorts. In becoming adapted to the work carried out by a particular organ the cells of which it is composed are changed in one way or another but, essentially and in principle, all the cells of all organisms are alike. Micro-organisms differ only in that their whole bodies consist of a single cell. This similarity in principle between all organisms confirms the idea, now generally accepted in scientific circles, that all living things on the

Earth are connected with one another; they form, so to speak, a family of blood relations. The more complicated organisms arose from simpler ones which gradually changed and grew more efficient. Thus, it is only necessary to explain how some very simple organism could have been formed in order to be able to understand the origin of all plants and animals.

However, as has already been said, even the simplest creatures, consisting of only one cell, are extremely complicated structures. Their main component, called protoplasm, is a semi-liquid, ductile, gelatinous substance, permeated with water but not water soluble. Protoplasm is made up of a large number of extremely complicated chemical substances (mainly proteins and their derivatives) which are never found except in organisms. These substances are not simply mixed but are in a special state which has not yet received much study. Owing to this the protoplasm has an extremely fine structure which is poorly differentiated even under the microscope, but is extraordinarily complicated. The idea that such a complicated structure with a completely determinate fine organization could arise spontaneously in the course of a few hours in structureless solutions such as broths or infusions is as wild as the idea that frogs could be formed from the May dew or mice from corn.

The extreme complexity of the structure of even the simplest organisms struck some scientists so forcibly that they were sure that an impassable abyss existed between the living and the dead. The transition from dead to living or organized seemed impossible, either in the present or in the past. 'The impossibility of spontaneous generation at any time whatever', writes the well-known English physicist W. Thomson, 'must be considered as firmly established as the law of universal gravitation.'

How then did life appear on the Earth ? There was a time when, according to the views now generally accepted among scientists, the Earth was a white-hot ball. Astronomy, geology, mineralogy and other exact sciences provide evidence for this and it is beyond doubt. This means that conditions on the Earth are such that it was unthinkable, not to say impossible, for life to exist on it. Only after the Earth had lost a considerable amount of its heat by dissipating it in the cold of interplanetary space, and the cooling had gone so far that the water vapour had formed the first hot seas, did the existence of organisms like those we now see become possible. In order to explain this contradiction a theory was evolved which bore the rather complicated name of 'panspermia'.

H. E. Richter was the originator of this theory. Starting from the hypothesis that everywhere in space there are small particles of solid material (cosmozoa) which have been cast off by celestial bodies, the author suggested that along with these particles and possibly attached to them, there were the germs of micro-organisms. Thus these germs could be carried from one celestial body which was inhabited by organisms to another which was not yet inhabited. If on this body the conditions of temperature and moisture were suitable for life, then the germs would begin to grow and develop and would be the original parents of all the organic creatures on the planet concerned.

This theory attracted many supporters in the world of science, among them such outstanding minds as those of Helmholtz and W. Thomson. Its proponents were mainly concerned to demonstrate scientifically the possibility of such a transfer of germs from one heavenly body to another with conservation of their viability. In fact, when all was said and done, the main question was whether or not spores could complete such a long and dangerous journey as a flight from one planet to another without being destroyed and while retaining the ability to grow and develop into a new organism. Let us examine closely what dangers the germ would meet on the way.

In the first place there is the cold of interplanetary space, about −220°C. Having been cast off from its home planet the germ would be doomed to spend long years, centuries or even millennia at such an appalling temperature before some lucky change would give it the possibility of arriving in a new world. The question inevitably arises as to whether the germ could survive such an ordeal. To solve this problem experiments were made on the resistance to cold of present-day spores. Experiments along these lines showed that spores tolerate cold excellently. They remain viable even after having been kept at −200°C for six months. Of course, six months is not 1,000 years but, all the same, this experiment gives us reason to suppose that at least some spores could stand the severe cold of interplanetary space.

A far more severe danger to germs is presented by their complete lack of protection from light rays. Their path between the planets is penetrated by the rays of the Sun which are destructive to most microbes. Some bacteria are destroyed by the action of direct sunshine after only a few hours while others are more stable, but all, without exception, are unfavourably affected by very strong irradiation. This unfavourable effect is, however, considerably mitigated when there is no oxygen in the atmosphere, and we know that there is no air in space so we have some reason to suppose that germs of life could survive even this ordeal.

But let us assume that a lucky chance has given our germ the opportunity of falling into the gravitational field of some planet on which the conditions of temperature and moisture are favourable for the development of life. It remains only for the wanderer to submit to the gravitational force and to fall on its new land. But here, at once, just as it is reaching its destination in space, a terrible danger awaits it. Before this germ had been in air-free space, but before it can arrive at the surface of the planet it must pass through a fairly thick layer of air which surrounds its planet on every side.

The phenomenon of 'falling stars' (or meteorites) must be well known to everyone. Scientists now explain this phenomenon as follows: there are, floating in interplanetary space, lumps of material of greater or lesser size, fragments of planets or comets which have flown into our solar system from distant parts of the Universe. When they pass close to the Earth they are attracted to it, but before they can fall on it they have to pass through the atmosphere. Owing to the friction of the air the quickly falling meteorite becomes heated to white heat and becomes visible against the dark vault of the sky. Only a few of these meteorites reach the Earth in the form of red-hot stones from the sky. Most of them get burnt up by the intense heat long before they reach its surface.

A similar fate must befall the germs. However, various considerations indicate that destruction of this sort need not necessarily occur. There is reason to suppose that at least some of the germs reaching the atmosphere of any planet would reach its surface in a viable state.

Futhermore, there is no need to dwell on colossal, astronomical periods of time during which the Earth could have been sown with germs of life from other worlds. These periods are reckoned in millions of years. If, during this period, even one of the thousands of millions of germs could have reached the surface of the Earth satisfactorily and found there conditions suitable for its development, then this would be enough for the formation of the whole organic world. All the same, in the present state of scientific knowledge, this possibility, though conceivable, seems not to be very probable.* In any case we have no facts which would directly contradict it.

However, the theory of panspermia is only the answer to the problem of the origin of earthly life, and not in any way to that of the origin of life in general.

Carus Sterne writes: 'If this hypothesis merely pushes the beginning of life backwards to the time of the first appearance in the world of celestial space, then, from the philosophical point of view it is a completely useless labour because whatever could have happened in the first world was also possible in the second and third and would be the act of creation or spontaneous generation.'

'There are two possibilities,' said Helmholtz, 'organic life either began (came into being) at some time or else it has always existed.' If we accept the former view, then the theory of panspermia loses all logical point, for, if life could originate at some point in the Universe, there is no reason to suppose that it could not originate on the Earth as well. The partisans of the theory under discussion therefore accept the theory of the eternity of life. They recognize that 'life can only change its form, it can never be created from dead material'. Thus they put an end once and for all to any further study of the problem of the origin of life. They try to set up an impassable barrier between the living and the dead and to impose a limit on the efforts of the human mind and on the boundless generalizations to which exact science leads it.

But do we have any logical right to accept the fundamental difference between the living and the dead? Are there any facts in the world around us which convince us that life has existed for ever and that it has so little in common with dead matter that it could never, under any circumstances have been formed or derived from it? Can we recognize organisms as formations which are entirely and essentially different from all the rest of the world?

We shall try to give an answer to these questions in the next chapter.

*Even such an out-and-out defender of the theory under discussion as Helmholtz says in this connection, 'I cannot deny it if anyone thinks that this hypothesis is not very probable or even in the highest degree improbable.' And he adds further that he is forced to take up the point of view of the theory of panspermia because of the complete impossibility of explaining the origin of life scientifically in any other way.

The World of the Living and the World of the Dead

The first and most eye-catching difference between organisms and the rest of the (mineral) world is their chemical composition. The bodies of animals, plants and microbes are composed of very complicated, so-called organic substances. With very few and insignificant exceptions we do not find these substances anywhere in the mineral world. Their main peculiarity consists in the fact that at high temperatures in the presence of air they burn while in the absence of air they carbonize. This shows that they contain carbon. This element is present in all organisms without exception. It forms the basis of all those substances of which protoplasm is made up. At the same time, however, it is no stranger to the mineral world.

The diamond is certainly more or less well known to everyone. It is a precious stone which, in the cut form, is much used as an adornment under the name of brilliant.

As early as 1690 the English scientific genius Newton put forward the idea that the diamond, the hardest and brightest body on the Earth, must contain combustible material. This suggestion was quickly confirmed. A diamond was submitted to trial by fire at the focal point of a burning mirror. It did not survive the test but became covered with cracks, glowed and burnt away before everyone's eyes. Later and more accurate experiments showed beyond doubt that the shining diamond is the blood-brother of the unattractive graphite, the substance from which pencils are made. Both consist entirely of carbon. They have no component other than this element.

However, carbon is met with in the mineral world, not only in the pure state, but also in combination with other elements. Chalk, marble, soda, potash and other compounds all contain carbon. In general, one of the most valuable properties of carbon is its tendency to form the most varied compounds with other elements. In the substances which make up the bodies of organisms the carbon is always found combined with hydrogen, oxygen, nitrogen, sulphur and phosphorus and often with several other elements. All these elements are widely distributed in the non-living world. In combination with oxygen, hydrogen forms water; the air around us consists of a mixture of oxygen and nitrogen; sulphur and phosphorus are found in many of the minerals which make up rock formations. Thus we see that all the elements which enter into organic compounds are also found in abundance in the mineral world. Even this alone gives reason to doubt the existence of any essential difference between the world of the living and that of the dead.

It may be, however, that this difference concerns not so much the elementary compositions as the actual compounds made up from the elements.

It has been considered that the substances of the organic and inorganic (mineral) world are so unlike one another that the former could not be obtained from the latter by artificial means under any circumstances. Some chemists of that time even maintained that it was impossible to obtain organic bodies simply because these substances could only be formed within the living organism by the action of a special 'vital force'. However, as early as 1828 the chemist Wöhler succeeded in preparing an organic substance artificially and so to cast doubt on the importance of the famous 'vital force'.

Since then the study of organic compounds has been advancing with rapid strides, and the further it has gone the clearer has become the falsity of the idea that there is a fundamental difference between these substances and inorganic bodies. Starting from the simplest inorganic substances, chemists can now prepare artificially almost all the substances which are encountered in organisms. Although some of these substances have not yet been obtained there is no doubt that they can be obtained in the near future. The structure of these substances has been studied in extreme detail. No special means of combination of the individual elements has been found in them. They obey the same physico-chemical laws with the same constancy as inorganic compounds.

The essential similarity between organic and inorganic substances has now become so obvious that not a single serious natural scientist would deny it and the protagonists of the view that there is a fundamental difference between the living and the dead have already stopped. They assert that organic compounds whether prepared artificially or isolated from organisms are just as dead as minerals. Life may be recognized only in bodies which have particular special characteristics. These characteristics are peculiar to living things and are not seen in the world of the dead.

What are these characteristics? In the first place there is the definite structure or organization. Then there is the ability of organisms to metabolize to reproduce others like themselves and also their response to stimulation.

Let us go over each of these characteristics and see whether it is really present only in the living organism or whether it is not, in some form or another, also found in the mineral world.

The most important and essential characteristic of organisms is, as is demonstrated by their very name, their organization, their particular form or structure. The bodies of all living things, beginning with the smallest bacteria and algae and ending with man, are constructed according to a definite plan in which the greatest importance attaches, not to the external visible organization but to the fine structure of the protoplasm of the cells which make up the organism. This structure is the same in general for all members of the animal and vegetable kingdoms. Unfortunately it has still not been studied very much. Various investigators have seen different structural formations in the semi-liquid protoplasm in the form of fibres, networks and alveoli. However, as these formations are so extremely small it is very hard to see them, even with the best microscopes now available. All the same it is certain that protoplasm has a definite structure and is not a homogeneous lump of slime. This structure holds the secret of life. Destroy it and there will remain in your hands a lifeless mixture of organic compounds.

Some scientists believe that this structure itself, this organization, could not have been formed spontaneously from structureless and, according to them, lifeless substances. Following the ancient Greek philosopher Empedocles, they repeat, in one form or another, the idea that the organization is closely bound up with the spirit which both constructs the body and is destroyed or flies away when the particular form is annihilated. However, if we adopt this position we must agree with another philosopher, Thales, in ascribing a spirit to a magnet since this extremely simple spirit which expresses itself in the attraction which iron holds for the magnet, depends on the structure of the magnet, that is to say on the arrangement of the particles in it. This structure has only to be disturbed, as when a magnetic stone is ground up in a mortar, and it will lose its spirit just like an organism which has been cut into pieces.

The world of the dead or mineral world is certainly not lacking in definite forms, it is not structureless. It is a property of most chemical substances that they try to take up particular forms, that is to form crystals. We now know that the finest particles of the substance forming the crystals are not just arranged anyhow in them, but are arranged according to a definite plan depending on the chemical composition and the conditions under which the crystals are formed by separation from a solution or from the molten state. It is here that a transition takes place from the hitherto formless, structureless substance to the organized body. In a solution the smallest particles of the substance are in disorder; the same is true even when a substance is melted. However, when a crystal separates out, the particles arrange themselves relative to one another in a strict order, like soldiers forming straight ranks on the command 'Attention!' The form of the crystal and the whole range of its other properties depend upon this arrangement. If the crystal is destroyed by disturbing the arrangement of the particles all these properties disappear.

There is thus no doubt that even the simplest crystals have a definite arrangement. At first glance this arrangement seems to be extremely simple. However, we shall immediately put this idea aside if we remember those marvellous 'ice flowers' which appear on the panes of our windows on a frosty day. In their delicacy, complexity, beauty and variety these 'ice flowers' may even look like tropical vegetation while all the time being nothing at all but water, the simplest compound we know. In small droplets of it the particles were scattered in disorder but they were cooled, the wind blew, the temperature fell below zero and these particles, complying with the eternal laws of nature, which are the same for both the living and the dead, arranged themselves in a definite order and, on the simple window pane, they produce pictures of fabulous gardens, glistening in the sunshine with all the colours of the rainbow.

If we leave the simplest compounds aside and go on to look at those forms which produce more complicated substances an even more varied and involved picture will be presented to our gaze which, in its complexity, is in no way inferior to the most detailed picture of the structures of organisms. However, before we enter on this review we must make a small excursion which, though perhaps rather tedious, is necessary for our further discussion into the realms of the comparatively young science of colloid chemistry which has already acquired great importance.

As early as 1861, the English scientist, Graham, divided all the chemical bodies known at that time into two main classes, crystalloids and colloids. To the first class belonged such substances as various salts, sugar, organic acids and so forth. The substances in this group formed crystals easily and, when dissolved, gave clear and completely transparent solutions. If such a solution is poured into a bag made of vegetable parchment or the bladder of an animal and the bag is placed in pure water, the dissolved substance will pass through the walls of the bag and be washed out of it by the water.

Colloids present a completely contrary picture. They very seldom crystallize and then only with great difficulty. Their solutions are usually cloudy and they cannot pass through vegetable or animal membranes. Graham assigned such substances as starch, proteins, gums and mucus to this group.

It turned out later that such an assignment of all substances between two groups was not altogether correct since the same chemical compound could turn up both as a colloid and as a crystalloid according to the conditions of the solution. Thus 'colloidness' was not a property of a particular substance but of its particular state.

There was, however, a considerable element of truth in the classification made by Graham since substances which have very large and complicated particles very often and easily give rise to colloidal solutions. The study of the colloidal state is therefore of special importance since the vast majority, if not all, of the substances of which protoplasm is made up have very large and complicated particles and therefore must give colloidal solutions.

As we have already said, colloidal substances do not give crystals, but still they are fairly easily precipitated as clots or lumps of mucus or jelly. We may take as our example of such formations the protein of eggs which is precipitated on boiling and the gelatin which sets on cooling (well known to everyone as a jelly). The separating out of such coagulates or precipitates from previously uniform solutions sometimes takes place amazingly easily for apparently insignificant reasons.

The coagulates or gels obtained by precipitating colloids are, at first sight, quite structureless. If we examine a lump of jelly under the microscope, even at a high magnification, it appears to us quite uniform. However, scientists have now invented very effective instruments with which they have succeeded in revealing the complicated structure of coagulates. Here it is not a question of straight lines and planes such as we meet in crystals, for here we have a whole network, a whole skein of fine threads which are interlaced, separating from one another and coming together again in a definite, complicated order. Sometimes these threads are very fine, on the other hand, sometimes they are thickened, fusing with one another to form small enclosed bubbles or alveoli.

The structure of coagulates is strikingly reminiscent of that of protoplasm. Unfortunately this structure has not yet been sufficiently well studied for us to be able to say anything conclusive about this resemblance. However, there can be no doubt that we are dealing here with phenomena of the same order. There is no essential difference between the structure of coagula and that of protoplasm.

It may be, however, that the difference between living and dead does not lie in the organization which, as we have seen, is present in both worlds, but in the other features which we mentioned, the ability of living organisms to metabolize to reproduce themselves and to respond to stimuli.

The shapes of crystals are unalterable, they are formed once and for all, while an organism may be compared with a waterfall which keeps its general shape constant although its composition is changing all the time and new particles of water are continually passing through it. The composition of the living body changes in just the same way. The organism takes in different substances from its environment; after a number of chemical changes it assimilates these substances, transforming what had been foreign compounds into parts of its own body. The organism grows and develops at the expense of these substances. However, just as a factory requires a certain amount of fuel to carry on its work, so, if the organism is to carry on its unceasing activities it should consume, that is to say break down, at least part of the material which it has assimilated, and this is what actually happens. In the process of respiration or fermentation the organism breaks down substances which it has already taken in and the products of their degradation or decomposition are given off into the environment. Thus life consists of continual absorption, construction and destruction.

However, if we make a detailed analysis and simply contemplate the phenomenon of metabolism which has been described, we shall not find, even here, anything specifically characteristic of the living world. In fact, the phenomenon of feeding, the assimilation of substances from the environment, is, of course, found in its simplest form even in crystals. Thus, a crystal of common salt, which is well known to everyone, will, if it is immersed in a supersaturated solution of the same substance, increase its size and grow by absorbing individual particles of substance from its environment (the solution) and making them part of its body. Even here, in this simple phenomenon, we have before us all the characteristic features of the phenomenon of nutrition. There is even more similarity between this phenomenon and the processes which occur in colloidal coagula. A lump of such material has the ability to extract from solution and absorb the most varied substances such as dyes. These latter do not just remain on the surface of the lump but penetrate deeply into it, some of them simply adhering to the tangled threads which constitute the lump while others enter into chemical reactions with these threads and combine firmly with them, forming component parts of the whole lump.

Study of the process of feeding of living protoplasm shows that this process too takes place in exactly the same way as has just been described. Solutions of different chemical substances penetrate into the protoplasm as a result of the action of comparatively simple and thoroughly studied physical forces, just the same forces which operate in colloidal coagula. Having entered into the protoplasm, one substance will quickly pass out of it while others will enter into a chemical reaction with it, combine with it and become parts of it. And here, all in all, we have a simple chemical reaction and not anything mysterious such as could only be accomplished by a 'vital force'.

Thus, when various chemicals are absorbed by lifeless coagula we are dealing with processes which take place in a way which is completely analogous with the first stage of metabolism, that is to say, feeding.

The following example shows that, even in the world of the dead, we can find processes which are essentially just the same as metabolism as a whole. If we take a small piece of so-called spongy platinum (platinum is one of the 'noble' metals which can be obtained by special methods not as sheets or solid lumps, but as a very delicate sponge with very fine holes and delicate walls) and throw it into a solution of hydrogen peroxide in water* then

---

*Hydrogen peroxide is a chemical compound which, like water, is composed of hydrogen and oxygen only but there is twice as much oxygen in it as in water. This compound is fairly stable and can usually stand for a very long time without breaking down. When it breaks down for any reason it gives rise to oxygen and water.

bubbles of oxygen will immediately begin to form on the surface of the lump. They are formed by the breakdown of the hydrogen peroxide and the process goes on quite rapidly and only stops when all the hydrogen peroxide has been broken down to oxygen and water. If we then remove, dry and weigh our piece of platinum we shall find its weight has remained just as it was before. The same piece may again be thrown into a new amount of hydrogen peroxide and will again decompose it quickly while itself remaining unaffected. Thus, a comparatively small piece of spongy platinum can decompose an unlimited quantity of hydrogen peroxide.

Chemists have been interested in the mechanism of this process for a long time, and as a result of many investigations we now know for certain that the decomposition of hydrogen peroxide by platinum takes place in the following way. First the peroxide is adsorbed on the platinum. As the peroxide cannot penetrate into the metal this adsorption only takes place on its surface. That is why it is important to use spongy platinum for this experiment as it has a very large surface at which the metal is in contact with the liquid. The particles of peroxide do not simply adhere to the surface of the platinum but form a chemical compound with it, namely the hydrate of platinum peroxide. Thus the piece of metal, like the living organism, extracts particles of hydrogen peroxide from the water and assimilates them into its body. However, it does not end at that. After a short time the hydrate of platinum peroxide on the surface of the metal breaks down to platinum, water and oxygen, the last being given off in the form of bubbles of gas. The reduced platinum can combine with a new portion of hydrogen peroxide and again break it down, the products of the decomposition being oxygen and water. The process is repeated until there is no hydrogen peroxide left in the solution surrounding the metal.

In the example given we have the simplest but still complete prototype of metabolism. It contains all the important elements of this process. The absorption of substances from the surrounding medium, their assimilation and breakdown and the giving off of the products of their decomposition. Just the same process takes place in any living organism, for example in any bacteria cultivated in a solution of nutritive substances. The bacteria absorb the substances from the solution, assimilate them and then break them down, giving off to the outside the products of their decomposition. Thus, a simple piece of metal behaves in just the same way as a living organism.

In this connection it must be pointed out that both phenomena (the metabolism of organisms and the decomposition of hydrogen peroxide by platinum) are not only similar, in their external form, but the actual mechanism of the process is similar in both cases. In all organisms, without exception, metabolism is brought about by means of so-called enzymes. This name is given to substances, the chemical nature of which is still only poorly understood, but which can comparatively easily be isolated from any animal or plant in solution in water or as a powder which is easily dissolved in water. All enzymes now known have the power to act on substances forming part of the living body in a very remarkable way. They alter these substances in one way or another (either by breaking them down or by causing them to combine with one another) while themselves remaining completely intact. A detailed study of this phenomenon has shown that enzymes act on the different organic compounds of the living body in just the same way as does platinum on a solution of hydrogen peroxide. In fact metabolism, in its most important aspects, does not consist of anything but a long series of successive enzymic processes following one another and related to one another like the links of an unbroken chain.

At present one of the most extensive sections of physiology, the science which studies the functioning of living organisms, is devoted exclusively to the problem of metabolism. The further the study of this complicated process goes on, the more closely and accurately we get to know the essential features of the processes which are carried out in the living cell, the more strongly we become convinced that there is nothing peculiar or mysterious about them, nothing that cannot be explained in terms of the general laws of physics and chemistry.

Thus, even the ability to metabolize cannot be taken as a special characteristic peculiar to living organisms.

We still have two 'peculiarities of life' left to discuss, namely, the capacity for self-reproduction and response to stimuli.

In what does the capacity of organisms for self-reproduction consist? In the simplest case it amounts to this: the elementary organism, the cell, divides itself into two halves each of which then grows into a new daughter cell in which the structure of the mother cell is reproduced down to the finest detail. This property, however, does not belong only to organisms, but to all bodies possessing a definite structure, without exception. Let us take the simplest case as our example. If we take a crystal of any substance such as alum, break it into two halves and place them in a supersaturated solution of the same substance what will happen will be that the halves of the crystal which had been placed in the solutions will quite quickly replace their missing faces, angles and edges at the expense of particles which had previously been floating freely in the solution. Before growing larger they take a form which reproduces in the finest details that of their mother, the original crystal.

The question may, however, arise that in the example which we have given we forcibly broke the crystal whereas the division of the cell apparently takes place spontaneously. Is that not the fundamental difference between the two phenomena ? The fact is, however, that it only seems to us that the division of cells takes place spontaneously, it really takes place under the influence of definite physical forces (capillary attraction, surface tension) which, though they certainly have not yet received much study, still are of just the same kind, in principle, as all the other physical forces.

An even more interesting phenomenon is that of the 'seeding' of supersaturated solutions. It occurs as follows. In some cases it is possible to concentrate a solution of a particular substance very strongly without that substance separating out in the solid form. However, if the most minute crystal or grain of the substance in question is dropped into the solution crystals will immediately begin to separate out of the solution, sometimes in such quantities that the whole mass becomes crystalline.

This shows that a crystal can cause the formation of bodies like itself which would otherwise not have been formed. If the particles which are scattered at random throughout the solution are to arrange themselves according to a definite plan to give a definite organization or form, that form must already be present.

Here we have the occurrence of the most amazing phenomena which may serve as a key to the understanding of other extremely complicated phenomena of the same order. 'Let us take the example of sulphur,' as Carus Sterne says, 'this is known to be a simple substance yet it depends on the temperature at which it changes from the liquid to the solid state which of two very different forms it will take, octahedral or prismatic*. If we place two such crystals on fine platinum wires in a supersaturated solution of sulphur in benzine then, in the neighbourhood of the prismatic crystal new prisms will be formed while octahedra will be formed near the octahedral crystal. When the two armies of crystals approach one another, the latter form is victorious at the first clash. Here is an example of the struggle for existence in the realm of crystals! '

Let us now go on to take a look at the last of the peculiarities of living things which we mentioned, that is, at responsiveness to stimuli. In all living things without exception we meet with a property which in its most general form may be described as follows. In an organism external and internal stimuli will cause something of the nature of a discharge and will induce the performance by it of some definite action (e.g. movement, etc.) which will carry out in a particular way according to its structure and the means at its disposal. It is a very characteristic feature of responsiveness that there should be a quantitative disparity between the energy, that is to say the forces, required to excite or bring about the response, and the work which is the response of the organism to the stimulus in question. Thus, for example, a relatively slight touch can be enough to induce the organism to move from one place to another or to carry out some other work requiring the expenditure of much force. The organism draws from within itself the forces (energy) required for this work.

Some scientists believe that responsiveness is a specially characteristic feature of organisms. However, if we take this view we shall have to regard a railway locomotive with the steam up as a living thing. In fact it is only necessary to apply a slight stimulus by shifting a lever and the locomotive will start to move, carrying out a very considerable amount of work at the expense of the fuel which is burnt in its boiler. This work is many times greater than that expended in moving the lever and is carried out in complete accordance with the structure of the locomotive.

Responsiveness is to be found, not only in organisms, but also in any

---

*On crystallizing, different substances take on regular geometrical shapes. The crystals are characterized by these forms. Some substances may crystallize in different forms under different conditions. In particular, sulphur can give crystals in the form of octahedra (two four-sided pyramids joined together by their bases) or hexagonal prisms.

physical body which has any noteworthy store of hidden (potential) energy. Comparatively insignificant causes may lead to the discharge of this energy which will lead to the carrying out of some particular work. A landslide caused by a comparatively slight movement of the air, the explosion of a powder magazine caused by a spark which happened to fall in it – these are very simple cases of the phenomenon of responsiveness.

With this we will finish our short review of the main feature of living organisms. We have seen that not one of these can be held to be inherent only in living things. But if this is so we have no reason to think of life as being something which is completely different in principle from the rest of the world. If life had always existed and had not arisen by generation bodily from the rest of the world, if it had not separated itself or, crystallized out at some time from this world, then it would inevitably have had characteristics peculiar to itself. But this is not so. The specific peculiarity of living organisms is only that in them there have been collected and integrated an extremely complicated combination of a large number of properties and characteristics which are present in isolation in various dead, inorganic bodies. Life is not characterized by any special properties but by a definite, specific combination of these properties.

In the course of the colossal length of time during which our planet, the Earth, has existed, the appropriate conditions must certainly have arisen in which there could have been the conjunction of properties which were formerly disjoined to form the combination which is characteristic of living organisms. To discover these conditions would be to explain the origin of life.

From Uncombined Elements to Organic Compounds

Astronomers tell us that very, very long ago, millions, perhaps even thousands of millions of years ago, the Earth existed in the form of an enormous cloud of incandescent gas. We cannot yet form any accurate idea of the extreme heat which prevailed in that cloud which represented the Earth of the future. In any case it was considerably higher than the highest which we can yet obtain artificially, temperatures at which metals such as iron boil, turning into vapour like water when it falls on a red-hot stove. Under these conditions there could be no question of the occurrence of any chemical compounds. Even at lower temperatures all the compounds we know breakdown into their simplest components, the elements. These elements exist as very small particles of matter (atoms) which were distributed freely and haphazardly in space and constituted the earliest cloud from which the Earth was later to arise. This was the primaeval chaos in which, it seems, everything was in disorderly, unrelated and irrational movement.

We can form some idea of such a state of matter from a study of the gases or vapours obtained by raising a great variety of materials to white heat. In the puffs or clouds of these vapours the individual particles of matter are also in constant motion, just as were the particles in the original cloud.

A study of matter in the vapour state has led scientists to the belief that even in the primaeval cloud there was not absolute chaos, but even there, affected by the eternally acting forces of physics, a certain order slowly but surely began to establish itself. In the first place the particles of the different elements of which the cloud was composed began to distribute themselves within it in some sort of order. Under the influence of that same force of gravity which attracts the falling stone towards the centre of the Earth, the heavier atoms in the cloud began to sink towards its centre while the lighter ones remained at its surface. Thus, the gases forming the cloud were more or less separated out into layers in it, the heavier ones being below, at the centre, while the lighter ones were on the surface.

As the cloud cooled, the material in it became denser and at a particular moment it went over from the vapour to the liquid state. In the centre of the cloud a red-hot liquid nucleus was formed, surrounded on all sides by an immense envelope of incandescent gas. It is quite reasonable that the nucleus should have been composed for the most part of the heaviest elements while the lighter ones remained in the gaseous atmosphere enclosing this nucleus.

According to the views of D. I. Mendeleev, the first nucleus of the Earth must have been composed of heavy metals, mainly iron, while further from the centre there were the alkali and the alkaline earth metals,* the metalloids and, finally, the outermost part of the atmosphere was composed of the lightest gases, mainly hydrogen. Carbon, the element which interests us

---

*The commonest alkali metals are potassium and sodium. Calcium, magnesium, barium and strontium are alkaline earth metals. The vapour density of these metals is considerably less than that of the heavy metals iron, copper, lead, mercury, silver, etc.

most, should, by this scheme, be present in the primaeval nucleus of the Earth in very close association with iron.

The idea is not the result of some theoretical speculations, but is firmly based on facts. It is easy to convince ourselves of this by going back to a study of present-day conditions. In fact, if the heaviest elements were at some time in the centre of the Earth they should still be there. That is to say, if the idea put forward above were true the heaviest elements should be found in the depths of the Earth even today.

Unfortunately, only a thin skin on the surface of the Earth is accessible to study by Man. We can only penetrate a few thousand feet into the depths of the Earth and this means that we cannot see with our own eyes what exists in its interior. However, scientists have succeeded in obtaining a very clear picture of it in other ways. They have managed, though this may seem strange at first sight, to weigh the Earth as a whole. Of course its weight is extremely great. In round figures it is 4 quadrillion kg. By direct measurements of the surface they have also determined its volume. What they found was that the weight was nearly six times that of an equal volume of water. We know, of course, that the crust of the Earth does not consist of water alone, but even in its densest parts it does not have a mean specific gravity as great as five to six times that of water. The specific gravity of the crust does not exceed two and a half. Thus, the Earth as a whole has a far higher specific gravity than has the outer part of it which is accessible for us to study. It is so heavy that its weight can only be explained on the supposition that it has a nucleus of heavy metal inside it.

This view is also supported by study of the masses of lava which are extruded from the depths on to the surface of the Earth by volcanoes. The older the mass and the deeper the region from which it poured on to the surface, the greater the amount of heavy metals it contains, especially iron.

Thus, the evidence which we can obtain by studying the state of affairs on the Earth now provides very definite confirmation of the ideas which we have put forward above. The results of studying the heavenly bodies are even more instructive.

Scientists now believe that each of the stars which we see originated, in the same way as we described for the Earth, from a gaseous cloud, by gradual cooling and increasing density of the vapours of which it was formed.

Depending on a number of factors (mainly on the size of the cloud) the process of solidification might occur quickly or slowly. The different heavenly bodies are now, therefore, at different stages of development. Some of these stars shine with a white or bluish light and are in the earliest stage of development, others, which have developed further, are yellowish and our Sun is one of these. Finally, the stars which have cooled most and are already going out shine with a red light. A further stage of cooling is represented by the planets which can no longer shine with their own light. Our Earth is one of these. Thus, a study of the different heavenly bodies gives us an idea of the different stages of cooling of our own planet.

This study has made great strides forward since Kirchhoff developed what is called the spectral method of analysis. The essence of the method is as follows. If a ray of white light is passed through a glass prism it is divided into its component parts and gives rise to a strip consisting of all the colours of the rainbow (spectrum). If we use the Sun (daylight passing through an opening in a shutter) or an electric light or any solid or liquid body heated to white heat as the source of light for our experiments, we shall obtain a continuous strip like a rainbow in which one colour gradually merges into the next. A different picture is obtained by passing light from incandescent gases through the prism. In this case the spectrum will not consist of a continuous strip but of individual lines, coloured by a particular colour and separated from one another by black spaces. In this connection it is especially interesting that each element has its own particular, completely characteristic distribution of bands in the spectrum. Thus, if we have an incandescent gas consisting of a mixture of elements we can, by studying the spectrum, find out just what elements are present in it. It is thus possible to analyse the mixture in this way.

This method of spectral analysis is so convenient and accurate that it has found wide application in the study and differentiation of many substances on the Earth, but its most important use has certainly been in astronomy as a method for studying the chemical composition of heavenly bodies.

From the furthest ends of the Universe innumerable heavenly bodies send their light to the Earth. By studying this light with a spectroscope we can determine with accuracy which elements are present in the incandescent gaseous envelope of the star in question.

Specially interesting information is obtained by spectral analysis of the gaseous envelope of the Sun. The Sun, as we have already said, is one of the yellow stars. In the centre of it there is a huge, red-hot liquid nucleus, the spectrum of which takes the form of a continuous, rainbow-coloured strip. This nucleus is surrounded on all sides by an envelope of incandescent gas, the atmosphere. Spectroscopic studies of the lowest layers of this atmosphere have shown that it is composed of the incandescent vapours of heavy metals. Here the spectroscope reveals the presence of iron, nickel, cobalt, calcium and manganese, and Roland has even succeeded in finding carbon here. A little higher, further from the centre of the Sun, there is a layer of the atmosphere in which it is easy to demonstrate the presence of the vapours of the lighter alkaline metals, potassium and sodium. Even higher, helium is predominant: this is an element which was first discovered on the Sun.* Helium is a very light gas. Of all the materials known on the Earth hydrogen is the only one which is lighter.

Hydrogen, too, is found in the solar atmosphere. It lies just above the helium. Finally, above the hydrogen there is a layer of what must be an even lighter gas, coronia, which has not yet been found on the Earth.

Thus, the idea which we put forward about the arrangement of the elements in gaseous masses by layers and which was based on purely theoretical considerations has been completely vindicated by observations on the arrangement of the elements in the atmosphere of the Sun.

There was a time when the Earth, too, was passing through the same stage of development as the Sun, namely that of being a yellow star. Later, as it gradually radiated its heat outwards into the cold interplanetary space, it became cooler and cooler. It turned from a yellow star into a red one, its light became dimmer and dimmer and finally went out altogether. The Earth became a dark planet.

What was happening during these changes to the carbon and other elements which now enter into the composition of living organisms?

We have seen that, at the time when the Earth was a yellow star the carbon was present partly in the form of incandescent vapours in the lowest layer of

*It was considerably later that the well-known British chemist Ramsay succeeded in isolating this element from an extremely rare mineral. Thus it was possible to confirm, by chemical methods, that helium had the properties which had for a long time been assigned to it on the basis of spectral analyses.

the atmosphere and partly in the molten state in the hot, liquid, central nucleus. In both places it was mixed with heavy metals, mainly iron. The temperature prevailing on the Earth at that time was still too high for any chemical compounds to be formed. The elements, though mixed with one another, remained free and did not combine with one another.

However, the Earth gradually cooled and the time must certainly have come when combination took place between the free elements which were mixed with one another. Spectroscopic studies of the red stars lead us to the conclusion that this must first have taken place when the Earth passed from the stage of being a yellow star to that of being a red one.

The first compounds to exist on the Earth must have been extremely stable to high temperatures because only compounds of this sort could have existed for long in the heat which then prevailed. The most thermostable compounds of carbon now known are its compounds with heavy metals, known as 'carbides'. The commonly known representative of this class of compounds is iron carbide.

At the period of existence of the Earth which we are considering carbon vapour was, as we have seen, mixed with the vapours of metals. There is, therefore, good reason to suppose that carbides themselves were the first compounds to arise on the Earth. The existence of other compounds at this time was unthinkable simply because of the temperature.

Only when the Earth had dissipated some of its heat and cooled still further did there arise the possibility of the formation of other compounds and they began to come into being, gradually clothing the central nucleus in a more or less thick envelope. This envelope, consisting of the substances which later gave rise to the rock formations, thus separated the carbides lying beneath them from the atmosphere of that time. However, the envelope was not very stable and in many places there must have been eruption of the internal material on to the surface of the Earth. At the same time, of course, the Earth was gradually cooling. The central nucleus shrank while its outer layers were forming a solid envelope. The latter could not follow the contraction of the underlying material and this gave rise to folds and cracks through which the red-hot carbides of metal poured out and erupted on to the surface of the Earth.

Such phenomena may be seen even now though, of course, on a much smaller scale. The eruption of volcanoes and the processes by which mountains are formed are explained by geologists in terms of just the same causes as we have given, namely the contraction of the Earth which is still continuing. It is true that the cracks which are formed now are, comparatively speaking, not very deep and the lava which erupts from volcanoes comes from a layer a long way above the central core, but it could not be otherwise, because the crust of the Earth has increased in thickness many times over since the time of the formation of the first envelope. Studies of volcanoes and masses of lava which have arisen at different periods of the existence of the Earth show that the most ancient volcanoes derived the products which they erupted from the deepest layers of the Earth.

Thus, at the period of existence of the Earth which we are considering, when it was a red star about to become extinguished, masses of carbides of iron and other metals which had formerly been concealed in its depths were being extruded on to its surface through cracks formed in its crust. Here, on the surface, they encountered the atmosphere of that time which differed in many respects from that of today. Water vapour was specially abundant in it. All the water in all the seas and oceans now on the Earth then existed in the form of superheated steam in the atmosphere. The carbides which flowed out on to the surface encountered this steam.

If we treat carbides of metals with superheated steam we obtain what are known as hydrocarbons, that is to say compounds consisting of carbon and hydrogen. These compounds must also have arisen when the carbides and steam met on the surface of the Earth. Of course some of these must immediately have been burnt, being oxidized by the oxygen of the air. However, under the conditions then prevailing this combustion must have been far from complete. Only a certain part (and a comparatively small one) of the hydrocarbons were fully oxidized, being converted to carbonic acid and water. A further part, owing to incomplete oxidation, gave rise to carbon monoxide and oxygen derivatives of hydrocarbons, while, finally, a certain proportion of the hydrocarbons completely escaped oxidation and was given off into the upper, cooler layers of the atmosphere without any alteration. The more the Earth cooled the lower became the temperature at which the interaction between the carbides and the water vapour took place, and less carbonic acid and more unoxidized hydrocarbons were formed.

Thus, the theoretical considerations put forward above lead us to the belief that at a particular time in the existence of the Earth compounds of carbon with hydrogen and oxygen were formed in its atmosphere. Let us see whether there are not facts in our natural surroundings supporting this idea.

Since the period under discussion is that during which the Earth passed through the stage of being a red star it would be quite appropriate to consider first what we know about these stars. Spectroscopic studies of the darkest red stars which are about to go out, carried out by the astronomer Vogel, led him to the conclusion that the atmospheres of these stars contain hydrocarbons. This fact was soon confirmed by several other workers.

Hydrocarbon lines have also been found in the spectra of comets, those heavenly bodies which from time to time pass through our solar system from interplanetary space. Furthermore, thanks to the studies of several scientists it has been found that cyan* (a compound of carbon and nitrogen) and carbon monoxide are present in the gases which form the tails of comets.

By origin comets are related to a further class of heavenly bodies, the meteorites. We have already discussed these in an earlier chapter. Meteorites are specially interesting because, in falling, some of them reach the surface of our Earth in a more or less undamaged state in the form of red-hot stones from the sky. Thus they are accessible to direct chemical examination. They are, so to speak, lumps of matter, samples reaching us from the boundless region of interstellar space. Analysis of meteorites and study of their composition gives us the opportunity of getting a direct knowledge of some of the materials of which the stars are made. Most meteorites consist of native iron, partly combined with carbon and sometimes containing carbon in such quantities that it has been possible to isolate it from certain falling stars in the form of diamond dust. This composition of the meteorites is an extra confirmation of the correctness of the view that carbon exists on heavenly bodies in the form of mixtures or chemical compounds with metals.

In meteorites, however, other carbon compounds have also been found. By causing samples from meteorites to incandesce by means of an electric spark, scientists have managed to show that the hydrocarbon lines are certainly present in their spectra. It has even proved possible to isolate from

---

* An obsolete term for nitrogen carbide, CN, now known to correspond to the gas C_2N_2 (J.D.B.).

some meteorites a considerable amount of hydrocarbons and to establish their nature by chemical studies.

Thus, we can demonstrate beyond doubt the presence of hydrocarbons on a number of heavenly bodies. This fact gives full support to the conclusions we had already drawn. There came a time in the life of the Earth at which the carbon which had been set free from its combination with metal and had combined with hydrogen formed a number of hydrocarbons. These were the first 'organic' compounds on the Earth.

Although only two elements, carbon and hydrogen, enter into the composition of these compounds these elements can join together in the most varied combinations and give rise to the most varied hydrocarbons. Organic chemists can now list a very large number of such compounds.

As the properties of hydrocarbons have been studied in great detail, a study of the conditions prevailing on the Earth when these compounds came into being makes it possible to put forward some suggestions as to which hydrocarbons were in fact formed. Without going into details we can only say that everything points to the view that it was the 'unsaturated' (free radical) hydrocarbons which were formed first, that is to say the most unstable members of the class we are discussing, having very large stores of chemical energy and great chemical potentialities, compounds which combine very easily both with one another and with other elements.

If these compounds could avoid oxidation at the time of their formation, then, during their stay in the hot, wet atmosphere of the Earth, they must certainly have combined with oxygen and given rise to the most varied substances composed of carbon, hydrogen and oxygen in various proportions (alcohols, aldehydes, ketones and organic acids).

Thus, all the considerations and facts which we have put forward above convince us that, even if not all the carbon, at least a great part of it, first appeared on the surface of the Earth, not in the form of the chemically inert carbon dioxide as had been thought, but in the form of unstable organic compounds capable of further transformation.

Let us leave these compounds for a while and take a look at what happened, during the period in the existence of the Earth which we are discussing, to the fourth element which enters into the composition of living things, namely nitrogen. At high temperatures nitrogen can form compounds with oxygen (the technical production of nitric acid depends on this). We are therefore justified in expecting the appearance of these compounds in the atmosphere of the Earth where the two elements involved were mixed. However, oxides of nitrogen are somewhat unstable. At temperatures of about 1,000°C these compounds break down and give off free nitrogen. Compounds of nitrogen with metals obtained industrially under conditions of white heat are far more stable. Such compounds must also have been formed in the atmosphere of the Earth by interaction between nitrogen and the incandescent vapour of the lighter metals. Later these compounds of metals with nitrogen, of a similar nature to carbides, were submitted to the action of superheated steam and formed ammonia,* which is a compound of nitrogen and hydrogen. Ammonia could also have been formed primarily at a far earlier stage in the existence of the Earth by the condensation of hydrogen and nitrogen in the upper layers of the atmosphere.

Furthermore, we cannot exclude the possibility of the formation of compounds of nitrogen and carbon. In this case the material obtained would be cyan, a substance with which we have already become acquainted when we were discussing the spectra of comets. Its presence in the gases surrounding these heavenly bodies confirms the possibility that it might also have been formed on the Earth. Thus, the atmosphere of the Earth at a certain period of its existence must have contained compounds of nitrogen in the form of ammonia and cyan as well as oxygen derivatives of hydrocarbons.

Although, from our point of view, cyan can hardly have played any important part in the further transformations of organic substances, we mention it because it is interesting to us in another way. It forms the starting-point of the extremely far-reaching and well-thought-out theory of the origin of life put forward by the well-known German physiologist Pflüger. We shall try to summarize it here. Pflüger says that there is an essential difference between dead protein, such as we find in the whites of hen's eggs, and the living protein of which living material, protoplasm, is made. This consists in the presence of cyan groups in the particles of living protein. However, cyan is only formed at red heat, when nitrogen-containing compounds react with incandescent carbon. 'For this reason nothing can be more clear than the possibility of the formation of cyan compounds at the time when the

*As everyone knows, spirits of hartshorn is a solution of ammonia in water.

Earth, as a whole or in part, was in a red-hot or white-hot state.' Only then was it possible for cyan to be formed that 'compound in which the beginning of life resides'. 'Therefore life arises from fire and its basis is derived from the time when the world was still a white-hot, incandescent globe.'

Pflüger also refers to the immeasurable length of the time during which the cooling of the Earth was taking place. 'Cyan and other organic compounds therefore had enough time and opportunities to indulge their great tendency to transformation and the formation of polymers.'* As a result of these transformations they gave rise to 'that self-transforming protein which constitutes living matter.'

The most recent studies do not confirm Pflüger's ideas as to the predominant part played by cyan. The occurrence of cyan groups in 'living proteins' has now come under grave doubt and even the concept of 'living protein' itself has become rather out of date. Protein does not form protoplasm, it only enters as a component part into this chemically and physically complicated formation. The 'self-transforming protein' as Pflüger described it, certainly does not exist. The ability of protoplasm to transform itself is inherent, not only in the chemical substances of which it is composed, but also in its physical structure or organization.

Nevertheless, Pflüger's theory has retained its importance until now. The fundamental proposition that 'life arose from fire' remains unshaken. Only in fire, only in incandescent heat could the substances which later gave rise to life have been formed. Whether it was cyan or whether it was hydrocarbons is not, in the last analysis, very important. What is important is that these substances had a colossal reserve of chemical energy which gave them the possibility of developing further and increasing their complexity. They contain, hidden within themselves, particles of that fire, that heat, that energy which the Earth so generously and prodigally scattered into the cold region of interstellar space. The hidden fire, this energy, served as the basis for the life that was to come.

---

*Polymers is a name for substances having the same elementary composition but a different weight of their particles. Polymers with large particles are usually formed by the combination of two or more smaller particles.

From the Organic Substance to the Living Thing

In the last section we left off our discussion of the Earth at the time when it was gradually cooling and going over from being a red star to being a dark planet. Finally the time came when the temperature of the surface layers of the Earth fell to 100°C. It became possible for water to exist in the form of liquid drops. Continuous downpours of rain fell upon the surface of the Earth from the moist atmosphere. They inundated it and formed a cover of water in the form of the original boiling ocean.

The first organic substances which had hitherto remained in the atmosphere were now dissolved in the water and fell to the ground with it. What were these substances ?

We have already remarked on their main property at the end of the last section. They were substances having a large store of chemical energy and possessing great chemical potentialities. While still in the terrestrial atmosphere they had begun to combine with one another to give rise to very complicated compounds. In addition, they combined with oxygen and ammonia to give hydroxy and amino-derivatives of hydrocarbons (i.e. compounds of hydrocarbons with oxygen and nitrogen respectively).

When these substances fell from the atmosphere into the primaeval ocean they did not stop interacting with one another. Individual components of organic substances floating in the water met and combined with one another. Thus ever larger and more complicated particles were formed.

We can easily create a fairly accurate picture for ourselves of this process of aggregation (polymerization) of organic substances on the Earth by studying it in our chemical laboratories. In fact, the conditions in which organic substances existed in the stage of development of the Earth which we are dealing with can be achieved comparatively easily in our present-day laboratories. If we submit such substances as hydrocarbon radicals to the conditions described above and leave them to themselves we shall find the whole chain of reactions set out above taking place. The hydrocarbon radicals will be oxidized at the expense of the oxygen in the water and air to give the greatest variety of derivatives (alcohols, aldehydes, acids, etc.). This process takes place specially quickly at high temperatures and in the presence of iron and other metals.

Oxidized hydrocarbons readily combine with one another to form more complicated compounds. Many of these substances can also combine with ammonia and give rise to the development of the most varied nitrogen derivatives.

The process of aggregation of organic substances usually occurs rather slowly it is true. However, this is not very important. Whether it takes several months or several years, we still get, as a result of these processes, a mixture of various substances having a very complicated structure.

In these mixtures we may even find, among others, compounds of the nature of carbohydrates* and proteins. Both of these types of compounds play an important part in the structure of living material. We find them in all animals and plants without exception. In combination with other and yet more complicated substances they are, as it were, the foundation of life.

Of course, the substances which we produce artificially are not exactly the ones which can be isolated from living organisms. However, they are, if we may express it so, related to these compounds. The elementary composition, the structure of the particles and the chemical properties are almost the same in the one as in the other. The difference is only in detail.

The substances obtained by the method described above can serve as good nutrient material. They are specially nutritious for micro-organisms such as bacteria and moulds. This fact is specially important and we shall give a little more time to it.

One of the main objections brought against the possibility of the spontaneous generation of life in the distant past has been put in its general aspects as follows:

'If we assume', says W. Preyer, one of the opponents of the theory of spontaneous generation, 'that at some time during the development of the Earth living material arose by primary generation from non-living material then we must suppose that this is still possible. However, the failure of numerous attempts directed towards finding out how to do this has shown that it is unlikely to the highest degree. If, on the other hand, those studying

*Carbohydrates are organic substances composed of carbon, hydrogen and oxygen, the hydrogen and oxygen being present in the same proportions as in water. The various sugars such as glucose, sucrose and fructose as well as starch and cellulose are typical examples of carbohydrates.

the first emergence of life assume that it was only possible at some time in the distant past, but now cannot take place; this is also improbable since the conditions required for pursuing life exist now and in fact must also have existed at the time when it is presumed that living material originated from inorganic substances, otherwise the product of the first origin could not have remained alive for long. It is therefore hard to see exactly what is lacking at the present time when primary generation is impossible.'

We have already seen in the last section, that at present what are lacking above all are those substances containing much chemical energy which are the only things from which life could develop and which, themselves, could only be formed at extremely high temperatures. However, even if such substances were formed now in some place on the Earth, they could not proceed far in their development. At a certain stage of that development they would be eaten, one after the other. Destroyed by the ubiquitous bacteria which inhabit our soil, water and air.

Matters were different in that distant period of the existence of the Earth when organic substances first arose, when, as we believe, the Earth was barren and sterile. There were no bacteria nor any other micro-organisms on it, and the organic substances were perfectly free to indulge their tendency to undergo transformations for many, many thousands of years.

It is, of course, hard to say what these transformations were and what sort of substances resulted from them. The only thing that is certain is that these transformations were mainly directed towards an aggregation of material and the formation of more and more complicated and larger and larger particles.

However, we have seen in one of the preceding sections that substances with large and complicated particles have a great tendency to form colloidal solutions in water. Sooner or later such colloidal solutions of organic substances must have come into being in the watery covering of the Earth and once they had arisen they continued to exist, their molecules becoming more complicated and larger as time went on.

The state of colloidal solution is not, however, stable. For various, sometimes extremely slight causes, the dissolved substances come out of the colloidal solution in the form of precipitates, coagula or gels. It is impossible, incredible, to suppose that in the course of many hundreds or even thousands of years during which the terrestrial globe existed, the conditions did not arise 'by chance' somewhere which would lead to the formation of a gel in a colloidal solution. Such formation of aggregated pieces of organic material floating freely in the boundless watery spaces of the ocean which gave rise to them must certainly have occurred at some time in the existence of the Earth.

The moment when the gel was precipitated or the first coagulum formed, marked an extremely important stage in the process of the spontaneous generation of life. At this moment material which had formerly been structureless first acquired a structure and the transformation of organic compounds into an organic body took place. Not only this, but at the same time the body became an individual. Before this it had been inseparably fused with all the rest of the world, dissolved in it. Now, however, it separated itself out, though still very imperfectly, from that world and set itself apart from the environment surrounding it.

With certain reservations we can even consider that first piece of organic slime which came into being on the Earth as being the first organism. In fact it must have had many of those features which we now consider characteristic of life. It was composed of organic substances, it had a definite and complicated structure which was completely characteristic of it. It had a considerable store of chemical energy enabling it to undergo further transformations. Finally, even if it could not metabolize in the full sense of the word, it must certainly have had the ability to nourish itself, to absorb and assimilate substances from its environment, for this is present in every organic gel.

It is hard to say precisely how the further development of this first organism went on, but still it is quite possible to establish the general direction of that development. Let us assume that in one of the corners of the Earth, in the turbulent waves of the ocean, there were formed, either at the same time or one after the other, two bits of gel. Even if they separated out from the same solution they could not have been exactly alike. In one way or another they must have differed, for absolute identity does not exist on the Earth. Both bits were formed and floated in something that was not just water. They were immersed, so to speak, in a nutrient mixture, in a solution, though a very weak one, of different substances, among which there were various organic compounds. And each of these bits of slime absorbed these substances from the medium which surrounded it. Each grew at the expense of these substances, but as each bit had a different structure from the other they assimilated the material from the environment at different rates, one faster, the other slower. The one with the physico-chemical organization which made it possible to carry out the process of assimilation of hitherto foreign substances from the environment more quickly also grew faster than its weaker, less well-organized comrade. The more it grew and the larger its surface became the wider became this difference in the rate of growth.

As this went on the danger of the piece losing its wholeness by breaking or being broken up into larger or smaller parts also increased. This must have happened to different pieces in quite different ways for purely mechanical reasons, such as the breaking of waves or surf, or it may have been due to surface tension. All the same, in one way or another, sooner or later it must have happened. The bit of gel could not go on growing for ever as a continuous mass. It must have broken down and given rise to new pieces, new 'primitive organisms'. These latter were constructed or organized just like their parent though, of course, they only constituted parts of its body, and therefore their structure was inherited by them from the gel from which they were formed.

This structure, however, was not something immutable or constant. Naturally it depended to some extent on the chemical composition of the gel, but this was changing all the time. New substances were continually entering it from the external medium, new compounds were always being incorporated into its body and its physical and chemical structure was continually changing.

The bits which were formed by the breaking up of the original gel were, of course, similar to one another at first, but after division each followed its own path. Each began to grow independently and the structure of each began to undergo changes which were peculiar to itself. This meant that even after quite a short time the sister fragments must have differed from one another in their structure. The old story must have been repeated. The more efficiently constructed ones began to grow faster and the less efficient lagged behind in their growth.

This was repeated for many, many years. The structure of one gel with all the changes which had arisen in it was acquired and inherited by all the bits which owed their origin to the break up of that gel. The newly formed bits grew more, their structure again underwent changes in one way or another and the changes were once more transmitted to their offspring.

However, in the course of this process of change, selection of the better organized bits of gel was always going on. It is true that the less well organized could grow alongside the more efficient but they must have soon stopped growing. Even when there was enough of the dissolved nutrient substances for all, the leading part was always played by the qualitatively better organized entities. The growth in mass of the gels followed a geometrical progression and therefore bodies which had even a relatively slight superiority soon outstripped their less efficient comrades in regard to their growth and development. Thus, slowly but surely, from generation to generation, over many thousands of years, there took place an improvement of the physico-chemical structure of the gels, an improvement mainly directed towards increasing the efficiency of the apparatus for absorption and assimilation of nutrient compounds. On this basis a whole series of new properties must have arisen which had been absent from the original gel, among others the ability to metabolize.

Present-day organisms burn up in their bodies a part of the material which they absorb from their environment. This is inevitable as it is only by means of the energy obtained by this burning that the further growth of their protoplasm and further assimilation of nutrients can take place. In just the same way, too, the original organisms, when they had used up a considerable amount of the energy concealed in them, had to resort to a process of respiration or fermentation to acquire the energy they needed for their further growth and development. Only those among them which, during the preceding transformations, had developed within themselves an apparatus enabling them to burn or ferment, more or less quickly, a part of the nutrient substances absorbed by them, could grow and develop further. The rest must have halted in their development.

However, there arose among those fortunate ones which had developed this power of metabolism, a fierce struggle for existence, a fight to the death. The amount of nutrient organic material in the surrounding medium was getting less. Part of it had already been absorbed by the organisms, while part was broken down, burnt up in the process of respiration or fermentation. Only the most complicated and efficient could grow and develop, all the rest either ceased to develop or perished. The further life progressed the less nutrient substances were available to the organisms and the more strongly and bitterly the struggle for existence was waged and the stricter and stricter became 'natural selection', rejecting all that was weak or backward and allowing only the most efficient to live.

At last the time came when all, or almost all, of the organic substances which had hitherto served as the only food of the original living things had disappeared. Now only those organisms which could adapt themselves to the new conditions of life were able to maintain and prolong their existence. For this purpose there were only two paths open to them: either they could continue to use their old means of nutrition, acquiring the organic substances which they needed for their nutrition by eating their weaker comrades, or they could turn in a new direction and develop, create within themselves, an apparatus which would enable them to nourish themselves on very simple inorganic compounds.

Only those living things which followed one of these courses could preserve themselves for further life. Having developed and perfected themselves further they finally gave rise to all the forms of organisms which we can now observe.

If we turn to the study of these modern forms, investigating their internal structure and getting to know their means of nutrition, we shall see that the few facts which we have in this field are in complete accord with the hypothesis as to the origin of life. The internal structure of the cells of modern organisms is always changing, becoming more efficient and more complicated. It is only because of the imperfection of our methods of study that we cannot observe this directly and that this structure appears to us as something constant, cast in a definite and final form. This is not, in fact, true. If we compare the internal structure of the cells of higher and lower organisms we shall find a considerable difference between them. The cells of bacteria or blue-green algae have a considerably simpler internal organization than those of higher animals and plants. This is because these micro-organisms which stand on the lowest step of the systematic ladder, are the direct descendants of the most ancient classes of organisms. They have halted their development and retained unchanged all the features of the structure of their distant ancestors while all other living things, by continually altering and improving themselves, have attained a more complicated form of cell-structure.

Thus, the direct study of modern organisms convinces us that those forms which have a very complicated internal organization of their cells have arisen from simpler forms by successive changes and improvements. If this is so, then we have no reason to deny that these comparatively simple forms arose, in their turn, from beings which had an even simpler organization, one which even approached that of a colloidal gel. It is true that no trace of these primitive living things now remains on the Earth, but this is no proof that they never existed. It should not be forgotten that at a certain period of the existence of the Earth they must have been completely wiped out by their more highly organized comrades.

We obtain even more interesting evidence from the study of the means of nutrition of our modern organisms. A considerable proportion of organisms – bacteria, fungi and animals – can only feed on organic substances. It must be pointed out that this is the method of nutrition used by the least highly organized living things such as the rhizomastigina and protomastigina which are regarded by all present-day systematists as representatives of the kinds of organisms which were the ancestors of all the living things on the Earth. This fact fully supports the idea that the method of consuming organic substances is the most ancient means of nutrition. The power of independent, 'autotrophic' feeding could only develop later as a result of a number of internal transformations and changes in their physico-chemical structure. It could not, however, develop all at once. Our knowledge of the means of nutrition of modern lower and higher organisms leads us to the conclusion that living things underwent many changes and tried out many possibilities before they could achieve the best form of independent nourishment or inorganic substances. Among the lowest organisms, the bacteria, we find alongside forms which are nourished solely on organic materials, other forms which have set up within themselves an apparatus which enables them to feed in another way. Here we can observe an amazing variety of modes of nutrition, means by which organisms try to extract the energy they need for life from the inorganic medium surrounding them. One of them obtains this energy by converting hydrogen sulphide dissolved in the water into sulphuric acid. For the same purpose another converts ammonia into nitrous and nitric acids while a third transforms reduced iron salts into oxidized ones. Whether we like it or not, we get the impression that all these various forms of nutrition have been devised because the organisms were forced to find some way out, something which would enable them to exist in the absence of dissolved organic materials.

However, not one of these forms of independent nutrition became widely used among organisms. All these methods of obtaining the energy required for life from terrestrial sources were found to be inefficient. A far more efficient method was one based on the absorption and use of the energy which the Sun sends to us on the Earth in the form of rays of light. All present-day green plants, from the tiny single-celled algae of our ponds to the mighty giant trees of the tropical forests use just this means of nutrition. By means of a very complicated apparatus, in which the green pigment chlorophyll plays an essential part, these organisms trap the energy of the Sun's rays which fall on them and, with its help, break down the abundant carbon dioxide in the air. As a result of this breakdown the plants acquire the possibility of using the carbon in the carbon dioxide to build up new organic substances which also serve as nourishment for them. The form of nutrition described is very efficient but also very complicated. There can be no doubt that the extremely complicated physico-chemical apparatus needed for it could only have been created as a result of a long series of transformations and alterations of the living cell. We must therefore regard it as the latest and newest form of nutrition.

With this we end our discussion of the origin of life. In our minds we have followed a long path from the incandescent atoms of carbon of the earliest nebula to the living things of our times. We have seen how it is possible to explain the origin of life while basing our ideas all the time on scientificially established facts. Of course, the explanation given here is only one of those possible. We still have very few facts available to enable us to maintain with complete certainty that the process under discussion took place in just this way and not somehow else. We still know very, very little about the structure of colloidal gels and even less about the physico-chemical structure of protoplasm. But this ignorance of ours is certainly only temporary. What we do not know today we shall know tomorrow. A whole army of biologists is studying the structure and organization of living matter, while a no less number of physicists and chemists are daily revealing to us new properties of dead things. Like two parties of workers boring from the two opposite ends of a tunnel, they are working towards the same goal. The work has already gone a long way and very, very soon the last barriers between the living and the dead will crumble under the attack of patient work and powerful scientific thought.

The Origin of Life

J. B. S. Haldane

UNTIL about 150 years ago it was generally believed that living beings were constantly arising out of dead matter. Maggots were supposed to be generated spontaneously in decaying meat. In 1668 Redi showed that this did not happen provided insects were carefully excluded. And in 1860 Pasteur extended the proof to the bacteria which he had shown were the cause of putrefaction. It seemed fairly clear that all the living beings known to us originate from other living beings. At the same time Darwin gave a new emotional interest to the problem. It had appeared unimportant that a few worms should originate from mud. But if man was descended from worms such spontaneous generation acquired a new significance. The origin of life on the Earth would have been as casual an affair as the evolution of monkeys into man. Even if the latter stages of man's history were due to natural causes, pride clung to a supernatural, or at least surprising, mode of origin for his ultimate ancestors. So it was with a sigh of relief that a good many men, whom Darwin's arguments had convinced, accepted the conclusion of Pasteur that life can originate only from life. It was possible either to suppose that life had been supernaturally created on Earth some millions of years ago, or that it had been brought to Earth by a meteorite or by microorganisms floating through interstellar space. But a large number, perhaps the majority, of biologists, believed, in spite of Pasteur, that at some time in the remote past life had originated on Earth from dead matter as the result of natural processes.

The more ardent materialists tried to fill in the details of this process, but without complete success. Oddly enough, the few scientific men who professed idealism agreed with them. For if one can find evidences of mind (in

Reprinted from J. D. Bernal, *Origin of Life*. London: Weidenfeld & Nicolson, 1967, pp. 242-249.

religious terminology the finger of God) in the most ordinary events, even those which go on in the chemical laboratory, one can without much difficulty believe in the origin of life from such processes. Pasteur's work therefore appealed most strongly to those who desired to stress the contrast between mind and matter. For a variety of obscure historical reasons, the Christian Churches have taken this latter point of view. But it should never be forgotten that the early Christians held many views which are now regarded as materialistic. They believed in the resurrection of the body, not the immortality of the soul. St Paul seems to have attributed consciousness and will to the body. He used a phrase translated in the revised version as 'the mind of the flesh', and credited the flesh with a capacity for hatred, wrath, and other mental functions. Many modern physiologists hold similar beliefs. But, perhaps unfortunately for Christianity, the Church was captured by a group of very inferior Greek philosophers in the third and fourth centuries AD. Since that date views as to the relation between mind and body which St Paul, at least, did not hold, have been regarded as part of Christianity, and have retarded the progress of science.

It is hard to believe that any lapse of time will dim the glory of Pasteur's positive achievements. He published singularly few experimental results. It has even been suggested by a cynic that his entire work would not gain a Doctorate of Philosophy today! But every experiment was final. I have never heard of any one who has repeated any experiment of Pasteur's with a result different from that of the master. Yet his deductions from these experiments were sometimes too sweeping. It is perhaps not quite irrelevant that he worked in his latter years with half a brain. His right cerebral hemisphere had been extensively wrecked by the bursting of an artery when he was only forty-five years old; and the united brain-power of the microbiologists who succeeded him has barely compensated for that accident. Even during his lifetime some of the conclusions which he had drawn from his experimental work were disproved. He had said that alcoholic fermentation was impossible without life. Buchner obtained it with a cell-free and dead extract of yeast. And since his death the gap between life and matter has been greatly narrowed.

When Darwin deduced the animal origin of man, a search began for a 'missing link' between ourselves and the apes. When Dubois found the bones of Pithecanthropus some comparative anatomists at once proclaimed that they were of animal origin, while others were equally convinced that they were parts of a human skeleton. It is now generally recognized that either party was right, according to the definition of humanity adopted. Pithecanthropus was a creature which might legitimately be described either as a man or an ape, and its existence showed that the distinction between the two was not absolute.

Now the recent study of ultramicroscopic beings has brought up at least one parallel case, that of the bacteriophage, discovered by d'Herelle, who had been to some extent anticipated by Twort. This is the case of a disease, or, at any rate, abnormality of bacteria. Before the size of the atom was known there was no reason to doubt that

> Big fleas have little fleas
> Upon their backs to bite 'em;
> The little ones have lesser ones,
> And so ad infinitum.

But we now know that this is impossible. Roughly speaking, from the point of view of size, the bacillus is the flea's flea, the bacteriophage the bacillus' flea; but the bacteriophage's flea would be of the dimensions of an atom, and atoms do not behave like fleas. In other words, there are only about as many atoms in a cell as cells in a man. The link between living and dead matter is therefore somewhere between a cell and an atom.

D'Herelle found that certain cultures of bacteria began to swell up and burst until all had disappeared. If such cultures were passed through a filter fine enough to keep out all bacteria, the filtrate could infect fresh bacteria, and so on indefinitely. Though the infective agents cannot be seen with a microscope, they can be counted as follows. If an active filtrate containing bacteriophage be poured over a colony of bacteria on a jelly, the bacteria will all, or almost all, disappear. If it be diluted many thousand times, a few islands of living bacteria survive for some time. If it be diluted about ten million fold, the bacteria are destroyed round only a few isolated spots, each representing a single particle of bacteriophage.

Since the bacteriophage multiplies, d'Herelle believes it to be a living organism. Bordet and others have taken an opposite view. It will survive heating and other insults which kill the large majority of organisms, and will multiply only in presence of living bacteria, though it can break up dead ones. Except perhaps in presence of bacteria, it does not use oxygen or display any other signs of life. Bordet and his school therefore regard it as a ferment which breaks up bacteria as our own digestive ferments break up our food, at the same time inducing the disintegrating bacteria to produce more of the same ferment. This is not as fantastic as it sounds, for most cells while dying liberate or activate ferments which digest themselves. But these ferments are certainly feeble when compared with the bacteriophage.

Clearly we are in doubt as to the proper criterion of life. D'Herelle says that the bacteriophage is alive, because, like the flea or the tiger, it can multiply indefinitely at the cost of living beings. His opponents say that it can multiply only as long as its food is alive, whereas the tiger certainly, and the flea probably, can live on dead products of life. They suggest that the bacteriophage is like a book or a work of art, which is constantly being copied by living beings, and is therefore only metaphorically alive, its real life being in its copiers.

The American geneticist Muller has, however, suggested an intermediate view. He compares the bacteriophage to a gene – that is to say, one of the units concerned in heredity. A fully coloured and a spotted dog differ because the latter has in each of its cells one or two of a certain gene, which we know is too small for the microscope to see. Before a cell of a dog divides this gene divides also, so that each of the daughter-cells has one, two, or none according with the number in the parent cell. The ordinary spotted dog is healthy, but a gene common among German dogs causes a roan colour when one is present, while two make the dog nearly white, wall-eyed and generally deaf, blind or both. Most of such dogs die young, and the analogy to the bacteriophage is fairly close. The main difference between such a lethal gene, of which many are known, and the bacteriophage, is that the one is only known inside the cell, the other outside. In the present state of our ignorance we may regard the gene either as a tiny organism which can divide in the environment provided by the rest of the cell; or as a bit of machinery which the 'living' cell copies at each division. The truth is probably somewhere in between these two hypotheses.

Unless a living creature is a piece of dead matter plus a soul (a view which finds little support in modern biology) something of the following kind must be true. A simple organism must consist of parts A, B, C, D and so on, each of which can multiply only in presence of all, or almost all, of the others. Among these parts are genes, and the bacteriophage is such a part which has got loose. This hypothesis becomes more plausible if we believe in the work of Hauduroy, who finds that the ultramicroscopic particles into which the bacteria have been broken up, and which pass through filters that can stop the bacteria, occasionally grow up again into bacteria after a lapse of several months. He brings evidence to show that such fragments of bacteria may cause disease, and d'Herelle and Peyre claim to have found the ultramicroscopic form of a common staphylococcus, along with bacteriophage, in cancers, and suspects that this combination may be the cause of that disease.

On this view the bacteriophage is a cog, as it were, in the wheel of a life-cycle of many bacteria. The same bacteriophage can act on different species and is thus, so to say, a spare part which can be fitted into a number of different machines, just as a human diabetic can remain in health when provided with insulin manufactured by a pig. A great many kinds of molecule have been got from cells, and many of them are very efficient when removed from it. One can separate from yeast one of the many tools which it uses in alcoholic fermentation, an enzyme called invertase, and this will break up six times its weight of cane-sugar per second for an indefinite time without wearing out. As it does not form alcohol from the sugar, but only a sticky mixture of other sugars, its use is permitted in the United States in the manufacture of confectionery and cake-icing. But such fragments do not reproduce themselves, though they take part in the assimilation of food by the living cell. No one supposes that they are alive. The bacteriophage is a step beyond the enzyme on the road to life, but it is perhaps an exaggeration to call it fully alive. At about the same stage on the road are the viruses which cause such diseases as smallpox, herpes, and hydrophobia. They can multiply only in living tissue, and pass through filters which stop bacteria.

With these facts in mind we may, I think, legitimately speculate on the origin of life on this planet. Within a few thousand years from its origin it probably cooled down so far as to develop a fairly permanent solid crust. For a long time, however, this crust must have been above the boiling-point of water, which condensed only gradually. The primitive atmosphere probably contained little or no oxygen, for our present supply of that gas is only about enough to burn all the coal and other organic remains found below and on the Earth's surface. On the other hand, almost all the carbon of these organic substances, and much of the carbon now combined in chalk, limestone, and dolomite, were in the atmosphere as carbon dioxide. Probably a good deal of the nitrogen now in the air was combined with metals as nitride in the Earth's crust, so that ammonia was constantly being formed by the action of water. The Sun was perhaps slightly brighter than it is now, and as there was no oxygen in the atmosphere the chemically active ultra-violet rays from the Sun were not, as they now are, mainly stopped by ozone (a modified form of oxygen) in the upper atmosphere, and oxygen itself lower down. They penetrated to the surface of the land and sea, or at least to the clouds.

Now, when ultra-violet light acts on a mixture of water, carbon dioxide, and ammonia, a vast variety of organic substances are made, including sugars and apparently some of the materials from which proteins are built up. This fact has been demonstrated in the laboratory by Baly of Liverpool and his colleagues. In this present world, such substances, if left about, decay – that is to say, they are destroyed by micro-organisms. But before the origin of life they must have accumulated till the primitive oceans reached the consistency of hot dilute soup. Today an organism must trust to luck, skill, or strength to obtain its food. The first precursors of life found food available in considerable quantities, and had no competitors in the struggle for existence. As the primitive atmosphere contained little or no oxygen, they must have obtained the energy which they needed for growth by some other process than oxidation – in fact, by fermentation. For, as Pasteur put it, fermentation is life without oxygen. If this was so, we should expect that high organisms like ourselves would start life as anaerobic beings, just as we start as single cells. This is the case. Embryo chicks for the first two or three days after fertilization use very little oxygen, but obtain the energy which they need for growth by fermenting sugar into lactic acid, like the bacteria which turns milk sour. So do various embryo mammals, and in all probability you and I lived mainly by fermentation during the first week of our pre-natal life. The cancer cell behaves in the same way. Warburg has shown that with its embryonic habit of unrestricted growth there goes an embryonic habit of fermentation.

The first living or half-living things were probably large molecules synthesized under the influence of the Sun's radiation, and only capable of reproduction in the particularly favourable medium in which they originated. Each presumably required a variety of highly specialized molecules before it could reproduce itself, and it depended on chance for a supply of them. This is the case today with most viruses, including the bacteriophage, which can grow only in presence of the complicated assortment of molecules found in a living cell.

The unicellular organisms, including bacteria, which were the simplest living things known a generation ago, are far more complicated. They are organisms – that is to say, systems whose parts co-operate. Each part is specialized to a particular chemical function, and prepares chemical molecules suitable for the growth of the other parts. In consequence, the cell as a whole can usually subsist on a few types of molecule, which are transformed within it into the more complex substances needed for the growth of the parts.

The cell consists of numerous half-living chemical molecules suspended in water and enclosed in an oily film. When the whole sea was a vast chemical laboratory the conditions for the formation of such films must have been relatively favourable; but for all that life may have remained in the virus stage for many millions of years before a suitable assemblage of elementary units was brought together in the first cell. There must have been many failures, but the first successful cell had plenty of food, and an immense advantage over its competitors.

It is probable that all organisms now alive are descended from one ancestor, for the following reason. Most of our structural molecules are asymmetrical, as shown by the fact that they rotate the plane of polarized light, and often form asymmetrical crystals. But of the two possible types of any such molecule, related to one another like a right and left boot, only one is found throughout living nature. The apparent exceptions to this rule are all small molecules which are not used in the building of the large structures which display the phenomena of life. There is nothing, so far as we can see, in the nature of things to prevent the existence of looking-glass organisms built from molecules which are, so to say, the mirror-images of those in our own bodies. Many of the requisite molecules have already been made in the laboratory. If life had originated independently on several occasions, such organisms would probably exist. As they do not, this event probably occurred only once, or, more probably, the descendants of the first living organism rapidly evolved far enough to overwhelm any later competitors when these arrived on the scene.

As the primitive organisms used up the foodstuffs available in the sea some of them began to perform in their own bodies the synthesis formerly performed haphazardly by the sunlight, thus ensuring a liberal supply of food. The first plants thus came into existence, living near the surface of the ocean, and making food with the aid of sunlight as do their descendants today. It is thought by many biologists that we animals are descended from them. Among the molecules in our own bodies are a number whose structure resembles that of chlorophyll, the green pigment with which the plants have harnessed the sunlight to their needs. We use them for other purposes than the plants – for example, for carrying oxygen – and we do not, of course, know whether they are, so to speak, descendants of chlorophyll or merely cousins. But since the oxygen liberated by the first plants must have killed off most of the other organisms, the former view is the more plausible.

The above conclusions are speculative. They will remain so until living creatures have been synthesized in the biochemical laboratory. We are a long way from that goal. It was only this year that Pictel for the first time made cane-sugar artificially. It is doubtful whether any enzyme has been obtained quite pure. Nevertheless I hope to live to see one made artificially. I do not think I shall behold the synthesis of anything so nearly alive as a bacteriophage or a virus, and I do not suppose that a self-contained organism will be made for centuries. Until that is done the origin of life will remain a subject for speculation. But such speculation is not idle, because it is susceptible of experimental proof or disproof.

Some people will consider it a sufficient refutation of the above theories to say that they are materialistic, and that materialism can be refuted on philosophical grounds. They are no doubt compatible with materialism, but also with other philosophical tenets. The facts are, after all, fairly plain. Just as we know of sight only in connection with a particular kind of material system called the eye, so we know only of life in connection with certain arrangements of matter, of which the biochemist can give a good, but far from complete, account. The question at issue is: 'How did the first such system on this planet originate?' This is a historical problem to which I have given a very tentative answer on the not unreasonable hypothesis that a thousand million years ago matter obeyed the same laws that it does today.

This answer is compatible, for example, with the view that pre-existent mind or spirit can associate itself with certain kinds of matter. If so, we are left with the mystery as to why mind has so marked a preference for a particular type of colloidal organic substances. Personally I regard all attempts to describe the relation of mind to matter as rather clumsy metaphors. The biochemist knows no more, and no less, about this question than anyone else. His ignorance disqualifies him no more than the historian or the geologist from attempting to solve a historical problem.

From The Rationalist Annual, 1929.

ON THE EARLY CHEMICAL HISTORY OF THE EARTH AND THE ORIGIN OF LIFE

By Harold C. Urey

Institute for Nuclear Studies, University of Chicago

Communicated January 26, 1952

In the course of an extended study on the origin of the planets[1] I have come to certain definite conclusions relative to the early chemical conditions on the earth and their bearing on the origin of life. Oparin[2] has presented the arguments for the origin of life under anaerobic conditions which seem to me to be very convincing, but in a recent paper Garrison, Morrison, Hamilton, Benson and Calvin,[3] while referring to Oparin, completely ignore his arguments and describe experiments for the reduction of carbon dioxide by 40 m. e. v. helium particles from the Berkeley 60-inch cyclotron. As I believe these experiments, as well as many previous ones using ultra-violet light to reduce carbon dioxide and water and giving similar results to theirs, are quite irrelevant to the problem of the origin of life, I wish to present my views.

During the past years a number of discussions on the spontaneous origin of life have appeared in addition to that by Oparin. One of the most extensive and also the most exact from the standpoint of physical chemistry is that by Blum.[3a]   It seems to me that his discussion meets its greatest difficulty in accounting for organic compounds from inorganic sources. This problem practically disappears if Oparin's assumptions in regard to the early reducing character of the atmosphere are adopted.

In order to estimate the early conditions of the earth, it is necessary to ask and answer the questions of how the earth originated, and how the primitive earth developed into the present earth.   The common assumption is that the earth and its atmosphere have always been as they are now, but if this is assumed it is necessary to account for the present highly oxidized condition by some processes taking place early in the earth's history. Briefly, the highly oxidized condition is rare in the cosmos and exists in the surface regions of the earth and probably only in the surface regions of Venus and Mars.   Beyond these we know of no highly oxidized regions at all, though undoubtedly other localized regions of this kind exist.   This is essentially the argument of Oparin.

The surface of the moon gives us the most direct evidence relative to the origin of the earth.   Gilbert[4] called attention to the great system of ridges and grooves radiating from Mare Imbrium, and Baldwin[5] has explained this as due to a colliding planetesimal some hundred kilometers in radius. My own studies show that the object contained metallic iron objects and and silicate materials and that water as such or as hydrated silicates arrived with such objects.   This collision occurred during the terminal stage of the moon's formation.   Some five other similar objects left their marks on the moon's surface and they all fell within a time span of some 10^5 years. Their temperatures were not high; probably not appreciably higher than present terrestrial temperatures.   The arguments for these conclusions are long and detailed and cannot be repeated here.

At the time such objects were falling on the moon similar objects fell on the earth.   The conditions were different because of the greater energy of such objects, 22 times as great if they fell from a great distance and 11 times as great if they fell from the circum surface orbit, and because of the presence of a substantial atmosphere on the earth.   The energy was sufficient to completely volatilize the colliding planetesimals and raise the gas to greater than $10,000°K$.   An object similar to the Imbrium planetesimal would have distributed its material over a region several thousand kilometers in linear dimensions and the explosion cloud would have risen far above the atmosphere.   Its materials would have fallen through the atmosphere in the form of iron and silicate rains and would have reacted with the atmosphere in the process.   (H. H. Nininger recently showed me spherical iron-nickel objects collected near Meteor Crater, Ariz., which were formed in such a rain.)   The objects contained metallic iron-nickel alloy, silicates, graphite, iron carbide, water or water of crystallization, ammonium salts and nitrides, that is substances which would supply the volatile and non-volatile constituents of the earth. The temperatures produced in these collisions were very high, but unless the accumulation was very rapid indeed the general temperature of the planet was not excessively high. That such objects fell on the moon and earth at the terminal stage of their formation I regard as certain, and it is difficult in this subject to be certain about anything. But regardless of the detailed arguments, those who postulate oxidizing conditions as the initial state of the earth should present some similar argument to justify their assumption.

The reactions taking place at that time of interest to us here are:

$$FeO + H_2 = Fe + H_2O; \qquad K_{298} = 1.7 \times 10^{-3}, \; K_{1200} = 0.97$$
$$Fe_3O_4 + H_2 = 3FeO + H_2O; \qquad K_{298} = 2.5 \times 10^{-2}, \; K_{1200} = 1.0$$
$$C + H_2O = CO + H_2; \qquad K_{298} = 10^{-16}, \qquad K_{1200} = 3.8 \times 10^{-6}$$
$$C + 2H_2 = CH_4; \qquad K_{298} = 7.8 \times 10^8, \; K_{1200} = 1.6 \times 10^{-2}$$
$$Fe_3C + 2H_2 = Fe + CH_4; \qquad K_{298} = 3.2 \times 10^{11}, \; K_{1200} = 5.9 \times 10^{-3}$$
$$NH_3 = \tfrac{1}{2}N_2 + \tfrac{3}{2}H_2; \qquad K_{289} = 1.2 \times 10^{-3}.$$

From these equilibrium constants one sees that hydrogen was a prominent constituent of the primitive atmosphere and hence that methane was as well. Nitrogen was present as nitrogen gas at high temperatures but may have been present as ammonia or ammonium salts at low temperatures. At high temperatures hydrogen would escape from the planet very rapidly, but if the temperatures were high objects must have arrived rapidly to replenish the lost hydrogen. If the objects arrived slowly then the temperatures were low and hydrogen did not escape rapidly. Thus it is very difficult to see how the primitive atmosphere of the earth contained more than trace amounts of other compounds of carbon, nitrogen, oxygen and hydrogen than CH_4, H_2O, NH_3 (or N_2) and H_2.

We now consider what could reasonably be expected to convert this atmosphere into the present one existing on the earth.* If there was a large

* Poole, J. H. J. (*Proc. Roy. Soc. Dublin*, **22**, 345 (1941)), Harteck, P., and Jensen, J. H. D. (*Z. Naturforschung*, **3a**, 581 (1948)), and Dole, M. (*Science*, **109**, 77 (1949)) have reviewed the modern evidence for the photochemical origin of free oxygen and the evidence against the photosynthetic origin. Their conclusions are accepted here. In a recent paper, Poole, J. H. J. (*Sci. Proc. Roy. Dublin Acad.*, **25**, 201 (1951)) definitely concludes that methane and ammonia were not present in the primitive atmosphere and that it consisted of H_2O, CO_2 and N_2 but contained no oxygen. I cannot accept this conclusion for carbon dioxide and nitrogen are almost as difficult to understand as free oxygen. The interior of the earth and the lavas which reach its surface are highly reducing and are not a likely source of highly oxidized materials. Poole assumes that carbon dioxide was present in large quantities in the primitive atmosphere. How was it produced from methane, graphite or iron carbide? He also assumes that it would remain in the atmosphere, but as is shown in the text and the references cited, carbon dioxide in the presence of water should react rapidly with silicates until its partial pressure reaches values in the neighborhood of those on the earth and Mars.

amount of hydrogen, the outer parts of the atmosphere beyond the convection zone would become highly enriched in hydrogen. Hydrogen would absorb light from the sun in the far ultra-violet[6] and since it does not radiate in the infra-red would become a high-temperature atmosphere just as exists on the earth at present, and hydrogen would be lost very rapidly. As hydrogen was depleted, the atmosphere would become a methane one and since methane and its photochemical disintegration products absorb a wide band of energy in the ultra-violet and radiate in the infra-red, the temperature of the high atmosphere would fall far below the present temperature of 1500° or 2000°C.[7] In fact, it might well approach present terrestrial surface or even lower temperatures. The loss of hydrogen would be decreased, but since an oxidized atmosphere is present on the earth it can be assumed that it escaped at some appropriate rate. The net process was the dissociation of water into hydrogen which escaped, and into oxygen which oxidized reduced carbon compounds to carbon dioxide, ammonia to nitrogen, and reduced iron to more oxidized states. When the methane and ammonia were oxidized free oxygen appeared and the present atmospheric conditions were established. As carbon dioxide was formed it reacted with silicates to form limestones, i.e.,

$$CaSiO_3 + CO_2 = CaCO_3 + SiO_2, \qquad K_{298} = 10^8.$$

Of course the silicates may have been a variety of minerals but the pressure of CO_2 was always kept at a low level by this reaction or similar reactions just as it is now. Plutonic activities reverse the reaction from time to time, but on the average the reaction probably proceeds to the right as carbon compounds come from the earth's interior,[8] and in fact no evidence for the deposition of calcium silicate in sediments seems to exist.†

The histories of Mars and Venus should be similar to that of the earth. Mars has no mountains higher than 750 meters. Thus the initial lunar type mountains were probably eroded by water and no folded mountains have been formed. The oxidation of methane to carbon dioxide and the

† The alternative to this course of events would be the production of carbon dioxide from the earth's interior and there is evidence that at least some oxidized carbon is so produced. Gases escaping from lava lakes of Hawaiian volcanoes are highly oxidized, so much so that it is difficult to account for the high states of oxidation (Day, A. L., and Shepherd, E. S., *J. Wash. Acad. Sci.*, **3**, 457 (1913); Shepherd, E. S., *Bull. Hawaiian Volcano Obs.*, **VII**, 97 (1919); *Ibid.*, **VIII**, 65 (1920).) It seems probable to the writer that atmospheric oxygen directly or indirectly is responsible for the oxidation. Observers report burning gases as escaping from lavas, thus indicating the escape of reduced gases from the lava. Also surface oxidation of the lava pool must occur, as is shown by the high temperatures in the surfaces of such pools. Ferric oxide is probably produced and this in turn would oxidize carbon within the magma. (See *Bull. Hawaiian Volcano Obs.*, **IX**, 113 (1921).) If atmospheric oxygen is the source, then the carbon dioxide does not represent oxidized carbon being supplied to the atmosphere, but on the contrary it represents reduced carbon being so supplied.

formation of limestone proceeded as on earth, but oxygen atoms or water molecules were lost from the planet. In an atmosphere containing oxygen and nitrogen, a high temperature should have existed on Mars just as exists on the earth, and due to the smaller gravitational field atoms of atomic weight, 16, should escape if atoms of atomic weight, 4, i.e. He, escape from the earth now, as they do. Finally, there results a desert planet with very small amounts of water and a pressure of carbon dioxide in its atmosphere about equal to that on the earth,[9] the excess carbon dioxide having reacted with the silicates to form limestones until its partial pressure was reduced to a low value. Since mountains are absent, volcanic activity must be small or non-existent and carbon dioxide has not been generated from limestone and silicon dioxide.

Venus started with a reduced atmosphere, which was oxidized to carbon dioxide and limestone by photochemical action. It cannot lose water, however, and the absence of water means that much less water was present initially than in the case of the earth, probably due to having been formed nearer to the sun and thus at a higher temperature. Assuming plutonic activity on Venus, the carbon dioxide has been regenerated by processes similar to those of the earth. In the absence of water the reaction of carbon dioxide with silicates is very slow and hence a dense atmosphere of carbon dioxide is possible. Thus a reasonable course of events can be postulated for this planet.‡

The Origin of Life.—The problem of the origin of life involves three separate questions in our present discussion: (1) the spontaneous formation of the chemical compounds which form the physical bodies of living organisms; (2) the evolution of the complex chemical reactions which are the dynamic basis of life; and (3) the source of free energy which alone can maintain the chemical reactions and synthesize the chemical compounds. It is only the first and third questions which will be discussed here. At present the source of free energy is sunlight through photosynthesis, but how were primitive living organisms maintained through a long enough period of time for the evolution of photosynthesis to occur?

It is suggested here that life evolved during the period of oxidation of highly reduced compounds to the highly oxidized ones of today. During

‡ The discussion up to this point, together with the suggestion that life originated during the period of oxidation of reduced carbon compounds to oxidized ones, was presented before the Geological Society of America, Washington, November, 1950. It was thought that there might be some condition of pressure, temperature and composition such that organic compounds became stable, thus making the synthesis of complex compounds possible. The general ideas were discussed with Dr. H. E. Suess, who was a Fellow at the Institute for Nuclear Studies. He made some studies relative to this problem which did not appear to be promising of positive results. The present paper follows a somewhat different approach to the problem by assuming the synthesis of organic compounds by means of ultra-violet light in the high atmosphere.

this period compounds of carbon, oxygen, hydrogen, and nitrogen were present in substantial amounts. It is also suggested that the source of free energy was the absorption of ultra-violet light in the high atmosphere by methane and water and other compounds produced from them by this photochemical action. This process also protected primitive organisms from ultra-violet light in the absence of an ozone layer as exists now.

Organic compounds are generally unstable relative to completely reduced or completely oxidized compounds throughout the entire range of hydrogen and oxygen pressures in chemical equilibrium with water at ordinary temperatures, but photochemical processes should produce such compounds. The exact conditions obtaining in a methane atmosphere can only be roughly estimated. Convection to higher altitudes than now occur on the earth should have been present, since a methane atmosphere radiates in the infra-red while an oxygen-nitrogen atmosphere does not. The atmosphere should have been cooler at high altitudes so that convection extended to higher levels than now, and water vapor should have been carried to high altitudes and photochemical products of the high atmosphere should have been moved rapidly downward.

The photochemical processes in a pure methane atmosphere can be estimated qualitatively. Methane absorbs in the ultra-violet below 1450 A. The total energy of the present solar spectrum below this wavelength is about 5×10^{-6} of the total energy. Methane dissociates into methyl and atomic hydrogen. Methyl probably absorbs at much longer wave-lengths and probably repulsive states exist resulting in the formation of methylene. This compound likewise would be dissociated into CH. Thus the reactions

$$CH_4 + h\nu \ (\lambda < 1500) = CH_3 + H$$
$$CH_3 + h\nu \ (\lambda < 2800) = CH_2 + H$$
$$CH_2 + h\nu \ (\lambda < 2800) = CH + H$$
$$CH \ + h\nu \ (\lambda < 2800) = C + H$$

will occur and with their reversal and the reaction of the primary products with each other and with secondary products a steady state of great complexity will be established, the details of which cannot be estimated because of many unknown factors. The absorption spectra of CH_3 and CH_2 and the kinetics of the back reactions and other reactions are unknown. The fraction of the sun's energy below 2000 A is 3.3×10^{-4} and below 2500 A is 2.2×10^{-3}, so that very appreciable dissociation of CH_3 and CH_2 may be expected. The energies per year cm.$^{-2}$ of the earth's surface for complete absorption of the sun's spectrum at the earth below 1500 A, 2000 A and 2500 A are:

	Cal. yr.$^{-1}$ cm.$^{-2}$
1500	1.6
2000	85
2500	570

Not much methyl would be produced directly but secondary reactions such as $CH + CH_4 = CH_3 + CH_2$ could be expected to produce more of this radical. Altogether a very considerable absorption of solar energy would be expected.

In a pure water atmosphere dissociation of water into hydroxyl and atomic hydrogen would occur by absorption of light below 1900 A. Also dissociation of hydroxyl should occur giving atomic oxygen. Secondary reactions should produce O_2, H_2O_2, HO_2 and O_3 in amounts very difficult to estimate.

A combined methane and water atmosphere quite obviously would give a great variety of compounds of carbon, hydrogen and oxygen. In particular the reaction

$$CH_4 + OH = CH_3 + H_2O$$

would occur, thus producing larger quantities of methyl. These compounds would move by convection to lower levels in some quantities, dissolve in rain water and produce solutions of organic compounds in the oceans. Here ammonium salts should have been present (see below), and and the formation of nitrogen-containing compounds would occur. Given time, some natural catalysts and very slow destruction of organic compounds because of the absence of living organisms, a large number of organic compounds would be expected. If all the present surface carbon were dissolved in the present oceans as organic compounds, the oceans would become approximately a one per cent solution of these compounds. Thus compounds suitable for living organisms were possible and probably abundant.

Though the conditions postulated above are not approximated in any past experiments so far as I have been able to determine, the extensive studies on photochemistry, free radicals produced by various methods and the effects of electrical discharges on chemical substances[10] leave no doubt that many compounds would be formed due to the absorption of ultra-violet light.

A free energy supply for primitive living organisms is necessary, for only in this way can an active metabolism be supported and in the absence of such metabolism only dead and not living organic substance is possible. Rabinowitch[11] estimates that the present annual energy from photosynthesis is 600 cal. cm.$^{-2}$ of the earth's surface. The following table shows the standard free energies for three types of reactions of carbon-hydrogen-oxygen compounds of different oxidation states of the carbon atom.

COMPOUND	DISPROPORTION TO CH_4, CO_2 AND H_2O	OXIDATION TO CO_2 AND H_2O	REDUCTION TO CH_4 AND H_2O
CH_4	0	-195.50	0
CH_3OH	-22.31	-168.94	-30.13
CH_2O	-27.0	-124.75	-42.63
CH_2O_2	-21.83	-70.71	-45.28
CO_2	0	0	-31.26
$^1/_6C_6H_{12}O_6$	-17.23	-114.97	-32.85

It is apparent from this table that substantial quantities of free energy are available from the disproportion and reduction reactions of organic compounds, i.e., reactions possible under reducing conditions, though the free energies of the oxidation reactions possible in the present oxidizing atmosphere are much larger. It is not intended to infer that the reactions listed are necessarily the ones used by primitive life, but only that they indicate the order of magnitude of the free energies available from similar reactions. (Yeast uses a disproportion reaction,

$$^1/_6C_6H_{12}O_6(aq.) = {}^1/_3C_2H_5OH + {}^1/_3CO_2, \qquad \Delta F^\circ_{298} = -8.68$$

as a source of its free energy.) The high energy photochemical reactions of the reducing atmosphere at high altitudes could not be highly effective because of back reactions and because of the high energy used for the elementary processes ($\sim$150,000 cal.). Hence, the free energy supply for the primitive life processes was much less than that of the present time probably not more than 10^{-3} or 10^{-4} as much. However, experimentation on metabolic processes was possible and probably proceeded on a substantial scale. Also, a great advantage accrued to the mutations producing photosynthesis, thus ensuring survival of these processes.

Porphyrins probably appeared during the reducing period as important constituents of enzymes. Also during this time, photosynthesis evolved, and as oxidizing conditions were established green plants became the fundamental, even if they are not the dominant, type of life. In this way the evolution of photosynthesis was possible before the free energy due to it was available to living organisms.

Time and Conditions of Transition Period.—The order in which reduced substances were oxidized depends on the free energies of the oxidation reactions,

$$^1/_2CH_4(g.) + O_2 = {}^1/_2CO_2 + H_2O, \qquad \Delta F^\circ_{298} \text{ kcal.} = -97.75$$
$$^4/_3NH_3(g.) + O_2 = {}^2/_3N_2 + 2H_2O, \qquad \Delta F^\circ_{298} \text{ kcal.} = -110.73$$
$$^1/_2H_2S + O_2 = {}^1/_2H_2SO_4 \text{ (1 molar)}, \qquad \Delta F^\circ_{298} \text{ kcal.} = -84.72$$
$$^1/_2FeS + O_2 = {}^1/_2 FeSO_4 \text{ (aq.)}, \qquad \Delta F^\circ_{298} \text{ kcal.} = -86.7.$$

Thus ammonia should oxidize first, then methane, and hydrogen sulfide third. However, if ammonium salts of organic acids are possible the stability of reduced nitrogen is greatly increased, and thus methane and ammonium ion would be oxidized more or less together. This is shown by the reaction,

$$^4/_3NH_4OOCCH_3 + O_2 = {}^2/_3N_2 + 2H_2O + {}^4/_3CH_3COOH, \quad \Delta F^o_{298} = -97.83.$$

The last two reactions show that sulfide sulfur will be oxidized after the methane and ammonium ion have been oxidized. This conclusion is supported by other oxidation-reduction reactions of sulfur and carbon.

The amount of carbon on the earth's surface is about 350 g. atoms cm.$^{-2}$ equivalent to 1.4×10^3 moles cm.$^{-2}$ or 2.5×10^4 g. cm.$^{-2}$ of water, if all carbon was initially present as CH_4 and all had been oxidized to CO_2. But carbon is produced from the earth's interior and its state of reduction is less than that of methane and part of the surface carbon is not now oxidized to carbon dioxide. Hence the total amount of water which has been decomposed in order to oxidize the carbon is more nearly half of the above value, or 700 moles cm.$^{-2}$ of water. Using the energy per year absorbed by the atmosphere in wave-lengths below 1500 A, i.e., 1.6 cal. yr.$^{-1}$ cm.$^{-2}$, and assuming that every quantum produces a hydrogen atom with 200,000 cal. per/ gram atom and that the hydrogen atom escaped from the earth, 10^8 years would be required for the oxidation process. If effectively all the energy below 2000 A is utilized in this way, the time would be 2.5×10^6 years. Neither assumption is realistic and no estimate can be made in this way except that the time might be either very long or comparatively short.

The hydrogen must escape in order for an oxidized atmosphere to be established, and if the methane-water atmosphere is cold, escape will be difficult. Interpolating from Spitzer's calculations[12] the escape of the required amount of hydrogen would require about 2×10^9 years, if atomic hydrogen escaped at 325°K. with an effective surface partial pressure of 10^{-3} atmosphere and 2×10^6 years if the surface pressure was one atmosphere. In the latter case the escape formula is not a good approximation but the time would be short nevertheless. If escape was by molecular hydrogen the temperature must be 650°K. for the same times of escape. The temperature of the methane water atmosphere at high altitudes was probably less than 325°K. and hence a long time for the escape of hydrogen from the reducing atmosphere is indicated.

Thode[13] and his coworkers have found that the ratios of the sulfur isotopes in the sulfides, elementary sulfur and sulfates are closely the same as this ratio in meteoritic sulfur until about 8×10^8 years ago, and after this time the sulfur and sulfides contain increasing amounts of S^{32} with time while the sulfates contain less amounts of this isotope. They ascribe this to the action of living organisms in promoting the oxidation and reduction of sulfur compounds, thus leading to a progressive separation of the isotopes, and suggest that life evolved about 8×10^8 years ago. This is a very interesting suggestion and may be a correct conclusion. It does not give a very long time for the evolution of the very comlpex organisms whose remains are found in the Cambrian rocks.

On the other hand, this date might mark the transition from the reducing to the oxidizing atmospheric conditions. The oxidation of sulfur and sulfides to sulfates would probably not occur to any large extent until free oxygen appeared or until photosynthesis was well developed. But limestones were deposited in large quantities early in the earth's history and graphite was not. The two reactions,

$$CaCO_3 + SiO_2 + 4H_2(g.) = CH_4 + 2H_2O + CaSiO_3, \qquad K = 3 \times 10^{16}$$

and

$$C + 2H_2(g.) = CH_4(g.), \qquad K = 8 \times 10^8,$$

make possible an estimate of the pressure of hydrogen and methane that would make these two events possible at the same time. The pressure of methane would only be 5 atmospheres if all the present surface carbon were methane and if part of this carbon were dissolved as organic compounds in the oceans, the partial pressure might well be about one atmosphere. Then, if limestones were deposited, the hydrogen partial pressure was less than 10^{-4} atmosphere if equilibrium existed, but was probably higher since complete equilibrium cannot be expected. The second equation shows that graphite would not be stable under these conditions. As the methane was consumed the hydrogen pressure must have decreased. The partial pressure of carbon dioxide was 10^{-8} atmosphere if calcium carbonate was present, and may have been higher than this as it is today. If the hydrogen pressure fluctuated and for brief periods exceeded some critical pressure, massive deposits of limestones would be possible, but organisms which experimented with calcareous shells would have had great difficulty in preventing the dissolution of their shells during these periods and the extinction of their species, and indeed no certain calcareous fossils have been found in the Precambrian.

It seems just barely possible that reducing conditions were maintained until some 8×10^8 years ago. Limestones could be deposited, graphite need not have been formed, living organisms might find some 10^{-3} atmosphere or even less of hydrogen with photochemically oxidized organic compounds sufficient for their metabolic processes, and the methane pressure could not have been above 5 atmospheres and was maintained at lower pressures by the solubility of the oxidized organic compounds in water. The precipitation of limestones in great quantities presents a difficulty to the hypothesis of a long period during which a reducing atmosphere was present. The presence of highly oxidized iron in "red beds" and hematite (Fe_2O_3) iron ores are justly regarded as evidence for atmospheric oxygen. Red beds apparently are unknown earlier than the late Precambrian.

Most of the great bodies of iron ore were laid down in the late Precambrian (Huronian) or were extensively eroded during this time. The iron ore of the Vermillion range of Minnesota is much earlier (Keewatin) and thus oxidation of ferrous iron to ferric oxide took place early in the earth's history.[14] It should be noted that ferrous oxide should be oxidized to magnetic iron oxide by water if the temperature is sufficiently high to make the reaction fast enough with respect to the time available. However, magnetic iron oxide cannot be oxidized to ferric oxide by water unless the hydrogen is removed. The relations are shown by the reactions,

$$3FeO \ + H_2O = \ Fe_3O_4 + H_2; \qquad K_{298} \ = \ 10^8, \ K_{500} \ = \ 10^4$$
$$2Fe_3O_4 + H_2O = 3Fe_2O_3 + H_2; \qquad K_{298} \ = \ 10^{-6}, K_{500} \ = \ 10^{-5}$$

Thus circulating hot water could produce ferric oxide even in the absence of free oxygen, but it would probably be a rare event. The magnetic iron oxide is formed when iron is corroded by water in boilers. The conditions of deposition of these ores of the Precambrian are not well understood, though as stated above the presence of highly oxidized iron justifies a strong presumption of an oxidizing atmosphere. Throughout the calculations it has been assumed that thermodynamic equilibria will be attained except for photochemical effects. This need not be the case and the presence of living organisms almost certainly would lead to important deviations from such equilibrium.

The red bacteria and some species of algae are able to use hydrogen and carbon dioxide in photosynthesis. This ability to use hydrogen is especially interesting because they do not find hydrogen available in their natural habitats. They appear to be living fossils from some former time and would live under conditions outlined above, though they prefer higher pressures of hydrogen than 10^{-3} atmosphere. Incidentally, modern plants prefer higher concentrations of carbon dioxide than those available in nature.§ If the present atmosphere should slowly change to a reducing one, it is certain that a substantial flora and fauna would survive. The flora would surely include many green plants and the fauna most of the principal orders of animals with the exception of the mammals and birds, i.e., the warm-blooded animals, for whom the reduced free energy supply would probably be fatal. A few aerobes would probably survive wherever photosynthesis was very intense. Aerobic organisms must naturally be most abundant under aerobic conditions, but mutations would surely supply anaerobic ones for life in a reducing atmosphere.[15]

Poole[16] thinks that oxygen may have been absent from the earth's atmosphere for some 10^9 years of the earth's history, but according to the evidence given here his model of the primitive atmosphere is not correct and hence his conclusion does not substantiate the present work. He shows that Tammann's thermal dissociation is not correct and that photochemical dissociation and the escape of hydrogen are necessary for the formation of free oxygen.   It is contended here that this mechanism is necessary to account for carbon dioxide as well.   Lane[17] has argued that free oxygen did not appear until late Precambrian times because of the reduced condition of the Keewatin Greenstone schists.   MacGregor[18] comes to similar conclusions from the Precambrian rocks of Rhodesia.   He suggests that the Precambrian iron deposits were concentrated from igneous surface rocks by solution of iron in the absence of free oxygen as ferrous carbonate which was precipitated as ferric oxide by the action of green plants in a lake or sea into which the rivers ran.   The plants may have been diatoms and hence the well-known banded structure of hematite and jasper may have been produced.   As indicated above, the origin of these deposits is not well understood and therefore these suggestions, while worthy of consideration, cannot be regarded as conclusive.

The general course of events and the favorable condition for the origin of life outlined in this paper in no way depend on the time of transition from reducing to oxidizing conditions being exactly some 8×10^8 years ago.   However, the evolution from inanimate systems of biochemical compounds, e.g., the proteins, carbohydrates, enzymes and many others, of the intricate systems of reactions characteristic of living organisms, and of the truly remarkable ability of molecules to reproduce themselves seems to those most expert in the field to be almost impossible.   Thus a time from the beginning to photosynthesis of two billion years may help many to accept the hypothesis of the spontaneous generation of life.   On the other hand, our judgment of an approximate time for the origin of life certainly is not so precise that we can say that 2×10^9 years are sufficient but 2×10^8 years are not.

It seems to me that experimentation on the production of organic compounds from water and methane in the presence of ultra-violet light of approximately the spectral distribution estimated for sunlight would be most profitable.   The investigation of possible effects of electric discharges on the reactions should also be tried since electric storms in the reducing atmosphere can be postulated reasonably.

Also theoretical investigations on hydrogen and methane-water atmospheres would be most helpful in estimating the time of transition from the reducing to the oxidizing atmosphere.   Most interesting in this connection would be more experimental data such as those of Dr. Thode and his coworkers on the abundance of the sulfur isotopes.   The time of transition should be recorded in the rocks, and some such indication as that observed, by Thode, or some change in the state of oxidation of some elements should occur and should be detectable providing the time of transition was not too early in the earth's history.

I have profited greatly from discussions of this subject with Professor

James Franck, who has often pointed out to me and others that complex organic compounds, even porphyrins, may have originated under approximately the conditions outlined in this paper.

Note added in proof: Since this paper was written an interesting paper by J. D. Bernal on the Physical Basis of Life (London, Routledge and Kegan Paul, 1951) has come to my attention, in which very similar suggestions have been made, but there are differences in details and the arguments used. His paper is worthy of serious study.

§ I am indebted to Dr. H. Gaffron for information relative to these bacteria and algae.

[1] Urey, H. C., *Geochim. et Cosmochim. Acta*, **1**, 209 (1951). This has been revised and extended in *The Planets: Their Origin and Development*, Yale University Press, 1952.

[2] Oparin, A. I., *Origin of Life*, Macmillan Co., New York, 1938.

[3] Garrison, W. M., Morrison, D. C., Hamilton, J. G., Benson, A. A. and Calvin, M., *Science*, **114**, 416 (1951).

[3a] Blum, H. F., *Time's Arrow and Evolution*, Princeton University Press, 1951.

[4] Gilbert, G. K., *Bull. Phil. Soc. Wash.*, **12**, 241 (1893).

[5] Baldwin, R. B., *The Face of the Moon*, University of Chicago Press, 1949.

[6] See Greenstein, J. L., *Atmospheres of the Earth and Planets*, edited by G. P. Kuiper, University of Chicago Press, 1949, pp. 117 ff.

[7] See Spitzer, L., Jr., *Ibid.*, edited by G. P. Kuiper, University of Chicago Press, 1949.

[8] Rubey, W. W., *Bull. Geol. Soc. Am.*, **62**, 1111 (1951).

[9] Kuiper, G. P., *Atmospheres of the Earth and Planets*, University of Chicago Press, 1949, p. 304.

[10] See, for example, Rice, F. O., and Rice, K. K., *The Aliphatic Free Radicals*, The Johns Hopkins Press, 1935; *Free Radical Mechanisms*, Steacie, E. W. R., Reinhold, 1946; Noyes, W. A., Jr., and Leighton, P. A., *The Photochemistry of Gases*, Reinhold, 1941; Rollefson, G. K., and Burton, M., *Photochemistry*, Prentice-Hall, 1939; Glockle G., and Lind. S. C., *Electrochemistry of Gases and Other Dielectrics*, John Wiley, 1939.

[11] Rabinowitch, E. I., *Photosynthesis and Related Processes*, Vol. I, New York, 1945.

[12] Spitzer, Lyman, Jr., *Atmospheres of the Earth and Planets*, edited by G. P. Kuiper, University of Chicago Press, 1949.

[13] Thode, H. G., Am. Chem. Soc. Meeting, New York, September, 1951. See Tudge, A. P., and Thode, H. G., *Can. J. Res.*, **28**, 567 (1950).

[14] Leigh, C. K., Lund, R. J., and Leigh, Andrew, *Geol. Survey Professional Paper* 184 (1935).

[15] See Deevey, E. S., Jr., *Scientific American*, **185**, No. 4, 68 (1951) for an interesting mention of Chironomius and of other anaerobic animals.

[16] *Loc. cit.*

[17] Lane, A. C., *Am. J. Sci.*, **43**, 42 (1917).

[18] MacGregor, A., *South African J. Sci.*, **24**, 155 (1927).

THE PREBIOLOGICAL PALEOATMOSPHERE:
STABILITY AND COMPOSITION*

JOEL S. LEVINE and TOMMY R. AUGUSTSSON**

Atmospheric Sciences Division, NASA Langley Research Center, Hampton, Virginia 23665, U.S.A.

and

MURALI NATARAJAN

Systems and Applied Sciences Corporation, Hampton, Virginia 23666, U.S.A.

(Received 10 December, 1981; in revised form 22 April, 1982)

Abstract. In the past, it was generally assumed that the early atmosphere of the Earth contained appreciable quantities of methane (CH_4) and ammonia (NH_3). This was the type of atmosphere believed to be the most suitable environment for chemical evolution, the nonbiological formation of complex organic molecules, the precursors of living systems. Photochemical considerations suggest that a CH_4–NH_3 dominated early atmosphere was probably very short-lived, if it ever existed at all. Instead, an early atmosphere of carbon dioxide (CO_2) and nitrogen (N_2) is favored by photochemical as well as geological and geochemical considerations. Photochemical calculations also indicate that the total oxygen column density of the prebiological paleoatmosphere did not exceed 10^{-7} of the present atmospheric level.

1. The Stability of a Possible Methane-Ammonia Paleoatmosphere

In the past, it was generally assumed that the early paleoatmosphere contained appreciable quantities of methane (CH_4) and ammonia (NH_3). Different lines of reasoning led to this conclusion which has been reviewed by McGovern (1969) and is briefly summarized here. It was believed that the primordial atmosphere was a remnant of the solar nebula that gave birth to the solar system some 4.5 billion years ago. Since hydrogen was the overwhelming species in the solar nebula, it was natural to assume that the primordial atmosphere should have been dominated by hydrogen compounds. In addition, the atmospheres of the outer planets, which are thought to have undergone little evolution over geological time, contain appreciable quantities of CH_4 and NH_3. Finally, the early experiments on chemical evolution indicated a CH_4 – NH_3 dominated atmosphere was the most suitable environment for initiating the development of complex organic molecules, the precursors of living systems (Miller, 1953; Miller and Urey, 1959).

An alternative theory, more widely accepted at the present time, for the origin of the atmosphere is that it is secondary, resulting from the release of gases originally trapped in the interior of the Earth, as opposed to it being primordial, a remnant of the solar nebula. The gases originally trapped in the interior which outgassed to form the atmosphere included H_2O, CO_2, chlorine (Cl), nitrogen (N), and sulfur (S). Rubey (1955) coined the term 'excess volatiles' as the total amount of a volatile in the Earth/atmosphere system that is not supplied by the weathering of igneous rocks. Recent estimates for the total amount of 'excess volatiles' are H_2O: 1.6×10^{24} g; CO_2: 2.3×10^{23} g; Cl: 3.0×10^{22} g;

* Paper presented at the 6th College Park Colloquium, October 1981.
** NRC-NASA Research Associate.

N: 4.9×10^{21} g; and S: 4.4×10^{21} g (Walker, 1977). The composition of the atmosphere does not reflect the composition of the outgassed volatiles. Almost all of the outgassed H_2O condensed out of the atmosphere and formed the oceans, while the bulk of the outgassed CO_2 formed carbonates in sedimentary rocks (Walker, 1977; Walker, 1978a; Levine, 1978). The outgassed nitrogen accumulated in the atmosphere to become the most abundant atmospheric species. Abelson (1966) reported that laboratory mixtures of N_2, H_2, and CO exposed to a discharge resulted in the formation of hydrogen cyanide (HCN), CO_2, and H_2O. HCN is a reactive intermediate species which can eventually form amino acids, purines, and pyrimidines (see for example, Ferris *et al.*, 1978). The laboratory experiments of Abelson (1966) indicated that chemical evolution was also possible in the type of outgassed paleoatmosphere suggested by Rubey (1955).

Other possible sources for prebiological CH_4 and NH_3 have been proposed. Holland (1962) suggested that CH_4 and NH_3 may have outgassed from the solid planet prior to the migration of metallic iron from the mantle to the core. However, current thinking holds that the iron core formed at the time that the Earth formed (Walker, 1976, 1977). Another possible source of CH_4 may have resulted from the breakdown of hydrocarbons originally trapped in the solid Earth when it formed from the solar nebula (Gold, 1979). According to this hypothesis, deep mantle carbon may be released in the form of CH_4. However, recent stable carbon isotope measurements revealed that most of the hydrocarbons, including CH_4, found in geochemical emanations result from the decomposition of organic matter in sedimentary rocks (Des Marais, 1981). This observation suggests that most of the CH_4 in present-day geothermal emanations required the preexistence of life (Des Marais, 1981). Another possible source of carbon for the prebiological paleoatmosphere may have been cometary material (Oró, 1961; Chang, 1979; Lazcano-Araujo and Oró, 1981; Levine *et al.*, 1981). A recently suggested localized source of NH_3 in the prebiological paleoatmosphere may have been the fixation of atmospheric N_2 to form NH_3 in the presence of naturally occurring titanium dioxide (TiO_2) (Henderson-Sellers and Schwartz, 1980).

Hence, while there is general agreement that the prebiological paleoatmosphere was reducing in nature, there is uncertainty whether it was strongly reducing (CH_4 and NH_3) one atmosphere of CH_4 (McGovern, 1969. Estimates of NH_3 in the early atmosphere range from a few parts per million (Sagan and Mullen, 1972) to as much as 25% of the one atmosphere of CH_4 (McGovern, 1969). Esimates of NH_3 in the early atmosphere range from a few parts per million (Sagan and Mullen, 1972) to as much as 25% of the total mass of the early atmosphere (Hart, 1978, 1979). In a recent paper, Hart (1979) suggested that the prebiological paleoatmosphere was strongly reducing, containing 2×10^{21} g of CH_4 plus 4×10^{21} g of NH_3 plus 6×10^{21} g CO plus 3×10^{21} g of hydrogen sulfide (H_2S), or the equivalent in any combination. For comparison, the mass of the present atmosphere is about 5.3×10^{21} g. The present atmosphere contains both CH_4 (with a mixing ratio of about 1.6 ppm) and NH_3 (about one part per billion, ppb). These levels of CH_4 and NH_3 in the present atmosphere are a result of continuous biogenic production of these gases at the surface. There are also anthropogenic sources of CH_4 and NH_3 (Levine and Schryer, 1978). However, there are no known chemical processes that lead to the formation of CH_4 or NH_3 in the atmosphere. Hence, any CH_4 or NH_3 that may have existed in the prebiological paleoatmosphere must have resulted as a remnant of the solar nebula; or have outgassed from the interior prior to the migration of metallic iron from the mantle to the core; in the case of CH_4, have been of cometary origin; or, in the case of NH_3, have been produced catalytically from N_2 by the action of titanium dioxide.

NH_3 and CH_4 are photochemically and chemically destroyed via photolysis by solar ultraviolet radiation and by reaction with the hydroxyl radical (OH). In addition, since NH_3 is very water-soluble, it can be lost from the atmosphere due to rainout. The photochemical/chemical loss mechanisms for NH_3 and CH_4 can be represented as:

$$NH_3 + h\nu \rightarrow NH_2 + H; \quad \lambda \leqslant 230 \text{ nm}; \tag{1}$$
$$NH_3 + OH \rightarrow NH_2 + H_2O; \quad k_2 = 2.3 \times 10^{-12} \exp(-3300/T); \tag{2}$$
$$\text{Rainout of } NH_3; \tag{3}$$
$$CH_4 + h\nu \rightarrow CH_2 + H_2; \quad \lambda \leqslant 145 \text{ nm}; \tag{4}$$
$$CH_4 + OH \rightarrow CH_3 + H_2O; \quad k_5 = 2.4 \times 10^{-12} \exp(-1710/T). \tag{5}$$

Kuhn and Atreya (1979) investigated the lifetime of NH_3 in the paleoatmosphere against photolysis (reaction 1). They found that the lifetime of NH_3 is very short, e.g., less than a day for an NH_3 mixing ratio of 10^{-8} to about 10^4 days for a mixing ratio of 10^{-4}. Previously, Ferris and Nicodem (1972) investigated the lifetime of NH_3 in the paleoatmosphere against photodestruction. The calculations of Ferris and Nicodem indicated that the lifetime of NH_3 against photodestruction is very dependent on the levels of other trace gases that may protect NH_3 from photodestruction, such as hydrogen sulfide (H_2S), which has strong features shortward of 240 nm, the spectral region of NH_3 photolysis. Ferris and Nicodem found that for H_2S levels on the order of 10^3 times that of NH_3 (which may be unrealistically large), the lifetime of NH_3 against photodestruction may have been very long — on the order of 2×10^9 yr. Kuhn and Atreya did not assess the lifetime of NH_3 against its reaction with OH (reaction 2), but concluded that it was probably too small to influence their results. Levine *et al.* (1980) have investigated the lifetime of NH_3 against rainout in the present atmosphere by comparing measurements of the vertical distribution of NH_3 in the troposphere with theoretical calculations obtained with a photochemical model. They concluded that the lifetime of NH_3 against rainout is about 10 days. Hence, it is evident that in the absence of a source, the lifetime of NH_3 in the prebiological paleoatmosphere is very short.

McGovern (1969) investigated the photochemistry of a methane-dominated paleoatmosphere. He considered the photolysis of CH_4 as the main destruction mechanism for CH_4 (reaction 4). McGovern reported that the photolysis of CH_4 formed several hydrocarbons which are very efficient infrared radiators, resulting in a cool exospheric temperature (500–1000 K) for this CH_4-dominated paleoatmosphere. This low exospheric temperature prevented the rapid exospheric escape at atomis hydrogen, resulting in the stability of a CH_4-dominated paleoatmosphere of 5 to 10×10^8 yr. More recently, Shimizu (1976) reexamined the question of the exospheric temperature in a CH_4-dominated paleoatmosphere and concluded that it should have exceeded 1300 K, making such a paleoatmosphere very short-lived. Lasaga *et al.* (1971) also investigated the photochemistry of a CH_4-dominated paleoatmosphere. They found that methyl and methylene radicals produced by the ultraviolet photolysis of CH_4 (reaction 4) combined to form heavier hydrocarbons, which eventually could have precipitated out of the paleoatmosphere forming an oil slick with a thickness of 1–10 meters at the surface of the Earth. Their calculations indicated that the equivalent of one atmosphere of CH_4 could have been polymerized by solar ultraviolet radiation in 10^6–10^7 yr.

In the calculations of McGovern (1969) and Lasaga *et al.* (1971) the photolysis of CH_4 occurred high in the paleoatmosphere — above 100 km. In this paper we investigate the lifetime of CH_4 against photolysis (reaction 4) as a function of altitude in the paleoatmosphere. The photolytic destruction rate of CH_4, $J(CH_4)$ (reaction 4) can be expressed as:

$$J(CH_4) = \int_{110}^{145\ nm} I_\infty(\lambda)\,\sigma(CH_4)\,e^{-\tau(i)\,\sec\theta} \tag{6}$$

$I_\infty(\lambda)$ is the solar flux incident at the top of the atmosphere (Ackermann, 1971), and $\sigma(CH_4)$ is the absorption cross section for CH_4 (McGovern, 1969). The solar flux incident at the top of the atmosphere and the absorption cross-section for CH_4 as a function of wavelength are shown in Figure 1. Referring to Equation (6), θ is the solar zenith angle

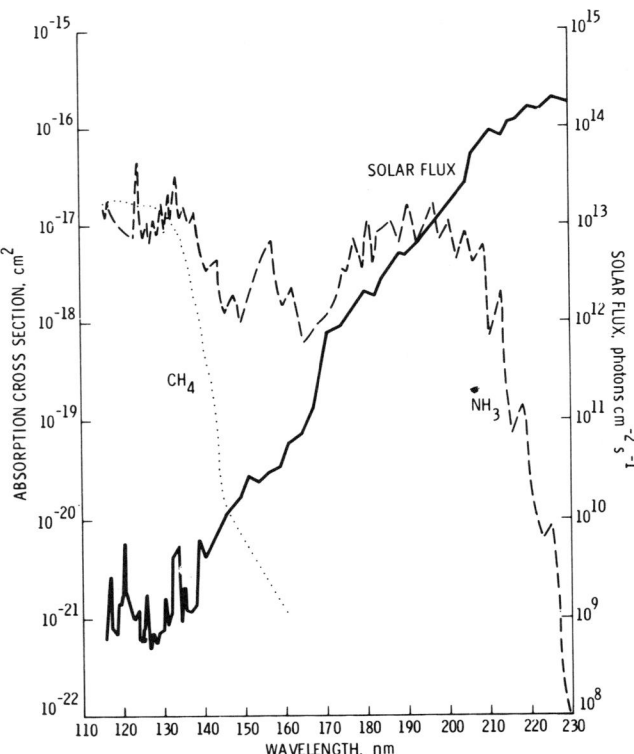

Fig. 1. Absorption cross-sections for CH_4 and NH_3 and solar flux incident at the top of the atmosphere (110–230 nm).

and $\tau(i)$ is the optical depth due to the various atmospheric species that absorb in the region of CH_4 photolysis, i.e., $\lambda \leqslant 145$ nm. These atmospheric species include CH_4, O_2, CO_2, and H_2O. $\tau(i)$ can be represented as:

$$\tau(i) = N(CH_4)\sigma(CH_4) + N(O_2)\sigma(O_2) + N(CO_2)\sigma(CO_2) + N(H_2O)\sigma(H_2O). \qquad (7)$$

$N(i)$ is the column density of the i-th species above a given altitude, z, and can be expressed as:

$$N(i) = \int_z^\infty n(i)\mathrm{d}z. \qquad (8)$$

$n(i)$ is the number density of the i-th species at an altitude of z. The absorption cross-sections for O_2, CO_2, and H_2O, as well as the solar flux incident at the top of the atmosphere, are shown in Figures 2 and 3, respectively. The absorption cross-sections are from Levine *et al.* (1979).

Once the value of $J(CH_4)$, the photolytic destruction rate of CH_4 given in Equation (6), is known, we can estimate the mean lifetime of CH_4 against photolytic destruction, $\tau(CH_4)$, at any altitude using the following expression:

$$\tau(CH_4) = \frac{1}{J(CH_4)}. \qquad (9)$$

To solve Equation (6) we must prescribe values of $n(i)$ for all of the absorbing gases in Equation (7). The calculation of the level of O_2 in the prebiological paleoatmosphere is

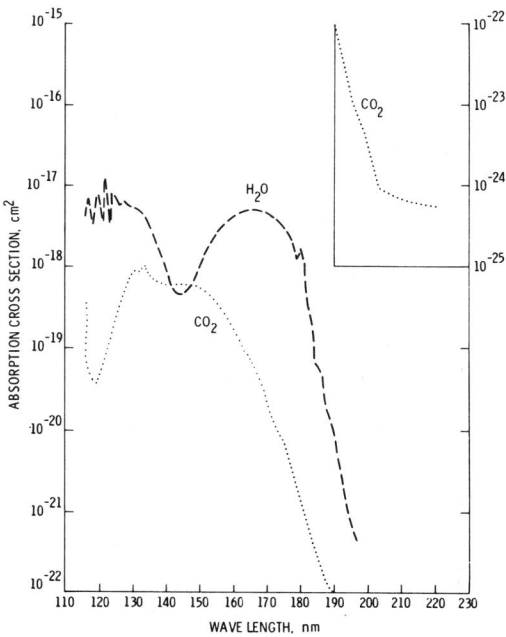

Fig. 2. Absorption cross-sections for CO_2 and H_2O (110–230 nm).

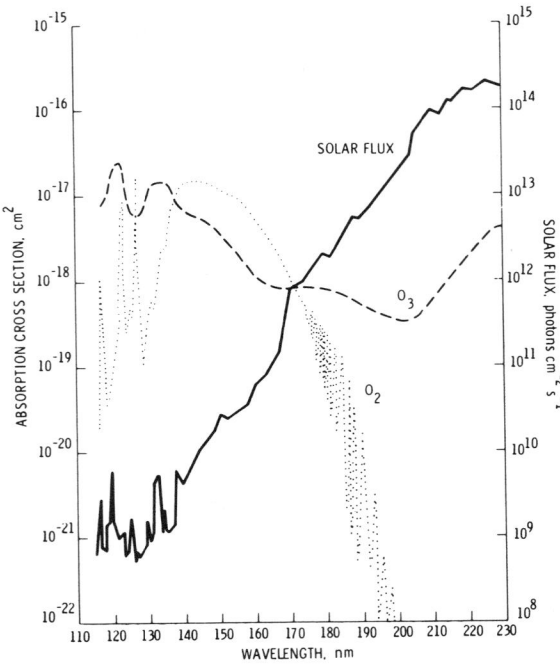

Fig. 3. Absorption corss-sections for O_2 and O_3 and solar flux incident at the top of the atmosphere (110–230 nm).

discussed in the following section. Our calculations indicate that the column density of O_2 in the prebiological paleoatmosphere was about 10^{-7} of the present atmospheric level (PAL). For simplicity we have further assumed that the prebiological paleoatmosphere contained pre-industrial levels of CO_2 (280 ppm) and present atmospheric levels of CH_4 (1.6 ppm) and H_2O (2% at the surface decreasing to a mixing ratio of 5.5×10^{-5} at the tropopause). Insertion of these gas concentrations in Equation (7) yields the following total atmospheric optical depths for each of these species: $\tau(O_2) =$ 2.3; $\tau(CH_4) = 2.5 \times 10^2$; $\tau(CO_2) = 3.0 \times 10^3$ and $\tau(H_2O) = 3.2 \times 10^5$. The calculated photolytic destruction rate of CH_4 at the top of the atmosphere ($z = 120$ km), $J_\infty(CH_4)$, is about 2.96×10^{-6} s^{-1}, which corresponds to a photolytic lifetime, $\tau_\infty(CH_4)$, of about 4 days. The calculated photolytic destruction rate of CH_4 at the surface of the Earth ($z = 0$), $J_0(CH_4)$, was found to approach zero, corresponding to an infinitely long lifetime of CH_4 against photolytic destruction at the surface. It is interesting to note that Kuhn and Atreya (1979) found that the photolytic destruction of NH_3 at the ground was very rapid, whereas in our calculations we found that the photolytic destruction of CH_4 at the ground was very, very slow. This can be explained by inspection of Figure 1. As we go from the photolysis threshold of $NH_3(\lambda = 230$ nm) to the photolysis threshold of $CH_4(\lambda = 145$ nm), the incident solar flux drops in intensity by more than four orders of magnitude. In addition to the decrease in available solar photons as we approach the photolysis threshold of CH_4, inspection of Figure 2 indicates that the absorption cross section of H_2O, which controls the photolysis of CH_4, increases by almost four orders of magnitude. The result is that at the surface there are no solar photons available for the photolysis of CH_4, and as a result, CH_4 has an infinitely long lifetime against photolysis.

We have also considered the lifetime of CH_4 at the surface against destruction by OH (reaction 5). The lifetime of CH_4 against the reaction with OH, $\tau(CH_4-OH)$, can be estimated using the following expression:

$$\tau(CH_4-OH) = \frac{1}{k_5\,[OH]} \tag{10}$$

To determine the lifetime of CH_4 against destruction via reaction with OH, we need to know levels of OH in the prebiological paleoatmosphere. In the present atmosphere OH is formed by the reaction of excited atomic oxygen ($O(^1D)$) and H_2O:

$$O(^1D) + H_2O \rightarrow 2\ OH. \tag{11}$$

It has been demonstrated experimentally (Ferris and Chen, 1975; Bossard and Toupance, 1981) and theoretically (Levine et al., 1979) that in the oxygen-deficient paleoatmosphere the photolysis of H_2O was an important source of OH via the reaction:

$$H_2O + h\nu \rightarrow OH + H; \quad \lambda \leqslant 240\ nm. \tag{12}$$

Calculations indicate that as the O_2 level decreases from the present atmospheric level (1 PAL) to 10^{-4} PAL, surface levels of OH increase from present atmospheric levels of about $10^6\ cm^{-3}$ to about $10^9\ cm^{-3}$. Walker (1978b) pointed out that very early in the Earth's history (corresponding to O_2 levels considerably below 10^{-4} PAL), molecular hydrogen (H_2) resulting from volcanic emissions readily reacted with OH:

$$H_2 + OH \rightarrow H_2O + H. \tag{13}$$

This reaction limited levels of OH in the early prebiological paleoatmosphere. For an assumed H_2 volcanic flux of about $1 \times 10^8\ cm^{-2}\ s^{-1}$, calculations indicate OH levels of about $10^5\ cm^{-3}$ (Kasting et al., 1979; Yung and McElroy, 1979). The variation of surface OH with level of O_2 in the paleoatmosphere is summarized in Figure 4. Using an OH

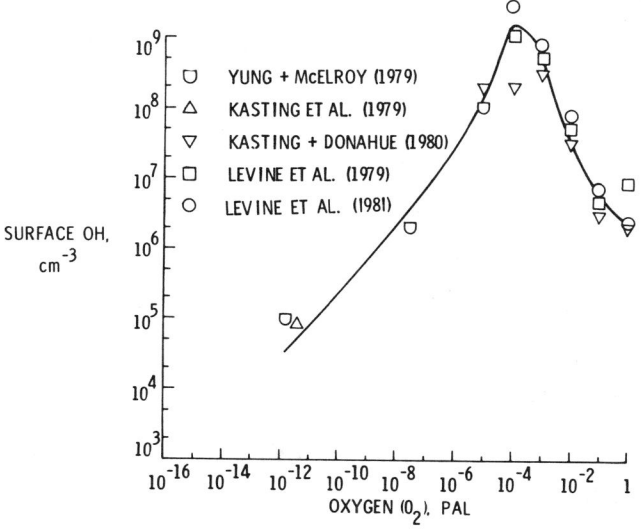

Fig. 4. Variation of surface concentration of OH as function of level of O_2 in paleoatmosphere.

value of about 10^5 cm^{-3} in Equation (10) indicates that the lifetime of CH_4 against loss due to OH is only about 50 years. Hence, photochemical considerations indicate that in the absence of a continuous source, the stability of a prebiological paleoatmosphere with appreciable quantities of CH_4 and NH_3 was extemely short-lived, if such an atmosphere ever existed at all. Abelson (1966) had previously argued that geochemical considerations do not support an early CH_4-NH_3 atmosphere. It is interesting to point out that very recent geological and geochemical evidence does not favor an early CH_4-NH_3 atmosphere (see for example, Kerr, 1980; Levine and Graedel, 1981).

2. Oxygen Levels in the Prebiological Paleoatmosphere

As shown in Equations (6) and (7), the photolytic destruction of CH_4 is controlled by the presence of gases that absorb in the spectral region of CH_4 photolysis. These gases include O_2, CO_2 and H_2O (see Figures 1–3). For our calculations we have assumed present atmospheric levels of H_2O and pre-industrial levels of CO_2. However, we must determine the level of O_2 in the prebiological paleoatmosphere. The physical and chemical processes that controlled the level of O_2 in the prebiological paleoatmosphere were discussed by Walker (1978b), Kasting *et al.* (1979), and Kasting and Walker (1981). In the prebiological paleoatmosphere the formation of O_2 was initiated by solar ultraviolet radiation, i.e., the photolysis of water vapor (H_2O), accompanied by the exospheric escape of atomic hydrogen (H), and the photolysis of carbon dioxide (CO_2). The photochemical reactions governing these processes are:

$$H_2O + h\nu \rightarrow OH + H; \quad \lambda \leqslant 240 \text{ nm, and,} \tag{12}$$
$$CO_2 + h\nu \rightarrow CO + O; \quad \lambda \leqslant 230 \text{ nm.} \tag{14}$$

Reaction (12) leads to the production of oxygen atoms (O) via:

$$OH + OH \rightarrow O + H_2O. \tag{15}$$

The oxygen atoms formed in reactions (14) and (15) form molecular oxygen (O_2) via the following reactions:

$$O + O + M \rightarrow O_2 + M, \quad \text{and,} \tag{16}$$
$$O + OH \rightarrow O_2 + H. \tag{17}$$

The destruction of O_2 is controlled by its photolysis and reaction with molecular hydrogen (H_2) resulting from volcanic emissions:

$$O_2 + h\nu \rightarrow O + O; \quad \lambda \leqslant 242 \text{ nm, and,} \tag{18}$$
$$2H_2 + O_2 \rightarrow 2H_2O. \tag{19}$$

Closely coupled to the origin and evolution of O_2 is the origin and evolution of O_3. O_3 is photochemically produced via the reaction:

$$O + O_2 + M \rightarrow O_3 + M. \tag{20}$$

O_3 is photochemically destroyed via photolysis and reaction with O:

$$O_3 + h\nu \to O_2 + O; \quad \lambda \leqslant 1100 \text{ nm, and,} \tag{21}$$
$$O_3 + O \to 2O_2 \tag{22}$$

In addition, O_3 is photochemically destroyed by a series of catalytic cycles involving the oxides of nitrogen ($NO + NO_2$), hydrogen ($OH + HO_2$), and chlorine ($Cl + ClO$).

To study the levels of O_2 and O_3 in the prebiological paleoatmosphere, the photochemical model of the paleoatmosphere of Levine et al. (1981) was modified. This model, which was originally developed to study the effect of anthropogenic perturbations on future levels of O_3, has been used to study the origin and evolution of O_3 for specified levels of paleoatmospheric O_2 ranging from 10^{-4} of the present atmospheric level (PAL) to the present atmospheric level (1 PAL) (Levine et al., 1981). The model was modified such that O_2, CO_2, and H_2 which were previously specified as input parameters are now considered as chemically-active species, whose vertical profiles are calculated using coupled species continuity/flux equations, which contain both chemical production and loss terms and vertical eddy transport. The inclusion of O_2, CO_2, and H_2 as chemically-active species brings to a total of 31 species whose vertical profiles are solved simultaneously by coupled species continuity/flux equations. The other calculated species are: O_3, O, N_2O, N, NO, NO_2, NO_3, N_2O_5, HNO_2, HNO_3, H_2O, H, OH, HO_2, H_2O_2, CH_4, CO, CH_3OOH, CH_2O, CH_3, HCO, CH_3O_2, CH_3O, Cl, ClO, HCl, CH_3Cl, and $ClNO_3$. Very short-lived species assumed to be in instantaneous photochemical equilibrium include: $O(^1D)$, ClO_2, and Cl_2. The model now includes 25 photochemical processes listed in Table I, and 77 chemical reactions listed in Table II.

The U.S. Standard Atmosphere Mid-Latitude Spring/Fall temperature profile is specified in the troposphere. A primordial temperature profile that decreases linearly from the tropopause to the mesopause is used. This profile is based on coupled photochemical/radiative equilibrium temperature calculations in the O_3-deficient paleoatmosphere (Levine and Boughner, 1979). Water soluble species, e.g. HNO_3, H_2O_2, CH_3OOH, HCl, ClO, and $ClNO_3$ are lost via rainout with a rainout coefficient of 1.0×10^{-6} s^{-1}. The model extends from the surface up to 53.5 km with the tropopause at a height of 14.5 km. Photodissociation rates are diurnally averaged for a latitude of $30°$ and solar declination of $0°$. The model includes the flux of radiation from the present Sun between 110–735 nm (Ackermann, 1971). As already noted, the complete photochemical model contains 31 species and 102 photochemical/chemical processes listed in Tables I and II. This model has been adapted for photochemical studies of the prebiological paleoatmosphere by appropriate choices for the boundary conditions for various species. In these calculations, the prebiological paleoatmosphere consists of present-day levels of nitrogen (N_2) and water vapor (H_2O), and pre-industrial levels (280 parts per million by volume — ppmv) of carbon dioxide (CO_2). The H_2O profile based on the U.S. Standard Atmosphere Mid-Latitude Spring/Fall is specified in the troposphere, and is calculated using a H_2O continuity/flux equation in the stratosphere. The surface flux of gases resulting from biological activity, e.g. CH_4, N_2O, and CH_3Cl, and those resulting from anthropogenic activity, e.g. NO_x ($NO + NO_2$), CCl_4, CH_3CCl_3 are set equal to zero in these calculations.

TABLE I
Photodissociation reactions and diurnally — averaged rates at 53.5 km for a latitude of 30° (equinoctial conditions) (for present atmosphere)

Reaction number	Reaction	Destruction (cm^{-3} s^{-1})
J1	$O_2 + h\nu \to O + O$	1.41×10^6
J2:	$O_3 + h\nu \to O + O_2$	4.40×10^6
J3:	$O_3 + h\nu \to O(^1D) + O_2$	7.07×10^7
J4:	$H_2O + h\nu \to OH + H$	5.32×10^2
J5:	$N_2O + h\nu \to O(^1D) + N_2$	1.29×10^2
J6:	$HNO_3 + h\nu \to OH + NO_2$	2.63×10^0
J7:	$NO_2 + h\nu \to O + NO$	1.33×10^4
J8:	$H_2O_2 + h\nu \to OH + OH$	2.52×10^2
J9:	$HNO_2 + h\nu \to OH + NO$	4.31×10^1
J10:	$NO_3 + h\nu \to NO + O_2$	7.85×10^{-1}
J11:	$NO_3 + h\nu \to NO_2 + O$	7.85×10^0
J12:	$N_2O_5 + h\nu \to NO_2 + NO_3$	5.21×10^{-4}
J13:	$HCl + h\nu \to H + Cl$	1.47×10^0
J14:	$ClO_2 + h\nu \to ClO + O$	6.41×10^{-3}
J15:	$ClO + h\nu \to Cl + O$	1.45×10^2
J16:	$Cl_2 + h\nu \to Cl + Cl$	1.43×10^{-6}
J17:	$ClNO_3 + h\nu \to ClO + NO_2$	4.77×10^{-4}
J18:	$CCl_4 + h\nu \to 2Cl + products$	1.27×10^{-5}
J19:	$CH_3Cl + h\nu \to CH_3 + Cl$	1.09×10^{-2}
J20:	$CH_2O + h\nu \to H + HCO$	4.00×10^1
J21:	$CH_2O + h\nu \to H_2 + CO$	4.15×10^1
J22:	$CH_3CCl_3 + h\nu \to Cl + products$	1.20×10^{-5}
J23:	$CH_3OOH + h\nu \to CH_3O + OH$	1.82×10^0
J24:	$NO + h\nu \to N + O$	9.12×10^1
J25:	$CO_2 + h\nu \to CO + O$	4.51×10^{-1}

TABLE II
Chemical reactions

No.	Reaction	Rate Constant (cm^3 s^{-1} or cm^6 s^{-1})
1	$O + O_2 + M \to O_3 + M$	$1.1 \times 10^{-34} \exp(510/T)$
2	$O + O_3 \to 2O_2$	$1.5 \times 10^{-11} \exp(-2218/T)$
3	$O(^1D) + O_2 \to O + O_2$	$3.2 \times 10^{-11} \exp(67/T)$
4	$O(^1D) + N_2 \to O + N_2$	$1.8 \times 10^{-11} \exp(107/T)$
5	$N_2O + O(^1D) \to 2NO$	6.6×10^{-11}
6	$N_2O + O(^1D) \to N_2 + O_2$	5.1×10^{-11}
7	$NO + O + M \to NO_2 + M$	$1.6 \times 10^{-32} \exp(584/T)$
8	$NO + O_3 \to NO_2 + O_2$	$2.3 \times 10^{-12} \exp(-1450/T)$
9	$NO_2 + O \to O_2 + NO$	9.3×10^{-12}
10	$NO_2 + O_3 \to NO_3 + O_2$	$1.2 \times 10^{-13} \exp(-2450/T)$
11	$NO + HO_2 \to NO_2 + OH$	$3.5 \times 10^{-12} \exp(250/T)$
12	$NO_2 + OH + M \to HNO_3 + M$	* See JPL Publication 81-3 (1981)
13	$HNO_3 + OH \to NO_3 + H_2O$	$1.5 \times 10^{-14} \exp(650/T)$
14	$H_2O + O(^1D) \to 2OH$	2.2×10^{-10}

Table II (Continued)

No.	Reaction	Rate Constant
		$(\text{cm}^3 \text{ s}^{-1}$ or $\text{cm}^6 \text{ s}^{-1})$
15	$H + O_2 + M \rightarrow HO_2 + M$	$2.1 \times 10^{-32} \exp(290/T)$
16	$H + O_3 \rightarrow OH + O_2$	$1.4 \times 10^{-10} \exp(-470/T)$
17	$OH + O \rightarrow H + O_2$	$2.3 \times 10^{-11} \exp(110/T)$
18	$OH + O_3 \rightarrow HO_2 + O_2$	$1.6 \times 10^{-12} \exp(-940/T)$
19	$OH + OH \rightarrow H_2O + O$	$4.5 \times 10^{-12} \exp(-275/T)$
20	$HO_2 + O \rightarrow OH + O_2$	4.0×10^{-11}
21	$HO_2 + O_3 \rightarrow OH + 2O_2$	$1.1 \times 10^{-14} \exp(-580/T)$
22	$HO_2 + OH \rightarrow H_2O + O_2$	4×10^{-11}
23	$HO_2 + HO_2 \rightarrow H_2O_2 + O_2$	2.5×10^{-12}
24	$H_2O_2 + OH \rightarrow HO_2 + H_2O$	$2.7 \times 10^{-12} \exp(-145/T)$
25	$OH + NO + M \rightarrow HNO_2 + M$	See JPL Publication 81–3 (1981)
26	$NO + NO_3 \rightarrow 2NO_2$	2.0×10^{-11}
27	$O(^1D) + N_2 + M \rightarrow N_2O + M$	3.5×10^{-37}
28	$O(^1D) + H_2 \rightarrow OH + H$	9.9×10^{-11}
29	$O(^1D) + CH_4 \rightarrow OH + CH_3$	1.4×10^{-10}
30	$NO_2 + O + M \rightarrow NO_3 + M$	1.0×10^{-31}
31	$NO_2 + NO_3 \rightarrow N_2O_5$	$1.5 \times 10^{-13} \exp(861/T)$
32	$N_2O_5 \rightarrow NO_2 + NO_3$	$1.2 \times 10^{14} \exp(-10319/T)$
33	$N_2O_5 + H_2O \rightarrow 2HNO_3$	1.0×10^{-20}
34	$N_2O_5 + O \rightarrow 2NO_2 + O_2$	3.0×10^{-16}
35	$N + NO_2 \rightarrow N_2O + O$	$2.1 \times 10^{-11} \exp(-800/T)$
36	$N + O_2 \rightarrow NO + O$	$4.4 \times 10^{-12} \exp(-3220/T)$
37	$N + NO \rightarrow N_2 + O$	3.4×10^{-11}
38	$N + O_3 \rightarrow NO + O_2$	1.0×10^{-15}
39	$Cl + O_3 \rightarrow ClO + O_2$	$2.8 \times 10^{-11} \exp(-257/T)$
40	$ClO + O \rightarrow Cl + O_2$	$7.7 \times 10^{-11} \exp(-130/T)$
41	$ClO + NO \rightarrow Cl + NO_2$	$6.5 \times 10^{-12} \exp(280/T)$
42	$Cl + CH_4 \rightarrow HCl + CH_3$	$9.6 \times 10^{-12} \exp(-1350/T)$
43	$Cl + H_2 \rightarrow HCl + H$	$3.5 \times 10^{-11} \exp(-2290/T)$
44	$Cl + HO_2 \rightarrow HCl + O_2$	4.8×10^{-11}
45	$Cl + H_2O_2 \rightarrow HCl + HO_2$	$1.1 \times 10^{-12} \exp(-980/T)$
46	$Cl + HNO_3 \rightarrow HCl + NO_3$	$1.0 \times 10^{-11} \exp(-2170/T)$
47	$Cl + CH_2O \rightarrow HCl + HCO$	$9.2 \times 10^{-11} \exp(-68/T)$
48	$HCl + OH \rightarrow Cl + H_2O$	$2.8 \times 10^{-12} \exp(-425/T)$
49	$HCl + O \rightarrow Cl + OH$	$1.1 \times 10^{-11} \exp(-3370/T)$
50	$Cl + O_2 + M \rightarrow ClO_2 + M$	See JPL Publication 81–3 (1981)
51	$ClO_2 + M \rightarrow Cl + O_2 + M$	$2.7 \times 10^{-9} \exp(-2650/T)$
52	$Cl + ClO_2 \rightarrow 2ClO$	8.0×10^{-12}
53	$Cl + ClO_2 \rightarrow Cl_2 + O_2$	1.4×10^{-10}
54	$OH + OH + M \rightarrow H_2O_2 + M$	See JPL Publication 81–3 (1981)
55	$H_2O_2 + O \rightarrow OH + HO_2$	$2.8 \times 10^{-12} \exp(-2125/T)$
56	$OH + CH_4 \rightarrow CH_3 + H_2O$	$2.4 \times 10^{-12} \exp(-1710/T)$
57	$ClO + NO_2 + M \rightarrow ClNO_3 + M$	See JPL Publication 81–3 (1981)
58	$O + ClNO_3 \rightarrow ClO + NO_3$	$3.0 \times 10^{-12} \exp(-808/T)$
59	$O(^1D) + HCl \rightarrow OH + Cl$	1.4×10^{-10}
60	$H_2 + OH \rightarrow H_2O + H$	$1.2 \times 10^{-11} \exp(-2200/T)$
61	$CH_3 + O_2 + M \rightarrow CH_3O_2 + M$	See JPL Publication 81–3 (1981)
62	$CH_3O_2 + HO_2 \rightarrow CH_3OOH + O_2$	$7.7 \times 10^{-14} \exp(1300/T)$
63	$CH_3O_2 + NO \rightarrow CH_3O + NO_2$	7.4×10^{-12}

Table II (Continued)

No.	Reaction	Rate Constant
		$(cm^3\ s^{-1}\ $ or $cm^6\ s^{-1})$
64	$CH_3O + O_2 \rightarrow CH_2O + HO_2$	$9.2 \times 10^{-13}\ exp(-2200/T)$
65	$CH_2O + OH \rightarrow HCO + H_2O$	1.0×10^{-11}
66	$CH_2O + O \rightarrow OH + HCO$	$3.0 \times 10^{-11}\ exp(-1550/T)$
67	$HCO + O_2 \rightarrow CO + HO_2$	5.0×10^{-12}
68	$CO + OH \rightarrow H + CO_2$	$1.35 \times 10^{-13}\ (1 + p_{atm})$
69	$CH_3Cl + OH \rightarrow Cl + products$	$1.8 \times 10^{-12}\ exp(-1112/T)$
70	$CH_3OOH + OH \rightarrow CH_3O_2 + H_2O$	$2.1 \times 10^{-12}\ exp(-145/T)$
71	$OH + CH_3CCl_3 \rightarrow Cl + products$	$5.4 \times 10^{-12}\ exp(-1820/T)$
72	$O + O + M \rightarrow O_2 + M$	$2.8 \times 10^{-34}\ exp(710/T)$
73	$H + H + M \rightarrow H_2 + M$	8.3×10^{-33}
74	$H_2 + O \rightarrow OH + H$	$3.0 \times 10^{-14}\ exp(-4480/T)$
75	$H + HO_2 \rightarrow O_2 + H_2$	$4.7 \times 10^{-11}\ (\times 0.29)$
76	$H + HO_2 \rightarrow H_2O + O$	$4.7 \times 10^{-11}\ (\times 0.02)$
77	$H + HO_2 \rightarrow OH + OH$	$4.7 \times 10^{-11}\ (\times 0.69)$

* *Chemical Kinetic and Photochemical Data for Use in Stratospheric Modelling*, JPL Publication 81–3, Jet Propulsion Laboratory, California Institute of Technology, California Institute of Technology, Pasadena, California, Jan. 15, 1981, 124 pages.

The surface flux of CO assumed to result from volcanic activity is set equal to 3.2×10^8 $cm^{-2}\ s^{-1}$ (Walker, 1977; Kasting *et al.*, 1979). The surface flux of HCl assumed to result from sea salt spray and volcanic activity is set equal to $1 \times 10^{10}\ cm^{-2}\ s^{-1}$ (Levine *et al.*, 1981). Oxygen (O_2) is considered as a chemically-active species and has a specified zero flux lower boundary condition, while molecular hydrogen (H_2) has a prescribed surface mixing ratio of 17 ppmv, both identical to the boundary conditions of Kasting and Walker (1981).

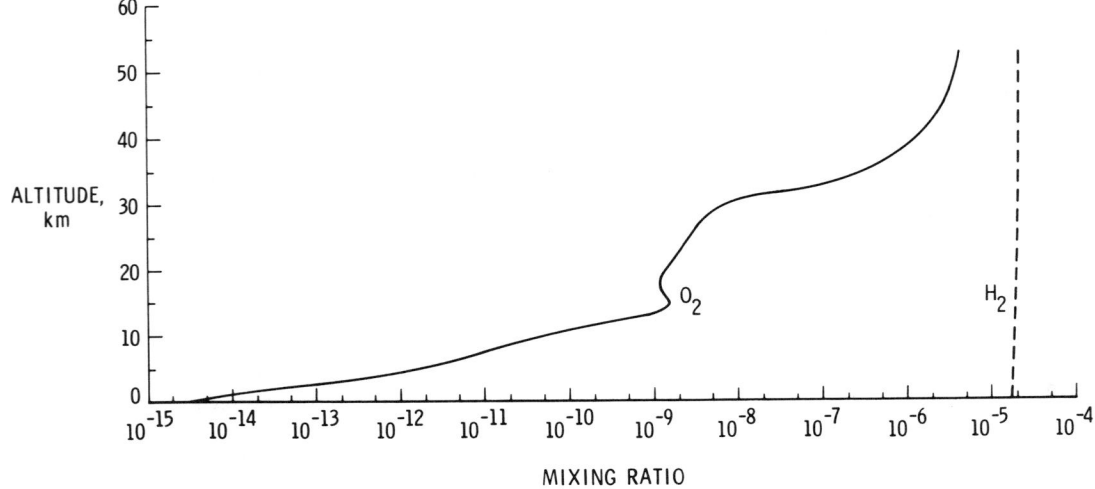

Fig. 5.   The vertical distribution of oxygen (O_2) and hydrogen (H_2) in the prebiological paleoatmosphere.

Calculated profiles of O_2 and H_2 in the prebiological paleoatmosphere are shown in Figure 5. The O_2 profile exhibits a minimum mixing ratio of about 10^{-15} at the surface, and a maximum mixing ratio approaching 10^{-5} at about 50 km, similar to the calculations of Kasting and Walker (1981). The O_2 maximum near 50 km results from the photolysis of CO_2 which is an important source of O atoms (reaction (14)). The O_2 profile in Figure 5 corresponds to an O_2 column density of about 10^{-7} of the present atmospheric O_2 column, and was the value of $N(O_2)$ used in Equation (7). The mixing ratio of H_2 remains constant (about 17 ppmv) between the surface and 53.5 km, as shown in Figure 5. Calculated profiles of O_3 and O in the prebiological paleoatmosphere are shown in Figure 6. The O_3 profile in Figure 6 corresponds to an O_3 column density

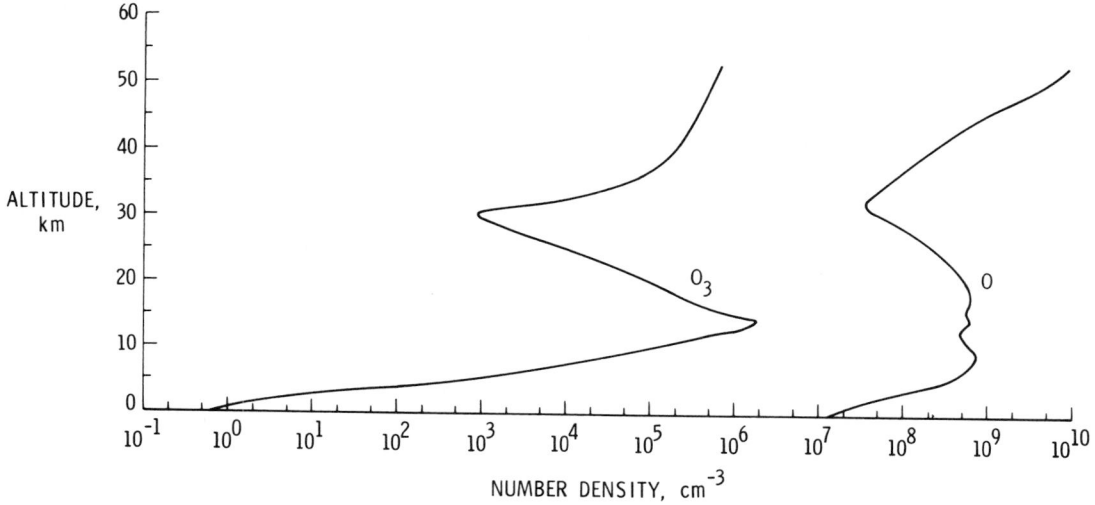

Fig. 6. The vertical distribution of atomic oxygen (O) and ozone (O_3) in the prebiological paleoatmosphere.

of about 1×10^{12} molecules cm^{-2}. For comparison, O_3 column densities of about 5×10^{15}, 7×10^{16}, 2×10^{18}, and 6×10^{18} molecules cm^{-2} were previously found for O_2 levels of 10^{-4}, 10^{-3}, 10^{-2} and 10^{-1} PAL, respectively (see Table 3 in Levine *et al.*, 1981). These calculations indicate that prebiological levels of O_3 offered no protection at the Earth's surface from lethal solar ultraviolet radiation. The O profile shown in Figure 6 reflects the photochemical/chemical production (reactions (14) and (15)) and loss (reactions (16) and (17)) of O. The O maximum above 40 km results from the photolysis of CO_2 (reaction (14)) and is responsible for the maximum in the distributions of both O_2 and O_3 (Figures 5 and 6, respectively). These calculations indicate that in the prebiological paleoatmosphere the oxygen species, i.e., O_2, O, and O_3 were extremely deficient. It was not until the origin, evolution, and explosion of photosynthetic organisms that O_2 became an important atmospheric gas. The transition from a weakly reducing atmosphere to a strongly oxidizing atmosphere, due to oxygen resulting from photosynthetic activity, was probably the major air pollution event in the history of our planet. In conclusion, photochemical calculations suggest that in the absence of a continuous source, a possible CH_4-NH_3 dominated paleoatmosphere was probably very short-lived, if such an atmosphere ever existed at all. Instead, a more mildly reducing prebiological paleoatmosphere of CO_2 and N_2 is favored.

References

Abelson, P. H.: 1966, *Proc. Natl. Acad. Sci. (U.S.)* 55, 1365–1372.

Ackermann, M.: 1971, 'Ultraviolet solar radiation related to mesospheric processes', in G. Fiocco (ed.), *Mesospheric Models and Related Experiments*, D. Reidel Publ. Co., Dordrecht, Holland, pp. 149–159.

Bossard, A. and Toupance, G.: 1981, 'Far UV photolysis of methane-water gaseous mixtures and the prebiotic synthesis of aldehydes', in Y. Wolman (ed.), *Origin of Life*, Proceedings of the 3rd ISSOL Meeting, D. Reidel Publ. Co., Dordrecht, Holland, pp. 93–100.

Chang, S.: 1979, 'Comets: Cosmic connections with carbonaceous meteorites, interstellar molecules and the origin of life,' in M. Neugebauer, D. K. Yeomens, J. C. Brandt, and R. W. Hobbs (eds.), *Space Missions to Comets*, NASA Reference Publication 2089, National Technical Information Service, Springfield, Virginia, 59–112.

Des Marais, D.: 1981, 'Stable isotope measurements of geothermal emanations', in *Research and Technology Annual Report 1981*, NASA TM–81333, Ames Research Center, Moffett, California, 6.

Ferris, J. P. and Chen, C. T.: 1975, *J. Amer. Chem. Soc.* 97, 2962–2967.

Ferris, J. P. and Nicodem, D. E.: 1972, *Nature* 238, 268–269.

Ferris, J. P., Joshi, P. C., Edelson, E. H., and Lawless, J. G.: 1978, *J. Mol. Evol.* 11, 293–311.

Gold, T.: 1979, *J. Petrol. Geol.* 1, 3–19.

Hart, M. H.: 1978, *Icarus* 33, 23–39.

Hart, M. H.: 1979, *Origins of Life* 9, 261–266.

Henderson-Sellers, A. and Schwartz, A. W.: 1980, *Nature* 287, 526–528.

Holland, H. D.: 1962, 'Model for the evolution of the Earth's atmosphere', in A. E. J. Engle, H. L. James, and B. F. Leonard (eds.), *Petrologic Studies: A Volume in Honor of A. F. Buddington*, Geological Society of America, N.Y., 447–477.

Kasting, J. F. and Donahue, T. M.: 1980, *J. Geophys. Res.* 85, 3255, 3263.

Kasting, J. F. and Walker, J. C. G.: 1981, *J. Geophys. Res.* 86, 1147–1158.

Kasting, J. F., Liu, S. C., and Donahue, T. M.: 1979, *J. Geophys. Res.* 84, 3097–3107.

Kerr, R. A.: 1980, *Science* 210, 42–43.

Kuhn, W. R. and Atreya, S. K.: 1979, *Icarus* 37, 207–213.

Lasaga, A. C., Holland, H. D., and Dwyer, M. J.: 1971, *Science* 174, 53–55.

Lazcano-Araujo, A. and Oró, J.: 1981, 'Cometary material and the origins of life on Earth', in C. Ponnamperuma (ed.), *Comets and the Origin of Life*, D. Reidel Publ. Co., Dordrecht, Holland, pp. 191–225.

Levine, J. S.: 1978, 'The evolution of H_2O and CO_2 on Earth and Mars', in C. Ponnamperuma (ed.), *Comparatice Planetology*, Academic Press, N.Y., 165–182.

Levine, J. S. and Boughner, R. E.: 1979, *Icarus* 39, 310–314.

Levine, J. S. and Graedel, T. E.: 1981, *Trans. Amer. Geophys. Union* 62, 1177–1181.

Levine, J. S. and Schryer, D. S. (eds.): 1978, *Man's Impact on the Troposphere*, NASA Reference Publication 1022, National Technical Information Service, Springfield, Virginia, 376 pp.

Levine, J. S., Augustsson, T. R., and Hoell, J. M.: 1980, *Geophys. Res. Lett.* 7, 317–320.

Levine, J. S., Hays, P. B., and Walker, J. C. G.: 1979, *Icarus* 39, 295–309.

Levine, J. S., Augustsson, T. R., Boughner, R. E., Natarajan, M., and Sacks, L. J.: 1981, 'Comets and the photochemistry of the paleoatmosphere', in C. Ponnamperuma (ed.), *Comets and the Origin of Life*, D. Reidel Publ. Co., Dordrecht, Holland, pp. 161–190.

McGovern, W. E.: 1969, *J. Atmos. Sci.* 26, 623–635.

Miller, S. L.: 1953, *Science* 117, 528–529.

Miller, S. L. and Urey, H. C.: 1959, *Science* 130, 245–251.

Oró, J.: 1961, *Nature* 190, 389–390.

Rubey, W. W.: 1955, Geol. Soc. Amer., Spec. Paper 62, *Crust of the Earth*, 631–650.

Sagan, C. and Mullen, G.: 1972, *Science* 177, 52–56.

Shimizu, M.: 1976, *Precamb. Res.* 3, 463–470.

Walker, J. C. G.: 1976, 'Implications for atmospheric evolution of the inhomogeneous accretion model of the origin of the Earth', in B. F. Windley (ed.), *The Early History of the Earth*, John Wiley and Sons, N.Y., pp. 537–546.

Walker, J. C. G.: 1977, *Evolution of the Atmosphere*, MacMillan, New York, 318 pp.

Walker, J. C. G.: 1978a, 'Atmospheric evolution on the inner planets', in C. Ponnamperuma (ed.), *Comparative Planetology*, Academic Press, N.Y., pp. 141–164.

Walker, J. C. G.: 1978b, *Pure Appl. Geophys.* 116, 222–231.

Yung, Y. L. and McElroy, M. B.: 1979, *Science* 203, 1002–1004.

Possible limits on the composition of the Archaean ocean

James C. G. Walker

Space Physics Research Laboratory, Department of Atmospheric and Oceanic Science, The University of Michigan, Ann Arbor, Michigan 48109, USA

It has been suggested that the partial pressure of carbon dioxide in the terrestrial atmosphere was larger in the past than it is at present[1-4]. In particular, partial pressures of 100–1,000 times those of the present have been invoked for the early Earth as a means of insuring equable climates at a time when the Sun was significantly less luminous[5-7]. While the climatic argument is not conclusive[8], it is worth considering whether the geological record is in any way inconsistent with the proposed high partial pressures of carbon dioxide. I now examine the potential impact of high carbon dioxide partial pressure on ocean chemistry and ask what constraints are imposed by the known record of chemical sedimentation through time. The evidence consists of the persistence throughout almost the entire sedimentary rock record of calcium carbonate and sulphate precipitation. I adopt a uniformitarian point of view that assumes no very great change in the conditions for the deposition of these chemical sediments. The methods of Holland[9] are used to set limits on the composition of the water from which precipitation occurred. I find no inconsistencies between the sedimentary rock record and presumed higher partial pressure of carbon dioxide early in Earth history, provided that high partial pressure was accompanied by a generally lower pH for seawater, higher concentrations of calcium and bicarbonate ions, and lower concentrations of carbonate and sulphate ions.

For specified values of the pH and carbon dioxide partial pressure the concentrations of carbonate and bicarbonate ions in solution are determined by equilibrium relationships among the carbon-bearing species[10]. At a temperature of 25 °C and pressure of a few atmospheres the relationships are

$$[H^+][HCO_3^-] = 10^{-7.5}P \qquad (1)$$

$$[H^+]^2[CO_3^{2-}] = 10^{-16.6}P \qquad (2)$$

where $[X]$ denotes the concentration of dissolved species X in mol l^{-1} and P denotes the partial pressure of carbon dioxide in atmospheres[11]. I ignore the weak dependence of the equilibrium constants on pressure and temperature. Constant carbonate and bicarbonate ion concentrations are shown as solid and dashed lines in Fig. 1.

Deposits of calcium carbonate are present in sedimentary rocks of all ages extending back to the oldest yet discovered[12]. From the persistence of calcium carbonate precipitation in the sedimentary rock record I estimate that the ocean has always been close to saturation with respect to calcium carbonate, as it is today. With the assumption that the product of calcium and carbonate ion concentration has not changed markedly with time, then,

$$[Ca^{2+}][CO_3^{2-}] = 10^{-6} \qquad (3)$$

where values for the present day oceanic average are as quoted by Broecker[13]. In Fig. 1, therefore, the lines of constant carbonate ion concentration apply also to the calcium ion concentration, with the values shown.

Sedimentary sulphate evaporites are known from rocks on all the contents except Antarctica and of all ages extending back to 3,500 Myr ago[14]. These sulphate evaporites include massive deposits tens of metres thick and covering thousands of square kilometres in horizontal extent. Many of them contain evidence of having been initially precipitated as gypsum ($GaSO_4.2H_2O$). With allowance for the ease with which evaporite deposits may have been destroyed by weathering

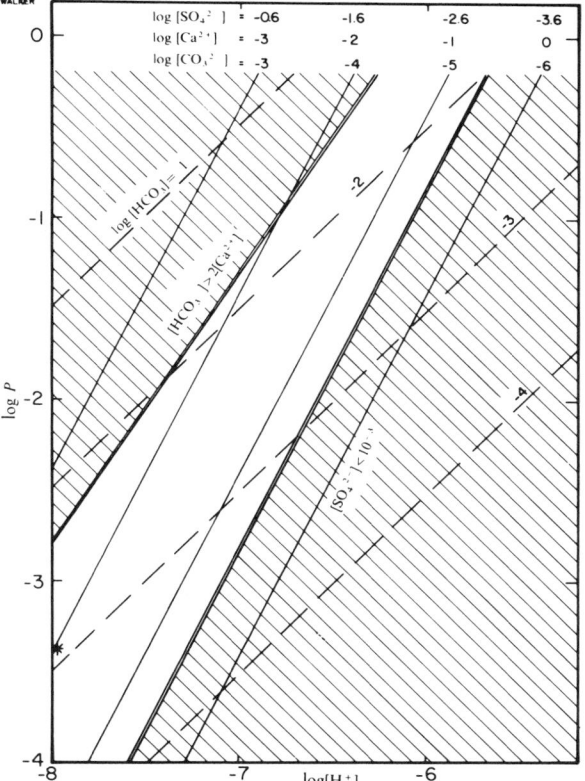

Fig. 1 Concentrations of dissolved species in seawater as functions of the carbon dioxide partial pressure and the concentration of H$^+$ ions. Concentrations are expressed in mol l^{-1} and the partial pressure in atmospheres. Equilibrium in the carbonate system yields the dashed lines for bicarbonate concentration and the solid lines for carbonate concentration. The values for calcium and sulphate concentrations are derived by assuming a constant product of calcium and carbonate concentrations and of calcium and sulphate concentrations. The star indicates the composition of present day seawater. The cross-hatched regions correspond to compositions that are not consistent with the sedimentary rock record.

processes, the record suggests that the mineralogy of the initial stages of evaporite formation has not changed significantly with time. The record is at least consistent with the uniformitarian assumption that gypsum was the first abundant mineral, after carbonate, to precipitate from evaporating seawater in the Precambrian as in the Phanerozoic[15]. The record is not sufficiently good to determine whether the proportions of carbonate and gypsum have changed with time. It is therefore conservative to assume that the product of calcium and sulphate ion concentrations has not changed with time. Then

$$[Ca^{2+}][SO_4^{2-}] = 10^{-3.6} \qquad (4)$$

and the lines of constant carbonate concentration in Fig. 1 apply also to sulphate, with values as shown. The oxidation state of sulphur in the Archaean ocean and atmosphere has been discussed by Brimblecombe and Walker[16,17]. Even in the absence of free oxygen it seems likely that photochemical reactions in the atmosphere and surface ocean would have yielded abundant sulphate from volcanic emanations of hydrogen sulphide and sulphur dioxide. The banded iron-formations of the Archaean and early Proterozoic provide evidence for abundant ferrous ions in solution[18-21], which argues against high sulphide concentrations because of the low solubility of iron sulphides[22].

Some constraints on the composition of seawater follow from the occurrence of massive deposits of evaporitic sulphate minerals some tens of metres thick. An approximate lower limit on the sulphate concentration can be derived from the requirement that such a deposit accumulate in a reasonable period of time. A generous upper limit on the evaporation rate in arid terrestrial environments is $10 \, m \, yr^{-1}$ (ref. 23). At this rate and at a sulphate concentration of $10^{-3} \, mol \, l^{-1}$ it would take 13,500 yr to accumulate a layer of gypsum 1 m thick. The stability of evaporitic environments for times longer than 100,000 yr seems unlikely, so sulphate concentrations $<10^{-3} \, mol \, l^{-1}$ are probably not consistent with the geological record. The excluded compositions are indicated by shading on the right of Fig. 1. A lower limit on carbonate ion concentration and upper limit on calcium ion concentration correspond to the lower limit on sulphate ion concentration.

If evaporating seawater is to precipitate first calcium carbonate and then calcium sulphate then the calcium ion concentration must exceed one half the bicarbonate ion concentration[9]. If this were not the case, precipitation of calcium carbonate would exhaust the calcium ions in seawater leaving none to enter gypsum. The persistence of calcium sulphate evaporites in the sedimentary rock record therefore sets an upper limit on the bicarbonate ion concentration indicated by the shaded area on the left of Fig. 1. The clear band in the middle of Fig. 1 depicts compositions of seawater that are consistent with the sedimentary rock record in the sense that they would have yielded no obvious secular changes in the minerals deposited during the initial stages of evaporite formation. The allowed concentrations are summarized as functions of carbon dioxide partial pressure in Fig. 2. Higher carbon dioxide partial pressure on the early Earth would have corresponded to generally higher concentrations of calcium, bicarbonate and hydrogen ions and generally lower concentrations of carbonate and sulphate ions. Cameron[24] has argued from sulphur isotope data that sulphate concentrations were $<10^{-3} \, mol \, l^{-1}$ in the Archaean ocean.

Modern seawater is close to saturation with respect to the mineral barite ($BaSO_4$) (ref. 10). The concentrations of barium ions shown in Fig. 2 are derived assuming that the product of barium and sulphate ion concentrations has not changed with time. The generally higher barium concentrations that correspond to higher carbon dioxide partial pressures are consistent with the deduction of Veizer *et al.*[25] that the concentration of barium ions was enhanced by about a factor of 10 in the Archaean ocean. The concentrations of iron ions are those that yield saturation with respect to the mineral siderite ($FeCO_3$), calculated in accordance with the factors described by Holland[19]. High concentrations (exceeding $10^{-4} \, mol \, l^{-1}$) of dissolved ferrous iron in Archaean and early Proterozoic seawater[26] are favoured as a source of the chemically precipitated iron in banded iron formations[18-21].

In the absence of silica-precipitating organisms, the oceans were presumably approximately saturated with respect to amorphous silica. Holland[9] estimates a concentration of dissolved silica of about 100 p.p.m. This high concentration of dissolved silica is presumably responsible for the abundance of chemically precipitated siliceous sediments in formations of Archaean and early Proterozoic age. In the absence of extensive continental land masses[27] there were presumably few large deposits of evaporitic halite ($NaCl$). Enhanced concentrations of sodium and chloride ions in Archaean seawater are a possibility. The oceans would be far from saturation with respect to halite even if all known evaporite deposits were dissolved[15].

I have not discussed possible reasons for the postulated high pressures of carbon dioxide on the early Earth. Mechanisms that relate to the rate of weathering of continental rocks[6,28,29] cannot be invoked for the Archaean era, when oceanic composition was dominated by interactions with the mantle rather than with continental rocks[25]. In general terms, the carbon dioxide partial pressure would have been determined by the partitioning of carbon between the fluid (ocean plus atmosphere) and solid (rock) phases. Higher internal temperatures and more tectonic activity on the early Earth may have promoted the remobilization of carbon from the solid to the fluid phases[1,30]. In the conditions described here, the fraction of total terrestrial carbon

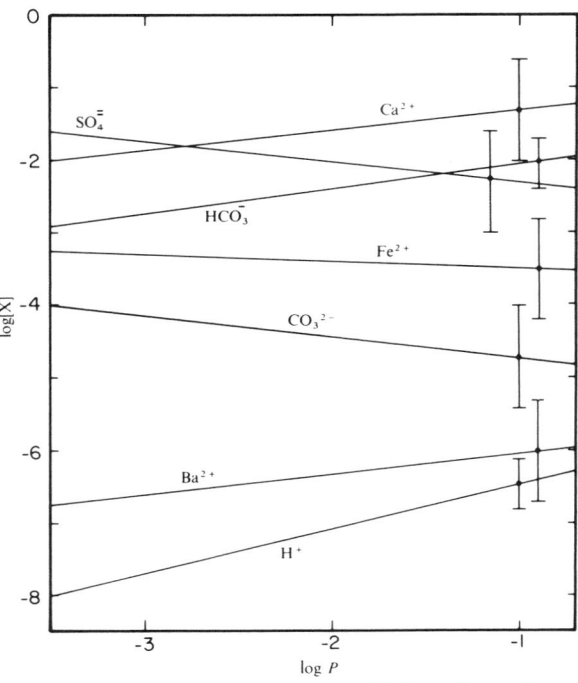

Fig. 2 Seawater composition consistent with the rock record as a function of carbon dioxide partial pressure. The lines correspond to the middle of the allowed region of Fig. 1. The bars represent the range of allowed concentrations. The barium ion concentration corresponds to the assumption of constant product of barium and sulphate concentrations. The iron concentration corresponds to an ocean saturated with respect to siderite.

in the fluid phase is still small. A more precise understanding of the controls on the composition of the Archaean ocean and atmosphere undoubtedly involves the details of the interaction of seawater with mantle rocks[25,27,31].

This paper describes research supported in part by NASA under grant NAGW-176 to the University of Michigan. It is a contribution of the Precambrian Paleobiology Research Group.

Received 26 November 1982; accepted 20 January 1983.

2. Hart, M. H. *Icarus* **33**, 23–39 (1978).
3. Henderson-Sellers, A. & Cogley, J. G. *Nature* **298**, 832–835 (1982).
4. Veizer, J. *Contr. Miner. Petrol.* **38**, 261–278 (1973).
5. Owen, T., Cess, R. D. & Ramanathan, V. *Nature* **277**, 640–642 (1979).
6. Walker, J. C. G., Hays, P. B. & Kasting, J. F. *J. geophys. Res.* **86**, 9776–9782 (1981).
7. Walker, J. C. G. *Paleogeogr. Paleoclimatol. Paleoecol.* **40**, 1–11 (1982).
8. Rossow, W. B., Henderson-Sellers, A. & Weinreich, S. K. *Science* **217**, 1245–1247 (1982).
9. Holland, H. D. *Geochim. cosmoschim. Acta* **36**, 637–651 (1972).
10. Riley, J. P. & Chester, R. *Introduction to Marine Chemistry* (Academic, New York, 1971).
11. Walker, J. C. G. *Evolution of the Atmosphere* (Macmillan, New York, 1977).
12. Allaart, J. H. in *The Early History of the Earth* (ed. Windley, B. F.) 177–189 (Wiley, London, 1976).
13. Broecker, W. S. *Chemical Oceanography* (Harcourt Brace Jovanovich, New York, 1974).
14. Walker, J. C. G. *et al.* in *The Earth's Earliest Biosphere: Its Origin and Evolution* (ed. Schopf, J. W.) (Princeton University Press, in the press).
15. Holland, H. D. *The Chemistry of the Atmosphere and Oceans* (Wiley-Interscience, New York, 1978).
16. Brimblecombe, P. & Walker, J. C. G. *J. geophys. Res.* (submitted).
17. Brimblecombe, P. & Walker, J. C. G. *J. geophys. Res.* (submitted 1982).
18. Drever, J. I. *Bull geol. Soc. Am.* **85**, 1099–1106 (1974).
19. Holland, H. D. *Econ. Geol.* **68**, 1169–1172 (1973).
20. Ewers, W. E. in *4th int. Symp. Environmental Biogeochemistry* (eds Trudinger, P. A. & Walter, M. R.) (Springer, New York, 1980).
21. Button, A. *et al.* in *Mineral Deposits and the Evolution of the Biosphere* (eds Holland, H. D. & Schidlowski, M.) 259–273 (Springer, Berlin, 1982).
22. Walker, J. C. G. *Life Sci. Space Res.* **8**, 89–100 (1980).
23. Sellers, W. D. *Physical Climatology* (University of Chicago Press, 1965).
24. Cameron, E. M. *Nature* **296**, 145–148 (1982).
25. Veizer, J., Compston, W., Hoefs, J. & Nielsen, H. *Naturwissenschaften* **69**, 173–180 (1982).
26. Veizer, J. in *The Earth's Earliest Biosphere: Its Origin and Evolution* (ed. Schopf, J. W.) (Princeton University Press, in the press).
27. Lovelock, J. E. & Whitfield, M. *Nature* **296**, 561–563 (1982).
28. Lovelock, J. E. & Watson, A. J. *Planet. Space Sci.* **30**, 795–802 (1982).
29. Walker, J. C. G. in *Comparative Planetology* (ed. Ponnamperuma, C.) 141–163 (Academic, New York, 1978).
30. Fryer, B. J., Fyfe, W. S. & Kerrich, R. *Chem. Geol.* **24**, 25–33 (1979).

1. Holland, H. D. in *The Early History of the Earth* (ed. Windley, B. F.) 559–567 (Wiley, London, 1976).

Climatic Consequences of Very High Carbon Dioxide Levels in the Earth's Early Atmosphere

JAMES F. KASTING AND THOMAS P. ACKERMAN

The possible consequences of very high carbon dioxide concentrations in the earth's early atmosphere have been investigated with a radiative-convective climate model. The early atmosphere would apparently have been stable against the onset of a runaway greenhouse (that is, the complete evaporation of the oceans) for carbon dioxide pressures up to at least 100 bars. A 10- to 20-bar carbon dioxide atmosphere, such as may have existed during the first several hundred million years of the earth's history, would have had a surface temperature of approximately 85° to 110°C. The early stratosphere should have been dry, thereby precluding the possibility of an oxygenic prebiotic atmosphere caused by photodissociation of water vapor followed by escape of hydrogen to space. Earth's present atmosphere also appears to be stable against a carbon dioxide–induced runaway greenhouse.

EARTH HAS APPROXIMATELY 60 BARS of carbon dioxide tied up in carbonate rocks, roughly two-thirds the amount present in the atmosphere of Venus (1, 2). This carbon, along with other volatile elements, was presumably brought to the earth during accretion as a component of infalling planetesimals. A substantial fraction of these volatile compounds should have been released upon impact (3–5). Carbon may have been degassed as CO_2 or as some more reduced gas (CO or CH_4), depending on the oxidation state of the infalling material and of the upper mantle. Once in the atmosphere, however, any reduced carbon species should have been oxidized to CO_2 by OH radicals produced from water vapor photolysis (6). Consequently, the earliest atmosphere may have contained large amounts of CO_2—up to one-third of the earth's total inventory, or 20 bars, according to Holland's estimate (3). If the fraction of the earth's surface occupied by continents was initially small, carbonate formation would have been inhibited and sea-floor carbonate sediments would have been rapidly recycled; thus, a dense (approximately 10 bar) CO_2 atmosphere could conceivably have persisted for several hundred million years (7).

To explore the possible climatic consequences of high CO_2 concentrations in the early atmosphere, we made a series of calculations using a one-dimensional radiative-convective climate model. The primary goal of these calculations was to determine whether a runaway greenhouse could have occurred on the early earth. A runaway greenhouse is here defined as an atmosphere in which water is present entirely as steam or clouds; no oceans or lakes are present at the surface. We concern ourselves only with times subsequent to the accretion period, when the earth was heated solely by absorption of solar radiation. The possibility of a runaway greenhouse during accretion will be considered elsewhere. A second purpose of this study was to determine the stability of a high CO_2 primitive atmosphere against water loss through photodissociation of water vapor followed by escape of hydrogen to space. An understanding of this latter question is needed in order to estimate the earth's initial water inventory and to predict the oxidation state of the early atmosphere. An interesting by-product of our calculation is an estimate of the stability of the earth's current atmosphere to large CO_2 increases.

The radiative-convective model employed here is based on one used in previous studies of the earth's climate system (8, 9). It has, however, been updated to include new absorption coefficients for H_2O and CO_2 (10) along with a self-consistent calculation of solar energy deposition (11). The band model coefficients used to define gaseous

absorption were derived for pressures of 0.1 and 1 bar; calculated transmission functions are not expected to be accurate at higher pressures. This should have little effect on our results, since the dominant mode of energy transport at these higher pressures is convection.

The most important physical assumptions made in the model are related to our treatment of tropospheric lapse rate, relative humidity, and clouds. The lapse rate was set equal to its moist adiabatic value, following Ingersoll's formulation (12), which is valid for large water vapor amounts. The use of the moist adiabatic lapse rate causes the surface temperature T_s to increase much more slowly with increasing CO_2 than it would in a fixed lapse-rate model (8, 13, 14) because the temperature of the upper troposphere increases more rapidly than does T_s.

Relative humidity cannot be calculated self-consistently with a one-dimensional model because it is determined by three-dimensional dynamical processes. Since our primary goal is to calculate upper limits on surface temperature, we wish to ensure that the troposphere is nearly saturated with water vapor at high CO_2 concentrations. At low CO_2 concentrations, however, we want the troposphere to revert to its present unsaturated state. To effect such a transition the tropospheric relative humidity was assumed to increase as the fractional amount of water vapor at the surface increased (15). This assumption is consistent with the idea that the behavior of atmospheric water vapor is related to its mixing ratio (12). Our parameterization has no rigorous theoretical justification, however, and may be regarded simply as an artifice for connecting unsaturated low T_s solutions to nearly saturated high T_s solutions.

The stratospheric water vapor content was estimated by allowing relative humidity to increase to unity above the convective region, provided that the H_2O volume mixing ratio, $f(H_2O)$, did not increase with altitude. This approximate cold-trapping mechanism tends to overestimate the H_2O content of the stratosphere because it ignores latitudinal variations in tropopause temperature. This is acceptable for our purposes because we wish to derive upper limits on surface temperature and hydrogen escape rate.

Clouds were not included explicitly in our model because we do not know how they would vary as a function of CO_2 concentration. Their effect on climate was included implicitly by adopting a high surface albedo ($A_s = 0.22$). This value of A_s was chosen because it allows the model to reproduce the

Space Science Division, NASA Ames Research Center, Moffett Field, CA 94035.

Reprinted with permission from *Science*, Volume 234, pp. 1383-1385. © 1986 by the American Association for the Advancement of Science.

observed mean value of T_s (288 K) for the present earth, given the present solar insolation. By holding A_s fixed, we assume zero cloud feedback at higher CO_2 levels. This parameterization ignores the effects of clouds on the infrared radiation budget. In reality, clouds do absorb infrared radiation; they also reflect more sunlight back to space than assumed in our model. Under certain (pathological) conditions—the development of a widespread optically thin cirrus layer, for instance—cloud feedbacks could cause surface temperatures at elevated CO_2 concentrations to be considerably higher than calculated here. For other possible changes, such as an increase in areal coverage by optically thick, low-altitude clouds, the increase in planetary albedo would probably more than offset any increase in the greenhouse effect, and a net cooling would result. If, on the other hand, fractional cloud cover remained constant and the cloud tops remained at roughly a constant temperature,

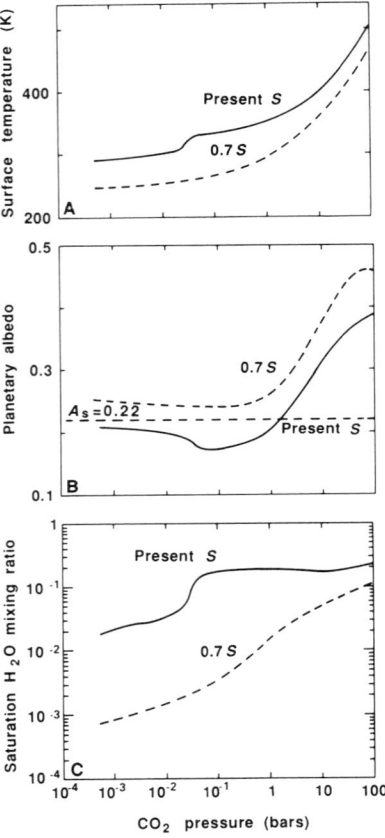

Fig. 1. Effect of increased CO_2 pressure on (A) surface temperature, (B) planetary albedo, and (C) saturation H_2O volume mixing ratio at the surface. Solid curves are for present solar luminosity S and atmospheric composition; dashed curves represent 30% reduced luminosity and no O_2 or O_3.

T_s should be virtually unaffected. Without a physical model for predicting which of these effects is most plausible, it seems fruitless to speculate as to what the actual effect of clouds would be.

Calculations were performed for both early and present earth. For the early earth model, the solar luminosity S was set at 70% of its present value (*16*), and a 0.8-bar N_2 background atmosphere was assumed (*17*). For the present earth model, the modern value of solar luminosity was used and O_2 and O_3 were included in the model atmosphere. The present earth calculations, besides being interesting from an academic standpoint, help to bracket some of the uncertainties in the problem by providing a reasonable upper limit on surface temperatures for the early earth. Even if several of our assumptions (for example, neglect of cloud feedback) turned out to be invalid, it seems unlikely that their effect on surface temperature could be as large as that of a 40% solar flux increase.

The results of increasing the CO_2 pressure up to 100 bars are shown in Fig. 1. The solid curves represent the present earth and the dashed curves represent the early earth. Both cases are apparently stable against the development of a runaway greenhouse. Surface temperature for the present earth model with 100 bars of CO_2 is 506 K, or 233°C (Fig. 1A). The saturation H_2O vapor pressure is 29 bars, so that the total surface pressure is 130 bars. At this pressure the oceans would boil only if T_s were about 100° hotter. Thus, the bulk of the earth's water (1.4×10^{24} g, or 270 bars) would reside in the ocean. For an early earth with 10 to 20 bars of CO_2, the oceans should have been even more stable; T_s would have been 85° to 110°C, with a vapor pressure of 0.6 to 1.5 bars.

An important factor in stabilizing the climate at high CO_2 concentrations is the increase in planetary albedo (A_p) caused by Rayleigh scattering. For the early earth model A_p remains near 0.25 at low CO_2 pressures, but rises to about 0.4 at CO_2 pressures of 10 to 20 bars (Fig. 1B). For the present earth model, A_p is lower because of increased absorption of solar energy by H_2O and O_3.

In addition to being stable against a runaway greenhouse, our early earth atmosphere is also stable against loss of water through photodissociation followed by escape of hydrogen to space. The maximum (diffusion-limited) escape rate is proportional to the total hydrogen mixing ratio in the stratosphere in all of its chemical forms (*18*). The stratospheric H_2O content remains low as long as water is a minor atmospheric constituent (*12*). For the early earth model,

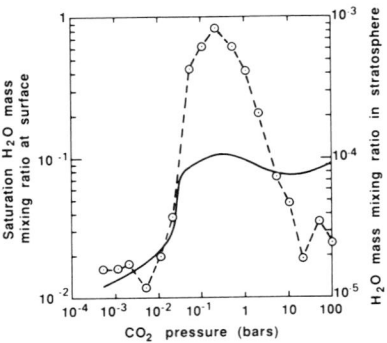

Fig. 2. Saturation mass mixing ratio of H_2O at the ground [solid curve, $c_{sat}(H_2O)$] and in the stratosphere (dashed curve, $c_0(H_2O)$) for the present earth model.

H_2O makes up less than 10% of the atmosphere (Fig. 1C), and the stratospheric H_2O mixing ratio is less than 10^{-7} at all CO_2 pressures. The associated hydrogen escape rate is $<5 \times 10^6$ H atoms per square centimeter per second. This rate is too small to affect the amount of water in the ocean and has a negligible impact on the atmospheric hydrogen budget. Thus, our results would appear to rule out the possibility of an oxygenic prebiotic atmosphere, which has been suggested by some workers (*19, 20*).

The sharp increase in T_s and $f_{sat}(H_2O)$ at CO_2 pressures near 0.03 bar in the present earth model is a consequence of our relative humidity parameterization (*15*). Above this CO_2 pressure, water vapor constitutes about 20% of the lower atmosphere, so that the troposphere is 80% saturated. This abrupt increase probably has little physical significance; a different choice of parameterizations could either shift the increase to a different CO_2 pressure or eliminate it entirely. Models that assume a constant relative humidity (*9*) do not exhibit such a feature. Neither, however, are they useful for determining upper limits on surface temperature or stratospheric water vapor abundances. The surface H_2O mixing ratio in the present earth model actually decreases slightly at CO_2 pressures between 1 and 10 bars. The reason is that CO_2 is not that efficient as a greenhouse gas; thus, the increase in surface pressure caused by the addition of CO_2 outstrips the increase in H_2O vapor pressure caused by higher T_s.

This behavior is seen more clearly if one plots the H_2O mass mixing ratio at the surface [$c_{sat}(H_2O)$] as a function of CO_2 pressure (Fig. 2). The maximum value of $c_{sat}(H_2O)$ occurs at a CO_2 pressure of 0.2 bar, at which point the greenhouse effect of CO_2 is large but the surface pressure is still relatively low. Corresponding to this maximum in $c_{sat}(H_2O)$ is a pronounced peak in stratospheric H_2O mass mixing ratio $[c_0(H_2O)]$ (Fig. 2). This peak is a consequence of the decreased efficiency of the cold trap at high H_2O levels (8, 12). At the high point of this curve the oceans could be depleted of water in about 9 billion years if the diffusion-limited hydrogen escape flux was achieved. A 30° increase in T_s, which could result from a mere 10% increase in solar flux according to our model, would cause $c_0(H_2O)$ to rise to about 0.1 at this CO_2 pressure. The oceans would then disappear in only a few hundred million years, as may have happened on Venus (8). The present atmosphere might therefore be said to be marginally stable against water loss at CO_2 pressures of 0.1 to 1 bar. At still higher CO_2 pressures, the stratosphere should once again be dry and the oceans should be more stable.

Our calculations do not prove that the early atmosphere must have been hot. The question of how CO_2 was originally partitioned between the atmosphere, ocean, and solid planet is complex, and it is possible that the early atmosphere was much less massive than we have assumed. We have shown, however, that a dense CO_2 atmosphere could have existed on the early earth without violating any known constraints on the planet's subsequent evolution.

REFERENCES AND NOTES

1. A. B. Ronov and A. A. Yaroshevsky, *Geochemistry* **11**, 1041 (1967).
2. H. D. Holland, *The Chemistry of the Atmosphere and Oceans* (Wiley, New York, 1978).
3. _____, *The Chemical Evolution of the Atmosphere and Oceans* (Princeton Univ. Press, Princeton, NJ, 1984).
4. B. M. Jakosky and T. J. Ahrens, in *Proceedings of the 10th Lunar and Planetary Science Conference* (Pergamon, New York, 1979), p. 2727.
5. T. Matsui and Y. Abe, *Nature (London)* **319**, 303 (1986).
6. J. F. Kasting, K. J. Zahnle, J. C. G. Walker, *Precambrian Res.* **20**, 121 (1983).
7. J. C. G. Walker, *Origins Life* **16**, 117 (1986).
8. J. F. Kasting, J. B. Pollack, T. P. Ackerman, *Icarus* **57**, 335 (1984).
9. J. F. Kasting, J. B. Pollack, D. Crisp, *J. Atmos. Chem.* **1**, 403 (1984).
10. Infrared absorption by H_2O and CO_2 was calculated in 55 spectral intervals by fitting synthetic spectra computed from the Air Force Geophysical Laboratories (AFGL) tape [R. A. McClatchey *et al.*, *Air Force Geophys. Lab. Tech. Rep. AFCRL-71-0279* (Bedford, MA, 1971)]. The calculations were performed at pressures of 1 and 0.1 bar and at temperatures of 200, 300, and 600 K. H_2O was assumed to obey the Fels-Goody model [S. B. Fels, *Appl. Opt.* **18**, 2634 (1979)] and CO_2 was assumed to follow the Malkmus model. Self-broadening efficiencies for H_2O and CO_2 were enhanced by factors of 5 and 1.3, respectively, and broadening of H_2O lines by CO_2 was enhanced by a factor of 2 [J. R. Izatt *et al.*, *J. Opt. Soc. Am.* **59**, 19 (1969); P. Varanasi, *J. Quant. Spectrosc. Radiat. Transfer* **11**, 223 (1971)]. The fitting method used here avoids errors that would otherwise be incurred by applying such band models over relatively broad (50 to 100 cm^{-1}) spectral intervals [J. T. Kiehl and V. Ramanathan, *J. Geophys. Res.* **88**, 5191 (1983)]. H_2O continuum absorption and pressure-induced CO_2 absorption were treated as described by Ingersoll (12). Absorption in the visible and near infrared was computed in each of 38 spectral intervals ranging from 0.2 to 4.4 μm. At wavelengths longer than 0.67 μm, exponential sum coefficients for H_2O and CO_2 were derived from the AFGL tape. The coefficients were computed for pressures of 10^{-3} to 10 bars and a temperature of 300 K. At wavelengths between 0.54 and 0.67 μm, H_2O absorption was estimated from tabulated lines in the solar spectrum [C. E. Moore *et al.*, *Natl. Bur. Stand.* **61** (Washington, DC, 1966)]. CO_2 absorption between 1.04 and 0.71 μm was estimated from laboratory spectra [G. Herzberg and L. Herzberg, *J. Opt. Soc. Am.* **43**, 1037 (1953)].
11. Rayleigh scattering by CO_2 is particularly important because its scattering cross section is about 2.5 times as high as that of air [(H. C. Van de Hulst, in *The Atmospheres of the Earth and Planets*, G. P. Kuiper, Ed. (Univ. of Chicago Press, Chicago, 1957), p. 49]. Our cross sections are the same as those used in another recent model of the climatic effects of large CO_2 increases [I. M. Vardavas and J. H. Carver, *Planet. Space Sci.* **32**, 1307 (1984)].
12. A. P. Ingersoll, *J. Atmos. Sci.* **26**, 1191 (1969).
13. R. S. Lindzen, A. Y. Hou, B. F. Farrell, *ibid.* **39**, 1189 (1982).
14. M. Lal and V. Ramanathan, *ibid.* **41**, 2238 (1984).
15. Tropospheric relative humidity was assumed to obey the relation

$$r = r_0\left[\frac{p/p_s - 0.02}{1 - 0.02}\right]^\Omega \qquad (1)$$

where p is pressure, p_s is surface pressure, and r_0 (= 0.8) is the surface relative humidity. The parameter Ω allows the relative humidity of the upper troposphere to vary. When $\Omega = 1$, Eq. 1 reduces to the empirical formula of Manabe and Wetherald [*J. Atmos. Sci.* **24**, 241 (1967)], which is used in many one-dimensional climate models. When $\Omega = 0$, the troposphere is 80% saturated throughout. This formulation follows a suggestion by R. D. Cess [*ibid.* **33**, 1831 (1976)] which he made on the basis of observed latitudinal gradients in T_s and r. He proposed that $\Omega = 1 - 0.03(T_s - 288 \text{ K})$. We have replaced this formula with

$$\Omega = 1 - \frac{f_{sat}(H_2O) - f_p}{0.1 - f_p}$$

with $0 \leq \Omega \leq 1$ and where $f_{sat}(H_2O)$ [$= p_{sat}(H_2O)/p_s$] is the saturation H_2O mixing ratio at the surface and f_p (= 0.0166) is the value of $f_{sat}(H_2O)$ for the present atmosphere ($T_s = 288$ K). Ω is assumed to be unity for $f_{sat}(H_2O) < f_p$ and 0 for $f_{sat}(H_2O) > 0.1$. We used this equation instead of Cess's formula because we feel that the behavior of water vapor in an atmosphere is more closely related to its mixing ratio than to surface temperature. It gives approximately the same result as Cess's formula for a 1-bar atmosphere, but yields a slower increase in r at higher surface pressures.
16. D. O. Gough, *Sol. Phys.* **74**, 21 (1981).
17. The surface pressure of the model atmosphere is given by $p_s = p_n + pCO_2 + p_{sat}(H_2O)$, where p_n is the pressure of gases other than CO_2 and H_2O. The CO_2 pressure so defined is the pressure that would be exerted by a pure CO_2 atmosphere containing a given amount of gas. It differs from the actual CO_2 partial pressure at the surface by a factor of $44/M$, where M is the mean molecular weight of the atmosphere. The CO_2 pressure used here is linearly proportional to CO_2 column content, whereas the actual CO_2 partial pressure is not.
18. D. M. Hunten, *J. Atmos. Sci.* **30**, 1481 (1973).
19. J. H. Carver, *Nature (London)* **292**, 136 (1981).
20. K. M. Towe, *Precambrian Res.* **20**, 161 (1983).

30 June 1986; accepted 6 October 1986

Annihilation of ecosystems by large asteroid impacts on the early Earth

Norman H. Sleep[*], Kevin J. Zahnle[†], James F. Kasting[‡] & Harold J. Morowitz[§]

[*] Department of Geophysics, Stanford University, Stanford, California 94305, USA
[†] NASA Ames Research Center, Moffett Field, California 94035, USA
[‡] Department of Geosciences, Pennsylvania State University, University Park, Pennsylvania 16802, USA
[§] Biology Department, George Mason University, Fairfax, Virginia 22030-4444, USA

Large asteroid impacts produced globally lethal conditions by evaporating large volumes of ocean water on the early Earth. The Earth may have been continuously habitable by ecosystems that did not depend on photosynthesis as early as 4.44 Gyr BP (before present). Only a brief interval after 3.8 Gyr exists between the time when obligate photosynthetic organisms could continuously evolve and the time when the palaeontological record indicates highly evolved photosynthetic ecosystems.

THE early Earth was bombarded by large projectiles at the same time that large impact basins formed on the Moon. It is sometimes asserted (for example, ref. 1) that these impacts would have precluded the continued existence of life before 3.8 Gyr, but little has been done until recently to quantify the untoward effects of impacts on life. Maher and Stevenson[2] considered the possibility that frequent impacts could have prevented the origin of life, particularly for sites at the surface. They did so by comparing the time between impacts at a given location on the Earth's surface with the time required for biogenesis. In this paper, we concern ourselves with a related, yet inherently different problem, namely: When did the last planet-sterilizing impact occur? That is, we assume that life originated in some unspecified manner, and then consider what it would have taken to have wiped it out. This approach is better because the timescale for originating life is very poorly constrained, whereas early ecosystems with well defined organisms are broadly constrained by analogy with modern ones. In addition, extant organisms and the palaeontological record relate only to continuous evolution after the last impact to sterilize the planet.

We argue that the major way in which the planet can be sterilized is by evaporation of ocean water. This process affects some organisms more strongly than others, depending on how much water is evaporated and what type of ecosystem is envisioned. Certain classes of ecosystems may be able to withstand thermal perturbations lasting hundreds or thousands of years. We first discuss the flux of impacting objects, then the effects of impacts, and finally their effects on plausible early ecosystems.

Flux of impacting objects on the early Earth

The impact of a Mars-sized object, which is believed to have formed the Moon, would have either melted or vaporized much of the Earth[3]. Neither living cells nor complex organic molecules would have survived this blow. It is the flux of lesser later projectiles on the early Earth that is relevant to biology. No direct record of the earliest impacts exists on the Earth. By contrast, the lunar record is well preserved and well studied. Thus, the lunar crust is a reasonable starting point for our study. Our approach follows that in refs 4–7. We represent impact fluxes in terms of the equivalent thickness that the projectiles would add to a planet if all the material were retained. This measure is independent of planetary size so long as gravitational focusing and the statistics of the small number of largest objects can be neglected.

Geochemical estimates of flux. A minimum estimate of the material that hit the moon is obtained from the meteoritic component retained in lunar samples. Rocks exposed in the lunar highlands, which are believed to be typical of the lunar upper crust, are breccias composed of pristine rocks, impact melts and still older breccias. It is generally agreed that Ir abundances indicate a meteoritic component of 1–2%. It is controversial whether Ni abundances indicate a comparable meteoritic component that lacks Ir (ref. 8) or an internal component that was reduced to metal in the crust[9]. Studies of lunar meteorites indicate that the excess nickel is local to the nearside landing sites and that the meteoritic component is between 1–2%[10]. The total thickness of lunar crust that is contaminated with siderophiles is not well constrained. Granulites believed to be deeply buried breccias as old as 4.26 Gyr (ref. 11) and 4-Gyr old near-surface breccias are both enriched in siderophiles, indicating that the meteoritic component is mixed in much of the upper crust. The megaregolith thickness is one estimate of the mixing depth. Spudis and Davis[12] give half the thickness of the lunar crust, 35 km. If the meteoritic component is between 1% and 4%, this implies a total meteoritic thickness between 0.35 km and 1.4 km, with 0.7 km being our preferred estimate.

The time of these impacts is also constrained. If a magma ocean existed on the Moon, the upper crust should have frozen from the top down while being stirred by impacts[7]. The date at which a mostly solid crust existed at any depth defines the start of the time interval for Ir and Ni retention in the lunar regolith. The oldest lunar rocks approach 4.5 Gyr and indicate that a crust existed early in the history of the planet[13]. The upper part of the lunar crust, ferroan anorthosite, solidified as early as 4.44 Gyr (ref. 14). The lunar crust was largely solid by 4.36 Gyr (the model age of KREEP (potassium rare-earth elements phosphorus) basalt), which is believed to be the date of the last solidification of the base of the crust[15]. Much of the accretion was over by 4.26 Gyr, the age of the oldest regolith sample.

Energy and mass of estimates of total projectiles. The lunar crust is so mildly churned that this fact by itself sets a useful upper limit on the flux of large projectiles. Large impacts were sufficiently rare that regional heterogeneities, primary stratification, and even local igneous bodies have not been obliterated[16,17]. Impacts similar to or larger than Imbrium seem to have been rare after the upper crust froze at 4.44 Gyr, as the crustal stratification remained intact over much of the lunar surface. In particular, impacts larger than Imbrium should have excavated mantle, but no lunar mantle samples have ever been found despite careful search. Similarly troctolite, an expected constituent of the lower crust which was excavated by Imbrium, is quite rare[18]. The total amount of material from impacts occurring since 4.44 Gyr should be less than the 1.88-km equivalent layer needed to saturate the surface with Imbrium-sized craters[4].

The size of the largest projectiles is relevant in this regard (Fig. 1). The energy of Orientale (diameter, 930 km), the youngest (3.8 Gyr) and best preserved of the basins, is obtained from the amount of buried heat by examining the thermal contraction of the centre of the basin[19]. If we assume that 25%

Reprinted with permission from *Nature*,
Volume 342, pp. 139–142. © 1989 by Macmillan Magazines Ltd.

of the energy is deeply buried, the impact energy is between 4×10^{25} J and 3×10^{26} J with the middle of the range 1.2×10^{26} J being our preferred estimate. This corresponds to an object of mass 1.4×10^{18} with a diameter of 93 km (assuming a projectile density of 3,300 kg m^{-3} and a velocity of 13 km s^{-1}). The ejecta mass[20] of 2.4×10^{19} kg implies a ratio of ejected mass to projectile mass of 17. The largest confirmed basin, Imbrium, ejected 3.6×10^{19} kg (ref. 21) or 1% of the 35-km-thick megaregolith layer. The impactor mass was 2×10^{18} kg (ref. 6) Baldwin[4] gives an equivalent thickness of 0.2 km of accretion after 4.30 Gyr, equal to the mass of 11 Imbrium projectiles.

Extrapolating to the Earth. The Earth should have been hit with many more projectiles than the moon as it has more surface area and larger gravity. Statistically, the Earth is also likely to get hit with larger objects. For the impact velocities we assume below, the Moon is hit by $\frac{1}{24}$ of the objects and the Earth the remaining $\frac{23}{24}$. The probability that the moon is not hit by any of the 16 largest objects is over 0.5, because $(\frac{23}{24})^{16} > 0.5$. Thus, if Imbrium is the largest lunar projectile after 4.30 Gyr, 16 larger objects are expected to have hit the Earth in the same period. The time and size of these impacts are difficult to constrain because of the small numbers involved. For example, the late impact of the Imbrium object on the Moon could be a statistical fluke. There is no convincing evidence for a 'spike' of late impacts between 3.8 Gyr and 3.9 Gyr and no obvious physical mechanism for producing them[6]. Conversely, it can be concluded that the number of huge objects that hit the Earth was not large.

Theoretical size–number distributions are one way to infer the size of the large objects that hit the Earth from the lunar record. For fragmentation, such mass distributions are power laws. For example, the cumulative number of objects with a mass greater than m is proportional to m^{1-q}. We can visualize the properties of such a distribution by 'binning' the objects in logarithmic intervals of mass. For $q = 2$, there is equal mass in every bin. For $q = \frac{5}{3}$, there is equal surface area in each bin, and most of the mass is concentrated in the largest objects. The largest is expected to be $2 - q$ of the total mass of the ensemble. With some algebra, it can be shown that the median fraction of the total mass hitting the moon is $(\frac{1}{23})^{1/1-q}$. The size distribution is maintained in a quasi-steady state by a fragmentation cascade. Theoretically, expected values of q range from $\frac{5}{3}$ for mild collisions to 2 for violent collisions[22].

The most relevant surviving class of objects for impacting the early Earth are asteroids with diameters between 130 and 260 km which are believed to be fragments of larger bodies. Such bodies would enter Earth-crossing orbits after the fragmentation event. The lower end of the range is comparable to the Imbrium object from which our extrapolation starts. For this size range, $q \approx 2$ (refs 22, 23). If this distribution applies, the largest objects hitting the Earth need not have exceeded 260 km.

The larger gravity of the Earth increases the probability and energy of impacts[24], by a factor of 1.74 assuming an approach velocity of 13 km s^{-1}, which corresponds to an impact velocity of 13 km s^{-1} on the Moon and 17 km s^{-1} on the Earth. The approach velocity (weighted for probability of impact) of present Earth-crossing asteroids, 16.8 km s^{-1} (calculated from ref. 25, Table 8.5.1), is higher for large early asteroids by an unknown amount because the sample includes small asteroids that began as comets[26] and small asteroid fragments ejected by collisions[27].

Global sterilization by large impacts

We are concerned here with global effects of large impacts which would sterilize the entire planet. Boiling of the ocean by the heat of rock vapour produced by the impact is the most obvious such effect. Other global effects, including pressure changes in the ocean from large tsunami waves, fouling the ocean with meteorite debris and ejecta, and salinity changes as the ocean is boiled off and later as the water rains out have no obvious planet-sterilizing effects and are not considered further. The crater itself is of only regional importance. For example, according to the scaling relations compiled by Maher and Stevenson[2], an asteroid of 500-km diameter moving at 17 km s^{-1} would create a crater 1,500 km in diameter with 0.1-km-thick ejecta extending over a region 4,000 km in diameter. The ejecta at the rim of this crater would be 3 km thick.

Energy considerations. The estimates of the energy necessary to boil the ocean or the photic zone are insensitive to whether the final state is considered to be isothermal, adiabatic or on the vapour-saturation curve. For water that is initially at 0 °C, isobaric boiling requires 2.5×10^6 J kg^{-1} at 1 bar and 2.1×10^6 J kg^{-1} at the critical pressure. Thus, $\sim 4 \times 10^{26}$ J of energy are needed to boil the 200-m-thick photic zone of the ocean, and 5×10^{27} J is needed to evaporate the entire ocean (1.4×10^{21} kg) and raise the temperature of the water vapour above the critical point. To further raise the surface temperature to the melting point of typical silicate rocks requires about half again as much energy.

An impact of an ocean-evaporating scale corresponds to a 440-km diameter (an object of mass 1.3×10^{20} kg)—roughly the size of the large asteroids Vesta and Pallas—hitting at 17 km s^{-1}. A 190-km-diameter (mass 1.1×10^{19} kg) object would evaporate the photic zone. We estimate that $\sim 25\%$ of the impactor energy goes into evaporating sea water, 25% is radiated to space, and 50% buried near the impact site. For impacts of this size, most of the material that leaves the crater is either melted or vaporized[24]. Rock vapour is produced directly by the impact and secondarily when fine particles of ejecta return to the atmosphere. Thus, the planet is quickly enveloped by about 100 bar of hot rock vapour and suspended droplets. The rock-vapour atmosphere radiates upwards to space and downwards onto the ocean with an effective temperature of 2,000 K. For an impact on an ocean-vaporizing scale, the radiative cooling time for rock vapour to condense is a few months. The precise radiating temperature is not too important, because roughly half of the rock vapour's energy is radiated to space, and roughly half is absorbed by the ocean.

To continue with the events following impact, the high opacity of sea water to infrared radiation concentrates boiling to a thin surface layer. Radiation from the rock vapour ablates the surface of the ocean, leaving the ocean depths cool. Convective mixing of the ocean is strongly inhibited by the temperature gradient. As water vapour is somewhat transparent to 2,000-K blackbody radiation, the oceans are not effectively shielded from the hot rock above, yet the water vapour is sufficiently opaque that it is quickly heated to the 2,000-K atmospheric temperature[28]. During this period the atmosphere is kept roughly isothermal around 2,000 K, the condensation temperature of silicates. Water clouds would not form under these conditions.

Once the rock vapour has condensed, a steady-state water-vapour atmosphere above a liquid ocean can radiate to space no faster than a certain threshold level that is only some 30% higher than Earth's present infrared flux[28]. This is precisely the 'runaway greenhouse' threshold often encountered in the comparative planetology of Earth and Venus[29]. This rate is determined by the transmission of infrared radiation through a moist, convective, H_2O-dominated atmosphere. The presence of clouds can only reduce the rate of thermal emission. Faster radiative cooling demands very high surface temperatures, $\sim 1,500$ K, to exploit the lower opacity of water vapour to visible and near-infrared radiation. The runaway greenhouse threshold also controls the maximum rate at which a water-vapour atmosphere can cool while water condenses.

For the minimal ocean-vaporizing impact, about half the ocean remains as liquid water at the time when the rock vapour has rained out. (About 300 m of rock raindrops would sit on the ocean floor.) At this point the atmosphere would consist of about half an ocean (140 bars) of hot ($> 1,500$ K) steam. Shortly afterward, the top of the steam atmosphere cools sufficiently to form a moist, convective, optically thick upper layer. Thereafter, the planet radiates to space or below at the runaway greenhouse threshold, which is negligible compared with the rate at which energy is transferred to the ocean. In the minimal ocean-vaporizing impact, the last droplet of ocean is evaporated as the first droplet of rain reaches the ground. The surface temperature is then near the critical temperature, 647 K.

The lifetime of the steam atmosphere can be estimated by dividing the energy invested in steam by the effective cooling rate of 150 W m^{-2}, this being the difference between the runaway greenhouse threshold and total absorbed solar irradiation around 4 Gyr BP. The cooling time is on the order of 2,000 years for a minimal ocean-vaporizing impact. (The cooling rate is twice as fast if the clouds are sufficiently opaque that no solar radiation is absorbed; this difference is unimportant here.) An impact ~50% more energetic than the minimum ocean-vaporizing impact will leave a 1,500-K steam atmosphere when the ocean evaporates. This takes 3,000 years to cool. Cooling times for still larger impacts may not be greatly longer, because the runaway greenhouse threshold does not apply for very hot surfaces. But it is clear that extremely adverse conditions are expected for at least 3,000 years after any impact even modestly larger than the minimal ocean-vaporizing impact.

By contrast, adverse conditions in the atmosphere and the shallow ocean persist for about 300 years for an impact which evaporates only the photic zone. Bottom-dwelling photosynthetic organisms would be killed directly as the ocean boiled away above them. Floating photosynthetic organisms might, if they were lucky, be continually mixed downward as the ocean surface boiled. Survivors would, however, be killed when they were mixed back up to the surface and exposed to hot, fresh rainwater.

Discussion and uncertainties. Given that 25% of the impact energy is used to evaporate sea water, a total impact energy of 2×10^{28} J is needed to evaporate the entire ocean. An object of mass 1.3×10^{20} kg (440-km) hitting the Earth at 17 km s^{-1} would suffice. Ocean vaporization is plausible at much later times given a scale-invariant impactor distribution. Using the inference that

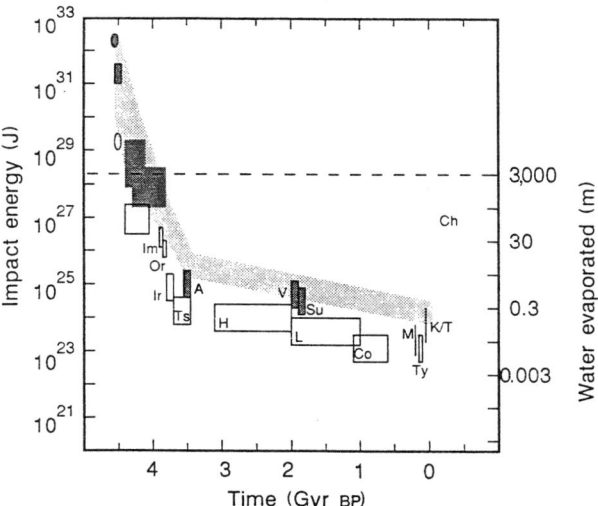

FIG. 1 The largest impacts on Earth and Moon. Open boxes are lunar, filled boxes terrestrial. Lunar craters are Tycho, Copernicus, Langrenus, Hausen, Tsiolkovski, Iridum, Orientale and Imbrium. Terrestrial events are the K/T impact, Manicougan, Sudbury, Vredevort and an impact energy corresponding to the thickness of Archaean spherule beds. Ovals are self energies of formation; the early box refers to a possible Moon-forming impact. Impact estimates between 3.8 and 4.4 Gyr are discussed in the text. The stippled region for Earth is inferred from these data. The depth of ocean vaporized by the impact is also given; the dashed line corresponds to an ocean-vaporizing impact. A possible but extremely unlikely collision with Chiron is placed safely in the future.

the Imbrium projectile is $\frac{1}{11}$ of the mass colliding with the Moon after 4.3 Gyr to obtain $q = 1.91$, the largest object inferred to hit the Earth since 4.44 Gyr is expected to account for 9% of the total accreted mass, or 3×10^{20} kg. Between 3.8 and 4.3 Gyr, an object 31 times as massive as the Imbrium projectile, or 6×10^{19} kg, is also statistically expected (Fig. 1). Given the uncertainties inherent in such a scale-invariant distribution, the last ocean-vaporizing impact may have occurred as early as 4.44 Gyr or as late as 3.8 Gyr. Impact vaporization of the photic zone as late as 3.8 Gyr is probable, however, whatever the size distribution, as the larger fragmental asteroids would suffice. A long time interval between the last impact to vaporize the ocean and the last to have evaporated the photic zone is thus statistically likely.

Survival of ecosystems

Biological systems require an energy source. The present biosphere receives its enthalpic buildup largely from solar energy; a second minor component in the deep sea is derived from the oxidation of reduced components at hydrothermal vents. The latter depends on oxidants being produced in the photic zone. It is likely that both these energy sources were available on the early Earth. The detailed nature of the earliest ecosystems is unknown because no sedimentary rocks older than 3.8 Gyr have been found. Life is unambiguously present at 3.56 Gyr in the Warrawoona Group in Australia; the organisms exhibited wall formation, some kind of motility, phototaxis and mat formation[30]. This complex and sophisticated set of properties should have required a substantial evolutionary history. In addition, carbon isotopes in the highly metamorphosed sediments of the 3.77-Gyr Isua Group of Greenland indicate that the ratio of organic carbon to carbonate in sediments was similar to the present implying that photosynthetic life was already abundant by Isua time[31].

Types of ecosystems. Even if individual organisms survive, it is necessary to reestablish a sustainable ecosystem immediately following the impact. The key issue becomes that of the primary producers.

Ecosystems where all the primary productivity is by obligate photosynthetic autotrophs have the primary producers in the top 200 m of the ocean where there is sufficient light. Impacts which sterilized this zone would destroy the global ecosystem and reset biogenesis.

Ecosystems where the primary producers are chemo-autotrophs, especially those at hydrothermal vents, would have been much better protected. However, there is still a problem of supplying oxidized material. The present ecosystem at hydrothermal vents ultimately depends on photosynthesis for its oxidants[32]. A primitive system of that type would die out when oxidants were exhausted. An ecosystem where oxidants are produced inorganically, by oxidation of volcanic SO_2 (ref. 33), or by photolysis of dissolved ferrous iron[34], would survive.

A system with photosynthetic primary producers that could act as faculative anaerobic heterotrophs using organic matter and oxidants is also difficult to sterilize. Such reserves are easily exhausted in a well mixed system, such as modern oil spills or the mid-depths of the ancient ocean stirred up by an impact. The oxidant and reductant reservoirs are heterogeneously distributed on the modern Earth, however, and were probably also heterogeneously distributed on the ancient Earth. The structure of the ancient ecosystem would depend on whether sulphates (or sulphites) were present in the ocean.

If significant sulphate (or sulphite) existed, the situation in the global ocean was basically similar to modern closed basins in that excess organic matter existed in the sediments and excess mobile oxidant, the sulphate, existed in the sea water. Heterotrophs could then exist on the interface between the two reservoirs, that is on the sea floor and in very shallow sediments. In the absence of sulphate or sulphite, the main oxidant would be ferric iron, which is less mobile than organic matter and tends to sink to the sea floor. Excess ferric oxide would accumulate on the sea floor beneath regions of high organic productivity[35]. In either case, two reservoirs are not easily exhausted as they are utilized by only a small mass of organisms living at the sea-floor interface.

Additional interfaces within the sedimentary column might escape the surficial heating. The thermal diffusion distance, $\sqrt{\text{diffusivity time}}$, is about a few hundred metres. Deep habitable interfaces, for example, might exist between sedimentary calcium sulphate deposits (or sediments with dissolved sulphate) and sediments with organic matter. By analogy, modern oil-field bacteria exist at the interface of oil and sulphate (or oxygen)-rich water[36]. Although such deep ecosystems could persist for millions of years, some survivors would need to evolve photosynthesis to sustain life.

Implications. The window in time for the emergence of extant life is defined by a palaeontological considerations on the lower end and by geophysical and planetological considerations on the upper. The necessity for a long evolutionary pre-history to evolve the Warrawoona fossils and the Isua carbon isotopes suggests the existence of life as early as 3.8 Gyr. Here we have been concerned with defining the upper limit. This limit depends on what type of early ecosystem is envisioned and, hence, what one believes is the earliest life form. Did photosynthetic prokary-

otes give rise to organisms that could utilize redox gradients or vice versa? Is it easier for life to originate in surficial environments or the deep ocean? Rather than attempting to resolve these biological issues, we examine the broader implications of our physical inferences.

If one postulates primitive photosynthetic biota as the sole primary producers, then annihilating events in the photic zone probably occurred as late as 3.8 Gyr, and our estimate for the time of the origin of continuous life converges on this date. The situation is less clear if one postulates early deep-marine life. (This situation is also less probable, as no evidence for such organisms exists in either the fossil record or in the evolutionary systematics of advanced life.) In this case, the upper limit may be as great as 4.44 Gyr if the maximum size of impactors was similar to fragmental asteroids or if the few largest objects in a scale-invariant distribution happened to hit the Earth early on. However, size and frequency of impacts should statistically increase as one moves further back in time. Thus, early organisms and ecosystems would have had to be increasingly sophisticated to survive. As this is counter to the usual course of evolution, the upper limit for continuous evolution would be around 4.0 Gyr, with considerable uncertainty. Conversely, a long continuity of life before 3.8 Gyr would imply occupation of the deep sea.   □

Received 5 July; accepted 18 September 1989.

1. Cloud, P. *Scient. Am.* **249** (March), 176–189 (1983).
2. Maher, K. A. & Stevenson, D. J. *Nature* **331**, 612–614 (1988).
3. Newsom, H. E. & Taylor, S. R. *Nature* **338**, 29–34 (1988).
4. Baldwin, R. B. in *Multi-Ring Basins, Proc. lunar planet. Sci. Conf.* **12A**, 19–28 (1981).
5. Baldwin, R. B. *Icarus* **71**, 1–18 (1987).
6. Baldwin, R. B. *Icarus* **71**, 19–29 (1987).
7. Hartmann, W. K. in *Proc. Conf. lunar Highlands Crust* (eds Papike, J. J. & Merrill, R. B.) 155–173 (Pergamon, New York, 1980).
8. Korotev, R. L. *J. geophys. Res.* **92**, E447–E461 (1987).
9. Ringwood, A. E. & Seifert, S. in *Origin of the Moon* (eds Hartman, W. K., Phillips, R. J. & Taylor, G. J.) 331–358 (Lunar and Planetary Institute, Houston, 1986).
10. Warren, P. H., Jerde, E. A. & Kallemeyn, G. W. *Earth planet. Sci. Lett.* **91**, 245–260 (1989).
11. Lindstrom, M. M. & Lindstrom, D. J. *J. geophys. Res.* **91**, D263–D276 (1986).
12. Spudis, P. D. & Davis, P. A. *J. geophys. Res.* **91**, E84–E90 (1986).
13. Swindle, T. D., Caffee, M. W., Hohenberg, C. M. & Taylor, S. R. in *Origin of the Moon* (eds Hartman, W. K., Phillips, R. J. & Taylor, G. J.) 331–357 (Lunar and Planetary Institute, Houston, 1986).
14. Carlson, R. W. & Lugmair, G. W. *Earth planet Sci. Lett.* **90**, 119–130 (1988).
15. Carlson, R. W. & Lugmair, G. W. *Earth planet Sci. Lett.* **45**, 123–132 (1979).
16. Pieters, C. M. *Rev. Geophys.* **24**, 557–578 (1986).
17. Davis, P. A. & Spudis, P. *J. geophys. Res.* **92**, E387–E395 (1987).
18. Marvin, U. B., Carey, J. W. & Lindstrom, M. M. *Science* **243**, 925–931 (1989).
19. Bratt, S. R., Solomon, S. C. & Head, J. W. *J. geophys. Res.* **90**, 12.415–12.433 (1985).
20. Bratt, S. R., Solomon, S. C., Head, J. W. & Thuber, C. H. *J. geophys. Res.* **90**, 3049–13064 (1985).
21. Spudis, P. D., Hawke, B. R. & Lucey, P. G. *Proc. lunar planet. Sci. Conf.* **18**, 155–168 (1988).
22. Hughes, D. W. *Mon. Not. R. astr. Soc.* **199**, 1149–1157 (1982).
23. Donnison, J. R. & Sugden, R. A. *Mon. Not. R. astr. Soc.* **210**, 673–682 (1984).
24. Melosh, H. J. *Impact Cratering: A Geological Process* (Oxford University Press, New York, 1989).
25. Basaltic Volcanism Project *Basaltic Volcanism on the Terrestrial Planets* (Pergamon, New York, 1981).
26. Hartmann, W. K., Tholen, D. J. & Cruikshank, D. P. *Icarus* **69**, 33–50 (1987).
27. Wetherill, G. W. *Icarus* **75**, 552–565 (1988).
28. Kasting, J. F. *Icarus* **74**, 472–494 (1988).
29. Ingersoll, A. P. *J. atmos. Sci.* **26**, 1191–1198 (1969).
30. Schopf, J. W. & Walter, M. R. in *Earth's Earliest Biosphere* (ed. Schopf, J. W.) 214–239 (Princeton University Press, 1983).
31. Schidlowski, M. *Nature* **333**, 313–318 (1988).
32. Jannasch, H. W. in *Hydrothermal Processes at Seafloor Spreading Centres* (eds Rona, P. A., Bostöm, K., Laubier, L. & Smith, K. L. Jr) 677–709 (Plenum, New York, 1983).
33. Walker, J. C. G. *Nature* **329**, 710–712 (1987).
34. Braterman, P. S., Cairns-Smith, A. G. & Sloper, R. W. *Nature* **303**, 163–164 (1983).
35. Walker, J. C. G. & Brimblecombe, P. *Precambrian Res.* **28**, 205–22 (1985).
36. Connan, J. *Adv. Petrol. Geochem.* **1**, 299–335 (1984).

ACKNOWLEDGEMENTS. We thank Chris Chyba for his review and H. J. Melosh for his review and for discussions. J.K. acknowledges support from NASAs Exobiology Program. N.S. was supported in part by the NSF. K.Z. and N.S. were supported by the NASA Ames University Consortium.

Endogenous production, exogenous delivery and impact-shock synthesis of organic molecules: an inventory for the origins of life

Christopher Chyba & Carl Sagan

Sources of organic molecules on the early Earth divide into three categories: delivery by extraterrestrial objects; organic synthesis driven by impact shocks; and organic synthesis by other energy sources (such as ultraviolet light or electrical discharges). Estimates of these sources for plausible end-member oxidation states of the early terrestrial atmosphere suggest that the heavy bombardment before 3.5 Gyr ago either produced or delivered quantities of organics comparable to those produced by other energy sources.

MICROSCOPIC fossils[1] and fossil stromatolites[2] indicate life originated on Earth more than 3.5 Gyr ago. Evidence for biologically mediated carbon isotope fractionation[3] suggests that life may already have existed by 3.8 Gyr ago. The terrestrial origins of life must therefore have coincided with the final stages of the heavy bombardment of the inner Solar System, during which those planetesimals 'left over' from planetary formation were largely swept up or scattered. This bombardment[4,5], known from radioactive dating of cratered surfaces on the Moon, and by comparisons of the lunar, martian and mercurian cratering records, declined in intensity through orders of magnitude, reaching its present comparatively low level by ~3.5 Gyr ago.

The heavy bombardment seems to have had important consequences for the origins of life, some deleterious and some favourable. Sufficiently large and fast impacts can erode planetary atmospheres. This effect, although possibly critical for Mars[6], was probably of little importance for the Earth[7]. The largest impactors could have led to an 'impact frustration' of life's origins[8-10], through the creation of a globe-encircling rock vapour atmosphere and evaporation of the euphotic zone, or even the entire terrestrial ocean[10].

During the heavy bombardment, volatile-rich impactors would have been delivering essential 'biogenic' elements to the terrestrial surface[7,11-13]. Moreover, comets, carbonaceous asteroids, and interplanetary dust particles (IDPs) are rich in organic molecules[14-16], so may have contributed directly to terrestrial prebiotic inventories. Impacts would also have shock-synthesized organics in the atmosphere[17,18]. Here we focus on these last two effects, comparing them quantitatively with the principal non-heavy-bombardment sources of prebiotic organics.

Delivery of intact exogenous organic matter

Exogenous sources deliver organic molecules more or less intact to Earth today. These include[14] those interplanetary dust particles small enough to be gently decelerated by the atmosphere, and meteorites large enough to avoid complete ablation, but small enough to be substantially decelerated during their fall. Some impactors catastrophically fragment during atmospheric passage, as seems to have happened[19,20] to a comet or comet fragment over Tunguska, Siberia, in 1908. Fragmentation of a CI carbonaceous chondrite took place over Revelstoke, Canada, in 1965; photomicrographs of recovered millimetre-sized fragments reveal unheated interiors[21], within which organics should have survived. (An investigation, probably requiring the examination of carbon isotope ratios, of whether exogenous organics are in fact present in the Revelstoke fragments needs to be done.) Finally, the discovery of apparently extraterrestrial amino acids in Cretaceous/Tertiary (K/T) boundary sediments at

Stevns Klint, Denmark, has been taken to suggest that a large fraction of cometary organics might in fact survive giant impacts[22], although both xenon measurements at the K/T boundary[14] and hydrodynamic simulations of organic pyrolysis in impacts[15] argue strongly otherwise. Zahnle and Grinspoon[23] have invoked the accretion of cometary dust as an explanation; other explanations may also be possible[24].

Each of these exogenous sources of organics should have been present on early Earth. Here we estimate their quantitative importance, scaling by the lunar impact record. We give our results as an exogenous organic mass flux through time, to allow the influx to be readily determined for whatever epoch a particular model for the origin of life suggests is appropriate. Such an approach involves several uncertainties, about which we are explicit; we also argue, however, that these are no worse than many encountered in typical estimates of endogenous sources of prebiotic organics.

Attempts to estimate the impact environment of early Earth often begin with analytical fits to the lunar cratering record[6-8], intended to minimize the model-dependence of the conclusions. In practice, however, this procedure faces numerous difficulties, which have often been disregarded. The oldest lunar province for which a radiometric date actually exists (the Apollo 16 and 17 uplands) is only 3.85–4.25 Gyr old; the ages of more heavily cratered provinces can at present only be estimated[4]. The entire interpretation of the heavy bombardment as representing exponentially decaying remnants of planetary formation is occasionally questioned by those favouring a lunar cataclysm[25]. We have previously shown[7] that different choices[6,8] for the decay rates fitted to lunar cratering data[4,5] can lead to substantially different conclusions about terrestrial mass influx during the heavy bombardment. More recently, we have argued that the more extreme of these choices can be excluded, as in contradiction with lunar and terrestrial geochemical data on meteoritic input[26]. Here we employ a model lunar bombardment history that we have demonstrated is in good agreement with the geochemical constraints[26]. Both the model and the constraints are summarized in Box 1.

Asteroid and comet impacts. The model developed in Box 1 for terrestrial mass accretion during the heavy bombardment is based on counts for lunar craters larger than 4 km in diameter. By equation (2), these correspond to impactors of masses greater than ~10^{10} kg, or radii above ~100 m. What is the fate of organics in such objects incident on Earth? Some might catastrophically fragment ('airburst') while traversing the atmosphere; this case is treated below. We have demonstrated that such objects that do not airburst are insufficiently decelerated ('aerobraked') by a 1-bar terrestrial atmosphere for their organics to survive the heat of impact[15]. But some models suggest that Earth may have had a dense, ~10-bar CO_2 atmosphere some time before 3.8 Gyr ago[29,30]. Under these conditions, comets with radii as large as ~100 m would have been sufficiently aerobraked (after substantial ablation during atmospheric passage) for most remaining organics to have survived impact with the terrestrial ocean[15]. Carbonaceous asteroids of similar size would not have been sufficiently aerobraked. Our earlier work on this topic[15] considered a broad range of possible early terrestrial impact environments, because of uncertainties in lunar cratering data[7]. Since then (see Box 1), we claim to have reduced these uncertainties by appealing to geochemical constraints[26].

Following Delsemme's analysis of comet Halley data[16], we take comets to be ~14% organic carbon by mass. The mass flux of cometary organic carbon surviving impact at time t is given by

$$\dot{m}(t) = \dot{m}(0)[\dot{n}(t)/\dot{n}(0)] = \dot{m}(0)f(t) \qquad (8)$$

where $\dot{n}(t)$ is given by equation (7), and, if we use the results of Chyba *et al.*[15] for cometary ablation, deceleration and resulting organic pyrolysis, $\dot{m}(0) = 6.6 \times 10^2(\psi/0.1)$ kg yr^{-1}. Here ψ is the mass fraction of the ancient impacting flux in the relevant size range (here, radii $\leqslant$ 100 m) that was cometary. We review the data pertaining to the value of ψ elsewhere[13] and find it inconclusive. Certain models[31,32] of outer planet formation imply comet fluxes through the inner Solar System that require the bulk of the heavy bombardment to have been cometary. But disparities between the cratering records of the terrestrial planets (the planets nearest the Sun) and those of some outer planet satellites suggest that comets could not have been the main component of the heavy bombardment population[33]. Admittedly, as long as only the lunar cratering record can be assigned absolute ages, such an objection remains unconfirmed. For now,

it is best to treat ψ as a free parameter. Our results, labelled 'comet impacts' in Fig. 1, take $\psi = 0.1$, and scale linearly, so the effect of a different choice for ψ is evident. More recent numerical modelling, extending earlier two-dimensional hydrodynamic impact models[15] to a full three dimensions, suggests[34] the two-dimensional results used here may underestimate organic delivery from cometary impacts by a factor of ~3.

Catastrophic airbursts. Photometric observations of Earth-crossing asteroids imply that ~$\frac{1}{3}$ of the current asteroid flux at Earth is C-type[35]. Of 17 terrestrial craters for which impactor type may be identified, only two seem to have had a carbonaceous chondritic composition[36]. Similarly, carbonaceous chondrites constitute only ~5–6% of stony meteorite falls[14], although they seem to represent over a third of 10^2–10^6 g Prairie Network fireballs (with another third seemingly having cometary origins)[37]. It therefore appears that carbonaceous chondrites have material strengths so low that they are typically unable to survive atmospheric passage without breakup[38]. Evidence from meteorite falls[20,38] seems consistent with a compressive strength for carbonaceous chondrites of $\sigma_c \approx 10^6$ Pa (10 bars), orders of magnitude below typical rock strengths. In this case, the catastrophic fragmentation of the Revelstoke object[21] may be a typical fate for carbonaceous impactors. Because of the increased surface-area-to-volume ratio of the resulting fragments, such airbursts might then provide an efficient mechanism for exogenous organics to reach Earth[15,24].

We estimate the magnitude of this source as follows. Melosh[39] has calculated the critical radius R_c of a meteoroid, greater than which—even should the object fragment—the fragments will have insufficient time to accelerate away from one another and follow independent trajectories before striking the ground. Only objects smaller than R_c are therefore able to deposit the bulk of their kinetic energy explosively in the atmosphere (that is, to

---

BOX 1   Impact statistics on the early Earth

Data[4] for the cumulative surface density of lunar craters with diameter $> D$, as a function of surface age t in billions of years (Gyr), are well modelled by an equation of the form[6,8]

$$N(t, D) = \alpha[t + \beta(e^{t/\tau} - 1)](D/4,000 \text{ m})^{-1.8} \text{ km}^{-2} \qquad (1)$$

We take $\tau = 144$ Myr, corresponding to a 100-Myr decay half-life for the impactor population, $\alpha = 3.5 \times 10^{-5}$, and $\beta = 2.3 \times 10^{-11}$ (refs 7, 26). Relying on the recent modelling of post-excavation crater collapse by McKinnon *et al.*[27], we have related[26] crater diameter D (in m) observed on the Moon to the mass m (in kg) and collision velocity v (in m s^{-1}) of an impactor by

$$m = 0.54 \gamma v^{-1.67} D_c^{0.44} D^{3.36} \qquad (2)$$

where $D_c \approx 1.1 \times 10^4$ m is a transition diameter at which lunar craters change morphology from simple to complex forms[27], and $v = 1.2 \times 10^4$ m s^{-1} for typical impacts on the Moon. The constant $\gamma = 1.4 \times 10^3$ kg s$^{-1.67}$ m$^{-2.13}$ depends on target and impactor densities, surface gravity, and impactor incidence angle. Equations (1) and (2) may be combined to give the number of objects of mass $> m$ that have struck the Moon as a function of time t:

$$n(>m, t) = 3.1 \times 10^8[t + 2.3 \times 10^{-11}(e^{t/\tau} - 1)]m^{-b} \text{ kg}^b \qquad (3)$$

where $b = 0.54$. The total mass, $M(t)$, incident on the Moon after some time t in impactors with masses in the range m_{min} to m_{max} is given by the integral:

$$M(t) = \int_{m_{max}}^{m_{min}} m[\partial n(>m, t)/\partial m]dm \qquad (4)$$

which yields

$$M(t) = 3.7 \times 10^8[t + 2.3 \times 10^{-11}(e^{t/\tau} - 1)](m_{max}^{1-b} - m_{min}^{1-b}) \text{ kg}^b \qquad (5)$$

To scale to the Earth, $M(t)$ must be multiplied by $\xi \approx 24$, the ratio of the terrestrial to the lunar gravitational cross sections at typical asteroid approach velocities[26].

Taking m_{max} to equal the mass of the lunar impactor that excavated the South Pole-Aitken basin ($D \approx 2,200$ km), or ~1.4×10^{19} kg by

equation (2), and m_{min} to be negligible, equation (5) yields a total mass of 1.0×10^{20} kg incident on the Moon subsequent to the solidification of the lunar crust ~4.4 Gyr ago. About half this mass would actually have been retained by the Moon. This result is in good agreement with the geochemical estimates of Sleep *et al.*[10] of the meteoritic component mixed into the lunar crust, which yield $(0.4–1.5) \times 10^{20}$ kg. Similarly, scaling equation (5) to Earth, and taking proper account of the statistical probability that the largest impactors incident on Earth were more massive than the largest incident on the Moon, gives an estimate of 1.5×10^{22} kg of material accumulated by Earth after 4.4 Gyr ago. This result is in good accord with geochemical estimates of post-core-formation meteoritic input. These estimates, based on chondritic abundances of highly siderophile elements in the terrestrial mantle, lie in the range $(1–4) \times 10^{22}$ kg (refs 4, 26, 28). Certain proposed fits[8] to other lunar cratering data sets[5] seem to predict siderophile abundances two orders of magnitude in excess of those observed; such fits can probably be excluded as models for the early terrestrial impact environment. On the other hand, the close agreement of equation (5) with both the lunar cratering record and the available lunar and terrestrial geochemical constraints suggests that it provides a reasonable, albeit procrustean, model for post-core formation terrestrial mass accretion during the heavy bombardment.

Equation (5), multiplied by ξ, gives the total mass incident on Earth in impactors within a certain size range, after some time t. A more useful quantity is the mass flux (kg yr^{-1}) at a particular time in the Earth's past. Combined with estimates of the organic mass fraction surviving delivery, this allows a quantitative comparison of these sources with photochemical or other *in situ* production rates of prebiotic organics as a function of time. To this end, we define $\dot{M}(t) \equiv \xi[\partial M(t)/\partial t]$, the terrestrial mass flux from objects within a given mass range (m_{min} to m_{max}) being accreted by Earth at a time t. From equation (5),

$$\dot{M}(t) = 8.9 f(t)(m_{max}^{1-b} - m_{min}^{1-b}) \text{ kg}^b \text{ yr}^{-1} \qquad (6)$$

where $f(t) = (1 + 1.6 \times 10^{-10} e^{t/\tau})$. Our discussion will also require $\dot{n}(t) \equiv \xi[\partial n(>m, t)/\partial t]$, the number flux of objects within a given mass range being accreted by Earth as a function of time. From equation (3),

$$\dot{n}(t) = 7.4 f(t)(m_{max}^{-b} - m_{min}^{-b}) \text{ kg}^b \text{ yr}^{-1} \qquad (7)$$

airburst). For an incidence angle of 45°, the critical radii for icy and stony meteoroids in a 1-bar terrestrial atmosphere are 407 m and 235 m, respectively.

These results are consistent with the few relevant observations that are available. The Tunguska object is typically estimated[19,20] to have had a mass of several billion (10^9) kilograms, corresponding to a comet ~100 m in radius. It seems to have exploded at an altitude of 6-9 km (ref. 19). The Lappajärvi crater, apparently the result of an impactor with a carbonaceous chondritic composition[36], required an object with a diameter of ~1 km; this impactor evidently did not airburst.

An approximate upper limit to organic delivery from airbursting asteroids and comets may therefore be obtained by taking m_{max} in equation (6) to be that appropriate to 235-m chondrites and 407-m comets. We take 10% of the impact object mass (for objects below ~0.5 km in radius) to be cometary, with the remainder split evenly between carbonaceous and non-carbonaceous objects (which average together to give a 1.3% organic carbon content[14]). Results are readily scaled to other assumptions about impactor compositions. Equation (6) then takes the form of equation (8), with $\dot{m}(0) = 3.6 \times 10^4$ kg yr^{-1}, and is shown in Fig. 1. Clearly this value is an upper limit, as it ignores organic destruction in the airburst event itself (the Tunguska object exploded with an energy of ~10^{16} J, or several megatons of high explosive[20]) or in subsequent ablation of the resulting fragments. Moreover, not all objects with radii $< R_c$ will airburst. If the early terrestrial atmosphere were as thick as ~10 bar, larger objects could airburst[39], and the upper limit derived here would increase by a factor of about 30.

Interplanetary dust particles (IDPs). Anders[14] has estimated the flux of intact organic matter reaching the contemporary Earth in IDPs and meteorites. The bulk of the mass in IDPs is in the ~100-μm radius range, or, for a typical initial density[40] ~1 g cm^{-3}, in particles with masses ~10^{-5} g. The mass scaling in equation (6) leads one to expect most incident mass to lie in the largest particles, but in fact this scaling breaks down for particles with masses $\leq 10^2$ g, probably because of dust production from larger bodies[41]. The global mass flux in particles below 10^2 g is observed[42,43] to increase with decreasing size until it peaks at ~10^{-5} g. This peak reaches a flux ~$10^5 - 10^6$ times higher than would be the case if the IDP mass spectrum did not deviate from power-law scaling.

Anders takes IDPs to be ~10% organic carbon by mass, a mean carbon content determined from 30 IDPs by X-ray analysis[44]. He takes IDPs in the $10^{-12} - 10^{-6}$ g (~0.6-60 μm radius) range to be sufficiently gently decelerated during atmospheric entry to deliver their organics intact; his 60-μm upper limit is in good agreement with results of both theoretical modelling[40] and direct examination of IDPs[45]. His lower limit is based on an estimate of the size below which IDP organics would be destroyed by ultraviolet photolysis; this choice could vary considerably without significant quantitative effects on the final result. It remains unclear whether the more fragile organic species would survive IDP atmospheric passage[24], especially at the high velocities appropriate to cometary-derived particles[40]; there is probably a kind of deceleration-heating natural selection of the thermally most stable species.

Earth is currently accreting[41] ~3.2×10^6 kg yr^{-1} of $10^{-12}-10^{-6}$ g IDPs, or $\dot{m}(0) \approx 3.2 \times 10^5$ kg yr^{-1} of intact organics[14]. It is unclear how we should scale this back in time through the heavy bombardment: would the mass influx peak at ~10^{-5} g have persisted? There are some suggestive, although less than compelling, data relevant to this question.

The current terrestrial mass influx from IDPs found by Hughes[42] from meteor observations agrees well with accretion rates inferred from terrestrial and lunar data. As Table 1 shows, these data rely on a number of different sampling methods, and show a remarkably constant net IDP mass flux over the past ~3.6 Gyr, suggesting that the current IDP mass flux is at least not a recent anomaly. Most of these data, however, take the

form of an integral over the IDP size distribution, so cannot prove that the shape of the flux curve has remained unchanged with time. In any case, they do not extend further back than the end of the heavy bombardment.

Most IDPs currently collected in the stratosphere belong to one of three main classes; on the basis of particle heating and compositional data, these have been tentatively identified as corresponding to origins in main-belt asteroids, comets with perihelia outside 1.2 AU, and Earth-crossing asteroids or comets[46]. The first two sources seem to be the most common. Main-belt asteroids seem unlikely to be a source representative of 'typical' heavy bombardment sources for IDPs, but Earth-crossing objects and comets with perihelia beyond 1.2 AU might well be.

Clearly we cannot say with confidence how to scale the IDP flux back beyond 3.6 Gyr ago. If IDPs are the product of cometary evaporation or of asteroid-asteroid collisions, one might expect their number to scale linearly, or as the square, respectively, of the number of such objects in the inner Solar System. Loss mechanisms, however, must also be considered. Whipple[47] has argued that, for dust whose loss rate is collisionally dominated, terrestrial accretion will scale as the square root of the cometary flux. But those IDPs actually able to contribute organics intact to Earth ($m \leq 10^{-6}$ g) have lifetimes dominated by Poynting-Robertson drag, not by collisions[43]. The relative importance of these two loss mechanisms could have changed in an early Solar System in which the IDP flux was greater. Production of $\leq 10^{-6}$ g particles (whose collisional production at present greatly exceeds their collisional destruction) would, however, also have increased as the square of the number of larger colliding particles. Relative production and destruction rates would have to be simultaneously modelled. To our knowledge, no such model currently exists.

Here we scale the IDP flux linearly with the lunar impact record, using equation (8) with $\dot{m}(0) \approx 3.2 \times 10^5$ kg yr^{-1} of organic carbon. A reasonable lower limit for the IDP flux on

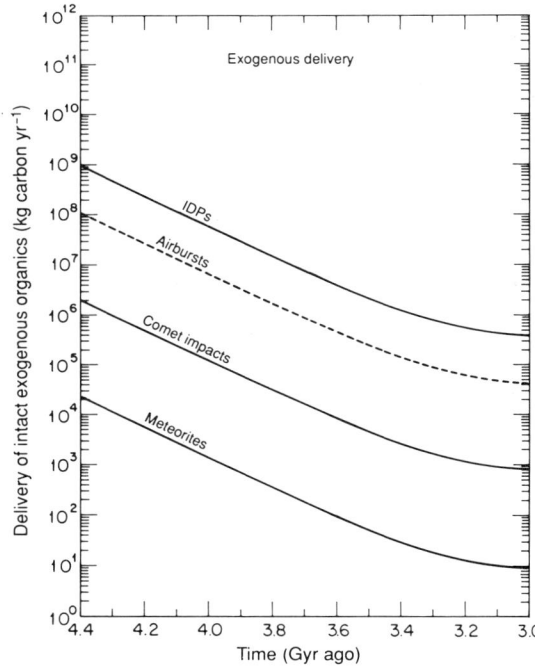

FIG. 1 Exogenous organic carbon delivered to the Earth as a function of time. Labels correspond to the appropriate sections of the text, where uncertainties are discussed. The dashed line indicates an upper bound.

Earth during the heavy bombardment is the current value. At 4 Gyr ago, this leads to an uncertainty of a factor of $\sim 10^2$.

Meteorites. Anders[14] has estimated contemporary terrestrial accretion of intact organic matter from meteorites, finding 7.6 kg carbon yr^{-1}. Taking this value as $\dot{m}(0)$ in equation (8) gives the meteoritic organic carbon input through time, as shown in Fig. 1. Despite uncertainties, this source is clearly negligible compared with IDPs.

Accretion of interstellar dust. An additional source of exogenous organics on early Earth, independent of the heavy bombardment, would have been the terrestrial accretion of dust as the Solar System passed through interstellar clouds. Greenberg[48] has estimated that during its first 7×10^8 years, Earth should have passed through ~ 4–5 such clouds, accreting organic molecules during each $\sim 6 \times 10^5$ yr passage at a rate 10^6–10^7 kg yr^{-1}. Even during cloud passage, this source would have been one to two orders of magnitude smaller than that due to IDPs.

Shock synthesis of organic molecules

Gilvarry and Hochstim suggested that shock waves from meteoroids traversing the terrestrial atmosphere may have synthesized organics on early Earth[49]. Bar-Nun et al. demonstrated that shock heating of reducing gas mixtures in the laboratory yielded amino acids in high yield[17], although more recent results[50] suggest that the yields first reported[17] for amino acid synthesis in $CH_4/C_2H_6/NH_3/H_2O$ atmospheres were overestimated by a factor of ~ 30. Nevertheless, atmospheric shock synthesis may have been an important source of terrestrial organics[17,18]. Estimates of both the early terrestrial impact environment and the coupling efficiency of an impactor's energy with the atmosphere may now be updated.

The mass of organics shock-synthesized in the atmosphere by an impactor is proportional to η, the organic synthesis efficiency (kilograms organic carbon produced per joule of shock energy): η is strongly dependent on atmospheric composition. Earlier models for a primordial reducing terrestrial atmosphere rich in CH_4 and NH_3 are now less favoured than that of a neutral atmosphere rich in CO_2 and N_2. (Geochemical and photochemical arguments for this conclusion have been summarized in, for example, refs 15 and 29.) Nevertheless, both because this recent preference is not beyond doubt, and because many atmospheric compositions intermediate between the two extremes have been considered, we will treat two oxidation state endmembers of a continuum of possible early-Earth atmospheres, one rich in CH_4, the other rich in CO_2.

A thermochemical model of shock synthesis in reducing $CH_4/N_2/H_2O$ atmospheres[51] gives an HCN production efficiency $\sim 10^{17.5}$ molecules J^{-1}, in excellent agreement with laboratory results[50,52] of $\sim 2 \times 10^{17}$ molecules J^{-1}. These experiments also show simultaneous formation of simple hydrocarbons, such as C_2H_2 and C_2H_4, as well as carbon soot[52]. Although these results are temperature-dependent, typical organic C yields, ignoring soot, are ~ 3 times that of HCN alone[52].

Because of the strength of the N_2 bond, we might expect efficiencies to increase in atmospheres where nitrogen is present as NH_3, rather than N_2. In fact, this effect is unimportant at the high shock temperatures appropriate here. Both high-power laser simulation, and theoretical high-temperature-equilibrium modelling, of shocks in NH_3/CH_4 atmospheres give HCN production yields[53] of $\sim 3 \times 10^{17}$ molecules J^{-1}. Hydrocarbons at least up to pentane (C_5H_{12}) are also produced. Using the experimental yields reported for all C-containing organic species gives a total efficiency for organic C production $\eta \approx 1.2 \times 10^{-8}$ kg J^{-1} for a reducing atmosphere[53], a result virtually identical with that found for N_2/CH_4 atmospheres[50-52]. Efficiencies for HCN production in $CO_2/N_2/H_2O$ atmospheres, are $\sim 10^{7.5}$ times smaller[51]. In this case, formaldehyde (H_2CO) production is comparable[18] to that of HCN, giving $\eta \approx 2.5 \times 10^{-16}$ kg J^{-1} for CO_2 atmospheres.

Shocks from meteors. Roughly 1.6×10^7 kg yr^{-1} strikes the Earth's atmosphere in meteors of mass 10^{-14}–10^2 g (refs 14, 41). These objects lose 100% of their kinetic energy to the atmosphere. Assuming an average velocity of 15 km s^{-1}, about midway between those typical for asteroidal and cometary IDPs[40], meteors deposit 1.8×10^{15} J yr^{-1} into the present atmosphere. Some fraction, e_c, of this energy is converted into atmospheric shock waves. Pollack et al.[54,55] have selected a value $e_c \approx 0.3$ for the efficiency of converting impactor kinetic energy into atmospheric shock heating, based on analyses of meteorite ablation in the contemporary atmosphere[56]. With this value, total organic production in the atmosphere by meteors (Fig. 2) is given by $\dot{m}(t) = \eta (6 \times 10^{14}$ J $yr^{-1}) f(t)$. Again, this result assumes that scaling with the lunar impact record is appropriate.

Shocks from airbursts. Objects as large as ~ 200–400 m in radius may airburst (see above), thereby coupling their entire kinetic energy to the atmosphere[57]. Equation (6) can be used to put an upper limit on the importance of this effect. If we assume that all objects with radii $\leqslant 300$ m will airburst, then Equation (6), for this choice of m_{max} and multiplied by $v^2/2$ (we take $v = 15$ km s^{-1}, the median terrestrial collision velocity of Earth-crossing asteroids[26]), gives $\dot{m}(t) = \eta (1.5 \times 10^{14}$ J $yr^{-1}) f(t)$. This upper limit is shown in Fig. 2.

Shocks from post-impact vapour plumes. For large impactors, the total energy imparted to the atmosphere by the rapidly expanding post-impact plume is much greater than that lost by the impactor during atmospheric passage[58,59]. Moreover, for a large impact, much of the atmosphere shocked by the incoming object will be shocked again by ejecta, largely pyrolysing products just synthesized during atmospheric passage[60]. Meteors (see above) deliver about ten times as much net shock energy to the atmosphere than does the atmospheric passage of all impactors

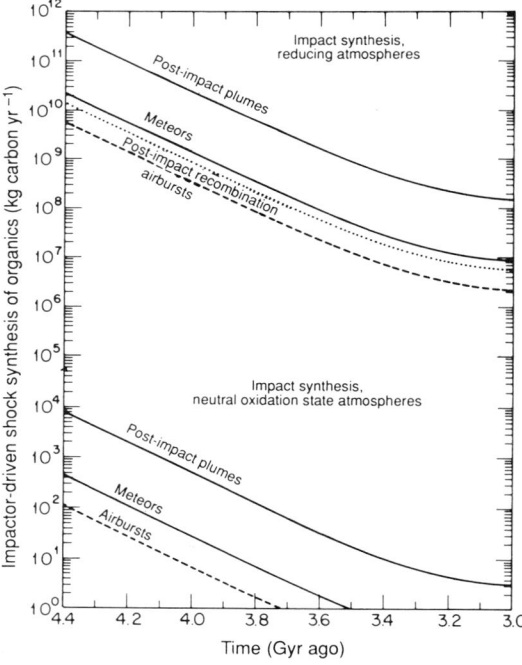

FIG. 2 Impactor-driven shock synthesis of organics in the terrestrial atmosphere as a function of time. Labels correspond to the appropriate sections of the text, where uncertainties are discussed. The upper curves are for $CH_4 + (N_2$ or $NH_3) + H_2O$ reducing atmospheres, and the lower curves are for $CO_2 + N_2 + H_2O$ neutral oxidation state atmospheres. Dashed lines indicate upper bounds, and the dotted line denotes an estimate that is especially poorly understood, although perhaps largely independent of the oxidation state of the atmosphere.)

TABLE 1   Estimates of terrestrial meteoric mass accretion

Sampling method	Effective sampling time	Mass influx (10^7 kg yr^{-1})
Atmospheric[86] Satellite, radio, and visual observations[14,42,43]	~1 yr	0.6–1.1
Pleistocene abyssal clays[87]	past few decades	0.3–8.0*
Cretaceous and Tertiary abyssal clays[41]	past ~10^6 yr	3–9†‡
Micrometeorite component in lunar soils[48,49]	33–67 Myr ago	1–5‡
	past ~3.6 Gyr	1.8–2.8§

* Hughes' preferred value[42] is 1.6×10^7 kg yr^{-1}, with an uncertainty (adopted here) of 'at least half an order of magnitude'.

† As corrected by Kyte and Wasson[41].

‡ After subtracting contribution from larger objects, following Kyte and Wasson[41].

§ Scaled by a factor of 24 from the Moon to the larger gravitational cross section of Earth[26].

with masses $> 10^2$ g combined[61]. Therefore we neglect atmospheric passage of large impactors, and concentrate on shock processing due to post-impact (plume) effects.

The simplest treatment of this problem would be first to multiply Equation (6) by $v^2/2$, to calculate the net kinetic energy flux at the terrestrial surface through time. Multiplying this result by η, and by ε, the fraction of an impactor's kinetic energy at surface impact returned to the atmosphere as shock energy, would then give an atmospheric organic production rate. We find $\dot{m}(t) = \eta\varepsilon(6 \times 10^{17}$ J yr$^{-1})f(t)$. Typical published values[59] for ε are ~1/8.

This approach, however, ignores several important effects. Most important, the plumes arising from the largest impacts are poorly matched to the atmosphere; they rise far above the atmosphere and encounter only a small fraction of it[60]. The result is that the shock processing yield per unit of impact energy decreases as the energy of the impact increases. (Failing to take such effects into account leads, following the procedure in the preceding paragraph, to the absurd conclusion that the largest objects (South Pole–Aitken size) striking the Earth would have shock-processed ~50% of an early reducing atmosphere.) Zahnle[60] has modelled these effects in detail for the case of impact production of nitric oxide in the present atmosphere. For an impact energy of 2×10^{25} J, he finds a NO yield ~1/40 that calculated by Prinn and Fegley[59] with the assumption of a constant NO production yield (independent of impact energy). Kasting[62] has integrated Zahnle's results over an impactor mass (hence, kinetic energy) distribution extending from 10^{13} to 10^{27} J, and finds an average NO production yield of 3.3×10^{14} molecule per joule of impact energy. (Kasting's choice of upper limit for impact energy is nearly identical to the one that we take for the South Pole–Aitken impactor; moreover, production efficiency depends only very weakly on the exact choice of exponent for the impactor mass distribution. We may therefore adopt his integration, an approximate treatment in any case, for our purposes here.) Prinn and Fegley[59] use a constant production yield of 2×10^{16} molecule per joule of energy imparted to the atmosphere. (Taking only 1/8 of an impactor's kinetic energy to be so imparted, their production yield is really 2.5×10^{15} molecule per joule of impact energy.) In laboratory shock-tube experiments, η measures production per joule imparted to the atmosphere, so Kasting's result implies that η for shock production of organics should analogously be scaled down by a factor $(3.3 \times 10^{14})/(2 \times 10^{16}) \approx 1/60$ for production by impact plumes. We therefore have $\dot{m}(t) = \eta(1/60)(6 \times 10^{17}$ J yr$^{-1})f(t)$, as shown in Fig. 2.

The resulting timescale at 4.4 Gyr ago for complete shock processing of a 1-bar CH$_4$ atmosphere is ~10^8 yr. A 1-bar CH$_4$ atmosphere on early Earth could therefore only have been

sustained if terrestrial volcanism or exogenous sources resupplied CH$_4$ to the atmosphere at a rate ~$2 \times 10^7 f(t)$ kg yr^{-1}, or 5×10^{10} kg yr^{-1} 4.4 Gyr ago. Terrestrial volcanoes are estimated[63] to release as much as ~4×10^{10} kg yr^{-1} of carbon into the atmosphere, today mostly in the form of CO$_2$. If carbon were released on a reducing primitive Earth as CH$_4$, early terrestrial volcanism would need to have been no more intense than today to maintain an early reducing atmosphere against impactor shock-processing.

Finally, we note that the effects of erosion of the atmosphere because of especially energetic impacts are unlikely to have substantially altered the results found here. Only ~10% of asteroid, and ~50% of comet, collisions with the Earth occur at velocities high enough to cause erosion[26], effects small compared with other uncertainties in the problem (unless comets made up the bulk of heavy bombardment impactors).

Post-impact recombination. Several authors have suggested that organic molecules may have formed on early Earth by recombination from reducing mixtures resulting from the shock vaporization of bolides on impact[64–66]. Others have suggested organic molecules might be similarly produced by impact shocks of terrestrial rocks[64,67]. McKay et al. have attempted to simulate the former process using a thermochemical model[65]. Their preliminary results show evidence for post-impact organic shock synthesis. But these simulations assumed as initial conditions a reducing mixture equivalent to the elemental composition of comets. In an actual impact, as much target material (oxygen-rich surface rocks) as impactor would be incorporated into the expanding and cooling vapour plume[6], and background atmosphere, possibly CO$_2$-rich, might also be entrained. The applicability of results for an initially reducing vapour mixture ([O] < [H$_2$]) is therefore unclear. Kasting[62] argues that organic carbon in the hot rock vapour plume resulting from an impact would have been nearly entirely converted to CO.

At the same time, there are some relevant experimental data. Barak and Bar-Nun have demonstrated amino acid shock synthesis even when initial gas mixtures contain large quantities of H$_2$O and air[68]. Mukhin et al. have stimulated shock vaporization and organic recombination using laser-pulse heating of a variety of terrestrial rocks and meteorite samples, and report a wide range of products of varying oxidation states[64]. Unfortunately, their experiments vaporized only a fraction of the target, so organics may have been released from melted and heated target material, as well as synthesized in post-vaporization recombination. Experiments in which the entire target is vaporized would be informative.

In the Mukhin et al. experiment, most carbon was incorporated into CO and CO$_2$, but typically several per cent went into hydrocarbons, and smaller amounts into HCN and aldehydes. The resulting ratio of (CO + CO$_2$) to hydrocarbons was roughly constant, even as the initial wt% carbon of the sample varied over more than an order of magnitude; this seems to be consistent with post-shock recombination, rather than escape of unshocked organics. Hydrocarbon production (like that of CO and CO$_2$) scaled with the carbon content of the sample. We will consider the result of directly scaling these experiments up to impacts of comets and asteroids with the early Earth. First, we note that organic production from surface rocks is unimportant (~10%) compared with that for the carbon-rich impactors, so we henceforth ignore the former. Taking 10% of the impactor mass to be cometary (~17% C), of which 50% is blown out into space[26], and the remainder asteroidal (~1.3% C), and taking the Mukhin et al. result that ~4% of impactor carbon is incorporated into organics, we find post-impact recombination produced organic carbon at a rate $\dot{m}(t) = (4.6 \times 10^6$ kg yr$^{-1})f(t)$ on the early Earth. To the extent that recombination occurs without significant entrainment of atmosphere, this result is independent of terrestrial atmosphere oxidation state. We emphasize the great uncertainties in the preceding calculation by displaying post-impact recombination results as a dotted line in Fig. 2.

TABLE 2　Inventory of main sources of prebiotic organics on Earth 4 Gyr ago

Source	Energy dissipation (J yr⁻¹)	Production efficiency, reducing atmosphere (kg J⁻¹)	Production efficiency, neutral* atmosphere (kg J⁻¹)	Organic production, reducing atmosphere (kg yr⁻¹)	Organic production, neutral* atmosphere (kg yr⁻¹)
Lightning	1 (18)†	3 (−9)	3 (−11)	3 (9)	3 (7)
Coronal discharge	5 (17)	3 (−10)	3 (−12)	2 (8)	2 (6)
Ultraviolet light ($\lambda < 2{,}700$ Å)‡	1 (22)	2 (−11)	—	2 (11)	—
Ultraviolet light ($\lambda < 2{,}300$ Å)§	5 (21)	—	5 (−14)	—	3 (8)
Ultraviolet light ($\lambda < 2{,}000$ Å)‖	6 (20)	5 (−12)	—	3 (9)	—
Atmospheric shocks (meteors)	1 (17)	1 (−8)	3 (−16)	1 (9)	3 (1)
Atmospheric shocks (post-impact plumes)	1 (20)	2 (−10)¶	4 (−18)	2 (10)	4 (2)
IDPs	—	—	—	6 (7)	6 (7)
Totals	—	—	—	2 (11)	4 (8)

* $[H_2]/[CO_2] \approx 0.1$.
† 1(18) should be read 1×10^{18}.
‡ Appropriate to absorption by H_2S (ref. 79).
§ Appropriate to absorption by CO_2 (ref. 74).
‖ Appropriate to absorption by H_2O in H_2O/CH_4 atmospheres (ref. 78).
¶ Following Kasting[62], averaged over Zahnle's[60] results for impactor energies from 10^{13} J to 10^{27} J.

Endogenous sources of prebiotic organics

At the time of the origins of life ~4 Gyr ago, IDPs may have been delivering ~10^8 kg yr⁻¹ of organic carbon, regardless of the oxidation state of the terrestrial atmosphere. In the case of an early reducing atmosphere, shocks by post-impact plumes would have been producing organic carbon at ~10^{10} kg yr⁻¹. How important would these heavy bombardment sources have been? We can put such rates into context by comparing them to the main endogenous sources of organics on the early Earth. Our list of possible terrestrial sources is not meant to be exhaustive. We do, however, explicitly cite those endogenous sources that have traditionally been considered to be the most important (Table 2). In the case of sources producing organics in the terrestrial atmosphere, Table 2 lists estimates of energy dissipation (J yr⁻¹), and experimentally derived values of production efficiency of organics (kg C J⁻¹) for reducing and neutral atmospheres. Organic production rates (kg yr⁻¹) are then derived.

As our standard 'neutral' or intermediate oxidation state atmosphere, we take one in which $[H_2]/[CO_2] \approx 0.1$. This choice is by no means the least reducing plausible neutral atmosphere. Kasting has recently considered an atmospheric H_2 mixture ratio of ~10^{-3}, at the upper end of the range used in many photochemical models for the early Earth, and based on H_2 outgassing rate near the upper limit of the flux likely to have been produced by early volcanoes or by photostimulated reduction of ferrous iron in the surface ocean[62]. (On the other hand, large impacts may have enhanced the atmospheric CO/CO_2 ratio[62].) We list $[H_2]/[CO_2] \approx 0.1$ in Table 1, as it is that neutral atmosphere in which organic delivery by IDPs just begins to match endogenous production. It has been experimentally demonstrated[69,70] that organic production rates drop precipitously by orders of magnitude as the ratio $[H_2]/[CO_2]$ falls below ~1, so that lower $[H_2]/[CO_2]$ ratios will strongly favour IDPs as the dominant early source of organics.

In a review[71] of contemporary data for terrestrial lightning and coronal energy dissipation, we recommended values of 1×10^{18} J yr⁻¹ and 5×10^{17} J yr⁻¹, respectively. (These values are factors of about 20 and 120 times smaller than those originally suggested by Miller and Urey[72] in their compilation of sources of organics on the prebiotic Earth. The greatest uncertainty in these estimates, which is difficult to quantify, may be the extrapolation from contemporary rates of electrical energy discharge back to the primitive Earth.) We reviewed experimental work on organic production efficiencies for these sources, finding coronal discharge efficiencies to be about an order of magnitude less than those for lightning[71]. Schlesinger and Miller have found[70] that organic yields drop by ~10^2 for H_2CO, HCN, and amino acids[69] as $[H_2]/[CO_2]$ drops from several (for which values organic production is roughly equal to that in CH_4 atmospheres) to ~0.1. We have therefore scaled production efficiencies for the latter atmosphere in Table 2 down by 10^2 relative to the reducing case.

Apart from electrical energy sources, a principal source of energy for organic production on the early Earth was ultraviolet light from the Sun. In the past decade it has been recognized, from observations of young solar analogue stars, that the early ultraviolet luminosity of the Sun would have been greater than is the case today (despite lower net solar luminosity)[73,74]. In Fig. 3, we use the results of Zahnle and Walker[73] for the evolution of the solar spectrum in 50-Å wavelength intervals (their Fig. 10) to derive the evolution of solar ultraviolet energy at the Earth as a function of time. Their model includes the evolution of important solar spectral lines; we follow it from a 10^8-yr-old Sun (4.5 Gyr ago) until the present. Our values for the current (4.6-Gyr-old) solar ultraviolet fluxes are drawn from Heroux

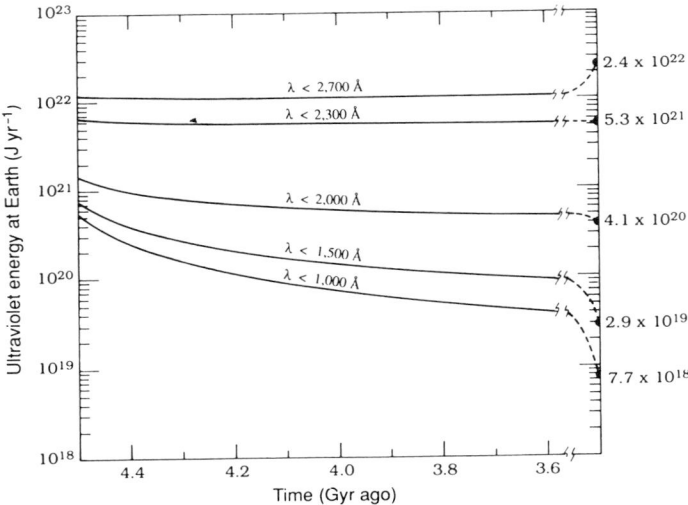

FIG. 3　Evolution of solar ultraviolet luminosity through time, beginning with a ~10^8-Myr-old Sun at 4.5 Gyr ago, and extending to the present. Dashed lines beyond 3.6 Gyr ago indicate qualitative evolution to current values (labelled numerically).

and Hinteregger[75] for wavelengths below 2,000 Å, and from Arvesen et al.[76] from 2,000 Å to 2,700 Å.

Figure 3 shows the evolution of solar energy at the Earth for wavelengths below some selected value. The values shown are chosen as upper limits to the photodissociation continuum of gases that may have been of importance in the early atmosphere. For example, CH_4 is transparent at wavelengths above $\sim$1,500 Å (ref. 77), H_2O above $\sim$2,000 Å (ref. 78), CO_2 above $\sim$2,300 Å (ref. 74), and H_2S above $\sim$2,700 Å (ref. 79). Table 2 lists the net energy available on Earth below the latter three wavelengths at 4 Gyr ago (taking into account cosine incidence and diurnal effects).

In Table 2, the production efficiency for reducing (CH_4/H_2O) atmospheres (at wavelength $\lambda < 2,000$ Å) is taken from experimental work by Ferris and Chen[78], and that for atmospheres containing a long-wavelength ultraviolet photon acceptor generating superthermal hydrogen atoms, such as H_2S (or H_2CO), is taken from Sagan and Khare[79] and Hong et al[80] (the latter authors demonstrate amino acid production using only CH_4 as a carbon source). These authors cite quantum yields for amino acids only; here we assume total organic carbon produced is $\sim$10 times that incorporated into amino acids, a typical ratio[69,70]. We assume that production efficiencies by ultraviolet acting on neutral atmospheres are $\sim$100 times lower than those appropriate for CH_4 atmospheres, again following ref. 70. (This scaling agrees with photochemical modelling results[81] for ultraviolet production of H_2CO in atmospheres where $[H_2]/[CO_2]$ varies from $\sim$3 to $\sim$0.3.) For all these results it was assumed that production in a given candidate atmosphere was limited by the flux of ultraviolet light, not by the abundance of reactants—an assumption which may or may not have held for a given early Earth atmosphere.

An inventory of organics on early Earth

Our results for exogenous delivery, impact-shock synthesis and endogenous production of organic molecules 4 Gyr ago are summarized in Table 2. As indicated by Miller and Urey[72], more energy is potentially available for organic synthesis from long-wavelength ultraviolet radiation than from any other source; and as indicated by Bar-Nun et al.[17], shock synthesis has the highest specific efficiency. The total productivities listed in Table 2 show $\sim$10^{11} kg yr^{-1} for reducing atmospheres 4 Gyr ago, and $\sim$10^8 kg yr^{-1} for mainly neutral CO_2 atmospheres with 10% H_2. With lower $[H_2]$, the endogenous productivities will be considerably less. Other atmospheres may have special mechanisms: for example, $[CH_4] \approx 10^{-4}$ in an N_2 atmosphere will give[82] $\sim$10^{10} kg yr^{-1} of HCN through short-ultraviolet photolysis. For both oxidation state end-member models, the heavy bombardment either generated or delivered quantities of organics roughly comparable to those from endogenous sources.

If all organic products were fully soluble in oceans of contemporary extent and depth, and had a mean lifetime of $\sim$10^7 yr against thermal degradation in mid-ocean vents or subducted plates, the steady-state organic abundance in the oceans 4 Gyr ago would have been $\sim$10^{-3} g per g for a reducing atmosphere and 10^{-6} g per g for our intermediate oxidation state atmosphere. Although freeze–thaw and other concentration mechanisms doubtless existed, reducing atmospheres seem to be more favourable for the origin of life by three orders of magnitude or more; the entire ocean would then have been, almost literally, a 'dilute soup' of organic matter.

Qualitative as well as quantitative differences are likely. For example, IDPs are radiation-hardened from their stay in interplanetary space and may be rich in amorphous carbon and polycyclic aromatic hydrocarbons, which are less interesting for the origin of life than the amino acids and nucleotide bases produced by endogenous and impact processes. But a pathway from such hydrocarbons to amino acids has been proposed[83] (and disputed[84]), and exogenous organics may be preferentially rich in amphiphilic vesicles, conceivably relevant to the origin

of the cell[85].

To summarize, which sources of prebiotic organics were quantitatively dominant depends strongly on the composition of the early terrestrial atmosphere. In the event of an early strongly reducing atmosphere, production by atmospheric shocks seems to have dominated that due to electrical discharges. Organic synthesis by ultraviolet light may, in turn, have dominated shock production, but only if a long-wavelength absorber such as H_2S were supplied to the atmosphere at a rate sufficient for synthesis to have been limited by ultraviolet flux, rather than by reactant abundance. In the apparently more likely case of an early terrestrial atmosphere of intermediate oxidation state, atmospheric shocks were probably of little importance for direct organic production. For $[H_2]/[CO_2]$ ratios of $\sim$0.1, net organic production was some three orders of magnitude lower than for reducing atmospheres, with delivery of intact exogenous organics in IDPs and ultraviolet production being the most important sources. At still lower $[H_2]/[CO_2]$ ratios, IDPs may have been the dominant source of prebiotic organics on the early Earth. Endogenous, exogenous and impact-shock sources of organics could each have made a significant contribution to the origins of life.   □

Christopher Chyba and Carl Sagan are at the Laboratory for Planetary Studies, Cornell University, Ithaca, New York 14853, USA. Christopher Chyba's present address is the Space Science Division, NASA Ames Research Center, Moffett Field, California 94035, USA.

1. Schopf, J. W. & Walter, M. R. in *Earth's Earliest Biosphere* (ed. Schopf, J. W.) 214–239 (Princeton University Press, Princeton, 1983).
2. Walter, M. R. in *Earth's Earliest Biosphere* (ed. Schopf, J. W.) 240–259 (Princeton University press, Princeton, 1983).
3. Schidlowski, M. *Nature* **333**, 313–318 (1988).
4. Basaltic Volcanism Study Project, *Basaltic Volcanism on the Terrestrial Planets* (Pergamon, New York, 1981).
5. Wilhelms, D. E. in *The Geology of the Terrestrial Planets*, NASA SP-469, 107–205 (1984).
6. Melosh, H. J. & Vickery, A. M. *Nature* **338**, 487–489 (1989).
7. Chyba, C. F. *Nature* **343**, 129–133 (1990).
8. Maher, K. A. & Stevenson, D. J. *Nature* **331**, 612–614 (1988).
9. Oberbeck, V. R. & Fogleman, G. *Nature* **339**, 434 (1989).
10. Sleep, N. H., Zahnle, K. J., Kasting, J. F. & Morowitz, H. J. *Nature* **342**, 139–142 (1989).
11. Oró, J. *Nature* **190**, 389–390 (1961).
12. Anders, E. & Owen, T. *Science* **198**, 453–465 (1977).
13. Chyba, C. F. *Nature* **330**, 632–635 (1987).
14. Anders, E. *Nature* **342**, 255–257 (1989).
15. Chyba, C. F., Thomas, P. J., Brookshaw, L. & Sagan, C. *Science* **249**, 366–373 (1990).
16. Delsemme, A. H. in *Comets in the Post-Halley Era* (eds Newburn, R. L., Neugebauer, M. & Rahe, J.) 377–428 (Kluwer, Dordrecht, 1991).
17. Bar-Nun, A., Bar-Nun, N., Bauer, S. H. & Sagan, C. *Science* **168**, 470–473 (1970).
18. Fegley, B., Prinn, R. G., Hartman, H. & Watkins, G. H. *Nature* **319**, 305–308 (1986).
19. Turco, R. P. *et al. Icarus* **50**, 1–52 (1982).
20. Levin, B. Yu. & Bronshten, V. A. *Meteoritics* **21**, 199–215 (1986).
21. Folinsbee, R. E., Douglas, J. A. V. & Maxwell, J. A. *Geochim. cosmochim. Acta* **31**, 1625–1635 (1967).
22. Zhao, M. & Bada, J. L. *Nature* **339**, 463–465 (1989).
23. Zahnle, K. & Grinspoon, D. *Nature* **348**, 157–160 (1990).
24. Chyba, C. F. *Nature* **348**, 113–114 (1990).
25. Ryder, G. *Eos* **71**, 313, 322–323 (1990).
26. Chyba, C. F. *Icarus* **92**, 217–233 (1991).
27. McKinnon, W. B., Chapman, C. R. & Housen, K. R. in *Uranus* (eds Bergstrahl, J. T., Miner, E. D. & Matthews, M. S.) (University of Arizona Press, Tucson, in the press).
28. Dreibus, G. & Wänke, H. in *Origin and Evolution of Planetary and Satellite Atmospheres* (eds Atreya, S. K., Pollack, J. B. & Matthews, M. S.) 268–288 (University of Arizona Press, Tucson, 1989).
29. Walker, J. C. G. *Origins Life* **16**, 117–127 (1986).
30. Kasting, J. F. & Ackerman, T. P. *Science* **234**, 1383–1385 (1986).
31. Fernàndez, J. A. & Ip, W.-H. *Icarus* **54**, 377–387 (1983).
32. Shoemaker, E. M. & Wolfe, R. F. *Proc. lunar planet. Sci. Conf.* **15**, 780–781 (1984).
33. Strom, R. G. *Icarus* **70**, 517–535 (1987).
34. Thomas, P. J., Brookshaw, L., Chyba, C. F. & Sagan, C. *Eos* **71**, 1429 (1990).
35. Shoemaker, E. M., Wolfe, R. F. & Shoemaker, C. S. in *Global Catastrophes in Earth History* (eds Sharpton, V. L. & Ward, P. D.), ESA SP-247, 155–170 (1990).
36. Grieve, R. A. F. *Geol. Soc. Am. Spec. Pap.* **190**, 25–37 (1982).
37. Ceplecha, Z. in *Comets, Asteroids, and Meteorites* (ed. Delsemme, A.) 143–152 (University of Toledo, Toledo, 1977).
38. Baldwin, B. & Sheaffer, Y. *J. geophys. Res.* **76**, 4653–4668 (1971).
39. Melosh, H. J. in *Multi-ring Basins, Proc. Lunar Planet. Sci.* (eds Schultz, P. H. & Merrill, R. B.) 29–35 (Pergamon, New York, 1981).
40. Love, S. G. & Brownlee, D. E. *Icarus* **89**, 26–43 (1991).
41. Kyte, F. T. & Wasson, J. T. *Science* **232**, 1225–1229 (1986).
42. Hughes, D. W. in *Cosmic Dust* (ed. McDonnell, J. A. M.), 123–185 (J Wiley, New York, 1978).
43. Grün, E., Zook, H. A., Fechtig, H. & Giese, R. H. *Icarus* **62**, 244–272 (1985).
44. Schramm, L. S., Brownlee, D. E. & Wheelock, M. M. *Meteoritics* **24**, 99–112 (1989).
45. Brownlee, D. E. *Annu. Rev. Earth planet. Sci.* **13**, 189–215 (1985).
46. Sanford, S. A. & Bradley, J. P. *Icarus* **82**, 146–166 (1989).
47. Whipple, F. L. *Mém. Soc. R. Liège 6eme Série* **9**, 101–111 (1976).
48. Greenberg, J. M. in *Comets and the Origin of Life* (ed. Ponnamperuma, C.) 111–127 (Reidel, Dordrecht, 1981).
49. Gilvarry, J. J. & Hochstim, A. R. *Nature* **197**, 624–625 (1963).

50. Bar-Nun, A. & Shaviv, A. *Icarus* **24,** 197–210 (1975).
51. Chameides, W. L. & Walker, J. C. G. *Origins Life* **11,** 291–302 (1981).
52. Rao, V. V., Mackay, D. & Trass, O. *Can. J. chem. Eng.* **45,** 61–66 (1967).
53. Scattergood *et al. Icarus* **81,** 413–428 (1989).
54. McKay, C. P., Scattergood, T. W., Pollack, J. B., Borucki, W. J. & Van Ghyseghem, H. T. *Nature* **332,** 520–522 (1988).
55. Pollack, J. B., Podolak, M., Bodenheimer, P. & Christofferson, B. *Icarus* **67,** 409–443 (1986).
56. Bronshten, V. A. *Physics of Meteoric Phenomena,* (Reidel, Dordrecht, The Netherlands, 1983).
57. Lewis, J. S., Watkins, G. H., Hartman, H. & Prinn, R. G. *Geol. Soc. Am. Spec. Pap.* **190,** 215–221 (1982).
58. O'Keefe, J. D. & Ahrens, T. J. *Geol. Soc. Am. Spec. Pap.* **190,** 103–120 (1982).
59. Prinn, R. G. & Fegley, B. *Earth planet. Sci. Lett.* **83,** 1–15 (1987).
60. Zahnle, K. J. in *Global Catastrophes in Earth History* (eds Sharpton, V. L. & Ward, P. D.), GSA SP-247, 271–288 (1990).
61. Chyba, C. F. thesis, Cornell Univ. (1991).
62. Kasting, J. F. *Origins Life* **20,** 199–231 (1990).
63. Walker, J. C. G. *Evolution of the Atmosphere* (Macmillan, New York, 1977).
64. Mukhin, L. M., Gerasimov, M. V. & Safonova, E. N. *Nature* **340,** 46–48 (1989).
65. McKay, C. P., Borucki, W. R., Kujiro, D. R. & Church, F. *Lunar planet. Sci. Conf. XX,* 671–672 (1989).
66. Oberbeck, V. R., McKay, C. P., Scattergood, T. W., Carle, G. C. & Valentin, J. R. *Origins Life* **19,** 39–55 (1989).
67. Freund, F., Dickinson, J. T., Becker, C. H., Freund, M. M. & Chang, S. *Origins Life* **16,** 291–292 (1986).
68. Barak, I. & Bar-Nun, A. *Origins Life* **6,** 483–506 (1975).
69. Schlesinger, G. & Miller, S. L. *J. molec. Evol.* **19,** 376–382 (1983).

70. Schlesinger, G. & Miller, S. L. *J. molec. Evol.* **19,** 383–390 (1983).
71. Chyba, C. & Sagan, C. *Origins Life* **21,** 3–17 (1991).
72. Miller, S. L. & Urey, H. C. *Science* **130,** 245–251 (1959).
73. Zahnle, K. J. & Walker, J. C. G. *Rev. Geophys. Space Phys.* **20,** 280–292 (1982).
74. Canuto, V. M., Levine, J. S., Augustsson, T. R. & Imhoff, C. L. *Nature* **296,** 816–820 (1982).
75. Heroux, L. & Hinteregger, H. E. *J. geophys. Res.* **83,** 5305–5308 (1978).
76. Arvesen, J. C., Griffin, R. N. & Pearson, B. D. *Appl. Opt.* **8,** 2215–2232 (1969).
77. Mount, G. H., Warden, E. S. & Moos, H. W. *Astrophys. J.* **214,** L47–49 (1977).
78. Ferris, J. P. & Chen, C. T. *J. Am. chem. Soc.* **97,** 2962–2967 (1975).
79. Sagan, C. & Khare, B. N. *Science* **173,** 417–420 (1971).
80. Hong, K.-Y., Hong, J.-H. & Becker, R. S. *Science* **184,** 984–987 (1974).
81. Pinto, J. P., Galdstone, G. R. & Yung, Y. L. *Science* **210,** 183–185 (1980).
82. Zahnle, K. J. *J. geophys. Res.* **91,** 2819–2834 (1986).
83. Shock, E. O. & Schulte, M. D. *Nature* **343,** 728–731 (1990).
84. Miller, S. L. & Bada, J. L. *Nature* **350,** 388–389 (1991).
85. Deamer, D. W. *Nature* **317,** 792–794 (1985).
86. Tuncel, G. & Zoller, W. H. *Nature* **329,** 703–705 (1987).
87. Barker, J. L. & Anders, E. *Geochim. cosmochim. Acta* **32,** 627–645 (1968).
88. Anders, E., Ganapathy, R., Krähenbühl, U. & Morgan, J. W. *Moon* **8,** 3–24 (1973).
89. Dohnanyi, J. S. *Science* **173,** 558 (1971).

ACKNOWLEDGEMENTS. We thank P. J. Thomas and L. Brookshaw for discussions, and K. Zahnle and H. J. Melosh for reviews of an earlier version of this manuscript. This work was supported by the Kenneth T. and Eileen L. Norris Foundation and NASA.

Section II

PREBIOTIC CHEMISTRY

PREBIOTIC CHEMISTRY

Now, when ultra-violet light acts on a mixture of water, carbon dioxide, and ammonia, a vast variety of organic substances are made, including sugars and apparently some of the materials from which proteins are built up. This fact has been demonstrated in the laboratory of Baly of Liverpool and his colleagues. In this present world, such substances, if left about, decay—that is to say, they are destroyed by micro-organisms. But before the origin of life they must have accumulated till the primitive oceans reached the consistency of hot dilute soup.

J. B. S. Haldane, 1929

What were the sources of organic compounds?

Since life processes are based on chemical principles, the source and composition of organic compounds on the prebiotic Earth has been and remains a central problem for research on biogenesis. Haldane posed this issue in his 1929 essay, citing earlier results from Baly's laboratory to support the idea that light energy, acting upon the primitive atmosphere, could generate more complex molecules. Although Baly's photoreduction of carbon dioxide could not be repeated by later workers, he was clearly on the right experimental track. The ultimate source of the Earth's carbon was extraterrestrial, since carbon would have been one of many elements delivered during accretion. The question, then, concerns the mechanism by which organic carbon entered the prebiotic environment.

There are three possibilities: (1) relatively simple organic compounds like hydrogen cyanide (HCN) and formaldehyde (HCHO) were synthesized in the early atmosphere or hydrosphere and then reacted to form more complex organic molecules like amino acids and carbohydrates, (2) relatively simple organic compounds like HCN and HCHO were delivered or produced by late accretionary infall and then underwent further chemical transformation into more complex molecules, or (3) complex organic carbon compounds were delivered intact by comets and meteorites and were then available to undergo further chemical reactions on the Earth's surface. These are not mutually exclusive pathways, of course, and it is plausible that some groups of compounds were synthesized primarily through terrestrial chemistry while others were brought to the Earth by cometary and meteoritic infall.

Sources and sinks of organics

From the principles of thermodynamics and kinetics, we know that complex chemical structures have finite lifetimes: any structure that is maintained away from equilibrium must have a source from which its components are replenished. Without such a source, the structure is degraded by one or more pathways to simpler molecules. In considering the presence of organic material in the prebiotic environment, we must take into account both its *sources* and its *sinks*. Several sources of organic compounds were described earlier. Examples of sinks include photochemical degradation of organic compounds exposed to ultraviolet light, hydrolytic degradation of organic solutes dissolved in aqueous phases, and thermal degradation by volcanic activity.

This picture would have changed drastically with the appearance of the first metabolic systems (cells). Because living systems are capable of capturing energy and nutrients from the environment and of dispersing the altered end products into the environment, life's origin would have produced a new and different kind of global source and sink for organics. Life's earliest autotrophic microorganisms (using light or inorganic compounds such as H_2S, CH_4, or NH_3 as sources of energy and using atmospheric CO_2 as a source of carbon) would have provided the first *biological source* of organic material, possibly a photosynthetic product resembling carbohydrates. Life's earliest heterotrophic microorganisms (using organic material as its source both of energy and of carbon) would have provided a *biological sink*.

Monomers include amino acids, simple carbohydrates, lipid like molecules, purines, and pyrimidines

The reductionistic research approach is powerful, and, over the past century, it has led to a clearer understanding of the basic molecules involved in cellular processes. Cells are bounded energetic systems—assemblies composed chiefly of macromolecules, large polymeric molecules that are, in turn, composed of much simpler monomers.

Amino acids are the monomers of proteins; simple sugars like glucose are the monomers of carbohydrates such as starch, cellulose, and glycogen. Nucleotides are the monomers of the nucleic acids, and they are composed of pentose sugars, phosphate, and organic bases called purines and pyrimidines, linked through phosphodiester bonds in the polymers we call DNA and RNA.

Lipids are composed of hydrophobic hydrocarbons linked to hydrophilic groups such as glycerol and phosphate. Structures composed of lipids are not polymeric by definition because they are stabilized by nonpolar physical forces rather than covalent chemical bonds. One example of such a structure is the lipid bilayer providing the indispensible membraneous boundary that delineates all cells.

Although the mass and boundary structure of cells are assembled from monomeric and polymeric components, none of these monomers or polymers is, in any sense, "alive." As a result of the circularity of metabolic pathways that define them as bounded systems, cells maintain internal potentials different from those in their environment. It is the maintenance of this particular energetic relationship of the bounded system to its environment that is the attribute of the living state (Swenson, 1991).

Organic synthesis can occur under simulated prebiotic environments

Arguing from the principle of continuity, it is reasonable to search for chemical pathways that plausibly lead from simple to complex molecules, particularly from the monomers described above. The scientific community was startled in 1953 by the publication of the first modern paper on a simulated prebiotic synthesis. That paper, reprinted here, clearly demonstrated that amino acids were synthesized when free energy in the form of electrical discharge was provided to a mixture of simple gases—hydrogen, methane, ammonia, and water. These results suggested immediately that amino acids were plausible constituents of the prebiotic environment, leading to the expectation that simple combinations of these monomers into proteinlike polymers might represent the first step toward the living state.

Within a few years, other monomers were added to this list. Oró (1961) and Oró and Kimball (1961) showed that purine synthesis occurred very readily when hydrogen cyanide molecules condensed into molecules such as adenine, a pentamer of hydrogen cyanide. Other investigators proposed that the formose reaction (base-catalyzed condensa-

tion of formaldehyde) could provide a rich source of carbohydrates, including pentoses and hexoses. Anders et al. (1973) suggested that hydrocarbons, the essential component of lipidlike molecules, could be synthesized under Fischer-Tropsch reaction conditions. From these early successes, it was easy to imagine that life's beginning involved little more than assembling a natural pot of hydrogen cyanide, formaldehyde, and hydrocarbons, stirring in phosphate in the presence of an energy source, and waiting for results: given enough time and the entire Earth as a reaction vessel, the monomeric and polymeric components of living systems must surely arise.

The source of some monomers is problematic

Although we know plausible pathways leading to amino acids, purines, and simple carbohydrates, there are also some surprising gaps, particularly as regards nucleic acids. For instance, pentose sugars such as ribose are essential components of the nucleic acids, yet it is not easy to understand how ribose could have been specifically selected from the mixture of sugars produced by the formose reaction. Purines and pyrimidines are not difficult to imagine as abundant prebiotic end products of hydrogen cyanide condensation, but no robust synthetic pathways for nucleotides have yet been discovered.

Lipids, and especially their hydrocarbon chains, are also relatively difficult to synthesize under plausible prebiotic conditions. While hydrocarbon chains are produced under Fischer-Tropsch conditions where water vapor and a gaseous form of carbon (such as carbon monoxide) are passed over a hot iron-powder catalyst, it is not easy to imagine how such specialized reactions might have occurred on the prebiotic Earth. Furthermore, the hydrocarbon chains must be oxidized to produce fatty acids of approximately 12 carbons or more, and the fatty acids must in turn be linked into more complex molecules like phospholipids, which are necessary for the self-assembly of stable membranes. This seems an increasingly improbable sequence of events. Of course, the seeming improbability may simply reflect our present state of knowledge: it was once also improbable that electrical discharge into a simulated prebiotic atmosphere might produce an abundant yield of amino acids, yet this is now a well-established reaction.

Phosphate itself lacks an obvious source. It is an essential but rare component of the contemporary biosphere and is often a limiting growth factor in lakes and seas. There is no reason to think that phosphate was more abundant on the prebiotic Earth than it is today. Even if all known phosphate deposits were entirely dissolved in the oceans, the phosphate concentration of seawater would increase to a level only slightly above its present 3-micromolar range. Such low concentrations presumably could not have supported robust prebiotic reactions, suggesting that phosphate was a limiting factor in certain prebiotic chemical reactions as well.

Organic compounds are present in comets and meteorites

Another potential source of reduced organic material is meteoritic and cometary infall during late accretion. We do not yet have samples of comets, but they are well known to contain organic carbon compounds, particularly hydrogen cyanide and formaldehyde. In fact, hydrogen cyanide was the first extraterrestrial organic compound to be reported, having been observed in the tail of Halley's comet in 1906.

Access to meteoritic material is provided by relatively common falls occurring every few years. Meteorites, therefore, are the most direct link we have between the planet we live on and its history as a member of the solar system. Based on radioactive decay products of meteorites, the age of the solar system (and, by extension, of the Earth) is estimated to be 4.6 billion years. Because of the violence of events accompanying the accretion process, the oldest geological formations on Earth are only 4 billion years old.

In September 1969, a bolide fireball exploded in the sky above Murchison, Australia, about one hundred miles north of Melbourne. Meteoritic stones rained down over an area of five square miles, and, within days, many kilograms of samples had been collected by scientists. The Murchison fall is an invaluable resource: it is relatively recent and provides a large quantity of material that has undergone minimal handling. Most of the organic carbon is present as a poorly defined keragenlike polymer. A small fraction of the organic material (measured in parts per million) contains compounds relevant to cellular life: amino acids, hydrocarbons, organic acids, and traces of purines.

One might ask how we can be so certain that such compounds are not contaminants. There are several lines of convincing evidence in this regard. First, the amino acids are racemic mixtures, whereas one would expect only L-amino acids had they come from terrestrial contamination. Furthermore, many of the meteoritic amino acids are not found in the proteins of terrestrial life-forms. Instead, they resemble the composition of amino acids produced by the Miller-Urey synthesis (Miller et al., 1976). Probably the most convincing evidence for an extraterrestrial origin is the stable isotope composition of the meteoritic compounds: the deuterium/hydrogen ratios in the organic polymer and in the amino acids is far from that expected of terrestrial contamination (Kerridge, 1982; Epstein et al., 1987).

Meteoritic organics confirm earlier conjectures about prebiotic chemistry

The discovery of extraterrestrial organic compounds in meteorites added weight to Urey's and Miller's observation that synthesis of relatively complex organic substances occurs abiotically and thereby provides a source of building blocks for early forms of life. One of the most significant questions at present concerns the role of extraterrestrial organics in prebiotic evolution. A minimal conclusion is that the presence of organic compounds in meteorites demonstrates the possibility of such reactions occurring in the early solar system. A broader possibility is that meteoritic material may have delivered significant amounts of complex organic carbon to the Earth's surface. Chyba et al. (1990) and Anders (1989) address this question, showing that both comets and meteorites could deliver up to 10^{20}g of organic carbon over the several hundred million years of late accretion following the primary accretion process.

The organic carbon added by this mechanism is in the same range as estimates of formaldehyde rainout rates (Pinto et al., 1980) and is several orders of magnitude greater than the current organic carbon inventory for living organisms, estimated to be 6–9 X 10^{17}g in land biota (Krumbein and Schellnhuber, 1990). Cometary and meteoritic infalls are therefore likely to have been significant sources of organic carbon in the prebiotic environment.

What compounds might be present in a mixture of organic substances derived from carbonaceous meteorites? The most abundant material in such meteorites is a complex kerogenlike polymer containing aromatic groups, followed by monocarboxylic acids and then aliphatic and aromatic hydrocarbons. Amino acids and purines are present in smaller quantities. The major fraction of cometary and meteoritic infall would have entered the oceans and presumably released organic content over a period of time. Water-soluble compounds such as amino acids would have dissolved, while relatively low-density hydrocarbons and their derivatives would have accumulated at the ocean surface, forming a kind of prebiotic oil slick. The effects of wind and tide would have concentrated the films at intertidal zones, just as they do today.

Because all organic compounds are degraded by several kinds of reactions, particularly photo-chemical damage caused by ultraviolet light, it is uncertain what concentrations of organics might have built up in the oceans. We have little idea of the relative amounts of material added by infall and terrestrial synthesis, nor do we know the rate at which the compounds were degraded. On a global scale, organic concentrations were probably very dilute. It has been calculated that, even under the most favorable conditions, amino acid concentrations would reach little more than millimolar ranges (Stribling and Miller, 1987). Local conditions capable of concentrating organic compounds were, therefore, probably essential to the origin of cells.

Polymers are condensation products

Given the presence of some source of monomers, the next step would involve polymerization chemistry. That is, reaction pathways must be discovered that allow monomers to form covalent bonds with one another to produce larger molecules.

Only a few chemical bonds commonly contribute to polymerization chemistry. These include ester bonds, peptide bonds, and glycoside bonds. *Ester bonds* form between an acid and an alcohol group. Examples of esters include the bonds linking phosphate and pentose sugars (phosphodiester linkages) in strands of nucleic acid and the linking of fatty acids to glycerol in the phospholipids of membranes. *Peptide bonds*, formed between the amino and carboxyl groups of amino acids, link fifty to several hundred amino acids in the macromolecules we call proteins. *Glycoside bonds* link simple sugars like glucose into polysaccharides such as starch and cellulose.

It is important to understand that all of the above bonds form during condensation reactions, or *dehydrolysis*, when the elements of water are removed from between the two functional groups. (The reverse reaction, *hydrolysis*, adds the elements of water to such linking bonds and thereby breaks them during digestion.) Furthermore, polymerization reactions require an input of energy; securing this input is the function of cell metabolism in all life today. We may ask, then, what kinds of linking reactions were plausible on the prebiotic Earth, and what were plausible sources of energy? The simplest sources of energy are heat and dehydration. Fox and Harada (1958) applied both to amino acid mixtures and found that polymers were produced with 100 or so amino acids coupled through a variety of bonds, including some peptide linkages. In later work, Usher (1977) showed that relatively low temperatures could drive the formation of phosphodiester bonds between two previously existing oligonucleotides. Hargreaves et al. (1977) and Oró and his coworkers (1978) demonstrated that simple phospholipids could be synthesized when fatty acids were heated and dried in the presence of glycerol and phosphate.

Yet these early successes were achieved under defined conditions and with purified compounds. It is not easy to imagine how simple drying and heating cycles could drive appropriate syntheses of polymers in the mixtures of organic compounds that must have been present in the prebiotic environment. Imagine that a dilute seawater solution of amino acids, carbohydrates, purines, phosphate, and fatty acids was dried and heated by the sun on a lava surface in an intertidal zone. As drying occurred, salt (sodium chloride) crystals would form, divalent cations would precipitate organic anions, and an immense variety of covalent bonds would be produced by condensation reactions between all possible combinations of functional groups. How might self-assembly processes have led to bounded systems of macromolecules capable of primitive metabolic pathways and growth processes? This, of course, is a central question of the origin of life, and only now are we beginning to appreciate the complex dimensions of the problem and the myriad of reactions that could have been possible.

Could mineral surfaces have been involved?

Several investigators have proposed that various minerals might have been involved in the process described above, perhaps providing an organizing substrate that helped sort specific assemblies and metabolic pathways from the complex mixtures in the prebiotic environment. The best known of these substrates is clay, first suggested by Bernal (1951) and later made central to a full hypothesis by Cairns-Smith (1982) in his book, *Genetic Takeover and the Mineral Origins of Life*. Cairns-Smith holds not only that mineral surfaces such as clays could act as templates for organizing organic matter but, further, that certain clay minerals contain a kind of structural information that once served as a primitive genetic material before nucleic acids appeared and took over the genetic function. Although there is no clear evidence in support of a genetic takeover, several investigators have provided convincing evidence that clay minerals can, indeed, act as catalysts, particularly in driving significant polymerization reactions related to nucleotide chemistry (Ferris et al., 1988).

A recent conjecture (Wächtershäuser, 1990) attracting considerable attention is the incorporation of certain iron sulfide minerals into a general scheme for prebiotic chemical evolution. This conceptual model is built on the fact that iron sulfides such as pyrite contain free energy as reduced (ferrous) iron. If that energy could be coupled to one or more primitive metabolic pathways, it would provide an energy source when in close contact with mineral surfaces capable of concentrating potential substrates.

Reprinted papers on prebiotic chemistry

The papers reprinted here show the development of our understanding of prebiotic chemistry. The first paper, by Groth and Suess (1938), demonstrates the remarkable foresight occasionally found in isolated papers that are ahead of their time. (It is published here in English translation for the first time.) In this work, the authors consider clearly the question of how reduced forms of carbon might have been produced on the early Earth, and they discuss a plausible photochemical synthesis. Groth and Suess refer to an earlier paper by Baly, Heilbron and Barker (1921), which claims that ultraviolet light can drive the photoreduction of carbon dioxide to formaldehyde. Chittenden and Schwartz (1981) have published a critical review of the history of this and other such work, arguing convincingly that claims for photoreduction of carbon dioxide are not well founded.

The second paper, by Garrison et al. (1951), describes the use of ionizing radiation in an attempt to "fix" carbon dioxide in a more reduced state, producing formic acid in yields up to 25 percent. The conditions chosen by Garrison et al. happened to be a weakly reducing chemical environment, with ferrous sulfate present in the solution as a source of reducing power. If the authors had chosen more highly reducing conditions, with hydrogen and ammonia present, it is likely that they would have found more complex organic molecules. The third and fourth papers, by Miller (1953) and by Miller and Urey (1959), show how free-energy sources (electrical discharge, ultraviolet light, ionizing radiation, and heat) can drive the synthesis of more complex molecules such as amino acids. The Urey-Miller results are generally considered to mark the beginning of our current understanding of reactions possible on the prebiotic Earth.

The fifth paper, by Oró (1961), added purines to the growing list of monomers that could be formed under simulated prebiotic conditions. The sixth paper, by Ferris et al. (1978), provides a chemical mechanism for the oligomerization of hydrogen cyanide (HCN) into the primary monomeric biomolecules (purines, pyrimidines, and amino acids). Together with HCN, formaldehyde is an essential component of many reactions leading to more complex organic compounds. We have therefore included a short paper by Pinto et al. (1980) that shows how significant amounts of formaldehyde could have been synthesized photochemically in the early atmosphere.

The next three papers concern reactions in which monomers form polymers. Because of the central role of phosphate in polymerization and metabolism, we have included a paper by Westheimer (1987) discussing why nature would have "chosen" phosphate. We have also included a paper by Fox and Harada (1958), the first to demonstrate that simple heating-drying is sufficient to drive the condensation of amino acids into polymerized compounds. This paper was a significant advance in showing that complex polymers were readily formed under plausible prebiotic conditions, but it has turned out to be much more difficult to produce polymers resembling amino acids. We include a paper by Ferris et al. (1988) that discusses this problem and shows how a clay—montmorillonite—can act as a catalyst in phosphate-ester formation. The synthesis of membrane-forming lipids under simulated prebiotic conditions was first demonstrated by Hargreaves et al. (1977) and by Oró et al. (1978), and the Hargreaves paper is reprinted here.

The last three papers concern extraterrestrial organic compounds and their possible delivery to the early Earth. The paper by Kvenvolden et al. (1970) was the first convincing demonstration that the amino acids known to be present in carbonaceous meteorites were, in fact, extraterrestrial in origin rather than contaminants. The papers by Chyba et al. (1990) and by Anders (1989) discuss how much of the organic content of comets and meteorites could be expected to survive atmospheric entry and impact and be delivered intact to the Earth's surface.

References

Anders, E. (1989) Pre-biotic organic matter from comets and asteroids. *Nature* 342:255-257.

Anders, E., Hayatsu, R. and Studier, M. H. (1973) Organic compounds in meteorites. *Science* 182: 781-790.

Baly, E. Ch. C., Heilbron, I. M. and Barker, W. F. (1921) Photocatalysis. Part I: The synthesis of formaldehyde and carbohydrates from carbon dioxide and water. *J. Chem. Soc.* 119:1025-1035.

Bernal, J. D. (1951) *The Physical Basis of Life*. London: Routledge and Kegan Paul.

Cairns-Smith, A. G. (1982) *Genetic Takeover and the Mineral Origins of Life*. Cambridge, U.K.: The Cambridge University Press.

Chittenden, G. J. F. and Schwartz, A. W. (1981) Prebiotic photosynthetic reactions. *BioSystems* 14:15-32.

Chyba, C. F., Thomas, P. J., Brookshaw, L. and Sagan, C. (1990) Cometary delivery of organic molecules to the early Earth. *Science* 249:366-373.

Epstein, S., Krishnamurthy, R. V., Cronin, J. R., Pizzarello, S. and Yuen, G.U. (1987) Unusual stable isotope ratios in amino acid and carboxylic acid extracts from the Murchison meteorite. *Nature* 326:477-479.

Ferris, J. P., Huang, C.-H. and Hagan, W. J. (1988) Montmorillonite: A multifunctional mineral catalyst for the prebiological formation of phosphate esters. *Orig. Life Evol. Biosphere* 18:121-133.

Ferris, J. P., Joshi, P. C., Edelson, E. H. and Lawless, J. G. (1978) HCN: A plausible source of purines, pyrimidines and amino acids on the primitive Earth. *J. Mol. Evol.* 11:293-311.

Fox, S. W. and Harada, K. (1958) Thermal copolymerization of amino acids to a product resembling protein. *Science* 128:1214.

Garrison, W. M., Morrison, D. C., Hamilton, J. G., Benson, A. A. and Calvin, M. (1951) Reduction of carbon dioxide in aqueous solutions by ionizing radiation. *Science* 114:416-418.

Groth, W. and Suess, H. (1938) Bemerkungen zür Photochemie der Erdatmosphäre.*Naturwissenschaften* 26:77.

Haldane, J. B. S. (1929) The origin of life. Reprinted in: J. D. Bernal, *The Origin of Life* (1967). London: Weidenfeld and Nicolson, pp. 242-249.

Hargreaves, W. R., Mulvihill, S. and Deamer, D. W. (1977) Synthesis of phospholipids and membranes in prebiotic conditions. *Nature* 266:78-80.

Kerridge, J. F. (1982) Isotopic composition of C, H, N in carbonaceous chondrite polymer using stepwise combustion. *Proc. Lunar Planet. Sci. Conf.* 13:381-382.

Krumbein, W. E. and Schellnhuber, H.-J. (1990) Geophysiology of carbonates as a function of bioplanets. In: *Facets of Modern Biogeochemistry – Festschrift for F. T. Degens*, V. Ittekkot, S. Kempe, W. Michaelif and A. Spitzy, eds., Berlin: Springer-Verlag, pp. 5-22.

Kvenvolden, K. A., Lawless, J., Pering, K., Peterson, E., Flores, J., Ponnamperuma, C., Kaplan, I. R. and Moore, C. (1970) Evidence for extraterrestrial amino-acids and hydrocarbons in the Murchison meteorite. *Nature* 228:923-926.

Miller, S. L. (1953) Production of amino acids under possible primitive Earth conditions. *Science* 117:528-529.

Miller, S. L. and Urey, H. C. (1959) Organic compound synthesis on the primitive Earth. *Science* 130:245-251.

Miller, S. L., Urey, H. C. and Oró, J. (1976) Origin of organic compounds on the primitive Earth and in meteorites. *J. Mol. Evol.* 9:59-72.

Oró, J. (1961) Mechanism of synthesis of adenine from hydrogen cyanide under possible primitive Earth conditions. *Nature* 191:1193-1194.

Oró, J. and Kimball, A. P. (1961) Synthesis of purines under possible primitive Earth conditions. I: Adenine from hydrogen cyanide. *Arch. Biochem. Biophys.* 94:217-227.

Oró, J., Sherwood, E., Eichberg, J. and Epps, D. E. (1978) Formation of phospholipids under primitive Earth conditions and the role of membranes in prebiological evolution. In: *Light-Transducing Membranes: Structure, Function and Evolution*, D. W. Deamer, ed., New York: Academic Press, pp. 1-19.

Pinto, J. P., Gladstone, G. R. and Yung, Y. L. (1980) Photochemical production of formaldehyde in Earth's primitive atmosphere. *Science* **210**:183-185.

Stribling, R. and Miller, S. L. (1987) Energy yields for hydrogen cyanide and formaldehyde syntheses: The HCN and amino acid concentration in the primitive ocean. *Origins of Life* **17**:261-273.

Swenson, R. (1991) Order, evolution, and natural law: Fundamental relations in complex system theory. In: *Cybernetics and Applied Systems*, C. Negoita, ed., New York: Marcel Dekker, pp. 125-148.

Usher, D. A. (1977) Early chemical evolution of nucleic acids: A theoretical model. *Science* **196**: 311-313.

Wächtershäuser, G. (1990) Evolution of the first metabolic cycles. *Proc. Natl. Acad. Sci. USA* **87**:200-204.

Westheimer, F. H. (1987) Why nature chose phosphates. *Science* **235**:1173-1178.

Remarks on the photochemistry of the Earth's atmosphere.

W. Groth and H. Suess
Institut für physikalische Chemie, Hamburg.
Naturwissenschaften **26** : 77 (1938).

Based on the composition of the Earth's crust and that of volcanic exhalations, geologists conclude that the Earth's primitive atmosphere contained no free oxygen[1]. In discussions concerning the formation of free oxygen, it has been repeatedly pointed out that photochemical processes, and more specifically photodissociation of water vapor abundant in the primitive atmosphere, could have led to its formation.

It will be shown in the following that the formation of free oxygen and so-called organic compounds at a time prior to the existence of organisms can be explained with laboratory experiments in photochemistry. Water vapor and carbon dioxide are presumed major constituents of the primitive atmosphere: both absorb significantly at 1800 Ångstroms. According to the law of radiation, such short-wave radiation is certainly contained in the solar spectrum; the strong absorption of oxygen and ozone in the ultraviolet, however, prevents its penetration of the atmosphere. Further, astrophysical considerations[2] make it likely that the absolute intensity of the extreme ultraviolet portion of solar radiation significantly exceeds that of a black body. This fact is substantiated by demonstration of the spectral lines of excited nitrogen atoms in night sky radiation[3] which can be due only to absorption of extremely short-wave radiation of about 600 Ångstroms.

With the aid of Harteck's xenon arc lamp, which emits resonance lines of xenon both at 1470 Ångstroms and 1295 Ångstroms, one of us has investigated the photodissociation of carbon dioxide[4] and water vapor[5]. The following reactions: CO_2 + light = CO + O, and H_2O + light = OH + H occur, as expected[7] on spectrographic grounds[6]. Furthermore, the experiments by Frankenburger[8] and one of us[9] have demonstrated that H-atoms react with CO to form formaldehyde or glyoxal, according to the equation: H + CO (+M) = HCO, and $2HCO$ = H_2CO + CO (or HCO-HCO).

When we illuminated mixtures of carbon dioxide and water vapor with the xenon arc lamp, both reactions occurred simultaneously. Aldehydes, formed from the CO produced and the dissociation products of water vapor, can be demonstrated by Schryver's reaction, and under certain circumstance their condensation products are deposited on the lamp window[10]. This process, which may have played an important role in the Earth's atmosphere, is similar to the assimilation reaction of plants: under the influence of light, oxygen and carbon compounds are formed from carbon dioxide and water. There is thus an explanation, first, for the initial appearance of free oxygen in the primitive atmosphere, and second, for the formation of certain carbon compounds considered prerequisite for the evolution of organisms. That these processes still take place in the present atmosphere, and that they may play a role in the CO_2-O_2 budget, is suggested by the investigations of Dhar and Ram[11], who demonstrated up to 1 mg/liter formaldehyde in rainwater.

1. See, e.g., Schwinner, *Lehrbuch der physikalischen Geologie* 1, 199. Berlin, 1936.
2. A. Unsöld, Physik der Sternatmosphären. Berlin, 1937. pp. 420-426.
3. M.N. Saha, *Proc. Roy. Soc. Lond.* (A) **160**, 155 (1937).
4. W. Groth, *Z. physik. Chem.* (B) **37**, 315 (1937)
5. W. Groth, *Z. physik. Chem.* (In press)
6. See H. Sponer, *Molekülspektren*. Berlin, 1935.
7. The carbon monoxide produced can be split further into C- and O- atoms by even shorter-wavelength radiation.
8. W. Frankenburger, *Z. Elektrochem.* **36**, 757 (1930).
9. W. Groth, *loc. cit.*
10. In the light of an ordinary mercury lamp, reactions occur only if the illuminated mixture contains mercury as a sensitizer. In the experiments of E. Ch. Baly, J.M. Heilbron and W. F. Baker (*J. Am. Chem. Soc.* **119** (1921) 1025) carbon dioxide-containing water was irradiated in liquid phase and formaldehyde formation was observed.
11. N.R. Dhar and A. Ram, Nature (London) 132, 819 (1933); *J. Physic. Chem.* **37**, 525 (1933).

Translation by J. Seeler of 'Bemerkungen zür Photochemie der Erdatmosphäre' from *Naturwissenschaften*, Volume 26 (1938), p.77, published with permission from Springer-Verlag Heidelberg.

Reduction of Carbon Dioxide in Aqueous Solutions by Ionizing Radiation[1]

W. M. Garrison, D. C. Morrison, J. G. Hamilton, A. A. Benson, and M. Calvin[2]

Crocker Laboratory, Radiation Laboratory,
and *Department of Chemistry,*
University of California, Berkeley

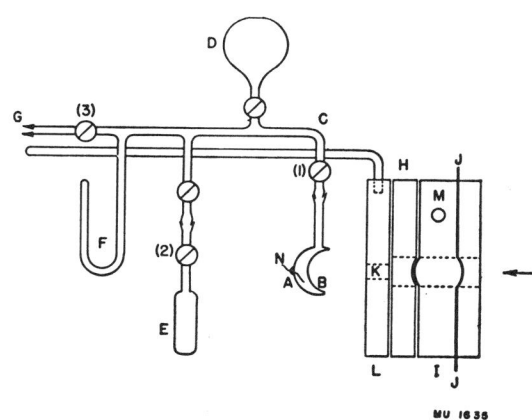

Fig. 1.

The question of the conditions under which living matter originated on the surface of the earth is still a subject limited largely to speculation. The speculation has a greater chance of approaching the truth when it includes and is based upon the ever wider variety of established scientific fact. One of the purposes of the observation reported herein is to add another fact that might have some bearing upon this interesting question.

One of the most popular current conceptions is that life originated in an organic milieu (1–5). The problem to which we are addressed is the origin of that organic milieu in the absence of any life. It appeared to us that one source—if not the only one—of reduced carbon compounds in complex arrangements might be the interaction of various high-energy radiations with aqueous solutions of inorganic materials, particularly carbon dioxide, and nitrogenous compounds such as ammonia and nitrogen, since it appears that these compounds were the commoner forms in which the essential elements were to be found on the primordial earth (6, 7).

Although it has long been known that high-energy radiations can cause organic decomposition and oxidation, it seemed useful to us to demonstrate that conditions could be found in which high-energy radiations could induce the reduction with water of carbon dioxide and the ultimate creation of polyatomic molecules (other than simple polymerization of monomers) of carbon, oxygen, hydrogen, and nitrogen.

The general technique employed was to bombard air-free aqueous solutions of C^{14}-labeled CO_2 in a closed system with and without the addition of ferrous sulfate. The bombardments were made using the 40-mev helium ion beam of the 60-inch cyclotron at Crocker Laboratory. To detect the amount and nature of the reduction products, chemical separations were made on the bombarded solution after the addition of carrier amounts of formic acid, formaldehyde, and methyl alcohol. These were separated as solid derivatives and assayed for C^{14} activity. In most of the bombardments 1 mc of 5–9% C^{14}-labeled CO_2 was used. This made it possible to detect the reduction of approximately one part in 10^6.

A diagram of the target assembly is shown in Fig. 1. The aqueous solutions were bombarded in an all-glass target cell (A) which consisted essentially of a 50-ml Pyrex flask, one side of which was drawn in to give a window (B) having an average thickness of approximately 5 mil over the bombarded area. The cell had a volume of 12 ml. It was connected to a glass manifold (C), which, in turn, was connected through stopcocks to a 100-ml product gas storage bulb (D) to a 25-ml CO_2 reservoir (E) to a mercury manometer (F), and to an outlet (G) through which the entire system could be evacuated. The assembly was supported on a bracket (H), which was fastened to the bell-jar–type target (I). The helium ion beam was brought out of the cyclotron vacuum through a 1.5-mil aluminum foil (J) and was delimited in cross section by the aperture (K) in plate (L). The target window was cooled by means of an air stream, which entered at (M) and emerged through the aperture (K). The beam current was monitored through the electrode (N). With the all-glass target cell it was necessary, because of the nonuniform thickness of the window, to calculate the number of ion-pairs produced from the amount of Fe^{+2} oxidation, assuming the same ion-pair yield for this reaction in the glass cell as was obtained in the cell having the platinum window. With the latter tar-

[1] The work described in this paper was sponsored by the U. S. Atomic Energy Commission.

[2] We wish to thank W. M. Latimer for helpful discussions and the crew of the 60-inch cyclotron at the Crocker Laboratory for the bombardments.

143

TABLE 1

Bombardment	(1)	(2)	(3)	(4)	(5)
Cell window	Glass	Glass	Glass	Platinum	Glass
Solute	$C^{14}O_2 + FeSO_4$	$C^{14}O_2 + FeSO_4$	$C^{14}O_2$	$CO_2 + FeSO_4$	$CO_2 + FeSO_4$
Vol of solution (ml)	12.0	12.0	12.0	12.0	12.0
Gas vol, manifold + product gas bulb (ml)	145.	145.	145.	145.	145.
Initial conc $C^{14}O_2$ in solution (M)	6.8×10^{-5}	8.2×10^{-5}	1.0×10^{-5}	8.2×10^{-5}	$\smile 8 \times 10^{-5}$
No. CO_2 molecules dissolved in H_2O phase	5.5×10^{17}	6.7×10^{17}	8.3×10^{17}	6.2×10^{17}	$\smile 7 \times 10^{17}$
Partial pressure of $C^{14}O_2$ in gas space (mm Hg)	2.4	2.9	3.6	2.7	
C^{14} activity (mc)	1	1	1	None	—
C^{14} in CO_2 (%)	9.0	6.0	4.8	—	—
Energy of emerging helium ions (mev)	40.	40.	40.	40.	40.
Bombardment current (μa)	0.5	0.5	0.5	0.5	0.5
Total bombardment (μa-hr)	0.75	0.042	0.042	0.13	0.13
No. ion-pairs produced in solution (assuming 32.5 eu/ip)	7.8×10^{21}	4.4×10^{20}	4.4×10^{20}	3.1×10^{21}	1.3×10^{21}
Hydrogen pressure after bombardment (mm Hg)	208.	8.	*	*	*
No. hydrogen molecules produced	1.1×10^{21}	$.43 \times 10^{20}$	—	—	—
Initial conc Fe^{+2} (M)	0.80	*	None	0.77	—
Conc Fe^{+3} after bombardment	0.23	—	—	0.10	—
No. of Fe^{+3} atoms formed	1.5×10^{21}	—	—	6.2×10^{20}	—
Ion-pair yield for Fe^{+3} formation	0.20	—	—	0.20	—
Total C^{14} activity in the HCOOH fraction (μc)	0.21	1.32	8.7×10^{-3}	Inactive	Inactive
No. CO_2 molecules reduced to HCOOH	2.2×10^{16}	1.5×10^{17}	1.2×10^{15}	—	—
Fraction of dissolved CO_2 reduced to HCOOH	0.04	0.22	1.4×10^{-3}	—	—
Ion-pair yield for HCOOH formation	2.9×10^{-6}	3.4×10^{-4}	2.8×10^{-6}	—	—
Total C^{14} activity in the HCHO fraction (μc)	5.7×10^{-3}	8×10^{-3}	Inactive	Inactive	Inactive
No. CO_2 molecules reduced to HCHO	6.4×10^{14}	9×10^{14}	—	—	—
Ion-pair yield for HCHO formation	0.82×10^{-7}	2.1×10^{-6}	—	—	—

* Not determined.

get cell, it was possible to estimate within a few per cent the energy loss of the helium ions in penetrating the 1.5-mm aluminum foil, 10 cm of air path, and the 1-mil platinum window.

The target cell was first flushed with nitrogen, then filled with triple-distilled deaerated water, or deaerated 1 M ferrous sulfate solution at a pH of approximately 3.5. The water was deaerated by boiling and then allowed to cool in a glass-stoppered vessel filled with nitrogen gas. The 1 M ferrous sulfate solutions were prepared by adding a known weight of ferrous sulfate. After the target cell was filled, it was immediately connected to the manifold, which was then evacuated until roughly 5% of the target solution had been evaporated. Stopcock (1) was then closed, and the manifold was evacuated, including the product gas storage bulb and that portion of the manifold stopcock (2) which was connected to the CO_2 reservoir containing approximately 1 mc of 5–9% C^{14}-labeled CO_2. After the evacuation was complete, the manifold was isolated by closing stopcock (3). Stopcocks (1) and (2) were then opened, and the CO_2 was allowed to equilibrate with the target solution. The target cell was then bombarded with a 0.5 μa beam of 40-mev helium ions. Bombardment data for each of the experiments are summarized in Table 1.

After bombardment, the target cell was allowed to stand for 1–2 hr to permit the induced radioactivity to decay out. Stopcocks (1) and (2) were closed, and the target cell was removed from the manifold. The solution was then treated with sulfuric acid to dissolve the ferric hydroxide, and adjusted to pH 1. The unreacted $C^{14}O_2$ was stripped with nitrogen and recovered in sodium hydroxide solution. After most of the high specific activity $C^{14}O_2$ had been removed, the solution was flushed with tank CO_2, which was discarded. A sample of the solution was withdrawn at this point for ferric ion analysis.

To the remainder of the solution formic acid, formaldehyde, and methyl alcohol carriers were added in amounts to give 100 mg of the isolated product—i.e., barium formate, methone derivative of formaldehyde, and barium carbonate prepared from the CO_2 formed on oxidation of the methyl alcohol fraction.

The pH of the solution was then adjusted to 7, and the formaldehyde and methyl alcohol were distilled *in vacuo*. The distillate was treated with methone solution in 50% excess and acidified. This precipitated the methone-formaldehyde derivative, and the methyl alcohol was separated from this mixture by a second vacuum distillation. The methyl alcohol distillate was wet-oxidized with a chromium trioxide-sulfuric acid mixture containing potassium iodate, and the evolved CO_2 was recovered as barium carbonate. In none of the bombardments was this barium carbonate fraction active.

The methone-formaldehyde precipitate was filtered off, washed, and dissolved in sodium hydroxide. The solution was acidified and the precipitate centrifuged, washed, redissolved in sodium hydroxide, and reprecipitated. This procedure was repeated, and then the methone-formaldehyde reaction product was recrystallized twice from acetone-water. A sample of the purified methone-formaldehyde product was counted.

The residue from the first distillation containing the formic acid was acidified to pH 1 and distilled *in vacuo*. The distillate was titrated to phenolphthalein end point with a saturated barium hydroxide solution after flushing with CO_2 followed by nitrogen. The precipitate of barium carbonate which formed was centrifuged off. The supernatant containing the barium formate was evaporated to approximately 0.5 ml and while warm was treated with absolute ethyl alcohol, which precipitated crystalline barium formate. This was redissolved in water and recrystallized in this manner four times.

A fraction of original solution, which had been removed for iron determination, was acidified with 6 *N* sulfuric acid and titrated with standard solution of potassium permanganate. A second fraction of this solution was reduced with sulfur dioxide and titrated with potassium permangante after the excess sulfur dioxide was removed by boiling. The Fe^{+3} concentration in the target solution was calculated from the difference in titer. In Table 1, Bombardments 1, 2, and 3 were made using the all-glass cell. Bombardment 4 was made using the cell having a 1-mil platinum window. With this cell, the helium ion beam incident on the solution had an energy of 35.8 mev. The number of ion-pairs produced in Bombardments 1, 2, and 3 were calculated, assuming that the ion-pair yield for ferric ion oxidation obtained in Bombardment 4 was also obtained using the all-glass target cells. This assumption is considered reasonable since the energy losses in the glass and platinum windows were of the same order of magnitude.

To insure that $HC^{14}OOH$ and $HC^{14}HO$ were actually produced by helium ion bombardment, the following additional control experiments were performed: (1) A sample of the original unbombarded target solution containing $C^{14}O_2$ and $FeSO_4$ was retained at approximately 30° C for 1 week and then processed in a manner identical with that used in separating the $HCOOH$, $HCHO$ and CH_3OH fractions in the bombarded samples. No C^{14} activity could be detected in these fractions from the unbombarded solution, indicating that reduced C^{14} compounds were not present in the original solution or formed by a metabolic process involving mold or other organisms. (2) A blank bombardment (#5) was made without added $C^{14}O_2$; the isolated $HCOOH$ and $HCHO$ carriers were inactive. (3) Mass absorption curves run on active barium formate produced in the radiation reduction of $C^{14}O_2$ were identical with those obtained using known samples of active barium formate prepared chemically and having the same specific activity and counting geometry. (4) No decay could be detected in the activity of the radiation produced $HC^{14}OOH$ and $HC^{14}HO$.

An examination of Table 1 demonstrates unequivocally that it is quite possible to reduce appreciable quantities of carbon dioxide to formic acid by means of water through the agency of radiation. In fact, it appears that approximately one fourth of the dissolved carbon dioxide was reduced in Expt 2. Whether the formic acid is further reduced to formaldehyde or whether the formaldehyde has its origin in a direct reduction of carbon dioxide still remains to be demonstrated, but formaldehyde can also be produced from carbon dioxide and water under the influence of radiation.

The actual ion-pair yield is certainly not optimal even in Expt 2 in view of the large excess of the number of ion-pairs produced over the number of molecules of carbon dioxide in the solution. Presumably this reduction is achieved by means of the secondary hydrogen atoms resulting from the ionization. The actual amount of reduction observed is clearly still only the resultant of the reduction and oxidation reactions. The oxidation reaction is presumably minimized by the destruction of the hydroxyl radicals by their reaction with ferrous ion (*8–10*).

Whether carbon—carbon bonds and carbon—nitrogen bonds can be formed and more highly organized structures created under the influence of high-energy radiations is at present under investigation.

References

1. OPARIN, A. I. *Origin of Life* (Trans. by S. Morgulis). New York: Macmillan, 271 (1938).
2. HOROWITZ, N. H. *Proc. Natl. Acad. Sci. U. S.*, **31**, 153 (1945).
3. SIMPSON, G. G. *The Meaning of Evolution*. New Haven. Conn.: Yale Univ. Press (1950).
4. VAN NIEL, C. B. *Photosynthesis in Plants*. Ames, Iowa: Iowa State Coll. Press, 437–95 (1949).
5. HALDANE, J. B. S. *The Origin of Life in Fact and Faith*. Baltimore, Md.: Pelican (1932).
6. LATIMER, W. M. *Science*, **112**, 101 (1950).
7. WARTENBURG, H. *Naturwissenschaften*, **18**, 400 (1950).
8. WEISS, J., et al. *J. Am. Chem. Soc.*, 3241, 3245, 3254, 3256 (1949); 2704, 2709 (1950); 25 (1951).
9. BURTON, M. *J. Phys. & Colloid Chem.*, **51**, 611 (1947).
10. ALLEN, A. O. *Ibid.*, **52**, 479 (1948).

A Production of Amino Acids Under Possible Primitive Earth Conditions

Stanley L. Miller[1, 2]

G. H. Jones Chemical Laboratory,
University of Chicago, Chicago, Illinois

The idea that the organic compounds that serve as the basis of life were formed when the earth had an atmosphere of methane, ammonia, water, and hydrogen instead of carbon dioxide, nitrogen, oxygen, and water was suggested by Oparin (*1*) and has been given emphasis recently by Urey (*2*) and Bernal (*3*).

In order to test this hypothesis, an apparatus was built to circulate CH_4, NH_3, H_2O, and H_2 past an electric discharge. The resulting mixture has been tested for amino acids by paper chromatography. Electrical discharge was used to form free radicals instead of ultraviolet light, because quartz absorbs wavelengths short enough to cause photo-dissociation of the gases. Electrical discharge may have played a significant role in the formation of compounds in the primitive atmosphere.

The apparatus used is shown in Fig. 1. Water is boiled in the flask, mixes with the gases in the 5-l flask, circulates past the electrodes, condenses and empties back into the boiling flask. The U-tube prevents circulation in the opposite direction. The acids and amino acids formed in the discharge, not being volatile, accumulate in the water phase. The circulation of the gases is quite slow, but this seems to be an asset, because production was less in a different apparatus with an aspirator arrangement to promote circulation. The discharge, a small corona, was provided by an induction coil designed for detection of leaks in vacuum apparatus.

The experimental procedure was to seal off the opening in the boiling flask after adding 200 ml of water, evacuate the air, add 10 cm pressure of H_2, 20 cm of CH_4, and 20 cm of NH_3. The water in the flask was boiled, and the discharge was run continuously for a week.

[1] National Science Foundation Fellow, 1952–53.
[2] Thanks are due Harold C. Urey for many helpful suggestions and guidance in the course of this investigation.

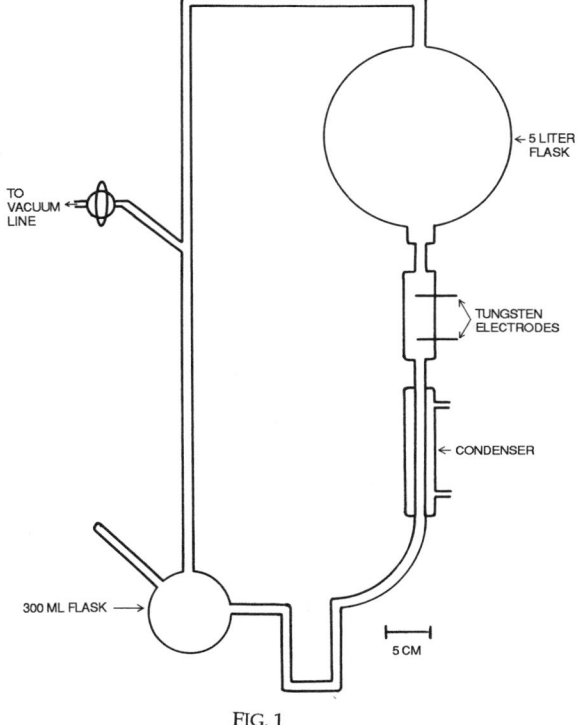

TO VACUUM LINE

← 5 LITER FLASK

TUNGSTEN ELECTRODES

← CONDENSER

300 ML FLASK →

5 CM

FIG. 1

During the run the water in the flask became noticeably pink after the first day, and by the end of the week the solution was deep red and turbid. Most of the turbidity was due to colloidal silica from the glass. The red color is due to organic compounds adsorbed on the silica. Also present are yellow organic compounds, of which only a small fraction can be extracted with ether, and which form a continuous streak tapering off at the bottom on a one-dimensional chromatogram run in butanol-acetic acid. These substances are being investigated further.

At the end of the run the solution in the boiling flask was removed and 1 ml of saturated $HgCl_2$ was added to prevent the growth of living organisms. The ampholytes were separated from the rest of the constituents by adding $Ba(OH)_2$ and evaporating *in vacuo* to remove amines, adding H_2SO_4 and evaporat-

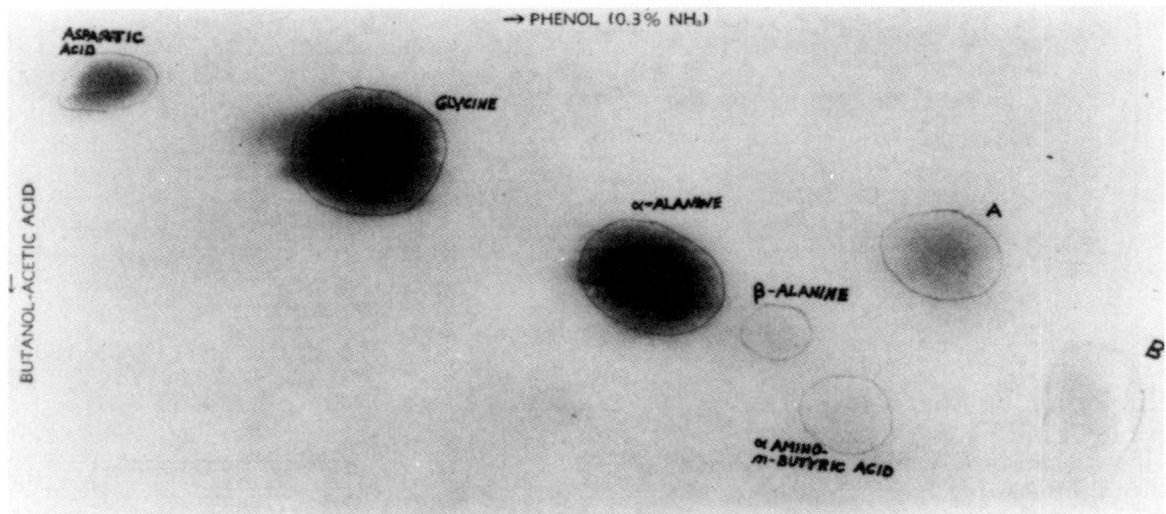

FIG. 2.

ing to remove the acids, neutralizing with $Ba(OH)_2$, filtering and concentrating *in vacuo.*

The amino acids are not due to living organisms because their growth would be prevented by the boiling water during the run, and by the $HgCl_2$, $Ba(OH)_2$, H_2SO_4 during the analysis.

In Fig. 2 is shown a paper chromatogram run in *n*-butanol-acetic acid-water mixture followed by water-saturated phenol, and spraying with ninhydrin. Identification of an amino acid was made when the R_f value (the ratio of the distance traveled by the amino acid to the distance traveled by the solvent front), the shape, and the color of the spot were the same on a known, unknown, and mixture of the known and unknown; and when consistent results were obtained with chromatograms using phenol and 77% ethanol.

On this basis glycine, α-alanine and β-alanine are identified. The identification of the aspartic acid and α-amino-*n*-butyric acid is less certain because the spots are quite weak. The spots marked A and B are unidentified as yet, but may be beta and gamma amino acids. These are the main amino acids present, and others are undoubtedly present but in smaller amounts. It is estimated that the total yield of amino acids was in the milligram range.

In this apparatus an attempt was made to duplicate a primitive atmosphere of the earth, and not to obtain the optimum conditions for the formation of amino acids. Although in this case the total yield was small for the energy expended, it is possible that, with more efficient apparatus (such as mixing of the free radicals in a flow system, use of higher hydrocarbons from natural gas or petroleum, carbon dioxide, etc., and optimum ratios of gases), this type of process would be a way of commercially producing amino acids.

A more complete analysis of the amino acids and other products of the discharge is now being performed and will be reported in detail shortly.

References

1. OPARIN, A. I. The Origin of Life. New York: Macmillan (1938).
2. UREY, H. C. *Proc. Natl. Acad. Sci. U. S.,* **38**, 351 (1952) ; *The Planets.* New Haven: Yale Univ. Press Chap. 4 (1952).
3. BERNAL, J. D. *Proc. Phys. Soc. (London),* **62A**, 537 (1949) ; **62B**, 597 (1949) ; *Physical Basis of Life.* London: Routledge and Kegan Paul (1951).

Manuscript received February 13, 1953.

Organic Compound Synthesis on the Primitive Earth

Several questions about the origin of life have been answered, but much remains to be studied.

Stanley L. Miller and Harold C. Urey

Since the demonstration by Pasteur that life does not arise spontaneously at the present time, the problem of the origin of life has been one of determining how the first forms of life arose, from which all the present species have evolved. This problem has received considerable attention in recent years, but there is disagreement on many points. We shall discuss the present status of the problem, mainly with respect to the early chemical history of, and the synthesis of organic compounds on, the primitive earth.

Many of our modern ideas on the origin of life stem from Oparin (1), who argued that the spontaneous generation of the first living organism might reasonably have taken place if large quantities of organic compounds had been present in the oceans of the primitive earth. Oparin further proposed that the atmosphere was reducing in character and that organic compounds might be synthesized under these conditions. This hypothesis implied that the first organisms were heterotrophic—that is, that they obtained their basic constituents from the environment instead of synthesizing them from carbon dioxide and water. Horowitz (2) discussed this point further and outlined how a simple heterotrophic organism could develop the ability to synthesize various cell constituents and thereby evolve into autotrophic organisms.

In spite of the argument by Oparin,

numerous attempts were made to synthesize organic compounds under the oxidizing conditions now present on the earth (3). Various sources of energy acting on carbon dioxide and water failed to give reduced carbon compounds except when contaminating reducing agents were present. The one exception to this was the synthesis of formic acid and formaldehyde in very small yield (10^{-7} H_2CO molecules per ion pair) by the use of 40-million-electron-volt helium ions from a 60-inch cyclotron (4). While the simplest organic compounds were indeed synthesized, the yields were so small that this experiment can best be interpreted to mean that it would not have been possible to synthesize organic compounds nonbiologically as long as oxidizing conditions were present on the earth. This experiment is important in that it induced a reexamination of Oparin's hypothesis of the reducing atmosphere (5).

The Primitive Atmosphere

Our discussion is based on the assumption that conditions on the primitive earth were favorable for the production of the organic compounds which make up life as we know it. There are many sets of conditions under which organic compounds could have been produced. All these conditions are more or less reducing. However, before accepting a set

of conditions for the primitive earth, one must show that reactions known to take place will not rapidly change the atmosphere to another type. The proposed set of conditions must also be consistent with the known laws for the escape of hydrogen.

Cosmic dust clouds, from which the earth is believed to have been formed, contain a great excess of hydrogen. The planets Jupiter, Saturn, Uranus, and Neptune are known to have atmospheres of methane and ammonia. There has not been sufficient time for hydrogen to escape from these planets, because of their lower temperatures and higher gravitational fields. It is reasonable to expect that the earth and the other minor planets also started out with reducing atmospheres and that these atmospheres became oxidizing, due to the escape of hydrogen.

The meteorites are the closest approximation we have to the solid material from which the earth was formed. They are observed to be highly reduced—the iron mostly as metallic iron with some ferrous sulfide, the carbon as elemental carbon or iron carbide, and the phosphorus as phosphides.

The atmosphere under these reducing conditions would contain some hydrogen, methane, nitrogen, and ammonia; smaller amounts of carbon dioxide and carbon monoxide; and possibly small amounts of other substances such as higher hydrocarbons, hydrogen sulfide, and phosphine. These substances were probably not present in equilibrium concentrations, but compounds which are thermodynamically very unstable in this highly reducing atmosphere—such as oxygen, oxides of nitrogen, and oxides of sulfur—could not have been present in more than a few parts per million. This is true of compounds which are unstable in the present oxidizing atmosphere of the earth, such as hydrogen, ozone, methane, and nitrous oxide.

The over-all chemical change has been the oxidation of the reducing atmosphere to the present oxidizing atmosphere. This

Dr. Miller is a member of the staff of the department of biochemistry, College of Physicians and Surgeons, Columbia University, New York. Dr. Urey is on the staff of the University of California, La Jolla, Calif.

is caused by the loss of hydrogen, which results in the production of nitrogen, nitrate, sulfate, free oxygen, and ferric iron. As is discussed below, many complex organic compounds would have been formed during the course of this over-all change, thereby presenting a favorable environment for the formation of life. Whether the surface carbon of the present earth was all part of the initial atmosphere or whether it has been escaping from the earth's interior in a somewhat reduced condition is not important to the over-all picture.

Escape of Hydrogen

We have learned in recent years that the temperature of the high atmosphere is 2000°K or more, and there is no reason to suppose that the same temperature was not present in the past. One might expect that a reducing atmosphere would be cooler than an oxidizing atmosphere because methane and ammonia can emit infrared radiation while the diatomic molecules, nitrogen and oxygen, cannot. Curtis and Goody (6) have shown that carbon dioxide is ineffective in emitting infrared radiation in the high atmosphere. This is due to the low efficiency of energy transfer from the translational and rotational to the vibrational degrees of freedom, and it seems likely that this would apply to methane as well.

The loss of hydrogen from the earth is now believed to be limited by the diffusion of H_2 to the high atmosphere, since almost all the water is frozen out before it reaches the high atmosphere. Urey (7) has discussed this problem and finds that the loss is entirely due to these effects and not to the Jeans escape formula.

The present rate of escape is 10^7 atoms of hydrogen per square centimeter per second, and it is proportional to the concentration of molecular hydrogen in the atmosphere, which is now 10^{-6} atm at the earth's surface. This rate would result in escape of hydrogen equivalent to 20 g of water per square centimeter in the last 4.5×10^9 years. This rate is not sufficient to account for the oxygen in the atmosphere (230 g/cm²).

In addition, we must account for the oxidation of the carbon, ammonia, and ferrous iron to their present states of oxidation. The oxidation of the 3000 g of surface carbon per square centimeter present on the earth from the 0 to the +4 valence state (that is, from C or

H_2CO to CO_2) would require the loss of 1000 g of hydrogen per square centimeter. At the present rate of escape this would require 2.5×10^{12} years. In order for this escape to be accomplished in 2.5×10^9 years (that is, between 4.5×10^9 and 2.0×10^9 years ago), a pressure of hydrogen at the surface of the earth of 0.7×10^{-3} atm would have been required. In order for the nitrogen, sulfur, and iron also to be oxidized, even larger losses and a higher pressure of hydrogen would have been needed. We use a figure of 1.5×10^{-3} atm for the hydrogen pressure in the primitive atmosphere.

These calculations are greatly oversimplified, since methane and other volatile hydrogen compounds would be decomposed in the high atmosphere and therefore a higher concentration of hydrogen might exist in the high atmosphere than is indicated by surface partial pressures. However, the results of the calculation would be qualitatively the same for hydrogen pressures different from the chosen value by an order of magnitude.

Equilibria of Carbon Compounds

The partial pressure of CO_2 in the atmosphere is kept low by two buffer systems. The first system, which is rapid, is the absorption in the sea to form HCO_3^- and H_2CO_3. The second, which is slow, is the reaction of carbon dioxide with silicates; for example

$$CaSiO_3 + CO_2 \rightarrow$$
$$CaCO_3 + SiO_2 \qquad K_{eq} = 10^6$$

The partial pressure of CO_2 at sea level (3.3×10^{-4} atm) is somewhat higher than the equilibrium pressure (10^{-8} atm), but very much lower than would be the case without the formation of limestones ($CaCO_3$).

The equilibrium constant at 25°C in the presence of liquid water for the reaction

$$CO_2 + 4H_2 \rightarrow CH_4 + 2H_2O$$

is 8×10^{22}. Assuming that equilibrium was attained, and using partial pressures $P_{CO_2} = 10^{-8}$ atm and $P_{H_2} = 1.5 \times 10^{-3}$ atm, we find that the pressure of CH_4 would be 4×10^3 atm. In order to have a reasonable pressure of CH_4, the partial pressure of CO_2 would have to be less than 10^{-8} atm, and limestones would not form.

Complete thermodynamic equilibrium could not exist in a reducing atmosphere

because of the dependence of the equilibrium proportions of compounds on pressure and hence on altitude. It is more likely that the steady-state concentrations of CO_2 and CH_4 would be determined not by the equilibrium at sea level but rather by the equilibrium at higher altitude, where the ultraviolet light would provide the activation energy to bring about rapid equilibrium. Under these conditions water would be a gas, and the equilibrium constant would be 10^{20}, so

$$K_{eq} = 10^{20} = P_{CH_4}P^2_{H_2O}/P_{CO_2}P^4_{H_2}$$
$$= X_{CH_4}X^2_{H_2O}/X_{CO_2}X^4_{H_2} \quad P^{-2}$$

where the X's are the mole fractions and P is the total pressure. If the surface partial pressures were $P_{CH_4} = 1$, $P_{CO_2} = 3.3 \times 10^{-4}$ (the present value), and $P_{H_2} = 1.5 \times 10^{-3}$, the X's would be equal to these partial pressures. We use $X_{H_2O} = 10^{-6}$, which is the present value for H_2O above the tropopause. Equilibrium will be established under these conditions where $P = 2.5 \times 10^{-9}$ atm—the present atmospheric pressure at about 180 km. It is reasonable to assume that equilibrium was established at some high altitude; therefore, carbon dioxide and hydrogen could both have been present at small partial pressures and methane could have been present at a moderate partial pressure in a reducing atmosphere where the pressure of hydrogen was 1.5×10^{-3} atm.

Carbon monoxide should not have been an important constituent of the atmosphere, as can be seen from the following reaction

$$CO_2 + H_2 \rightarrow CO + H_2O_{(l)} \quad K_{eq} = 3.2 \times 10^{-4}$$
$$P_{CO}/P_{CO_2} = 3.2 \times 10^{-4}P_{H_2}$$

Using $P_{H_2} = 1.5 \times 10^{-3}$, we have the ratio $P_{CO}/P_{CO_2} = 5 \times 10^{-7}$, which is independent of pressure. Furthermore, carbon monoxide is a relatively reactive compound, and should any significant quantities appear in the atmosphere, it would react rather rapidly to give organic compounds, carbon dioxide and hydrogen, and formate.

Rubey (8) and Abelson (9) have argued that the surface carbon and nitrogen have come from the outgassing of the interior of the earth instead of from the remaining gases of the cosmic dust cloud from which the earth was formed. The carbon from the outgassing of the earth is a mixture of CO_2, CO, and CH_4, and hydrogen may be present. While outgassing may have been a significant process on the primitive earth, this does not mean that the atmosphere was necessarily composed of CO_2 and CO. The thermodynamic considerations discussed above would still apply. The carbon dioxide would dissolve in the ocean to form bicarbonate, and $CaCO_3$ would be deposited, and the CO would be unstable, as is demonstrated above.

Many writers quote "authorities" in regard to these questions without understanding what is fact and what is opinion. The thermodynamic properties of C, CO, CO_2, CH_4, N_2, NH_3, O_2, H_2O, and other similar substances are all well known, and the equilibrium mixtures can be calculated for any given composition without question. The only point open to argument is whether equilibrium was approximated or whether a nonequilibrium mixture was present. A mixture of hydrogen and carbon monoxide or hydrogen and carbon dioxide is very unstable at $25°C$, but does not explode or react detectably in years. But would such mixtures remain in an atmosphere for millions of years subject to energetic radiation in the high atmosphere? We believe the answer is "No." These mixtures would react even without such radiation in geologic times. Hydrogen and oxygen will remain together at low temperatures for long times without detectable reaction by ordinary methods. The use of radioactive tracers shows that a reaction is proceeding at ordinary temperatures nonetheless.

The buffer systems of the ocean and the calcium silicate–calcium carbonate equilibrium were of sufficient capacity to keep the partial pressure of the carbon dioxide in the atmosphere at a low value; hence, the principal species of carbon in the atmosphere would have been methane, even though the fraction of surface carbon in the oxidation state of carbon dioxide was continuously increasing. This would have been true until the pressure of H_2 fell below about 10^{-6} atm. It is likely that shortly after this, significant quantities of molecular oxygen would have appeared in the atmosphere.

Equilibria of Nitrogen Compounds

The equilibrium concentrations of ammonia can be discussed by considering the reaction

$$\frac{1}{2}N_2 + \frac{3}{2}H_2 \rightarrow NH_3 \qquad K_{25} = 7.6 \times 10^2$$

Using $P_{H_2} = 1.5 \times 10^{-3}$, we have $P_{NH_3}/P_{N_2}^{1/2} = 0.04$.

Ammonia is very soluble in water and

Table 1. Present sources of energy averaged over the earth.

Source	Energy (cal cm^{-2} yr^{-1})
Total radiation from sun	260,000
Ultraviolet light	
$\lambda < 2500$ A	570
$\lambda < 2000$ A	85
$\lambda < 1500$ A	3.5*
Electric discharges	4†
Cosmic rays	0.0015
Radioactivity (to 1.0 km depth)	0.8‡
Volcanoes	0.13§

* Includes the 1.9 cal cm^{-2} yr^{-1} from the Lyman α at 1216 A (39). † Includes 0.9 cal cm^{-2} yr^{-1} from lightning and about 3 cal cm^{-2} yr^{-1} due to corona discharges from pointed objects (40). ‡ The value, 4×10^9 years ago, was 2.8 cal cm^{-2} yr^{-1} (41). § Calculated on the assumption of an emission of lava of 1 km^3 ($C_p = 0.25$ cal/g, $P = 3.0$ g/cm^3) per year at $1000°C$.

therefore would displace the above reaction toward the right, giving

$$\frac{1}{2}N_2 + \frac{3}{2}H_2 + H^+ \rightarrow NH_4^+$$

$$(NH_4^+)/P_{N_2}^{1/2}P_{H_2}^{3/2} = 8.0 \times 10^{13}(H^+)$$

which is valid for pH's less than 9. At $pH = 8$ and $P_{H_2} = 1.5 \times 10^{-3}$, we have

$$(NH_4^+)/P_{N_2}^{1/2} = 47$$

which shows that most of the ammonia would have been in the ocean instead of in the atmosphere. The ammonia in the ocean would have been largely decomposed when the pressure of hydrogen fell below 10^{-5} atm, assuming that the pH of the ocean was 8, its present value. A higher pH would have made the ammonia less stable; the converse is true for a lower pH.

All the oxides of nitrogen would have been unstable and therefore rare. Hydrogen sulfide would have been present in the atmosphere only as a trace constituent because it would have precipitated as ferrous and other sulfides. Sulfur would have been reduced to hydrogen sulfide by the reaction

$$H_2 + S \rightarrow H_2S \qquad K = 6 \times 10^4$$

It is evident that the calculations do not have a quantitative validity because of many uncertainties with respect to temperature, the processes by which equilibrium could be approached, the atmospheric level at which such processes would be effective, and the partial pressure of hydrogen required to provide the necessary rate of escape. In view of these uncertainties, further calculations are unprofitable at the present time. However, we can conclude from this dis-

cussion that a reducing atmosphere containing low partial pressures of hydrogen and ammonia and a moderate pressure of methane and nitrogen constitutes a reasonable atmosphere for the primitive earth. That this was the case is not *proved* by our arguments, but we maintain that atmospheres containing large quantities of carbon monoxide and carbon dioxide are not stable and cannot account for the loss of hydrogen from the earth.

Synthesis of Organic Compounds

At the present time the direct or indirect source of free energy for all living organisms is the sunlight utilized by photosynthetic organisms. But before the evolution of photosynthesis other sources of free energy must have been used. It is of interest to consider the sources of such free energy as well as the origin of the appropriate chemical compounds containing excess free energy which supplied the energy for chemical evolution prior to the existence of what should be called living organisms, and before the evolution of photosynthesis.

Table 1 gives a summary of the sources of energy in the terrestrial surface regions. It is evident that sunlight is the principal source of energy, but only a small fraction of this is in the wavelengths below 2000 A which can be absorbed by CH_4, H_2O, NH_3, CO_2, and so on. If more complex molecules are formed, the absorption can move to the 2500-A region or to longer wavelengths where a substantial amount of energy is available. With the appearance of porphyrins and other pigments, absorption in the visible spectrum becomes possible.

Although it is probable, it is not certain that the large amount of energy from ultraviolet light would have made the principal contribution to the synthesis of organic compounds. Most of the photochemical reactions at these low wavelengths would have taken place in the upper atmosphere. The compounds so formed would have absorbed at longer wavelengths and therefore might have been decomposed by this ultraviolet light before reaching the oceans. The question is whether the rate of decomposition in the atmosphere was greater or less than the rate of transport to the oceans.

Next in importance as a source of energy are electric discharges, such as lightning and corona discharges from pointed objects, which occur closer to the earth's surface and hence would have effected more efficient transfer to the oceans.

Cosmic-ray energy is negligible at present, and there is no reason to assume it was greater in the past. The radioactive disintegration of uranium, thorium, and potassium was more important 4.5×10^9 years ago than it is now, but still the energy was largely expended on the interior of solid materials such as the rocks, and only a very small fraction of the total energy was available in the oceans and atmosphere. Volcanic energy is not only small but its availability is very limited. A continuous source of energy is needed. It contributes little to the evolutionary process to have a lava flow in one part of the earth at one time and to have another flow on the opposite side of the earth years later. For a brief time heat is available at the surface of the lava, but the surface cools and heat flows slowly from the interior for years, making the surface slightly warm. Only a very small contribution to the evolutionary process could be contributed by these energy sources.

Electric Discharges

While ultraviolet light is a greater source of energy than electric discharges, the greatest progress in the synthesis of organic compounds under primitive conditions has been made with electric discharges. The apparatus used by Miller in these experiments was a closed system of glass, except for tungsten electrodes. The water is boiled in a 500-ml flask which mixes the water vapor and gases in a 5-lit. flask where the spark is located. The products of the discharge are condensed and flow through a U-tube back into the 500-ml flask. The first report (10) showed that when methane, ammonia, water, and hydrogen were subjected to a high-frequency spark for a week, milligram quantities of glycine, alanine, and α-amino-*n*-butyric acid were produced.

A more complete analysis (11, 12) of the products gave the results shown in Table 2. The compounds in the table account for 15 percent of the carbon added as methane, with the yield of glycine alone being 2.1 percent. Indirect evidence indicated that polyhydroxyl compounds (possibly sugars) were synthesized. These compounds were probably formed from condensations of the formaldehyde that was produced by the

electric discharge. The alanine was demonstrated to be racemic, as would be expected in a system which contained no asymmetric reagents. It was shown that the syntheses were not due to bacterial contamination. The addition of ferrous ammonium sulfate did not change the results, and the substitution of N_2 for the NH_3 changed only the relative yields of the compounds produced.

This experiment has been repeated and confirmed by Abelson (13), by Pavlovskaya and Passynsky (14), and by Heyns, Walter, and Meyer (15). Abelson worked with various mixtures of H_2, CH_4, CO, CO_2, NH_3, N_2, H_2O, and O_2. As long as the conditions were reducing conditions—that is, as long as either H_2, CH_4, CO, or NH_3 was present in excess—amino acids were synthesized. The products were the same and the yields as large in many of these mixtures as they were with methane, ammonia, and water. If the conditions were oxidizing, no amino acids were synthesized. These experiments have confirmed the hypothesis that reducing atmospheres are required for the formation of organic compounds in appreciable quantites. However, several of these mixtures of gases are highly unstable. Hence the synthesis of amino acids in these mixtures does not imply that such atmospheres were present on the primitive earth.

Heyns, Walter, and Meyer also performed experiments with different mixtures of gases, with results similar to Abelson's. These workers also used CH_4, NH_3, H_2O, and H_2S. They obtained ammonium thiocyanate, thiourea, and thioacetamide as well as compounds formed when H_2S was absent.

The mechanism of synthesis of the amino acids is of interest if we are to extrapolate the results in these simple systems to the primitive earth. Two alternative proposals were made for the synthesis of the amino and hydroxy acids in the spark discharge system. (i) Aldehydes and hydrogen cyanide are synthesized in the gas phase by the spark. These aldehydes and the hydrogen cyanide react in the aqueous phase of the system to give amino and hydroxy nitriles, which are hydrolyzed to amino and hydroxy acids. This mechanism is essentially a Strecker synthesis. (ii) The amino and hydroxy acids are synthesized in the gas phase from the ions and radicals that are produced in the electric discharge.

It was shown that most, if not all, of the amino acids were synthesized accord-

ing to the first hypothesis, since the rate of production of aldehydes and hydrogen cyanide by the spark and the rate of hydrolysis of the amino nitriles were sufficient to account for the total yield of amino acids (12).

This mechanism accounts for the fact that most of the amino acids were α-amino acids, the ones which occur in proteins. The β-alanine was formed not by this mechanism but probably by the addition of ammonia to acrylonitrile (or acrylamide or acrylic acid), followed by hydrolysis to β-alanine.

The experiments on the mechanism of the electric discharge synthesis of amino acids indicate that a special set of conditions or type of electric discharge is not required to obtain amino acids. Any process or combination of processes that yielded both aldehydes and hydrogen cyanide would have contributed to the amount of α-amino acids in the oceans of the primitive earth. Therefore, whether the aldehydes and hydrogen cyanide came from ultraviolet light or from electric discharges is not a fundamental question, since both processes would have contributed to the α-amino acid content. It may be that electric discharges were the principal source of hydrogen cyanide and that ultraviolet light was the principal source of aldehydes, and that the two processes complemented each other.

Ultraviolet Light

It is clear from Table 1 that the greatest source of energy would be ultraviolet light. The effective wavelengths would be $CH_4 < 1450$ A, $H_2O < 1850$ A, $NH_3 < 2250$ A, CO < 1545 A, $CO_2 < 1690$ A, $N_2 < 1100$ A, and $H_2 < 900$ A. It is more difficult to work with ultraviolet light than with electric discharges because of the small wavelengths involved.

The action of the 1849-A Hg line on a mixture of methane, ammonia, water, and hydrogen produced only a very small yield of amino acids (16). Only NH_3 and H_2O absorb at this wavelength, but apparently the radical reactions formed active carbon intermediates. The limiting factor seemed to be the synthesis of hydrogen cyanide. Groth (17) found that no amino acids were produced by the 1849-A line of mercury with a mixture of methane, ammonia, and water, but that amines and amino acids were formed when the 1470-A and

1295-A lines of xenon were used. The 1849-A line produced amines and amino acids with a mixture of ethane, ammonia, and water. The mechanism of this synthesis was not determined. Terenin (18) has also obtained amino acids by the action of the xenon lines on methane, ammonia, and water.

We can expect that a considerable amount of ultraviolet light of wavelengths greater than 2000 A would be absorbed in the oceans, even though there would be considerable absorption of this radiation by the small quantities of organic compounds in the atmosphere. Only a few experiments have been performed which simulate these conditions.

In a most promising experiment, Ellenbogen (19) used a suspension of ferrous sulfide in aqueous ammonium chloride through which methane was bubbled. The action of ultraviolet light from a mercury lamp gave small quantities of a substance with peptide frequencies in the infrared. Paper chromatography of a hydrolyzate of this substance gave a number of spots with Ninhydrin, of which phenylalanine, methionine, and valine were tentatively identified.

Bahadur (20) has reported the synthesis of serine, aspartic acid, asparagine, and several other amino acids by the action of sunlight on paraformaldehyde solutions containing ferric chloride and nitrate or ammonia. Pavlovskaya and Passynsky (21) have also synthesized a number of amino acids by the action of ultraviolet light on a 2.5-percent solution of formaldehyde containing ammonium chloride or nitrate. These high concentrations of formaldehyde would not have occurred on the primitive earth. It would be interesting to see if similar results could be obtained with $10^{-4}M$ or $10^{-5}M$ formaldehyde. This type of experiment deserves further investigation.

Radioactivity and Cosmic Rays

Because of the small amount of energy available, it is highly unlikely that high-energy radiation could have been very important in the synthesis of organic compounds on the primitive earth. However, a good deal of work has been done in which this type of energy has been used, and some of it has been interpreted as bearing on the problem of the origin of life.

Dose and Rajewsky (22) produced

Table 2. Yields from sparking a mixture of CH_4, NH_3, H_2O, and H_2; 710 mg of carbon was added as CH_4.

Compound	Yield [moles ($\times 10^5$)]
Glycine	63.
Glycolic acid	56.
Sarcosine	5.
Alanine	34.
Lactic acid	31.
N-Methylalanine	1.
α-Amino-n-butyric acid	5.
α-Aminoisobutyric acid	0.1
α-Hydroxybutyric acid	5.
β-Alanine	15.
Succinic acid	4.
Aspartic acid	0.4
Glutamic acid	0.6
Iminodiacetic acid	5.5
Iminoacetic-propionic acid	1.5
Formic acid	233.
Acetic acid	15.
Propionic acid	13.
Urea	2.0
N-Methyl urea	1.5

amines and amino acids through the action of x-rays on various mixtures of CH_4, CO_2, NH_3, N_2, H_2O, and H_2. A small yield of amino acids was obtained through the action of 2 Mev electrons on a mixture of CH_4, NH_3, and H_2O (23).

The formation of formic acid and formaldehyde from carbon dioxide and water by 40 Mev helium ions was mentioned previously. These experiments were extended by using aqueous formic acid (24). The yield per ion pair was only 6×10^{-4} for formaldehyde and 0.03 for oxalic acid. Higher yields of oxalic acid were obtained from $Ca(HCO_3)_2$ and NH_4HCO_3 by Hasselstrom and Henry (25). The helium ion irradiation of aqueous acetic acid solutions gave succinic and tricarbolic acid along with some malonic, malic, and citric acids (26).

The irradiation of 0.1- and 0.25-percent aqueous ammonium acetate by 2 Mev electrons gave glycine and aspartic acid (27). The yields were very small. Massive doses of gamma rays on solid ammonium carbonate yielded formic acid and very small quantities of glycine and possibly some alanine (28).

The concentrations of carbon compounds and the dose rates used in these experiments are, in all probability, very much larger than could be expected on the primitive earth, and the products and yields may depend markedly on

these factors, as well as on the effect of radical scavengers such as HS^- and Fe^{2+}. It is difficult to exclude high-energy radiations entirely, but if one is to make any interpretations from laboratory work, the experiments should be performed with much lower dose rates and concentrations of carbon sources.

Thermal Energy

The older theories of the formation of the earth involved a molten earth during its formation and early stages. These theories have been largely abandoned, since the available evidence indicates that the solar system was formed from a cold cloud of cosmic dust. The mechanisms for heating the earth are the gravitational energy released during the condensation of the dust to form the earth and the energy released from the decay of the radioactive elements. It is not known whether the earth was molten at any period during its formation, but it is clear that the crust of the earth would not have remained molten for any length of time.

Studies on the concentration of some elements in the crust of the earth indicate that the temperature was less than 150°C during this lengthy fractionation, and that it was probably close to present terrestrial temperatures (29).

Fox (30) has maintained that organic compounds were synthesized on the earth by heat. When heated to 150°C, malic acid and urea were converted to aspartic acid and ureidosuccinic acid, and some of the aspartic acid was decarboxylated to α- and β-alanine. The difficulty with these experiments is the source of the malic acid and urea on the primitive earth—a question not discussed by Fox. Fox has also synthesized peptides by the well-known reaction (31) of heating amino acids at 150° to 180°C, and the yield of peptides has been increased by using an excess of aspartic or glutamic acid (32). There is a difficulty connected with heating amino acids and other organic compounds to high temperatures. Geological conditions can heat amino acids to temperatures above 100°C over long periods of time, but it is not likely that this could occur over short periods. Abelson (33) has shown that alanine, one of the more stable amino acids, decarboxylates to methylamine and carbon dioxide. The mean life of alanine is 10^{11} years at 25°C but only 30 years at 150°C. Therefore, any extensive heating of amino acids will result in their destruction, and the same is true for most organic compounds. In the light of this, and since the surface of the primitive earth was probably cool, it is difficult to see how the processes advocated by Fox could have been important in the synthesis of organic compounds.

Surface Reactions, Organic Phosphates, and Porphyrins

It is likely that many reactions were catalyzed by adsorption on clay and mineral surfaces. An example is the polymerization of aminoacetonitrile to glycine peptides in the presence of acid clays, by Akabori and his co-workers (34). Formaldehyde and acetaldehyde were shown to react with polyglycine adsorbed on kaolinite to give serine and threonine peptides. This field offers many possibilities for research.

Gulick (35) has pointed out that the synthesis of organic phosphates presents a difficult problem because phosphate precipitates as calcium and other phosphates under present earth conditions, and that the scarcity of phosphate often limits the growth of plants, especially in the oceans. He proposes that the presence of hypophosphites, which are more soluble, would account for higher concentrations of phosphorus compounds when the atmosphere was reducing. Thermodynamic calculations show that *all* lower oxidation states of phosphorus are unstable under the pressures of hydrogen assumed in this article. It is possible that stronger reducing agents than hydrogen reduced the phosphate or that some process other than reduction solubilized the calcium phosphate. This problem deserves careful attention.

The synthesis of porphyrins is considered by many authors to be a necessary step for the origin of life. Porphyrins are not necessary for living processes if the organism obtains its energy requirements from fermentation of sugars or other energy-yielding organic reactions. According to the heterotrophic theory of the origin of life, the first organisms would derive their energy requirements from fermentations. The metabolism of sulfate, iron, N_2, hydrogen, and oxygen appears to require porphyrins as well as photosynthesis. Therefore, porphyrins probably would have to be synthesized before free energy could be derived from these compounds. While porphyrins may have been present in the environment

before life arose, this is apparently not a necessity, and porphyrins may have arisen during the evolution of primitive organisms.

Intermediate Stages in Chemical Evolution

The major problems remaining for an understanding of the origin of life are (i) the synthesis of peptides, (ii) the synthesis of purines and pyrimidines, (iii) a mechanism by which "high-energy" phosphate or other types of bonds could be synthesized continuously, (iv) the synthesis of nucleotides and polynucleotides, (v) the synthesis of polypeptides with catalytic activity (enzymes), and (vi) the development of polynucleotides and the associated enzymes which are capable of self-duplication.

This list of problems is based on the assumption that the first living organisms were similar in chemical composition and metabolism to the simplest living organisms still on the earth. That this may not be so is obvious, but the hypothesis of similarity allows us to perform experiments to test it. The surprisingly large yields of aliphatic, hydroxy, and amino acids—α-amino acids rather than the other isomers—in the electric-discharge experiments, plus the arguments that such syntheses would have been effective on the primitive earth, offer support for this hypothesis. Further support can be obtained by demonstrating mechanisms by which other types of biologically important compounds could be synthesized.

Oparin (1) does not view the first organism as a polynucleotide capable of self-duplication but, rather, as a coacervate colloid which accumulates proteins and other compounds from the environment, grows in size, and then splits into two or more fragments, which repeat the process. The coacervate would presumably develop the ability to split into fragments which are very similar in composition and structure, and eventually a genetic apparatus would be incorporated which would make very accurate duplicates.

These two hypotheses for the steps in the formation of the first living organism differ mainly in whether the duplication first involved the relatively accurate duplication of nucleic acids, followed by the development of cytoplasm duplication, or whether the steps occurred in

the reverse order. Other sequences could be enumerated, but it is far too early to discuss profitably the exact nature of the first living organism.

It was probably necessary for the primitive organisms to concentrate organic and inorganic nutrients from their environment. This could be accomplished by means of a membrane or by absorption on rocks or clays (36). The development of optical activity in living organisms is another important problem. This has been discussed by many authors and is not taken up here.

Life on Other Planets

Life as we know it—and we know of no other variety of life than that existing on the earth—requires the presence of water for its chemical processes. We know enough about the chemistry of other systems, such as those of silicon, ammonia, and hydrogen fluoride, to realize that no highly complex system of chemical reactions similar to that which we call "living" would be possible in such media. Also, much living matter exists and grows actively on the earth in the absence of oxygen, so oxygen is *not* necessary for life, although the contrary is often stated. Moreover, the protecting layer of ozone in the earth's atmosphere is not necessary for life, since ultraviolet light does not penetrate deeply into natural waters and also because many carbon compounds capable of absorbing the ultraviolet light would be present in a reducing atmosphere.

It is possible for life to exist on the earth and to grow actively at temperatures ranging from $0°C$, or perhaps a little lower, to about $70°C$. It seems likely that if hot springs were not so temporary, many plants and possibly animals would evolve which could live in such temperatures. Plants are able to produce and accumulate substances which lower the freezing point of water, and hence they can live at temperatures below $0°C$. At much lower temperatures the reactions would probably be too slow to proceed in reasonable periods of time. At temperatures much above $120°C$, reaction velocities would probably be so great that the nicely balanced reactions characteristic of living things would be impossible. In addition, it is doubtful whether the organic polymers necessary for living organisms would be stable much above $120°C$; this is probably true even when allowance is made for the amazing stability of the enzymes of thermophilic bacteria and algae.

Only Mars, Earth, and Venus conform to the general requirements so far as temperatures are concerned. Mars is known to be very cold and Venus may be too hot. Observations of the black-body emission of radio waves from Venus indicate surface temperatures of 290° to 350°C (37). The clouds of Venus have the polarization of water droplets. Clearing of the clouds occurs, and this indicates that the clouds are composed of some volatile substance, for nonvolatile dust could hardly settle out locally. However, no infrared bands of water have been observed. It is possible that this is due to a very dry, high atmosphere, such as is characteristic of the earth, and to a cloud level that rises to very near the tropopause, so that there is little water vapor above the reflecting layer.

Mars is known to be very cold, with surface temperatures of +30°C to −60°C during the day. The colors of Mars have been observed for many years by many people. The planet exhibits seasonal changes in color—green or bluish in the spring and brown and reddish in the autumn. Sinton (38) has observed an absorption at 3.5 μ in the reflected light of Mars. This corresponds to the C-H stretching frequency of most organic compounds, but many inorganic compounds have absorptions at this wavelength. The changing colors of Mars and the 3.5 μ absorption are the best evidence, however poor it may be, for the existence of life on the planet. One thing that can be stated with confidence is that if life exists there, then liquid water must have been present on the planet in the past, since it is difficult to

believe that life could have evolved in its absence. If this was so, water must have escaped from the planet, as very little water remains there now and no liquid water has been observed. Hence, oxygen atoms must escape from the planet. This is possible if the high atmosphere has a temperature of 2000°K, and this may well be the case in view of the high temperatures in the high atmosphere of the earth.

Surely one of the most marvelous feats of 20th-century science would be the firm proof that life exists on another planet. All the projected space flights and the high costs of such developments would be fully justified if they were able to establish the existence of life on either Mars or Venus. In that case, the thesis that life develops spontaneously when the conditions are favorable would be far more firmly established, and our whole view of the problem of the origin of life would be confirmed (42).

References

1. A. I. Oparin, *The Origin of Life* (Macmillan, New York, 1938; Academic Press, New York, ed. 3, 1957).
2. N. H. Horowitz, *Proc. Natl. Acad. Sci. U.S.* 31, 153 (1945).
3. E. I. Rabinowitch, *Photosynthesis* (Interscience, New York, 1945), vol. I, p. 81.
4. W. M. Garrison, D. C. Morrison, J. G. Hamilton, A. A. Benson, M. Calvin, *Science* 114, 416 (1951).
5. H. C. Urey, *Proc. Natl. Acad. Sci. U.S.* 38, 351 (1952); *The Planets* (Yale Univ. Press, New Haven, Conn., 1952), p. 149.
6. A. R. Curtis and R. M. Goody, *Proc. Roy. Soc. (London)* 236A, 193 (1956).
7. H. C. Urey, in *Handbuch der Physik*, S. Flügge, Ed. (Springer, Berlin, 1958), vol. 52.
8. W. W. Rubey, *Geol. Soc. Am. Spec. Paper No. 62* (1955), p. 631.
9. P. H. Abelson, *Carnegie Inst. Wash. Year Book No. 56* (1957), p. 179.
10. S. L. Miller, *Science* 117, 528 (1953).
11. ——, *J. Am. Chem. Soc.* 77, 2351 (1955).
12. ——, *Biochim. et Biophys. Acta* 23, 480 (1957).
13. P. H. Abelson, *Science* 124, 935 (1956); *Carnegie Inst. Wash. Year Book No. 55* (1956), p. 171.
14. T. E. Pavlovskaya and A. G. Passynsky, *Reports of the Moscow Symposium on the Origin of Life* (Aug. 1957), p. 98.
15. K. Heyns, W. Walter, E. Meyer, *Naturwissenschaften* 44, 385 (1957).
16. S. L. Miller, *Ann. N. Y. Acad. Sci.* 69, 260 (1957).
17. W. Groth, *Angew. Chem.* 69, 681 (1957); —— and H. von Weyssenhoff, *Naturwissenschaften* 44, 510 (1957).
18. A. N. Terenin, in *Reports of the Moscow Symposium on the Origin of Life* (Aug. 1957), p. 97.
19. E. Ellenbogen, *Abstr. Am. Chem. Soc. Meeting, Chicago* (1958), p. 47C.
20. K. Bahadur, *Nature* 173, 1141 (1954); in *Reports of the Moscow Symposium on the Origin of Life* (Aug. 1957), p. 86; *Nature* 182, 1668 (1958).
21. T. E. Pavlovskaya and A. G. Passynsky, *Intern. Congr. Biochem. 4th Congr. Abstr. Communs.* (1958), p. 12.
22. K. Dose and B. Rajewsky, *Biochim. et Biophys. Acta* 25, 225 (1957).
23. S. L. Miller, unpublished experiments.
24. W. M. Garrison et al., *J. Am. Chem. Soc.* 74, 4216 (1952).
25. T. Hasselstrom and M. C. Henry, *Science* 123, 1038 (1956).
26. W. M. Garrison et al., *J. Am. Chem. Soc.* 75, 2459 (1953).
27. T. Hasselstrom, M. C. Henry, B. Murr, *Science* 125, 350 (1957).
28. R. Paschke, R. Chang, D. Young, *ibid.* 125, 881 (1957).
29. H. C. Urey, *Proc. Roy. Soc. (London)* 219A, 281 (1953).
30. S. W. Fox, J. E. Johnson, A. Vegotsky, *Science* 124, 923 (1956); *Ann. N.Y. Acad. Sci.* 69, 328 (1957); *J. Chem. Educ.* 34, 472 (1957).
31. For a review, see E. Katchalski, *Advances in Protein Chem.* 6, 123 (1951).
32. S. W. Fox et al., *J. Am. Chem. Soc.* 80, 2694, 3361 (1958); —— and K. Harada, *Science* 128, 1214 (1958).
33. P. H. Abelson, *Ann. N.Y. Acad. Sci.* 69, 276 (1957).
34. S. Akabori, in *Reports of the Moscow Symposium on the Origin of Life* (Aug. 1957), p. 117; *Bull. Chem. Soc. Japan* 29, 608 (1956).
35. A. Gulick, *Am. Scientist* 43, 479 (1955); *Ann. N.Y. Acad. Sci.* 69, 309 (1957).
36. J. D. Bernal, *Proc. Phys. Soc. (London)* 62A, 537 (1949); *ibid.* 62B, 597 (1949); *The Physical Basis of Life* (Routledge and Kegan Paul, London, 1951).
37. G. H. Mayer, T. P. McCullough, R. M. Sloanaker, *Astrophys. J.* 127, 1 (1958).
38. W. Sinton, *ibid.* 126, 231 (1957).
39. W. A. Rense, *Phys. Rev.* 91, 299 (1953).
40. B. Schonland, *Atmospheric Electricity* (Methuen, London, 1953), pp. 42, 63.
41. E. Bullard, in *The Earth as a Planet*, G. P. Kuiper, Ed. (Univ. of Chicago Press, Chicago, 1954), p. 110.
42. One of the authors (S. M.) is supported by a grant from the National Science Foundation.

Mechanism of Synthesis of Adenine from Hydrogen Cyanide under Possible Primitive Earth Conditions

IT has recently been shown that adenine and the purine precursors 4-aminoimidazole-5-carboxamidine and formamidine are formed spontaneously from hydrogen cyanide in water–ammonia systems under conditions assumed to have existed on the primitive Earth[1].

I wish to propose now a mechanism for the formation of adenine from hydrogen cyanide under the conditions of the above experiments which is in line with previous observations[2]. Adenine formation from hydrogen cyanide is thought to be initiated by base catalysis and to require the participation of 4-aminoimidazole-5-carboxamidine and formamidine as the key intermediates. The different steps involved in this synthesis are summarized in Fig. 1 and briefly described below. Direct experimental evidence for the last and most important step in this synthesis is also given below.

Reaction 1. In the presence of ammonium hydroxide, cyanide ion is formed which condenses with hydrogen cyanide to form iminoacetonitrile. Indirect evidence for the formation of iminoacetonitrile as the primary product of the base-catalysed hydrogen cyanide polycondensation has been obtained from kinetic studies on the formation of polymeric hydrogen cyanide[3].

Reaction 2. As soon as iminoacetonitrile is formed cyanide ion condenses with this compound to form aminomalonodinitrile. The transformation of hydrogen cyanide into a hydrogen cyanide trimer, presumed to be aminomalonodinitrile, although not completely elucidated, was claimed a long time ago by several investigators[4], and this trimer of hydrogen cyanide, of its isomer α,β-diiminopropionitrile, has also been considered as an intermediate in the formation of tetrameric hydrogen cyanide[5].

Reactions 3 and 4. Hydrogen cyanide as well as aminomalonodinitrile are ammonolysed into the corresponding amidines, formamidine and aminomalondiamidine. These reactions deserve no special comment, since the ammonolysis of nitriles is well known, and is in fact one of the main methods used in the preparation of amidines.

Reaction 5. Aminomalondiamidine condenses with formamidine to yield 4-aminoimidazole-5-carboxamidine. Indirect evidence for this reaction can be found in the very similar syntheses of 4-amino-5-substituted imidazoles carried out by other workers[6]. In the experiments carried out in my laboratory the appearance of 4-aminoimidazole-5-carboxamidine as well as formamidine among the reaction products was observed very early during the course of the hydrogen cyanide condensation reaction. In addition to 4-aminoimidazole-5-carboxamidine, 4-aminoimidazole-5-carboxamide was also isolated

from the reaction mixture. The latter compound was observed to be formed by hydrolysis of the former, as was to be expected from the strongly basic conditions of the reaction mixture.

Reaction 6. In the final step, 4-aminoimidazole-5-carboxamidine condenses with another molar equivalent of formamidine to yield adenine. In support of this reaction, preliminary kinetic studies on the formation of formamidine and 4-aminoimidazole-5-carboxamidine have shown that these compounds make their appearance and disappearance in the reaction mixture as if they were intermediates of adenine. The cyclization of 4-aminoimidazole-5-carboxamidine with formamidine into adenine was thus considered the key step in the elucidation of the mechanism of adenine synthesis. For this reason, and because the reactants were easily available, an independent experimental demonstration of this reaction was undertaken.

This cyclization reaction was studied under different conditions of pH and carried out as follows: About 50 mgm. of 4-aminoimidazole-5-carboxamidine and 90 mgm. of formamidine hydrochloride were dissolved in 1 ml. of distilled water. Exactly 0·2 ml. aliquots of this solution were mixed with an equal volume of either water, 1 M disodium hydrogen phosphate, or 7 N ammonium hydroxide. The pH of the resulting reaction mixtures was approximately 1, 7 and 12 respectively. These three mixtures were heated in closed vessels at a temperature of 130° C. for 12 hr. At the end of the reaction time, a very small amount of residue was separated by centrifugation from each reaction product and each of the supernatants was analysed by paper chromatography and ultra-violet

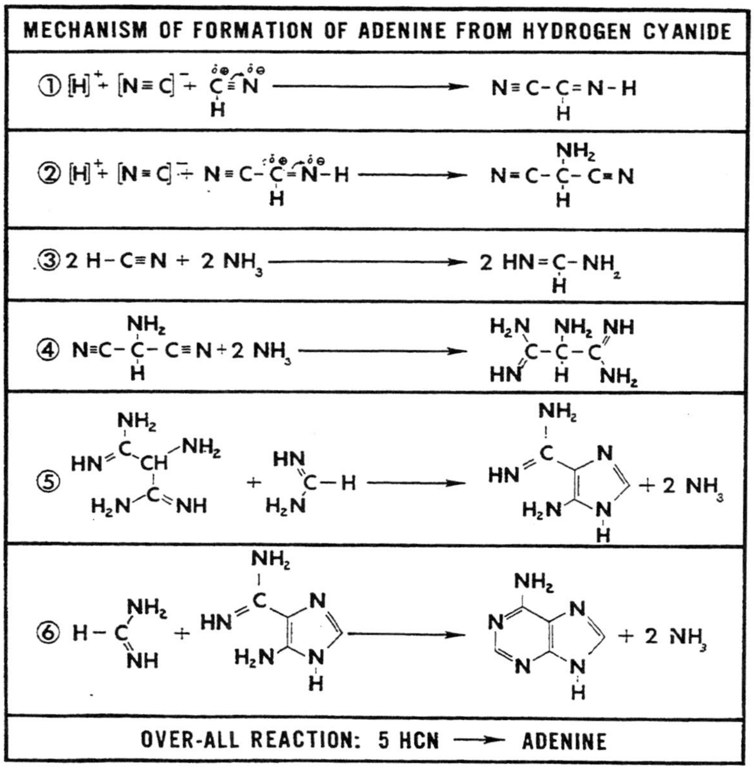

Fig. 1

Reprinted with permission from *Nature,*
Volume 191, pp. 1193-1194. © 1961 by Macmillan Magazines Ltd.

spectrophotometry as described previously[1]. In all cases adenine was formed in high yield. Some unreacted 4-aminoimidazole-5-carboxamidine was observed in the acid and neutral reaction mixtures, but none in the basic reaction mixture. The experiment has been repeated recently at lower temperatures, and similar results have been obtained.

Although alternative mechanisms cannot yet be ruled out, the evidence obtained in my laboratory as well as recent evidence on the so-called 'one-step' synthesis of purines[7] supports the imidazole pathway outlined in Fig. 1 as a mechanism for the synthesis of adenine from hydrogen cyanide under possible primitive Earth conditions.

Work is being carried out at present on the isolation of the three-carbon compounds assumed to be the precursors of the 4-amino-5-substituted imidazoles isolated, and also on the synthesis of other purines from hydrogen cyanide.

This work was supported in part by research grant No. G–13117 from the National Science Foundation.

J. Oró

Chemistry Department,
University of Houston,
Houston 4,
Texas.

[1] Oró, J., *Biochem. Biophys. Research Comm.*, 2, 407 (1960); Oró, J., *Nature*, 190, 389 (1961); Oró, J., and Kimball, A. P., *Arch. Biochem. Biophys.* (in the press).

[2] Grossman, F., Cristopher, J., and Canavan, G., in *Proc. Nat. Sci. Found. Summer Inst. for High School Juniors*, 8 (University of Houston, 1960). Oró, J., *Fed. Proc.*, 20, 352 (1961); *Proc. Fifth Intern. Cong. Biochem., Moscow* (in the press).

[3] Völker, T., *Angew. Chem.*, 62, 728 (1957); *Angew. Chem.*, 72, 379 (1960).

[4] Lange, O., *Ber.*, 6, 99 (1873); Wippermann, R., *Ber.*, 7, 767 (1874). See also: Long, D. A., and George, W. D., *Proc. Chem. Soc.*, 285 (1960).

[5] Bredereck, H., Schmötzer, G., and Becher, H. S., *Ann.*, 600, 87 (1956).

[6] Richter, E., Loeffler, J. E., and Taylor, E. C., *J. Amer. Chem. Soc.*, 82, 3144 (1960).

[7] Taylor, E. C., and Cheng, C. C., *Tetrahedron Letters*, No. 12, 9 (1959).

HCN: A Plausible Source of Purines, Pyrimidines and Amino Acids on the Primitive Earth*

J.P. Ferris[1], P.C. Joshi[1], E.H. Edelson[1], and J.G. Lawless[2]

[1]Department of Chemistry, Rensselaer Polytechnic Institute, Troy, NY 12181, USA
[2]NASA Ames Research Center, Moffett Field, CA 94035, USA

Summary. Dilute (0.1 M) solutions of HCN condense to oligomers at pH 9.2. Hydrolysis of these oligomers yields 4,5-dihydroxypyrimidine, orotic acid, 5-hydroxyuracil, adenine, 4-aminoimidazole-5-carboxamide and amino acids. These results, together with the earlier data, demonstrate that the three main classes of nitrogen-containing biomolecules, purines, pyrimidines and amino acids may have originated from HCN on the primitive earth. The observation of orotic acid and 4-aminoimidazole-5-carboxamide suggests that the contemporary biosynthetic pathways for nucleotides may have evolved from the compounds released on hydrolysis of HCN oligomers.

Key words: HCN — Cyanide — HCN oligomers — 4,5-Dihydroxypyrimidine — Orotic Acid — 5-Hydroxyuracil — Adenine — 4-Aminoimidazole-5-carboxamide — Prebiotic — Primitive earth

Introduction

HCN is considered to have been an important source of biological molecules on the primitive earth. Oro and Kimball (1961, 1962) first demonstrated the syntheses of the purine adenine (6) and the amino acids glycine, alanine and aspartic acid from HCN. These results have been confirmed and extended by several groups and it is now known that adenine, 4-aminoimidazole-5-carboxamide (4), hydantoins and at least seven amino acids are formed with HCN as the only reactant. (Oro and Kamat, 1961; Lowe et al., 1967; Matthews and Moser, 1967; Ferris et al., 1973a; Ferris et al., 1974a, 1974b; Labadie et al., 1967, 1968).

Aqueous, pH 9.2 solutions of cyanide of concentration greater than 0.01 M condense to give a tetramer (2) which reacts further to give oligomers (Sanchez et al., 1967). These oligomers are compounds of unknown structure of molecular weight

---

*Chemical Evolution XXX. For the previous paper see J.P. Ferris, P.C. Joshi and J.G. Lawless, Biosystems, *9*, 81 (1977)

500—1000 which yield biological molecules such as amino acids and purines upon hydrolysis (Scheme 1) (Ferris et al., 1972, 1973b). Similar solutions in which the cyanide concentrations are 0.01 M or less do not form oligomers but instead the cyanide hydrolyzes in a stepwise fashion to yield formamide and formic acid. Oligomer formation does take place in preference to hydrolysis when 0.01 M solutions of cyanide are cooled to -23.4^{o}C. A eutectic containing 74.5 mole per cent cyanide is formed at this temperature. The HCN concentration is so great that oligomer formation proceeds at an appreciable rate even though the temperature is -23.4^{o}C (Sanchez et al., 1966).

In Oro's initial investigations (1961, 1962), adenine was found as a reaction product when 1—11 M solutions of HCN were used. Indirect evidence suggested that this synthesis proceeds via the condensation of formamidine with a trimer of HCN (aminomalononitrile, *1*) and with 4-aminoimidazole-5-carbonitrile (*5*) (Scheme 1) (Sanchez et al., 1967; Ferris and Orgel, 1965, 1966a,b). This synthesis would not be plausible in dilute solution because of the competing hydrolysis of formamidine to formamide (Sanchez et al., 1967). Previously we found that diaminomaleonitrile undergoes a very efficient photochemical rearrangement to 4-aminoimidazole-5-carbonitrile, a compound which readily yields a wide variety of purines under conditions which may have existed on the primitive earth (Ferris and Orgel, 1966a,b; Ferris et al., 1969; Ferris and Kuder, 1970; Sanchez et al., 1968). The latter is a more plausible purine synthesis since it does not require a high steady state concentration of formamidine.

Scheme 1.

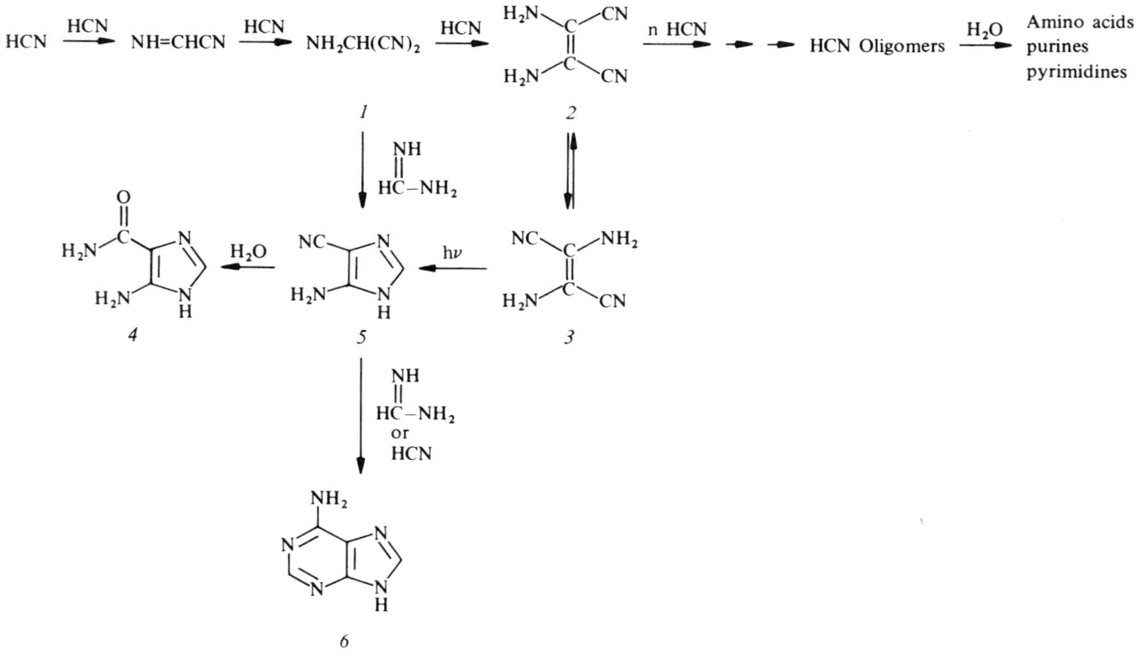

Experimental

General Procedures. UV spectra were determined on a Unicam SP 800 and IR spectra were determined on a Perkin-Elmer Model 137 spectrophotometer. Paper chromatography was performed by the ascending technique for 18—24 h using Whatman 3MM paper and the following solvent systems (1) 1-propanol/NH_4OH (17%) (3:1); (2) 1-butanol/acetic acid/H_2O (12:3:5); (3) *t*-butyl alcohol/methyl ethyl ketone/formic acid/H_2O (40:30:15:15); (4) 1-butanol/water (86:14); (5) isopropyl alcohol/H_2O/conc. HCl (56:18.4:16.6); (6) 1-butanol/acetic acid/water (2:1:1); (7) isopropyl alcohol/NH_4OH/H_2O (20:1:2); (8) benzene/acetic acid/H_2O (125:72:3); (9) 1-butanol/acetic acid/water (4:1:1); (10) ethyl acetate/formic acid/H_2O (70:20:10); (11) ethyl acetate/formic acid/H_2O (60:5:35 — upper layer); (12) t-butyl alcohol/methyl ethyl ketone/H_2O/formic acid (44:44:11:0.26). The Dowex 1-X8 (200 g) was washed batchwise before use with 4 ℓ of distilled water then packed in a glass column and washed with 1 ℓ of distilled water, 1 ℓ of 4N HCl, distilled water, 1 ℓ of 4N NH_4OH and finally with distilled water. The Dowex 50W-X8 (200 g) was washed in the same way except the 4N NH_4OH wash preceeded the 4N HCl wash.

Gas Chromatography and Combination GC/MS. Gas chromatographic procedures similar to those of Butts (1972) were used. Gas chromatography was performed on 6 ft. 0.2 mm i.d. glass column packed with either 3% OV-1 or 3% OV-17 on 80/100 mesh Chromosorb W-HP. The column temperature was maintained initially at 70° or 90° for 3 or 4 min and was then programmed to 275° at 10°/min. The injection port was maintained at 240° and the flame detector at 300°. The gas flow rates were: helium or nitrogen, 40 ml/min; hydrogen, 30 ml/min; and air, 300 ml/min. Standard samples of n-alkanes were run under the same conditions to give retention times relative to the internal standards, expressed as retention indices. The analyses were performed on either a Varian 2400 or Hewlett-Packard 7620A chromatograph equipped with flame ionization detectors. The GC/MS analyses were performed on a Perkin-Elmer 990 gas chromatograph interfaced with the ion source of a CEC 21-491 mass spectrometer. The samples passed through a membrane separator from the gas chromatograph to the mass spectrometer (Folsome et al., 1973). Data processing was performed on a Decision GC/MS computer system.

Samples were converted to volatile trimethylsilyl derivatives for gas chromatography unless noted otherwise. The samples (0.5—4 mg) were dissolved in 50 μl of pyridine and to this was added 50 μl of Regisil ^{TM} which is a mixture of 99% bis(trimethylsilyl)trifluoroacetamide and 1% trimethylchlorosilane. The samples were sonicated to facilitate solution, sealed in ampuls and heated at 150° for 2.5 h or at 60° for 18—24 h. The cooled reaction solution was injected directly into the gas chromatograph.

Oligomerization and Fractionation of HCN. A 0.1 M solution of HCN was adjusted to pH 9.2 by the addition of conc. NH_4OH and the solution was stored in the dark in glass-stoppered bottles at room temperature for 4—12 months. The concentration of cyanide as measured by the method of Scoggins (1972), decreased by about one half after six months reaction time. The pH of the solution was still 9.2 when the oligomerization was terminated. One liter of the oligomerization mixture was filtered to remove a small amount of black insolubles and it was then added to a column of 200 g of Dowex 1 (OH$^-$). The column was eluted with 2 ℓ of distilled water and the combined eluates were then chromatographed on 200 g of Dowex 50 (H$^+$). The Dowex 50 column was eluted with 1--2 ℓ of distilled water and the eluate was concentrated using a rotary evaporator to give the neutral fraction. The Dowex-1 column was then washed with 3—4 ℓ of 3 N HCl and the eluate was concentrated to yield 430 mg of acidic and amphoteric oligomers. This fraction was heated at 70° in vacuo for 7—10 days to sublime the oxalic acid and NH$_4$Cl present leaving 325 mg of acidic and amphoteric oligomers. The Dowex 50 column was washed with 2 ℓ of 3 N NH$_4$OH and concentrated to give 25 mg of basic material.

To prepare HCN oligomers in the absence of oxygen, the 0.1 M cyanide solution was adjusted to pH 9.2 with NH$_4$OH. This solution was cooled in a liquid nitrogen bath and subjected to four freeze-pump-thaw cycles. Nitrogen gas was admitted to the flask on the last cycle. The flask was maintained in an atmosphere of nitrogen for 7 months before the oligomerization was terminated and the products were fractionated by ion exchange as described above. The yield of acidic and amphoteric oligomers before sublimation was 584 mg per liter of 0.1 M cyanide.

Hydrolysis and Fractionation of the Hydrolysate of the HCN Oligomers. The oligomer hydrolyses were performed for 24 h at 110° in sealed ampuls in either 6 N HCl or in aqueous solution which was adjusted to pH 8.5 with conc. NaOH. A final pH of 8.3—8.5 was observed after the pH 8.5 hydrolyses were terminated. The hydrolysate solutions were concentrated to dryness using a rotary evaporator and then dissolved in distilled water for analysis. Oligomers prepared in the absence of oxygen were degassed by 4 freeze-pump-thaw cycles after dissolution in 6 N HCl or at pH 8.5 before hydrolysis at 110°.

About 10 ml of an aqueous solution of the hydrolysate was passed through a column of 25—30 g of Dowex 50 (H$^+$) and the acidic and neutral fraction was eluted with water. The basic and amphoteric compounds were eluted with 4 N NH$_4$OH.

GC/MS Analysis of the HCN Oligomer Hydrolysate. The basic and amphoteric fraction of the 6 N HCl hydrolysate (Chart 2) was silylated and chromatographed on an OV-1 column. The presence of the following compounds was established by the comparison of their mass spectrum with that of a silylated authentic sample: Glycine, 4,5-dihydroxypyrimidine, *meso-* and *dl-*diaminosuccinic acid and β-alanine (Fig.1). The relative yields of the silylated derivatives of glycine and 4,5-dihydroxypyrimidine were calculated to be 1:0.85 respectively from their relative ion monitor responses on the mass spectrometer. Several unidentified compounds were also observed.

The GC trace of the pH 8.5 hydrolysate was similar to that of the 6 N HCl hydrolysate with the exception that no 4,5-dihydroxypyrimidine was observed and lower yields of the other biomolecules were found when the same hydrolysis time period was used. The presence of glycine and the *meso-* and *dl-*diaminosuccinic acid was confirmed from the mass spectra. No significant differences were observed in the nature of the biomolecules formed from oligomers prepared in the absence or presence of oxygen.

The basic and amphoteric fraction resulting from the acid hydrolysis of the basic HCN oligomers (Chart 1) contained α-aminoisobutyric acid, glycine, 4,5-dihydroxypyrimidine and β-alanine as shown by the GC/MS analysis of the silylated sample.

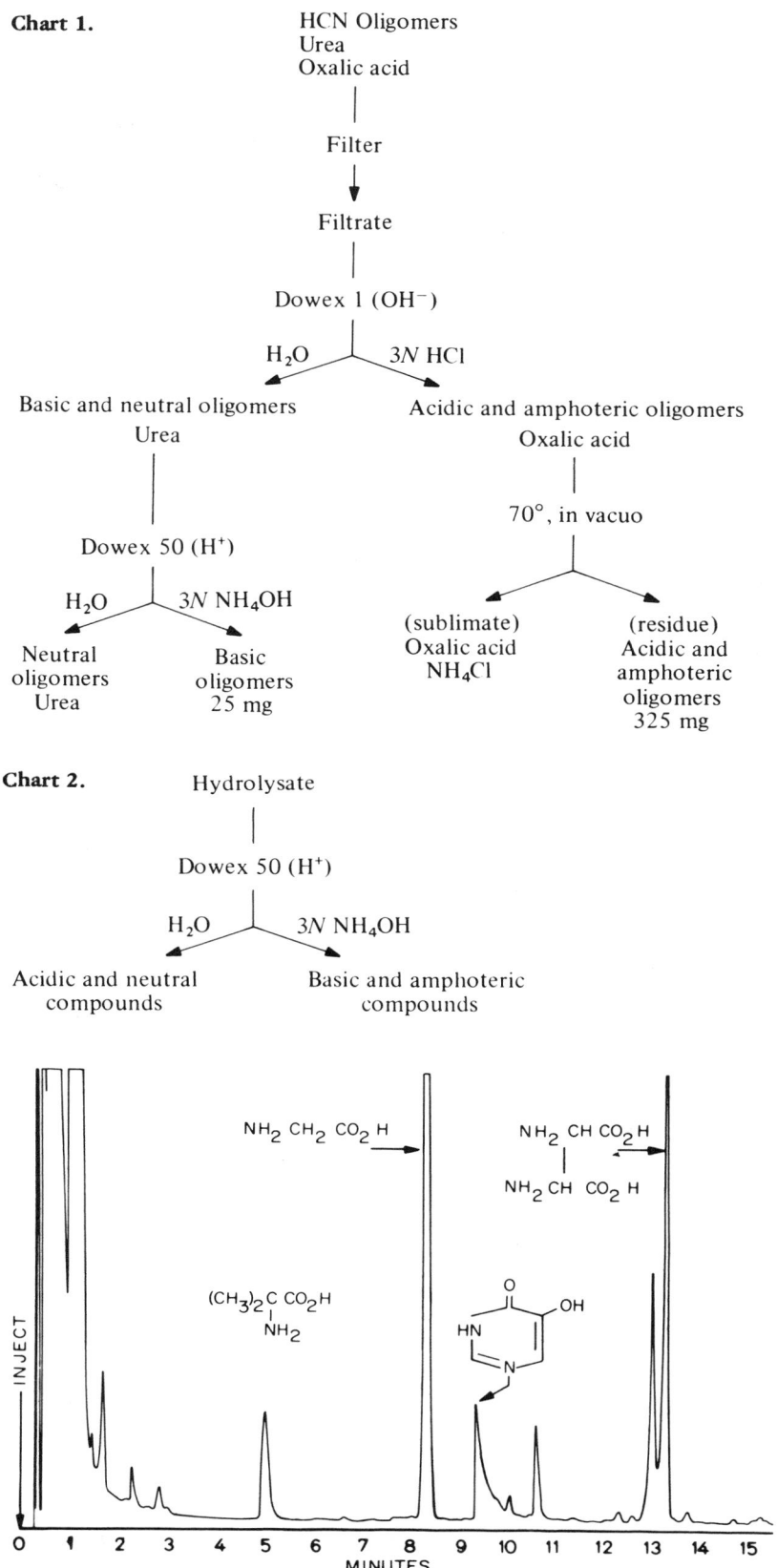

Chart 1.

HCN Oligomers
Urea
Oxalic acid
Filter
Filtrate
Dowex 1 (OH⁻)
H₂O
3N HCl
Basic and neutral oligomers
Urea
Acidic and amphoteric oligomers
Oxalic acid
70°, in vacuo
Dowex 50 (H⁺)
H₂O
3N NH₄OH
Neutral oligomers Urea
Basic oligomers 25 mg
(sublimate) Oxalic acid NH₄Cl
(residue) Acidic and amphoteric oligomers 325 mg

Chart 2.

Hydrolysate
Dowex 50 (H⁺)
H₂O
3N NH₄OH
Acidic and neutral compounds
Basic and amphoteric compounds

NH₂ CH₂ CO₂ H
NH₂ CH CO₂ H
NH₂ CH CO₂ H
(CH₃)₂C CO₂H
NH₂
O
OH
HN
N
INJECT
0 1 2 3 4 5 6 7 8 9 10 11 12 13 14 15
MINUTES

Fig. 1. Gas chromatographic trace of the trimethylsilyl derivatives of the basic and amphoteric compounds formed on acid hydrolysis of the HCN oligomers

The silylated acid and neutral fraction (Chart 2) obtained by hydrolysis of the acidic and amphoteric oligomers (Chart 1) with 6 N HCl or at pH 8.5 did not give reproducible GC/MS results. Apparently the silyl derivatives decomposed after elution from the GC column but before reaching the ion source of the mass spectrometer. Oxalic acid was consistently observed as the major hydrolysis product. The presence of 5-hydroxyuracil and orotic acid was suggested by one GC/MS run but this could not be reproduced. The presence of these compounds was confirmed after the substances were purified as described later.

4,5- Dihydroxypyrimidine. This substance is a major product in the basic and amphoteric fraction of the acid hydrolysate (Chart 2). It gives an intense UV-absorbing spot on paper chromatography of this fraction which gives an orange color with diazotized sulfanilic acid (Grimmett and Richards, 1975). This compound was not observed when the hydrolysis was performed at pH 8.5 but another pyrimdine is formed which is converted to 4,5-dihydroxypyrimidine on acid hydrolysis. The 4,5-dihydroxypyrimidine exhibits the following R_f values: system (R_f); (1) 0.25; (2) 0.52; (4) 0.35; (7) 0.12; (8) 0.68. The following UV maxima were observed: λ_{max} 255 nm (0.1 M HCl) 260 and 287 nm (0.1 M NaOH) and 268 nm (distilled water). It exhibited a retention time of 9.75 min (retention index = 13.70) on OV-1 and 9.45 min (retention index = 14.54) on OV-17. The quantitative yield of 4,5-dihydroxypyrimidine in the acid hydrolysate of the HCN oligomers was found to be 17.5 mg per liter of the oligomerization mixture.

Combination GC/MS of the silylated derivative gave a molecular ion at m/e 256 and an M-15 at m/e 241. High resolution GC/MS, performed in the laboratory of Dr. A. Burlingame (Space Sciences Laboratory, University of California at Berkeley) gave molecular weights of 256.1057 and 241.0832 respectively for these peaks. The molecular formula calculated for $C_{10}H_{20}N_2O_2Si_2$ is 256.1063 and for $C_9H_{17}N_2O_2Si_2$ is 241.0829. These data require an empirical formula of $C_4H_4N_2O_2$ for the parent compound.

An authentic sample of 4,5-dihydroxypyrimidine was prepared by the acid hydrolysis of 4,5-diaminopyrimidine (Aldrich) using a procedure similar to that of McOmie and Turner (1963): mp about 274° (start dec. at 265°) lit mp 268-269° decomp. (McOmie and Turner, 1963) λ_{max} (1 N HCl) 255 nm, ϵ 9180; 1 N NaOH 260 nm, ϵ 7620 amd 287 nm, ϵ 7060). The synthetic sample exhibited UV and mass spectral data identical with those of the compound isolated from the HCN oligomers.

Orotic Acid. Orotic acid was detected in the acid and amphoteric fraction of the oligomer hydrolysate (Chart 2) if the hydrolysis was performed at pH 8.5 or with 6 N HCl followed by hydrolysis with 1 N NaOH. No orotic acid was detected if the oligomers were not hydrolyzed or if only 6 N HCl was used for hydrolysis. The presence of orotic acid was established using the following methods:

a) Specific Color Test. A color test (Adachi et al., 1963; Kesner et al., 1975) was used to establish the presence of orotic acid in the total hydrolysate before it was fractionated on an ion exchange resin. The results are given in Table 1. The yields were determined by comparison with standard orotic acid solutions measured at 480 nm. The absorbance at 480 nm was corrected with a blank containing the colorimetric reagents but not orotic acid. It was determined by paper chromatography that barbituric acid, a substance which gives the same color test with this reagent, was not present in the hydrolysate.

Table 1. Orotic acid yield (μg)[a]

	Acid/Base Hydrolysis	pH 8.5 Hydrolysis
100 ml of oligomerization mixture	5.5	nd
50 mg of acidic and amphoteric oligomers	8.3[b]	43
20 mg of acidic and amphoteric oligomers prepared in the absence of oxygen	nd	10[c]

[a]Yields determined colorimetrically as described in the text, nd = not determined

[b]This yield is equivalent to 5.4 μg per 100 ml of the oligomerization mixture since this volume contains 30–35 mg of acidic and amphoteric oligomers

[c]This yield corresponds to 45 μg of orotic acid per 50 mg of the oligomers prepared in the presence of oxygen. (584 mg/ℓ of acidic and amphoteric oligomers are obtained in the absence of oxygen as compared to 325 mg/ℓ in the presence of oxygen). This corresponds to a total yield of 280 μg/ℓ orotic acid from the oligomerization mixture

b) Paper Chromatography. Orotic acid was observed as a UV-absorbing substance which had the same R_f as an authentic sample in systems 1, 3, 5, and 6. This UV-absorbing area gave a positive color test for orotic acid (section a). Orotic acid was also detected by paper chromatography and ultraviolet spectroscopy in the hydrolysate of the oligomers prepared in the absence of oxygen.

c) Gas Chromatography/Mass Spectrometry. A sample of the acid and neutral hydrolysate (Chart 2) from the acidic and amphoteric oligomers (Chart 1) prepared by pH 8.5 hydrolysis was partially purified by paper chromatography in system 1. The UV absorbing area with the same R_f as orotic acid was eluted, silylated and subjected to gas chromatography on OV-1 and OV-17. The predominant peak had the same retention time as orotic acid and exhibited the same mass spectrum as an authentic sample of silylated orotic acid.

5-Hydroxyuracil. The acidic and neutral fraction (Chart 2) from the acid hydrolysis of the acidic and amphoteric oligomers (Chart 1) was chromatographed in system 10 and the UV-absorbing material with the same R_f as 5-hydroxyuracil was eluted and rechromatographed in system 2. The UV spectra of the authentic 5-hydroxyuracil and the material from the HCN oligomers were identical when measured in acidic and neutral solutions (λ_{max} 275 nm) and were similar in basic medium (λ_{max} 298 nm and 295 nm respectively). The silyl derivative of the substance isolated from the HCN oligomers gave the same gas chromatographic retention time on OV-1 and OV-17 and the same mass spectrum as a standard sample. The mass spectrum was identical with a published spectrum. (Markey et al., 1974). The yield of 5-hydroxyuracil, estimated by comparison of the intensity of the UV absorption on the paper chromatogram with varying concentrations of standards applied to the same paper, was 80-120 μg/l of the total oligomerization mixture.

Unidentified Pyrimidine Released by pH 8.5 Hydrolysis of HCN Oligomers. Chromatography in system 1 of the basic and amphoteric fraction (CHart 2) of the pH 8.5 hydrolysate revealed a large array of UV-absorbing and fluorescent spots, several of which gave a blue color when sprayed with the Pauly Reagent (diazotized sulfanilic acid) (Grimmett and Richards, 1965). The substance which exhibited an R_f value of 0.1-0.2 in system 1 as well as UV absorption and a blue Pauly color was eluted and rechromatographed in system 10. This material exhibited an R_f of 0.1, it yielded 4,5-dihydroxypyrimidine on acid hydrolysis and it exhibited the following UV maxima: 280 nm (aqueous solution and 1 N HCl) and 290 nm (1 N NaOH). This compound is insoluble in the silylation mixture and as a consequence it has not been possible to obtain mass spectral data. Our observation of a UV spectrum, a positive Pauly test and its hydrolysis to 4,5-dihydroxypyrimidine constitutes strong evidence for the presence of a 4,5-substituted pyrimidine ring system in the compound.

Attempted Detection of 4-Hydroxypyrimidine. The detection of 4-hydroxypyrimidine in the Murchison meteorite (Folsome et al., 1973), prompted our search for it among the pyrimidines formed by hydrolysis of the HCN oligomers. It could not be detected in the basic and amphoteric fraction of the acid hydrolysate (Chart 2) when it was chromatographed in system 1. The UV absorbing material with the same R_f as 4-hydroxypyrimidine from this chromatogram was eluted and rechromatographed in system 10. No UV absorbing spot with the same R_f as 4-hydroxypyrimidine was observed within the limit of detection of 2.5-5 μg.

Adenine. Adenine was detected in the basic and amphoteric fraction of the hydrolysate (Chart 2) by comparison of the R_f of a UV absorbing substance in systems 1 through 5 with that of an authentic sample. This substance also gave a positive test for adenine with the silver nitrate - bromophenol blue reagent (Hais and Macek, 1963) after chromatography in system 1. The UV-absorbing area corresponding in R_f to adenine in system 1 exhibited the same UV spectrum as that of an authentic sample when measured in 0.1 M HCl (262.5 nm), 0.1 M NaOH (268.5 nm) and distilled water (261 nm). Optimal UV spectral data were obtained when the paper chromatographic sheets were washed with 1% oxalic acid solutions prior to use (Hanes and Isherwood, 1949). The adenine yield was 1 mg/l of the original oligomerization mixture. Adenine was also released by acid hydrolysis and pH 8.5 hydrolysis from oligomers prepared in the absence of oxygen.

The identity of adenine in the acid hydrolysate was confirmed by silylation of a sample which had been purified by chromatography in system 1 followed by elution from Dowex 1 with water. The major product exhibited the same retention time as an authentic sample when the silylated derivative was chromatographed on OV-1 and OV-17 (retention indices of 18.47 and 20.61 respectively)

Adenine Analysis Before Oligomer Hydrolysis. A 15 mg sample of the basic HCN oligomers (Chart 1) was chromatographed in system 1 with an authentic sample of adenine. Although no well-defined UV absorbing spot was obtained in the region which had the same R_f as authentic adenine this region was eluted with water and rechromatographed, in system 10. No spot corresponding to adenine was detected.

The limit of detection of adenine is 3-5μg. Since 15 mg of the basic oligomers corresponds to 0.5 l of the oligomerization mixture this means there is less than 6-10 μg
of adenine in 1 l of the oligomerization mixture. Adenine was detected in the same
sample after acid hydrolysis by the comparison of the R_f value (system 1) and the
UV spectrum of an eluted sample with that of an authentic sample.

4-Aminoimidazole-5-Carboxamide. The basic and amphoteric hydrolysate (Chart 2)
obtained by the acid hydrolysis of the acidic and amphoteric oligomers (Chart 1) was
chromatographed using systems 1, 4 and 9. Several compounds which gave a positive
test with diazotized sulfanilic acid (Grimmett and Richards, 1965) were observed. One
of these was a faint spot which gave the same blue color and R_f value as an authentic
sample of 4-aminoimidazole-5-carboxamide (AICA). A portion of the acid hydrolysate
was chromatographed in system 1 by developing the paper twice. The paper was dried
between developments. Elution of the material with the same R_f as AICA yielded a
substance which exhibited a UV spectrum which was identical with that of AICA;
λ_{max} 266.5 nm (0.1 M HCl), 277 nm (0.1 M NaOH) and 266 nm (water).

No AICA was detected initially when the corresponding oligomer fraction prepared
in the absence of oxygen was hydrolyzed with 6 N HCl or at pH 8.5 for 24 h. Control
experiments established that 2 mg of AICA is destroyed under these hydrolysis conditions. AICA was detected after a 1.5 h hydrolysis by its R_f value and color test using
system 1. The substance produced by acid hydrolysis when eluted from the paper
chromatogram gave the same UV spectrum as an authentic sample when measured in
acidic, basic and neutral solution.

Hydrolysis of HCN Oligomers Without Prior Ion Exchange Fractionation. Since the
possibility existed that the biomolecules released on hydrolysis of the HCN oligomers
resulted from transformations taking place in the presence of the ion exchange resins,
qualitative analyses were performed to see if the same compounds were released on
direct hydrolysis of the oligomerization mixture. Amino acids, adenine, 4,5-dihydro-
oxypyrimidine, AICA and orotic acid were all detected as hydrolysis products.

Biomolecules from HCN Oligomers Prepared from 1 M Cyanide Without Added NH$_3$.
The pH of a 1 N NaCN solution was adjusted to 9.2 with conc. HCl and the solution
was allowed to stand in a stoppered vessel in the dark for 1 year. The amphoteric and
acidic HCN oligomers were isolated in the usual way (Chart 1). Acid hydrolysis of
this fraction yielded adenine and 4,5-dihydroxypyridine. These substances were purified by paper chromatography as described previously. The adenine was identified by
comparison of its UV spectrum in acidic and basic solution with that of an authentic
sample. The 4,5-dihydroxypyrimidine was identified by comparison of the color
produced with diazotized sulfanilic acid and diazotized p-nitroaniline (McOmie and
Turner, 1965) with those colors produced using an authentic sample chromatographed
on the same paper.

Guanidine. Paper chromatographic analysis of the basic and amphoteric hydrolysate
(Chart 2) formed by 6 N HCl and pH 8.5 hydrolysis of the acidic and amphoteric oligomers (Chart 1) revealed the presence of guanidine. It was detected by positive color
tests with the pentacyanoferrate and the diacetyl reagents (Smith, 1969). A yield of
3.3-4.4 mg/l of the oligomerization mixture was established by comparison of the

intensity of the guanidine spot, obtained by acid hydrolysis of 5 mg of the acid and amphoteric oligomers, with varying amounts of authentic guanidine applied to the same paper chromatogram. A guanidine yield of 3.5-4.5 mg/l and 2.5-3 mg/l was obtained by 6 N HCl hydrolysis and pH 8.5 hydrolysis, respectively, of oligomers formed in the absence of oxygen.

Guanidinoacetic Acid. Guanidinoacetic acid was detected in the basic and amphoteric hydrolysate (Chart 2) of the acidic and amphoteric oligomers (Chart 1) by the identity of R_f value with an authentic sample when measured in systems 1, 2 and 7. The compound was detected using the pentacyanoferrate and diacetyl reagents (Smith, 1969). The yield of guanidinoacetic acid was found to be 0.8 mg/l of the oligomerization mixture after acid hydrolysis and 0.4 mg/l after pH 8.5 hydrolysis.

Identification of Amino Acids by GC/MS of the (+)-2-Butyl Ester Trifluoroacetamide Derivatives

A. *Preparation of (+)-2-Butyl-N-trifluoroacetamides of Amino Acids.* To a 1-2 mg sample of the amino acid mixture was added 0.5 ml of 3 N HCl in (+) 2-butanol. The mixture was sonicated for a minute to facilitate dissolution of the sample and then heated in a sealed tube for 2 h at 100-150°C. The solution was concentrated to dryness with warming to 60°C, 1 ml of CH_2Cl_2 was added and the solution was again concentrated to dryness with warming. The residue was taken up in 0.8 ml of CH_2Cl_2 and 0.3 ml of trifluoracetic anhydride was added. The mixture was sonicated for one minute and allowed to stand at room temperature for 1 h. The mixture was concentrated to dryness using a rotary evaporator and a room temperature water bath. CH_2Cl_2 (0.2 ml) was added and the mixture was again concentrated and the residue was dissolved in 0.1 ml of CH_2Cl_2 for gas chromatography.

B. *Gas Chromatography/Mass Spectrometry of the (+)-2-Butyl trifluoroacetamide Derivatives of Amino Acids.* Samples (2 μl) were chromatographed on a 150 ft x 0.02 in. Ucon 75-H-90,000 capillary column using a model 7620A Hewlett-Packard Gas Chromatograph (Gil-Av et al., 1965; Pollock et al., 1965). The column was held for 8 min at 100° and programmed at 10/min. to 160°. The helium flow rate was 6 ml/min, make up helium was 47 ml/min, air 300 ml/min and hydrogen 30 ml/min. This column gave fair (not baseline) separation of the diastereomeric derivatives of valine, alanine, isoleucine, leucine, proline, phenylalanine and glutamic acid but not aspartic acid.

The acidic and amphoteric oligomers and the basic oligomers (Chart 1) were hydrolyzed and the basic and amphoteric hydrolysate (Chart 2) was analyzed (Table 2). With the exception of the isomers of diaminosuccinic acid the same amino acids were detected for each preparation. The yield of amino acids was much greater with the acid hydrolysis than with pH 8.5 hydrolysis. N-methylglycine and β-alanine were not observed in the previous study which utilized the l-butyl-N-trifluoroacetamide derivatives (Ferris et al., 1974a). No glutamic acid or isoleucine, previously reported compounds, were detected in the present work. Previously the glutamic acid was only observed on hydrolysis of the neutral fraction of the HCN oligomers, a fraction which was not investigated in this study. Isoleucine was observed irregularly and in very small yield in the previous study (Ferris et al., 1974a).

Table 2. GC/MS analysis of D-2-butyl TFA amino acids from HCN oligomers[a]

Amino Acid	Sample[b]		
	Acid Hydrolysis Acidic and Amphoteric Oligomers	Acid Hydrolysis Basic Oligomers	pH 8.5 Hydrolysis Deoxygenated Acidic and Amphoteric Oligomers
α-aminoisobutyric	M	M	M
alanine	P	P	—
N-methylglycine	P	P	—
ammonia	T	M	M
glycine	S	S	S
β-alanine	M	M	P
aspartic	M	M	M
diaminosuccinic	M	—	P
diaminosuccinic	M	—	P

[a]Identity established by mass spectral analysis with the exception of the pH 8.5 hydrolysis of the deoxygenated oligomers where only the GC retention time was used. The GC trace in the latter case was similar to the other runs

[b]S = strong, M = medium, T = trace, P = possibly present but not in sufficient amount to establish unequivocally

Results and Discussion

The present study was performed to investigate:

1) Which Biomolecules are Formed from Dilute (0.1M) Cyanide Solutions. Since it is unlikely that cyanide concentrations of 1-11M, that had been used in previous studies, were ever present on the primitive earth, we investigated 0.1 M cyanide solutions. The choice of 0.1 M cyanide is a compromise since it represents an absolute upper limit for primitive earth cyanide ion concentrations and a practical lower limit for oligomer formation in the absence of the concentration of the HCN by eutectic formation (Sanchez et al., 1966).

2) The Effect of Molecular Oxygen on the Formation of Biomolecules from HCN. Oxygen had not been excluded in most of the studies carried out previously because it was reported to have no effect on the oligomerization of cyanide (Völker, 1960). Recently it was demonstrated that diaminomaleonitrile (2) is readily air oxidized (Ferris and Ryan, 1973). This result opened the question of a possible role for molecular oxygen in the formation of biomolecules from HCN since there was little or no oxygen in the atmosphere of the primitive earth (Miller and Orgel, 1974).

3) Variation in the Structures of the Biomolecules Released from the HCN Oligomers when Hydrolyses are Performed at pH 8.5. Previously all the hydrolyses were performed with 6 N HCl. Neutral hydrolysis conditions are most reasonable since the pH of the primitive earth oceans was probably pH 8 ± 1 (Miller and Orgel, 1974).

In our experimental approach we allowed 0.1 M cyanide to condense to oligomers under plausible primitive earth conditions. The acidic and amphoteric oligomers, isolated by ion exchange techniques (Ferris et al., 1974a) were then subjected to hydrolysis at 110°C. The pH 8.5 hydrolysis of these oligomers simulates their slow breakdown on the primitive earth with the release of the requisite biomolecules essential for the origins of life. It is necessary to do these hydrolyses at 110° in the laboratory so that they proceed at a reasonable rate. It is assumed that neither the nature of the reaction products nor their relative yields is changed by the use of 110° as compared with the longer term, lower temperature hydrolyses that probably occurred on the primitive earth.

The detection of 4,5-dihydroxypyrimidine (9), orotic acid (8) and 5-hydroxyuracil (10) is the first observation of the formation of pyrimidines using HCN as the only starting material (Table 3). The yield of 4,5-dihydroxypyrimidine is especially impressive since it is similar to that of glycine and greater than adenine. The structure of 4,5-dihydroxypyrimidine was deduced by the low resolution and high resolution GC/MS of the bis(trimethylsilyl) derivative (Folsome et al., 1973). This structure was confirmed by a direct mass spectral comparison with an authentic sample prepared by the acid hydrolysis of 4,5-diaminopyrimidine (McOmie and Turner, 1963).

Orotic acid (8) is only observed when the oligomer hydrolysis is performed at pH 8.5. This is probably because 5-carboxymethylidine hydantoin (7), a precursor to orotic acid, is formed by acid hydrolysis and this in turn rearranges to orotic acid only in neutral or basic solution (Ferris et al., 1974b). Presumably these reactions proceed sequentially at pH 8.5 to yield orotic acid.

The adenine yield we observed from oligomers prepared using 0.1 M HCN is one-tenth to one-fifth the optimum yields obtained using 2-11 M HCN solutions (Oro and Kimball, 1961). This yield is at least ten times higher than expected if the first step in the adenine synthesis is the slow bimolecular reaction of aminomalononitrile (1) with formamidine to give 4-aminoimidazole-5-carbonitrile (5). The concentration of each of these reactants should be decreased at least ten fold in 0.1 M HCN as compared to 1 M or 10 M HCN solutions and therefore the adenine yield should be about one hundredth to one thousandth that observed by Oro and Kimball (1961). Since these results and other results, which will be cited later, demonstrate that formamidine is not a likely reaction intermediate, the principal pathway for adenine synthesis must involve initial formation of HCN oligomers which release adenine on hydrolysis (Scheme 1).

The extent of HCN oligomer formation is a function of the concentrations of cyanide and diaminomaleonitrile. This was demonstrated by the observation that the rate of oligomer formation increased when 2 is added to freshly prepared cyanide solutions (Sanchez et al., 1967). The steady state concentration of 2 is a complex function of cyanide concentration as evidenced by the observation that its steady state concentration from 0.1 M cyanide is greater than that from 5 M cyanide after a 5 day reaction time (Sanchez et al., 1967). Since the adenine yield does not decrease by the square of the concentration as the cyanide level is decreased from 11 M to 0.1 M, a bimolecular reaction between aminomalononitrile and formamidine is probably not involved in the synthesis of adenine from 0.1 M cyanide. A pathway via diaminomaleonitrile and HCN oligomers is consistent with these data.

Further support for the proposal that formamidine is not involved in adenine synthesis is our observation that both adenine and 4,5-dihydroxypyrimidine are released from oligomers that were prepared in the absence of added NH_3. The ammonia concentrations (from the hydrolysis of HCN) must be very low and the corresponding concentration of formamidine must be essentially zero. Consequently it is very unlikely that formamidine is involved in the formation of adenine.

Formamidine does have a central role in the HCN oligomerizations performed under anhydrous conditions (Wakamatsu et al., 1966; Yamada et al., 1968, 1969). In the absence of water the HCN oligomerization proceeds at a slower rate and there is no loss of formamidine by hydrolysis. Hence, the higher concentration of formamidine and the slower oligomerization result in the more efficient conversion of aminomalono-nitrile to adenine by the route shown in Scheme 1 when anhydrous conditions are used.

No adenine was detected when the oligomeric mixture from 0.1 M cyanide was analyzed before hydrolysis. The absence of free adenine is also consistent with a synthetic pathway via HCN oligomers which release adenine upon hydrolysis. This is also the predominant reaction pathway with concentrated solutions of cyanide as evidenced by the observation that only 16% of the adenine formed from 11.1 M cyanide is free in solution (Oro and Kimball, 1961).

The pH 8.5 hydrolyses reported in Table 3 were performed on oligomers which were prepared and hydrolyzed in the absence of oxygen. The same reaction products were obtained when a small amount of molecular oxygen was present during the oligomeriza-tion and hydrolysis reactions. Although it has been shown that diaminomaleonitrile (2) is readily oxidized by oxygen (Ferris and Ryan, 1973), these data show that the presence of molecular oxygen does not result in any significant changes in the reaction products. This is probably because all the non-degassed reaction solutions were stop-pered and were not agitated during the olgomerization reaction. As a consequence, the oxygen which is initially in solution is scavanged by the initially formed diamino-maleonitrile. Since very little additional oxygen diffuses into the stoppered flasks a steady state concentration of 2 is built up and the oligomerization proceeds. There is a lower steady state concentration of 2 and correspondingly smaller yields of HCN oligomers if the solutions are saturated with oxygen.

A unified pathway for the formation of the heterocyclic compounds listed in Table 3 is given in scheme 2. This Scheme is an oversimplification since these compounds are not found free in solution but are incorporated into HCN oligomers. The final products (6, 8, 9, 10) are obtained by hydrolysis of the oligomers. A pathway similar to that proposed for the formation of 4,5-hydroxypryrimidine (9) and 5-hydroxyuracil has been proposed for the synthesis of 2-substituted-4,5-diamino-6-cyanopyrimidines from diaminomaleonitrile (2) (Begland, 1975). As discussed previously, only HCN is used as a reactant (Scheme 2) with the exception of the proposed role of guanidine or its equivalant in the formation of 5-hydroxyuracil (10). Guanidine is a plausible reactant since it is formed in appreciable yield from HCN (Table 3). The decarboxyl-ation reactions postulated in the formation of 4,5-dihydroxypyrimidine (9) and 5-hydroxyuracil (10) are analogous to the decarboxylation of β-ketoacids (Ferris and Miller, 1966). The validity of the proposed decarboxylation step was shown by our conversion of 5-aminoorotic acid (11) to 5-hydroxyuracil (10) in 29% yield when heated at 110° for 18 h in 6 N HCl.

The previous report of the release of amino acids on acidic and pH 8 hydrolysis of the HCN oligomers (Ferris et al., 1974a) was confirmed in this study. The presence

Scheme 2.

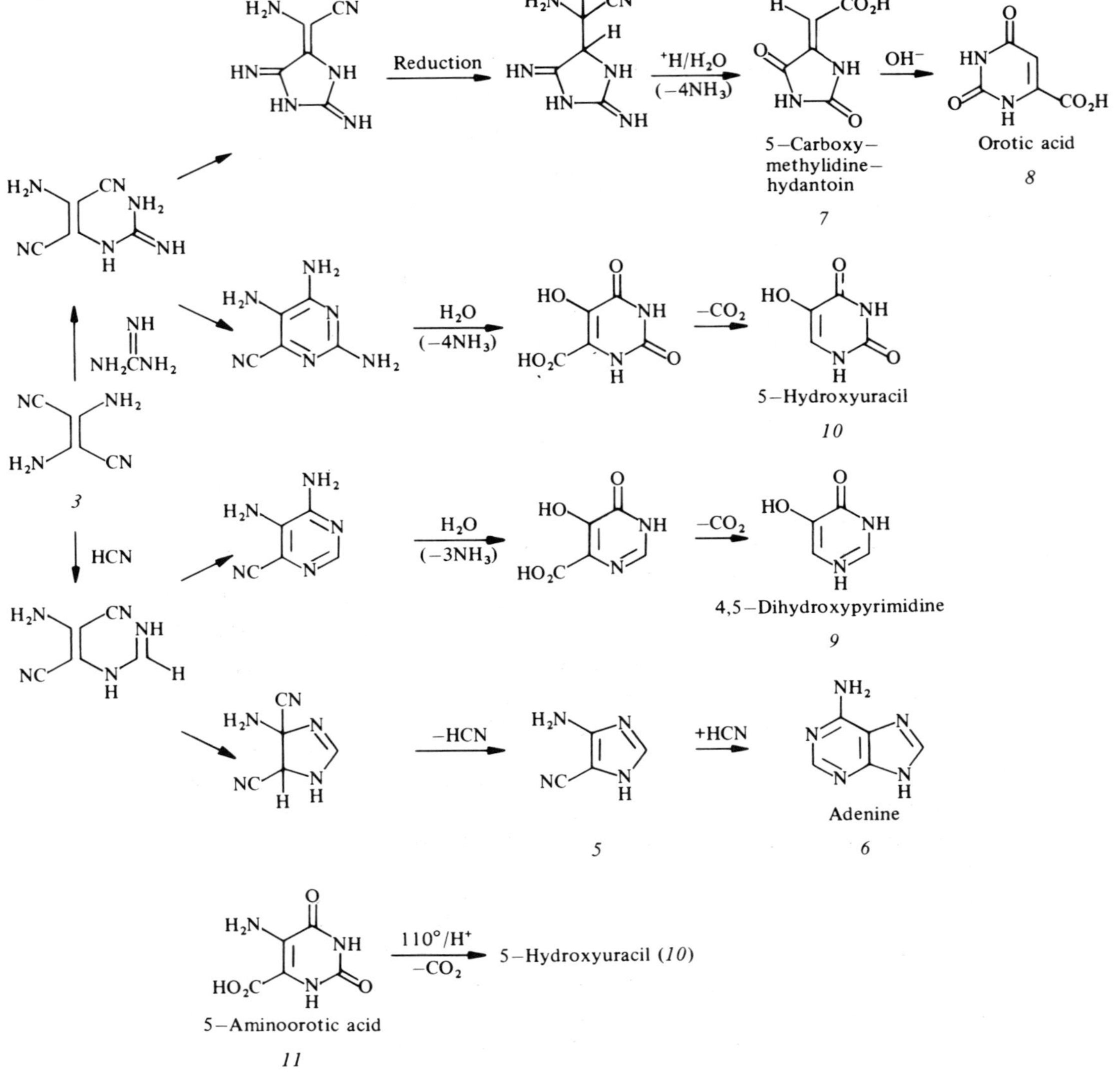
Reduction
$^{+}H/H_2O$
$(-4NH_3)$
OH^-
5–Carboxy–methylidine–hydantoin
7
Orotic acid
8
3
HCN
H_2O
$(-4NH_3)$
$-CO_2$
5–Hydroxyuracil
10
H_2O
$(-3NH_3)$
$-CO_2$
4,5–Dihydroxypyrimidine
9
–HCN
+HCN
Adenine
5
6
$110°/H^+$
$-CO_2$
5–Hydroxyuracil (10)
5–Aminoorotic acid
11

of glycine, *meso-* and *dl*-diaminosuccinic acid, α-aminoisobutyric acid and β-alanine was established by GC/MS analysis of their silylated derivatives (Butts, 1972). Alanine, N-methylglycine and aspartic acid were also identified along with glycine, β-alanine, α-aminoisobutyric acid and the diaminosuccinic acid isomers by the GC/MS analysis of the (+)-2-Butyltrifluoracetyl derivatives (Table 2) (Gil-Av et al., 1965; Pollock et al., 1965). The quantitative yields of glycine and diaminosuccinic acid isomers given in Table 3 were obtained using an amino acid analyzer.

Table 3. Biomolecules from hydrolysis of HCN oligomers[a]

	6 N HCl	pH 8.5[b]
4,5-dihydroxypyrimidine	0.7 —0.9%	—
orotic acid (8)	—	0.009%
5-hydroxyuracil (10)	0.003%	_d
pyrimidine (structure unknown)[b]	—	+
adenine (6)	0.03—0.04%	+
4-aminoimidazole-5-carboxamide (4)	+	+[e]
glycine	0.6%	0.1%
diaminosuccinic acid	0.1%[c]	+
aspartic acid	+	+
alanine	+	+
β-alanine	+	+
α-aminoisobutyric acid	+	+
guanidinoacetic acid	0.03%	+
guanidine	0.2%	0.1%

[a]HCN oligomers were prepared and hydrolyzed as described in the Experimental section. The yields shown are based on starting HCN. Conversions are approximately 2x as great since only about 50% of the cyanide is consumed. It is assumed that 5 moles of HCN are required to form each mole of heterocyclic compound and 4 moles are required to form each mole of glycine. The identity of each compound was established by comparison of the mass spectrum of a volatile derivative with that of an authentic sample except for the following which were purified by chromatography and identified by their UV spectra and/or specific color tests: adenine, 4-aminoimidazole-5-carboxamide, guanidinoacetic acid and guanidine. The presence of the compound is indicated by (+) and absence by (−). When quantitative analyses were not performed no analysis is indicated by (nd). The presence of guanine, xanthine and hypoxanthine is suggested by UV data but these were not reported because their identity has not been confirmed by mass spectral data.

[†]Products from oligomers prepared from degassed HCN solutions. Hydrolyses were performed in evacuated ampuls.

[b]This compound gives a blue color with diazotized sulfanilic acid and it is converted to 4,5-dihydroxypyrimidine by hydrolysis with 6 N HCl.

[c]Sum of the yields of meso- and dl-diaminosuccinic acid. They are formed in approximately a 2:1 ratio respectively.

[d]Control experiments demonstrated that this compound is destroyed by the hydrolysis conditions used in this study.

[e]4-aminoimidazole-5-carboxamide decomposes under acid hydrolysis conditions and when hydrolyzed at pH 8.5 for 24 h. It can be detected in the hydrolysate of the HCN oligomers formed in the absence of oxygen when a 1.5 h hydrolysis time is used.

HCN and Chemical Evolution

Representatives of the three major classes of nitrogen containing biomolecules, purines, pyrimidines and amino acids are formed in comparable amounts by the hydrolysis of the oligomers formed from dilute aqueous solutions of HCN. Since these biomolecules are released from HCN oligomers when either acidic or mildly basic hydrolysis conditions are used it is likely that they would have been formed from the HCN oligomers even if there were large differences in the pH in different locales on the primitive earth. HCN is produced under a variety of plausible primitive earth conditions so it seems very likely that it condensed to these oligomers on the primitive earth. Although these biomolecules can be formed from other reactants (Miller and Orgel, 1974; Ferris et al., 1974c; Chittenden et al., 1976; Miller, 1970) it is particularly impressive that all three structural types are also formed directly from HCN. This primitive earth synthesis is more compelling than those which require different reactants and reaction conditions for each structural type. Although we do not maintain that HCN was the sole source of biomolecules, it is not nescessary, starting from HCN, to postulate a variety of environments on the primitive earth in order to account for the formation of each different structural type of biological molecule.

Two of the pyrimidines obtained from HCN oligomers are present in contemporary biological systems. Orotic acid is a central intermediate in the biosynthesis of pyrimidine nucleotides (Lieberman et al., 1955). 5-Hydroxyuracil is a minor component of yeast RNA (Lis and Passarge, 1966; Hayes and Lis, 1973) and its derivative, uracil-5-oxyacetic acid is present in the anticodons of several transfer RNA's (Yanio and Barrell, 1969; Murao et al., 1970; Weiss, 1973). The biochemical similarity of uridine and 5-hydroxyuridine was demonstrated by the observation that 5-hydroxyuridine undergoes most of the metabolic transformations of uridine in Ehrlich ascites cells (Smith and Vissar, 1965) and is hydrolyzed with yeast nucleoside hydrolase (Lis and McLaughlin, 1971). These findings suggest that 5-hydroxyuracil is compatible with contemporary biochemical processes and consequently it may have been utilized in primitive nucleic acids.

Although the attempted prebiotic syntheses of pyrimidine nucleosides starting from cytosine, uracil or thymine has not been successful (Fuller et al., 1972a,b) we feel that such a synthesis is likely if 4,5-dihydroxypyrimidine or 5-hydroxyuracil is used in their place. The previous lack of success probably reflects the limited nucleophilicity of the non-basic heterocyclic nitrogen atoms present in these pyrimidines. These would be poor nucleophiles and would not effectively displace groups present in the 1'-position or ribose. The successful synthesis of purine nucleosides under the same reaction conditions can be explained by the presence of the nucleophilic imidazole nucleus (Fuller et al., 1972a,b). This prebiotic purine synthesis is very similar to the "salvage" pathway for the biosynthesis of purine nucleotides in which the purine base displaces the pyrophosphate grouping from the 1-position of 5-phosphorylribose-1-pyro—pyrophosphate (Lehninger, 1975). It seems likely that 4,5-dihydroxypyrimidine, which contains a basic heterocyclic nitrogen, would react with ribose derivatives to give the corresponding nucleoside. A similar argument can be made for 5-hydoxyuracil and the metal chelates of orotic acid.

Intermediates in the contemporary biosynthesis of purines and pyrimidines are also formed from dilute cyanide solutions. Orotic acid (8) and 4-aminoimidazole-5-carboxamide (4), which as the ribotides serve as precursors to pyrimidines and purines respectively, are both present. Their formation from HCN and their central role in contemporary biosynthetic pathways suggests that when the supply of preformed purines and pyrimidines became limiting, only those primitive life forms which could synthesize purines and pyrimidines survived (Horowitz, 1945). A provocative corollary to this hypothesis is that some of the fundamental metabolic pathways in contemporary life may have been established several billion years ago as a consequence of the reaction products formed from dilute aqueous solutions of HCN.

The present study demonstrated that formamidine is not essential for the formation of purines in the absence of UV light. Previously it was believed that formamidine was an essential intermediate in the synthesis of adenine (Sanchez et al., 1967). It was felt that this constituted a serious constraint of adenine synthesis from HCN because formamidine is rapidly hydrolyzed under primitive earth conditions. As a consequence a route to adenine was proposed which required the photochemical rearrangement of 2 (Scheme 1) (Sanchez et al., 1967; Ferris and Orgel, 1966a,b; Ferris et al., 1969; Ferris and Kuder, 1970). We now consider the photochemical pathway and the non-photochemical pathway via HCN oligomers to be plausible complementary routes for purine synthesis on the primitive earth.

A primitive earth scenario for the synthesis of these biomolecules would be the constant formation of HCN by the action of electric discharges (Sanchez et al., 1967) shock waves (Bar-Nun and Tauber, 1972) and ultraviolet light (Hubbard et al., 1975; Ferris and Chen, 1975) on the primitive atmosphere. The HCN would be dissolved in rain droplets and carried to the earth's surface where it would oligomerize in lakes and oceans (Sanchez et al., 1967; Miller and Orgel, 1974). These oligomers slowly hydrolyze to yield the biomolecules which eventually organized into the primitive forms of life.

Appreciable amounts of biomolecules could have formed on the primitive earth from the HCN dissolved in lakes and rivers. A one liter solution of $0.1\,M$ cyanide yields, in one year, sufficient HCN oligomers to give 7.5 μmol adenine, 130 μmol 4,5-dihydroxypyrimidine and 160 μmol glycine after hydrolysis. A small primitive lake the size of Lake George, New York (volume 2 km^3 or 2×10^{12} liters) could yield sufficient HCN oligomers annually to give about 10^7 mol of adenine and 10^8 mol of 4,5-dihydroxypyrimidine and glycine if it contained $0.1\,M$ cyanide. A lake the size of Lake Superior would have the potential of producing 6000 times as much of these biomolecules each year. The quantities of these compounds would increase more than a 1000 fold over $10^3 - 10^6$ years. While it is recognized that there are many assumptions lurking in these simple calculations (e.g. inefficient synthesis of cyanide or its complexation with metal ions or hydrolysis of the cyanide) which could change the final values by several orders of magnitude, the results demonstrate that significant quantities of all three classes of nitrogen-containing biomolecules may have been formed exclusively from HCN on the primitive earth.

Acknowledgment. The authors thank Dr. A Burlingame and M.F. Walls of Space Sciences Laboratory at U.C. Berkley for the high resolution GC/MS data and Dr. J. Evans, Finnigan Corporation for some of the low resolution GC/MS data. Dr. A. Lobo, N.Y. State Health Laboratories supplied most of the amino acid analyzer data. A portion of this research was performed by J.P.F. at the NASA Ames Research Center while supported by a National Research Countil Resident Research Associateship. The research at RPI was supported by NSF Grants CHE 76–11000 and MPS 7304352 and a NASA Grant NGR 30–018–148.

References

Adachi, T., Tanimura, A., Asahina, M. (1963). J. Vitaminol. **9**, 217–226
Bar-Nun, A., Tauber, M.E. (1972). Space Life Sciences **3**, 254–259

Begland, R.W. (1975). U.S. Patent 3, 883, 532

Chittenden, G.J.F., Schwartz, A.W. (1976). Nature, 263, 350−351

Ferris, J.P. Chen, C.T. (1975). Nature 258, 587−588

Ferris, J.P., Wos, J.D., Nonner, D.W., Oro, J. (1974a). J. Mol. Evol. 3, 225−231

Ferris, J.P., Wos, J.D., Lobo, A.P. (1974b). J. Mol. Evol. 3, 311−316

Ferris, J.P., Zamek, O.S., Altbuck, A.M., Freiman, H. (1974c). J. Mol. Evol. 3, 301−309

Ferris, J.P., Ryan, T.J. (1973). J. Org. Chem. 38, 3302−3307

Ferris, J.P., Donner, D.B., Lobo, A.P. (1973a). J. Mol. Biol. 74, 499−510

Ferris, J.P., Donner, D.B., Lobo, A.P. (1973b). J. Mol. Biol. 74, 511−518

Ferris, J.P., Donner, D.B., Lotz, W. (1972). J. Amer. Chem. Soc. 94, 6968−6974

Ferris, J.P., Kuder, J.E. (1970). J. Amer. Chem. Soc. 92, 2527−2533

Ferris, J.P., Kuder, J.E., Catalano, A. (1969). Science 166, 765−766

Ferris, J.P., Miller, N.C. (1966). J. Amer. Chem. Soc. 88, 3522−3527

Ferris, J.P., Orgel, L.E. (1966a). J. Amer. Chem. Soc. 88, 3829−3831

Ferris, J.P., Orgel, L.E. (1966b). J. Amer. Chem. Soc. 88, 1074

Ferris, J.P., Orgel, L.E. (1965). J. Amer. Chem. Soc. 87, 4976−4977

Folsome, C.E., Lawless, J.G., Romiez, M., Ponnamperuma, C. (1973). Geochim. Cosmochim. Acta, 37, 455−465

Fuller, W.D., Sanchez, R.A., Orgel, L.E. (1972a). J. Mol. Biol. 67, 25−33

Fuller, W.D., Sanchez, R.A., Orgel, L.E. (1972b). J. Mol. Evol. 1, 249−257

Gil-Av, E., Charles, G., Fisher, R., (1965). J. Chromatog. 17, 408−410

Grimmett, M.R., Richards, E.L. (1965). J. Chromatog. 18, 605−608

Hais, I.M., Macek, K., Paper Chromatography 1963, Czechoslovak. Academy of Sciences, p. 814 (Prague)

Hanes, C.S., Isherwood, F.A. (1949). Nature 164, 1107−1112

Hayes, S.J., Lis, A.W. (1973). Physiol. Chem. and Physics 5, 87−107

Horowitz, N.H. (1945). Proc. Nat. Acad. Sci. U.S. 31, 153−157

Hubbard, J.S., Voecks, G.E., Hobby, G.L., Ferris, J.P., Williams, E.A., Nicodem, D.E. (1975). J. Mol. Evol. 5, 223−241

Kesner, L., Aronson, F.L., Iverman, M.S., Chan, P.C. (1975). Clin. Chem. 21, 353−355

Labadie, M., Jensen, R., Neuzil, E. (1968). Biochem. Biophys. Acta 165, 525−533

Labadie, M., Jensen, R., Neuzil, E. (1967). Bull. Soc. Chim. Biol. 49, 673−682

Lehninger, A.L. (1975). Biochemistry, p. 742, New York: Worth

Lieberman, I., Kornberg, A., Simms, E.S. (1955). J. Biol. Chem. 215, 403−415

Lis, A.W., McLaughlin, D.I. (1971). Physiol. Chem. and Physics 3, 168−174

Lis, A.W., Passarge, W.E. (1966). Arch. Biochem. Biophys. 114, 593−595

Lowe, C.U., Rees, M.W., Markham, R. (1967). Nature 199, 219−222

Markey, S.P., Urban, W.G., Levine, S.P. (1974). Mass Spectra Compounds of Biological Interest, U.S. Atomic Energy Commission Report TID-26553 (available from National Technical Information Service, U.S. Dept. of Commerce, Washington, D.C.).

Matthews, C.N., Moser, R.E. (1967). Nature 215, 1230−1234

McOmie, J.F.W., Turner, A.B. (1963). J. Chem. Soc., 5590−5593

Miller, S.L., Orgel, L.E. (1974). The Origins of Life on Earth, Englewood Cliffs: Prentice-Hall

Miller, S.L. (1970). Origins of Life, Vol. I, L. Margulis, ed. p. 214, New York: Gordon and Breach

Murao, K., Saneyoshi, M., Harada, R., Nishimura, S. (1970). Biochem. Biophys. Res. Commun. **38**, 657–662

Oro, J., Kimball, A.P. (1962). Arch. Biochem. Biophys. **96**, 293–313

Oro, J., Kimball, A.P. (1961). Arch Biochem. Biophys. **94**, 217–226

Oro, J., Kamat, J.S. (1961), Nature **190**, 442–443

Pollock, G.E., Oyama, U.I., Johnson, R.O. (1965). J. Gas Chromatog. **3**, 174–176

Sanchez, R.A., Ferris, J.P., Orgel, L.E. (1968). J. Mol. Biol. **38**, 121–128

Sanchez, R.A., Ferris, J.P., Orgel, L.E. (1967). J. Mol. Biol. **30**, 223–253

Sanchez, R.A., Ferris, J.P., Orgel, L.E. (1966). Science **153**, 72–73

Scoggins, M.W. (1972). Anal. Chem. **44**, 1294–1296

Smith, D.A., Visser, D.W. (1965). J. Biol. Chem. **240**, 446–453

Smith, T. (1969). Chromatographic and Electrophoretic Techniques, 3rd Edition, p. 289 New York: Interscience

Visher, E., Chargaff, E. (1948). J. Biol. Chem. **176**, 703–714

Volker. T. (1960). Angew. Chem. **176**, 379–384

Wakamatsu, H., Yamada, Y., Saito, T., Kumashiro, I., Takenishi, T. (1966). J. Org. Chem. **31**, 2035–2036

Weiss, C.B. (1973). J. Mol. Evol. **2**, 199–204

Yamada, Y., Noda, I., Kumashiro, I., Takenishi, T. (1969). Bull Chem. Soc. Japan **42**, 1454–1456

Yamada, Y., Kumashiro, I., Takenishi, T. (1968). J. Org. Chem. **33**, 642–647

Yanio, M., Barrell, B.G. (1969). Nature **222**, 278–279

Received September 22, 1977; Revised December 5, 1977

Photochemical Production of Formaldehyde in Earth's Primitive Atmosphere

Joseph P. Pinto, G. Randall Gladstone, and Yuk Ling Yung

Abstract. *Formaldehyde could have been produced by photochemical reactions in Earth's primitive atmosphere, at a time when it consisted mainly of molecular nitrogen, water vapor, carbon dioxide, and trace amounts of molecular hydrogen and carbon monoxide. Removal of formaldehyde from the atmosphere by precipitation can provide a source of organic carbon to the oceans at the rate of 10^{11} moles per year. Subsequent reactions of formaldehyde in primeval aquatic environments would have implications for the abiotic synthesis of complex organic molecules and the origin of life.*

The photochemical oxidation of methane (CH_4) is known to be a major source of formaldehyde (H_2CO) in Earth's atmosphere (*1*). The presence of CH_4 and other reduced gases, in abundances greatly exceeding those expected in thermochemical equilibrium, arises from biological activity (*2*). This report is concerned with the production of H_2CO, by photochemical reactions in the atmosphere, at an early stage in Earth's history before life has originated; H_2CO is one of the simple molecules that might have played a key role in the abiotic synthesis of complex organic molecules on the primitive Earth.

The laboratory synthesis of H_2CO and other important organic molecules has been carried out by a number of workers. The initial mixtures used to simulate the composition of the primitive atmosphere have ranged from highly reducing (*3*) to slightly oxidizing (*4*) materials. Although the oxidation state of Earth's atmosphere before life had originated is still highly uncertain, it is generally agreed that it would have been determined by the outgassing history and the composition of volcanic gases. There is strong indirect evidence for a geologically rapid initial degassing of volatiles, rather than a continuous release over the history of the terrestrial planets. Supporting arguments for this view have been presented by Fanale (*5*) for Earth, by Walker (*6*) for Earth and Venus, and by McElroy *et al.* (*7*) for Mars. It is also unlikely that free iron would have been present in the upper mantle for more than 5×10^8 years after Earth had formed (*8*). Once native iron had been removed from the upper mantle by differentiation of the core and mantle, vol-

canic emissions of CH_4 would have been negligible and the composition of volcanic gases would not differ significantly from that of the present (*8*). However, the outgassed volatiles may have contained a small fraction (a few percent) of reduced compounds such as H_2 and CO, in amounts slightly greater than those observed today (*6*). It has been implicitly assumed by earlier workers that the reduced materials was lost solely through the escape of H_2 from the exosphere. Our study raises the possibility of the production of reduced organic carbon from this material.

The major atmospheric constituents after Earth had differentiated should be CO_2, N_2, and H_2O (*8*), with smaller amounts of reduced gases such as H_2 and CO. For the purposes of this investigation, we have assumed that the abundances of CO_2, N_2, and H_2O are the same as today's and that the abundance of H_2 was governed by the balance between volcanic release, the escape of hydrogen atoms, and photochemical reactions. The mixing ratio of H_2 in the atmosphere

was calculated to be of the order of 10^{-3} (*9*). Except for the absence of a stratospheric thermal inversion, the thermal structure was also taken to be the same as the present. The reactions and rate coefficients used in the photochemical model are presented in Table 1 (*10*).

In the absence of shielding by O_2, CO_2 and H_2O are photolyzed in the troposphere according to the reactions

$$CO_2 + h\nu \rightarrow CO + O \qquad (J1)$$
$$H_2O + h\nu \rightarrow H + OH \qquad (J2)$$

Hydrogen atoms, formed in reaction J2 react with CO to yield formyl radicals (HCO) by

$$H + CO + M \rightarrow HCO + M \qquad (R7)$$

The HCO radicals are removed by the following paths (*11*)

$$HCO + H \rightarrow H_2 + CO \qquad (R8)$$
$$HCO + h\nu \rightarrow H + CO \qquad (J3)$$
$$HCO + HCO \rightarrow H_2CO + CO \qquad (R9)$$

The production of H_2CO from CO_2 is described by the following reaction sequence

$$2(CO_2 + h\nu \rightarrow CO + O) \qquad (J1)$$
$$2(H_2O + h\nu \rightarrow H + OH) \qquad (J2)$$
$$2(H + CO + M \rightarrow HCO + M) \qquad (R7)$$
$$HCO + HCO \rightarrow H_2CO + CO \qquad (R9)$$
$$CO + OH \rightarrow CO_2 + H \qquad (R5)$$
$$2(O + H_2 \rightarrow OH + H) \qquad (R2)$$
$$3(OH + H_2 \rightarrow H_2O + H) \qquad (R1)$$
$$3(H + H + M \rightarrow H_2 + M) \qquad (R6)$$

which may be summarized as

$$CO_2 + 2H_2 \rightarrow H_2CO + H_2O$$

In our standard photochemical model, the abundances of the major atmospheric gases are as follows: N_2, 0.8 bar; H_2O, 0.012 bar; CO_2, 2.4×10^{-4} bar; CO, 2.4×10^{-7} bar; and H_2, 8.0×10^{-4} bar (*12*). Vertical profiles of important minor

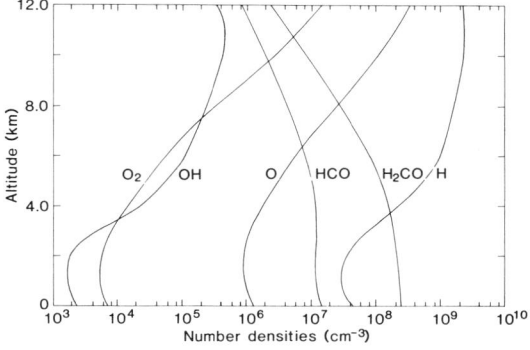

Fig. 1. Concentrations of important minor species in the standard model. The surface mixing ratios of N_2, H_2O, CO_2, CO, and H_2 are 1, 1.5×10^{-2}, 3×10^{-4}, 3×10^{-7}, and 1×10^{-3}, respectively. Except for H_2O, these gases are well mixed throughout the troposphere. The height-independent eddy diffusion coefficient is 10^6 cm^2 sec^{-1}.

constituents are shown in Fig. 1. Production and loss rates of HCO radicals and H_2CO are given in Fig. 2. Most of the H_2CO ($\approx$ 99 percent) is destroyed by photolysis (reactions J4 and J5). However, a small fraction is incorporated into rain droplets and delivered to the oceans. We calculated the rate of removal of H_2CO from the atmosphere by precipitation, using the scavenging coefficients of Wofsy [13] for H_2CO in the present atmosphere. For the standard model the rainout rate of H_2CO was 2.8×10^8 molecule $cm^{-2} sec^{-1}$. A variety of models were also constructed to study the sensitivity of our results to the choice of the essential input parameters [14].

The atmospheric abundance of CO_2 was controlled by a number of processes, including volcanic release, the weathering of surface rocks, and dissolution in the oceans. It is most likely [15] that the partial pressure of CO_2 in the primitive

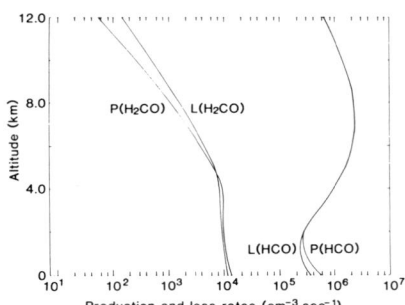

Fig. 2. Production (P) and loss (L) rates of HCO and H_2CO in the standard model.

atmosphere was buffered at about 2×10^{-4} bar by the latter two processes. However, recent work [16] suggests a possibly higher concentration of CO_2 in the primitive atmosphere. In this case, the production and subsequent rainout rate of H_2CO could greatly exceed that

calculated in our standard model [14]. Our results are also somewhat sensitive to the thermal structure. Knauth and Epstein [17] have suggested that surface temperatures could have reached 340 K during the early Archean (3.7×10^9 years ago). In this calculation, the distribution of relative humidity was held fixed at 50 percent [14].

For the case of an H_2 mixing ratio of 10^{-3}, the concentration of O_2 is exceedingly small. The tropospheric profile of O_2 is governed, in this case, by photochemical equilibrium. Later evolutionary stages may be represented by the models, in which the H_2 abundance has decreased. In addition, the emergence of O_2, perhaps because of photosynthesis [18], would have curtailed the production of H_2CO. This would have occurred through reactions such as

$$O + HCO \rightarrow OH + CO \quad (R12)$$

$$O_2 + HCO \rightarrow HO_2 + CO \quad (R13)$$

The value given for the wet removal of H_2CO, in the standard model, is sufficient to fill the oceans, at their present volume, to a $10^{-3}M$ solution in 10×10^6 years. At this concentration, H_2CO in solution may polymerize [19]. The results of Ponnamperuma [20] indicate that, in the presence of ultraviolet radiation of wavelengths near 2800 Å, polymerization may take place in a $3 \times 10^{-4}M$ solution of H_2CO. Other processes may also have occurred [4]. However, this does not alter the conclusion that simple organic molecules could have been delivered to the oceans from photochemical reactions in the atmosphere under weakly reducing conditions. Further laboratory studies should be performed to assess the factors governing the production and stability of H_2CO in conditions appropriate to the primitive Earth.

JOSEPH P. PINTO
Goddard Institute for Space Studies,
New York 10025

G. RANDALL GLADSTONE
YUK LING YUNG
California Institute of Technology,
Pasadena 91125

Table 1. Reactions and rate coefficients used in the photochemical model. Units for two- and three-body rate constants are cubic centimeters per second and centimeters raised to the sixth power per second, respectively. Globally averaged photodissociation rate constants (per second) are given for 0 and 12 km, respectively. Numbers are values used in the standard model; T, absolute temperature; P_{atm}, atmospheric pressure.

Reaction	Rate constant		Source in [10]
	0 km	12 km	
J1 $CO_2 + h\nu \rightarrow CO + O$	2.3×10^{-12}	3.5×10^{-11}	a
J2 $H_2O + h\nu \rightarrow H + OH$	9.3×10^{-15}	1.4×10^{-9}	a
J3 $HCO + h\nu \rightarrow H + CO$	1.0×10^{-2}	1.0×10^{-2}	b
J4 $H_2CO + h\nu \rightarrow H + HCO$	1.8×10^{-5}	1.8×10^{-5}	c
J5 $H_2CO + h\nu \rightarrow H_2 + CO$	2.5×10^{-5}	2.5×10^{-5}	c
J6 $H_2O_2 + h\nu \rightarrow OH + OH$	5.5×10^{-5}	5.7×10^{-5}	a
J7 $O_2 + h\nu \rightarrow O + O$	4.1×10^{-10}	4.7×10^{-10}	a
J8 $O_3 + h\nu \rightarrow O_2 + O$	2.3×10^{-4}	2.3×10^{-4}	a
R1 $H_2 + OH \rightarrow H_2O + H$	$1.2 \times 10^{-11} \exp(-2200/T)$		d
R2 $H_2 + O \rightarrow H + OH$	$7.0 \times 10^{-11} \exp(-5000/T)$		e
R3 $O + OH \rightarrow O_2 + H$	1.7×10^{-11}		f
R4 $OH + OH \rightarrow H_2O + O$	$1.0 \times 10^{-11} \exp(-500/T)$		d
R5 $CO + OH \rightarrow CO_2 + H$	$1.3 \times 10^{-13}(1 + P_{atm})$		d
R6 $H + H + M \rightarrow H_2 + M$	$2.6 \times 10^{-33} \exp(375/T)$		g
R7 $H + CO + M \rightarrow HCO + M$	$2.0 \times 10^{-33} \exp(-850/T)$		h
R8 $H + HCO \rightarrow H_2 + CO$	3.0×10^{-10}		i
R9 $HCO + HCO \rightarrow H_2CO + CO$	6.3×10^{-11}		i
R10 $OH + HCO \rightarrow H_2O + CO$	5.0×10^{-11}		h
R11 $O + HCO \rightarrow H + CO_2$	1.0×10^{-10}		j
R12 $O + HCO \rightarrow OH + CO$	1.0×10^{-10}		j
R13 $O_2 + HCO \rightarrow HO_2 + CO$	5.0×10^{-12}		d
R14 $HO_2 + HCO \rightarrow H_2O_2 + CO$	1.0×10^{-11}		k
R15 $H + H_2CO \rightarrow H_2 + HCO$	$2.8 \times 10^{-11} \exp(-1540/T)$		d
R16 $OH + H_2CO \rightarrow H_2O + HCO$	$1.7 \times 10^{-11} \exp(-100/T)$		d
R17 $O + O + M \rightarrow O_2 + M$	$9.6 \times 10^{-34} \exp(480/T)$		e
R18 $O + O_2 + M \rightarrow O_3 + M$	$9.9 \times 10^{-29} T^{-2.1}$		d
R19 $O + O_3 \rightarrow O_2 + O_2$	$1.5 \times 10^{-11} \exp(-2218/T)$		d
R20 $H + O_2 + M \rightarrow HO_2 + M$	$1.6 \times 10^{-28} T^{-1.4}$		d
R21 $O + HO_2 \rightarrow OH + O_2$	3.5×10^{-11}		d
R22 $H + HO_2 \rightarrow H_2 + O_2$	1.4×10^{-11}		ℓ
R23 $H + HO_2 \rightarrow OH + OH$	3.2×10^{-11}		ℓ
R24 $H + HO_2 \rightarrow H_2O + O$	9.4×10^{-13}		ℓ
R25 $OH + HO_2 \rightarrow H_2O + O_2$	4.0×10^{-11}		d
R26 $HO_2 + HO_2 \rightarrow H_2O_2 + O_2$	2.5×10^{-12}		d
R27 $H_2O_2 + O \rightarrow OH + HO_2$	$2.8 \times 10^{-12} \exp(-2125/T)$		d
R28 $H_2O_2 + OH \rightarrow H_2O + HO_2$	$1.0 \times 10^{-11} \exp(-750/T)$		d
R29 $H + O_3 \rightarrow OH + O_2$	$1.4 \times 10^{-10} \exp(-470/T)$		d
R30 $OH + O_3 \rightarrow HO_2 + O_2$	$1.6 \times 10^{-12} \exp(-940/T)$		d
R31 $HO_2 + O_3 \rightarrow OH + 2O_2$	$1.1 \times 10^{-14} \exp(-580/T)$		d

References and Notes

1. S. C. Wofsy, J. C. McConnell, M. B. McElroy, *J. Geophys. Res.* **77**, 447 (1972).
2. J. E. Lovelock and L. Margulis, *Tellus* **26**, 2 (1974).
3. S. L. Miller and H. C. Urey, *Science* **130**, 245 (1959).
4. J. S. Hubbard, J. P. Hardy, N. H. Horowitz, *Proc. Natl. Acad. Sci. U.S.A.* **68**, 574 (1971); P. H. Abelson, *ibid.* **54**, 1490 (1965).
5. F. P. Fanale, *Chem. Geol.* **8**, 279 (1971).
6. J. C. G. Walker, *Evolution of the Atmosphere* (Macmillan, New York, 1977).
7. M. B. McElroy, T. Y. Kong, Y. L. Yung, *J. Geophys. Res.* **82**, 4379 (1977).
8. H. D. Holland, in *Petrologic Studies—A Volume in Honor of A. G. Buddington*, A. E. J.

Engle, H. L. James, B. F. Leonard, Eds. (Geological Society of America, Boulder, Colo., 1962), p. 447. However, according to the preliminary results of J. S. Lewis (personal communication), the composition of degassed volatiles could have been the following: H_2O, 20 percent; CO_2, 50 percent; CO, 20 percent; and H_2, 10 percent by volume. This composition is much more reducing than that used in the calculation.

9. M. B. McElroy, *Chemical Processes in the Solar System: A Kinetic Perspective* (Butterworths, London, 1975); Y. L. Yung and M. B. McElroy, *Science* **203**, 1002 (1979). Analyses of the ratio of deuterium to hydrogen by Y. L. Yung and M. B. McElroy [*Bull. Am. Astron. Soc.* **9**, 497 (1977)] and H. Craig and J. E. Lupton [*Earth Planet Sci. Lett.* **31**, 369 (1976)] indicate that large quantitites of H_2 could have escaped from Earth's atmosphere in the past. These results are consistent with an H_2 mixing ratio of $\sim 10^{-3}$ in the primitive atmosphere. The production and subsequent removal of H_2CO represents an additional loss of H_2. However, it is unlikely that such a calculation would yield results outside our current range of estimates for the H_2 mixing ratio.

10. Sources of rate coefficient data are as follows: (a) See Yung and McElroy (9). (b) Estimated. A a mean cross section of 3×10^{-19} cm^2 was derived for the visible band from the theoretical study of P. J. Bruna, R. J. Buenker, S. D. Peyerimhoff [*J. Mol. Struct.* **32**, 217 (1976)]. A. A. Borisov, V. T. Galochkin, S. A. Mulenko, A. N. Oraevskii, E. F. Starodubtsev, A. F. Suchkov [*Sov. J. Quantum Electron.* **8**, 1094 (1978)] have estimated mean cross section of 1×10^{-20} cm^2 for the same band. The values of Bruna *et al.* have been used here. (c) J. G. Calvert, J. A. Kerr, K. L. Demerjian, R. D. McQuigg, *Science* **175**, 751 (1972). (d) W. B. DeMore, *Jet Propulsion Lab. Publ. 79-27* (1979). (e) J. A. Logan, M. J. Prather, S. C. Wofsy, *Philos. Trans. R. Soc. London* **290**, 187 (1978). (f) R. T. Watson, personal communication. (g) S. C. Liu and T. M. Donahue, *J. Atmos. Sci.* **31**, 1118 (1974). (h) D. L. Baulch, D. D. Drysdale, J. Duxbury, S. J. Grant, *Evaluated Kinetic Data for High Temperature Reactions* (Butterworths, London,

1976), vol. 3. (i) J. P. Reilly, J. H. Clarke, C. B. Moore, G. C. Pimentel, *J. Chem. Phys.* **69**, 4381 (1978). (j) R. F. Hampson and D. Garvin, Eds., *Reaction Rate and Photochemical Data for Atmospheric Chemistry 1977* (Special Publication 513, National Bureau of Standards, Washington, D.C., 1978). (k) Estimated. (l) W. Hack, H. G. Wagner, K. Hoyermann, *Ber. Bunsenges. Phys. Chem.* **82**, 713 (1978); W. Hack, A. W. Preuss, H. G. Wagner, K. Hoyermann, *ibid.* **83**, 212 (1979).

11. Note that R7 and R8 constitute an effective cycle for the catalytic recombination of atomic hydrogen. This cycle also accounts for the difference in the computed atomic hydrogen densities at low altitudes between this model and that of J. F. Kasting, S. C. Liu, T. M. Donahue, *J. Geophys. Res.* **84**, 3097 (1979).

12. The one-dimensional model extends from 0 to 120 km with a vertical resolution of 2 km, based on the numerical scheme developed by M. Allen, Y. L. Yung, and J. W. Waters (in preparation). The equations of continuity are solved for all species. Vertical transport of long-lived species is described in terms of eddy diffusion. For the standard model, a uniform value of the eddy diffusion coefficient (K) of 10^6 cm^2 sec^{-1} was chosen. Concentrations of the major species H_2O, CO_2, and H_2 were fixed at the lower boundary, and a zero flux was specified for CO. Concentrations of radical species were determined on the basis of their photochemical production and loss rates. A flux of reduced material of $\sim 10^{10}$ cm^{-2} sec^{-1} (mostly as H_2) is required at the lower boundary, in order to maintain the steady state.

13. S. C. Wofsy, *Annu. Rev. Earth Planet. Sci.* **4**, 441 (1976).

14. For H_2 mixing ratios of 10^{-4}, 3×10^{-4}, and 3×10^{-3}, the H_2CO rainout rate is 1.7×10^7, 4.4×10^7, and 1.0×10^9 molecule cm^{-2} sec^{-1}, respectively. For CO_2 mixing ratios of 6×10^{-4}, 1.2×10^{-3}, and 3×10^{-3}, the rainout rate is 4.8×10^8, 1.2×10^9, and 3.1×10^9 cm^{-2} sec^{-1}. Adjusting the value of k_8 to 1×10^{-10}, 2×10^{-10}, and 5×10^{-10} leads to rainout rates of 1.9×10^9, 4.5×10^8, and 1.4×10^8 cm^{-2} sec^{-1}. Adopting a uniform value for K of 10^5 cm^2 sec^{-1} leads to a rain-

out rate of 2.3×10^8 cm^{-2} sec^{-1}, and raising the surface temperature by 10 percent to 317 K leads to a rainout rate of 8.8×10^7 cm^{-2} sec^{-1}.

15. F. P. Fanale, personal communication.

16. T. Owen, R. D. Cess, V. Ramanathan, *Nature (London)* **277**, 640 (1979).

17. L. Knauth and S. Epstein, *Geochim. Cosmochim. Acta* **40**, 1095 (1976).

18. L. V. Berkner and L. C. Marshall, *J. Atmos. Sci.* **22**, 225 (1965).

19. Perhaps the most important question concerning the fate of H_2CO, once it had been precipitated into the primitive oceans, is its interaction with solar ultraviolet radiation. Preliminary results of G. L. Kok (private communication) indicate that H_2CO in solution does not exhibit the same absorption spectrum as gaseous H_2CO. Photolysis for H_2CO in solution is therefore unlikely, as it undergoes a rapid and almost complete hydrolysis to methylene glycol, $CH_2(OH)_2$, and to other products by processes such as the Cannizzaro reaction [J. F. Walker, *Formaldehyde* (Reinhold, New York, 1964)]. These data imply a certain degree of stability for H_2CO in solution. This question should be investigated further with attention being paid to factors such as pH, the oxidation-reduction state, and the presence of other ultraviolet absorbers in the primitive oceans.

20. C. Ponnamperuma, in *Origin of Prebiological Systems and of Their Molecular Matrices*, S. W. Fox, Ed. (Academic Press, New York, 1965), p. 221.

21. One of us (J.P.P.) thanks J. C. G. Walker for many helpful discussions. We also thank M. Allen, F. P. Fanale, G. L. Kok, C. B. Moore, J. S. Lewis, and J. J. Morgan for their useful comments. J.P.P. is a NASA National Research Council resident research associate. This work was also supported in part by NASA contract NSG-7376 under the Planetary Atmospheres Program. Contribution 3357 of the Division of Geological and Planetary Sciences, California Institute of Technology.

7 February 1980; revised 17 June 1980

Why Nature Chose Phosphates

F. H. WESTHEIMER

Phosphate esters and anhydrides dominate the living world but are seldom used as intermediates by organic chemists. Phosphoric acid is specially adapted for its role in nucleic acids because it can link two nucleotides and still ionize; the resulting negative charge serves both to stabilize the diesters against hydrolysis and to retain the molecules within a lipid membrane. A similar explanation for stability and retention also holds for phosphates that are intermediary metabolites and for phosphates that serve as energy sources. Phosphates with multiple negative charges can react by way of the monomeric metaphosphate ion PO_3^- as an intermediate. No other residue appears to fulfill the multiple roles of phosphate in biochemistry. Stable, negatively charged phosphates react under catalysis by enzymes; organic chemists, who can only rarely use enzymatic catalysis for their reactions, need more highly reactive intermediates than phosphates.

PHOSPHATE ESTERS AND ANHYDRIDES DOMINATE THE LIVing world. The genetic materials DNA and RNA are phosphodiesters. Most of the coenzymes are esters of phosphoric or pyrophosphoric acid. The principal reservoirs of biochemical energy [adenosine triphosphate (ATP), creatine phosphate, and phosphoenolpyruvate] are phosphates. Many intermediary metabolites are phosphate esters, and phosphates or pyrophosphates are essential intermediates in biochemical syntheses and degradations. Synthetic organic chemists, however, preferentially use other groups for linking hydroxyl, carboxyl, and amino groups, and for activating them for reaction. Why were phosphates, and almost no other groups, selected by evolution for biochemical transformations? Why did nature choose phosphates? In this article, an attempt is made to answer those questions.

The Role of Phosphates

A few of the biologically important phosphates are listed in Table 1. Granted that phosphates are ubiquitous in biochemistry, what do they do? The answer is that they can do almost everything. In synthetic organic chemistry, nucleophilic displacement reactions at a carbon atom generally require a good "leaving group" [the anion of a strong acid or the conjugate base (for example, water) of a cationic acid (for example, hydronium ion)]. A large number of such leaving groups, such as chlorides, bromides, iodides, tosylates, and triflates, is in use, but phosphates are almost never used. In metabolic reactions, the leaving group is usually phosphate or pyrophosphate

The author is Morris Loeb Professor of Chemistry, Emeritus, at Harvard University, Cambridge, MA 02138.

(1), although sulfonium salts (such as S-methyl methionine) are also used. A typical example of the displacement of a pyrophosphate residue is shown below.

(In this article, inorganic phosphate ion is sometimes abbreviated as P_i and inorganic pyrophosphate as PP_i.)

Elimination reactions in synthetic organic chemistry also require good leaving groups, and here also chemists use groups such as chlorides and tosylates. They also use trialkylamines (which are released from quaternary ammonium salts), sulfoxides and selenoxides, and many other groups, but seldom do they use phosphates. In contrast, although biochemical eliminations of water from hydroxyl compounds often occur without isolable intermediates, when such intermediates are found, they are usually phosphates or pyrophosphates (2). Two examples, the first of which is a decarboxylative elimination, are shown below.

Although some biochemical processes proceed as S_N2 reactions, others follow the S_N1 pathway. The formation of terpenes and steroids depends on generating an allylic carbonium ion from dimethylallyl pyrophosphate (3):

The corresponding processes in synthetic organic chemistry also involve familiar leaving groups such as bromides and tosylates; the variety is large but seldom includes phosphates or pyrophosphates. A summary of the contrasts between the practices of organic chemists and the pathways in biochemistry is shown in Table 2.

The Importance of Being Ionized

A possible explanation for the role of phosphates begins with a paper published in 1958 by Davis entitled "The importance of being ionized" (4). Davis's thesis was that living organisms must conserve their metabolites within the cell membrane. If these compounds had diffused through the membranes of evolutionarily primitive organisms, they would have been lost by dilution in the water outside the cell. Most electrically neutral molecules will have some solubility in lipid and will pass through a membrane; most ionized molecules, that is, salts, are insoluble in lipids. More precisely, the pK of an acid should be less than 4 and that of a base greater than 10 to ensure that only a small fraction of the compound remains in the un-ionized form at physiological pH. This general rule is not absolute. Polyhydroxylated compounds such as sugars may be lipophobic without being ionized, and compounds that are extremely insoluble in water, such as steroids, either become part of the membranes (5) or are conserved within cells in special ways or both (6). But certainly molecules can be kept within membranes if they remain ionized. The first pK's of phosphoric acid and of phosphate mono- and diesters are about 2, so that phosphates are ionized at physiological pH and therefore are trapped within cells.

An interesting example of Davis's principle in action concerns the biosynthesis of histidine (7). The details of the transformations are irrelevant to this discussion; the important point is that all the molecules are ionized. The intermediates that occur early in the sequence are all monoesters of phosphoric or pyrophosphoric acid, and, although all phosphate groups are removed from the intermediates that occur late in the sequence, the last phosphate is not removed until other ionized groups have been added. Some of the charged groups carry a negative charge, and some a positive charge, but at no place in this synthesis of histidine are any of the intermediates uncharged. This is not an isolated example; Davis's generalization can be illustrated with a number of different reaction sequences (8). [Adapted with permission from (7).]

(Ribose - Ⓟ represents 5-phosphoribose; Ribose - Ⓟ - Ⓟ - Ⓟ represents the corresponding triphosphate)

Table 1. Examples of phosphates in biochemistry.

Phosphate	Acid derivative
DNA	Diester of phosphoric acid
RNA	Diester of phosphoric acid
ATP	Anhydride of phosphoric acid
Creatine phosphate	Amide of phosphoric acid
Phosphoenolpyruvate	Enol ester of phosphoric acid
Pyridoxal phosphate	Phenol ester of phosphoric acid
Nicotine adenine dinucleotide	Ester and anhydride of phosphoric acid
Fructose 1,6-diphosphate	Ester of phosphoric acid
Glucose-6-phosphate	Ester of phosphoric acid
Isopentenyl pyrophosphate	Ester of pyrophosphoric acid
Ribose-6-phosphate-1-pyro-phosphate	Ester of phosphoric and pyrophosphoric acids

Moreover, the electrostatic interaction of positive and negative charges constitutes the simplest and possibly the most primitive mode of interaction among molecules. The negative charges on phosphates are important in the binding of coenzymes to enzymes and in the "packaging" of nucleic acids.

A linking group for nucleic acids. In addition to the chemistry of intermediary metabolites, the linking of the nucleosides of DNA and RNA by phosphates must be explained. An argument can be advanced on the basis of the twin assumptions that the nucleotides must be connected so as to form a tape that contains a genetic code, and that the tape can later be taken apart, so as to reuse the nucleotides. The conservation of structures as complicated as nucleotides is certainly desirable for the economy of any organism. The best chemistry for taking a tape apart will probably be hydrolytic; if so, then the best connection will be an esterification reaction, or something similar. (A discussion of alternatives to esters occurs later in this article.) To make a tape from small molecules, the connecting agent must be at least bivalent in order to supply one connection for each of the nucleotides. Further, if the resulting molecules must also be ionized, the connecting agent must be at least trivalent. Phosphoric acid comes immediately to mind; sulfuric acid would not do, as it is only a dibasic acid. (Furthermore, sulfate esters are too reactive for a stable genetic material.) Of course, there are numerous molecules that can make two covalent bonds and remain charged; citric acid, glutamic acid, ethylene diamine monoacetic acid, arsenic acid, and silicic acid are examples. Nothing so far indicates that phosphoric acid is unique.

Rates of hydrolysis. Then why phosphoric acid? The answer depends on some basic principles of physical-organic chemistry and on an additional biological constraint on living systems. Living systems must be reasonably stable. Biomolecules not only must be confined within a bag defined by a lipid membrane; these molecules must survive in water for an appreciable time, preferably a long time. Metabolites, or at least some metabolites, can be short-lived, but not the genetic material.

What are the rates of hydrolysis of different kinds of compounds, and in particular what is the effect of structure on the rates of hydrolysis of esters? How can these rates be controlled? In the nearly neutral pH of the biological environment, esters such as ethyl acetate (9) can survive at ambient temperature for many months. But even a simple gene will have a thousand ester bonds, and most will have several thousand; further, even a simple organism must have a thousand genes or more. If a single ester bond in the genetic material is cleaved during the lifetime of the organism, it may fail to reproduce. Of course, with bacteria, a small sample may contain many billions of organisms, and only a few of them need to reproduce to maintain the species. Still, natural selection will favor a genetic material that, in the majority of the cells, will last for times comparable to the lifetime of the organism. Certainly higher organisms, whose numbers are limited, would have special difficulties with labile genetic material. If that material is to survive, no more than one in a million ester bonds should cleave during the lifetime of the organisms. At room temperature, and at neutral pH, one molecule of ethyl acetate in a million is hydrolyzed every hour (9), and at higher acidities or basicities the time required for hydrolysis is less. In the acidic environment of fruits (10), one out of a million molecules of a carboxylic acid ester will hydrolyze every few seconds. The pH during the early stages of evolution is uncertain, but it may have been far from neutrality. At best, esters of carboxylic acids would be marginal as a genetic material for rapidly dividing bacteria and totally inadequate for longer lived animals or plants. Some sterically hindered esters that are more stable might serve, but the system would become unnecessarily complicated.

The rate of hydrolysis of triesters of phosphoric acid is somewhat less than that for electrically neutral carboxylic acids such as ethyl acetate (11), and, although phosphate triesters might be somewhat better as a genetic material, they would still be inadequate. As explained above, however, metabolites should be charged to keep them within the cell membrane, and this charge must be negative, since a negative charge has an important second effect: it sharply diminishes the rate of nucleophilic attack on the ester. Nucleophiles, such as hydroxide ion, are repelled by negative charges and therefore react less rapidly with anions than with neutral substrates. Furthermore, the same effect is qualitatively present with respect to electrically neutral nucleophiles such as water. In this case, attack involves the pushing of a lone pair of electrons from the nucleophile into the electrophilic site of reaction. Because the phosphate ester groups in DNA are negatively charged, they are relatively resistant to hydrolysis and are therefore preferable to esters of carboxylic acids as a genetic material. On the other hand, a residual positive charge near the site of hydrolysis on the genetic tape would not work; such a charge would increase rather than decrease the rate of nucleophilic attack.

The effect of negative charge. Two questions immediately arise. First, how much slower is nucleophilic attack on a negatively charged substrate as compared with attack on a neutral species, and second, would not this effect also occur with diesters of other tribasic acids such as citric acid? The answers to these questions come from standard physical-organic chemistry and show the advantage of phosphates.

The effect of charge on ionization constants and on reaction rates was noted and explained by Bjerrum (12) in 1923; a crude model that gives moderately good quantitative results was published (13) in 1938. To what extent will a negative charge retard the approach of a nucleophile to a phosphorus or carbon atom, or indeed to the atom of any element, in an ester? Qualitatively, the effect will be comparable to the effect of a negative electric charge on the ionization of the corresponding polybasic acid. When a polybasic acid such as phosphoric acid or citric acid ionizes, the covalent bond between a proton and the oxygen atom to which it is attached must be broken, and the electrostatic work required to take the proton to infinity in the field of the negative charge of the incipient anion must be overcome. The work required to remove a second proton from a dibasic acid should be similar to that to remove the first proton, except that the second proton must be removed not only in the field of the incipient charge on the oxygen atom to which the proton was attached but also in the additional electrostatic field of the negative charge produced by the first ionization. This additional electrostatic work is reflected in the ratio of the first to the second ionization constants of the acid. (A statistical factor must also be taken into account in calculating the ratio of the first to the second ionization constants.)

Table 2. Contrasts between synthetic organic chemistry and biochemistry. The leaving groups shown are typical examples for the indicated type of reaction. The symbols Ar, R, and Ac stand for aryl, alkyl, and acyl groups, respectively.

Reaction type	Synthetic chemistry	Biochemistry
Leaving group in S_N2	Cl^-, Br^-, I^-, $ArSO_3^-$	OPO_3^{3-}, $P_2O_7^{4-}$
Leaving group in S_N1	Cl^-, Br^-, I^-, $ArSO_3^-$	OPO_3^{3-}, $P_2O_7^{4-}$
Leaving group in E2	Cl^-, Br^-, I^-, $ArSO_3^-$, NR_3, $-S-Ar$, H_2O	OPO_3^{3-}, H_2O
Driving force	Ac_2O, H_2SO_4, KOH, carbodiimides	ATP

The electrostatic work of ionization decreases with increasing distance between a proton that is undergoing ionization and the residual negative charge or charges in a molecule. The work decreases with increasing distance more than is predicted for the interactions of charges in a uniform dielectric medium because the medium is not uniform; the lines of force between the charges flow through both the molecule and the external dielectric. The result is that the electrostatic effect is astonishingly greater for small distances than for larger ones. This result is in accord with theory (13), and is apparent from the ratios of the successive ionization constants of phosphoric and citric acids, as shown in Table 3 (14). The successive ionization constants of phosphoric acid differ by factors of more than 10^5, whereas those for citric acid differ by factors of less than 50, even when the statistical factors are included.

The attack of hydroxide ion on esters will be similarly affected by charge. Specifically, the rate constants for the attack of hydroxide on dimethyl phosphate anion should be and are less than that for the attack of hydroxide ion on trimethyl phosphate by a factor of more than 10^5 (15). The experimental data, which have been extrapolated to 35°C, are shown in Table 4.

The rate for attack by hydroxide ion on trimethyl phosphate is less than that for attack on ethyl acetate by a factor of 25; the rate for attack of hydroxide on the anion of dimethyl phosphate is less than that for attack on ethyl acetate by a factor that approaches 5 million. (The statistical factors are sufficiently small by comparison with these numbers that they can be neglected.) The half-time for the hydrolysis of dimethyl phosphate in a solution of $1N$ alkali at 110°C is about a day (15). Such stability guarantees that genetic material will survive. Further questions related to hydrolysis are discussed later.

Possible Alternatives to Phosphate Esters

Citric acid. In this case the negative charge generated by the ionization of the first proton is distant from the second proton, and the electrostatic effect is correspondingly much less. The effect is reflected in the ionization constants of Table 3. The negative charge will then also cause only a relatively small change (about a factor of 20) in the rate of saponification of the unsymmetrical diester, as compared to that of the triester of citric acid (16). (In both cases, the central carboxyl residue is attacked.) The change is real but insufficient to stabilize genetic material. The negative charges on the diesters of phosphoric acid in DNA help retain the molecules within the cell membrane and impart to the nucleic acid the stability required for reproducible genetics.

Arsenic acid and silicic acid. Another compound that must be considered as a basis for a possible genetic material is arsenic acid, which is also tribasic. The poisonous effects, however, of compounds of arsenic probably cannot be avoided, since these effects are

Table 3. Ionization constants for orthophosphoric and citric acids.

Acid	K_1 (M)	K_2 (M)	K_3 (M)	K_1/K_2	K_2/K_3
Phosphoric	7.5×10^{-3}	6.2×10^{-8}	2.2×10^{-13}	1.2×10^5	2.8×10^5
Citric	6.0×10^{-4}	1.5×10^{-5}	4.0×10^{-7}	40	38

centered in the lower valence states of arsenic (*17*), and the reduction of pentavalent arsenic is much easier than that of pentavalent phosphorus. In any case, arsenic esters are totally unsuitable; the hydrolysis of esters of arsenic acid is remarkably fast. The triisopropyl ester in neutral water at room temperature is completely hydrolyzed in a couple of minutes (*18*). Apparently the hydrolysis of the diesters is even faster than that of the triesters. [Compare with the hydrolysis of phosphites (*19*).]

Silicic acid is more abundant in nature than phosphoric acid and is tetravalent, but it is also unsuitable. Its esters, like those of arsenic acid, hydrolyze far too rapidly (*20*) to survive, and furthermore silicic acid is too weak an acid; its first *p*K (*21*) is about 9.5. Its diesters would not ionize in neutral solution and would not carry a negative charge.

Amides. If esters hydrolyze too easily, why not use amides, which are much more resistant to hydrolysis? Proteins are obviously quite stable in aqueous media. If the system had been designed with amide bonds, presumably natural selection would have led to the synthesis of appropriate amines and of enzymes with sufficient specificity to avoid confusion between structural and enzymatic proteins on the one hand and the postulated proteinlike genetic tape on the other. There must be (and are) other compelling reasons that account for the choice of phosphates.

In general, a driving force is required for chemical and biochemical reactions, such as the formation of a genetic tape. The transfer of nucleotides from molecules such as ATP or guanosine triphosphate (GTP) in the synthesis of ribonucleic acids, or the transfer of nucleotides from molecules such as deoxyadenosine triphosphate or deoxyguanosine triphosphate in the synthesis of deoxyribonucleic acids, or the transfer of the PO_3^- residue from ATP in many phosphorylations all make use of the favorable properties of phosphoric anhydrides. In sharp contrast to carboxylic anhydrides, the phosphoric anhydrides are protected by their negative charges from rapid attack by water and other nucleophiles so that they can persist in an aqueous environment (*22*) even though they are thermodynamically unstable, and thus can drive chemical processes to completion in the presence of a suitable catalyst (enzyme). This remarkable combination of thermodynamic instability and kinetic stability was noted many years ago by Lippmann, who correctly ascribed the kinetic stability to the negative charges in ATP (*23*). A citric acid anhydride would not survive long in water and could not serve as a convenient source of chemical energy.

Prebiotic Chemistry

Despite the argument given above, the choice of phosphate esters rather than proteins for the genetic tape may lie in chemical history rather than in chemical kinetics. Recently, both Cech and co-workers (*24*) and Altman and co-workers (*25*) have found that RNA exhibits catalytic properties. Prior to these discoveries, several investigators (*26*) had speculated that RNA was both the original genetic and the original catalytic material; now Cech and co-workers (*27*) have strongly reinforced these early suggestions. This hypothesis is strengthened by the finding (*25, 28*) by Guerrier-Takada, Pace, Altman, and co-workers that ribonuclease-P consists of two subunits, one of which is a protein and the other RNA. Although the

catalytic subunit is surprisingly the RNA, the fact that a protein is associated with the nucleic acid and increases its catalytic efficiency suggests that ribonuclease-P is a biochemical fossil (*29*).

Perhaps the original enzymes were made of RNA and then later adsorbed peptides from the prebiotic soup, peptides that stabilized the RNA or enhanced its catalytic efficiency. Eventually the protein component, because of greater variety of catalytic groups, may have taken over the catalytic function and in all but a few cases shed the RNA. If this scenario is correct, then the reason why the genetic material consists of phosphate esters instead of amides is a historical one. Phosphate esters came first and worked well. Natural selection operated on the system and added deoxyribosides, but phosphates, which served well, were not changed. The question might be turned around to ask why proteins were necessary at all. The answer is implied in the discussion above; the greater structural variety of amino acids allowed better catalytic properties in protein enzymes than in those composed of RNA; further, a reasonable prebiotic pathway leads from one to the other.

Catalysis by RNA (*30*) was discovered in a study of the self-splicing of the nucleic acid; the reactions require an ester exchange between phosphate esters. Similar ester exchange is also of great biological importance in genetic recombination. Comparable reactions are rare with amides; the necessity for such ester exchange provides an additional reason to prefer phosphate esters over proteins for the genetic material.

Hydrolysis of RNA. Although RNA is a phosphodiester and carries a negative charge, it is relatively susceptible to hydrolysis; the rate of its spontaneous reaction with water, extrapolated to room temperature, is about 100 times greater than that of DNA (*31*). This occurs because the 2'-hydroxyl group of RNA acts as an internal nucleophile and leads to the formation of a 2',3'-cyclic phosphate with cleavage of the RNA chain (below) (*32*). Intramolecular reactions

(Where Py is either cytosine or uracil)

are generally enormously more rapid than the corresponding external ones (*33*). In this case, the cyclic phosphate that is formed is highly strained (*34*), but the cyclization proceeds nonetheless, presumably because of the large positive entropy of the cleavage reaction. The cyclic phosphate in turn readily undergoes hydrolysis. The thermodynamics of the hydrolysis of the cyclic ester depends on the strain energy that is released during cleavage of the ring, and the kinetics depend in part on that strain (*35*).

Since RNA is hydrolyzed relatively easily, it is not as well adapted as is DNA as a storage material for genetic material. The evolutionary advantage of DNA is apparent, although the evolutionary pathway to the difficult reduction process (*36*) is obscure. Some viruses contain RNA and not DNA. However, in the tight viral package, water is excluded, so that the hydrolytic instability of RNA is not relevant. Subsequently, the time for viral reproduction in a cell during which the viral RNA is exposed to water is only a few minutes. In addition, not all of the enormous number of viral particles need reproduce to preserve the organism.

Enzymatic hydrolysis. A reason then why nature prefers phosphate esters is that they are quite stable. Nevertheless, a biochemical system must not be so stable that it cannot be taken apart. DNA must be metabolized. In fact, a phosphate diester that undergoes spontaneous hydrolysis at a negligible rate can readily be cleaved by enzymatic catalysis, since enzymes can increase reaction rates by factors of 10^9 to 10^{12}. Understanding the mechanisms of enzyme action is among the important challenges to physical organic chemistry today, but from the point of view of this article the fact that enzymes exist that can hydrolyze the diesters of phosphoric acid is sufficient.

Dianions of monoesters of phosphoric acid. But another problem arises in connection with the chemistry of phosphates. The third ionization constant of phosphoric acid is smaller than the second by a factor of 10^5. Should not then the hydrolysis of the monoester dianions of phosphoric acid be slower than that of diesters by a similar factor? If dimethyl phosphate monoanion is hydrolyzed in 1N base at 110°C with a half-time of a day, will the hydrolysis of the dianion under the same conditions require 3000 years? Since, however, the second pK of phosphoric acid is about equal to physiological pH, a considerable quantity of monoanion will be present, and subject to nucleophilic attack.

Monomeric metaphosphates. Actually, hydrolysis and ester exchange for monoester dianions of phosphoric acid occur by way of a different mechanism. Monoester dianions can decompose in analogy to the S_N1 reaction of carbon chemistry to yield the monomeric metaphosphate ion PO_3^- (37). (Monomeric metaphosphoric acid, HPO_3, is an unstable chemical intermediate. The stable phosphoric acid, H_3PO_4, is called orthophosphoric acid.) In contrast to nitrate ion, which is its analog in the first row of the periodic table, PO_3^- is a powerful electrophile (38) and is unstable in water relative to dihydrogen orthophosphate by approximately 32 kcal/mol (39).

The maximum rate of hydrolysis of monoesters of phosphoric acid occurs at pH 4 (37, 40), where the major species present is the monoprotonated monoanion; the pathway postulated for its decomposition by way of monomeric metaphosphate is shown in Eq. 7.

$$CH_3OPO_3H^- \rightleftharpoons \overset{H}{\underset{+}{CH_3OPO_3^{2-}}} \rightarrow CH_3O^- + PO_3^-$$

$$PO_3^- + H_2O \rightarrow H_2PO_4^- \qquad (7)$$

This mechanism was first postulated (37) in the early 1950s, and much research has been performed in the three subsequent decades to test this idea (41). Despite some argument about details, the need for monomeric metaphosphate has been firmly established. The species has been identified in the gas phase (42) and in tertiary butyl alcohol (43), but a vigorous (if friendly) debate (44) has centered on whether the monomeric metaphosphate ion is ever truly free in water.

In phosphorylations, PO_3^- is transferred, for example, from ATP to various substrates. Nucleophilic attack has been inferred because many such reactions result in stereochemical inversion (45) at phosphorus; at present, only the solvolysis of *p*-nitrophenyl phosphate, made chiral at the phosphorus atom with ^{16}O, ^{17}O, and ^{18}O (43), is known to proceed with racemization. Paradoxically, however, the kinetics for some reactions that proceed with inversion at phosphorus resemble those of S_N1 processes (46). In the transition state for phosphorylations, metaphosphate is essentially completely formed, while the bond to the monomeric metaphosphate residue

Table 4. Rates of saponification at 35°C.

Ester	$k(M^{-1} sec^{-1})$	k, relative*
$(CH_3O)_2PO_2^-$	2.0×10^{-9}	1.0
$(CH_3O)_3P{=}O$	3.4×10^{-4}	2×10^5
$CH_3CO_2C_2H_5$	1.0×10^{-2}	5×10^6

*Rate relative to that of $(CH_3O)_2PO_2^-$.

from an incoming nucleophile has barely begun to form (44). In such a loose transition state, the monomeric metaphosphate ion is nearly, even if not completely, free. Protons are known only in the gas phase and are never free in water, yet mechanistic chemistry requires the concept of protons. The concept of monomeric metaphosphates is needed to explain how monoesters of phosphoric acid and ATP, for example, react in the absence of enzymes at measurable rates despite their negative charges, and react by enzymatic catalysis sufficiently rapidly to participate in metabolism.

Conclusions

In summary, the existence of a genetic material such as DNA requires a compound for a connecting link that is at least divalent. In order that the resulting material remain within a membrane, it should always be charged, and therefore the linking unit should have a third, ionizable group. The linkage is conveniently made by ester bonds, but, in order that the ester be hydrolytically stable, that charge should be negative and should be physically close to the ester groups. All of these conditions are met by phosphoric acid, and no alternative is obvious. Furthermore, phosphoric acid can form monoesters of organic compounds that can decompose by a mechanism other than normal nucleophilic attack, a mechanism that allows them sufficient reactivity to function in intermediary metabolism.

Finally, we can answer the question concerning the choices made by chemists and by natural selection. Chemists need to use reactive intermediates, or at least intermediates that will react at only moderately elevated temperatures. They cannot afford to use compounds as stable as the phosphate anions; phosphate dianions are poor leaving groups in S_N1 or S_N2 reactions, or in eliminations. On the other hand, biochemistry could not tolerate compounds as reactive as alkyl bromides or dialkyl sulfates; the metabolites would spontaneously hydrolyze much too rapidly. Furthermore, the reactivity of alkyl halides and sulfates is directly related to their toxicity; they alkylate and destroy essential metabolites and enzymes. Biochemistry uses phosphate ester anions that undergo slow hydrolysis in the absence of enzymes, and rapid hydrolysis in the presence of enzymes, and that are readily contained within lipid membranes. We can understand the choices made both by chemists and by the process of natural selection. They are both correct.

REFERENCES AND NOTES

1. J. Preiss and P. Handler, *J. Biol. Chem.* **233**, 493 (1958); J. G. Flaks, M. J. Erwin, J. M. Buchanan, *ibid.* **228**, 201 (1957); A. Kornberg, I. Lieberman, E. S. Simms, *ibid.* **215**, 417 (1955); S. C. Hartman and J. M. Buchanan, *ibid.* **233**, 451 (1958).
2. H. Morell, M. J. Clark, P. F. Knowles, D. B. Sprinson, *ibid.* **242**, 82 (1967).
3. K. Bloch, in *Ciba Foundation Symposium on the Biosynthesis of Isoprenoid Compounds*, G. E. W. Wolstenholm and M. O'Connor, Eds., (Churchill, London, 1959), p. 4; C. D. Poulter and H. C. Rilling, in *Biosynthesis of Isoprenoid Compounds*, J. W. Porter and S. L. Spurgeon, Eds. (Wiley, New York, 1981), vol. 1, p. 161.
4. B. D. Davis, *Arch. Biochem. Biophys.* **78**, 497 (1958).
5. K. Bloch, *Curr. Top. Cell Regul.* **18**, 289 (1981).
6. J. L. Goldstein and M. S. Brown, *J. Lipid Res.* **25**, 1450 (1984).
7. See, L. Stryer, *Biochemistry* (Freeman, San Francisco, ed. 2, 1981), p. 499, for example.
8. *Biochemical Pathways* (Boehringer, Mannheim, West Germany, ed. 3, 1978).
9. A. J. Kirby, in *Comprehensive Chemical Kinetics*, C. H. Banford and C. F. H. Tipper, Eds. (Elsevier, Amsterdam, 1972), vol. 10, p. 153.
10. The pH of the sap of a Macintosh apple is about 4.0.

11. J. R. Cox and O. B. Ramsay, *Chem. Rev.* **64**, 317 (1964).
12. N. Bjerrum, *Z. Phys. Chem.* **106**, 219 (1923).
13. J. G. Kirkwood and F. H. Westheimer, *J. Chem. Phys.* **6**, 506 (1938); F. H. Westheimer and J. G. Kirkwood, *ibid.*, p. 513; F. H. Westheimer and M. W. Shookhoff, *J. Am. Chem. Soc.* **61**, 555 (1939).
14. M. Kutake, Ed., *Constants of Organic Compounds* (Asakura, Tokyo, 1963).
15. J. Kumamoto and F. H. Westheimer, *J. Am. Chem. Soc.* **77**, 2515 (1955).
16. J. Pinnow, *Z. Electrochem.* **24**, 21 (1918).
17. F. C. Knowles and A. W. Benson, *Trends Biochem. Sci.* **8**, 178 (1983).
18. J.-M. Crafts, *Bull. Soc. Chim. France Ser. 2* **14**, 99 (1870); C. D. Baer, J. O. Edwards, P. H. Rieger, *Inorg. Chem.* **20**, 905 (1981).
19. V. E. Bel'skii and G. Z. Motygullin, *Bull. Acad. Sci. USSR* **1967**, 2427 (1967); E. T. Kaiser, M. Panar, F. H. Westheimer, *J. Am. Chem. Soc.* **85**, 602 (1963).
20. Ebelman, *Ann. Chem.* **57**, 334 (1846); R. Aelion, A. Loebel, F. Eirich, *Recl. Trav. Chim. Pays-Bas* **69**, 61 (1950); *J. Am. Chem. Soc.* **72**, 5705 (1950).
21. D. Barby, T. Griffiths, A. R. Jacques, D. Pawson, *Spec. Publ. Chem. Soc.* **31**, 320 (1977).
22. D. L. Miller and F. H. Westheimer, *J. Am. Chem. Soc.* **88**, 1507 (1966).
23. F. Lippmann, in *Phosphorus Metabolism*, W. D. McElroy and H. B. Glass, Eds. (Johns Hopkins Press, Baltimore, 1951), vol. 1, p. 521.
24. A. J. Zaug and T. R. Cech, *Science* **231**, 470 (1986).
25. C. Guerrier-Takada, K. Gardiner, T. Marsh, N. Pace, S. Altman, *Cell* **35**, 849 (1983).
26. H. B. White III, *J. Mol. Evol.* **7**, 101 (1976); C. M. Visser and R. M. Kellog, *ibid.* **11**, 163, (1978); *ibid.*, p. 171; see also L. E. Orgel, *J. Mol. Biol.* **38**, 381 (1968).
27. B. L. Bass and T. R. Cech, *Nature (London)* **308**, 820 (1984).
28. C. Guerrier-Takada and S. Altman, *Science* **223**, 285 (1984); S. Altman, *Cell* **36**, 237 (1984).
29. F. H. Westheimer, *Nature (London)* **319**, 534 (1986).
30. A. J. Zaug and T. R. Cech, *Cell* **19**, 331 (1980); *Biochemistry* **25**, 4478 (1986); T. R. Cech, *Sci. Am.* **255**, 64 (November 1986).
31. J. Eigner, H. Boedtker, G. Michaels, *Biochim. Biophys. Acta* **51**, 165 (1961).
32. F. M. Richards and H. W. Wyckoff, in *The Enzymes*, P. D. Boyer, Ed. (Academic Press, New York, ed. 3, 1971), vol. 4, p. 647.
33. A. J. Kirby, *Adv. Phys. Org. Chem.* **17**, 183 (1980).
34. J. M. Sturtevant, J. A. Gerlt, F. H. Westheimer, *J. Am. Chem. Soc.* **95**, 8168 (1973).
35. F. H. Westheimer, *Acc. Chem. Res.* **1**, 70 (1968); K. Taira, T. Fanni, D. Gorenstein, *J. Am. Chem. Soc.* **106**, 1521 (1984).
36. L. Thelander and P. Reichard, *Annu. Rev. Biochem.* **48**, 133 (1979); M. A. Ator and J.-A. Stubbe, *Biochemistry* **24**, 7214 (1985).
37. W. Butcher and F. H. Westheimer, *J. Am. Chem. Soc.* **77**, 2420 (1955); P. W. C. Barnard *et al.*, *Chem. Ind. (London)* **1955**, 760 (1955).
38. F. H. Westheimer, *Chem. Rev.* **81**, 313 (1981).
39. J. P. Guthrie, *J. Am. Chem. Soc.* **99**, 3991 (1977).
40. M. C. Bailly, *Bull. Soc. Chim. France Ser. 5* **9**, 314 (1942).
41. J. Rebek and F. Gavina, *J. Am. Chem. Soc.* **97**, 1591 (1975); D. Samuel and B. Silver, *J. Chem. Soc.* **1961**, 4321 (1961); A. J. Kirby and A. G. Varvoglis, *ibid.* **1968B**, 135 (1968); F. Ramirez and J. F. Maracek, *Tetrahedron* **35**, 1581 (1979); A. C. Satterthwait and F. H. Westheimer, *J. Am. Chem. Soc.* **103**, 1177 (1981); K. C. Calvo and F. H. Westheimer, *ibid.* **105**, 2827 (1983).
42. D. J. Harven *et al.*, *J. Am. Chem. Soc.* **101**, 7409 (1979); M. Henchman, A. A. Viggiano, J. F. Paulson, A. Freedman, J. Wormhoudt, *ibid.* **107**, 1453 (1985).
43. S. L. Buchwald, J. M. Friedman, J. R. Knowles, *ibid.* **106**, 4911 (1984); J. Friedman and J. R. Knowles, *ibid.* **107**, 6126 (1985); J. Knowles, personal communication.
44. W. P. Jencks, *Acc. Chem. Res.* **13**, 161 (1980).
45. J. R. Knowles, *Annu. Rev. Biochem.* **49**, 877 (1980); S. L. Buchwald, D. E. Hansen, A. Hassett, J. R. Knowles, *Methods Enzymol.* **87**, 279 (1982); K. C. Calvo, *J. Am. Chem. Soc.* **107**, 3690 (1985).
46. A. J. Kirby and A. G. Varvoglis, *J. Am. Chem. Soc.* **89**, 415 (1967); K. C. Calvo and F. H. Westheimer, *ibid.* **106**, 4205 (1984); K. C. Calvo and J. M. Berg, *ibid.*, p. 4202.
47. This article is based on the George Willard Wheland Lecture, presented at the University of Chicago, 14 May 1986. The author wishes to thank S. A. Benner, A. Eschenmoser, J. R. Knowles, and L. E. Orgel for their helpful suggestions.

Thermal Copolymerization of Amino Acids to a Product Resembling Protein

Attempts to produce a true proteinoid from all of the common amino acids (1) by concerted application of information now accumulated (2, 3) have yielded such materials. Chromatograms of acid hydrolyzates of the dialyzed products have displayed spots with R_f's of all of the amino acids except tryptophan, which was found in the unhydrolyzed polymer by Hopkins-Cole test. A critical feature of producing such proteinoids is employment of considerable molar excess of dicarboxylic amino acid (4).

To prepare the proteinoid, 2.0 g of L-glutamic acid was heated for 1 hr in an oil bath at 170°C, and into this melt was stirred a finely ground mixture of 2.0 g of DL-aspartic acid with 1.0 g of an amino acid mixture used for microbial assay (5). The mixture was heated for 3 hr under a blanket of CO_2 in the oil bath at 170°C. After being allowed to cool, the resultant glass was vigorously rubbed with 20 ml of water which converted the product to a granular precipitate. This was allowed to stand overnight and was then filtered and washed with 10 ml of water and 10 ml of ethanol. The solid was next washed by dialysis in a cellophane bag in an agitated water bath for 4 days. Yields, by weight, were usually much in excess of 15 percent. A chromatogram of a hydrolyzed sample of the clear soluble fraction

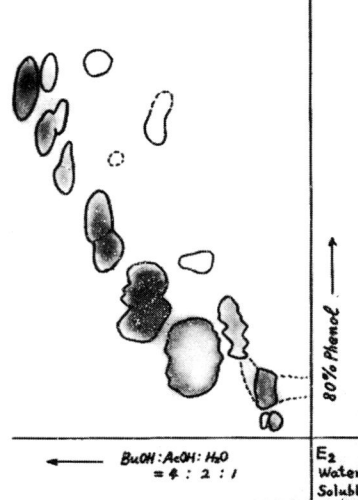

Fig. 1. Chromatogram of hydrolyzed sample of polymer of the common amino acids.

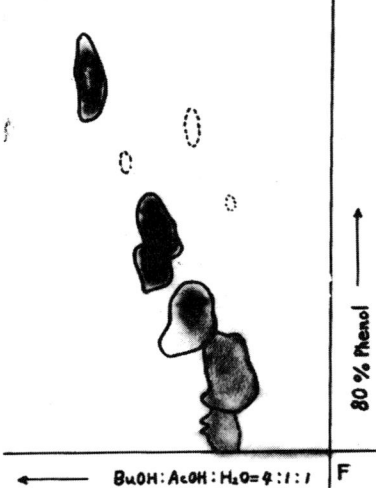

Fig. 2. Chromatogram of hydrolyzate of polymer of aspartic and glutamic acids, glycine, alanine, and leucine.

of nondiffusible proteinoid is shown in Fig. 1. Hydrolyzates of the solid had the same pattern on chromatography. The polypeptide nature of the polymer was substantiated by biuret tests and infrared analysis.

Variations on the synthesis have included replacement of glutamic acid by L-glutamine without preheating, and the added use of 85 percent phosphoric acid (6). In each of these, chromatograms similar to that shown in Fig. 1 were obtained.

The recovery of five amino acids, as shown in Fig. 2, from the same five reacted, precludes explanation of the results as bacterial contamination (7). The possibility of trapping of amino acids in the solid fraction was eliminated by a negative ninhydrin test and by chromatographic results of the hydrolyzed dialysis residue of the soluble sodium salt of the proteinoid; these chromatograms resembled that shown in Fig. 1. The possibility of difficultly soluble diketopiperazines was similarly eliminated (8). An unanswered question, however, is that of the extent to which all of the amino acids are present in each of the peptides of the presumed mixture obtained.

One of the first proteinoids showed a mean chain weight of 4900 and contained 15 percent of glutamic acid, 71 percent of aspartic acid, and 14 percent of the other amino acids by DNP assay (2). The N-terminal amino acids were 48 percent of glutamic acid, 13 percent of aspartic acid, and 39 percent of the

others; these analyses indicate a nonrandom arrangement of residues. Microbial assay by the Shankman Laboratories of Los Angeles reveals 0.2 to 2.0 percent of each of 13 amino acids, smaller proportions of serine, threonine, and cystine, and large proportions of the dicarboxylic amino acids. Studies of the effects of time, temperature, ratios of reactants, and phosphoric acid have revealed ways of increasing the percentage of basic and neutral amino acids (9).

Although these results do not prove that primordial protein was produced thermally, nor prove interpretations of biochemical origins suggested by unexpected results (10) from such experiments, the entire outlook appears worthy of more serious consideration because of these findings. It becomes possible to visualize in fuller detail an overlapping origin of anabolic reactions, enzymic protein, and genic nucleic acid (11). In the purely preparative realm, the scope for synthesis of peptides, including proteinoids, is enlarged by recognition of the possibilities from copolymerization of amino acids (12).

SIDNEY W. FOX
KAORU HARADA

Oceanographic Institute and
*Chemistry Department, Florida
State University, Tallahassee*

References and Notes

1. S. W. Fox, Am. Scientist **44**, 347 (1956).
2. K. Harada and S. W. Fox, J. Am. Chem. Soc. **80**, 2694 (1958).
3. G. D. Maier, M.S. thesis, Iowa State College (1956).
4. Probable need for molar excess of either aspartic acid or glutamic acid appeared partly in experiments by one of us (S.W.F.) at the Scripps Institution of Oceanography, University of California, during the summer of 1957. The hospitality of Dr. Denis L. Fox is gratefully acknowledged.
5. K. A. Kuiken, W. H. Norman, C. M. Lyman, F. Hale, L. Blotter, J. Biol. Chem. **151**, 615 (1943).
6. The technical assistance of Mrs. Donna Keith is acknowledged.
7. Mr. Allen Vegotsky has also carried a methanol-sterilized sample of polymer through to a result like that shown in Fig. 1.
8. We thank Mr. George W. Knight in the Laboratory of Dr. Karl Dittmer for repeating the essential synthesis and analysis from written directions.
9. A detailed description of such studies is in preparation.
10. S. W. Fox, J. E. Johnson, A. Vegotsky, Science **124**, 923 (1956); S. W. Fox, J. Chem. Educ. **34**, 472 (1957).
11. S. W. Fox, A. Vegotsky, K. Harada, P. D. Hoagland, Ann. N.Y. Acad. Sci. **69**, 328 (1957).
12. This project is supported by grant RG-4666 of the National Institutes of Health, U.S. Public Health Service, grant G-4566 of the National Science Foundation, and by the General Foods Corporation. This report is contribution No. 97 of the Oceanographic Institute, Florida State University.

20 February 1958

MONTMORILLONITE: A MULTIFUNCTIONAL MINERAL CATALYST FOR THE PREBIOLOGICAL FORMATION OF PHOSPHATE ESTERS

JAMES P. FERRIS, CHUN-HSIEN HUANG, and WILLIAM J. HAGAN, Jr.

Department of Chemistry, Rensselaer Polytechnic Institute, Troy, N.Y. 12180-3590, U.S.A.

(Received 9 July, 1987)

Abstract. Reaction of diiminosuccinonitrile (DISN) with 3'-AMP in the presence of alkali- and alkaline earth-montmorillonites results in the formation of 2',3'-cAMP in aqueous solution. Little or no 2',3'-cAMP is produced when metal ion concentrations equivalent to that of the metal ion associated with the homoionic clays are used instead of mobntmorillionite. Yields comparable to those obtained with DISN are obtained when diaminomaleonitrile (DAMN) is used in place of DISN as the condensing agent. DAMN, a compound which is more stable than DISN in aqueous solution, is oxidized to DISN on the surface of the clay by Fe^{+3} in the clay lattice. DISN, the true condensing agent, is thus generated in the presence of the bound 3'-AMP on the montmorillonite surface. The montmorillonite catalyzes the DISN-mediated formation of 2',3'-cAMP and this product, which binds much less strongly than does the 3'-AMP, is desorbed from the clay surface. This research established that the montmorillonite performs four different functions in its role as catalyst: (1) Binding one of the substrate molecules (3'-AMP) (2) Activating the second substrate (DAMN) (3) Catalyzing the formation of 2',3'-cAMP (4) Releasing the reaction product so another substrate molecules can bind to the montmorillonite.

1. Introduction

The laboratory demonstration of the oligomerization of unblocked monomers in aqueous solution using prebiological conditions is both a fundamental and recalcitrant problem in the field of chemical evolution. In none of the reported polymerizations does the proposed prebiological reaction proceed in the presence of water using reagents that may have been readily available on the primitive earth. For example, anhydrous reaction conditions are required to prepare oligopeptides from amino acids (Fox and Dose, 1972). In one approach to the synthesis of oligoribonucleotides, anhydrous heating in the presence of both a nucleotide triphosphate and the condensing agent cyanamide is required (Rao *et al.*, 1980). The condensation of activated ribonucleotides to oligonucleotides proceeds regiospecifically under mild conditions in aqueous solution, but the reaction requires the presence of a preformed RNA template and the 2-methylimidazolide derivative of the phosphate, reactants that may not have been present on the primitive earth (Inoue and Orgel, 1982). The condensation of some ribonucleotide diphosphates occurs in the absence of a template to give oligomers that are linked by the pyrophosphate bond. This reaction proceeds in aqueous solution and requires the the use of the imidazolide derivative of the phosphate grouping as the activating agent (Schwartz and Orgel, 1985; Schwartz, 1986).

In previous studies from this laboratory we reported that diiminosuccinonitrile (DISN), a compound which is formed readily by the the oxidation of a tetramer of

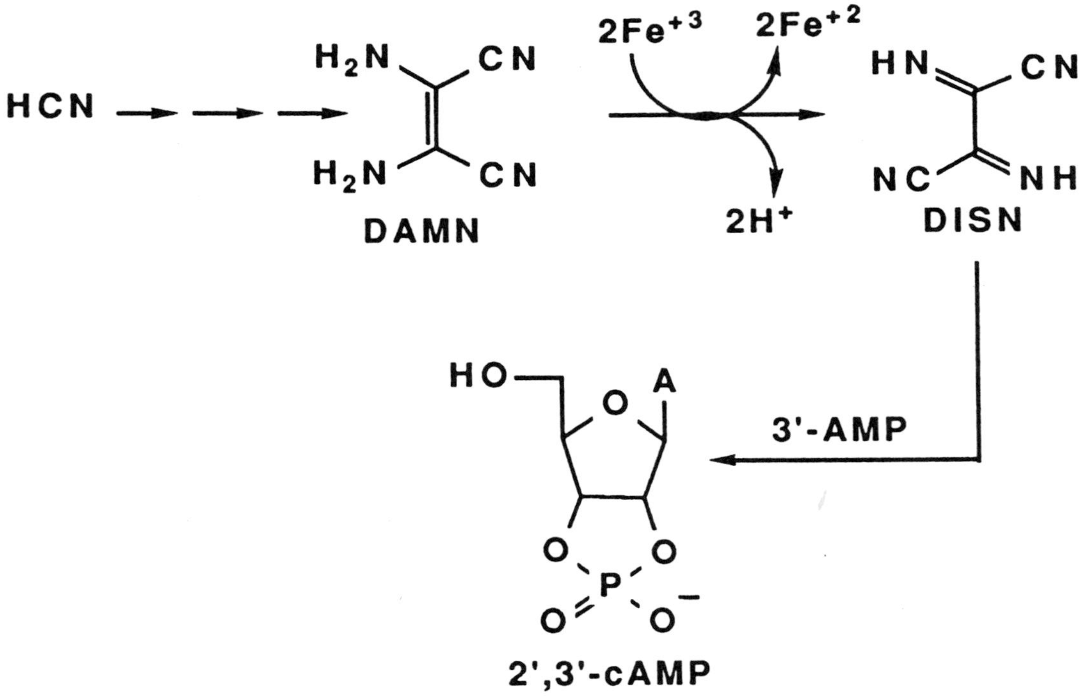

Fig. 1.   A chemical system for the formation of the phosphodiester bond in aqueous solution. HCN condenses to DAMN which in turn is oxidized to DISN. DISN is the condensing agent which effects the intramolecular formation of the phosphodiester bond in 3'-AMP to generate 2',3'-cAMP.

hydrogen cyanide (diaminomaleonitrile, DAMN) effects the cyclization of 3'-ribonucleotides to the corresponding 2',3'-cyclic nucleotides in aqueous solution (Ferris *et al.*, 1982; Ferris *et al.*, 1984) (Figure 1). This model system for the formation of the phosphodiester linkage proceeds in aqueous solution at 60°C and is catalyzed by divalent metal ions. Thus it fulfills the prerequisites of a generally applicable prebiological reaction in that it takes place in aqueous solution, it utilizes a plausible prebiotic condensing agent and the reaction proceeds at moderate temperatures(0–80°C). A major drawback of this approach is that the condensing agent (DISN) is hydrolyzed by water and consequently an excess is required to achieve modest yields of the 2',3'-cyclic nucleotide.

It was felt that montmorillonite clays may serve to bind and concentrate the nucleotides so that if it were possible to generate the DISN on the surface of the montmorillonite, it would react efficiently with the proximate bound nucleotide to generate a higher yield of the corresponding cyclic nucleotide (Ferris and Hagan, 1986). This possibility was explored using Zn^{+2}-montmorillonite since nucleotides bind to it strongly in aqueous solution (Lawless *et al.*, 1985; Ferris and Hagan, 1986). Higher yields of 2',3'-cAMP were observed when the reaction of 3'-AMP and DISN was performed in the presence of Zn^{+2}-montmorillonite, but the yields were only slightly higher than when a concentration of aqueous Zn^{+2} corresponding to that of the Zn^{+2} bound to the clay was used (24.3% versus 17.8%) (Ferris and Hagan, 1986).

If DAMN was substituted for DISN, low yields of 2′,3′-cAMP ($< 5\%$) were obtained when homoionic montmorillonites with transition metal ions as the exchangeable cation were investigated (Hagan, 1985). Since it has been reported that 5′-AMP binds to a series of homoionic montmorillonite clays (Lawless *et al.*, 1985; Banin *et al.*, 1985) it was anticipated that 3′-AMP would also bind to these clays. The effect of a series of homoionic clays on the cyclization of 3′-AMP was investigated in the present study and it was observed that the formation of 2′,3′-cyclic AMP proceeds efficiently in the presence of some alkali- and alkaline earth-montmorillonites under conditions where no reaction is observed in the absence of montmorillonite. A preliminary report of some of these findings has been published (Ferris *et al.*, 1986).

2. Experimental*

A. MATERIALS AND METHODS

The materials and methods used in this research were described in Ferris and Hagan (1986). All high performance liquid chromatography (HPLC) was performed with a Waters Microbondapak C-18 column.

B. REACTIONS

1. *General Procedure for the Formation of 2′,3′-cAMP*

A 2 ml solution of 0.05M 3′-AMP in 0.2M PIPES (pH 6.35) was added to a 8 ml screw cap test tube containing 200 mg of montmorillonite and the mixture was shaken on a Vortex mixer for 30 sec and allowed to stand at room temperature for 16−20 h. This mixture was centrifuged and 40 μl of the supernatant was diluted to 10 ml and the UV absorbance at 260 nm was measured. The extent of 3′-AMP binding to the montmorillonite was determined by the decrease in absorbance when compared with the absorbance of a comparable dilution of the stock solution. To the 3′-AMP and montmorillonite mixture was added 2 ml of either 12.2 mM DAMN (final concentration 0.0061 M) of 9.44 mM DISN (final concentration 0.0047 M) in 0.2M PIPES. The final concentration of 3′-AMP was 0.025 M. The mixture was shaken for 30 sec on a Vortex mixer and heated at 60°C for 3 h. The mixture was then cooled to room temperature, the pH measured, 4 ml of 0.2 M EDTA was added and then the tube was shaken on the Vortex mixer. After 20 min the mixture was centrifuged, a portion of the supernatant was filtered and 1 ml of the filtrate was diluted to 10 ml with distilled water. The diluted sample (40 μl) was used for HPLC analysis (Ferris *et al.*, 1984).

---

* Abbreviations: DAMN, diaminomaleonitrile; DISN, diiminosuccinonitrile; 3′-AMP, 3′-adenylic acid; 5′-AMP, 5-adenylic acid; 2′,3′-cAMP, 2′,3′-cyclic adenylic acid; ApAp, adenyl-3′, 5′-adenosine-3′-phosphate; (Ap$_4$), adenyl-3′, 5′-adenyl-3′, 5′-adenyl-3′, 5′-adenosine-3′-phosphate; PIPES; 1-4-piperazinebis(ethanesulfonic acid).

TABLE I

Decomposition of DAMN on native Montmorillonite[a]

Native Montmorillonite (mg)	DAMN		Loss	
	0.2M PIPES		No buffer	
	Final pH	(%)	Final pH	(%)
0	6.5	3.4	5.7	4.7
50	6.5	28.9	5.5	24.6
100	6.4	42.7	5.2	30.7
150	6.4	51.5	5.1	36.8
200	6.3	59.9	5.0	43.1

[a] The analysis of DAMN was performed by HPLC. The reaction conditions were the same as those described in the general experimental procedure except no 3'-AMP was used.

3. Results and Discussion

Reaction variables including the concentrations of 3'-AMP, DAMN and DISN, the quantity of clay and the presence or absence of 0.2 M PIPES buffer were varied in a systematic fashion to determine the factors which affect the yield of 2',3'-cAMP. Studies on the decomposition of DAMN by the Fe^{+3} located in the aluminosilicate lattice of the native montmorillonite, i.e., the mineral sample that is only processed by grinding and washing with distilled water before use, showed that the oxidation proceeded more efficiently in the presence of 0.2 M PIPES buffer than with no buffer (Table I). It has not been determined whether this phenomenon is due to the pH variation in the absence of buffer or whether PIPES increases the accessibility of interlamellar surfaces (Ferris and Hagan, 1986).

The yield of 2',3'-cAMP was found to be directly dependent on the amount of Na^+-montomorillonite used (Table II). It was decided to use 200 mg of montmorillonite because it gave a yield of 2',3'-cAMP that could be accurately measured by HPLC. Previous studies established that in the presence of 0.2 M PIPES 90 μmole of DAMN are oxidized per gram of native montmorillonite (Ferris and Hagan, 1986), so the 200 mg. of clay should oxidize 18 μmole of DAMN. The extent of DAMN oxidation is less than this value when 200 mg of montmorillonite was used, but it was greater than the expected values of 9 and 4.5 μmole when 100 and 50 mg. of clay were used respectively (Table II). A comparable amount of DISN was used (18.8 μmole) to facilitate the comparison between DAMN and DISN as condensing agents.

The yield of 2',3'-cAMP was not changed appreciably by the presence of 0.2 M PIPES buffer when DISN was used directly as the condensing agent (Table II). A 3.3-fold higher yield of the cyclic AMP was observed in the presence of PIPES when DAMN was used as the condensing agent. It can be seen from the data in Table I that the decomposition of DAMN is slightly greater in the presence of 0.2M PIPES but the difference is not a factor of 3.3 in magnitude.

The extent of the catalysis by various homoionic montmorillonites was then inves-

TABLE II

Variation in 2′,3′-cAMP yield with the amount of Na⁺-Montmorillonite[a]

Na⁺ Mont-morillonite (mg)	3′-AMP Adsorbed (μmol)	Final pH	DAMN Reacted (μmol)	DISN Reacted (μmol)	Yield 2′,3′-cAMP (%)[b]
50	4.4	6.4	9.7	–	2.3
50	3.7	6.4	–	18.8	5.1
100	8.0	6.4	12.5	–	4.5
100	7.8	6.4	–	18.8	9.0
200	13.9	6.3	15.9	–	9.8
200	14.8	6.3	–	18.8	16.3
200	8.2	6.0	15.3	–	3.0[c]
200	9.4	6.2	–	18.8	11.5[c]

[a] The reaction conditions are described in the general experimental procedure using 100 μmol of 3′AMP except the amount of Na⁺-montmorillonite was varied.
[b] Yield based on DAMN or DISN reacted.
[c] Reactions performed in the absence of PIPES.

tigated. The best yields of 2′3,′-cAMP were obtained using Na⁺-and Ca⁺²-montmorillonite (Table III). The transition metal montmorillonites bind 3′-AMP more strongly than the alkali and alkaline earth derivatives of the clay but give lower yields of 2′,3′-cAMP. The yield of 2′,3′-cAMP obtained with the native clay is comparable with that obtained using Na⁺-or Ca⁺²-montmorillonite, a finding consistent with the report that this smectite has Na⁺ and Ca⁺² as its principal exchangeable cations (American Petroleum Institute, 1951; Ferris et al., 1979). It was established that the formation of 2′,3′-cAMP was not catalyzed by the the exchangeable cation associated with the montmorillonite. In reactions performed under the same conditions except the soluble metal ion was used in place of the metal ion-montmorillonite complex (Table V), no 2′,3′-cAMP was detected except for the small yields observed using Ca⁺² and Cu⁺². The virtual absence of metal ion catalysis clearly established the catalytic role of the clay surface in the formation of 2′,3′-cAMP. This catalytic effect is much greater than was observed previously with Zn⁺²-montmorillonite (Ferris and Hagan, 1986).

The most extensive mineral catalysis was observed with homoionic montmorillonites in which Na⁺, K⁺ and Ca⁺² are the exchangeable cations. These clays are the least effective in binding the 3′-AMP. Up to this point in our research the strategy had been to equate potential catalysis with strong binding so minerals were screened on the basis of their affinity for organic reactants. The transition metal montmorillonites, which are the most effective in binding nucleotides, may indeed be good catalysts for the formation of the phosphate ester bond but they may also form complexes with DISN (Ferris et al., 1984) and catalyze the hydrolysis of DISN or the DISN-AMP adduct. This is one possible explanation of the low yields of 2′,3′-cAMP using transition metal montmorillonites.

The nature of the binding of 3′-AMP to montmorillonite is not clear. It was con-

TABLE III

Reaction of 3′-AMP and DISN in the presence of Montmorillonite[a]

Homoionic Clay	3′-AMP (μmol)	Final pH	3′-AMP Adsorbed (μmol)	2′,3′-cAMP[b] (%)
Native	10	6.25	6.1	1.1
	50	6.30	18.1	6.8
	50	6.30	18.2	6.6
	100	6.25	22.9	10.5
	100	6.25	22.4	11.7
	200	6.36	29.0	14.0
Na^+	10	6.25	6.9	1.6
	50	6.28	12.7	8.9
	100	6.30	14.8	16.3
	200	6.30	16.6	21.7
K^+	10	6.05	7.7	1.3
	100	6.10	20.3	16.3
Ca^{+2}	50	6.40	17.4	9.4
	100	6.36	24.3	15.7
	200	6.40	26.2	21.3
Fe^{+3}	50	6.20	35.3	3.3
	100	6.12	43.6	7.9
Cu^{+2}	100	6.14	52.7	3.0
Zn^{+2}	50	6.30	46.8	0.7
	100	6.36	68.5	6.2
	200	6.36	75.4	13.2

[a] The general procedure outlined in the Experimental Section was used except the quantity of 3′-AMP was varied.
[b] Yield based on DISN added, except when 10 μmol 3′-AMP was used and then it is based on 3′-AMP.

cluded that the binding of 5′-AMP to montmorillonite in the pH 4.5−5.5 range is by the interaction of the negatively charged phosphate group to the positive charges at the edges of the clay particles or by ligand exchange with coordinated water (Lawless *et al.*, 1985; Banin *et al.*, 1985). The nature of the binding to homoionic montmorillonites in the pH 6−7 range used in the present study was not investigated.

The actual mechanism of the montmorillonite catalysis of 2′,3′-cAMP formation has not been established but two postulates have been formulated on the basis of the available data. The first is based on the observation that the yields of 2′,3′-cAMP in the solution phase reactions are very low, suggesting that the rate of the bimolecular reaction between DISN and 3′-AMP is slow relative to the hydrolysis of DISN. Since the clay-mediated formation of 2′,3′-cAMP proceeds under conditions where no cyclic nucleotide can be detected in the absence of montmorillonite, the clay en-

TABLE IV

Reaction of 3'-AMP and DAMN in the presence of montmorillonite[a]

Montmorillonite	3'-AMP (μmol)	Final pH	3'-AMP Adsorbed (μmol)	DAMN Reacted[b] (μmol)	2'3'-cAMP[c] (%)
Native	10	6.10	7.5	17.4	0.6
	50	6.24	20.5	15.4	5.2
	50	6.24	20.5	15.4	5.3
	100	6.23	26.5	15.1	10.2
	100	6.22	26.9	15.1	9.5
Na$^+$	10	6.24	6.9	15.7	0.8
	50	6.26	12.6	16.0	4.8
	100	6.26	13.9	15.9	9.8
	200	6.26	16.0	14.0	20.3
K$^+$	10	6.05	7.6	15.5	0.5
	100	6.08	20.3	15.2	8.2
Ca^{+2}	10	5.90	6.9	d	0.8
	50	6.40	17.0	20.6	4.7
	100	6.35	23.7	18.6	8.3
	200	6.35	28.0	14.2	8.6
Fe^{+3}	50	6.20	35.4	13.7	4.5
	100	6.12	43.5	15.4	10.4
Cu^{+2}	100	6.15	52.4	26.0	4.4
Zn^{+2}	10	6.02	9.3	d	0
	50	6.28	48.3	17.4	0
	100	6.30	66.2	17.7	2.5
	200	6.32	73.0	17.8	5.6

[a] Reaction performed as described in the general procedure in the Experimental except the 3'-AMP concentration was varied.

[b] Unreacted DAMN determined by HPLC analysis.

[c] Yield based on DAMN reacted, except when 10 μmol 3'-AMP is used and then it is based on 3'-AMP.

[d] Not determined.

hances the reaction between DISN and 3'-AMP. The enhanced reaction rate may be due to the concentration of the reactants on the mineral surface. While the adsorption of 3'-AMP on montmorillonite has been measured, the corresponding binding of DISN has yet to be established.

Catalysis of the cyclization of the 3'-AMP-DISN adduct is a second explanation for the role of montmorillonite. The precedent for this explanation comes from our observation that divalent metal ions enhance the yield of 2',3'-cAMP in the reaction of DISN or BrCN with 3'-AMP (Ferris et al., 1984). It was proposed that the divalent cations bind to the 3'-AMP-DISN adduct thus shielding the negative charge on the phosphate group and enhancing its electrophilic character. The rate of the attack of the 2'-hydroxyl group is greater because of the increased electrophilicity of the phos-

TABLE V

Metal ion catalysis of the DISN-mediated formation of
2′,3′-cAMP[a]

Metal Ion[b]	Yield 2′,3′-cAMP (%)
–	0
Na^+	0
Ca^{+2}	3.4
K^+	0
Fe^{+3}	0[b]
Zn^{+2}	0[c]
Cu^{+2}	3.9[c]

[a] Yield based on DISN as the limiting reagent. An amount of metal ion equivalent to the amount of exchangeable cation associated with 200 mg of homoionic montmorillonite was used, 0.0175 M for M^{+2} and 0.035 M for M^{+1}. The reaction conditions were the same as those outlined in the general procedure for the formation of 2′,3′-cAMP.

[b] A precipitate formed which did not disssolve upon addition of EDTA. The recovery of 3′-AMP was 36% indicating loss by adsorption on the precipitate.

[c] A precipitate formed which dissolved on adding EDTA after termination of the reaction.

phate group. A similar enhanced rate of cyclization of 2′,3′-cAMP is possible at the edges of the montmorillonite where exposed Al^{+3} sites may serve as Lewis acid catalysts (Paecht-Horowitz, 1976). Thus the cyclization step will proceed more rapidly at the mineral edge than it does in solution. Although the two mechanistic proposals are not fully established and do not preclude the possibility of other reaction intermediates, these models do suggest two ways that clay minerals may have catalyzed the formation of the phosphate ester bond on the primitive earth.

The yields of 2′,3′-cAMP obtained when DAMN was used as the condensing agent generally parallel those obtained with DISN with the exception of Fe^{+3}-montmorillonite (Table IV). The montmorillonites with alkali metal exchangeable cations gave high yields of 2′,3′-cAMP. The comparable trend in the yields confirms our previous conclusion the DAMN is oxidized to DISN by the clay. DAMN may be subject to other reaction processes as well since the product yield is slightly less than when it is the condensing agent.

The more extensive catalysis observed with the Fe^{+3}-montmorillonite using DAMN as the condensing agent as compared with DISN (Tables III and IV) is difficult to understand. It is not due to the generation of larger amounts of DISN by the exchangeable Fe^{+3} because as can be seen in Table IV the amount of recovered DAMN is about the same with every montmorillonite except Cu^{+2}. From these data we conclude that the exchangeable Fe^{+3} is not reacting with the DAMN and the only oxidant is the Fe^{+3} bound in the lattice of the montmorillonite. This finding indicates that the oxidation potential of the exchangeable Fe^{+3} is less than

TABLE VI

AMP binding to Montmorillonite clays[a]

Homoionic Clay	AMP Adsorption (%)	Soluble 2',3'-cAMP (%)	Increase upon EDTA Wash (%)		
			2',3'-cAMP	2'-AMP	3'-AMP
none	–	100	–	–	–
Na$^+$	12	81	– 2.6	2.7	3.9
Native	17	78	1.3	1.9	3.2
Ca^{2+}	22	71	3.2	2.4	3.9
Zn^{2+}	36	60	– 1.4	11.1	16.7

[a] A mixture of 2',3'-cAMP (0.025 M) and montmorillonite (100 mg) in 1 ml of 0.2 M PIPES (pH 6.4) was allowed to stand at room temperature for 24 h and the mixture was then centrifuged. The extent of adsorption (column 2) represents the difference between the initial concentration of 2',3'-cAMP and the sum of the supernatant concentrations of 2',3'-cAMP, 2'-AMP and 3'-AMP determined by HPLC analysis of a 0.3 ml aliquot that was diluted with 0.3 ml of 0.2 M PIPES and 0.6 ml of 0.2 M EDTA. The percent 2',3'-cAMP in the supernatant is given in column 3. The clay and the supernatant were again mixed, 1 ml of 0.2 M PIPES buffer and 2 ml of 0.2 M EDTA were added, and the mixture was centrifuged. The amount of each AMP isomer was determined by HPLC and the increase after EDTA treatmet is given in columns 4–6. About 90% of the initial 2',3'-cAMP was accounted for after EDTA treatment.

that of the lattice-bound Fe^{+3}. This is not a surprising result because there are theoretical (Aronowitz et al., 1982) and experimental studies which are in agreement with the observation that lattice-bound Fe^{+3} is a stronger oxidizing agent than Fe^{+3} in solution or bound to an ion exchange resin. What is usual is the observation that Fe^{+3} which is cation-exchanged to montmorillonite is a weaker oxidant than soluble Fe^{+3} or Fe^{+3} bound to an ion-exchange resin. In our previous studies we demonstrated that both of the latter forms of Fe^{+3} oxidized DAMN to DISN (Ferris et al., 1982) but our present studies suggest that the exchangeable Fe^{+3} associated with montmorillonite does not. The reason for this diminished reactivity is not known but it may have been an important factor on the primitive earth where some organic sediments may not have been oxidized by the Fe^{+3} bound to minerals.

Binding studies established that 2',3'-cAMP is not adsorbed on to montmorillonite under the reaction conditions used in this research. The investigation of 2',3'-cAMP binding is complicated by its montmorillonite-catalyzed hydrolysis (Table VI). It is clear that 3'-AMP binds more strongly to Zn^{+2}-montmorillonite than does 2',3'-cAMP when the extent of adsorption in Tables III and VI respectively is compared. While the extent of adsorption of AMP isomers on Na$^+$-, Ca^{+2}-, and native montmorillonite given in Table VI is comparable to that for 3'-AMP given in Table III, the amount of 2',3'-cAMP bound to the clay is less than the total quantity of nucleotide adsorbed as shown by the elution of greater amounts of 2- and 3'-AMP on treatment with EDTA (Table VI).

The absence of 2'-AMP as a reaction product (yield less than 0.2% with Na$^+$, Ca^{+2}-, and native-montmorillonite) in the experiments reported in Table III is further evidence for the facile desorption of 2',3'-cAMP from the clay. The binding studies in Table VI demonstrate that montmorillonite catalyzes the hydrolysis of

2',3'-cAMP and that in the absence of montmorillonite no hydrolysis takes place. Hydrolysis of 2',3'-cAMP is known to give a 1:1.5 ratio of 2'- and 3'-AMP respectively (Usher and Yee, 1979) so the failure to observe 2'-AMP establishes that 2',3'-cAMP does not bind to montmorillonite. The stronger binding of 3'-AMP thus serves to protect the product from hydrolysis by blocking adsorption sites that would otherwise be available to catalyze the breakdown of the 2',3'-cAMP.

4. Minerals, HCN, and the Prebiotic Formation of Phosphate esters

The possible role of clays in processes leading to the origins of life has been a subject of conjecture since 1947 when Bernal first suggested in a lecture that clays may have served to concentrate and orient small molecules for subsequent reactions and then protect the intercalated reaction products from degradation (Bernal, 1949). This concept has been extended by Cairns-Smith (1982) in his postulate that life processes were originally effected by clays and that contemporary organic compounds took over the 'clay life' at a later evolutionary stage. There are no experimental reports of clay catalysis in prebiological systems that approach the chemical complexity proposed by Bernal and Cairns-Smith. The polymerization of alanine adenylate on sodium montmorillonite has been reported (Paecht-Horotwitz, 1976) but there has been one report where it was not possible to reproduce these findings (Brack, 1976). It has not been possible to repeat the synthesis of the amino acid adenylates using the zeolite Decalso F as catalyst (Warden *et al.*, 1974).

In the montmorillonite-catalyzed reaction reported herein the smectite exhibits many of the characteristics of an enzyme (Figure 2). It binds one of the substrate molecules (3'-AMP) (Step A), activates the masked form of the condensing agent (oxidizes DAMN to DISN) (Step B), it catalyzes the reaction (Step C) between DISN and 3'-AMP, and the product formed (2',3'-cAMP) is desorbed from the catalytic surface. These findings provide direct experimental support for the postulate that minerals may have served as prototypical enzymes on the primitive earth. It might be argued that this system does not model the catalysis observed with enzymes because the Fe^{+2} generated in the montmorillonite lattice is not converted back to Fe^{+3} in the course of these transformations. However, this conversion is likely to have occurred on the primitive earth by the photochemical expulsion of an electron from the Fe^{+2} (Braterman *et al.*, 1983) or by the oxidation of Fe^{+2} by photochemically generated radicals (Banin and Rishpon, 1979), thereby completing the reaction cycle.

It is significant that catalysis is observed with the unprocessed (native) montmorillonite. While it is not claimed that this particular montmorillonite was present on the primitive earth, it is reassuring to observe that the laboratory procedures used for the generation of homoionic species were not essential for catalysis. That catalysis was observed with Na^+- and Ca^{+2}-montmorillonite is also significant. Since nucleotides bind more strongly to montmorillonites in which transition metals are the exchangeable cations (Lawless *et al.*, 1985; Ferris and Hagan, 1986), it was assumed that such

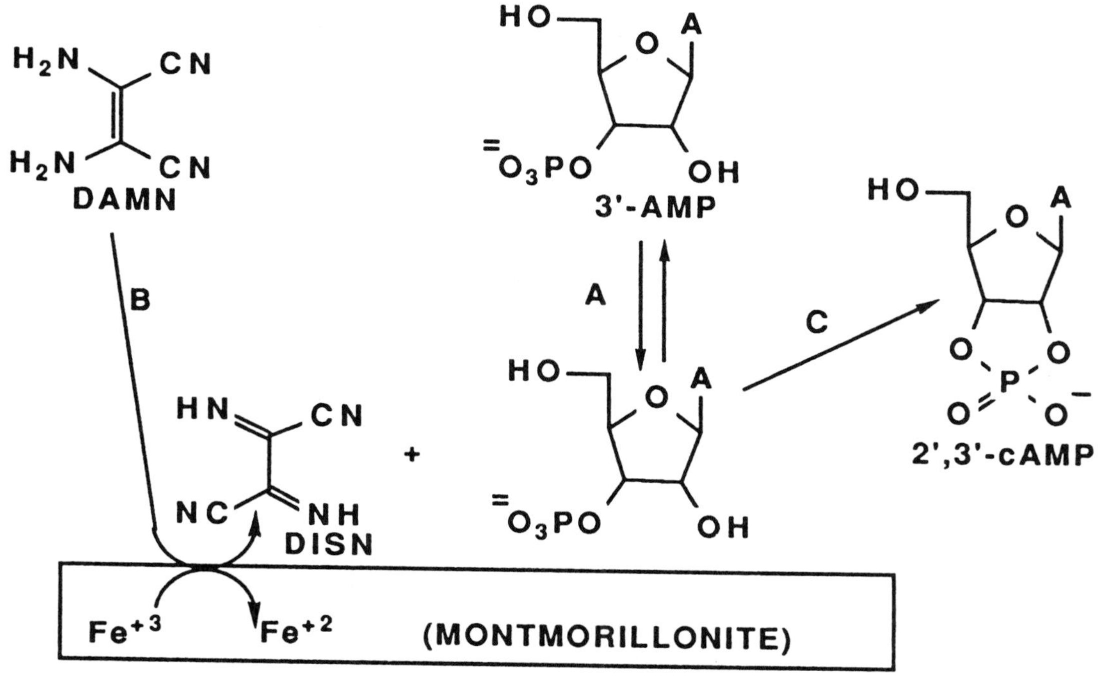

Fig. 2. Montmorillonite catalysis of the solution phase process shown in Figure 1. A. The reversible binding of 3'-AMP to montmorillonite. B. DAMN reacts with lattice-bound Fe^{+3} in montmorillonite to generate DISN proximate to 3'-AMP. C. Montmorillonite catalysis of the reaction of DISN and 3'-AMP to generate 2',3'-cAMP which is released from the montmorillonite surface.

clays would be better catalysts. While it is not difficult to imagine primitive earth environments where transition metal ions constituted an appreciable concentration of the exchangeable cations, the prevalence of these environments was undoubtedly much less than those where the alkali and alkaline earth cations predominated.

These studies do not establish that montmorillonites were the most efficient or the only mineral catalysts on the primitive earth. Insoluble phosphates (Miller and Parrish, 1964: Acevedo and Orgel, 1986), iron and manganese carbonates (Joe et al., 1986) and iron oxide-hydroxides (Holm, 1985) are a few of the other classes of minerals which are currently being investigated. But the present work does provide experimental verification for the Bernal postulate (1949) that minerals may have had a central role in primitive earth processes.

The use of DISN as a condensing agent follows directly from our studies on the prebiotic chemistry of HCN. In these earlier studies it was established that representatives of the three major classes of nitrogen-containing biological molecules, purines, pyrimidines and amino acids, may be produced directly from HCN in a 'one-pot' reaction (Ferris and Hagan, 1984 and references therein). The central intermediate in all this chemistry is DAMN, a compound which forms directly from HCN in mildly alkaline solutions (Sanchez et al., 1967). The oxidation of DAMN to DISN by Fe^{+3} and other oxidants is a well-established chemical reaction, so if it is assumed that HCN was the source of nitrogen-containing biomolecules on the primitive earth, then it follows that DISN was also formed in the presence of mild oxidants.

The principal thrust of the present research was to establish whether montmorillonite will catalyze the formation of the phosphate ester bond. However, the formation of the 2',3'-cAMP does constitute a pathway to RNA oligomers since it has been possible to effect the polymerization of this monomer (Verlander *et al.*, 1973; Verlander and Orgel, 1974). As the long-range goal of the research is to explore the clay-catalyzed oligomerization of nucleotides we investigated the polymerization of ApAp in the presence of Zn^{+2}-montmorillinote (C.-H. Huang, unpublished results from this laboratory). The objective of the study was to see if was possible to form higher oligomers using reaction conditions where there was an ApAp associated with each Zn^{+2} cation of the montmorillonite. It can be calculated from the charge: surface area ratio of the montmorillonite that each Zn^{+2} would be, on the average, separated by 14–22 Å. In the interlayer between two clay platelets, the Zn^{+2} cations associated with both platelets would be separated by 10–15.5 Å. Inspection of a molecular model of ApAp indicated that the distance between its 3'-phosphate and 5'-OH groupings is about 11 Å so some ApAp molecules bound to the montmorillonite may be in close proximity. However, no $(Ap)_4$ was detected upon addition of DISN or 1-(3-dimenthylaminopropyl)-3-ethylcarbodiimide to a solution of ApAp and a suspension of the nucleotide bound to montmorillonite. The only new product that was detected had an HPLC retention time close to that of ApAp and it was assumed to be the corresponding cyclic phosphate (ApA-2',3'-cAMP). The facile intramolecular cyclization of the 2'-hydroxyl group proximate to the phosphate probably proceeds much more rapidly than the bimolecular reaction with another nucleotide even though the nucleotide is bound on a adjacent site on the clay surface. Currently studies are in progress using 5'-AMP, since it is expected that the formation of 3',5'-cAMP will not be as facile as that of the formation of 2',3'-cAMP from 3'-AMP and thus it may be possible to observe the formation of oligomers of pA.

Acknowledgements

This research was supported by NSF grant CHE−85−06377. The HPLC equipment used was purchased with funds from NASA grant (NGR3−018−148).

References

Acevedo, O. and Orgel, L. E.: 1986, *Origins of Life* **16**, 441.

American Petroleum Institute: 1951, Clay Mineral Standards. American Petroleum Institute, Project 49, Preliminary Report 7B, Chemical Analysis, Columbia University, New York.

Aronowitz, S., Coyne, L., Lawless, J., and Rishpon, J.: 1982, *Inorg. Chem.* **21**, 3589.

Banin, A. and Rishpon, J.: 1979, *J. Mol. Evol.* **14**, 133.

Banin, A., Lawless, J. G., Mazzurco, J., Church, F. M., Margulies, L., and Orenberg, J. B.: 1985, *Origins of Life* **15**, 89.

Bernal, J. D.: 1949, *Proc. Roy. Soc.* **62A**, 537.

Brack, A. L.: 1976, *Clay Minerals* **11**, 117.

Braterman, P. S., Cairns-Smith, A. G., and Sloper, R.: 1983, *Nature* **303**, 163.

Cairns-Smith, A. G.: 1982, *'Genetic Takeover'*, Cambridge University Press, Cambridge, Great Britain.

Ferris, J. P., Edelson, E. H., Mount, N. M., and Sullivan, A. E.: 1979, *J. Mol. Evol.* **13**, 317.

Ferris, J. P., Hagan, W. J., Jr., Alwis, K. W., and McCrea, J.: 1982, *J. Mol. Evol.* **18**, 304.

Ferris, J. P. and Hagan, W. J.: 1984, *Tetrahedron* **40**, 1093.

Ferris, J. P., Yanagawa, H., Dudgeon, P. A., Hagan, W. J., Jr., and Mallare, T. E.: 1984, *Origins of Life* **15**, 29.

Ferris, J. P. and Hagan, W. J.: 1986, *Origins of Life* **17**, 69.

Ferris, J. P., Huang, C.-H., and Hagan, W. J., Jr.: 1986, *Origins of Life* **16**, 473.

Fox, S. W. and Dose, K.: 1972, *Molecular Evolution and the Origin of Life*, Marcel Dekker, New York.

Hagan, W. J., Jr.: 1985, Ph.D. Thesis, RPI.

Holm, N. G.: 1985, *Origins of Life* **15**, 131.

Inoue, T. and Orgel, L. E.: 1982, *J. Mol. Biol.* **162**, 201.

Joe, H., Kuma, K., Paplawsky, W., Rea, B., and Arrhenius, G.: 1986, *Origins of Life* **16**, 369.

Lawless, J. G., Banin, A., Church, F. M., Mazzurco, J., Huff, R., Kao, J., Cook, A., Lowe, T., Orenberg, J. B., and Edelson, E.: 1985, *Origins of Life* **15**, 77.

Miller, S. L. and Parris, M.: 1964, *Nature* **204**, 1248.

Paecht-Horowitz, M.: 1976, *Origins of Life* **7**, 369, and references therein.

Rao, M., Odom, D. G., and Oro, J.: 1980, *J. Mol. Evol.* **15**, 317.

Sanchez, R. A., Ferris, J. P., and Orgel, L. E.: 1967, *J. Mol. Biol.* **30**, 223.

Schwartz, A. W. and Orgel, L. E.: 1985, *J. Mol. Evol.* **21**, 299.

Schwartz, A. W.: 1986, *Origins of Life* **16**, 444.

Usher, D. A. and Yee, D.: 1979, *J. Mol. Evol.* **13**, 287.

Verlander, M. S., Lohrman, R., and Orgel, L. E.: 1973, *J. Mol. Evol.* **2**, 303.

Verlander, M. S. and Orgel, L. E.: 1974, *J. Mol. Evol.* **3**, 115.

Warden, J. T., McCullough, J. J., Lemmon, R. M., and Calvin, M.: 1974, *J. Mol. Evol.* **4**, 189.

Synthesis of phospholipids and membranes in prebiotic conditions

IT is generally agreed that stable membranes were prerequisite to the assembly of the earliest self-replicating systems[1-4]. Phospholipids, which are ubiquitous in biological membranes and which self-assemble in aqueous environments into stable lipid bilayers and vesicles[4], are obvious candidates for prebiotic membrane components. We report here the abiotic synthesis of various lipids, including membranogenic phospholipids.

Our rationale in establishing simulated primaeval conditions included considerations of possible precursors, energy sources, catalysts and the behaviour and stability of the reactants and products. Derivatives of 1,2-diacylglycerophosphate are common membrane lipids, suggesting fatty acids, glycerol and phosphate as precursors. Heat derived from the Sun was a major primaeval source of energy[5], and recent evidence indicates that average temperate zone surface temperatures may have reached 70 °C before 3×10^9 yr ago[6]. The abiotic synthesis of polypeptides (molecular weight > 1,600) at 65 °C, in the absence of water or catalysts, has been detected in amino acid mixtures after incubation times of 40 d (ref. 7); we selected similar conditions for our experiments. Although various clays such as kaolin (an aluminium silicate) have been found to promote condensation reactions in aqueous conditions[8,9], this mineral is rare in Precambrian sediments[10]. However, silicon oxide or silica is present in nearly all rocks[11], and silicate rocks have been shown to promote condensations in certain conditions[12]. We tested the catalytic properties of silicic acid (silica gel H) and kaolin as possible models for natural silica-containing sands and clays. (Since completing this work, we have learned that washed and ignited sea sand promotes polypeptide formation when heated with mixtures of amino acid[13].) Dicyanamide has been widely used as a possible prebiotic condensing agent, and as it has been shown to catalyse glycerol phosphorylation in acid solutions[5], it was included in some experiments. Our general rationale was that tide pools containing dilute solutions of glycerol, phosphate and cyanamide, and having surface films of various hydrocarbon derivatives, could evaporate and produce mixtures of hydrophobic and hydrophilic compounds within a hot sand or clay matrix.

To synthesise neutral lipids, ^{14}C-glycerol and various C12 lipids were mixed in duplicate vessels and the solvent was removed by vacuum or by evaporation during subsequent incubation at 65 °C. Relatively short chain, saturated lipids were used as precursors for glycerolipids because they are fluid at 20–46 °C and are not labile to oxidation within the chain. In some cases, prewashed kaolin or silicic acid was added before the removal of the solvent. After drying, all samples were incubated for 1 week. Lipids were extracted and chromatographed on thin-layer plates of silica gel, placed over X-ray film for autoradiography, and the radioactive spots were scraped for scintillation counting. Typical autoradiographic patterns are shown in Fig. 1. Glycerol incubated alone showed only a single spot at the origin. Monoglyceride was the major reaction product from fatty acid, and was also prominent in the fatty aldehyde samples. Also tentatively identified were 1,3- and 1,2-diglycerides, and small amounts of triglycerides and several unknown lipids, possibly long chain cyclic acetals of glycerol. Total yields of glycerolipids were 3.1% from fatty acid and 7.7% from fatty aldehyde. Fatty alcohol formed little, if any, monoglyceride or other glycerolipid in the absence of added agents. With kaolin, however, detectable amounts of labelled lipids (0.2% yield) were formed.

Figure 2 compares the net yield of labelled lipid in the presence and absence of kaolin and silica gel. Surprisingly, kaolin seems to inhibit the formation of labelled lipids from fatty acid, but has much less effect on the reactivity of

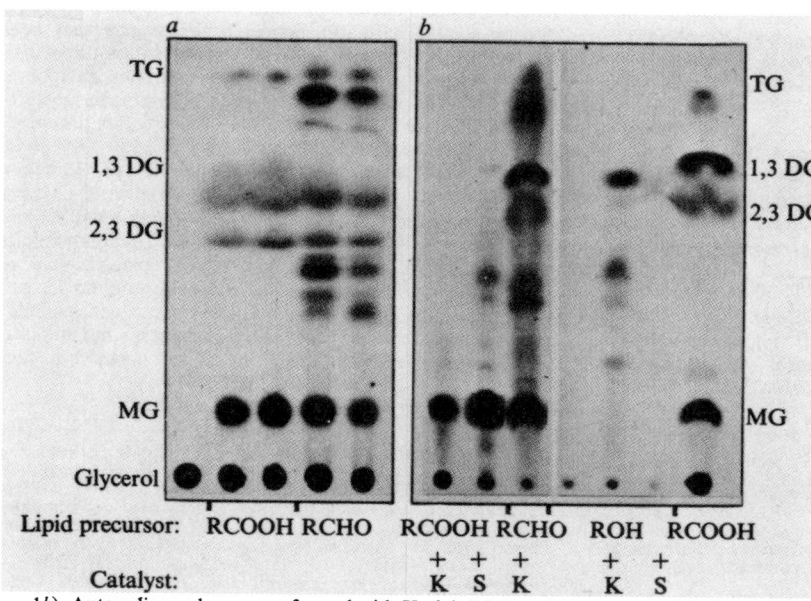

Lipid precursor: RCOOH RCHO RCOOH RCHO ROH RCOOH
Catalyst: + + + + +
  K S K K S

Fig. 1 Autoradiograms of ^{14}C-glycerolipids including mono- (MG), di- (DG), and triglycerides (TG). 50 nmol ^{14}C-glycerol (2×10^6 d.p.m., ICN, Irvine) and 250 nmol C12 acid (RCOOH, Sigma), alcohol (ROH, Sigma), or aldehyde (RCHO, Aldrich) were placed in autoclaved microreaction vessels with loose fitting screw tops. In some cases, 10 mg kaolin (K, J. T. Baker, Phillipsburg) or silica gel H (S, Sigma), prewashed with acetone and chloroform-methanol, was added to the vessel before the removal of the organic solvent by vacuum. Reaction vessels containing aldehyde were sealed individually in small plastic canisters flushed with argon and containing a few $CaCl_2$ pellets. In one experiment using fatty acid and fatty alcohol, open test tubes were used, without desiccant (reactants were suspended in H_2O which was evaporated during the first 2–3 d of incubation). Lipid yield from fatty acid was indistinguishable from the experiment using vacuum-evaporated microreaction vessels. After 1 week at 65 °C, total lipids were extracted in chloroform–methanol or ethyl ether and chromatographed on thin-layer plates of silica gel (F-254; EM Lab., Elmsford) twice in hexane–ethyl ether–acetate (100 : 200 : 1) (Fig. 1a), or once in a more complex solvent system[20] followed by the first solvent (Fig. 1b). Autoradiography was performed with Kodak RP-Royal X-omat film; exposure times of 3–7 d at −85 °C were followed by development in Diafine. One week exposure revealed 50–100 d.p.m. ^{14}C. Standard lipids were monolaurin (Sigma), dilaurin (Supelco, Bellefonte) and trilaurin (Sigma), run as internal standards in some samples and adjacent to unknowns in some cases. Identification of unknowns is tentative, as it is based solely on TLC R_fs in several solvent systems (some samples were eluted and rechromatographed). Glycerol and lipids were used as received, except the aldehyde, which was purified on silicic acid in hexane and stored under argon.

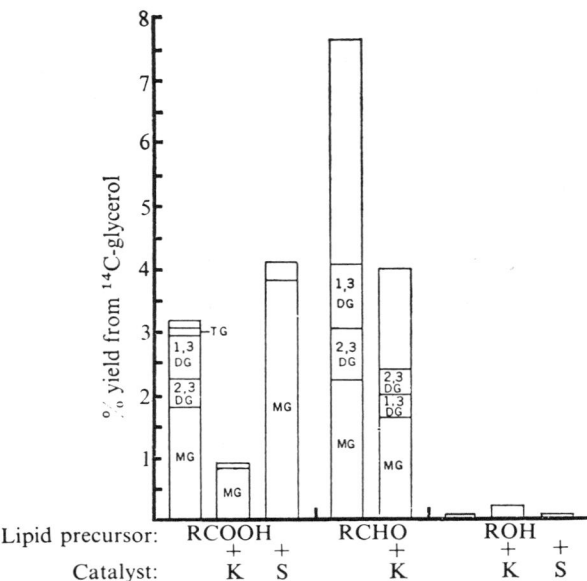

Fig. 2 Quantitation of ¹⁴C-glycerolipids from C12 acid (RCOOH), aldehyde (RCHO) and alcohol (ROH). Radioactive lipids detected by autoradiography were scraped from TLC plates into scintillation vials. One ml H_2O was added and after 1 h, 5 ml cocktail ('Tritosol' modified from Fricke[21]) was added. Percentage yield is based on estimated initial 1.8×10^6 c.p.m., though total c.p.m. recovered from samples free of K and S were substantially lower, due to vaporisation of glycerol and adsorption to containment canister when present. RCHO samples contained substantial amounts of apparent triglycerides, but it was not feasible to count these spots separately from large amounts of an unknown lipid running just below. Unlabelled sections of bars in all other cases represent unidentified compounds.

glycerol–fatty aldehyde. Silicic acid promotes the formation of monoglycerides from fatty acid (3.8% yield compared to 1.8% yield without catalyst), but inhibits the synthesis of more complex glycerolipids.

As initial experiments suggested that monoglycerides and more complex glycerolipids could have accumulated in evaporated tide pools, we selected an ether monoglyceride (chimyl alcohol) as a precursor for phospholipids. Ether monoglyceride is stable in the conditions used and avoided the requirement for large amounts of glycerol and fatty acid in the samples. Silicic acid, kaolin and dicyanamide were tested as possible condensing agents. Duplicate samples containing ether monoglyceride, dodecanoate and prewashed silicic acid and kaolin were mixed by vortexing in 2 ml H_2O containing phosphate (20 mM) and glycerol (6.5 mM), or sodium dicyanamide (50 mM), or both (Table 1). All tubes

were incubated sealed for 12 h at 65 °C, then all but one pair were allowed to evaporate to dryness in the vented oven. After 2 d the open tubes were completely dry in appearance and, after another 7 d, all tubes were extracted with chloroform–methanol–0.1 M KCl (ref. 14). Polar lipids were precipitated from acetone and chromatographed on thin-layer plates.

The polar lipid fraction from sample 2 (Table 1) contained phospholipids with R_fs similar to phosphatidic acid (PA), phosphatidyl glycerol (PG), and phosphatidyl glycerophosphate (PGP), and some residual neutral lipids. Total yield of lipid phosphorus was 0.2% of the phosphate in the reaction mixture. Samples 1 and 4 each yielded <0.015% lipid phosphorus, suggesting that dicyanamide and evaporation together enhanced phospholipid synthesis more than tenfold. Several polar lipid-like compounds (ninhydrin negative) were present in sample 3, but no phospholipids were found.

To determine if the partially purified phospholipids could form membranes, an aliquot from the polar fraction of sample 2 was dried in a small tube. The sample was suspended in 30 mM NaCl, 2 mM Na_2HPO_4 (pH 7.4) by warming to 50 °C in a water bath. After a few seconds, a slightly opalescent solution with no visible particles was obtained. Phase contrast microscopy revealed a heterogeneous suspension of apparent vesicles (Fig. 3a). Brief sonication with a probe sonicator at 50–55 °C resulted in a more homogeneous population of vesicles (negative stain electron micrograph, Fig. 3b). A concentrated suspension, made without sonication in the above buffer containing 25% glycerol, was prepared for freeze-fracture electron microscopy. This technique exposes the hydrophobic plane of lipid bilayer membranes and preserves the structure of lipid–water phases[15]. The presence of lipid vesicles of various sizes in this preparation (Fig. 3c) confirmed our light microscope and negative stain electron microscope observations.

This work has implications concerning the possible origin of life on Earth. Although considerable progress has been made in the abiotic synthesis of important biomolecules, including phospholipid precursors, in simulated primaeval conditions[5,16], the only similar studies of prebiotic glycerolipid synthesis are those of Degans et al.[8,9], who reported the synthesis of neutral lipids from glycerol and fatty acids in a water–kaolin mixture held at 85 °C for 2 weeks. Neither reagent purity nor control experiments were described, however, and yields were relatively small (about 0.1%). Our results show that both fatty acid and fatty aldehyde react rapidly with glycerol in the absence of catalysts to form the precursors of membrane lipids, and that phospholipids and lipid-membrane vesicles can assemble in possible prebiotic conditions. We suggest that silicates other than kaolin are likely to have promoted such syntheses.

Table 1 Synthesis of phospholipid. All samples contained ether monoglyceride (chimyl alcohol, 10 mg, Sigma), dodecanoate (10 mg), prewashed K (50 mg), prewashed S (50 mg), and sodium azide (0.01% initial concentration). When present, phosphate was 20 mM Na_2HPO_4, pH 7.8, glycerol was 6.5 mM, and sodium dicyanamide (Aldrich) was 50 mM. Lipids extracted from alkaline samples after addition of 0.1 M KCl (ref. 14) were only partly soluble in chloroform (C) or ethyl ether (E), but readily soluble in C–methanol (M) (2 : 1). Polar lipids were precipitated by adding 10 volumes of acetone to 1 volume of C–E (1 : 1) containing partially solubilised lipid and storing at −20 °C overnight. Insoluble (polar) lipids were collected after centrifugation in a clinical centrifuge at 4 °C. Phospholipid was identified by spray reagent[22] and by the method of Lowry et al.[23] after scraping spots from charred preparative plates of silica gel H run in C–M–H_2O (65 : 25 : 4). Compounds similar to PA, PGP and PG were present at greater than 0.065, 0.016, and 0.005 μmol lipid phosphate in the acetone fractions of sample 2. Samples 1 and 4 each contained less than 0.006 μmol total lipid phosphorus.

Sample	Treatment	Phosphate	Glycerol	Dicyanamide	Phospholipid detected
1	Aqueous	+	+	+	Trace
2	Evaporated to dryness	+	+	+	0.2% Yield
3	Evaporated to dryness	−	−	+	None
4	Evaporated to dryness	+	+	−	Trace

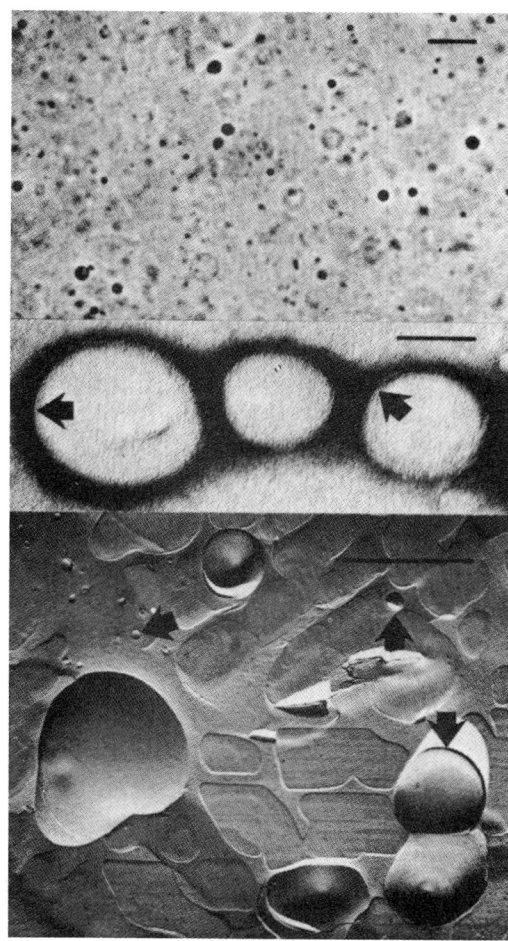

It is significant that phosphatidyl glycerol was apparently formed in our reaction mixture. This lipid, and its aminoacyl derivatives, are prominent in most prokaryotic membranes[17]. Our results show that membrane-forming lipids require only three precursors: glycerol, fatty acid or aldehyde, and phosphate. An alternative and well-studied model system for prebiotic membranes consists of proteinoid 'microspheres', prepared from concentrated solutions of amino acid polymers[18]. There is general agreement, however, that biological membranes require lipid bilayers as barriers to free diffusion of metal ions and substrate molecules. Because our results show that membranogenic lipids are readily produced in simulated prebiotic conditions, we concur with suggestions (Goldacre in ref. 4)[3,4,19] that lipid bilayer membranes could have been present before the accumulation of proteins, polynucleotides and other complex molecules.

This study was supported by a grant from the NSF. W.R.H. was supported by a UC Regents Fellowship and an Earle C. Anthony Fellowship in Zoology.

W. R. HARGREAVES

S. J. MULVIHILL

D. W. DEAMER

Department of Zoology,
University of California,
Davis, California 95616

Received 27 October; accepted 21 December 1976.

1 Oparin, A. I., Deborin, G. A. & Yanopolskaya, N. D. in *Molecular Evolution* (eds. Rohlfing, D. L. & Oparin, A. I.) 343–352 (Plenum, New York, 1972).
2 Miller, S. L. & Urey, H. C. *Science* 130, 245–251 (1959).
3 Weissmann, C. in *Cell Membranes: Biochemistry, Cell Biology and Pathology* (eds. Weissmann, G. & Claiborne, R.), xiii–xvi (H. P. Publ., New York, 1975).
4 Bangham, A. D. *Prog. Biophys. molec. Biol.* 18, 29–95 (1968).
5 Kenyon, D. H. & Steinman, G. *Biochemical Predestination* (McGraw-Hill, New York, 1969).
6 Knauth, L. P. & Epstein, S. *Geochim. cosmochim. Acta* 40, 1095–1108 (1976).
7 Rohlfing, D. L. *Science* 193, 68–69 (1976).
8 Matheja, J. & Degens, E. T. *Structural Molecular Biology of Phosphates* (Fischer, Stuttgart, 1971).
9 Harvey, G. R., Mopper, K. & Degens, E. T. *Chem. Geol.* 9, 79–87 (1972).
10 Weaver, C. E. *Geochim. cosmochim. Acta* 31, 2181–2196 (1967).
11 Ireland, H. A. in *Silica in Sediments* (ed. Ireland, H. A.) 1–3 (Spec. Publs Soc. econ. Paleont. Miner., Tulsa No. 7, 1959).
12 Degens, E. T. *Chem. Geol.* 13, 1–10 (1974).
13 Rohlfing, D. L. & McAlhaney, W. W. *Biosystems* (in the press).
14 Bligh, E. G. & Dyer, W. J. *Can. J. Biochem. Physiol.* 37, 911–917 (1959).
15 Deamer, D. W., Leonard, R., Tardieu, A. & Branton, D. *Biochim. biophys. Acta* 219, 47–60 (1970).
16 Nooner, D. W., Gibert, J. M., Gelpi, E. & Oro, J. *Geochim. cosmochim. Acta* 40, 915–925 (1976).
17 Mindich, L. in *Bacterial Membranes and Walls* (ed. Leive, L.), 1–36 (Dekker, New York, 1973).
18 Fox, S. W. & Dose, K. *Molecular Evolution and the Origin of Life* (Freeman, San Francisco, 1972).
19 Shah, D. O. in *Exobiology* (ed. Ponnamperuma, C.), 235–265 (North-Holland, Amsterdam, 1972).
20 Freeman, C. P. & West, D. *J. Lipid Res.* 7, 324–327 (1966).
21 Fricke, U. *Analyt. Biochem.* 63, 555–558 (1975).
22 Vaskovsky, V. E. & Kostetsky, E. Y. *J. Lipid Res.* 9, 396 (1968).
23 Lowry, O. H., Roberts, N. R., Leiner, K. Y., Wu, M–L. & Farr, A. L. *J. biol. Chem.* 207, 1–17 (1954).

Fig. 3 Vesicles from polar lipid extracts of sample 2. *a*, Phase contrast micrograph (bar is 10 μm) of lipid suspension (1 mg ml⁻¹, made by evaporating solvent from aliquot in small tube, adding 30 mM NaCl, 2mM Na₂HPO₄ (*p*H 7.4), and warming to 50–55 °C for 1 min. *b*, Negative stain electron micrograph (bar is 0.1 μm) of above preparation after probe sonication (Bronwill Biosonik, Will Corp., Rochester, New York) for 10 s at 53 °C. Vesicles were stained with uranyl acetate (1%) after sample was applied to carbon–formvar coated grid and blotted dry. Note several lamellae (arrows). *c*, Freeze-fracture electron micrograph (bar is 1.0 μm) of lipid suspension (5 mg ml⁻¹) made by warming lipid film as above with the buffer containing 25% glycerol as a cryoprotectant. Note vesicles of various sizes (arrows).

Evidence for Extraterrestrial Amino-acids and Hydrocarbons in the Murchison Meteorite

by

KEITH KVENVOLDEN
JAMES LAWLESS
KATHERINE PERING
ETTA PETERSON
JOSE FLORES
CYRIL PONNAMPERUMA

Exobiology Division,
Ames Research Center, NASA,
Moffett Field,
California 94035

I. R. KAPLAN

Department of Geology,
University of California,
Los Angeles,
California 90024

CARLETON MOORE

Center for Meteorite Studies,
Arizona State University,
Tempe, Arizona 85281

Organic molecules found in meteorites seem to have been formed before the meteorites reached Earth.

THE Oparin–Haldane[1,2] hypothesis of chemical evolution, postulating the formation of organic compounds before the appearance of life on Earth, has been substantiated by experiments in the laboratory[3-5]. Many of the molecules important to living processes can be synthesized in conditions which may have prevailed on a primitive planet[6,7]. The fundamental concept of chemical evolution would be further substantiated by the finding of compounds of biological significance in extraterrestrial conditions.

Recent studies of intergalactic space, using the techniques of radioastronomy, have shown the presence of water[8,9], ammonia[10], hydrogen cyanide[11], formaldehyde[12], carbon monoxide[13] and cyanoacetylene[14], which are considered to be precursors in the formation of many biochemicals. Some evidence has also been put forward for the possible presence of molecules such as porphyrins[15] and polyaromatic compounds[16,17] in the interstellar medium. The lunar samples from the Apollo 11 and Apollo 12 missions provided a further opportunity for the search for carbon compounds in extraterrestrial materials, but the analysis of the fines from Mare Tranquillitatis[18] and Oceanus Procellarum[19] revealed only minute traces of such compounds. No conclusive evidence is available for the presence of any biomolecules indigenous to the lunar surface.

On the other hand, meteorites have been analysed for organic compounds for over a century[20-22]. Berzelius[23] examined the Alais, Wöhler[24] the Kaba and Berthelot[25] the Orgueil, and reported the presence of substances of organic origin. These investigations have continued, and there is general agreement about the presence of polymeric organic matter in carbonaceous chondrites[20]. The inherent potentiality for contamination resulting from the ubiquitous distribution of biomolecules on Earth[26,27] leads us to believe that many of the results reporting the presence of organic compounds in meteorites are inconclusive, but the results of our present investigation seem to resolve some of these ambiguities and provide evidence for amino-acids and hydrocarbons of possible extraterrestrial origin.

The Murchison meteorite fell at about 11.00 a.m., local time, September 28, 1969, near Murchison, Victoria, Australia (lat. 36° 36′, long. 145° 12′)[28]. The parent object broke up during flight and scattered many fragments over an area of about 5 square miles. Most fractured surfaces on the individual pieces have fusion crusts. Notable and distinctive features of many of the individual stones are networks of deep cracks extending into their interiors. Several stones were picked up soon after the fall and many have been collected since that time particularly during February and March, 1970. Many of the stones broke on impact with the ground.

For this study we selected those stones which had the fewest cracks, the least exterior contamination, and which generally appeared to have a massive character. The samples we examined contained 2·0 weight per cent carbon and 0·16 weight per cent nitrogen. Chemically, the Murchison meteorite is a type II carbonaceous chondrite (C-2).

Amino-acids

An interior piece of the meteorite (10 g) was pulverized and refluxed with 50 ml. of triply distilled water for 20 h at about 110° C. The extract was recovered by centrifugation and the residue rinsed twice with 50 ml. of water. The extract and rinses were combined, evaporated to dryness, and hydrolysed with 50 ml. of 6 M HCl for 20 h at about 100° C. This hydrolysate was evaporated to dryness and desalted by ion exchange using Dowex 50 (H+) followed by Dowex 2 (OH-). Portions of the recovered product were analysed by conventional ion exchange chromatography (Beckman–Spinco 120), by capillary gas chromatography of amino-acid derivatives (Perkin Elmer 881), and by gas chromatography combined with mass spectrometry (CEC 21-491). The meteorite residue after water extraction was hydrolysed with 6 M HCl for 20 h at about 110° C. This hydrolysate was treated in the same manner as the water extract.

Conventional ion exchange chromatography of the hydrolysed water extract from the 10 g sample revealed a number of peaks having retention times similar to common amino-acids. Five abundant amino-acids suggested by the chromatograms were glycine (6 µg/g), alanine (3 µg/g), glutamic acid (3 µg/g), valine (2 µg/g) and proline (1 µg/g).

The presence of the amino-acids, suggested by ion exchange chromatography, and their enantiomeric distribution were established by gas chromatography of the N-trifluoroacetyl-D-2-butyl esters of the amino-acids[29-32]. In this procedure, amino-acids with an asymmetric centre were transformed to diastereomers and the enantiomeric distribution of these compounds was determined on two different capillary columns—'UCON 75-H-90,000' and 'XE 60'. Fig. 1 shows a representative chromatogram resulting from an analysis of amino-acid derivatives. Peaks of the common amino-acids identified on the chromatogram confirm the identification made by ion exchange chromatography. Moreover, both enantiomers of amino-acids with asymmetric centres are present. The percentage of D-amino-acid enantiomers present was calculated from the gas chromatograms to be about 50 per cent for alanine, 45 per cent for glutamic acid, 40–47 per cent for valine and 40–43 per cent for proline. The presence of 2-methylalanine and sarcosine, amino-

Reprinted with permission from *Nature*, Volume 228, pp. 923-926.

acids not commonly found in biological systems, was suggested by the gas chromatogram.

To obtain an unambiguous identification, the compounds eluted from the gas chromatograph were introduced through a membrane separator into the mass spectrometer. The mass spectra obtained were compared with the spectra of known standards. In this manner the identities of the amino-acids, glycine, alanine, valine, proline and glutamic acid, were confirmed. Furthermore, the identity of each enantiomeric derivative of alanine, valine and proline was established. The D enantiomeric derivative of glutamic acid was confirmed but because of interfering compounds the spectra of the L enantiomeric derivative was equivocal.

The identities of 2-methylalanine and sarcosine were also established by mass spectrometry. Fig. 2 shows the spectra obtained from a standard N-TFA–D-2-butyl ester of 2-methylalanine compared with the derivatized compound from the Murchison meteorite sample. Similarly, Fig. 3 illustrates the spectra of a derivatized standard sample of sarcosine and a derivatized compound from the meteorite sample. As in the mass spectra of the N-TFA-*n*-butyl esters of amino-acids[33] the key peaks in the spectra of the N-TFA-D-2-butyl esters of 2-methylalanine and sarcosine are the $(M - C_4H_9COO)^+$ peak and the $(M - C_4H_9O)^+$ peak.

Amino-acids have been found previously in other meteorites[20,34] but because the distributional patterns of these amino-acids are similar to amino-acids in fingerprints, the occurrence of amino-acids in meteorites has commonly been attributed to contamination[20]. The total concentration of glycine, alanine, valine, proline and glutamic acid obtained from Murchison meteorite was about 15 μg/g, which is about twice the total concentration of these same compounds found free in Murray, another type II carbonaceous chondrite[34]. In previous work serine has usually been reported as among the most abundant amino-acid in both meteorites and fingerprints[20]. In this study serine was tentatively identified only by ion exchange chromatography and its concentration was an order of magnitude less than the concentration of glycine. The unique distribution of amino-acids in the Murchison meteorite cannot be explained satisfactorily on the basis of contamination by human hands.

The distribution of the enantiomers of alanine, valine, proline and glutamic acid also supports the idea that the amino-acids from the Murchison meteorite are not recent biological contaminants. Amino-acids in living systems are of the L configuration except in rare instances. In recent sediments the L configuration dominates[32].

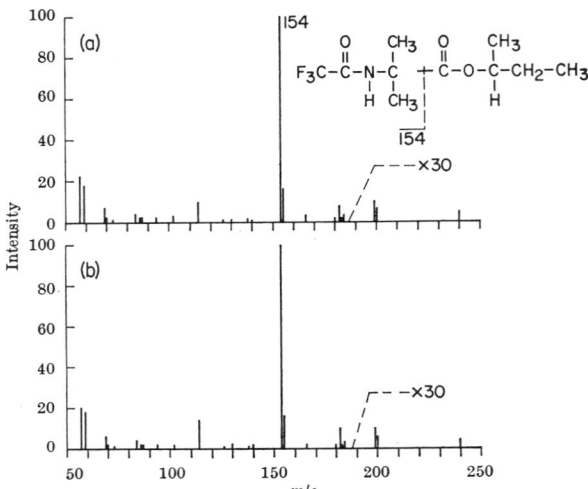

Fig. 2. Mass spectra of N-trifluoroacetyl-D-2-butyl ester of: *a*, 2-methylalanine standard; *b*, compound from the Murchison meteorite.

Studies of amino-acid configurations in other carbonaceous chondrites show that the distribution of enantiomers resembles those of recent sediments and soils[35]. The D and L enantiomers of amino-acids in the Murchison meteorite are almost equally abundant. Slight biological contamination might account for the small dominance of the L isomers of valine, proline and glutamic acid. The distribution of amino-acid enantiomers in Murchison can be accounted for either as a racemic modification of biologically produced amino-acids or as a racemic mixture of abiotically produced amino-acids. The discovery of 2-methylalanine and sarcosine, which are not commonly found in proteins, supports the hypothesis that the whole collection of amino-acids was produced abiotically. In chemical evolution experiments simulating the primitive atmosphere, both sarcosine and 2-methylalanine have been noted among the products[36].

Hydrocarbons

Ten g of the pulverized meteorite were extracted by sonication in a centrifuge tube three times successively, with about 20 ml. benzene–methanol (9 : 1). After centrifugation the extract was evaporated to dryness in a rotary evaporator. This residue was redissolved in 1–2 ml. hexane, which was applied to a 10 g activated silica gel column. The column was eluted with 15 ml. portions of hexane, benzene and benzene–methanol. Activated copper strips were placed in the hexane and benzene fractions and allowed to stand overnight to remove any free sulphur. After the removal of the copper the three fractions were evaporated to about 100 μl. About one-tenth of the hexane fraction was analysed by gas chromatography using a 50 foot 'OV-1' support coated open tubular column in a Perkin–Elmer '880' gas chromatograph. The remainder of the sample was analysed by gas chromatography combined mass spectrometry ('CEC 21-491'). Mass spectra were taken every 3 min during the chromatography.

A second sample (15 g) was similarly extracted. After the removal of sulphur the hexane fraction was analysed by thin layer chromatography using silica gel plates impregnated with silver nitrate to separate saturated alkanes from the unsaturated[37]. Saturated alkanes were eluted from the silica gel with hexane, and one-twenty-

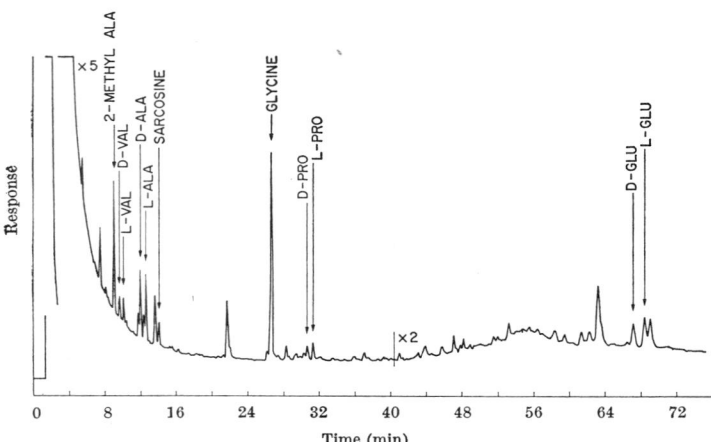

Fig. 1. Gas chromatogram of N-trifluoroacetyl-D-2-butyl esters of amino-acids in acid hydrolysed water extract of a 10 g sample of the Murchison meteorite. Gas chromatography: 0·02 inch × 150 feet capillary column coated with 'UCON 75-H-90,000'; temperature programmed from 100° to 150° C at 1° C/min.

fifth of the eluate was analysed by gas chromatography as before. A small fraction (approximately one-third) was volatilized directly into the source of a 'CEC 21-110' mass spectrometer for a series of high resolution photoplate spectra, and one-third was vaporized directly into the source of the 'CEC 21-110' mass spectrometer to obtain a series of low resolution spectra.

The chromatograms in Fig. 4 indicate that a very complex mixture of alkanes has been isolated from the two meteorite samples. A rough estimate of their abundance can be made by assuming that the arch shaped trace is an equilateral triangle composed of a series of peaks spaced at very close intervals. This estimate suggests that about 350 μg of alkanes were isolated from the 10 g sample and 175 μg from the 15 g sample.

Contamination seems to be limited to a small group of compounds eluted at 210°–249° C. These have been routinely observed in our laboratory and seem to be airborne contaminants.

The distribution pattern of the alkanes isolated from the 10 g sample is a symmetrical trace (Fig. 4) with a peak at 180° C, which corresponds to the elution temperature of n-C_{18}, with a shoulder in the 100°–140° C region. On this shoulder (130° C) is a peak with the retention time corresponding to a saturated n-C_{13} hydrocarbon, and represents about 0·1 μg of hydrocarbons. Other—much smaller—peaks on the chromatogram correspond, in retention time, to n-C_{18} and n-C_{22}-C_{23}. The mass spectra of the various GLC fractions taken at 3 min intervals were very similar and showed a predominance of fragment ions corresponding to the homologous series:

$$(C_nH_{2n-3})^+ > (C_nH_{2n-1})^+ > (C_nH_{2n-5})^+ > (C_nH_{2n+1})^+$$

Mass spectra of fractions eluting from the column at higher temperatures show an increasing predominance of the $(C_nH_{2n}-5)^+$ ion over the $(C_nH_{2n}-1)^+$ ion. The peak emerging at 130° C has not been completely characterized. The mass spectra of the species coinciding with the appearance of this peak on the gas chromatogram shows intense ions at m/e 127 and 182.

The gas chromatographic distribution pattern of the saturated alkanes isolated from the 15 g sample by thin

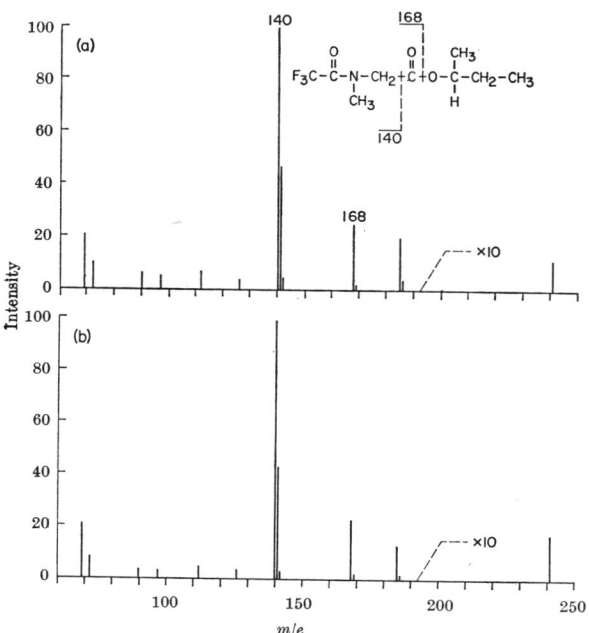

Fig. 3. Mass spectra of N-trifluoroacetyl-D-2-butyl ester of: a, standard of sarcosine; b, compound from the Murchison meteorite.

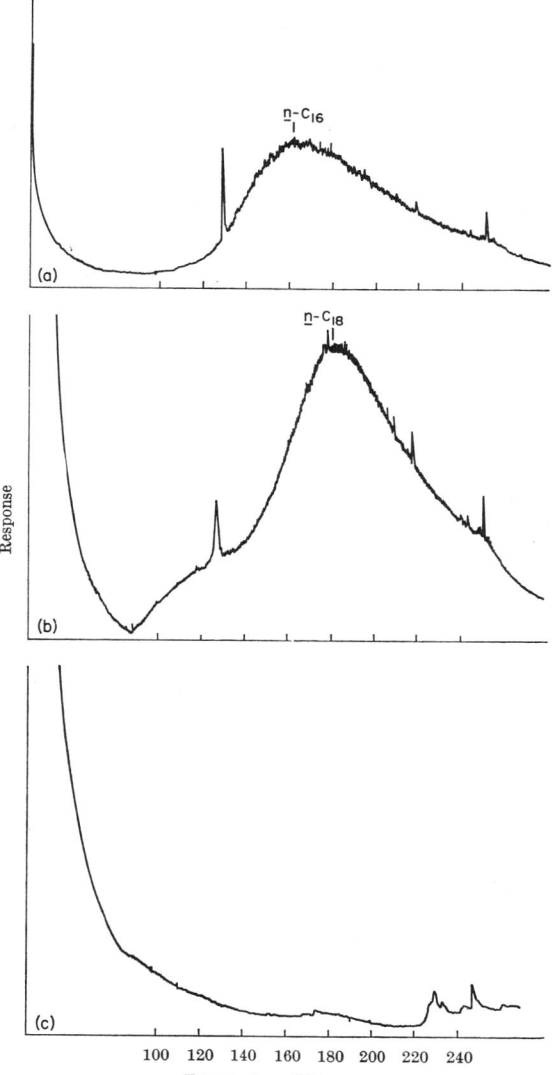

Fig. 4. Gas chromatograms of alkanes isolated from benzene–methanol extracts of two samples of the Murchison meteorite: a, 1/10 of alkane fraction isolated from 10 g sample (attenuation ×50); b, 1/25 of alkane fraction isolated from 15 g sample (attenuation ×20); c, chromatographic blank (attenuation ×20) gas chromatography: 50 feet support coated open tubular column with 'OV-1' liquid phase; temperature programmed from 100° to 240° C at 2° C/min.

layer chromatography is also a symmetrical trace (Fig. 4), but it differs from the first sample, in that it reaches a maximum at 160° C (n-C_{16} retention time) and does not have a shoulder in the 100°–140° C region. This may result from removal of small amounts of unsaturated isomers by AgNO₃ chromatography. The low resolution mass spectra of this alkane mixture resembled those from the first sample. Several predominant molecular ions are present (Fig. 5), suggesting that the hydrocarbons in this particular sample are relatively stable to fragmentation on electron bombardment. These molecular ions (m/e 278, 292, 306) correspond to the bicyclic, homologous series C_nH_{2n-4}. A series of relatively intense peaks at m/e 221, 235 and 263 suggests that the loss of hydrocarbon fragments, ranging from $-CH_3$ to $-C_4H_9$, is a favourable process. This evidence, and the relatively large intensity of the m/e 151 peak relative to m/e 137 and m/e 165, suggests that these hydrocarbons closely

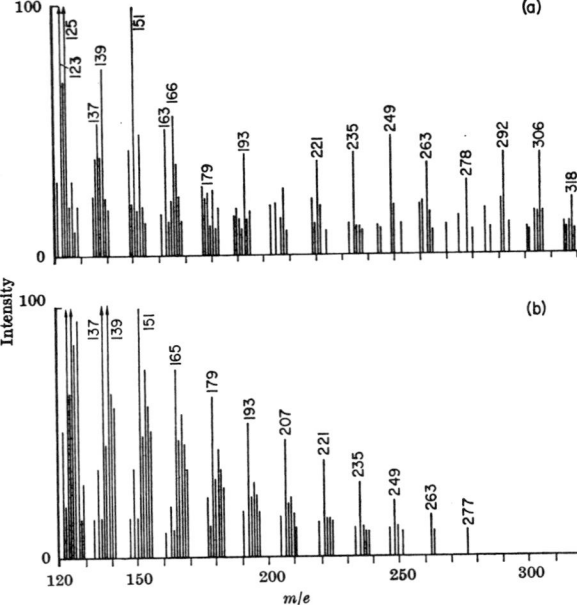

Fig. 5. Low resolution mass spectrum of alkane fractions obtained on volatilization at 125° C from the direct probe of a 'CEC 21–110' mass spectrometer. Ion intensities normalized to *m/e* 151. *a*, Fraction from 15 g sample of the Murchison meteorite; *b*, fraction from spark discharge in methane.

resemble decahydronaphthalenes containing short chain hydrocarbon substituents.

There is a marked resemblance between the gas chromatographic traces of the aliphatic hydrocarbons synthesized by the action of electrical discharges on methane[38] and those from the meteorite. The analysis of these samples by thin layer chromatography on silica gel plates impregnated with AgNO₃ (ref. 37) indicates that they consist largely of saturated alkanes. The mass spectral data reveal the same dominant homologous series in both samples (Fig. 5). The similarity between the aliphatic hydrocarbons from the meteorite and the spark discharge material, on these three counts, is suggestive of a possible abiogenic origin for the hydrocarbons in the Murchison meteorite.

Further investigations are in progress to determine the possible presence of aromatic and unsaturated hydrocarbons, carboxylic acids and porphyrins.

Carbon Isotope Distribution

Various carbon compounds were analysed for $^{13}C/^{12}C$ following the technique of Smith and Kaplan[39]. The results (expressed in the conventional manner as $\delta^{13}C$, relative to Peedee Belemnite) are given in Table 1. These results, together with the carbon content (2·5 per cent) place the meteorite as a type II carbonaceous chondrite[39]. The values obtained for the organic solvent extractable material are significantly more enriched in ^{13}C than any previous measurements on such compounds in other carbonaceous chondrites. Extracts of terrestrial sediments and petroleum fall in the range $\delta^{13}C = -25$ to -35 parts per thousand[40,41]. Results obtained from the meteorite confirm the indigenous nature of the carbon and indicate that the amount of terrestrial contamination is insignificantly low.

Table 1

Form of carbon analysed	$\delta^{13}C$ PDB (parts per thousand)
(i) Total carbon	−7·3, −7·1
(ii) Carbonate	+45·4
(iii) Benzene–methanol soluble compounds	+4·4, +4·8, +5·9
(iv) Insoluble carbon (after solvent extraction and demineralization of silicates with HF)	−10·6

The results of our investigation suggest the indigenous nature of amino-acids and hydrocarbons in the Murchison meteorite. The $\delta^{13}C$ values of +4·4 to +5·9 for the extractable organic material of the meteorite fall into a range widely different from terrestrial organic matter, and the presence of the amino-acids glycine, alanine, valine, proline, glutamic acid, 2-methylalanine and sarcosine was unequivocally established. The presence of almost equal amounts of the D and L enantiomers of valine, proline, alanine and glutamic acid minimizes the possibility of terrestrial contamination and suggests a possible extraterrestrial origin. The presence of the two amino-acids, 2-methylalanine and sarcosine, which are not generally found in biological systems, is indicative of a possible abiogenic synthesis. The gas chromatographic pattern of distribution of hydrocarbons and their mass spectrometry fragmentation pattern similar to that obtained by an abiotic synthesis in the laboratory support the contention that the organic molecules identified here are abiotic and possibly extraterrestrial in origin.

We thank Mr Michael Romiez and Miss Cheryl Boynton of the Ames Research Center and Mrs Chari Petrowski of the Department of Geology, University of California, Los Angeles, for their assistance.

Received September 29; revised October 19, 1970.

1 Oparin, A. I., *Proischogdenie Zhizni*. (Moscovsky Robotchii, Moscow, 1924).
2 Haldane, J. B. S., *Rationalist Ann.*, 148 (1928).
3 *The Origin of Life on Earth* (edit. by Oparin, A. I.) (Pergamon, Oxford, 1959).
4 *The Origins of Prebiological Systems and of Their Molecular Matrices* (edit. by Fox, S. W.) (Academic Press, 1965).
5 *Proc. Third Intern. Conf. on the Origin of Life, Pont-a-Mousson, France* (1970) (edit. by Buvet, R., and Ponnamperuma, C.) (North-Holland, Amsterdam, in the press).
6 Ponnamperuma, C., and Gabel, N. W., *Space Life Sci.*, 1, 64 (1968).
7 Lemmon, R. M., *Chem. Rev.*, 70, 95 (1970).
8 Cheung, A. C., Rank, D. M., Townes, C. H., Thornton, D. D., and Welch, W. J., *Nature*, 221, 626 (1969).
9 Snyder, L. E., and Buhl, D., *Astrophys. J.*, 155, L65 (1969).
10 Cheung, A. C., Rank, D. M., Townes, C. H., Thornton, D. D., and Welch, W. J., *Phys. Rev. Lett.*, 21, 1701 (1968).
11 Snyder, L. E., and Buhl, D., *IAU Circular*, No. 2251 (1970).
12 Snyder, L. E., Buhl, D., Zuckerman, B., and Palmer, P., *Phys. Rev. Lett.*, 22, 679 (1969).
13 Wilson, R. W., Jefferts, K. B., and Penzias, A. A., *Astrophys. J.*, 161, L43 (1970).
14 Turner, B. E., *IAU Circular*, 2268 (1970).
15 Johnson, F. M., *Proc. Regional Symp. on Planetary Geology and Geophysics* (Amer. Astron. Soc. and New England Academic Community, 1967).
16 Donn, B., *Astrophys. J.*, 152, L129 (1968).
17 Donn, B., in *Exobiology* (edit. by Ponnamperuma, C.) (North-Holland, Amsterdam, 1970).
18 Lunar Science Conference, *Science*, 167, 751 (1970).
19 Chang, S., Kvenvolden, K., Lawless, J., Ponnamperuma, C., and Kaplan, I. R. (in the press).
20 Hayes, J. M., *Geochim. Cosmochim. Acta*, 31, 1395 (1967).
21 Nagy, S., *Geol. for Förhandlingar* (Stockholm), 88, 235 (1966).
22 Studier, M. H., Hayatsu, R., and Anders, E., *Science*, 149, 1455 (1965).
23 Berzelius, J. J., *Ann. Phys. Chem.*, 33, 113 (1834).
24 Wöhler, M. F., and Hörnes, M., *Sitzber. Akad. Wiss. Wien, Math.-naturw. Kl.*, 34, 7 (1859).
25 Berthelot, M., *CR Acad. Sci.*, 67, 849 (1868).
26 Hamilton, P. B., *Nature*, 205, 284 (1965).
27 Oro, J., and Skewes, H. B., *Nature*, 207, 1042 (1965).
28 *Rep. Center for Shortlived Phenomena*, No. 779 (Smithsonian Inst., Cambridge, Mass., 1969).
29 Gil-Av, E., Charles, R., and Fischer, G., *J. Chromatog.*, 17, 408 (1965).
30 Pollock, G. E., Oyama, V. I., and Johnson, R. O., *J. Gas Chromatog.*, 3, 174 (1965).
31 Pollock, G. E., and Oyama, V. I., *J. Gas Chromatog.*, 4, 126 (1966).
32 Kvenvolden, K. A., Peterson, E., and Brown, F. S., *Science*, 169, 1079 (1970).
33 Gelpi, E., Koenig, W. A., Gibert, J., and Oro, J., *J. Chromatog. Sci.*, 7, 604 (1969).
34 Kaplan, I. R., Degens, E. T., and Reuter, J. H., *Geochim. Cosmochim. Acta*, 27, 805 (1963).
35 Nakaparksin, S., *Diss. Abst.*, 70-4495, 4016B (1970).
36 Miller, S. L., and Urey, H. C., *Science*, 130, 245 (1959).
37 Murphy, M. T. J., in *Organic Geochemistry, Methods and Results* (edit. by Eglinton and Murphy), 75 (Springer-Verlag, 1969).
38 Ponnamperuma, C., Woeller, F., Flores, J., Romiez, M., and Allen, W. V., *ACS Adv. Chem.*, 80, 280 (1969).
39 Smith, J. W., and Kaplan, I. R., *Science*, 167, 1367 (1970).
40 Degens, E. T., *Geochemistry of Sediments* (Prentice-Hall, Englewood Cliffs, 1965).
41 Silverman, S. R., in *Isotopic and Cosmic Chemistry* (edit. by Craig, H., Miller, S. L., and Warserburg, G. J.) (North-Holland, Amsterdam, 1964).

Cometary Delivery of Organic Molecules to the Early Earth

Christopher F. Chyba, Paul J. Thomas,* Leigh Brookshaw,† Carl Sagan

It has long been speculated that Earth accreted prebiotic organic molecules important for the origins of life from impacts of carbonaceous asteroids and comets during the period of heavy bombardment 4.5×10^9 to 3.8×10^9 years ago. A comprehensive treatment of comet-asteroid interaction with the atmosphere, surface impact, and resulting organic pyrolysis demonstrates that organics will not survive impacts at velocities greater than about 10 kilometers per second and that even comets and asteroids as small as 100 meters in radius cannot be aerobraked to below this velocity in 1-bar atmospheres. However, for plausible dense (10-bar carbon dioxide) early atmospheres, we find that 4.5×10^9 years ago Earth was accreting intact cometary organics at a rate of at least $\sim 10^6$ to 10^7 kilograms per year, a flux that thereafter declined with a half-life of $\sim 10^8$ years. These results may be put in context by comparison with terrestrial oceanic and total biomasses, $\sim 3 \times 10^{12}$ kilograms and $\sim 6 \times 10^{14}$ kilograms, respectively.

A SIGNIFICANT FRACTION OF EARTH'S VOLATILE INVENTORY may have been acquired as a late-accreting veneer from impacts of C-type asteroids (1) and comets (2, 3) during the period of heavy bombardment of the inner solar system 4.5×10^9 to 3.8×10^9 years ago. In addition to simple volatile molecules such as H_2O and short-chain hydrocarbons, C-type asteroids and comets are also rich in complex organics. In the case of asteroids, this is known from direct analysis of carbonaceous chondrite meteorites (4, 5), which are spectroscopically similar to C-type asteroids (6), their presumptive parent bodies. As long ago as 1908, this organic component in carbonaceous meteorites led Chamberlin and Chamberlin (7) to suggest that infalling "planetesimals" may have been an important source of prebiotic organic material on early Earth. In 1961, Oró (8) suggested, on the basis of spectroscopic observations of carbon- and nitrogen-containing radicals in cometary comae, that comets may have played a similar role. Since then, the possible relevance of cometary organics to the terrestrial origins of life has received considerable attention (9–11).

Over the last decade, this potential cometary source of prebiotic organics has taken on new importance, as an emerging consensus in planetary science has replaced earlier models of a primordial reducing terrestrial atmosphere rich in methane (CH_4) and ammonia (NH_3) with that of a neutral one rich in carbon dioxide (CO_2) and molecular nitrogen (N_2). Such early CO_2-rich atmospheres are implied by "hot" accretion scenarios for Earth, in which core formation takes place quickly, leaving the upper mantle (and hence, outgassed carbon) in an oxidized state (10, 12). The short photodissociation lifetimes of CH_4 and NH_3 in model paleoatmospheres reinforce this conclusion (13). If subduction-resistant continental platforms were largely or entirely absent on Earth before 3.8×10^9 years ago, the ~ 60 bars of CO_2 currently sequestered in continental platforms would have led to a dense, 10- to 20-bar CO_2 early terrestrial atmosphere (14), a conclusion that is consistent with solutions to the early faint sun "paradox," and with arguments based on comparative inner solar system planetology (15). Early Earth atmospheres with CO_2 pressures as high as 100 bars appear to be stable against a runaway greenhouse effect and are therefore consistent with oceans of liquid water (16).

Syntheses of key prebiotic molecules [such as hydrogen cyanide (HCN) and formaldehyde (H_2CO), precursors to purines, pyrimidines, and amino acids, and to sugars, respectively] would have been much more difficult in CO_2 atmospheres than in reducing ones (10). Experiments with CO_2 gas mixtures demonstrate that, relative to atmospheres rich in CH_4 or carbon monoxide (CO), yields of HCN and H_2CO, as well as amino acids themselves, drop precipitously as the H_2/CO_2 ratio falls below unity (17). However, although these results are suggestive, there is certainly at present no requirement to invoke impactor-delivered organics to account for the evolution of life on Earth. Many other speculative mechanisms for terrestrial prebiotic synthesis have been proposed, including production of HCN by photolysis of N_2 and CH_4 in a hypothesized weakly reducing atmosphere (18), reduction of CO_2 to CO by impactor-delivered iron (19), volcanic production of H_2 by H_2O decomposition, or reduction of CO_2 in solution or in the presence of clays [see (10) for a review]. The goal of investigators at present must be to quantify as well as possible prebiotic organic production from the various proposed sources, so that their comparative importance may be weighed. Our purpose here is to extend this compilation to the case of cometary delivery of organics.

Pyrolysis of Cometary Organics by Impact

A long-standing objection to cometary delivery of intact organic molecules to Earth is that these compounds would be totally dissociated by the heat of cometary atmospheric passage and the ensuing impact. This has led Clark to suggest "an improbable,

C. F. Chyba, P. J. Thomas, and C. Sagan are at the Laboratory for Planetary Studies, Cornell University, Ithaca, NY 14853. L. Brookshaw is in the Section of Applied Physics, Yale University, Box 2159, Yale Station, New Haven, CT 06520.

*Present address: Department of Physics and Astronomy, University of Wisconsin, Eau Claire, WI 54702.
†Present address: Lawrence Livermore National Laboratory, IGPP L-413, Post Office Box 808, Livermore, CA 94550.

fortuitous soft-landing of a cometary nucleus" as a mechanism for the origin of life (*11*, p. 209). Although such unlikely events cannot be ruled out, they are also extremely difficult to quantify. However, a closer examination reveals that the claim that cometary organics will be pyrolyzed at impact is uncertain. Consider a comet colliding with Earth at a velocity of 18 km s^{-1}. [Some ~25% of short-period (SP) comet-Earth collisions should occur at or below this velocity (*3*).] If we assume that upon impact all of the comet's kinetic energy goes into heating the comet, treat the comet as composed of water ice, and choose a specific heat of water appropriate to the resulting high temperatures and pressures [~4 kJ kg^{-1} K^{-1} (*20*)], we find the comet to be shocked to an average temperature of ~40,000 K. This temperature is certainly too high to permit any organics to survive. However, it is too high by only a factor of 20 to 40. Because much of the kinetic energy of a comet with a density of 1 g cm^{-3} will be partitioned not into heating the impactor but into kinetic energy of ejecta and target heating (*21*), it is possible that aerobraking (slowing by atmospheric drag) and uneven distribution of shock energy throughout the impacting projectile will conspire to yield some region of the comet for which temperatures remain low enough to allow at least the hardier organics to survive. This possibility can only be quantified by detailed numerical modeling of cometary atmospheric passage and surface collision. But before pursuing such models, we must first review what is known about the abundance and volatility of cometary organics.

Cometary Organics: Results from the Halley Apparition

Laboratory analysis of CI and CII carbonaceous chondrite meteorites reveals them to be 3 to 5% by mass organic heteropolymer (*4*). In situ mass spectroscopy in the coma of comet Halley by the Giotto and Vega spacecraft revealed Halley dust to be fully one-third organic by mass (*22*). The volatile (gas) fraction of the comet appears to be ~14% by mass organic (*23*); assigning a standard gas/dust mass ratio of 1:1 gives an overall cometary organic component of ~25% by mass. This may seem extraordinary, but it should be remembered that Halley has approximately cosmic abundances of carbon and nitrogen—elemental abundances higher by more than an order of magnitude than those of CI chondrites (*23*, *24*). Spectroscopic observations of comet Halley suggest that the production rate of H$_2$CO is about 2% by mass that of H$_2$O (*25*). In situ ion mass spectrometry for comet Halley (*26*) and spectroscopy of comet Wilson (*27*) suggest that CH$_4$ is present at 1 to 4% H$_2$O by mass. We therefore take H$_2$CO and CH$_4$ to account for 4 and 5% by mass of cometary organics, respectively. Perhaps ~7% by mass is HCN (*23*). Such estimates should be treated as rough at best.

Approximately 3% of the organic carbon in carbonaceous chondrites is in the form of amino acids (*5*). It is not known if comets contain comparable (or indeed any) amounts of such compounds. However, the discovery that amino acids from the Murchison meteorite are as a group highly enriched in deuterium (δD = 1370%) strongly suggests that the amino acids or their precursors were formed at low temperatures in interstellar clouds (*5*, *28*). If amino acids formed in interstellar clouds, they should be at least as abundant in comets as in carbonaceous chondrites. However, this conclusion need not hold if only amino acid precursors are present in interstellar dust. Despite dedicated searches, free amino acids remain undetected in interstellar clouds (*29*), although amino acids in dust would remain undetectable.

Finally, although we cannot entirely rule out the possibility that cometary organics are biological in origin (*30*), we believe that there is no persuasive evidence for this claim, and that much more

plausible alternatives exist (*31*). The problem we address here is whether abiotic cometary organics may have played a role in stocking the terrestrial prebiotic inventory. We do not concern ourselves in the following with panspermia theories.

Volatility of Cometary Organics

Most organic material in carbonaceous chondrites is in the form of high molecular weight, "intractable" heteropolymer (*4*, *5*). Laboratory experiments under conditions meant to simulate radiation processing of interstellar dust and cometary ices suggest that cometary organics are similar (*32*, *33*), a result supported by the success of such materials in modeling organic spectral features in a number of comets (*34*). As a well-investigated analog to such intractable organics, we use terrestrial kerogens, which are both compositionally similar to meteoritic carbonaceous material (*4*) and provide reasonable spectroscopic fits to certain asteroids and outer solar system objects (*35*). The optical constants of type II kerogens and an organic extract from the Murchison carbonaceous chondrite are also in good mutual agreement (*36*). Terrestrial kerogens are biological in origin; our use of them as structural analogs to [presumably abiotic (*31*)] cometary organics carries no implication of a biological origin for the latter.

Kinetic parameters from pyrolysis experiments with type I and II kerogens indicate that such material can survive temperatures of ~850 K for 1 s (*37*). (As described below, in the impacts of relevance to this discussion, 1 s is an overestimate for the duration of the most intense shock heating that cometary organics will endure. Well before this time, shock temperatures will have peaked and begun quickly to decline.) However, about 50% of the resulting kerogen pyrolyzate is composed of *n*-alkanes up to C$_{30}$, as well as such aromatics as benzene, toluene, and *m,p*-xylene (*38*). Therefore, to understand the ultimate fate of kerogen-like organics in the impactor, we must examine the fate of their components, for example, short-chain hydrocarbons and simple aromatics (and their oxygen and nitrogen analogs and substitutions). Kinetic parameters derived from shock-tube pyrolysis of such compounds (for example, ethane or benzene) at temperatures of ~1000 to 3000 K, the best laboratory analog we have to organic destruction in large impacts, show that these basic organic units will survive at ~1200 K for 1 s (*39*, *40*).

These temperatures are probably severe underestimates of the thermal stability of at least the hardiest of cometary organics. Thermogravimetric analyses of the involatile organic residue produced by electron, spark, or ultraviolet irradiation of cosmically abundant gases show that the residues are typically 50% stable at temperatures ranging from 600 to 1200 K; time scales for such analyses are far in excess of 1 s (*32*). Similar material produced in analogous ice irradiation experiments provides a good match to observed cometary organic emission features (*34*). Therefore, we expect that there is some (probably polycyclic) component of cometary organics that is much more thermally stable than the compounds being explicitly considered here.

Kinetic parameters from shock-tube experiments with CH$_4$ show that it survives temperatures of ~1500 K for 1 s (*41*). Numerous shock-tube experiments with H$_2$CO (*42*) and HCN (*43*) have been performed; we estimate from these results that these compounds will withstand shock temperatures of ~1200 and ~1800 K, respectively (*44*).

To our knowledge, kinetic parameters for the pyrolysis of amino acids have been measured only in solution. The results of such experiments are of questionable applicability to shock pyrolysis, but, until gas-phase results become available, we approximate impact survival of amino acids using these results. Alanine should withstand temperatures of ~700 K for 1 s (45), whereas other amino acids should withstand temperatures in the range of 600 to 800 K (46).

Impact Environment of Early Earth

The existence of complex terrestrial microorganisms by 3.5×10^9 years ago and evidence for biologically mediated carbon isotope fractionation in 3.8×10^9 year Isua metasediments (47) suggest choosing the period before 3.8×10^9 years ago to assess cometary delivery of organics. Attempts to estimate the impact environment of early Earth typically begin with an analytical fit to the lunar cratering record. The impactor flux thus derived can then be scaled to Earth's larger gravitational cross section. Following such a procedure for several published lunar cratering data sets and choices of time constant for the decay of the early impactor flux, Chyba (3) found three possible values, spanning the current uncertainties:

$$M(t) = 0.76[t + 4.57 \times 10^{-7} (e^{t/\tau_A} - 1)] (m_2^{1-b} - m_1^{1-b}) \text{ kg}^b \text{ km}^{-2}$$

(1)

$$M(t) = 0.99[t + 2.3 \times 10^{-11} (e^{t/\tau_B} - 1)] (m_2^{1-b} - m_1^{1-b}) \text{ kg}^b \text{ km}^{-2}$$

(2)

$$M(t) = 0.40[t + 5.6 \times 10^{-23} (e^{t/\tau_C} - 1)] (m_2^{1-b} - m_2^{1-b}) \text{ kg}^b \text{ km}^{-2}$$

(3)

for the total mass $M(t)$ that has been incident in impactors with masses in the range m_1 to m_2 on a lunar surface of age t. Here $b \approx 0.47$, t is in 10^9 years, $\tau_A = 220 \times 10^6$, $\tau_B = 144 \times 10^6$, and $\tau_C = 70 \times 10^6$ years. If we perform the appropriate gravitational scaling, Eqs. 1 through 3 yield a total cometary mass incident on Earth since 4.5×10^9 years ago of $M_{tot} = 1 \times 10^{21}$, 3×10^{21}, or $7 \times 10^{23} \alpha$ kg, respectively, where α is the fraction of the ancient impacting mass flux that was cometary (3). Nearly all of this mass was collected before 3.8×10^9 years ago, that is, during the period of heavy bombardment, which is also the time of interest for the origins of life.

The parameter α is not well constrained. Estimates of current asteroidal and cometary cratering rates suggest that comets account for 10 to 30% of the recent production of terrestrial impact craters >10 km in diameter, although "extinct" comets or possible comet showers could raise this fraction to over 50% (48). How to extrapolate this result back to the early solar system is not clear, however. Although none of the data pertaining to this question are conclusive, it seems likely that $\alpha \lesssim 10$ to 20% (2). For definiteness, we set $\alpha = 0.1$ in the following. Our results scale linearly, so the effect of a different choice of α will be evident.

Comets are ~18% carbon by mass (23), so Eqs. 1, 2, and 3 imply that Earth has collected 2×10^{19}, 6×10^{19}, or 1×10^{22} ($\alpha/0.1$) kg of cometary carbon, respectively, which may be compared to the total terrestrial surface (ocean, atmosphere, and sedimentry column) carbon inventory of ~9×10^{19} kg (14, 49). Clearly, there is a discrepancy between the terrestrial carbon inventory predicted by Eq. 3 and that actually present on Earth. A similar discrepancy exists for H_2O (3). Impact erosion by sufficiently large and fast-moving comets and asteroids could reduce this discrepancy for carbon (which may have been eroded as atmospheric CO_2) but appears incapable of doing so for condensed volatiles such as H_2O (3). It therefore appears that the volatile flux predicted by Eq. 3 for $\alpha = 0.1$ is too large by a factor of ~100.

Conversely, the mass flux predicted by Eqs. 1 and 2 is probably too small. Sleep et al. (50) have used lunar highland iridium and nickel abundances to calculate the total amount of material impact-ing the moon since the solidification of the upper lunar crust ~4.4 $\times 10^9$ years ago. Scaling to Earth yields a value $M_{tot} = 5 \times 10^{21} \alpha$ kg as a lower limit for the total cometary mass incident on Earth subsequent to 4.5×10^9 years ago (3).

Taking these limits into account and recalling that comets are about 25% organic by mass suggests that ~10^{20} kg of cometary organics have impacted Earth over geological time. Clearly, as many investigators have noted, Earth may have acquired a vast quantity of extraterrestrial organics. The key question, however, is what fraction of these organics actually survived atmospheric passage and impact shock heating. We now turn to models of these two processes.

Atmospheric Entry and Ablation

We model atmospheric entry by a finite difference numerical scheme that incorporates aerobraking and ablation. In this model, both atmospheric drag and ablation rate are functions of cross-sectional area, which in turn changes as a result of ablation (51, 52). As a conservative assumption, we treat all organic molecules in ablated material as completely pyrolyzed.

Impactor ablation is determined by the drag coefficient C_D, the heat of ablation ζ, and the heat transfer coefficient C_H. These parameters are functions of impactor shape, composition, mass, and velocity. From aerodynamic theory of hypervelocity impactors in the terrestrial atmosphere, we set $C_H = 0.01$ (21, 53) and $C_D = 0.92$ (21). These values are consistent with typical estimates (52). The exact value of ζ is a function of material type and the specific process of ablation. Passey and Melosh (54) suggested taking ζ for an iron meteorite to be the average of the heats of fusion and vaporization, 5 MJ kg^{-1}. The same procedure, if we use the heats of fusion and vaporization of ordinary chondrites (55), also yields 5 MJ kg^{-1}. Given this chondritic value for ζ, we scale to cometary and carbonaceous chondritic values using observed ablation coefficients $\sigma = C_H/C_D\zeta$ for cometary, carbonaceous chondritic, and ordinary chondritic meteors, taking C_H/C_D to be the same for all three types. [Values of σ for different meteor types have been determined from luminosity observations of some 3000 small meteors of varying compositions, densities, and orbital elements (56).] We find ζ for carbonaceous chondrites and comets to be 3.2 and 1.6 MJ kg^{-1}, respectively. This cometary value is within 25% of the average of the heats of vaporization and fusion of ice (57), a remarkable agreement between observations of cometary meteors and laboratory data. Even models with detailed self-consistent calculations of the above parameters (55) produce values that are nearly constant throughout atmospheric passage.

As initial conditions for atmospheric entry, we need typical cometary impact velocities with Earth. Statistics for SP comet velocities and impact probabilities may be used to calculate the percentage of cometary collisions with Earth that occur at a given velocity (3). Some 25% of SP cometary collisions with Earth occur with velocities at or below 18 km s^{-1}, and another 50% occur with velocities between 18 and 23 km s^{-1}. Because models of the early cometary bombardment of the inner solar system (58) indicate that the flux of comets scattered directly from the Uranus-Neptune region (that is, following SP-like orbits) dominated by several orders of magnitude the flux of those (long-period) comets first scattered out to the Oort cloud, we may use SP velocities as typical for cometary collisions with Earth during the heavy bombardment.

Calculations based on the use of our atmospheric entry model show that, for a cometary impactor (density $\delta = 1$ g cm^{-3}, an initial velocity of 23 km s^{-1}, and $\zeta = 1.6$ MJ kg^{-1}), even impactors of initial radius $r_0 \approx 100$ m have their speeds reduced $\lesssim 10\%$ by aerobraking, if we assume an atmospheric density similar to that of

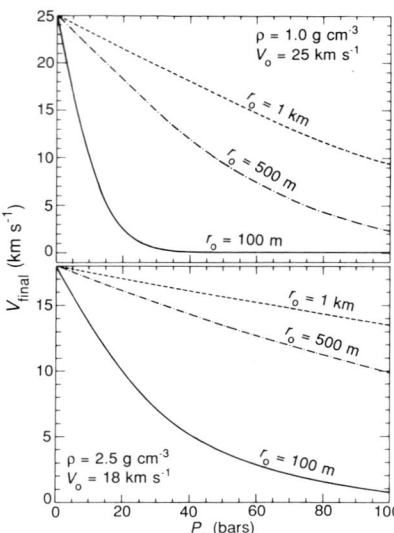

Fig. 1. Impact velocities as a function of surface atmospheric pressure for generic cometary and chondritic impactors at an incidence angle of 0° with radii of 100, 500, or 1000 m. In this illustrative model, the atmosphere is equivalent to the contemporary terrestrial one, with a surface pressure increased to P_0.

present-day Earth (pressure $P_0 = 1$ bar). Impactors with $\delta = 2.5$ g cm^{-3}, an initial velocity of 18 km s^{-1}, and $\zeta = 3.2$ MJ kg^{-1} [parameters appropriate for chondritic asteroids (3, 6)] are aerobraked less than cometary impactors of the same size, because of their greater density. Large objects, with $r_0 \gtrsim 1$ km, impact Earth's surface with their speeds essentially unchanged.

However, aerobraking in a primordial dense CO_2 terrestrial atmosphere may have been much more effective than in the present one. For a primordial atmosphere with $P_0 = 10$-bar CO_2 (14–16), with consequent (16) greenhouse-enhanced surface temperature and H_2O abundance (0.6 bar), 0.8-bar N_2, and resulting scale height (7.3 km), 100-m comets striking the atmosphere at a typical incidence angle of 45° with initial velocities V_0 of 23 or 18 km s^{-1} are aerobraked to final impact speeds of 5.9 km s^{-1}. (For comparison, at an incidence angle of 0°, a comet with $V_0 = 18$ km s^{-1} would be aerobraked under such conditions to a final velocity of 8.1 km s^{-1}.) Asteroidal impactors are not slowed to below ~10 km s^{-1}. Mass loss due to ablation is 65 to 83% for cometary impactors and ≤25% for chondrites with $r_0 = 100$ m (59). Larger objects will be less ablated, because of their smaller ratio of area to volume. In a primordial atmosphere with $P_0 = 20$-bar CO_2, final velocities for 100-m comets at 45° and 23 km s^{-1} are ~1 km s^{-1}. Although the results of these self-consistent 10- and 20-bar early atmospheres will be used for all subsequent calculations, we have used a simplified model in Fig. 1 to provide graphic intuition of cometary and chondritic aerobraking in terrestrial atmospheres ranging from 1 to 100 bars, for impactors with radii of 100, 500, and 1000 m.

Throughout the calculations presented here, we have taken the cometary density $\delta \equiv 1.0$ g cm^{-3}. Calculations of the density of comet Halley based on nongravitational forces suggest a large range of possible densities, 0.2 to 1.5 g cm^{-3} (60). How will a different choice for δ affect the above results? In the aerobraking model used here (51, 52), both deceleration and ablation in the atmosphere are, to a good approximation, proportional to $\rho/\delta r$, where ρ is the atmospheric density, and δ and r are the density and radius of the impactor, respectively. Thus, for a given r, the results found above for aerobraking a comet with a density of 1 g cm^{-3} in a 10-bar atmosphere will be approximately the same as those for a comet with a density of 0.5 g cm^{-3} in a 5-bar atmosphere. This approximation is imperfect, as a 5-bar CO_2 atmosphere will not have exactly the same greenhouse temperature or proportional H_2O abundance (and

hence scale height) as a 10-bar CO_2 one. Are radii of 0.1 to 1.0 km reasonable values for SP comets? Using a cometary radius–absolute magnitude (H_{10}) relation derived from comet Halley, Hughes (61) found that about 6, 45, and 75% of SP comets have radii $r < 0.1$, 0.5, and 1.0 km, respectively. Provided Halley's r-H_{10} relation is roughly representative, SP comets with radii of several hundred meters are not uncommon.

We have not attempted to model the effects of impactor fragmentation due to aerodynamic stresses. At present, our understanding of this process is incomplete, and existing models make many simplifying assumptions (54, 55). Neglecting this process renders our estimates of survivable organic material conservative, as fragmentation greatly increases the effective surface area of an impactor, enhancing aerobraking and reducing impact velocity. This effect, of importance only for impactors with radii ≤1 km, will be discussed further below. Catastrophic fragmentation of a comet or comet fragment appears to have occurred in Tunguska, Siberia, in 1908 (62). Fragmentation of a CI carbonaceous chondrite occurred over Revelstoke, Canada, in 1965; photomicrographs of resulting fragments up to 0.5 cm in size showed no heat damage within an exterior layer ≤0.7 mm thick (63), suggesting that such airbursts would indeed allow organic inclusions to survive. Fragmentation of "sun-grazing" and "Jupiter-grazing" comets, presumably due to tidal stresses, also suggests that some comets are loosely cohesive [tensile strengths of ≤10^3 to 10^5 dyne cm^{-2} (64)] and subject to fragmentation upon entry into Earth's atmosphere.

Hydrodynamic Simulation of Cometary Impact

Given appropriate post-aerobraking collision speeds with Earth, we then calculate resulting shock temperatures throughout the impactor, via a numerical scheme called the smoothed particle hydrodynamics (SPH) method (65). A hydrodynamic model is acceptable because typical shock pressures (~10^{12} to 10^{13} dyne cm^{-2} or ~10^2 to 10^3 GPa) are much higher than the material strengths of rock or ice (~1 GPa). The advantage of SPH in impact simulations is that motion of material is represented by motion of mass points; this avoids problems with complex or tangled grids, boundaries, or multiple materials. SPH allows us to calculate (via equations for continuity, momentum, and energy, applied at the mass points) the density, pressure, their derivatives, and the internal energy of the material, using known statistical methods based on smoothing kernels (66). As each point has a mass and a smoothed density, it may be viewed as a particle. The calculation is advanced forward in time by calculating pressure forces at each mass point and moving the points accordingly.

The SPH formulation allows one-, two-, or three-dimensional simulations to be done with almost equal facility. For constant resolution, however, the number of mass points increases as the power of the dimension. As a result of computing constraints, the SPH simulations reported here are two-dimensional, with a resolution of 8% of the radius of the comet.

An important problem in modeling shock transitions with particle schemes is particle interpenetration at the shock interface. Simple dimensional arguments show that particle interpenetration will occur when the relative velocity of particles is on the order of the local sound speed. To overcome this difficulty when modeling supersonic flows, we have added an artificial viscosity to the momentum equation. In the continuum limit, the artificial viscosity terms reduce to standard viscous terms in the Navier-Stokes equation. The effect of these terms is to broaden the shock front while reducing postshock oscillations. The magnitude of the viscous coefficients is reduced to the minimum consistent with ensuring that particle penetration will not exceed the resolution of the calculation (67).

With SPH we are able to calculate the density and internal energy of every particle at each time step. We ignore radiative heat transfer in our calculations, as it has a negligible effect on calculated temperatures ($\leq 0.01\%$). To calculate pressures and particle tem-

peratures, we need an equation of state (EOS) for each material. We use two independent methods to calculate temperature and pressure from density and internal energy, the Tillotson EOS (TEOS) (68) and the Los Alamos Sesame II EOS database (69).

The TEOS is based on a set of analytic equations that can be used over a wide range of pressures, densities, and materials. The parameters incorporated in the analytic equations are fixed for each material by experimental data. TEOS has been widely used (21, 70), although it is incapable of representing material strength (by negative pressures) and does not adequately represent two-phase media (70). The advantages of TEOS are its computational simplicity and its calculation of pressure as a function of density and internal energy, both, in turn, easily calculable in SPH. The major disadvantage of TEOS is that temperature is not explicitly determined. Fortunately, the specific heat of H_2O is known over a wide range of thermodynamic parameters (20). For our purposes, it is adequate to take the specific heat of water to be a constant [4 kJ $kg^{-1} K^{-1}$ (20)], as we will use TEOS only as a check on results given by the Sesame II database.

The latter is a library of some 100 materials. Pressures and thermal energies for each material are tabulated over a wide range of densities and temperatures; programs are supplied to invert and interpolate from these tables, so that pressure and temperature can be found as a function of density and thermal energy. We have used Sesame II together with our SPH code to trace the evolution of temperature throughout the bodies of comets and asteroids during impact on basalt and ocean surfaces. A wide variety of impact velocities and angles, as well as impactor sizes, has been investigated. Throughout, we have been guided by the conservative requirement that only those collisions that result in some fraction of the impactor remaining heated to a temperature $T \leq 1800$ K for ≤ 1 s are possible contributors to the terrestrial organic inventory. (For projectiles small enough to be sufficiently aerobraked for temperatures to remain below ~ 2000 K, shock temperatures peak and begin to decline in well under 1 s; in calculating organic survivability, we conservatively treat the peak temperature reached by any given particle to extend for this length of time.) For both chondrites and comets impacting basalt surfaces at all plausible impact speeds for bodies with $r_0 > 100$ m, none of the impactor experiences $T < 1800$ K. However, comets impacting the ocean surface at velocities of ≤ 10 km s^{-1} do have some fraction heated to a maximum temperature below this level. [For example, comets with impact speeds of 5.9 km s^{-1}, as found above for aerobraking in a 10-bar CO_2 atmosphere, have 100, 87, and 40% of their mass exposed to temperatures <1800, 1200, and 1000 K, respectively (71). Substantial quantities of HCN, CH_4, H_2CO, and even kerogen-like

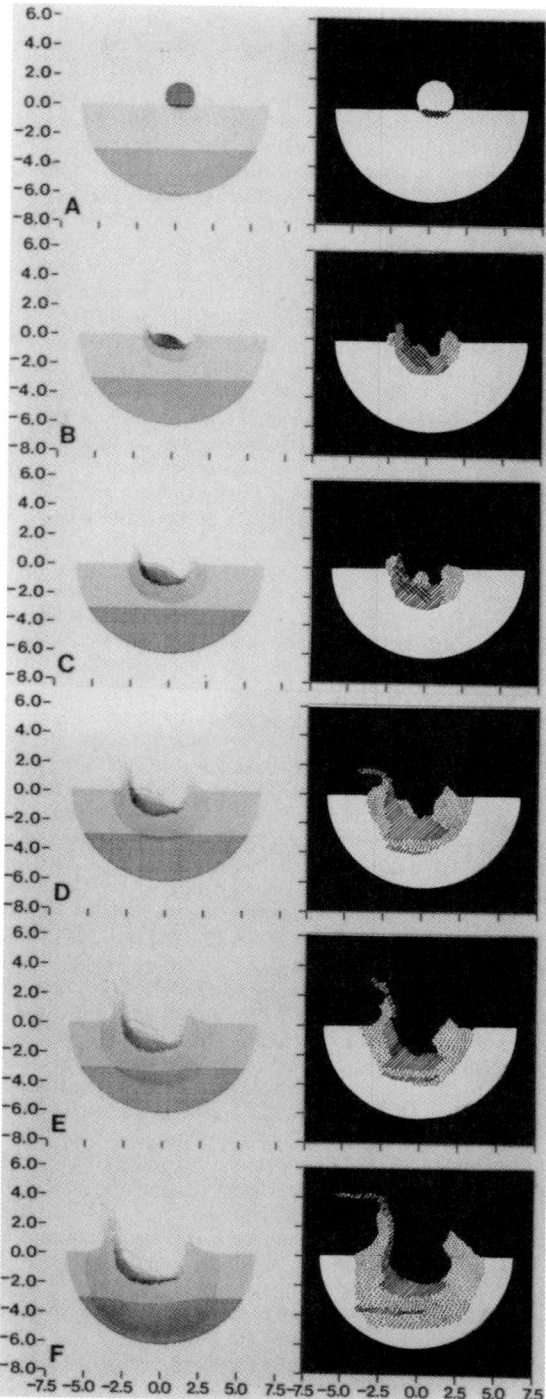

Fig. 2. Step-by-step evolution of an oblique cometary impact at 7.5 km s^{-1} onto an ocean 3 km deep underlain by a basaltic sea floor. To demonstrate the effects of an underlying seabed, we have given the comet a radius of 1 km, although aerobraking would only reduce impact velocity to this extent for a much smaller comet. In the images on the right, the "X" areas represent temperatures in the range 1800 to 2300 K, diagonal 1300 to 1800 K, speckled 800 to 1300 K, and white 300 to 800 K. The accompanying monochrome images identify particles from the comet, ocean, and basaltic sea floor. The images are spaced at intervals of 0.2 s. Note that, despite the oblique imact (**A**), the shock front is circular (**B** and **C**). Distortion of the shock after it reaches the sea floor (**D-F**) is due to the fact that the density of basalt is greater than that of water. A secondary shock is reflected off the seabed (E and F) but does not heat the comet significantly. The higher temperatures generated as the shock front passes into the sea floor are due to the fact that the specific heat capacity of basalt is less than that of water. The coolest fraction of the cometary material is found in the region surrounding that part of the comet opposite the point of impact and in the ejecta sheet. Our simulations show that this material is ejected from the vicinity of the impact without significant further heating.

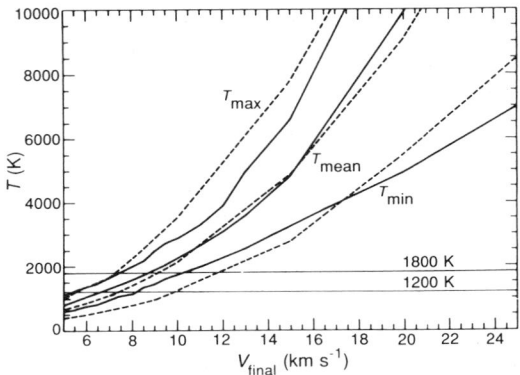

Fig. 3. Maximum, minimum, and mean temperatures for a comet impacting an ocean 3 km deep. The results here are independent of comet radius, provided the ocean is at least several cometary radii deep. The dotted lines denote temperatures obtained from the Tillotson EOS with an assumed constant specific heat capacity for ice. The solid lines represent temperatures from Sesame EOS. The temperatures 1800 and 1200 K are the temperatures for which HCN and simple organics (such as short-chain aliphatics, benzene, and H_2CO), respectively, survive shock heating for time scales comparable with those of impact. For impact velocities ≤ 10 km s^{-1}, a significant fraction of the organic inventory survives impact.

heteropolymers and amino acids would survive this impact intact.] This is so because the greater specific heat and lower density of water result in a smaller fraction of the kinetic energy of the impactor being converted into impactor thermal energy. In these calculations, neither the depth of the ocean nor the absolute size of the impactor is important, provided the ocean is at least several impactor radii deep. These results are essentially independent of impact angle for the cases calculated (up to 45° to the normal). Figure 2 illustrates the step-by-step evolution traced by our code for an ocean-impacting comet. Figure 3 shows the resulting maximum, mean, and minimum temperatures experienced as a function of impact velocity.

Cometary Delivery of Organics to Early Earth

To determine the flux of cometary organics delivered intact to early Earth, we must convolve the incident cometary mass flux, the cometary organic mass fraction, ablative loss during atmospheric passage, and pyrolysis of organics on impact. The relevant incident cometary mass flux is that corresponding to objects 100 to 200 m in size, because our simulations show that, in a 10-bar CO_2 atmosphere, only these will be aerobraked to impact velocities low enough to permit organics to survive.

Equations 1, 2 and 3 will yield the total mass incident on Earth in comets in a given size range since time t. A much more useful quantity is the mass flux (in kilograms per year) at a particular time in Earth's past, which both facilitates comparison with photochemical or other production rates of prebiotic organics (*10, 72–74*) and allows us to state our conclusions contingent on particular models for early terrestrial atmospheric evolution and the origin of life, for example, "if the time of interest for stocking the prebiotic inventory was 4.2×10^9 years ago and if the atmospheric density at that time was 10 bars, then the flux of intact cometary organics was $\dot{m}_{org}$ kilograms per year." [For example, it may be that delivery of organic molecules before some time t was useless to the origins of life, because of subsequent surface pyrolysis of organics by giant impacts (*50*).] In order to present our results in this form, we find

$\dot{m}(t) = \partial M(t)/\partial t$, with $M(t)$ taken from Eqs. 1 through 3, and scale to Earth. We calculate ablation, aerobraking, and subsequent shock-heating and pyrolysis over mass bins for m_1 and m_2 corresponding to comets with radii from r_1 and r_2, taking radial bin sizes of 50 m.

Equations 4 through 6 then give $\dot{m}_{org}(t)$, the rate of organic accretion by Earth, as a function of time:

$$\dot{m}_{org}(t) = 420\ (1 + 2.1 \times 10^{-6}e^{t/\tau_A})\ (\alpha/0.1)\ \text{kg year}^{-1} \quad (4)$$

$$\dot{m}_{org}(t) = 540\ (1 + 1.6 \times 10^{-10}e^{t/\tau_B})\ (\alpha/0.1)\ \text{kg year}^{-1} \quad (5)$$

$$\dot{m}_{org}(t) = 220\ (1 + 8.0 \times 10^{-22}e^{t/\tau_C})\ (\alpha/0.1)\ \text{kg year}^{-1} \quad (6)$$

These equations require a choice of α only for objects with radii ~ 100 m; because larger objects do not contribute to $\dot{m}_{org}(t)$, the cometary percentage of these impactors is irrelevant to our results. Equations 4 through 6 imply that, in a 10-bar CO_2 atmosphere 4.5×10^9 years ago, Earth was accreting organics at the rate of 7×10^5, 3×10^6, and 1×10^9 kg year^{-1}, respectively. (The 25% lowest velocity comets account for $\geq 50\%$ of these.) This flux thereafter declined with time constants τ_A, τ_B, or τ_C (Fig. 4). Using the empirical constraints on the impact record described above, we find that the most likely range of organic accretion is $\sim 10^6$ to 10^7 kg year^{-1}. (Of course, both ultraviolet photolysis of organics settling and raining out of the atmosphere and thermal and cosmic-ray decomposition of those accumulating on the surface would lower this total.) Intact heteropolymer and its pyrolyzates each represented $\sim 40\%$, and HCN, CH_4, and H_2CO represent about 9, 6, and 4%, respectively, of this surviving organic mass fraction. Amino acids, if present at carbonaceous chondrite abundances, were $\sim 1\%$ of the total. (There is a kind of impact selection of the thermodynamically most stable species, but significant amounts of comparatively labile organics also survive impact.) For a 20-bar atmosphere 4.5×10^9 years ago, Earth would have accreted only several times as many cometary organics, although these would have experienced considerably less shock heating. These results may be put in context by comparison with terrestrial oceanic and total biomasses, $\sim 3 \times 10^{12}$ kg and $\sim 6 \times 10^{14}$ kg, respectively (*49*).

We have not treated the possibility of the delivery of intact organics by efficient fragmentation or airburst. We estimate the magnitude of this effect in the following way. A bolide will undergo airburst when the differential pressure between its leading and trailing hemispheres exceeds its tensile strength. But for comets with radii $r_0 \gtrsim 1$ km, an atmospherically induced pressure wave would not have time to traverse the comet before terrestrial impact, so that only impactors with $r_0 \lesssim 1$ km may undergo airburst (*75*). From Eqs. 1 through 3, the incident mass ratio between comets in the ranges 200 to 1000 m and 50 to 200 m is about 13. If all comets in the range 200 to 1000 m were to undergo airburst and if all their organics were to reach the surface intact, then our previous estimates of terrestrial organic accretion would be too low by a factor of ~ 50 (*76*).

Zhao and Bada (*77*) have recently identified large quantities of apparently extraterrestrial nonprotein amino acids, α-aminoisobutyric acid (AIB) and racemic isovaline, in Cretaceous-Tertiary (K/T) boundary sediments. Because iridium abundances require the K/T bolide (assuming a single object was responsible) to have been ~ 10 km in radius, according to our results amino acids (or any other organics) could not have survived the resulting impact in a presumptive 1-bar atmosphere 65×10^6 years ago. As just argued, neither would a 10-km-diameter bolide have undergone airburst, so it appears that the claimed extraterrestrial amino acids cannot be the result of intact delivery by a comet or asteroid. It has been suggested that dust evolved from a large comet trapped in the inner solar system may have provided the source (*78*). Alternatively, the K/T amino acids may be the result of post-impact quench synthesis. Barak and Bar-Nun (*73*) have found efficient production of AIB (at

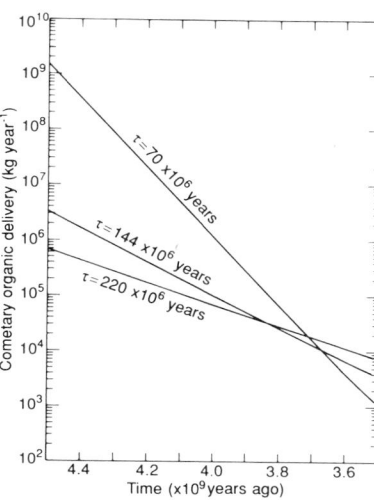

Fig. 4. Cometary delivery of organics to Earth as a function of time (from Eqs. 4 through 6), for a 10-bar CO_2 atmosphere and a typical cometary density δ of ~1 g cm^{-3}. If comets instead have a density of ~0.5 g cm^{-3}, the results shown would correspond approximately to 5-bar CO_2 atmospheres.

levels ~10% that of glycine and alanine) in shock tube experiments with initial mixtures of 10% ethane (C_2H_6) + 10% NH_3 in argon + air + H_2O. Production efficiency seems to be unaffected by the background air, at least up to O_2/C_2H_6 ratios of 2.5. Therefore, the quench synthesis explanation for K/T amino acids is not yet ruled out.

Anders (79) recently suggested that interplanetary dust may have been an important source of prebiotic organics for early Earth. He calculates the current infall rate of organic carbon in interplanetary dust to be $f \approx 3 \times 10^5$ kg year^{-1}. If we assume that f scales proportionally to the impactor flux in the inner solar system, the time dependencies of Eqs. 4 through 6 may be used to extrapolate f into the past. We then find a terrestrial infall rate 4.5 × 10^9 years ago of organic carbon from interplanetary dust of ~10^8 to 10^{10} kg year^{-1}.

Even in ~10-bar early terrestrial atmospheres, this rate exceeds the conservative estimates we have derived here for ~100-m comets. However, we have left almost entirely unquantified likely augmentations to the latter estimate, including those arising from ablated material, airbursts, lower temperatures implied by three-dimensional impact simulations [as opposed to the two-dimensional simulations used here (71)], a possible extremely thermally stable fraction of cometary organics, and a relaxation of our approximation that peak shock heating extends over the full 1 s of the impact simulation. Moreover, the role of (small) carbonaceous asteroids is yet to be determined. In any case, the nature of the organics provided by interplanetary dust and impacting comets would have been radically different. Comets could deliver extremely volatile molecules, such as CH_4 or nonpolymeric HCN and H_2CO, whereas the organics in interplanetary dust are almost certainly highly processed and refractory.

Finally, we emphasize that we have ignored a host of possible mechanisms for impact generation of organics, including production by shock waves in reducing or neutral atmospheres (72, 73) or even by vaporization and recombination of target material (80). Moreover, substantial organic recombination may take place by the quenching of the vaporized cometary projectile itself (81). However, efficient organic synthesis in dense background CO_2 atmospheres by means of such recombination is problematic.

Typical estimates of photochemical production rates for HCN and H_2CO in terrestrial CO_2 atmospheres (taking into account rainout efficiencies) are ~10^9 kg year^{-1} (74). But these rates drop precipitously as the H_2/CO_2 ratio falls below ~1 (17). It is

intriguing that it is exactly these dense CO_2 atmospheres, where photochemical production of organic molecules should be the most difficult, in which intact cometary organics would be delivered in large amounts.

REFERENCES AND NOTES

1. E. Anders and T. Owen, *Science* **198**, 453 (1977).
2. C. F. Chyba, *Nature* **330**, 632 (1987).
3. ———, *ibid.* **343**, 129 (1990).
4. L. L. Wilkening, *Naturwissenschaften* **65**, 73 (1978).
5. J. R. Cronin, S. Pizzarello, D. P. Cruikshank, in *Meteorites and the Early Solar System*, J. F. Kerridge and M. S. Matthews, Eds. (Univ. of Arizona Press, Tucson, 1988), pp. 819–857.
6. E. M. Shoemaker, J. G. Williams, E. F. Helin, R. F. Wolfe, in *Asteroids*, T. Gehrels, Ed. (Univ. of Arizona Press, Tucson, 1979), pp. 253–282.
7. T. C. Chamberlin and R. T. Chamberlin, *Science* **28**, 897 (1908).
8. J. Oró, *Nature* **190**, 389 (1961).
9. S. Chang, in *Space Missions to Comets*, M. Neugebauer, D. K. Yeomans, J. C. Brandt, Eds. (NASA Conference Publication 2089, National Aeronautics and Space Administration, Washington, DC, 1978), pp. 59–111; C. Ponnamperuma, *Comets and the Origin of Life* (Reidel, Boston, 1981); J. M. Greenberg, in *The Galaxy and the Solar System*, R. Smoluchowski, J. N. Bahcall, M. S. Matthews, Eds. (Univ. of Arizona Press, Tucson, 1986), pp. 103–115; P. J. Thomas, C. F. Chyba, L. Brookshaw, C. Sagan, *Lunar Planet. Sci. Conf.* **20**, 117 (1989); A. Delsemme, *Adv. Space Res.* **9** (no. 6), 25 (1989).
10. S. Chang, D. DesMarais, R. Mack, S. L. Miller, G. E. Strathearn, in *Earth's Earliest Biosphere*, J. W. Schopf, Ed. (Princeton Univ. Press, Princeton, NJ, 1983), pp. 53–92.
11. B. C. Clark, *Origins Life* **18**, 209 (1988).
12. J. C. G. Walker, *Evolution of the Atmosphere* (Macmillan, New York, 1977); D. J. Stevenson, in *Earth's Earliest Biosphere*, J. W. Schopf, Ed. (Princeton Univ. Press, Princeton, NJ, 1983), pp. 32–40.
13. W. R. Kuhn and S. K. Atreya, *Icarus* **37**, 207 (1979); J. S. Levine and T. R. Augustsson, *Origins Life* **15**, 299 (1985).
14. J. C. G. Walker, *Origins Life* **16**, 117 (1986).
15. J. F. Kasting, O. B. Toon, J. B. Pollack, *Sci. Am.* **258**, 90 (February 1988); J. F. Kasting and O. B. Toon, in *Origin and Evolution of Planetary and Satellite Atmospheres*, S. K. Atreya, J. B. Pollack, M. S. Matthews, Eds. (Univ. of Arizona Press, Tucson, 1989), pp. 423–449; R. Durham and J. W. Chamberlain, *Icarus* **77**, 59 (1989).
16. J. F. Kasting and T. P. Ackerman, *Science* **234**, 1383 (1986); J. F. Kasting, *Palaeogeogr. Palaeoclimatol. Palaeoecol.* **75**, 83 (1989).
17. G. Schlesinger and S. L. Miller, *J. Mol. Evol.* **19**, 376 (1983); *ibid.*, p. 383.
18. K. J. Zahnle, *J. Geophys. Res.* **91**, 2819 (1986).
19. J. F. Kasting, *Origins Life* **19**, 225 (1989); in preparation.
20. I. V. Shurshalov, *Izv. Akad. Nauk SSSR Mekh. Zhidk. Gaza* **2**, 184 (1967); *Fluid Dyn.* **2** (no. 6), 133 (1967) (English transl.).
21. J. D. O'Keefe and T. J. Ahrens, in *Geological Implications of Impacts of Large Asteroids and Comets on the Earth*, L. T. Silver and P. H. Schultz, Eds. (SP-190, Geological Society of America, Boulder, CO, 1982), pp. 103–120.
22. F. R. Krueger and J. Kissel, *Naturwissenschaften* **74**, 312 (1987).
23. A. H. Delsemme, *Philos. Trans. R. Soc. London Ser. A* **325**, 509 (1988).
24. E. Anders and N. Grevesse, *Geochim. Cosmochim. Acta* **53**, 197 (1989); E. K. Jessberger, A. Christoforidis, J. Kissel, *Nature* **332**, 691 (1988).
25. Production of H_2CO in comet Halley was observed from one of about four vents only, at ~4.5% that of H_2O by number. This corresponds to about 2% H_2O by mass overall. See M. J. Mumma and D. C. Reuter, *Astrophys. J.* **344**, 940 (1989); D. C. Reuter and M. J. Mumma, *Bull. Am. Astron. Soc.* **21**, 937 (1989).
26. M. Allen *et al.*, *Astron. Astrophys.* **187**, 502 (1987).
27. H. P. Larson, H. A. Weaver, M. J. Mumma, S. Drapatz, *Astrophys. J.* **338**, 1106 (1989).
28. S. Epstein, R. V. Krishnamurthy, J. R. Cronin, S. Pizzarello, G. U. Yuen, *Nature* **326**, 477 (1987).
29. L. E. Snyder *et al.*, *Astrophys. J.* **268**, 123 (1983), and references therein.
30. F. Hoyle and N. C. Wickramasinghe, *Nature* **328**, 117 (1987); *ibid.* **331**, 123 (1988); *ibid.*, p. 666.
31. C. F. Chyba and C. Sagan, *ibid.* **329**, 208 (1987); *ibid.* **332**, 592 (1988); in *Interstellar Dust: Contributed Papers*, A. G. G. M. Tielens and L. J. Allamandola, Eds. (NASA Conference Publication 3036, National Aeronautics and Space Administration, Washington, DC, 1989), pp. 433–435.
32. C. Sagan and B. N. Khare, *Nature* **277**, 102 (1979); ———, J. S. Lewis, in *Saturn*, T. Gehrels and M. S. Matthews, Eds. (Univ. of Arizona Press, Tucson, 1984), pp. 788–807.
33. J. M. Greenberg, in *Comets*, L. L. Wilkening, Ed. (Univ. of Arizona Press, Tucson, 1982), pp. 131–163; W. R. Thompson, B. G. J. P. T. Murray, B. N. Khare, C. Sagan, *J. Geophys. Res.* **92**, 14933 (1987); B. N. Khare *et al.*, *Icarus* **79**, 350 (1989).
34. C. Chyba and C. Sagan, *Nature* **330**, 350 (1987); ———, M. J. Mumma, *Icarus* **79**, 362 (1989).
35. J. Gradie and J. Veverka, *Nature* **283**, 840 (1980); D. P. Cruikshank *et al.*, *Icarus* **53**, 90 (1983).
36. E. T. Arakawa *et al.*, *Bull. Am. Astron. Soc.* **21**, 940 (1989).
37. We calculate 1-second survival temperatures from the Arrhenius equation in the form $T = (E_a/R)[\ln(\tau A)]^{-1}$, where temperature T is in kelvins, R is the gas constant, τ is the duration of shock heating (taken to be 1 s), and A (in s^{-1}) and E_a

are the pre-exponential and activation energy kinetic parameters. For type I kerogen, $A = 1.7 \times 10^{14}$ s^{-1} and $E_a = 55.19$ kcal mol^{-1}; for type II, $A = 6.7 \times 10^{12}$ s^{-1} and $E_a = 48.82$ kcal mol^{-1}. See A. K. Burnham, R. L. Braun, H. R. Gregg, A. M. Samoun, *Energy Fuels* **1**, 452 (1987).

38. C. Arnosti and P. J. Müller, *Org. Geochem.* **11**, 505 (1987).
39. For benzene, $A = 5 \times 10^{16}$ s^{-1} and $E_a = 88.2$ kcal mol^{-1}; see S. Khandelwal and G. Skinner, in *Shock Waves in Chemistry*, A. Lifshitz, Ed. (Dekker, New York, 1981), pp. 1–57.
40. For ethane, $A = 3.2 \times 10^{16}$ s^{-1} and $E_a = 88.0$ kcal mol^{-1}; see W. Tsang, *Int. J. Chem. Kinet.* **10**, 41 (1978).
41. Shock experiments with CH_4 yield $A = 1.3 \times 10^{15}$ s^{-1} and $E_a = 104$ kcal mol^{-1} (*39*).
42. H_2CO experimental results have been compiled by J. Warnatz, in *Combustion Chemistry*, W. C. Gardiner, Ed. (Springer-Verlag, New York, 1984), chap. 5.
43. Shock-tube experiments for HCN are summarized in A. Szekely, R. K. Hanson, C. T. Bowman, *J. Phys. Chem.* **88**, 666 (1984).
44. We know of no shock-tube pyrolysis experiments for HCN or H_2CO conducted at sufficiently high pressures to yield first-order rate constants. Shock-tube pyrolysis of these compounds at low (~1 bar) pressures have provided second-order rate constants (*42, 43*), giving decomposition half-lives dependent on the concentration of background gas (usually argon), which cannot be reliably extrapolated to cometary impacts. We take activation energies for HCN and H_2CO decomposition from those shock pyrolysis experiments with temperatures most appropriate for the impacts considered here: $E_a = 76.5$ kcal mol^{-1} for $H_2CO + Ar \rightarrow H + HCO + Ar$ with $1000\ K < T < 3000\ K$ (*42*), and $E_a = 117$ kcal mol^{-1} for $HCN + Ar \rightarrow H + CN + Ar$ with $2200 < T < 2700$ [P. Roth and Th. Just, *Ber. Busenges. Phys. Chem.* **80**, 171 (1976)]. First-order kinetic parameters for breaking C–H bonds in simple hydrocarbons give A in the range 2.0×10^{13} to 1.0×10^{15} s^{-1}, with typical values of ~10^{14} s^{-1}. We therefore adopt $A = 10^{14}$ s^{-1} for HCN and H_2CO. Altering this choice by a factor of ~10 results in a change in the 1-s survival temperature by only about 7%. We consider the choice of these parameters for HCN to be conservative, in the sense that they may underestimate HCN survival in impacts. HCN spectra consistent with temperatures as high as 2800 K have been observed in carbon stars; see, for example, J. H. Gobel *et al.*, *Astrophys. J.* **235**, 104 (1980).
45. Thermal degradation of alanine in dilute solution yields the first-order parameters $A = 2.2 \times 10^{11}$ s^{-1} and $E_a = 44$ kcal mol^{-1}. See P. H. Abelson, in *Researches in Geochemistry*, P. H. Abelson, Ed. (Wiley, New York, 1959), pp. 79–103.
46. This temperature range has as its end members the amino acids threonine, with $A = 2 \times 10^{12}$ s^{-1} and $E_a = 33.8$ kcal mol^{-1}, and phenylalanine, with $A = 1.7 \times 10^{8}$ s^{-1} and $E_a = 30.8$ kcal mol^{-1}. See J. R. Vallentyne, *Geochim. Cosmochim. Acta* **28**, 157 (1964). The major primary decomposition step for α-amino acids is decarboxylation [M. A. Ratcliff, E. E. Medley, P. G. Simmonds, *J. Org. Chem.* **39**, 1481 (1974)]; a vapor phase study of decarboxylation in 3-butenoic acid gives $A = 2.2 \times 10^{11}$ s^{-1} and $E_a = 40.6$ kcal mol^{-1}, or survival for 1 s at 780 K [G. G. Smith and S. E. Blau, *J. Phys. Chem.* **68**, 1231 (1964); S. W. Benson and H. E. O'Neal, *Kinetic Data on Gas Phase Unimolecular Reactions* (Publication NSRDS-NBS 21, U.S. Government Printing Office, Washington, DC, 1970)].
47. M. Schidlowski, *Nature* **333**, 313 (1988).
48. E. M. Shoemaker, *Annu. Rev. Earth Planet. Sci.* **11**, 461 (1983); ——— and R. F. Wolfe, in *The Galaxy and the Solar System*, R. Smoluchowski, J. N. Bahcall, M. S. Matthews, Eds. (Univ. of Arizona Press, Tucson, 1986), pp. 338–386.
49. J. M. Hayes, I. R. Kaplan, K. W. Wedeking, in *Earth's Earliest Biosphere*, J. W. Schopf, Ed. (Princeton, Univ. Press, Princeton, NJ, 1983), pp. 93–134.
50. N. H. Sleep, K. J. Zahnle, J. F. Kasting, H. J. Morowitz, *Nature* **342**, 139 (1989).
51. E. J. Öpik, *Physics of Meteor Flight in the Atmosphere* (Interscience, New York, 1958).
52. H. J. Melosh, *Impact Cratering* (Oxford Univ. Press, New York, 1989).
53. H. J. Allen, A. Seiff, W. Winovich, *NASA Technical Report TR R-185* (National Aeronautics and Space Administration, Washington DC, 1963); V. C. Liu, *Geophys. Res. Lett.* **5**, 309 (1978).
54. Q. R. Passey and H. J. Melosh, *Icarus* **42**, 211 (1980).
55. B. Baldwin and Y. Shaeffer, *J. Geophys. Res.* **76**, 4653 (1971).
56. Z. Ceplecha, in *Comets, Asteroids, and Meteorites*, A. H. Delsemme, Ed. (University of Toledo, Toledo, OH, 1977), chap. 5; V. A. Bronshten, *Physics of Meteoric Phenomena* (Reidel, Boston, 1983).
57. P. W. Atkins, *Physical Chemistry* (Freeman, New York, ed. 3, 1986).
58. J. A. Fernández and W.-H. Ip, *Icarus* **54**, 377 (1983); E. M. Shoemaker and R. F. Wolfe, *Lunar Planet. Sci. Conf.* **15**, 780 (1984).
59. Reducing ζ tends to reduce impact speed and increase ablation. For example, an impactor with $r_0 = 100$ m, $\delta = 1$ g cm^{-3}, an initial velocity of 23 km s^{-1}, and an impact angle of 45° in a 10-bar CO_2 atmosphere will impact with velocities of 5.9, 7.9, and 8.4 km s^{-1} and will experience 83, 56, and 40% ablation for values of ζ appropriate to comets, carbonaceous chondrites, and ordinary chondrites, respectively.
60. R. Z. Sagdeev, P. E. Elyasberg, V. I. Moroz, *Nature* **331**, 240 (1988).
61. D. W. Hughes, in *Symposium on the Diversity and Similarity of Comets* (European Space Agency, Brussels, 1987), pp. 43–48; *Icarus* **73**, 149 (1988).
62. B. Y. Levin and V. A. Bronshten, *Meteoritics* **21**, 199 (1986), and references therein.
63. R. E. Folinsbee, J. A. V. Douglas, J. A. Maxwell, *Geochim. Cosmochim. Acta* **31**, 1625 (1967).
64. Z. Sekanina, in *Comets*, L. L. Wilkening, Ed. (Univ. of Arizona Press, Tucson, 1982), pp. 251–287.
65. J. J. Monaghan, *Comput. Phys. Rep.* **3**, 71 (1985); H. Pongracic, thesis, Monash University, Clayton, Australia (1988).
66. SPH is based on the interpolation of a function f such that

$$\langle f(\mathbf{r},h)\rangle = \int_V f(\mathbf{r}')W(\mathbf{r} - \mathbf{r}',h)\,d\mathbf{r}' \qquad (7)$$

where $\mathbf{r} = \mathbf{r}(x,y,z)$, W is the interpolating, or "smoothing" kernel, h is the smoothing length, and V is the solution domain. The smoothing length defines the resolution of the calculation. Any feature in f that has a length scale $<< h$ will be smoothed out and will not appear in $\langle f \rangle$. This is an important consideration for impact calculations, as the width of the shock front can never be smaller than h and is normally ~$2h$. Derivatives of f can be estimated in a similar fashion. With the assumption $fW \rightarrow 0$ on the boundary (satisfied for most problems, since $W \rightarrow 0$ as $|\mathbf{r} - \mathbf{r}'| \rightarrow \infty$), we have

$$\left\langle \frac{\partial f}{\partial x} \right\rangle = -\int_V f \frac{\partial}{\partial x'} W(\mathbf{r} - \mathbf{r}',h)\,d\mathbf{r}' \qquad (8)$$

so that derivatives of f are replaced by derivatives of W. If f is known at the points $\mathbf{r}_i$, where $i = 1, 2, \ldots, N$, the equations above can be approximated by summations. The summation equations produced from the equations of motion of the particles are then easily solved by computer.
67. The need to restrict the interpenetration of particles at a shock interface places a lower limit on the magnitude of the artificial viscosity. All results quoted here use this lower limit. No constraint is placed on the upper limit of the artificial viscosity. Tests done to examine the effect of increasing the artificial viscosity show that variations of ~10% occur in the results, insignificant relative to other uncertainties in the calculation. In most problems the inclusion of an artificial viscosity gives satisfactory results. However, spurious heating occurs when materials collide supersonically. This effect (called wall heating) has been shown to be greatly reduced, and the shock fronts sharpened [W. H. Noh, *J. Comput. Phys.* **72**, 78 (1987)], by the inclusion of an artificial thermal diffusion. We include an artificial thermal diffusion along the lines of that of J. J. Monaghan, "SPH Meets the Shocks of Noh" (personal communication, 1988).
68. J. H. Tillotson, *General Atomic Corp. Rep. GA-3216* (1962).
69. "Sesame '83: Report on the Los Alamos Equation-of-State Library" (LALP-83-4, Los Alamos National Laboratory, Los Alamos, NM, 1983).
70. W. Benz, W. L. Slattery, A. G. W. Cameron, *Icarus* **71**, 30 (1987); *ibid.* **74**, 516 (1988); W. Benz, A. G. W. Cameron, H. J. Melosh, *ibid.* **81**, 113 (1989).
71. Preliminary three-dimensional models of cometary impacts indicate that our estimates of the fraction of the comet exposed to $T < 1200$ K may be underestimates, in some cases by as much as a factor of ~10. This is consistent with work reported by W. T. Brown, in *Shock Waves in Condensed Matter*, W. J. Nellis, L. Seman, R. A. Graham, Eds. (AIP Conference 78, American Institute of Physics, New York, 1981), pp. 529–533, who found pressure reductions by factors of ~2 for three-dimensional codes over two-dimensional approximations. These differences will be further explored in future work (P. J. Thomas, L. Brookshaw, C. F. Chyba, C. Sagan, in preparation).
72. A. Bar-Nun, N. Bar-Nun, S. H. Bauer, C. Sagan, *Science* **168**, 470 (1970); B. Fegley, R. G. Prinn, H. Hartman, G. H. Watkins, *Nature* **319**, 305 (1986).
73. I. Barak and A. Bar-Nun, *Origins Life* **6**, 483 (1975).
74. R. Stribling and S. L. Miller, *ibid.* **17**, 261 (1987); J. P. Pinto, G. R. Gladstone, Y. L. Yung, *Science* **210**, 183 (1980).
75. This does not require all comets with radii ≤1 km to undergo airburst, however. P. R. Weissman [*Nature* **314**, 517 (1985)] cited two terrestrial impact craters of diameters 8.5 and 14 km, for which impactor compositions are known to have been either carbonaceous chondritic or cometary. These craters are consistent with impacts by either intact bolides or tightly clustered fragments (*54*) but are certainly inconsistent with a Tunguska-like airburst. Applying a standard impactor mass-diameter scaling law (*3*) for either a C-chondrite or a comet impacting Earth at ~20 km s^{-1} yields radii of ~200 and 600 m for the two impactors.
76. A more careful examination would treat cometary fragments after an airburst according to the methods outlined in this article. One way to calculate the mass fraction of fragments in a given size range is to assume a power-law distribution of fragment sizes [D. L. Turcotte, *J. Geophys. Res.* **91**, 1921 (1986)]. One such distribution is $N(m) = Cm^{-D/3}$, where $N(m)$ is the number of fragments with mass $> m$, C is a constant fixed by the total mass, and D is a material-dependent constant. This equation can be used to calculate the mass fraction of fragments in the range 50 to 200 m, which may then be combined with Eqs. 1 through 3 to calculate the mass of surviving organics. For geological materials D is typically ~2.5; more fragile materials have smaller values. If D is in the range 2.2 to 2.8, we find that our neglect of fragmentation underestimates terrestrial organic accretion by a factor of 4 to 9.
77. M. Zhao and J. L. Bada, *Nature* **339**, 463 (1989).
78. K. Zahnle and D. Grinspoon, in preparation.
79. E. Anders, *Nature* **342**, 255 (1989).
80. L. M. Mukhin, M. V. Gerasimov, E. N. Safonova, *ibid.* **340**, 46 (1989).
81. C. P. McKay, W. R. Borucki, D. R. Kujiro, F. Church, *Lunar Planet. Sci. Conf.* **20**, 671 (1989); V. R. Oberbeck, C. P. McKay, T. W. Scattergood, G. C. Carle, J. R. Valentin, *Origins Life* **19**, 39 (1989).
82. We thank H. Pongracic for advice on using SPH in impact calculations and for allowing us to read draft copies of her thesis (*65*), A. Burnham and D. Camp for helpful discussions and preprints on kerogen pyrolysis, and S. Lyon for help in implementing the Los Alamos Sesame database. We are especially grateful to E. Anders, B. Clark, K. Zahnle, and two anonymous referees for their comments on an earlier version of this manuscript. We have benefited greatly from discussions with S. L. Miller, J. Oró, G. Hoffman, C. P. McKay, J. Melosh, J. Kasting, J. L. Bada, A. Bar-Nun, S. Chang, P. Gierasch, and P. Schultz. This work was supported by National Aeronautics and Space Administration grants NGR 33-010-220, NGR 33-010-101 (C.S.), NGT-50302 (C.F.C.), NEGW-1023 (P.J.T.), Office of Naval Research grant N00014-B3-K-0610 (L.B.), and a grant from the Kenneth T. and Eileen L. Norris Foundation.

Pre-biotic organic matter from comets and asteroids

Edward Anders

Enrico Fermi Institute and Department of Chemistry, University of Chicago, Chicago, Illinois 60637-1433, USA

SEVERAL authors[1-3] have suggested that comets or carbonaceous asteroids contributed large amounts of organic matter to the primitive Earth, and thus possibly played a vital role in the origin of life. But organic matter cannot survive the extremely high temperatures ($>10^4$ K) reached on impact, which atomize the projectile and break all chemical bonds. Only fragments small enough to be gently decelerated by the atmosphere—principally meteors of 10^{-12}–10^{-6} g—can deliver their organic matter intact[4]. The amount of such 'soft-landed' organic carbon can be estimated from data for the infall rate of meteoritic matter. At present rates, only ~0.006 g cm^{-2} intact organic carbon would accumulate in 10^8 yr, but at the higher rates of ~4×10^9 yr ago, about 20 g cm^{-2} may have accumulated in the few hundred million years between the last cataclysmic impact and the beginning of life. It may have included some biologically important compounds that did not form by abiotic synthesis on Earth.

Figure 1 shows the present estimates of the meteoritic mass influx on Earth[5], including meteors, meteorites and crater-forming bodies (asteroids and comets). The meteor data[6] cover only the past few decades, but as they agree[7] with noble-metal measurements in deep-sea sediments[5,8] and lunar soils[9], they should be representative of the past ~3.6×10^9 yr. The cratering curve should be similarly representative, being based in part on crater counts on lunar maria that are 3.3–3.8×10^9 yr old (ref. 10).

Now consider the size ranges in which organic matter can be preserved (shaded areas in Fig. 1). At very small masses, ultraviolet photolysis alters and destroys the organic matter. For the sake of definiteness I set the limit for such destruction at $<10^{-12}$ g. At higher masses, atmospheric friction heats the particle to increasingly higher temperatures. The nominal threshold for boiling of silicates is $>10^{-6}$ g, but it depends strongly on density, velocity, entry angle and fragmentation[6,11]. Well-preserved interplanetary dust particles (containing noble gases, nuclear tracks and organic matter) range up to 50 μm in size[11] and as many of these may be fragments of larger aggregates or particles, I adopt 10^{-6} g as the upper limit for survival of organic matter. In the next few intervals, much or all of the mass is lost by ablation[6,11], and any remainder is strongly heated by conduction. Ablation calculations and thermoluminescence data show that pyrolytic temperatures (>600 °C) generally extend <1 mm into the meteorite[12], and thus survival of organic matter—at least for low-velocity objects—becomes possible at larger masses, say 10 g. At still larger masses, atmospheric deceleration becomes progressively less effective, causing increasing destruc-

tion in the atmosphere[13] or on impact. A reasonable upper limit is 10^8 g, as meteorites of that mass usually reach the ground with a significant fraction of their cosmic velocity. (Also, the largest stony meteorite showers have masses of only a few tonnes.)

It is useful to estimate the accretion rate not only of organic C but also of total C. The first four lines of Table 1 give such data for the main types of meteoritic material. A few numbers require justification.

For meteors, I use the mean carbon value of ~10% for interplanetary dust particles[14] rather than the 24% reported for comet Halley dust[15], as the latter was measured near the comet nucleus where ices had not yet completely vapourized. This value should also apply to dead comets, which are far more likely to strike the Earth than live comets. (The luminous lifetime of an Earth-crossing short-period comet is typically ~10^3–10^4 revolutions, whereas its dynamical lifetime is ~10^7 revolutions.) For the crater-forming bodies, I assume equal numbers of comets, carbonaceous asteroids and non-carbonaceous asteroids. Several lines of evidence indicate a ratio of ~1:1 for the two asteroid classes (see below), but the fraction of comets is poorly constrained, with estimates[10,16] in the range 10–70%. For meteorites, we assume a carbon content of 1.3%, representing 50% ordinary chondrites (median ~0.1% C) and 50% carbonaceous chondrites (range 0.35–4.8% C). Although the latter constitute only 6% of observed stony meteorite falls, this rarity reflects preferential destruction in the atmosphere owing to their friable nature. Carbonaceous objects seem to account for about one-half of the photographic fireballs[17] with final velocities $\leqslant 8$ km s^{-1}, one-half of the asteroids near the 3:1 Kirkwood gap which are thought to be the principal source of meteorites[18], 13 out of 32 Apollo–Amor asteroids[19] and 3 out of 5 chondritic impactors chemically identified at terrestrial craters[20].

Next, I consider the unmelted material. Meteors pass through the atmosphere intact, but meteorites lose much of their mass by ablation. The ablation loss depends strongly on velocity, mass and crushing strength[13]. I shall use 90% for non-carbonaceous chondrites[21] and 99.4% for carbonaceous chondrites (the latter figure is chosen to reduce the proportion of C-chondrites from 50% to 6%).

Obviously, meteors supply most of the organic carbon. It will not be pristine, having been altered by solar ultraviolet and atmospheric heating, but this could make its chemistry more, rather than less, interesting. Meteorites supply only a negligible fraction (~2×10^{-5}) of the total.

The accretion rate of intact organic C, 3.2×10^8 g yr^{-1}, would give 3.2×10^{16} g, or 0.006 g cm^{-2}, in 10^8 yr. This amount is either large or small, depending on the frame of reference: although five times greater than the mass of the marine biosphere, it corresponds to a mean carbon concentration in the oceans of only 2×10^{-5} g l^{-1}. But the true value must be higher, because meteorite infall rates were much higher during the first ~10^9 yr of the Earth's history, when the inner planets swept up left-over

TABLE 1   Accretion rates on Earth

Sources	Mass range (g)	Mass accretion rate (Gg yr^{-1})	Carbon content (%)	Carbon accretion rate (Gg yr^{-1})
Total meteoritic matter	10^{-14}–10^{18}	78	—	4.2
Meteors	10^{-14}–10^{2}	16	10	1.6
Meteorites*	10^{1}–10^{8}	0.058	1.3	7.6×10^{-4}
Crater-forming bodies†	10^{8}–10^{18}	62	4.2	2.6
Unmelted material, contributing organic matter				
Meteors	10^{-12}–10^{-6}	3.2	10	0.32
Meteorites, noncarbonaceous‡	10^{1}–10^{8}	2.9×10^{-3}	0.1	2.9×10^{-6}
Meteorites, carbonaceous§	10^{1}–10^{8}	1.9×10^{-4}	2.5	4.7×10^{-6}

* Assumed pre-atmospheric abundances: 50% carbonaceous chondrites (2.5% C) and 50% noncarbonaceous meteorites (0.1% C).
† Assuming equal mass fractions of dead comets (10% C), carbonaceous asteroids (2.5% C), and noncarbonaceous asteroids (0.1% C).
‡ Assumed ablation loss 90%.
§ Assumed ablation loss 99.4%, to yield observed fraction of 6% carbonaceous chondrites among stony meteorites.

Reprinted with permission from *Nature*, Volume 342, pp. 255-257.
© 1989 by Macmillan Magazines Ltd.

planetesimals as well as asteroids and comets scattered by the giant planets. To assess the amount of organic carbon contributed by this 'early intense bombardment', we first consider the origin of the Earth's crustal carbon (1.4×10^{23} g; ref. 22).

Less than 10^{-3} of this carbon can come from meteorite infall at present rates over 4.5×10^9 yr. Even if we count all meteors in Fig. 1 and integrate the comet–asteroid distribution to $m = 8.7 \times 10^{20}$ g (one impact of this mass or larger is expected in 4.5×10^9 yr), the total mass is only 2.1×10^{21} g (including 7.4×10^{19} g meteors). For a bombarding population composed equally of comets, carbonaceous asteroids and non-carbonaceous asteroids, the mean C content is 4.2%, which gives a total carbon accretion of 9×10^{19} g in 4.5×10^9 yr, or only 7×10^{-4} of the Earth's crustal C.

There is good reason to believe that the remaining C (and N, H) was acquired as a 'late veneer' of carbonaceous material during the final stages of the Earth's accretion[23-25]. Not only C, N and H but also other volatile elements (Tl, Pb, B, In, Bi, Cl, Br, I, Ar and Kr) are present in the Earth's crust in approximately C1-chondrite proportions[23,25]. As these elements differ in volatility and atomic weight, it is unlikely that the observed, flat abundance pattern (Fig. 4 of ref. 25) could have arisen from 'impact volatilization'[26,27]—a popular but dubious concept for the final stages of Earth's accretion. It is more likely to reflect an original nebular condensation pattern that was preserved during formation of the crust. (These elements are incompatible with mantle minerals and would strongly concentrate in the crust during mantle differentiation.) The threshold of ~ 400 K for carbon condensation as complex organic matter[28] is still recognizable in the asteroid belt, where dark, carbonaceous asteroids become dominant at >2.5 AU (ref. 29). About 4.0×10^{24} g of C1-chondrite-like material (6.7×10^{-4} $M_\oplus$, or a 3.5-km layer) would suffice to account for the 1.4×10^{23} g carbon at the Earth's surface, and the amounts needed for N and H are rather similar (2.9×10^{-4} and 1.5×10^{-3} $M_\oplus$).

Two further lines of evidence indicate an infall of this order. First, the uniform enrichment of noble metals (Pd, Re, Os, Ir, Au) in the upper mantle, at $(7.32 \pm 0.99) \times 10^{-3}$ C1-chondrite abundances[30], apparently implies that highly oxidized, chondritic material fell on the Earth after core formation, leaving its siderophile elements stranded[30,31]. One mantle xenolith showing such enrichment, a sheared garnet lherzolite of particularly primitive composition[32], comes from >150 km depth and so the 'megaregolith' from the late bombardment presumably extends to at least this depth. This would imply $>5 \times 10^{-4}$ $M_\oplus$ of chondritic material in the upper mantle, which is remarkably consistent with the above values for C, N, H and noble metals (overall mean $= (7.4 \pm 2.7) \times 10^{-4}$ $M_\oplus$). Second, the size and age distribution of impact basins on the Moon indicates that the Earth accreted 1.3×10^{-4}–3.8×10^{-2} $M_\oplus$ of material between 3.8 and 4.5 Gyr ago[33]. Some authors have postulated a very large influx of cometary bodies from the region of the outer planets[34], but as only one of the last nine large bodies to strike the Moon[35,36] resembled C-chondrites[30], only part of the bombarding population was carbonaceous.

Although this late accretional component was $\sim 10^3$ times greater than the total meteoritic influx at present rates, it probably yielded a smaller proportion of intact organic matter. Like today, the most efficient vehicle for delivery of organic matter would be meteors in the range 10^{-12}–10^{-6} g (or higher if the atmosphere was denser). The meteor fraction probably was at least as large as the present one, that is, $\sim 20\%$ of the total (Table 1), as the disintegration rate of comets should have been similar and the collision rate in the asteroid belt was higher[37]. Much of the soft-landed organic matter would be destroyed or degraded, however, first by the high temperatures of the crust and atmosphere during the most intense stages of the bombardment, then by recurrent vapourization of the ocean and by burial. Material cycled through the >150-km megaregolith would be pyrolysed to elemental C and small molecules such as CO and CH_4, so only organics accumulated in the ocean and in the last

few kilometres of the regolith would escape serious degradation.

A reasonable guess—good to perhaps a factor of 3–5—is that only $\sim 1\%$ of the Earth's surface carbon ($\sim 1.4 \times 10^{21}$ g) accreted under conditions permitting survival of organic carbon. (One per cent of the accretable material ($\sim 5 \times 10^{22}$ g) would suffice to evaporate the ocean and to deposit several kilometres of ejecta, so accumulation of organic matter probably began only during the last few per cent of the accretion process, but at first with very low efficiency.) If the fraction of meteors, and hence intact organic carbon, was the same as in the most recent infall, that is $\sim 8\%$ (Table 1), this corresponds to 1.1×10^{20} g organic C, or 10^2 times the total infall in 4.5×10^9 yr at present rates. This amount is equivalent to 21 g cm^2 or an oceanic carbon concentration of 0.08 g l^{-1}.

Even 1.1×10^{20} g C is comparable to the 10^{20}–10^{21} g of HCN and HCHO expected from Miller–Urey reactions in the Earth's atmosphere ($H_2/CO_2 = 1$) over 600 Myr (ref. 38). Amounts aside, meteoritic and cometary organic matter may well have played an important role in pre-biotic evolution, as it formed by different processes (ion-molecule and photochemical reactions in interstellar clouds, grain catalysis in the solar nebula, aqueous chemistry in asteroids) and thus contains a different mix of compounds, including N-heterocyclics, aromatics, amphiphilic compounds, porphyrin-like pigments, and so on, that are not readily made by Miller–Urey reactions[28,39-41].

There are several uncertainties in this estimate: the fraction of carbon-rich objects in the final stages of bombardment, the fraction disintegrated to meteor-sized dust, the onset of safe conditions on Earth, the pressure of the atmosphere, and the net yield of organics from large, friable bodies. A higher pressure would extend aerobraking to higher masses[42], thus raising the soft-landed fraction in Fig. 1. Survival of dust from large, friable bodies would have the same effect. Öpik[43] has calculated that aerodynamic pressure would cause large stony meteorites to disintegrate into a flat sheet of dust and rubble, which would be decelerated to less than crater-forming velocities at masses $\leqslant 10^{12}$ g. Thomas et al.[42] have suggested that part of this debris might be decelerated so gently as to permit survival of organic matter.

At first sight, the empirical data are permissive, or even encouraging. No unmelted material has been found from the Tunguska object, but this may merely reflect its inconspicuous nature and lack of ferromagnetism. Very large amounts of two non-protein amino acids (AIB and ISOVAL) have been reported $+0.3$ m and -0.5 m from the Cretaceous/Tertiary (K/T) boundary at Stevns Klint, and have been attributed to soft-landed

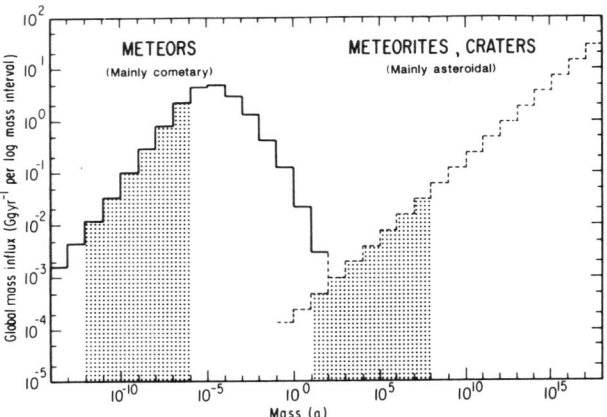

FIG. 1 Infall rate of meteoritic matter on Earth (adapted from ref. 5). Intervals where organic matter can survive passage through atmosphere are shaded. The curve on the right is based on the relation[5] $N = 0.54$ $r^{-2.1}$ ($N =$ number of impacts per Myr, $r =$ radius in km), for an assumed density of 3 g cm^{-3}. The corresponding mass accretion rate (Gg yr^{-1}) between r_1 and r_2 is 15.83 ($r_2^{0.9} - r_1^{0.9}$).

material from a comet[44]. However, these results are in sharp conflict with measurements of Xe in 9 K/T boundary samples from 3 sites, including 6 from Stevns Klint[45,46]. The AIB/Ir ratio is about five times the value in the Murchison C2-chondrite and implies essentially complete survival of the impactor, whereas the Xe/Ir values range from $\leqslant 6 \times 10^{-3}$ to 2×10^{-5} of the C-chondrite value, and imply essentially complete destruction. Differences in temperature sensitivity of AIB and Xe cannot explain this discrepancy, as amino acids are far more readily lost than is Xe (seconds at 1,000 K against hours at 1,200–1,500 K), and neither can differences in composition. Comets are ~10-fold enriched in C and N and hence perhaps in amino acids, but they are not likely to be significantly depleted in Ir, as judged from the C-chondrite-like pattern of the micrometeorite component in lunar soils[9] and from the relatively normal abundances of cosmochemically similar elements in comet Halley[15] (Fe = 0.62, Ni = 0.87, Ca = 1.10 of C1-chondrite values). Xenon actually is more abundant in IDPs[47] than in C1 chondrites, by a factor of about ten.

Furthermore, there are theoretical arguments against survival of unmelted material from comets the size of the K/T impactor (~10^{18} g). During atmospheric passage, such a body would lose only ~10^{-3} of its momentum or mass, unless it had a very small entry angle or very irregular (non-equant) shape[48]. Moreover, the 'ground track' a of the comet's atmospheric trajectory is $h/\tan \alpha$, where h is the altitude where significant fragmentation begins and α is the elevation angle. If $h = 50$ km (ref. 49) and $\alpha \geqslant 20°$, then a is only $\leqslant 137$ km, much smaller than the ~1,200-km radius of the fireball. Any debris spalled from the comet during atmospheric passage will be engulfed by the fireball and destroyed.

Survival of organic matter against this destruction mechanism becomes possible when the fireball radius is shorter than the ground track, that is at masses <10^{15} g. If a significant fraction of the spalled-off dust is gently decelerated[42] without melting or pyrolysis, then friable bodies of 10^8–10^{12} g may contribute some organic matter. However, at present it seems that the best vehicle for carrying organic matter to Earth is cometary dust formed before atmospheric entry. □

Received 7 July; accepted 6 October 1989.

1. Oró, J. Nature 190, 389–390 (1961).
2. Delsemme, A. H. Origins Life 14, 51–60 (1984).
3. Deamer, D. W., Brack, A., Morowitz, H., Weber, A. & Usher, D. Origins Life (in the press).
4. Anders, E., Hayatsu, R. & Studier, M. H. Science 182, 781–790 (1973).
5. Kyte, F. T. & Wasson, J. T. Science 232, 1225–1229 (1986).
6. Hughes, D. W. in Cosmic Dust (ed. McDonnell, J. A. M.) 123–185 (Wiley, New York, 1978).
7. Dohnanyi, J. S. Science 173, 558 (1971).
8. Barker, J. L. & Anders, E. Geochim. cosmochim. Acta 32, 627–645 (1968).
9. Anders, E., Ganapathy, R., Krähenbühl, U. & Morgan, J. W. The Moon 8, 3–24 (1973).
10. Shoemaker, E. M. A. Rev. Earth planet. Sci. 11, 461–494 (1983).
11. Brownlee, D. E. A. Rev. Earth planet. Sci. 13, 189–215 (1985).
12. Sears, D. W. Mod. Geol. 5, 155–164 (1975).
13. Baldwin, B. & Shaeffer, Y. J. geophys. Res. 76, 4653–4668 (1971).
14. Schramm, L. S., Brownlee, D. E. & Wheelock, M. M. Meteoritics 24, 99–112 (1989).
15. Jessberger, E. K., Christoforidis, A. & Kissel J. Nature 332, 691–695 (1988).
16. Weissman, P. R. Geol. Soc. Am. Spec. Pap. 190, 15–24 (1982).
17. Wetherill, G. W. & ReVelle, D. O. Icarus 48, 308–328 (1981).
18. Wetherill, G. W. Phil. Trans. R. Soc. A323, 323–337 (1987).
19. Wetherill, G. W. & Shoemaker, E. M. Geol. Soc. Am. Spec. Pap. 190, 1–13 (1982).
20. Palme, H., Grieve, R. A. F. & Wolf, R. Geochim. cosmochim. Acta 45, 2417–2424 (1981).
21. ReVelle, D. O. J. atmos. terr. Phys. 41, 453–473 (1979).
22. Mason, B. Principles of Geochemistry 3rd edn (Wiley, New York, 1966).
23. Anders, E. Acct. chem. Res. 1, 289–298 (1968).
24. Turekian, K. K. & Clark, S. P. Jr. Earth planet. Sci. Lett. 6, 346–348 (1969).
25. Anders, E. & Owen, T. Science 198, 453–465 (1977).
26. Matsui, T. & Abe, Y. Nature 322, 526–528 (1986).
27. Zahnle, K. J., Kasting, J. F. & Pollack, J. B. Icarus 74, 62–97 (1988).
28. Hayatsu, R. & Anders, E. Topics curr. Chem. 99, 1–37 (1981).
29. Gradie, J. & Tedesco, E. Science 216, 1405–1407 (1982).
30. Morgan, J. W. J. geophys. Res. 91, 12375–12387 (1986).
31. Kimura, K., Lewis, R. S. & Anders, E. Geochim. cosmochim. Acta 38, 683–701 (1974).
32. Morgan, J. W., Wandless, G. A., Petrie, R. K. & Irving, A. J. J. Tectonophys. 75, 47–67 (1981).
33. Chyba, C. F. Nature 330, 632–635 (1987).
34. Ip, W.-H. & Fernandez, J. A. Icarus 74, 47–61 (1988).
35. Gros, J., Takahashi, H., Hertogen, J., Morgan, J. W. & Anders, E. Proc. Lunar Sci. Conf. 7, 2403–2435 (1976).
36. Hertogen, J., Janssens M.-J., Takahashi, H., Palme, H. & Anders, E. Proc. Lunar Sci. Conf. 8, 17–45 (1977).
37. Chapman, C. R. & Davis, D. R. Science 190, 553–556 (1975).
38. Stribling, S. & Miller, S. L. Origins Life 17, 261–273 (1987).
39. Deamer, D. W. Nature 317, 792–794 (1985).
40. Mullie, F. & Reisse, J. Topics curr. Chem. 139, 83–117 (1987).
41. Cronin, J. R., Pizzarello, S. & Cruikshank, D. P. in Meteorites and the Early Solar System (eds Kerridge, J. F. & Matthews, M. S.) 819–857 (Univ. Arizona Press, 1988).
42. Thomas, P. J., Chyba, C. F., Brookshaw, L. & Sagan, C. Lunar planet. Sci. 20, 1117–1118 (1989).
43. Öpik, E. J. in Proc. Geophys. Lab. Lawrence Radiation Lab. Cratering Symp., Rep. UCRL-6438, 2, Paper S. 1–28 (Lawrence Radiation Lab., Berkeley, 1961).
44. Zhao, M. & Bada, J. L. Nature 339, 463–465 (1989).
45. Wolbach, W. S., Lewis, R. S. & Anders, E. Science 230, 167–170 (1985).
46. Lewis, R. S. & Wolbach, W. S. Meteoritics 21, 434–435 (1986).
47. Hudson, B., Flynn, G. J., Fraundorf, P., Hohenberg, C. M. & Shirck, J. Science 211, 383–386 (1981).
48. Emiliani, C., Kraus, E. B. & Shoemaker, E. M. Earth planet. Sci. Lett. 55, 317–334 (1981).
49. Turco, R. P. et al. Icarus 50, 1–52 (1982).

ACKNOWLEDGEMENT. I thank G. W. Wetherill for advice. This work was supported in part by NASA.

Section III

SELF-ASSEMBLY OF SUPRAMOLECULAR SYSTEMS

SELF-ASSEMBLY OF SUPRAMOLECULAR SYSTEMS

The appearance of life on this Earth was not so much a matter of change of substance for, in fact, life made use of all substances already occurring on Earth, particularly the universal medium of water. It was, rather, an enormous increase in the information and complexity with which these elements were combined both in spatial arrangement, forming the molecules of which life is built, and in dynamics, enabling them to reproduce the patterns through endless and varied generation.

J. D. Bernal, 1967

Any analysis of life's beginning must confront questions of how a prebiological system of molecules first assembled from a complex mix of precursors, how it maintained that assembly, and how it acquired the capacity to grow and reproduce.

How did such systems arise from the prebiotic milieu? There were no lipids, proteins, or nucleic acids, which means that the earliest systems must have self-assembled from whatever organic molecules were available, perhaps helped along by some inorganic substrate. Some of the organics presumably had properties relevant to life processes, such as the ability to absorb light energy and to transduce it into photochemical energy. Others may have been primitive catalysts or had the capacity to store information in some way. Some molecules must have had the ability to self-assemble and interact with one another, thereby forming supramolecular systems with properties arising from the whole system rather than from individual molecular components. In this section, we explore such processes of self-assembly.

Some self-assembly processes involve hydrogen bonding

A familiar self-assembly process is the crystallization of water into ice. At the molecular level, water can be understood as small clusters of individual molecules held together by physicochemical forces called *hydrogen bonds*. Because these forces are found in compounds containing hydrogen, oxygen, and nitrogen, we must consider hydrogen bonding as a potential organizing force in prebiotic self-assembly. The bonds arise because oxygen and nitrogen are relatively electronegative and draw electrons toward their nuclei, so that attached hydrogen atoms become relatively electropositive. The resulting structures, containing both negative and positive poles, are, therefore, called *dipoles*, and they have physical properties distinctly different from nonpolar molecules such as hydrocarbons.

In a dipole such as water, hydrogen bonds form between hydrogen and oxygen atoms of neighboring water molecules. As water cools, hydrogen bonds become increasingly common. At the freezing point of water, the bonds become pervasive so that individual water molecules "freeze" into ice crystals.

The reason for dwelling on this point is that hydrogen bonds also stabilize the self-assembly of many biological macromolecules. The two strands of the DNA double helix, for instance, are held together by hydrogen bonding between base pairs. Other examples include the *alpha helix* and *beta pleated sheet*, structures of the folded protein in which hydrogen bonds form between neighboring amino acids.

Chirality may have evolved through self-assembly processes

One of the most puzzling aspects of all contemporary life-forms is their incorporation of molecules with the structural property of *chirality* (from *cheír*, the Greek word for "hand"). Indeed, chiral molecules are said to have "handedness"—referring to their particular asymmetry, which is like the asymmetry of a pair of human hands. Such molecules exist in two forms that are mirror images of each other, just as the left hand is the mirror image of the right. The two mirrored forms are referred to as *stereoisomers*.

With a few rare exceptions, all amino acids of living systems are L-stereoisomers (from *levo-*, the Latin prefix for "left"), and the most biologically important carbohydrates are D-stereoisomers (from *dextro-*, the Latin prefix for "right"). Naturally occurring glucose—now referred to as D-glucose—was formerly called *dextrose* because of this property. D-glucose is the most abundant monosaccharide in nature and is the monomeric building block ot the most abundant polysaccharides, starch and cellulose.

It is reasonable to think that the universal chiral character of biomolecules may be an important clue to the origin of life. That is, because the mixture of prebiotic organic molecules available for self-assembly must have been *racemic* (containing both D- and L-forms of amino acids and various other monomers) if we were to understand how chiral molecules were selected from such a mixture, we might be closer to understanding how living systems began.

Several explanations for the chiral selectivity of biomolecules have been put forward, but, so far, none has been generally accepted. One suggestion takes into account the fact that stereoisomers interact differentially with polarized light (Wald, 1957). A solution of D-glucose, for example, rotates the plane of polarized light in a clockwise direction, while L-glucose rotates the plane counterclockwise. Thus, because sunlight passing through the atmosphere takes on a modest degree of polarization, it may have been that, on the early Earth, the polarized UV component of sunlight degraded D-stereoisomers of amino acids and left a solution more concentrated in the L-amino acids. Another suggestion, also by Wald (1957), is that, because certain mineral surfaces are chiral in character, one of the stereoisomers was selectively adsorbed to the surface and was concentrated thereby. Although both suggestions are interesting, neither has stood up to experimental tests. (See Kenyon and Steinman, 1969, for a review.)

A more likely possibility is that chiral selectivity first occurred in one or more self-assembly processes and was then amplified through an autocatalytic mechanism (Bonner et al., 1981; Fajszi and Czege, 1981). A good example of such selectivity was demonstrated by Joyce et al. (1984) using RNA templates and activated monomers. In this system, activated guanosine mononucleotides self-assemble on a poly-C RNA template through Watson-Crick base-pairing. The monomers then polymerize to form oligonucleotides. Under these conditions, Joyce and his colleagues found that the polymerization process was markedly inhibited by racemic mixtures of the activated monomers, while the appropriate stereoisomer readily produced polymers up to 20 nucleotides in length. The authors concluded that "there may have been a prebiotic system in which the optical specificity of monomer incorporation was very high. If so, such a system would have enjoyed a considerable advantage in the prebiotic environment."

Nonpolar forces also stabilize supramolecular assembly: phase separation

A variety of molecules dispersed in aqueous solutions do not simply dissolve but instead form aggregates in a predictable manner. A simple example of this is dispersion of hydrocarbon molecules (oil) in water: the hydrocarbons self-assemble into small droplets, aggregating the oil phase away from the water phase.

Why do nonpolar compounds typically form separate phases in aqueous solutions? The reason is that the presence of a hydrocarbon chain partially disrupts the extensive hydrogen bonding in liquid water. As a result, water molecules tend to fall into an orderly array around the chain, which decreases the overall entropy of the system. Because all systems tend to change in such a way that entropy increases, hydrocarbon chains dispersed in an aqueous phase will tend to associate with one another, thereby removing themselves from the water and allowing the water molecules to become more disorderly. It is now understood that this process, called the *hydrophobic effect* by Tanford (1978) in a paper reproduced here, is a significant stabilizing force in biological macromolecular systems (Morowitz, 1981).

Amphiphilic molecules form membranes

An important example of the hydrophobic effect is the ability of certain molecules to self-assemble into aggregates called bilayers. Such molecules have both polar and nonpolar groups. The polar groups tend to interact strongly with water, while the nonpolar groups tend to interact with nonpolar environments. For this reason, such molecules are called *amphiphilic*, because they have both "water-loving" (hydrophilic) and "oil-loving" (lipophilic) portions on the same molecule. (A synonymous term, arising from the same basic concept, is *amphipathic*.) Examples of amphiphiles include soap and commercial detergents, whose ability to self-assemble into bubbles is common knowledge.

Less well-known examples of amphiphiles are the lipids involved in all living cells. The three primary classes of lipids are fats, sterols (such as cholesterol), and phospholipids. The latter are the basic components of all contemporary membrane structures, forming lipid bilayers that provide a permeability barrier to dissolved solutes. Similar physical properties in primitive amphiphiles presumably stabilized the membraneous boundary structures necessary for the origin of cellular life.

Self-assembling structures

How can bounded microenvironments be established? This is a central question for origins-of-life investigations. Different kinds of organizing microenvironments have been proposed, including mineral surfaces such as clays, thin films on inorganic surfaces, and various forms of aggregates. Oparin (1924), for instance, proposed that colloidal coacervates were a possible model for the first organized structures:

> Sooner or later such colloidal solutions of organic substances must have come into being in the watery covering of the Earth and once they had arisen they continued to exist, their molecules becoming more complicated and larger as time went on.... [We] suppose that in the course of many hundreds or even thousands of years during which the terrestrial globe existed, the conditions... would lead to the formation of a gel in a colloidal solution. The moment when the gel was precipitated or the first coagulum formed, marked an extremely important stage in the process of the spontaneous generation of life. At this moment material which had formerly been structureless first acquired a structure and the transformation of organic compounds into an organic body took place. Not only this, but at the same time the body became an individual. Before this it had been inseparably fused with all the rest of the world, dissolved in it. Now, however, it separated itself out, though still very imperfectly, from that world and set itself apart from the environment surrounding it.... With certain reservations we can even consider that first piece of organic slime which came into being on the Earth as the first organism.

The proteinoid microspheres reported by Fox and Harada (1958) are similar to the original conjectures of Oparin, and Fox has argued that such microspheres are potential models for the earliest protocells.

In later years, with the advent of electron microscopy, it became apparent that living cells are not colloidal in nature but, instead, are bounded by membranes composed of both lipid and protein. The lipid is present as fluid lipid bilayers; this permeability barrier is penetrated by integral transmembrane proteins that act as channels, carriers, and ion pumps, forming a supramolecular structure now referred to as the fluid mosaic membrane (Singer and Nicolson, 1972). The ion gradients resulting at the bilayer surface are an energy source for multiple cell functions, including the production of ATP, excitability, and coupled transport of nutrients. It seems only reasonable, then, to include the assembly of lipid bilayer membranes as an essential step in the evolutionary process leading to the first cells (Goldacre, 1958; Deamer and Oró, 1980; Fleischaker, 1990).

Lipid bilayer membranes readily encapsulate large molecules

Given that primitive membranes must have self-assembled at some point to produce protocellular structures, we can ask how such structures might have encapsulated larger molecules involved in catalysis and information processing. After all, when a small volume is enclosed by the membraneous permeability barrier, the same membrane that encloses a microenvironment also tends to exclude macromolecules. There must be some reversible process by which the barrier can be broken, allowing entry of large molecules, and then resealed.

This turns out to be a characteristic property of model systems such as liposomes, defined as small vesicles composed of fluid lipid bilayers (Bangham, 1970). From studies of liposomes, it has been established that several simple processes provide a reversible breakage-resealing. The first is mechanical. That is, when liposomes are agitated in aqueous dispersion, they break and reseal, thus capturing any solutes in the surrounding solution. A second process involves cycles of drying and wetting. When liposomes are dried, they tend to fuse into multilayered structures that sandwich any solutes present during drying. Upon rehydration, the layers reseal to form vesicles with encapsulated macromolecules (Deamer and Barchfeld, 1982). It is easy to imagine similar drying-wetting cycles occurring in the prebiotic environment and large molecules being encapsulated by liposomal systems.

The third process involves osmotic gradients. It has been known for many years that erythrocytes (red blood cells) swell and burst when exposed to dilute hypotonic solutions. For example, the concentration of salt (sodium chloride) in blood is about 0.15 M. If fresh blood is mixed with a salt solution that is more dilute, say 0.015 M, a tenfold concentration gradient of salt is produced across the membrane. A concentration gradient of water across the membrane is also produced, with a higher concentration of water outside (in the dilute salt solution) than inside (in the concentrated salt solution). This means that water will diffuse across the membrane into the more concentrated salt solution within (a process called *osmosis)*. As a result, the red blood cells first swell and then rupture to release their content of salt and hemoglobin (a process called *hemolysis*.) The membranes then reseal, capturing any remaining solute molecules.

A similar process can be induced in liposomes by using osmotic concentration gradients to produce swelling and rupture. If this occurs in the presence of macromolecules, the large molecules leak inward across the membrane. When osmotic equilibrium is reached, the bilayer seals and thereby captures the macromolecules. Again, it is not difficult to imagine that osmotically driven rupturing and resealing of primitive membranes could have captured macromolecules in liposomal systems on the prebiotic Earth.

Permeability constraints on primitive cell functions

As noted in the previous discussion, even as a membrane-enclosed space can provide a micro-environment necessary for cell function, it also limits access to nutrients required for cell growth. How would an early cell have solved the problem of selective permeability?

First, it might be helpful to have some idea of what membrane permeability means. For example, if we compare the rates at which water and ions diffuse across a lipid bilayer, we observe that small, uncharged, water molecules undergo passive transport at a rate approximately one billion times faster than ions. This means that water can equilibrate across liposome membranes in microseconds, while a sodium-ion gradient would require hours to equilibrate. Because cell function depends on the maintenance of ion gradients, this immense barrier to ionized solutes is essential to all life processes.

Nutrients such as amino acids and phosphate are ionized as well, however, and their extremely low permeation rates would have presented a problem to early cells. For instance, modern bacterial cells can divide about once every twenty minutes, which means that they transport sufficient phosphate across their membranes to double their content of nucleic acid in that time interval. This remarkable expansion is possible because bacteria have carrier enzymes that capture phosphate and transport it across the bilayer barrier. If bacteria had to depend on passive phosphate transport such as that which occurs across typical lipid bilayer barriers, they could divide only once every year or so! This might not seem to matter over the geological timescales available for prebiotic evolution, but it seems improbable that a primitive cell could maintain structure or function if permeation rates were too low: the macromolecular assembly would have degraded and been dispersed long before.

The self-assembly of functioning protocells

We can now summarize the events that led to the self-assembly of functioning primitive cells. (See Cavalier-Smith, 1987, and Baeza et al., 1987, for more detailed discussions.) The postulated events represent only one way in which life could have begun; other pathways have been proposed. The following scheme is not a consensus view but, rather, is one plausible scenario that builds on what we know about self-assembly processes.

The first step would be the self-assembly of amphiphilic, lipidlike compounds into membraneous structures, forming vesicles with enclosed microenvironments. These boundary structures would have the capacity to make energy available in the form of ion gradients and to permit inward transport of nutrients. Furthermore, because membraneous vesicles can self-assemble, break, and reseal, they are, in principle, capable of growth and division. Several possible mechanisms by which the amphiphilic membrane components could be synthesized from nutrient precursors have been proposed (Morowitz et al., 1988).

Three other processes must have been incorporated into the protocellular machinery: a mechanism for active inward transport of nutrients from the environment, energy-transduction pathways capable of activating the nutrient molecules, and growth of the encapsulating membrane concurrent with growth of the encapsulated macromolecular system.

An additional ongoing process would be the self-assembly of catalytic polymers that could interact with other polymers able to store information in a sequence of monomers. That sequence in turn would, in some manner, determine the sequence of monomers in the catalyst. In contemporary cells, the polymers are protein enzymes and nucleic acids. However, in the protocell, both catalytic and information-containing sites could have been present in the same molecule, as suggested by recent studies of RNA enzymes, or ribozymes (to be discussed in Section V). Regardless of whether such polymers and their catalysts arose in open environments or evolved within membraneous boundary structures, what is important is that such molecules came to share the same microenvironment. This is not difficult to imagine, since macromolecules have been shown to be readily encapsulated by amphiphilic molecules under simulated prebiotic conditions, as discussed earlier.

Would such a system be considered a living cell? It clearly represents a minimal set of basic cellular processes: energy and nutrient capture from the local environment, growth through catalyzed production of its components, and replication of information-storage molecules, all encapsulated within a membraneous boundary structure. However, some important features of contemporary life processes are missing: a method for coding between a nucleotide sequence and a peptide sequence (*a genetic code*), a method for catalyzing the synthesis of peptides from activated amino acids (*translation*), and a mechanism to introduce changes into the genome so that lineages of populations can respond to their environment and change over time (*evolution*). Some means for selecting stereoisomers of amino acids and carbohydrates must also have been available. How these essential life processes were incorporated into living systems on the early Earth represents the end of our current knowledge of the origins of life.

Reprinted papers on self-assembly

The papers reprinted here were chosen to illustrate some of the main self-assembly properties of molecules related to formation of cell systems. The paper by Tanford (1978) was included for its discussion of the role of hydrophobicity in the self-assembly of amphiphilic molecules into complex structures. Similar processes must also have been involved in the original self-assembly of primitive cellular systems.

The paper by Bangham et al. (1965) is generally considered to be the first convincing demonstration of phospholipid self-assembly into bilayer vesicles that encapsulate ionic solutes. These liposomal vesicles provide useful models for investigations of protocellular structure and function.

The third paper (Deamer and Barchfeld, 1982) builds on the liposome concept, showing how macromolecules can be encapsulated by lipid bilayers under simulated prebiotic conditions. The fourth paper, by Morowitz et al. (1988), outlines the properties that must be fulfilled by the simplest protocell, describing how a system of self-assembling amphiphilic molecules could function to capture energy and nutrients. The fifth paper (Deamer and Pashley, 1989) reports the presence of amphiphilic molecules in the organic components of carbonaceous meteorites and demonstrates the capacity of those molecules to self-assemble into membrane structures.

References

Baeza, I., Ibanez, M., Lazcano, A., Santiago, C., Arguello, C., Wong, C. and Oró, J. (1987) Liposomes with polyribonucleotides as model of precellular systems. *Origins of Life* 17:321-331.

Bangham, A. D. (1970) The liposome as a membrane model. In: *Permeability and Function of Biological Membranes*, L. Bolis, A. Katchalsky, R. D. Keynes, W. R. Lowenstein and B. A. Pethica, eds., Amsterdam: North-Holland, pp. 195-206.

Bangham, A. D., Standish, M. M. and Watkins, J. B. (1965) Diffusion of univalent ions across the lamellae of swollen phospholipids. *J. Mol. Biol.* 13:238-252.

Bernal, J. D. (1967) *The Origin of Life*. London: Weidenfeld and Nicolson.

Bonner, W. A., Blair, N. E. and Dirbas, F. M. (1981) Experiments on the abiotic amplification of optical activity. *Origins of Life* 11:119-134.

Cavalier-Smith, T. (1987) The origin of cells: A symbiosis between genes, catalysts and membranes. *Cold Spring Harbor Symp. Quant. Biol.* 52:805-824.

Deamer, D. W. and Barchfeld, G. L. (1982) Encapsulation of macromolecules by lipid vesicles under simulated prebiotic conditions. *J. Mol. Evol.* 18:203-206.

Deamer, D. W. and Oró, J. (1980) Role of lipids in prebiotic structures. *BioSystems* 12:167-175.

Deamer, D. W. and Pashley, R. (1989) Amphiphilic molecules in organic extracts of the Murchison carbonaceous chondrite: Surface activity and membrane formation. *Orig. Life Evol. Biosphere* 19:21-38.

Fajszi, C. and Czege, J. (1981) Critical evaluation of mathematical models for the amplification of chirality. *Origins of Life* 11:143-162.

Fleischaker, G. R. (1990) Origins of life: An operational definition. *Orig. Life Evol. Biosphere* 20:127-137.

Fox, S. W. and Harada, K. (1958) Thermal copolymerization of amino acids to a product resembling protein. *Science* 128:1214.

Goldacre, R. J. (1958) Surface films: Their collapse on compression, the shapes and sizes of cells, and the origin of life. In: *Surface Phenomena in Biology and Chemistry*, J. F. Danielli, K. G. A. Pankhurst and A. C. Riddiford, eds., New York: Pergamon Press, pp. 12-27.

Joyce, G. F., Visser, G. M., van Boeckel, C. A. A., van Boom, J. H., Orgel, L. E. and van Westrenen, J. (1984)

Chiral selection in poly(C)-directed synthesis of oligo(G). *Nature* **310**:602-604.

Kenyon, D. H. and Steinman, G. (1969) *Biochemical Predestination.* New York: McGraw-Hill, pp. 214-217.

Morowitz, H. J. (1981) Phase separation, charge separation and biogenesis. *BioSystems* **14**:41-47.

Morowitz, H. J., Heinz, B. and Deamer, D. W. (1988) The chemical logic of a minimal protocell. *Orig. Life Evol. Biosphere* **18**:281-287.

Oparin, A. I. (1924) The Origin of Life [*Proiskhozhdenie zhizny.* Moscow: Izd. Moskovshii Rabochii]. English translation in: J. D. Bernal, *The Origin of Life* (1967). London: Weidenfeld and Nicolson, pp. 199-234.

Singer, S. J. and Nicolson, G. L. (1972) The fluid mosaic model of the structure of cell membranes. *Science* **175**:720-731.

Tanford, C. (1978) The hydrophobic effect and the organization of living matter. *Science* **200**:1012-1018.

Wald, G. (1957) Origin of optical activity. *Ann. New York Acad. Sci.* **69**:352-368.

The Hydrophobic Effect and the Organization of Living Matter

Charles Tanford

One of the principal objects of theoretical research in any department of knowledge is to find the point of view from which the subject appears in its greatest simplicity.
—J. WILLIARD GIBBS, in a letter to the American Academy of Arts and Sciences, January 1881

It is almost exactly 100 years since the publication of Gibbs' paper (*1*) "On the equilibrium of heterogeneous substances." The time now seems appropriate to use his method of thermodynamic analysis as a basis for thinking about the organization of the living cell and, beyond that, the assembly of cells into multicellular organisms. One should not be under the illusion that such an analysis will produce dramatic new advances, such as might result from the discovery of a hitherto unknown chemical substance or from the determination of the three-dimensional structure of an important biological macromolecule. But science does not progress by startling discoveries alone, and thermodynamic analysis serves its purpose by assimilating the new into the old and the part into the whole. It can create a simple unified conceptual framework for biology, as it has done for chemistry, and has the potential for clarifying research problems and indicating productive pathways for solving them.

It might at first seem an unrewarding task to apply equilibrium thermodynamics to biology at all, for life is surely the antithesis of an equilibrium state. In fact, however, the dynamic processes of life take place within an organized structural framework that turns over slowly or not at all. It is realistic to consider this framework (tentatively, as long as no compelling contrary evidence presents itself) as being essentially at equilibrium, subject only to spatial and temporal constraints that are usually self-evident. Such constraints, in the form of physical or kinetic barriers, exist whenever the laws of equilibrium thermodynamics are applied, and the constraints in living systems are no different from those encountered in simple chemical systems.

Thermodynamic Background

The great achievement of Gibbs was the definition of the chemical potential (*2*), which permits the state of equilibrium in a complex chemical system to be defined precisely, and often as simply as in a purely mechanical or electrical system. In this article we are interested in the assembly of molecules into organized structures, without chemical interconversions—that is, in the distribution of molecules between various phases or places in the biological system. The Gibbs equilibrium condition is that the chemical potential (μ_i) of each definable component must have the same value in each phase or place accessible to it; that is

$$\mu_{ia} = \mu_{ib} = \mu_{ic} = \ldots \quad (1)$$

where the subscripts a, b, c, . . . designate the possible locations of each component.

The phases with which we deal are solutions or mixtures, and the free energy includes a major contribution from the statistical entropy of mixing, which is nonspecific and does not depend on what substances are being mixed. Most workers have followed the procedure of Gurney (*3*), segregating this nonspecific statistical factor from the more interesting specific parts of the chemical potential. The procedure is based on an ideal expression for the entropy of mixing, and leads to $RT \ln X_i$ (where R is the gas constant, T is absolute temperature, and X_i represents mole fraction) as the purely statistical contribution to each μ_i. Equation 1, for any two environments, can then be rewritten as

$$\mu^0_{ia} + RT \ln X_{ia} = \mu^0_{ib} + RT \ln X_{ib} \quad (2)$$

where μ^0_i is the specific part of the chemical potential, called by Gurney the unitary potential. It includes the free energy of the inherent molecular motions of an isolated molecule of type i, plus the free energy arising from specific interactions with neighboring molecules, which are solvent molecules in a dilute solution or adjacent molecules of type i in a pure phase. Both factors are the sums of the enthalpy and entropy contributions; that is, $RT \ln X_i$ is not to be thought of as representing the total contribution of entropy to the chemical potential.

The thermodynamics of biological organization does not involve biosynthesis or other chemical transformations, but focuses solely on where molecules prefer to go after they have been synthesized. The properties of the isolated molecules on the two sides of Eq. 2 are the same, and the difference between μ^0_{ia} and μ^0_{ib} therefore comes exclusively from interactions with adjacent solvent or like molecules. It is moreover implicit in the use of the term "organization" that the distributions of interest are lopsided, greatly favoring one environment over another, so that one is always concerned with large differences in the unitary potential. Because of this, the possible error arising from use of an ideal expression for the entropy of mixing, which could in principle lead to incorpo-

Summary. Biological organization may be viewed as consisting of two stages: biosynthesis and assembly. The assembly process is largely under thermodynamic control; that is, as a first approximation it represents a search by each structural molecule for its state of lowest chemical potential. The hydrophobic effect is a unique organizing force, based on repulsion by the solvent instead of attractive forces at the site of organization. It is responsible for assembly of membranes of cells and intracellular compartments, and the absence of strong attractive forces makes the membranes fluid and deformable. The spontaneous folding of proteins, however, involves directed polar bonds, leading to more rigid structures. Intercellular organization probably involves polar bonds between cell surface proteins.

The author is James B. Duke Professor of Physical Biochemistry at Duke University Medical Center, Durham, North Carolina 27710. This article is based on a lecture which he gave at the 5th International Conference on Chemical Thermodynamics, Ronneby, Sweden, August 1977.

ration of part of the true mixing entropy into μ''_i, is probably unimportant (*4*). I conclude, therefore, that only the interactions of molecules of type *i* with their immediate environment are important in Eq. 2, and the equation can now be used in two ways. (i) Analytical data for well-defined model systems can be used to evaluate the potential differences $\mu^0_{ia} - \mu^0_{ib}$ for appropriate differences. (ii) Conversely, we can estimate differences in μ^0_i between states that involve environments similar to those of model systems and predict the relative concentrations X_{ia} and X_{ib}.

Hydrophobic Effect

Hydrophobic forces are not the only forces that can make large contributions to differences in unitary potential, even in the water-bathed environment of biological structures. They are not involved in the structure of bone and probably play only a minor role in the organization of the nucleus. They are important in the folding of proteins to produce compact three-dimensional structures, but other forces are equally important, especially in stabilizing a unique structure for each individual protein and making it relatively rigid and proof against deformation. I have chosen to emphasize hydrophobic forces in this article because they lead to structures that are not rigid and are thus uniquely suited for the first critical steps in the organization of living matter, where deformability is not only a virtue, but very likely a necessity.

The phenomenological definition of the hydrophobic effect (*5*) begins with the fact that hydrocarbon molecules have a much higher solubility in liquid hydrocarbons (or most other organic solvents) than they do in water. The unitary potential in the nonaqueous solvent (μ^0_{HC} if the solvent is liquid hydrocarbon) is more negative than that in water (μ^0_w) by several kilocalories per mole. The same phenomenon is observed with polar derivatives of hydrocarbons, where the difference in unitary potentials appears to be an additive function of the constituent groups. The contribution from the polar group favors the aqueous medium, but the hydrocarbon portion favors the nonaqueous medium, and its contribution to the difference between μ^0_{HC} and μ^0_w is essentially the same as it would be for a similar structural moiety within a pure hydrocarbon molecule. For linear alkyl chains the contribution to $\mu^0_{HC} - \mu^0_w$ is −850 calories per mole per CH_2 group and −2000 calories per mole per CH_3 group at 25°C (*5–9*). Nonlinear

Fig. 1. Schematic diagram of an amphiphilic solute in aqueous solution. The strong attractive forces between the polar end and the surrounding water molecules probably neutralize the CH_2 group immediately adjacent to the polar group.

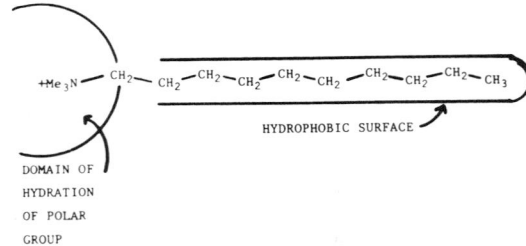

DOMAIN OF
HYDRATION
OF POLAR
GROUP

alkanes are somewhat less hydrophobic than their linear isomers because they are more compact and have a smaller area of contact with the solvent. Results for linear, branched, and cyclic alkyl chains can be accommodated by a single expression if the hydrophobic effect is considered as proportional to the surface area of the cavity created in the aqueous solvent by the hydrocarbon moiety (*10–13*).

For the purpose of the present discussion, the precise numbers are less important than the underlying mechanism from which the hydrophobic effect derives its name. The common mechanism for solvent preferences is based largely on solute-solvent attractive forces. Ionic and polar solutes are more soluble in water than in liquid hydrocarbon because there are strong attractive forces between the solute and water molecules, and polar substituents on alkyl chains make a positive contribution to $\mu^0_{HC} - \mu^0_w$ for the same reason. In the case of alkyl chains, however, solute-solvent attractive forces are weak both in the hydrocarbon environment and in the aqueous medium; in fact, the dipole–induced dipole attraction between H_2O and a CH_2 group may be slightly stronger than the attraction produced by dispersion forces between CH_2 groups. The large negative values of $\mu^0_{HC} - \mu^0_w$ for hydrocarbon solutes are due to the strong cohesive forces between water molecules and the fact that the network of H_2O-H_2O bonds is isotropic. Alkyl chains are literally squeezed out of the aqueous medium.

The basis for this distinction has long been known and has been explained eloquently in the past (*14–16*). However, because the absence of strong attractive forces between alkyl chains will be very important in the subsequent discussion, and because there has not been complete acceptance of the concept of hydrophobicity (*17*), I will note that the distinction is based on three separate lines of evidence.

1) Our knowledge of intermolecular forces: attractive forces between hydrocarbon groups or between them and other hydrophobic solutes, such as the inert

gases, are the weakest such forces known.

2) Thermodynamic parameters other than the chemical potential: the partial molal heat capacity is especially compelling, as it can be measured separately in the organic solvent and in water, and it is the value in water that is anomalous to an extraordinary degree.

3) The indifference of hydrocarbon solutes to most organic solvents; for example, μ^0_i values in CCl_4 are virtually identical to those in hydrocarbon solvents. Water has an almost unique place as an extremely poor solvent for hydrocarbon solutes.

Minor details. Numerical values for the hydrophobic free energy have a small measure of uncertainty because it is not possible to correct experimental data for the effects of nonideal mixing in the organic phase. The experimental results on which the numbers are based do not include data for large complex fused rings: a very careful study with highly purified cholesterol (*9*) gave an anomalous value for $\mu^0_{HC} - \mu^0_w$, apparently not as the result of nonideal mixing statistics, which suggests that fused rings may be somewhat less hydrophobic than other molecules with the same surface area. Another factor affecting numerical calculations is the likelihood that the CH_2 group proximal to a polar group in an amphiphilic molecule does not contribute to hydrophobicity because, as illustrated by Fig. 1, water molecules in its vicinity are preempted by the attractive forces for the polar group (*18*). None of these considerations have any effect on the conclusions reached in this article.

Organizational Drive

The organizational drive of the hydrophobic effect asserts itself whenever there is in an aqueous medium a sufficiently high concentration of amphiphilic molecules (or ions) containing a polar or charged group at one end, attached to a relatively large hydrocarbon moiety, as in Fig. 1. The opposing thermodynamic preferences of the two ends of such a molecule are most simply satisfied by self-association to form an aggregate with the hydrocarbon chains in the middle, avoiding contact with water as much as possible, and the hydrophilic polar groups at the surface, as illustrated by Fig. 2. The particle is called a micelle, and it typically contains on the order of 100 molecules per particle. Formation of a micelle is therefore a highly cooperative process, depending on a very high power of the monomer concentration. This means that there is a critical concentration (strictly speaking, a narrow concentration range) below which no micelles exist and above which virtually all added amphiphile enters the micellar state. The critical concentration depends chiefly on the size of the hydrophobic moiety, ranging from close to 1 molar for derivatives with hexyl chains to less than 10^{-9} molar for biological phospholipids.

The size and shape of micelles is determined by a combination of geometric and thermodynamic factors (19). One of the geometric factors is the surface/volume ratio, which for any reasonable assumed micelle shape has to decrease as the size increases; in other words, as the number of alkyl chains in the hydrocarbon core of the micelle becomes larger, the surface area per emerging chain has to become smaller. The second geometric factor, evident from Fig. 2, is that one dimension of a micelle core is restricted: it cannot be larger than the length of two alkyl chains, one approaching from each side. Moreover, the alkyl chains are normally in a liquid state; that is, their length is less than that of fully extended chains (20). The limited dimension calculated on this basis is about 24 angstroms for two dodecyl chains and about 30 angstroms for the chains in the C_{16} to C_{18} range normally found in phospholipids. These are small dimensions: if micelles were spherical with the diameter of the hydrophobic core equal to the limiting dimension, one could accommodate no more than 20 monomers with dodecyl chains in one micelle, or no more than 32 monomers with chains in the C_{16} to C_{18} range. Experimentally observed aggregation numbers are much larger than this.

Surface area calculations suggest why these small spherical micelles are thermodynamically unstable. Their surface/volume ratio would be large, and the area per emerging chain would be $> 100Å^2$. This means that there would be considerable open space between head groups with contact between water and hydrocarbon, so that the hydrophobic drive force would not be optimally satisfied. Any increase in size would decrease

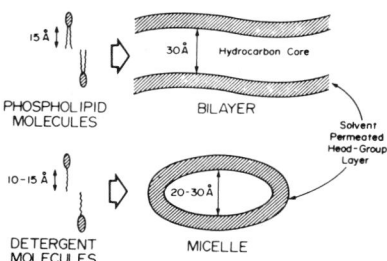

Fig. 2. Self-association of detergents and phospholipids to form micelles or planar bilayers, both shown in cross section. Most micelles are disklike and, being fluid, probably do not have a unique regular shape (22).

the area per emerging chain. Growth as a sphere being impossible, there are essentially two alternatives. One dimension can be kept fixed at the limited value and growth allowed in two perpendicular directions, forming an overall disklike structure; or two dimensions can be kept fixed and growth allowed along only one dimension, forming a cylindrical or rodlike particle. As shown in Table 1, a disklike structure represents the more efficient way to reduce the area and is therefore thermodynamically favored. Small micelles are, in fact, usually disklike, as demonstrated by measurements of their hydrodynamic properties (19, 21, 22).

It should be noted that the surface area of a micelle may not become too small because of repulsion between hydrated head groups. The optimal micelle size (on the order of 100 for dodecyl chains) seems to be reached in Table 1 when the area per head group is about 60 to 65 $Å^2$.

Table 1. Surface areas for micelles formed from dodecyl derivatives with a single alkyl chain per molecule as a function of aggregation number. The areas are calculated for micelles with a smooth hydrocarbon core (with no allowance for surface corrugations) at 1.5 Å from the core surface. The CH_2 group adjacent to the polar group is assumed to remain in the aqueous medium.

Assumed geometry	Aggregation number	Area per chain = area per head group ($Å^2$)
Largest possible sphere	20	106
Disk with rounded edge*	50	76
	100	61
	200	51
Oblate ellipsoid*	100	64
Cylinder with caps†	100	72
Prolate ellipsoid†	100	78

*Two different geometric models that may be used to estimate areas of disklike micelles. The oblate ellipsoid is a poor model for very large aggregation numbers. †Two different geometrical models that may be used to estimate areas of rodlike micelles.

Calculations in Table 2 for longer alkyl chains indicate that a higher aggregation number is needed to attain the same area, and this agrees with the effect of alkyl chain length on experimentally measured micelle size.

Thermodynamics of Micelle Formation

Rigorous thermodynamic equations (19, 23) can be obtained in either of two ways. One can consider micelles of aggregation number m, with m ranging from 2 to ∞, to be separate chemical species, and then work in terms of the chemical equilibria $mZ \rightleftarrows Z_m$, where Z represents a monomeric molecule (23). Alternatively, one can consider the micelles to be separate phases and apply the equilibrium condition that the chemical potential of an amphiphile molecule in a micelle of size m ($\mu_{mic,m}$) is the same in all micelles and equal to its chemical potential as a monomer in aqueous solution (μ_w). If the latter approach is used (both approaches give the same result), the expression used for the chemical potential in Eq. 2 is correct only for μ_w. The corresponding term for $\mu_{mic,m}$ must take account of the facts that it is whole micelles and not individual amphiphile molecules that are mixed with solvent and that $\mu_{mic,m}$ represents $1/m$ of the chemical potential of the micelle. One obtains

$$\mu_{mic,m} = \mu^0_{mic,m} + (RT/m) \ln(X_m/m) \quad (3)$$

and, for the equilibrium condition

$$RT \ln(X_m/m) =$$
$$m(\mu^0_w - \mu^0_{mic,m}) + mRT \ln X_w \quad (4)$$

where X_m is the mole fraction of amphiphile in micelles of size m, and Eqs. 3 and 4 apply separately to all possible values of m. If one can estimate the unitary potential difference ($\mu^0_{mic,m} - \mu^0_w$) as a function of m, one obtains a size distribution function as a function of X_w and from it the experimentally determinable average size and the critical concentration for micelle formation.

It is not possible to go into details of the calculation here. However, reasonable choices of the parameters that go into the calculation do lead to correct predictions of micelle size and critical micelle concentration for amphiphiles with different head groups, and they confirm the observation that a disklike shape is preferable to a rodlike shape and that a surface area of about 60 to 65 $Å^2$ per head group is appropriate for optimal stability (24). To illustrate what is meant by reasonable choices of parameters, I mention one parameter, the gain in uni-

Table 2. Surface areas for diacyl phospholipids with chains in the hexadecyl to oxadecyl range. The conditions are as described in the legend of Table 1.

Hydrocarbon chains per particle	Area (Å²)	
	Per alkyl chain	Per head group
Disk with rounded edge		
50	93	186
100	74	148
200	61	122
1,000	45	90
10,000	36	72
100,000	34	68
*Spherical vesicle**		
1,000	37	74
10,000	33	66

*Experimental aggregation numbers for single-walled vesicles formed spontaneously by sonication are about 2000, corresponding to 4000 hydrocarbon chains per vesicle.

tary potential from transferring an alkyl chain from water to the hydrocarbon interior of a micelle, which makes a major contribution to $\mu^o_{mic,m} - \mu^o_w$. The hydrocarbon core of a micelle cannot be regarded as equivalent to bulk liquid hydrocarbon because it is a two-dimensional rather than a three-dimensional liquid. Individual alkyl chains are flexible and can diffuse freely in two dimensions, but they cannot tumble. This restriction must make the unitary potential more positive in a micelle than in bulk hydrocarbon. To obtain agreement with experiment in the dependence of critical micelle concentration on alkyl chain lengths, for amphiphiles with a variety of head groups, one must set the contribution per CH_2 group at −710 cal/mole, a reasonable value in relation to the figure of −850 cal/mole given earlier for bulk hydrocarbon. (The terminal CH_3 group of an alkyl chain is presumably as free in a micelle as in bulk liquid, but its contribution is one of a number of invariant parameters, only the sum of which enters into the theoretical calculation.)

Phospholipids

Phospholipid molecules contain two hydrocarbon chains per head group instead of the single hydrocarbon chain of amphiphiles that form small micelles (Fig. 1). This difference alone accounts for the formation of extended phospholipid bilayers, as shown by the data in Table 2, which are for an alkyl chain length typical of those found in natural phospholipids. Suppose, for example, one were to form a small micelle containing 200 hydrocarbon chains, which would give a surface area that would be

in the optimal range if there were one head group per chain. A micelle of this size formed by diacyl phospholipids would, however, have only 100 head groups, and the area per head group would be twice the area per emerging chain. This is very far from the optimal area, and an area in the optimal range can be approached only if the disk-shaped micelle is grown to virtually infinite size, where it becomes effectively an extended bilayer, as illustrated in Fig. 2.

Another way to reduce the average area, which does not require as high a degree of aggregation, is to form a closed vesicle with an internal water-filled cavity, as illustrated by Fig. 3. Counting both inside and outside areas, the average area for a particular aggregation number is less for such a vesicle than for a circular disk with a rounded edge. Vesicles are, in fact, formed spontaneously whenever phospholipids are dispersed in an aqueous medium. They tend to be multiwalled rather than single-walled when the concentration of lipid is initially large, but single-walled vesicles are readily obtained under appropriate conditions, as when the lipid suspension is stirred vigorously by sonication (25–27).

Compartmentalization: The First Step in Biological Organization

The formation of vesicles is the first essential step in biological organization; it is, in effect, a beginning in the definition of a living cell. It occurs spontaneously, simply as a result of the drive toward thermodynamic equilibrium, once the appropriate molecules (phospholipids) have been synthesized. From the point of view of thermodynamics, however, there is an element of paradox. Because the vesicle wall contains a core of liquid hydrocarbon that is 30 Å thick, it is an effective barrier to ions, hydrophilic metabolites, proteins, nucleic acids, and so on. Communication between the inside and the outside of the vesicle is thus prevented, and there is no longer a pathway for equilibration between them. In other words, lipid molecules, in striving for their own thermodynamic equilibrium, have created the kind of structure necessary to prevent equilibration between most water-soluble substances. The chemical composition of the internal space need not be the same as that of the external environment, and the reactions that go on are, in principle, independent of the external environment as long as the vesicle walls remain intact.

A sealed-off compartment is not yet

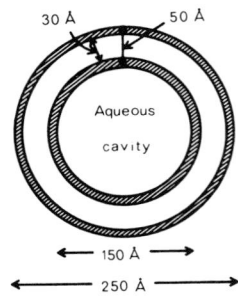

Fig. 3. Schematic diagram of a phospholipid vesicle. The dimensions given are close to the experimental dimensions of egg yolk lecithin vesicles formed by sonication (25).

representative of a cell or of the kind of intracellular compartments that exist within the cells of higher organisms. It is necessary to have controlled access for metabolites to provide energy within the compartment; building blocks to synthesize nucleic acids and proteins (and more phospholipid to permit cell growth and division); ion channels to conduct nerve impulses; recognition sites in the outer surface of cell membranes to organize multicellular structures or act as receptors for circulating hormones; mechanisms to maintain ion gradients and to secrete hormones, enzymes, and neurotransmitters; and sometimes channels to permit communication between adjacent cells. These additional structural elements have to have extraordinary specificity. One might think of them as analogous to stopcocks that provide access to a laboratory flask, but they need not only to have the mechanical ability to control flow but also to be the brains and hands of the experimenter, to determine precisely what reagent to add and when to do it. The only molecules known to be capable of such specificity are proteins, and membrane-bound proteins and glycoproteins that carry out these vital functions have been discovered in large number in recent years. In the following section I will show that the ultimate location of these proteins in membranes is probably also under thermodynamic control, and that the hydrophobic effect again plays a critical role in the process.

Protein Folding and Membrane-Bound Proteins

It is well established that the three-dimensional structure of proteins is determined by the sequence of amino acids alone: an unfolded linear polypeptide chain spontaneously folds to the same final structure in vitro as it does in the living cell (28, 29). The final structure is

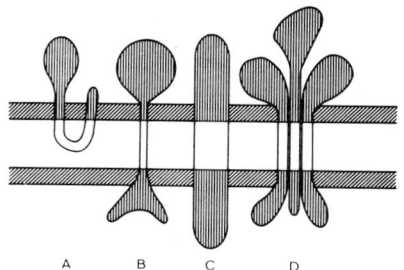

Fig. 4. Schematic representations of membrane-bound proteins. The shaded portions have hydrophilic surfaces; the unshaded portions have hydrophobic surfaces in contact with the hydrocarbon core of the bilayer. Proteins may lie entirely on one side of the membrane (A) or may have portions exposed on both sides (B to D). The hydrophobic part of the protein may represent only a small part of the whole (B). Membrane proteins, like other proteins, may be composed of more than one like or unlike subunit, held together by noncovalent forces (D).

largely under thermodynamic control, subject as usual to certain constraints, the major one being kinetic. The process begins with the folding of a particular portion of the polypeptide chain to its own thermodynamically stable state, and that portion may remain intact during subsequent stages of the process, especially if they are fast. Some structural states may thus be kinetically inaccessible during the search for a final equilibrium structure. How frequently (if ever) such kinetic constraints prevent attainment of a hypothetical true equilibrium state of lower free energy than the native state is not known.

The forces involved in protein folding are also well understood (16). Hydrophobic forces play an important role (30), but steric constraints, hydrogen bonds, and hydration energies of charged groups are equally important. For many water-soluble proteins the end result is a compact globular particle with charged groups invariably at the surface and hydrophobic groups mostly on the inside, avoiding contact with water as best they can. Hydrophobic groups are derived from amino acids with hydrocarbon side chains and are considerably smaller than the hydrocarbon chains of phospholipids or of micelle-forming amphiphiles. Furthermore, the amino acids with hydrophobic side chains are usually interspersed with hydrophilic amino acids in the sequence. The exact structures that are known (31) (for about 30 or 40 water-soluble proteins) thus do not contain conspicuous large internal domains consisting purely of hydrocarbon,

and the overall structures tend not to be significantly deformable. The peptide groups of the polypeptide backbone are intrinsically hydrophilic, but they appear to be stable away from contact with water if they can form hydrogen bonds with each other, such as occur in the well-known α-helix and the parallel or antiparallel β-sheet structures. The existence of such structures within the folded protein further enhances the structural rigidity.

How are membrane-bound proteins different? In discussing this I will consider only the so-called intrinsic or integral proteins (32, 33)—that is, proteins that traverse the phospholipid bilayer or are firmly anchored in it, with intimate contact between protein and hydrocarbon chains, as illustrated schematically by Fig. 4. All proteins that play the kind of functional role envisaged above are in this category. Amino acid compositions are known for many of them, and it is clear that they are constructed from the same amino acids as water-soluble proteins; moreover, they do not have a significantly different overall composition in terms of the fractions of the amino acids that are hydrophilic and strongly hydrophobic. In particular, they have about the same fraction of ionic side chains as water-soluble proteins, and these side chains have to be in close contact with water. Even in the absence of actual structural information, it is therefore certain that much of the protein surface must ordinarily project outside the bilayer into the adjacent medium, as in the schematic models of Fig. 4. The existence of these external regions is appropriate from a functional point of view, because all the functions envisaged involve recognition of extramembranous ions or water-soluble substrates, hormones, and so on.

There is substantial evidence that the location of membrane proteins in membranes is under thermodynamic control; proteins have been extracted in various ways, and normally functioning protein-containing membranes have been reconstituted. Moreover, most membrane proteins are virtually insoluble in aqueous salt solutions but can be solubilized in appropriate benign detergents without changes in structural parameters that can be measured (for instance, overall compactness and circular dichroism spectrum) (34). Any part of their function that can be assayed in solution, such as the specific ion-stimulated adenosinetriphosphatase activity of adenosine triphosphate–driven ion pumps, is generally preserved under these conditions. Since detergent micelles offer the same kind of local environment to the protein as a phospholipid bilayer—that is, a narrow layer of hydrocarbon bounded on both sides by an aqueous medium (Fig. 2)—it is reasonable to infer that this kind of environment is thermodynamically favored. If the principle that amino acid sequence determines three-dimensional structure is preserved (there is no reason to suppose that it is not), membrane proteins must differ from water-soluble proteins at the primary sequence level. Their amino acid sequences must direct spontaneous folding to three-dimensional structures that contain substantial hydrophobic surfaces, structures that would have a high chemical potential in an aqueous medium and a much lower one if the protein comicellizes with other substances, whether they are detergents forming small micelles or lipids forming more extended bilayers (35).

Only two or three membrane proteins have known amino acid sequences (none of them is involved in transmembrane communication), and the data support the idea that membrane association is the

+H₃N-Ala-Glu*-Gly-Asp*-Asp*-Pro-Ala-Lys*-Ala-Ala-PHE-Asp*-
  1   5   10

Ser*-LEU-Gln*-Ala-Ser*-Ala-Thr*-Glu*-TYR-ILE-Gly-TYR-Ala-
       15   20   25

TRP-Ala-Met-VAL-VAL-VAL-ILE-VAL-Gly-Ala-Thr*-ILE-Gly-
              30   35

ILE-Lys*-LEU-PHE-Lys*-Lys*-PHE-Thr*-Ser*-Lys*-Ala-Ser*-COO⁻
     40   45   50

Fig. 5. Amino acid sequence of the coat protein of bacteriophage fd (36, 47). Amino acids with hydrophobic side chains are shown by capital letters, and those with strongly hydrophilic side chains by an asterisk. The presumed membrane-bound region is enclosed by solid lines; it may include residues 40–42 because the lysine side chain is long enough to allow the α carbon of residue 40 to be within the hydrocarbon core, while the charged NH₃⁺ group lies beyond the next two hydrophobic side chains on the outside.

result of specialized sequences. One established sequence is for the coat protein of the filamentous bacteriophage of the fd type (*36*), shown in Fig. 5. It contains an unbroken sequence of about 20 hydrophobic residues, with a single threonine the only midly hydrophilic amino acid in the sequence. The protein is bound to the membrane of the host bacterium (*Escherichia coli*) during phage morphogenesis, and this segment of sequence is responsible for the binding and has been shown to be the membrane-bound portion in reconstituted lipid vesicles (*37*). Another established sequence is that of cytochrome b$_5$ from liver microsomal membranes. This protein lies entirely on one side of the membrane (as in Fig. 4A), and it has a three-dimensional structure composed of independently folded domains, which can be isolated as separate fragments after mild proteolysis (*38, 39*). The major fragment is water-soluble and contains the active site of the protein. A 40-residue fragment, relatively rich in hydrophobic amino acids, is responsible for association with the membrane. Instead of a single long hydrophobic sequence it has two shorter ones of eight residues apiece, separated by about ten polar residues (*40*), suggesting that attachment to the membrane is by two shallow U-shaped inserts.

Aqueous Channels Through Membranes

The short hydrophobic sequences just discussed are appropriate for forming solid inserts for anchoring receptors or recognition proteins to a membrane, but are not suited for transmembrane proteins involved in the transport of ions or polar molecules. These proteins must have more sophisticated structures that include provision for the passage of hydrophilic substances through the membrane. Figure 6 shows a purely speculative structure that could accomplish this. This structure might be formed by an alternating sequence in which there is a strongly hydrophilic residue at every third or fourth position in an otherwise hydrophobic stretch of amino acids. Such a sequence could form helices with a predominantly hydrophobic surface, but possessing one hydrophilic edge (*41, 42*). Six or more such helices could combine to form a structure with a hydrophobic surface, but with an aqueous channel down the middle. The individual helices could be parts of the same polypeptide chain, as in bacteriorhodopsin (*43*), or the overall structure could be an oligomer of identical chains, each contributing one helix.

Unique Fitness of Hydrophobic Forces

The structural organization I have described (solely in terms of the final equilibrium state; the means of arriving at this state represent a separate problem, considered later) consists of three stages: formation of phospholipid bilayers and vesicles, folding of proteins to their native structures, and insertion of appropriately structured proteins into bilayers. The thermodynamic driving force for the first and last stage is a repulsive force acting on the separated elements and not a preferential attraction. I believe this to be an essential feature of the process. Because of it, micelles and bilayers are fluid. There are no strong attractive forces between alkyl chains in the hydrocarbon core and no preferences for particular neighbors: micelles and bilayers are formed from mixtures of amphiphiles as readily as from pure ones. This indifference to local arrangement is crucial for detergency: soap and detergent micelles indiscriminately absorb any molecule that has a hydrophobic piece. Incorporation of membrane proteins into membranes can be thought of as a form of detergency that comes into play whenever a protein molecule with a hydrophobic piece has been synthesized in the cell.

Moreover, fluidity per se is vitally important for life processes (*33*). Cells have to be capable of deformation, as part of locomotion, for example, or when squeezed by neighboring cells. They have to be capable of instantly resealing if accidentally nicked at some point on the surface.

These essential properties of membranes—deformability and accommodation to appropriate insertions—would be virtually impossible to achieve if the permeability barriers that represent the first stage of compartmentalization were to

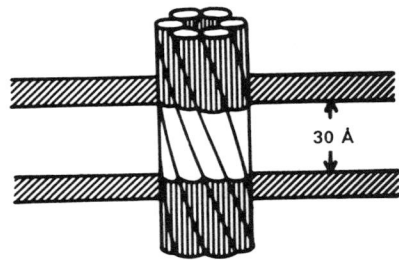

Fig. 6. Hypothetical model for the formation of an aqueous channel across a phospholipid bilayer. The twisted cylinders are helical segments of a polypeptide chain. Shaded portions of the surface are hydrophilic; unshaded portions are hydrophobic. [This is patterned after a proposal by Inouye (*42*) for a somewhat different kind of protein.]

be based on specific attractive forces, "bricks and mortar," that normally lead to rigid inflexible structures (illustrated by protein molecules, where internal hydrogen bonds play an important role). The fact that nature uses the hydrophobic force as the factor that creates compartments and the connectors between them is not an arbitrary choice. It is difficult to imagine it done in any other way.

Concluding Remarks

In this article I suggest that a reasonable initial approach to an understanding of the organization of living matter, even in the most complex multicellular organisms, is to divide the process into two parts: (i) biosynthesis of the appropriate molecules and its control and (ii) assembly of these molecules into organized structures. The second part of the process appears to be largely under thermodynamic control; that is, organized structures result simply from the search of each molecule for its position of lowest chemical potential.

I have focused on the creation of cell membranes and means of communication between the two sides, but there is every reason to believe that the same principle applies to the extension of cells in space, the creation of intercellular contacts, and other aspects of biological organization. That virus particles are self-assembling is already well established (*44*). The contribution of the hydrophobic effect to the chemical potential would, of course, be much less important for these other aspects of organization than it is for the formation of membranes. In fact, we can expect directed polar bonds to play a dominant role whenever the desired end result is a rigid structure, as exemplified by the role of rigid networks of hydrogen bonds in the formation of the ordered structures of proteins.

Thermodynamic analysis serves to simplify and clarify one's view of a subject. It answers some questions, but may be even more useful in defining problems for future investigation. In fact, in the development of physical chemistry, identification of an equilibrium state (and establishment that it is the equilibrium state) has usually been followed by investigation of the mechanism by which the equilibrium state is attained. The same must be true in biology, and an intriguing question, which is currently receiving much attention from cell biologists, concerns proteins that traverse membranes and have exposed hydrophilic portions on both sides (Fig. 4). The phospholipid bilayer, once formed, is impermeable to ionic molecules, but these proteins are synthesized on one side of an already existing bilayer, and some of their ionic amino acid side chains have to pass through the bilayer to reach the other side. The current status of work on this problem has been reviewed (45). A similar question can be asked with reference to the entry of nucleic acids into cells, such as occurs on infection by viruses, and to the transfer of newly synthesized phospholipid molecules to the external half of a bilayer (46).

It is noteworthy that these questions do not arise in relation to an unexpected failure to attain an equilibrium state. On the contrary, they seek to explain the existence of a state close to equilibrium where there are known physical constraints that would be expected to block the approach to equilibrium.

References and Notes

1. J. W. Gibbs, *Trans. Conn. Acad.* **3**, 108 (1876); *ibid.* **3**, 343 (1878).
2. Enthusiasm for the use of chemical potentials has not always been universal. For instance, in G. N. Lewis and M. Randall [*Thermodynamics and Free Energy of Chemical Substances* (McGraw-Hill, New York, 1923)] there is only one reference to chemical potentials (p. 158), in fine print.
3. R. W. Gurney, *Ionic Processes in Solution* (McGraw-Hill, New York, 1953; Dover, New York, 1962), p. 90.
4. J. H. Hildebrand *et al.*, *Regular and Related Solutions: The Solubility of Gases, Liquids, and Solids* (Van Nostrand Reinhold, New York, 1970), chap. 5. The data given in this reference and results from my own laboratory suggest that deviations from ideal mixing behavior, as when molecules of different sizes and shapes are mixed, are small where solvent-solute interactions are weak. Only the extreme case of high polymers would constitute an exception.
5. C. Tanford, *The Hydrophobic Effect: Formation of Micelles and Biological Membranes* (Wiley, New York, 1973).
6. C. McAuliffe, *J. Phys. Chem.* **70**, 1267 (1966).
7. S. S. Davis, T. Higuchi, J. H. Rytting, *J. Pharm. Pharmacol.* **24** (Suppl.), 30P (1972).
8. R. Smith and C. Tanford, *Proc. Natl. Acad. Sci. U.S.A.* **70**, 289 (1973).
9. D. B. Gilbert, C. Tanford, J. A. Reynolds, *Biochemistry* **14**, 444 (1975).
10. J. A. Reynolds, D. B. Gilbert, C. Tanford, *Proc. Natl. Acad. Sci. U.S.A.* **71**, 2925 (1974).
11. R. B. Hermann, *J. Phys. Chem.* **76**, 2754 (1972).
12. M. J. Harris, T. Higuchi, J. H. Rytting, *ibid.* **77**, 2694 (1973).
13. C. Chothia, *Nature (London)* **254**, 304 (1975).
14. G. S. Hartley, *Aqueous Solutions of Paraffin-Chain Salts* (Hermann, Paris, 1936), chap. 7.
15. H. S. Frank and M. W. Evans, *J. Chem. Phys.* **13**, 507 (1945).
16. W. Kauzmann, *Adv. Protein Chem.* **14**, 1 (1959).
17. For example, see J. H. Hildebrand [*J. Phys. Chem.* **72**, 1841 (1969)]. Arguments against the concept that invoke the vapor state [for example, R. Wolfenden and C. A. Lewis, Jr., *J. Theor. Biol.* **59**, 231 (1976)] neglect the cohesive energy gain that is involved in forming any condensed state. In effect, they compare no environment with an aqueous environment.
18. Direct evidence for one system is given by A. A. Ribeiro and E. A. Dennis, *J. Phys. Chem.* **81**, 957 (1977).
19. C. Tanford, *ibid.* **76**, 3020 (1972); *ibid.* **78**, 2469 (1974).
20. P. J. Flory, *Statistical Mechanics of Chain Molecules* (Wiley, New York, 1969), chap. 3.
21. R. J. Robson and E. A. Dennis, *J. Phys. Chem.* **81**, 1075 (1977).
22. C. Tanford, Y. Nozaki, M. F. Rohde, *ibid.*, p. 1555.
23. C. Tanford, in *Micellization, Solubilization and Microemulsions*, K. Mittal, Ed. (Plenum, New York, 1977), p. 119.
24. The optimal area depends on the strength of repulsive interaction relative to the hydrophobic free energy. Optimal areas are different for different head groups and do not remain exactly the same for a particular head group when the size of the hydrophobic moiety is changed. The argument in the text, based on a universal optimal range of 60 to 65 $Å^2$, is not intended to be the basis for quantitative calculations.
25. C. Huang, *Biochemistry* **8**, 344 (1969).
26. S. Batzri and E. D. Korn, *Biochim. Biophys. Acta* **298**, 1015 (1973).
27. J. Brunner, P. Skrabal, H. Hauser, *ibid.* **455**, 322 (1976).
28. C. B. Anfinsen, *Science* **181**, 223 (1973).
29. M. Levitt and A. Warshel, *Nature (London)*, **253**, 694 (1975).
30. C. Tanford, *J. Am. Chem. Soc.* **84**, 4260 (1962).
31. *Cold Spring Harbor Symp. Quant. Biol.* **36** (1971). This entire volume is devoted to the structures of proteins and viruses.
32. D. E. Green, *Ann. N.Y. Acad. Sci.* **195**, 150 (1972).
33. S. J. Singer and G. L. Nicolson, *Science* **175**, 720 (1972).
34. C. Tanford and J. A. Reynolds, *Biochim. Biophys. Acta* **457**, 133 (1976).
35. The two sides of a membrane in vivo are distinct, and proteins that traverse the membrane (Fig. 4, B to D) are always found inserted in the same direction. This means that each surface domain is at equilibrium with respect to location in an aqueous or hydrocarbon environment, but the hydrophilic domains on the two sides are presumably not at equilibrium, there being ordinarily no reason for a thermodynamic preference for either part to be on one side rather than the other. Sidedness is no longer present in detergent solution, and in at least some cases preference for one direction over the other is lost when previously solubilized proteins are reinserted into vesicles. See, for example, L. Packer, C. W. Mehard, G. Meissner, W. L. Zahler, S. Fleischer, *Biochim. Biophys. Acta* **363**, 159 (1974).
36. Y. Nakashima and W. Konigsberg, *J. Mol. Biol.* **88**, 598 (1974).
37. B. K. Chamberlain, Y. Nozaki, C. Tanford, R. E. Webster, in preparation.
38. L. Spatz and P. Strittmatter, *Proc. Natl. Acad. Sci. U.S.A.* **68**, 1042 (1971).
39. L. Visser, N. C. Robinson, C. Tanford, *Biochemistry* **14**, 1194 (1975).
40. D. Corcoran and P. Strittmatter, *Fed. Proc. Fed. Am. Soc. Exp. Biol.* **36**, 897 (1977).
41. The model shown is based on a model suggested by Inouye (42) for a bacterial lipoprotein [V. Braun, *Biochim. Biophys. Acta* **415**, 335 (1975)] of known sequence. The ratios of hydrophilic to hydrophobic amino acids are, however, reversed.
42. M. Inouye, *Proc. Natl. Acad. Sci. U.S.A.* **71**, 2396 (1974).
43. Bacteriorhodopsin occurs in the purple membrane of *Halobacter halobium* in a two-dimensional crystalline array, and its structure, containing seven transmembrane helices per molecule, has been determined to a resolution of 7 Å [R. Henderson and P. N. T. Unwin, *Nature (London)* **257**, 28 (1975)]. The amino acid sequence is not known.
44. A. Klug and A. C. H. Durham, in (*31*), p. 449; A. Kellenberger, *Ciba Found. Symp.* **7**, 189 (1972).
45. J. E. Rothman and J. Lenard, *Science* **195**, 743 (1977). The mechanism has to be vectorial because transmembrane proteins are always inserted in a unique direction (35), and it appears to be effective only at the actual time of biosynthesis of a polypeptide chain.
46. A more subtle problem that has not been actively pursued has to do with differences in acyl chain composition of lipids from different membranes. About half of them are generally unsaturated, which serves to prevent crystallization of the hydrocarbon core of the bilayer and keep it fluid. But no known purpose is served by such deviations from the norm as the presence of docosohexanoyl chains (six unconjugated double bonds) as the predominant unsaturated chains in retinal rod outer segment membranes. Similarly, there is no obvious functional need for the presence of cholesterol in the membranes of eukaryotic cells.
47. Abbreviations: Ala, alanine; Glu, glutamic acid; Gly, glycine; Asp, aspartic acid; Pro, proline; Lys, lysine; Phe, phenylalanine; Ser, serine; Leu, leucine; Gln, glutamine; Thr, threonine; Tyr, tyrosine; Ilc, isoleucine; Trp, tryptophan; Met, methionine; and Val, valine.
48. I am grateful to J. A. Reynolds for numerous helpful suggestions in the preparation of this article, and to the National Science Foundation for research support.

Diffusion of Univalent Ions across the Lamellae of Swollen Phospholipids

A. D. Bangham, M. M. Standish and J. C. Watkins

Agricultural Research Council Institute of Animal Physiology
Babraham, Cambridge, England

(*Received 16 March 1965, and in revised form 6 May 1965*)

The diffusion of univalent cations and anions out of spontaneously formed liquid crystals of lecithin is remarkably similar to the diffusion of such ions across biological membranes. If the unit structure of the liquid crystal is accepted as being that of a bimolecular leaflet, then these leaflets have been shown to be many orders of magnitude more permeable to anions than to cations. The diffusion rate for cations is very significantly controlled by the sign and magnitude of the surface charge at the water/lipid interface. There is a decrease of the diffusion rate for cations as the negative charge on the lipid structure decreases—which diminishes to zero for a positively charged membrane—the diffusion rate of anions remaining very high. The exchange diffusion rate of Cl^- and I^- was greater than that of F^-, NO_3^-, SO_4^{2-} and HPO_4^{2-} but no significant differences were detectable for the cation series, Li^+, Na^+, K^+, Rb^+ and choline. The membranes are very permeable to water.

Because the diffusion rate of cations is low, the phospholipid liquid crystalline structures appear to "bind" or "capture" cations. It is found that as the surface charge of the lipid lamellae is increased, the amount of cation per μmole of lipid increases. It is argued that if the cation is sequestered in aqueous compartments between the bimolecular leaflets, and if the thickness of the aqueous compartments is determined by the surface charge density of the lipid head groups and by the ionic strength of the aqueous phase in accordance with double-layer theory, the amount of cation trapped would also be expected to vary.

1. Introduction

The recognition that biological cells exploit the surface-active properties of lipids to define anatomical membranes has, in recent years, encouraged many workers to develop and study model systems based upon the orientation of lipids at interfaces (Bangham, 1963). A considerable advance was made when Mueller, Rudin, Ti Tien & Wescott (1962*a,b*, 1963,1964) and Haydon (quoted in Taylor, 1963), simultaneously and independently reported a technique for the preparation of isolated bimolecular lipid membranes separating two aqueous compartments. Such preparations, although somewhat fickle, have enabled a variety of physical parameters to be measured. The technique lends itself pre-eminently to electrical studies of a.c. and d.c. resistances and of capacitance (Huang, Wheeldon & Thompson, 1964; Hanai, Haydon & Taylor, 1964). The major criticism of the technique however, is that the precise composition of the "black" (bimolecular) membrane is in some doubt, since it has not been found possible to spread the membranes in the absence of a relatively large mole fraction of a "filler" hydrocarbon and of water-insoluble solvents. Indeed, according to Clements & Wilson (1962), if as little as 1% of the lipid mass in a membrane contains non-polar compounds, e.g. chloroform, the membrane may be considered to be in a fully anaesthetized state. A further difficulty is encountered when lipid mixtures analogous to those present in biological membranes fail to produce useful membranes.

The liquid crystal (smectic mesophase) is a preferred phase structure of many biological lipids in the presence of water or salt solutions. (See review by Dervichian, 1964.) It may be ascribed to the nature and heterogeneity of the hydrocarbon moieties and to the possession by the lipids of either polar, ionogenic or both types of head group. The precise geometry of the structures depends upon the relative concentration of the two principal components (lipid and water), the temperature, the composition of the lipid and the salt concentration of the aqueous phase (Bangham, 1963; Haydon & Taylor, 1963; Lawrence, 1961; Luzzati & Husson, 1962). Over a wide range of such variables, however, the commonest phase structure appears to be that of a layer lattice giving rise to spherulites and myelins, both composite structures consisting of many concentric bimolecular layers of lipid each separated by an aqueous compartment. For thermodynamic reasons (Haydon & Taylor, 1963), it is probable that at equilibrium each and every lipid bilayer forms an unbroken membrane—there being no exposed hydrocarbon/water interfaces—from which it follows that every aqueous compartment would be discreet and isolated from its neighbour, including a complete separation of the outermost aqueous compartment of the whole structure from the continuous aqueous phase in which it is suspended. Since these liquid crystalline structures form spontaneously when dry lipids are allowed to swell in aqueous salt solutions, it seemed reasonable to test for the integrity of unbroken membranes by measuring the amount of electrolyte solution trapped in the aqueous spaces, and also the rate of any subsequent leakage of ions into the bulk phase. The lipids were swollen in the presence of one salt species and then dialysed exhaustively against an iso-osmotic solution of another species; the amount and nature of the original salt remaining in the lipid structures were then determined. In the event, it was also possible to measure the relative leakage rates of both cations and anions out of the compound structure under a number of different conditions. Indeed, the final experimental arrangement closely resembled the pattern of experiment undertaken very widely for systems of biological interest where, for example, the diffusion of Na^+, K^+ or Cl^- from a red cell is measured as a function of an external environment. One of the advantages of this technique is that the composition of the lipids forming the membranes can be precisely varied, particularly with respect to the sign and magnitude of the surface charge.

2. Materials and Methods

(a) *Materials*

Ovolecithin was prepared by alumina and silicic acid chromatography (Dawson, 1963) and stored after phosphate analysis as a standardized solution in chloroform at $-20°C$. Purity checks were made by thin-layer chromatography and by microelectrophoresis of a suspension in 0.001 M-NaCl—isoelectric behaviour at pH 7.0 indicating the absence of the more common contaminating molecules or breakdown products. The long-chain compounds, dicetyl phosphoric acid, stearyl amine and docosanyl pyridinium bromide ($C_{22}Py^+Br^-$) were obtained commercially.

Salt solutions were prepared from analytical grade materials dissolved in water which had been glass distilled and subsequently passed over a deionizing column (Elgastat

Minor, type C403). The isotopically labelled salts, ^{42}KCl (0.145 M; initial activity of 20 mc/g K); $K^{36}Cl$ (obtained solid at a specific activity of 500 μc/g Cl) and ^3H_2O (200 mc/ml.) were obtained from the Radiochemical Centre, Amersham.

The dialysis tubing was $\frac{1}{4}$ in. Visking obtained from Hudes Merchandising Company Ltd. and Sephadex type G50 (coarse) was obtained from Pharmacia (Great Britain) Ltd.

(b) Methods

(i) Procedure for investigating the capture and leakage of cations

Enough (15 μmoles times the number of portions to be used) of the standard ovolecithin was pipetted into a small (50-ml.) round-bottomed flask; if the lecithin membranes were to be given a positive or negative charge, the requisite amount of the long-chain ion was added at this stage, in chloroform solution. The chloroform was then evaporated off under vacuum, and when the lipids were dry the swelling salt solution was added (usually 0.145 M, 1.0 ml./15 μmoles lipid). The flask was then gently agitated for about 1 hr until all the opaque, white lipid material had been freed from the bottom of the flask. At this stage, the dispersion could either be stored overnight at 5°C or used after a 2-hr period of equilibration at room temperature.

For use, 1-ml. portions of the thoroughly mixed dispersions were pipetted into wetted lengths of dialysis tubing tied off at their lower ends, through a throat made from two concentric, snugly fitting lengths of polythene which gripped the dialysis tubing. The bags were then sealed with tight-fitting polythene plugs. Up to 9 of these sealed bags were then clipped to a circular polythene frame and placed in a wide-mouthed polythene bottle of 1 litre capacity. 550 ml. of the isotonic dialysing salt solution were then added, the screw cap firmly replaced and the bottle clipped to a mixer rotating at 1 rev./min. By trial and error it was found that efficient dialysis could only be achieved when there was both thorough mixing of the dialysate—due to the tumbling action of the polythene frame—and gentle mixing of the dispersion due to the movement of air bubbles inside the dialysis bag. Further, it was found that 5 successive 30-min dialyses at room temperature against 550 ml. of fresh salt solution were necessary to reduce the concentration of the untrapped original swelling solution to acceptably low levels without significantly affecting the amount of trapped cation. The bags were then removed from their frame using either rinsed forceps or salt-free gloved fingers, and dropped into cleaned and steamed glass-stoppered 20-ml. test tubes containing a further 10 ml. of isotonic salt solution. From this point, the leakage of ions from individual lipid dispersions was measured under a variety of conditions.

In the early stages of this work, Na^+ and K^+ concentrations were measured by flame photometry, and it was then usual to dialyse dispersions containing Na^+ or K^+ into isotonic choline chloride which did not interfere with the estimation of either cation. K^+ could be measured quite accurately in the presence of 0.145 M-Na^+, and a number of experiments was carried out on such a system; however the reverse was not practicable. A low gas-pressure flame photometer (Evans Electroselenium MKII) was suitable for these measurements, since it combusted the highly concentrated choline chloride and also the lipid dispersions—thereby also enabling residual quantities of Na^+ or K^+ in the lipids to be estimated at the end of an experiment. It was thus possible to account for all the Na^+ and/or K^+ remaining in a dialysis bag after its initial dialysis.

Isotopes were measured in appropriate equipment as follows: $^{42}K^+$ and $^{36}Cl^-$ samples were measured as 10-ml. portions in a thin-glass liquid Geiger–Mueller tube (20th Century M6) and 0.25-ml. portions of ^3H-labelled H_2O, solubilized with 2.5 ml. ethanol and 5 ml. of toluene (A.R.) containing 0.3% PPO (2:5-diphenyloxazole) and 0.01% POPOP (1:4-bis-2-(5-phenyloxazolyl)-benzene (Nuclear Enterprises Ltd.) to form a clear solution, were counted in a glass container placed on a refrigerated photomultiplier, coupled to a scaler.

(ii) Procedure for investigating the capture and diffusion of anions

Preliminary experiments indicated quite clearly that Cl^- was diffusing out of the lipid dispersions several orders of magnitude faster than Na^+ or K^+ and it was therefore necessary to devise a more rapid method of exchanging the untrapped salt solution. For this purpose about 3 g of Sephadex G50 (coarse) were swollen in the dialysing solution (24 hr) and allowed to settle in a glass tube to make a column 8 to 10 cm high. 1 ml. of the original lipid dispersion was then gently placed on the moist surface of the Sephadex and chased through with a requisite volume of dialysing solution. The lipid dispersion passed rapidly over the Sephadex and emerged as a single, somewhat diluted fraction, followed later by the untrapped salt (Papahadjopoulos & Hanahan, 1964).

(iii) *Birefringence and electron microscopy*

About 10 μl. of a chloroform solution (600 μg P/ml.) of lecithin with or without added long-chain compound was allowed to dry on a cavity slide. When dry, the lipid was flooded with salt solution and a coverslip sealed on. After equilibration for 5 hr, photographs were taken with Polaroids in the crossed position and with a first-order red compensator in order that birefringent and non-birefringent material could be more easily seen. Electron microscopic preparations were carried out as previously described (Bangham & Horne, 1964).

(iv) *Measurement of zeta potentials*

Electrophoretic mobilities were measured in an apparatus of the type described by Bangham, Flemans, Heard & Seaman (1958). Measurements were made by direct observation of the dispersion in the stationary layer of the cylindrical apparatus. Mobilities were converted into zeta potentials using the appropriate equation for $\kappa a > 300$ and assuming negligible specific surface conductance:

$$\zeta = 12{\cdot}9\,\frac{v}{E} \tag{1}$$

where v equals velocity and E is the field strength (Davies & Rideal, 1961).

3. Results and Discussion

(a) *Macroscopic and microscopic appearance of lecithin dispersions*

Pure ovolecithin, which by gas–liquid chromatography (Huang *et al.*, 1964) has been shown to contain a heterogeneous mixture of saturated and unsaturated fatty acids, swells readily even in strong electrolyte to form a milky dispersion which does not froth. The absence of frothing upon shaking indicates the extreme paucity of free molecules in solution available to adsorb at the air/water interface, an observation that fits in well with the reputed low critical micelle concentration (Robinson, 1960). Admixture of up to 15 moles per cent of either a long-chain cation or long-chain anion does not noticeably alter the quality of the dispersion.

Microscopically, a typical dispersion presented complete heterogeneity of size and shape. Large, typical myelinic fragments could be seen and many smaller structures exhibiting Brownian movement. With crossed Polaroids, and depending upon the magnitude of the zeta potential, the fragments exhibited positive, negative or zero birefringence. The birefringent properties could be more easily demonstrated by simply allowing the lecithin mixtures to swell on a coverslip without perturbation (Plate I). The interpretation of the variation in birefringence as a function of zeta potential is dealt with elsewhere (Bangham, Haydon & Papahadjopoulos, manuscript in preparation) but, since its recognition and implication formed the starting point of the present investigation, a brief analysis is given here. The apparent birefringence is the sum of an intrinsic component, which may be positive or negative, and a form component (due to the parallel array of structures) which is negative. The intrinsic birefringence is a characteristic of the species and orientation of individual molecules, but the form component can vary in magnitude as a function of the relative thickness and refractive index of the two phases—in this case, the thickness of the lipid bimolecular

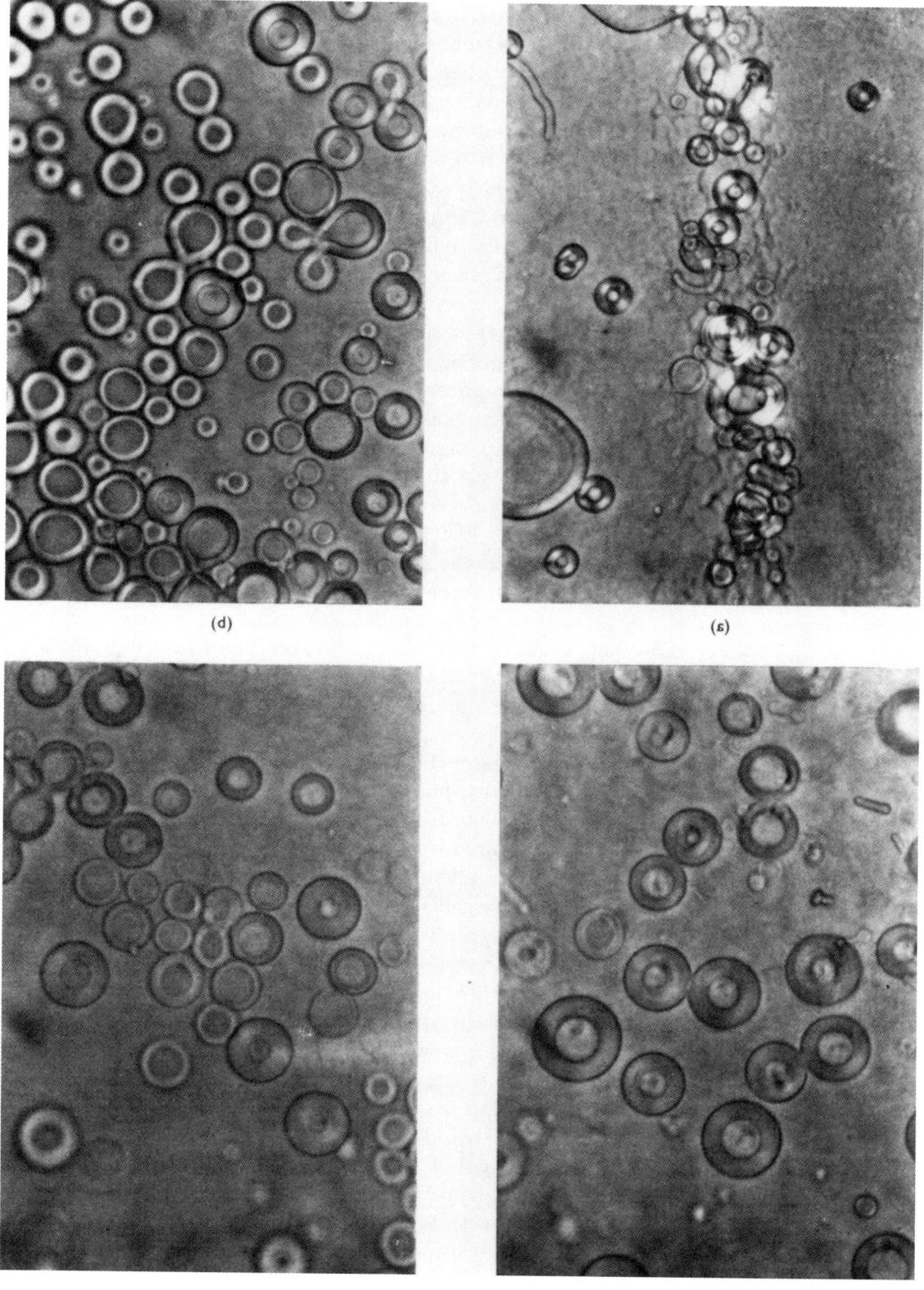

PLATE I. Light photomicrographs taken with crossed Polaroids and first-order red compensator.
(a) Lecithin only. (b) Lecithin/5% dicetyl phosphoric acid. (c) Lecithin/10% dicetyl phosphoric acid.
(d) Lecithin/15% dicetyl phosphoric acid. All preparations were swollen in 0·145 M-KCl. × 800.

[facing p. 240

layers and of the aqueous compartment. If the equilibrium distance between two adjacent lamellae depends upon the mechanical constraints of the swelling structure and the attractive and repulsive forces between adjacent lamellae (Verwey & Overbeek, 1948), absence of any net charge, as for example with pure lecithin, will result in close opposition of lamellae and strong positive birefringence. The presence of a net charge—whether positive or negative—will cause a separation of adjacent lamellae and an intensification of the negative form birefringence, ultimately resulting in an over-all loss or even reversal in the sign of the observed birefringence. Such a relationship depending on the degree of swelling of lipids in electrolytes of varying concentration was first pointed out by Palmer & Schmitt (1941) in X-ray diffraction studies of mixed brain lipids.

Ironically for Bangham & Horne (1964), it is these data which more than any other undermine the validity of the electron micrographs taken of charged lecithin dispersions negatively stained in potassium phosphotungstate. Plate II shows such a charged dispersion, where because of the lack of control of the ionic environment during drying, the dimensions of the apparent aqueous compartment do not truly represent the pre-stained system.

(b) *Capture of cation as a function of surface charge*

The birefringence data summarized in the preceding section, would indicate that ovolecithin swells to form aqueous compartments increasing in thickness as a direct function of the surface charge, a result to be predicted by a simple consideration of the Verwey–Overbeek theory (Verwey & Overbeek, 1948) relating to the equilibrium separation of charged surfaces in an electrolyte medium.

Figure 1 shows that the amount of cation, e.g. K^+, remaining in association with the phospholipid structure, even after prolonged dialysis, is finite for pure lecithin, and increases directly as a function of the surface charge increments whether positive or negative. This result suggests that cations taking part in the original swelling process are restricted in their subsequent diffusion out of the lipid, i.e. they are "captured", and that the amount captured is proportional to a physical property of the lipid structure rather than to any chemical ion-pairing or binding. That the amount of K^+ captured in positively charged structures, for a given surface charge, is not numerically equal to the amount captured in a negatively charged system, can be partly ascribed to the difference in the absolute amount and sign of gegen or screening ions required by electrical double-layer theory. The difference in the molecular volume of a double-chain anion and a single-chain cation has been neglected.

A further observation was that if the systems were swollen in equimolar mixtures of Na^+ and K^+ chloride, both cations were recovered in equivalent proportions and their sum equalled the amount trapped when one type of ion was used, a result which suggests that neither cation is being specifically bound to the membrane phosphate groups. It should also be noted (Fig. 1) that pure lecithin, although isoelectric over a wide pH range (Bangham, Pethica & Seaman, 1958; Anderson & Pethica, 1956), nevertheless, swells in electrolyte solutions to form liquid crystalline structures (Plate I) and captures some 0·1 μmole of cation per μmole of lecithin. In support of this minimal swelling, we have shown that the long-spacing of an X-ray diffraction pattern of uncharged lecithin dispersions in 0·145 M-KCl gives a value of 69 Å. By assigning an average thickness of 58 Å for a full bimolecular layer there remains an aqueous compartment equivalent to two to three water layers.

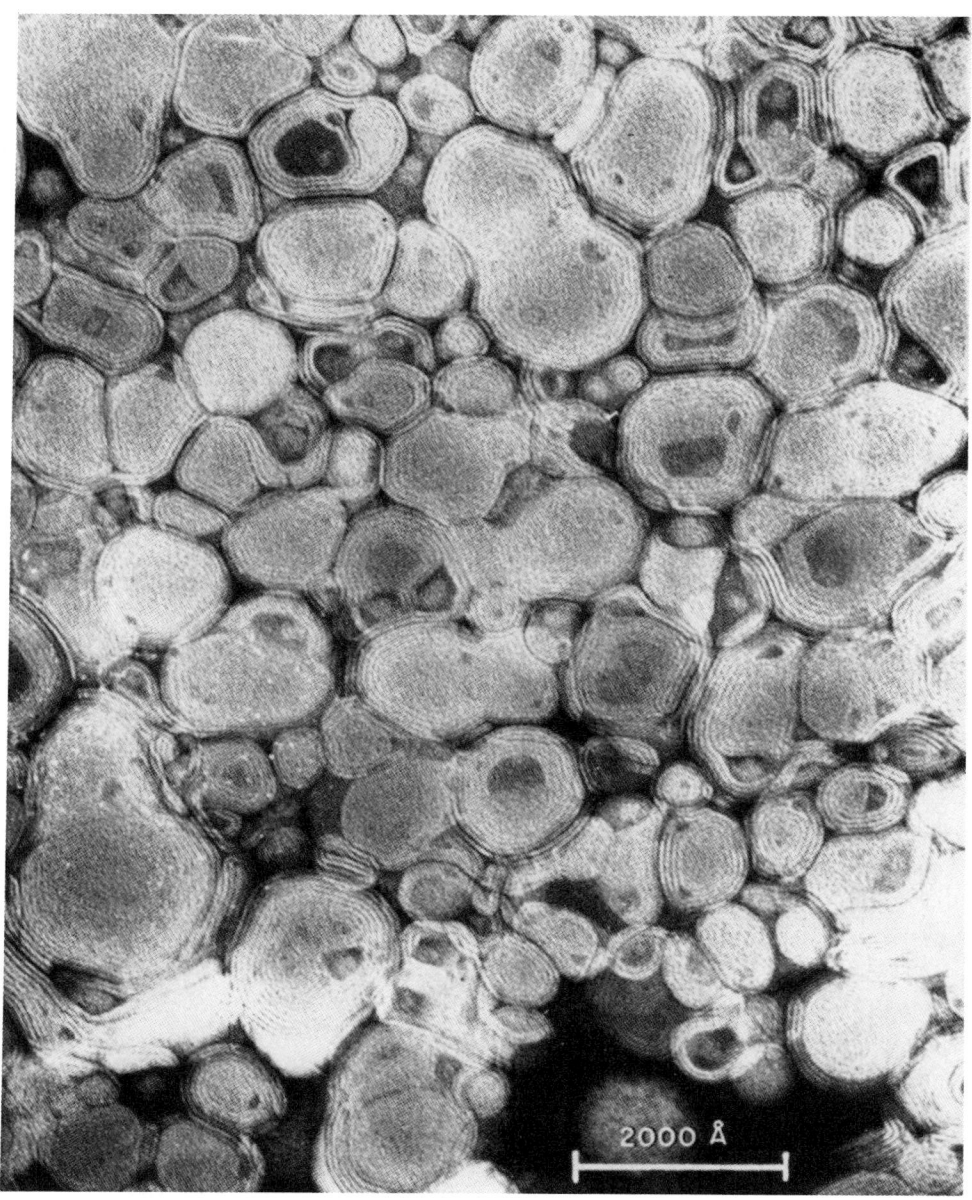

PLATE II. Lecithin/10% dicetyl phosphoric acid ultrasonically dispersed in 2% potassium phosphotungstate.

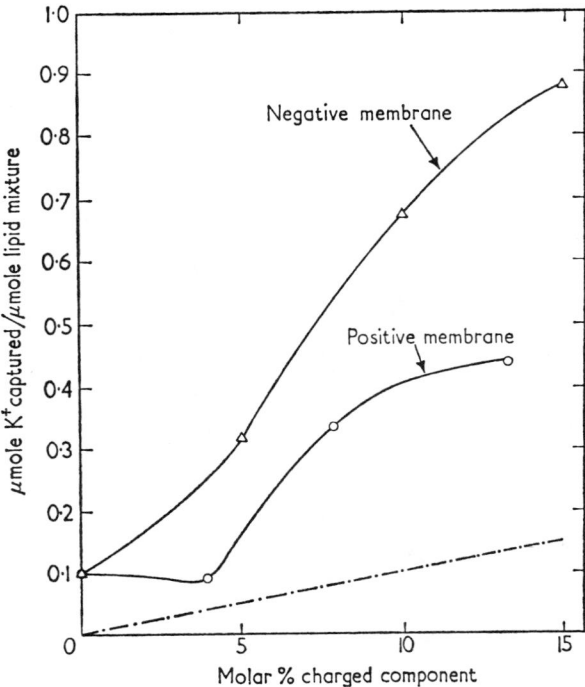

FIG. 1. Amount of K$^+$ captured by swollen lecithin structures as a function of surface charge. Lecithin plus dicetyl phosphoric acid, —△—△—; lecithin plus stearyl amine, —O—O—; fixed ion contribution (i.e. % dicetyl phosphoric acid added to lipid), ·—·—·—.

(c) *Diffusion of cations out of swollen phospholipid structures*

The results of the preceding section presuppose that during the initial dialysis there is no significant diffusion of captured cation out of the phospholipid structure. The lower capture values for positively charged systems for example, might be explicable on the grounds that captured cation was exchanging more rapidly with dialysate ions than in a negatively charged system. However, as will be shown in this section, the reverse is true.

As described in the Methods section, the exhaustively dialysed phospholipid mixture, contained inside a dialysis bag, was transferred into 10-ml. portions of isotonic tracer-free salt solution or choline chloride and the appearance of cation measured in the dialysate. By convention, this further diffusion process will be termed "leakage". If the leakage characteristics are to be attributed solely to the lipid system, then it is essential that the diffusion rate through the dialysis bag be many times greater than that across the lipid lamellae. Evidence for this is presented in Fig. 2 (bag only). When a dialysis bag of the same dimension as that used in the normal procedure is filled with 1·0 ml. of 0·145 M-KCl containing ^{42}K$^+$ and then placed in 10 ml. of 0·145 M-KCl (non-isotopic), the leakage of ^{42}K$^+$ monitored over the initial period showed an extremely rapid diffusion rate, almost complete equilibrium being reached in 30 minutes. This corresponded to a diffusion rate or total flux of $8·23 \times 10^{-5}$ mole min^{-1}. In comparison, the initial leakage rate of ^{42}K$^+$ associated with a well-dialysed dispersion of 15% dicetyl phosphoric acid/lecithin when contained in a similar dialysis bag gave an over-all diffusion rate of $1·38 \times 10^{-7}$ mole min^{-1}. Thus for ^{42}K$^+$, it would appear that the diffusion limiting process is attributable to some property of the lipids.

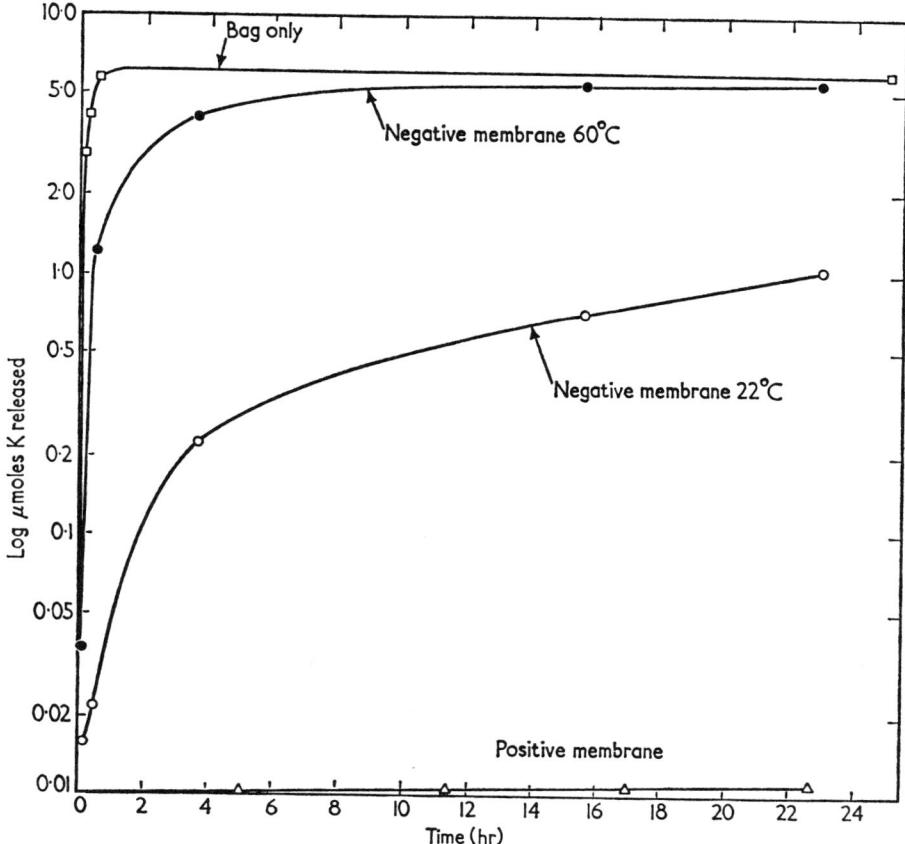

Fig. 2. Amount of $^{42}K^+$ in μmoles appearing outside a dialysis bag at 22°C. 6·0 μmoles of $^{42}K^+$ as 0·145 M-KCl inside the bag, —□—□—; 6·0 μmoles of $^{42}K^+$ as 0·145 M-KCl trapped inside 15 μmoles of a 5% dicetyl phosphoric acid/lecithin mixture,—○—○—; 6·0 μmoles of $^{42}K^+$ as 0·145 M-KCl trapped inside 15 μmoles of a 5% stearyl amine lecithin mixture, —△—△—; amount of $^{42}K^+$ in μmoles appearing outside a dialysis bag at 60°C when 6·0 μmoles of $^{42}K^+$ as 0·145 M-KCl are trapped inside 15 μmoles of a 5% dicetyl phosphoric acid/lecithin mixture, —●—●—.

At room temperature cations continue to diffuse out of negatively charged phospholipids for a period of some 24 hours at a rate which conforms to the calculated concentration gradient. For example, if a theoretical leakage-time curve is constructed, using Fick's law applied to the initial rates observed during the first 30 minutes, it can be calculated that the $^{42}K^+$ concentration during the subsequent 24 hours lies close to the experimental points.

That is to say

$$\frac{dc_2}{dt} = DA\,(c_1 - c_2) \tag{2}$$

where D is the diffusion coefficient, A is the unknown external surface area of the particles, and c_1 and c_2 are the internal and external concentrations respectively (μmoles/ml.). c_1 and c_2 are related by the relationship:

$$V_1 c_1 + V_2 c_2 = m$$

where V_1 is the volume of the electrolyte captured (ml.)

V_2 is the volume of the dialysate into which the K^+ is leaking (ml.)

m is the total quantity of K^+ captured by the lipid (μmoles)

and c_1 and c_2 are as defined above.

Rearranging equation (2) gives:

$$\int \frac{dc_2}{(c_1 - c_2)} = \int DA \, dt + \text{a constant.} \tag{3}$$

Now in order to integrate, c_1 must be expressed in terms of c_2 from the expression given above.

Integration then gives

$$c_2 = \frac{m}{V_2 + V_1} \left(1 - \exp \frac{V_2 + V_1}{V_1} DAt \right). \tag{4}$$

The value of DA is obtained by use of equation (2) applied to the initial leakage rate values given in Fig. 2 (negative membrane, 22°C).

Since it is proposed that the swollen phospholipid is a multi-layer system, the cation flux would be expected to obey Fick's law for a two-compartment system only for relatively small losses of internal K^+. After this stage, corresponding to depletion of the outer layers, the diffusion would have to take place across an in-series arrangement of lamellae and the diffusion kinetics would therefore be expected to resemble the analogous heat flux from a solid sphere (initially at uniform temperature) situated in a well-stirred medium. The results at 22°C shows that for only a 1/6th loss of cation the diffusion kinetics follow those indicated by Fick's law, whilst the results at 60°C deviate significantly from Fick's law after approximately a 1/6th loss. The results are not detailed enough to test whether the kinetics are those of the "solid sphere" (Carslaw & Jaeger, 1947), but the deviation would appear to represent the expected departure from two-compartment kinetics, and suggests some form of multi-layer system.

Justification for the use of the Fick diffusion law is found in Fig. 3, where it is seen that the initial leakages from negatively charged lipid systems are linear with time, and increase as the negative surface charge density increases, a relationship which might be predicted for two reasons. First, cations will concentrate in the electrical double-layer adjacent to a negatively charged surface, their concentration rising

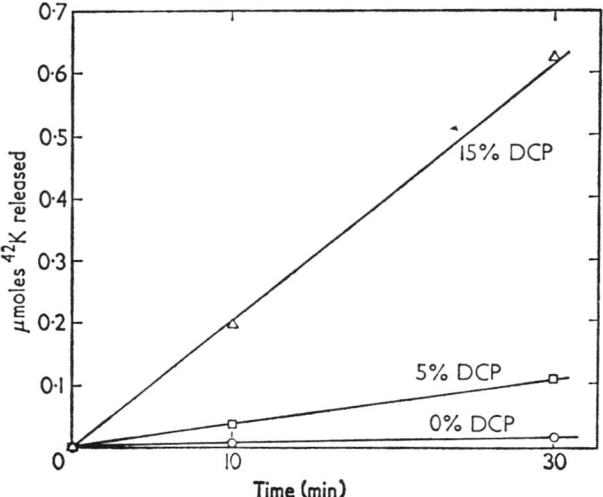

FIG. 3. Initial leakage rates of $^{42}K^+$ out of lecithin alone, —O—O—; lecithin/5% dicetyl phosphoric acid, —□—□—; and lecithin/15% dicetyl phosphoric acid into 0·145 M-KCl (non-isotopic), —△—△—. Temperature 37°C.

exponentially with surface charge density, and therefore by Fick's law the rate of diffusion will be proportionately increased. However, simple application of the Boltzmann distribution law to give the cation concentration adjacent to the interface would indicate leakage rates to be in the ratio $15\% : 10\% : 5\% : 0\% = 2 \cdot 72 : 2 \cdot 05 : 1 \cdot 44 : 1 \cdot 0$, which fails to account in itself for the much greater increase observed (Fig. 3). Second, an increase in the surface charge density would increase the intermolecular electrostatic energy of repulsion, giving rise to a time-average increase in pore size existing normal to the plane of the lipid lamellae. Any increase in the size of these pores, the effective diameter of which must be of the order of ionic radii, could significantly increase the diffusion rate. It should be noted, however, that since swelling accompanies the increase in charge density, an increase in surface area will also take place and this will increase the absolute leakage rate, whilst the diffusion rate per unit area may or may not increase.

The discrepancy between the observed rate of change of cation leakage and the expected increment if calculated by Boltzmann and Fick relationships is in many ways similar to the unexplained e-fold change in Na^+ conductance across a nerve membrane following a depolarizing potential of 4 to 6 mv. For a resting membrane potential of -70 mv, the external surface of the nerve membrane would carry an extra positive charge for every $2 \cdot 5 \times 10^4 Å^2$ of surface, which would decrease to one per $2 \cdot 75 \times 10^4 Å^2$ after depolarization. The reduction by 5 to 10% of the positive charge density may be considered to have taken place in a plane of fixed membrane anions, e.g. phosphatidylserine, triphosphoinositide or other phosphatides, and their removal would result in a net increase in the negative surface charge density. In the present system, a 10% increase in the negative surface charge density from 267 $Å^2/$ charge (that is 15% dicetyl phosphoric acid) to 293 $Å^2/$charge gives rise to an $e/20$-fold increase which represents an amplification of $1 \cdot 52$ of the expected increase for an inert membrane system obeying the Boltzmann distribution law and Fick's law of diffusion.

Whether Coulombic interaction between the charged head-groups would be sufficient to cause break-up of the bimolecular leaflets is a matter for speculation (Lucy, 1964). Myelins of pure lecithin have been shown to break up into spherical micelles (Bangham & Horne, 1964) as a result of penetration by a wedge-shaped molecule, e.g. lysolecithin. Long-chain surface-active ions would also tend to behave like wedge-shaped molecules, and on a time-average basis, give rise to transient foci of spherical micelles or larger holes bounded by a curved surface of polar head groups. Either structure would breach the bimolecular leaflet completely, permitting comparatively unrestricted diffusion to proceed. Micelle formation from a lamellar structure would involve an entropy increase, and it is perhaps not without significance that Keynes & Aubert (1964) find a cooling effect in electric eel electric organ immediately following depolarization, even when the discharge is allowed to occur on open circuit, i.e. the organ is triggered but no current flows.

Cation leakage is exhibited down to zero surface charge, i.e. pure lecithin, but with the addition of as little as 5 moles per cent of a long-chain cation, leakage is completely prevented. The relationship between the measured zeta potential and leakage of cation is shown in Fig. 4, and clearly illustrates the valve-like action of fixed surface charge on cation permeability. The effectiveness of this apparent charge barrier is demonstrated by the fact that no cation emerges from a positively charged lipid system even after several hours leakage at 60°C (cf. also Fig. 2, positive membrane).

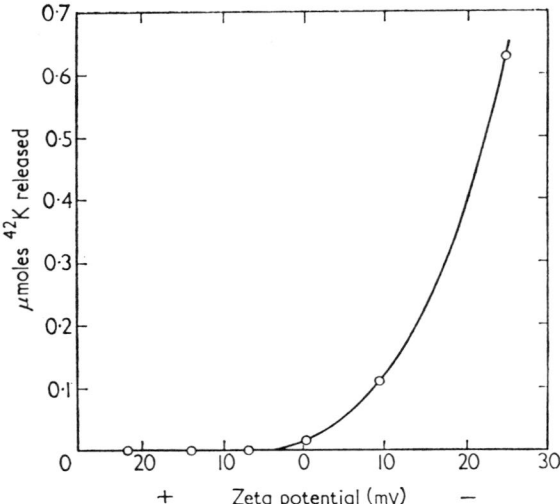

Fig. 4. Amount of $^{42}K^+$ released in 30 min at 37°C from lecithin/long-chain ion mixtures as a function of measured zeta potential (mv). Dialysis against 0·145 M-KCl (non-isotopic).

Such a result might again be predicted by a consideration of the Boltzmann distribution law, but this explanation is confounded by the finding that chloride ions diffuse almost as quickly out of both positively and negatively charged systems. Moreover, their diffusion rates are several orders of magnitude greater (see section (d)). The completeness with which such a small percentage of long-chain cation inhibits the diffusion of K^+ or Na^+ associated with the lipid structure lends further support to the conclusion that the leakage that is observed from negatively charged systems is not due to mechanical rupture of the enveloping membranes.

Bearing in mind the unique property of cation selectivity exhibited by many cell membranes, an attempt was made to determine whether the structures in the present experiments distinguished between any of the univalent alkali metal cation series. Experimentally, well-dialysed dispersions containing $^{42}K^+$ were transferred to a series of tubes containing 0·145 M solutions of the chlorides of Li^+, Na^+, K^+, Rb^+ and choline. After periods of 30 and 60 minutes, the amount of $^{42}K^+$ appearing in the dialysate was measured. For a series of surface charge densities (0 to 15% dicetyl phosphoric acid) no significant difference was detected. However, it is felt that before such a property is entirely discounted for these model systems, other fixed-charge molecules should be substituted, e.g. phosphatidyl serine, triphosphoinositide.

The effect of temperature on the initial leakage rate of cation out of negatively charged membrane is shown in Fig. 5. As in the earlier consideration of leakage rates where the concentration inside the lipid was assumed to be equal to that of the swelling solution, *viz* 0·145 M, and zero outside, initial slopes afford a convenient way of determining the specific diffusion rate. The specific diffusion rate is equal to DA and represents the rate of diffusion when unit concentration difference exists across the membrane.

At $\qquad\qquad\qquad t = 0 \qquad c_2 = 0$

equation (2) becomes $\qquad\qquad \dfrac{dc_2}{dt} = DAc_1 \qquad\qquad\qquad\qquad (5)$

where $DA (= k)$ can be regarded as the velocity constant of a unimolecular reaction. The velocity constant may be expressed in terms of the Arrhenius equation:

$$k = A' \exp \left(\frac{-E}{RT} \right) \tag{6}$$

where A' and E are practically independent of temperature. E is the energy of activation of the process and A' is a constant characteristic of the particular molecular species taking part.

Equation (6) may be rewritten as:

$$\ln k = - \frac{E}{RT} + \ln A' \tag{7}$$

and on plotting $\ln k$ *versus* $1/T$ a straight line of gradient $- E/R$ is obtained from which E can be calculated.

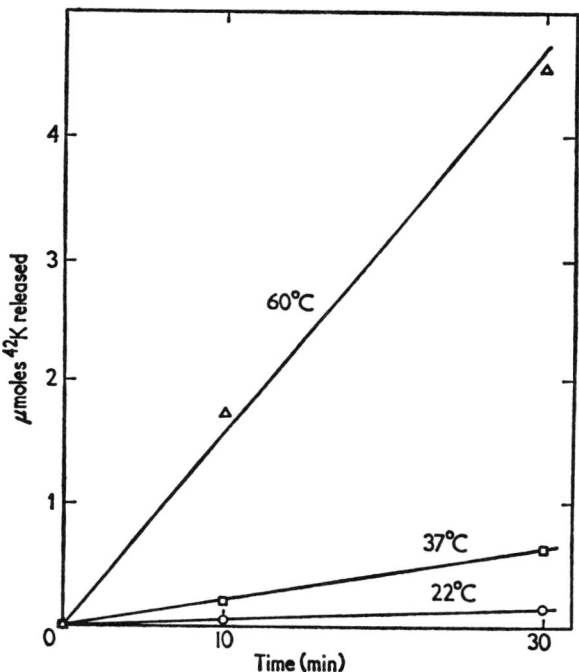

FIG. 5. Initial leakage of $^{42}K^+$ out of lecithin/15% dicetyl phosphoric acid mixtures at 22°C, —O—O—; 37°C, —□—□—; and 60°C, —△—△— into 0·145 M-KCl (non-isotopic).

This procedure was carried out using the data presented in Fig. 5, giving a value of E of 15 kcal./mole. Some measure of the energy barrier presented to the diffusing molecule may be obtained by comparing this value with those obtained for free diffusion of electrolyte in aqueous solutions of about 5 kcal./mole and of the values summarized by Glynn (1957) for ion diffusion across erythrocyte membranes of about 15 kcal./mole. It has already been mentioned that positively charged membranes remain impermeable even at raised temperatures.

(d) *Diffusion of anions*

When the experimental procedure which had been followed for the measurement of cation capture and leakage was applied to a study of $^{36}Cl^-$, it was immediately apparent that the method was inadequate. Preliminary results showed that well within the period of the first preliminary dialysis, $^{36}Cl^-$ had exchanged completely with Cl^- (non-isotopic). Without access to a flow tube apparatus (Hartridge & Roughton, 1923) and wishing to obtain at least a qualitative estimate of diffusion rate, it was decided to improvise with a method which involved passing the lipid dispersion, swollen by and suspended in $0·145$ M-$K^{36}Cl$, down a Sephadex column made up with a $0·145$ M salt solution (non-isotopic). It was found that the milky lipid dispersion passed rapidly through the Sephadex bed, followed by a clear fraction containing most of the radioactively labelled ions. If a Geiger–Mueller counter was placed at the lower end of the Sephadex column, the radioactivity of the emerging fractions could be recorded.

When for example, the Sephadex was swollen with Cl^- or I^- solutions, there was no peak of radioactivity associated with the lipid fraction after a passage of time through the column of five minutes. When, however, the Sephadex was made up with a SO_4^{2-}, NO_3^- or F^- salt, appreciable radioactivity was recorded even though the lipid had been in contact with the Sephadex for ten minutes (Fig. 6). Such qualitative

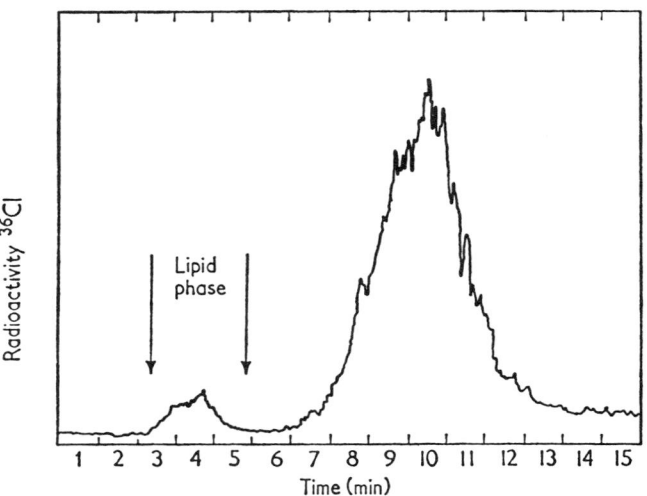

FIG. 6. A radioactivity profile of fractions emerging from a Sephadex column saturated with $0·145$ M-KNO_3. At zero time, 15 μmoles of a lecithin/5% dicetyl phosphoric acid dispersion in $0·145$ M-$K^{36}Cl$ was placed on the surface of the column. The lipids were washed through with KNO_3. When the Sephadex was saturated with Cl^- or I^- salts, no radioactivity was associated with the lipid fraction.

results were confirmed quantitatively by measuring the radioactivity in the lipid fraction and after phosphate analysis, expressing the amount of $^{36}Cl^-$ in μmoles recovered per μmole of lipid. Table 1 summarizes these results and would seem to be explained by the greater hydrated ionic radii of the SO_4^{2-}, NO_3^- and F^- ions; a principle which does not appear to apply to the cation diffusion.

(e) *Behaviour of water*

If tritiated water is used to label the aqueous phase of the swollen lipids which are subjected to the rapid dialysis method, only a trace of isotope emerges with the lipid fraction. Qualitatively, this result suggests that water is exchanging as fast if not

TABLE 1

Anion	μmoles/μmole lipid
Cl^-	0·0605
I^-	0·0598
F^-	0·1
NO_3^-	0·243
SO_4^{2-}	0·243
HPO_4^{2-}	0·35

Amount of $^{36}Cl^-$ (μmoles) associated per μmole of phospholipid swollen in 0·145 M-K^{36}Cl emerging from a Sephadex column saturated with 0·145 M-potassium solutions. Constant time of passage.

faster than chloride, a result which is compatible with the effect of hypotonic solution on lipid dispersions swollen and dialysed in strong salt solutions (Bangham, Standish & Weissmann, 1965) and which indicates that the membranes are susceptible to osmotic lysis. Bangham *et al.* (1965) have shown that both positively and negatively charged membranes release cations after being transferred to a hypotonic medium. The release from positively charged membranes indicates a physical rupture of the membranes, since it has already been shown (Fig. 4) that a simple increase in the positive zeta potential does not result in any cation leakage. The rapid accumulation of water in any inter-lipid compartments will have two effects; first an over-all volume change, and second a tendency to increase the electrostatic repulsion between the lamellae, due to the dilution of screening ions. Unless the lateral cohesion of the molecules forming the bimolecular membrane is sufficient to constrain such forces, rupture will result.

4. Conclusions

The most remarkable and perhaps least predictable result that emerges from these studies is the degree of selectivity of diffusion between cations and anions exhibited by spontaneously ordered charged or neutral phospholipid structures. If the light and electron microscopic evidence, coupled with the X-ray diffraction and birefringent properties have been interpreted correctly, then simple bimolecular leaflets of lecithin have been shown to present a diffusion barrier to cations several orders of magnitude greater than to anions. Combining such a property with the extreme degree of permeability to water and the ensuing osmotic sensitivity, the analogy to the behaviour of certain biological membranes (e.g. erythrocytes) becomes irresistible. The unexpected aspects of the experimental findings is that while the behaviour of cations can partly be reconciled with the sign and magnitude of the charge prevailing at the membrane surface, the anions appear to be relatively free to diffuse whether fixed charges of either sign are present or not. One might conclude therefore, that the mosaic arrangement of coplanar zwitterions which a pure lecithin bimolecular leaflet might present, is not perfectly crystalline but is free to form time-average positively charged surface clusters. Furthermore, with a negatively charged membrane, the anions must be finding such pores or clusters in areas of the membrane which are beyond the effective sphere of influence of the charge-contributing molecule; a suggestion that is not unreasonable since the charge densities employed are low. Although there are no direct measurements of the surface charge of biological lipid membranes, the details of their composition suggest that the net negative charge would also be low.

The control of cation diffusion by surface charge alone is decisive: from zero for positively charged membranes, and increasing as a direct function of the negative charge on the membrane. The amplification of the cation diffusion may be due to a decreased intermolecular cohesion of the bilayer superimposed upon an increase in the leakage dictated by the Boltzmann and Fick relationships. Transient foci of micelles, initiated by local concentration of the long-chain ions, may equally be responsible for this amplification and also for the very large increases observed in excitable membranes. In the experimental system which has been used, zeta potentials have been increased by incorporating increasing molar proportions of fixed long-chain anions, the ionic strength remaining constant. However, the possibility exists that biologically excitable membranes rely upon the equally effective method of increasing the negative zeta potential by decreasing the activity of positive screening ions in the region close to the surface.

Finally, there may be analogies to be drawn between the swelling properties of phospholipid liquid-crystalline structures, controlled as they appear to be by electrical double-layer phenomena, and the swelling behaviour of complete biological structures, e.g. mitochondria, composed as they so often are, of close packed lipid-containing membranes.

We wish to acknowledge the unstinting help of Nigel Miller in all aspects of this investigation. One of us (M. M. S.) was a visiting scientist from Unilever, Colworth House, Bedford, and another (J. C. W.) was a visiting scientist from the Department of Physiology, Australian National University, Canberra, Australia.

REFERENCES

Anderson, P. J. & Pethica, B. A. (1956). *Proc. 2nd Inter. Conf. Biochemical Problems of Lipids*, p. 24. London: Butterworths.

Bangham, A. D. (1963). In *Advances in Lipid Research*, ed. by R. Paoletti & D. Kritchevsky, vol. 1, New York: Academic Press.

Bangham, A. D. & Horne, R. W. (1964). *J. Mol. Biol.* 8, 660.

Bangham, A. D., Flemans, R., Heard, D. H. & Seaman, G. V. F. (1958). *Nature*, 182, 642.

Bangham, A. D., Pethica, B. A. & Seaman, G. V. F. (1958). *Biochem. J.* 69, 12.

Bangham, A. D., Standish, M. M. & Weissmann, G. (1965). *J. Mol. Biol.* 13, 253.

Carslaw, H. S. & Jaeger, J. C. (1947). *Conduction of Heat in Solids*. Oxford: University Press.

Clements, J. A. & Wilson, K. M. (1962). *Proc. Nat. Acad. Sci., Wash.* 48, 1008.

Davies, J. T. & Rideal, E. K. (1961). In *Interfacial Phenomena*, p. 134. London: Academic Press.

Dawson, R. M. C. (1963). *Biochem. J.* 88, 414.

Dervichian, D. G. (1964). *Progr. Biophys. Biophys. Chem.* 14, 265.

Glynn, I. M. (1957). *Progr. Biophys. Biophys. Chem.* 8, 241.

Hanai, T., Haydon, D. A. & Taylor, J. L. (1964). *Proc. Roy. Soc.* A, 281, 377.

Hartridge, H. & Roughton, F. J. W. (1923). *Proc. Roy. Soc.* A, 104, 376.

Haydon, D. A. & Taylor, J. L. (1963). *J. Theoret. Biol.* 4, 281.

Huang, C., Wheeldon, L. & Thompson, T. E. (1964). *J. Mol. Biol.* 8, 148.

Keynes, R. D. & Aubert, X. (1964). *Nature*, 203, 261.

Lawrence, A. S. C. (1961). In *Surface Activity and Detergency*, ed. by K. Durham, ch. 7. London: Macmillan & Co. Ltd.

Lucy, J. A. (1964). *J. Theoret. Biol.* **7**, 360.

Luzzati, V. & Husson, F. (1962). *J. Cell Biol.* **12**, 207.

Mueller, P., Rudin, D. O., Ti Tien, H. & Westcott, W. C. (1962a). *Nature*, **194**, 979.

Mueller, P., Rudin, D. O., Ti Tien, H. & Westcott, W. C. (1962b). *Circulation*, **26**, 1167.

Mueller, P., Rudin, D. O., Ti Tien, H. & Westcott, W. C. (1963). *J. Phys. Chem.* **67**, 534.

Mueller, P., Rudin, D. O., Ti Tien, H. & Westcott, W. C. (1964). *Recent Progress in Surface Science*, ed. by J. F. Danielli, K. G. Pankhurst & A. C. Riddiford. Oxford: Pergamon Press.

Palmer, K. J. & Schmitt, F. O. (1941). *J. Cell Comp. Physiol.* **17**, 385.

Papahadjopoulos, D. & Hanahan, D. J. (1964). *Biochim. biophys. Acta*, **90**, 436.

Robinson, N. (1960). *Trans. Faraday. Soc.* **56**, 1260.

Taylor, J. L. (1963). Thesis, Cambridge University.

Verwey, E. J. W. & Overbeek, J. Th. (1948). In *Theory of the Stability of Lyophobic Colloids*. Amsterdam: Elsevier Publishing Co.

Encapsulation of Macromolecules by Lipid Vesicles under Simulated Prebiotic Conditions

David W. Deamer and Gail L. Barchfeld

University of California, Davis, California 95616, USA

Summary. Phospholipid vesicles (liposomes) were subjected to dehydration-hydration cycles in the presence of 6-carboxyfluorescein or salmon sperm DNA. We found that the vesicles fused into multilamellar structures during dehydration with solutes trapped between the lamellae. Upon rehydration the lamellae swelled and formed large vesicular structures containing solute. This model can be used to study encapsulation of macromolecules by lipid membranes to form protocellular structures under prebiotic conditions.

Key words: Liposomes — Prebiotic evolution — Membrane fusion — Macromolecule encapsulation

1. Introduction

A variety of possible replicating systems have been proposed for early life forms. From knowledge of contemporary organisms, it is clear that at some point in time one or more such systems must have become encapsulated within a membranous structure containing lipid. Hargreaves et al. (1977, 1978) and Oro et al. (1977, 1979) have demonstrated that various kinds of lipid, including phospholipids, could have been present on the prebiotic earth, and that such lipids readily form closed vesicles. We now ask how lipid vesicles might have encapsulated macromolecules, and whether the membranous microenvironment could have contributed to the efficiency with which a potentially replicating macromolecule might act as a template for guiding the polymerization of certain monomers.

Usher (1977) has proposed that thermal dehydration cycles represent a likely source of free energy to drive polymerization reactions. We have adopted this model as a mechanism for lipid synthesis (Hargreaves et al. 1977) and report here that a dehydration-hydration cycle causes lipid dispersions to fuse and encapsulate various solutes, including macromolecules.

2. Experimental

In a typical experiment, dispersions of a variety of phospholipids were produced by brief sonication (3 min, Biosonic III, 40 watts nominal power) in distilled water at lipid concentrations of 1 mg/ml. Under these conditions the lipid forms liposomes which we consider to be a model for lipid dispersions available on the prebiotic earth. Results for phosphatidylcholine will be presented here, since synthesis of this lipid under simulated prebiotic conditions has been recently demonstrated by Rao et al. (1982). Aliquots of the liposomes were combined with the material to be encapsulated in varying mass ratios. (See legends for experimental details). The mixture was then dried under a stream of nitrogen at several temperature ranges. As will be described, these conditions cause fusion of the liposomes into lamellar arrays with material trapped between the lipid bilayers. The lipid was then rehydrated by sealing the tubes with a moist cotton plug for 2 h, followed by addition of distilled water to restore the original volume. The lipid lamellae swell during rehydration and are caused to form cell-sized vesicles by gentle agitation. A fraction of the material originally sandwiched between lipid layers is trapped in the vesicles. We used 6-carboxyfluorescein (6-CF) and salmon sperm DNA to demonstrate encapsulation, and monitored encapsulation efficiency either by fluorescence methods or by the absorption of DNA at 260 nm.

Offprint requests to: D.W. Deamer

3. Results

Fig. 1 A and B show liposomes before and after the dehydration-rehydration step. Note that the original dispersion cannot be resolved by light microscopy, but that after rehydration large unilamellar and multilamellar structures are readily observed. The most convincing demonstration of vesicle fusion and encapsulation is to perform the experiment with a fluorescent compound originally outside the vesicles, then to observe the resulting encapsulation by fluorescence microscopy and fluorometry. Therefore 6-CF was added to a final concentration of 0.1 mM in the liposome dispersion before dehydration. Fig. 1C and D show the

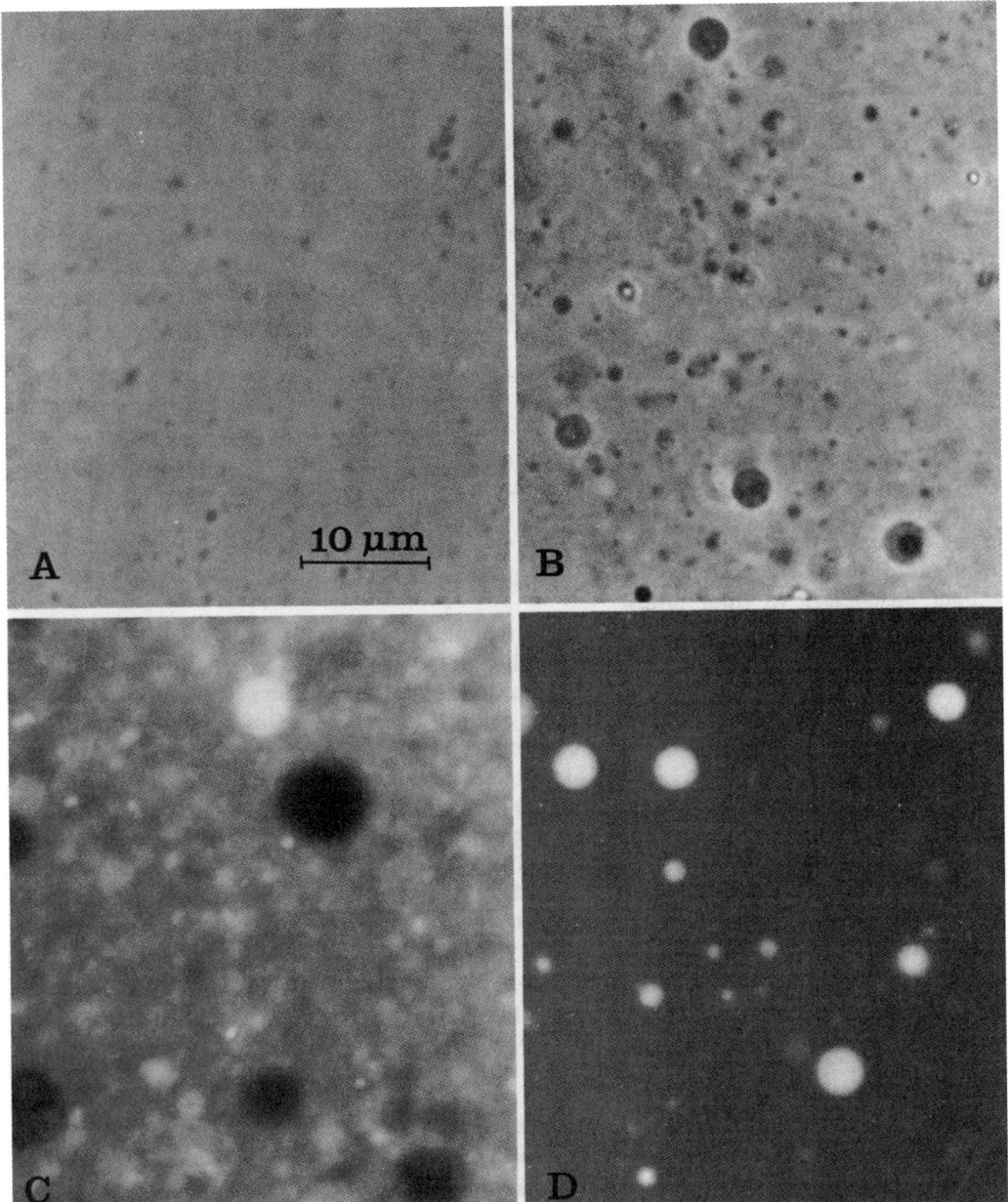

Fig. 1. Egg phosphatidylcholine (1 mg/ml) was probe sonicated until translucent (~3 min.) and examined by phase microscopy. Only a few particles at the limits of resolution could be observed (**A**). One ml of the suspension was then dried at 60°C for two hours under a nitrogen stream. The dried film was rehydrated for three hours by insertion of a moist cotton plug into the tube, followed by addition of 1 ml water and vortexing. Large vesicles were observed (**B**) indicating that the original liposomes had fused during rehydration, then resealed into larger structures during rehydration and vortexing. C shows a fluorescence micrograph of a similar preparation in which 6-carboxyfluorescein (0,1 mM) had been added after sonication but before dehydration, and **D** shows the same preparation after the external dye was removed by gel filtration (Sephadex G-50). The fusion induced by the dehydration-rehydration cycle permitted encapsulation of the carboxyfluorescein, as demonstrated by the fluorescence of the vesicles

post-hydration vesicles before and after gel filtration through Sephadex G-50. Note that the 6-CF is clearly encapsulated, and that this is most apparent after filtration, which removed the dye external to the vesicles. It is significant that before the filtration step some dark vesicles are seen which have apparently excluded dye, as will be discussed.

It was possible to calculate the fraction of the encapsulated 6-CF by measuring the amount of 6-CF fluorescence appearing in the liposome fraction after gel filtration, and comparing it with the total amount of 6-CF originally added. Values ranged from 3 to 16% encapsulation efficiency, depending on the amount of lipid originally present (Table 1). None of the 6-CF is adsorbed to liposome surfaces, because complete release occurs if the membranes were made permeable by addition of 0.1 mM Triton X-100, a non-solubilizing concentration.

Similar experiments were run with salmon sperm DNA, with two variations in procedure. After the rehydration step, the liposomes were suspended in a buffer system suitable for DNAse activity, and 20 units of pancreatic DNAse were added, followed by 1 h incubation at room temperature. This procedure digested any DNA that might be external to the liposomes or adhering to their surfaces. The liposomes were then centrifuged to remove them from the DNA solution (30 min, 10,000 x g), washed once in the same buffer, and the DNA was extracted by ethanol pre-

cipitation. Again, the encapsulation efficiency was dependent on the amount of lipid present (Table 1) and reached approximately 50% efficiency at 50:1 mass ratios of lipid/DNA. The DNA-containing liposomes under phase microscopy were similar to those shown in Fig. 1B.

4. Discussion

The results reported here show that phospholipid readily encapsulates both large and small molecules by a relatively simple process that would have been available on the prebiotic earth. We envisage the process described here to be analogous to the hydration-dehydration cycle of a fresh water tide pool containing dilute dispersions of membranogenic lipids and solutes. We note that other proposed protocells, for instance, the proteinoid microspheres described by Fox and Harada 1958, Fox and Dose 1977 are less amenable to such a direct encapsulation mechanism. We also suggest that the dehydration-fusion described here may have considerable value as an experimental tool for encapsulating compounds in liposomes or fusing lipids with biological membranes.

The mechanism of fusion under these conditions probably depends on the inability of lipid vesicles to maintain a stable bilayer structure during the dehydration cycle. As the vesicles are concentrated during drying, they must flatten. Fusion presumably occurs where vesicles come into contact, due to the inability of the bilayer to maintain a stable bilayer structure around the flattened edges. During fusion, external solutes would be sandwiched between the resulting lamellae. When rehydration occurs, the lamellae swell and disperse into larger vesicular structures (Fig. 2). It should be noted that according to this mechanism

Table 1

| | Phosphatidylcholine (mg) | | | | | |
	0.2	0.5	1.0	2.0	5.0	10
6-carboxyfluorescein (% encapsulated)	–	2.7	6.0	4.8	12.1	16
Deoxyribonucleic acid (% encapsulated)	1.1	5.0	9.8	24	45	39

Encapsulation was carried out by dehydration-rehydration as described in Figure 1, with varying amounts of phospholipid (egg phosphatidylcholine, Avanti Inc.). Carboxyfluorescein was added to 1 ml of the liposomes in water to a final concentration of 0.1 mM before drying. The fraction encapsulated was estimated by filtering the liposomes through Sephadex G–50 and measuring the carboxyfluorescein appearing in the liposome fraction by fluorometry. In the experiment with DNA, 100 μg salmon sperm DNA was used instead of 6-CF. Following rehydration, 1 ml of buffer was added (50 mM Tris-HCl, 5 mM MgCl$_2$, 0.2 mM β-mercaptoethanol) followed by addition of 20 units pancreatic DNAse (Sigma). After one hour, the liposomes were pelleted (10 kg, 60 min.) washed once in buffer, and the DNA content was precipitated by addition of 4 ml ethanol, followed by freezing in liquid nitrogen and centrifugation (30 min, 10,000 X g, -10°C). The DNA pellet was resuspended in 2 ml potassium phosphate buffer (0.1 M, pH 8.0) and scanned spectrophotometrically. The absorbance at 260 nm was then compared with that of the original DNA present and expressed as percent encapsulated. The data represent averages of values obtained from two separate experiments.

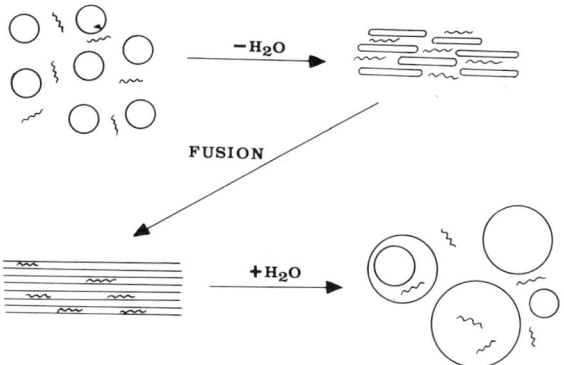

Fig. 2. Proposed encapsulation mechanism. During dehydration, liposomes touch and finally fuse, thereby trapping solutes between alternating lipid bilayers of multilamellar structures. Upon rehydration, the lamellae swell to form larger vesicles which have encapsulated a significant fraction of the original solute, in this case, the macromolecules indicated by zig-zag lines.

some of the unilamellar vesicles which form by fusion of the original internal volume will actually exclude dye. This is supported by the observation that some of the vesicles do in fact exclude 6-CF, as shown in Fig. 1. We also noted this effect with vesicles containing encapsulated DNA which were stained with acridine orange and observed by fluorescence microscopy.

There are two possible applications of these results to our understanding of prebiotic evolution. First, encapsulation of ions through dehydration-hydration cycles can provide ion concentration gradients by a non-enzymatic mechanism (Stoeckenius 1978), and it is conceivable that early life forms could take advantage of such gradients. Second, we suggest that if a template and monomers are encapsulated in vesicles, the enclosed microenvironment would have an ordering influence during dehydration. This would bring monomers into close contact with the template in a two-dimensional space and thereby enhance the efficiency with which monomers utilize the template to direct replication.

Acknowledgements. The authors thank Kenneth Turner for technical help with the DNA encapsulation studies.

References

Eichberg J, Sherwood E, Epps DE, Oro J (1977) Cyanamide mediated syntheses under plausible primitive earth conditions. IV. The synthesis of acylglycerols. J Mol Evol 10:221–230

Epps DE, Nooner DW, Eichberg J, Sherwood E, Oro J (1979) Cyanamide mediated syntheses under plausible prebiotic earth conditions VI. The synthesis of glycerol and glycerophosphate. J Mol Evol 14:235–241

Fox S, Dose K (1977) Molecular evolution and the origin of life. Marcel Dekker Inc., New York

Fox S, Harada K (1958) Thermal copolymerization of amino acids to a product resembling protein. Science 128:1214

Hargreaves WR, Deamer DW (1978) Liposomes from ionic, single-chain amphiphiles. Biochemistry 17:3759–3768

Hargreaves WR, Mulvihill S, Deamer DW (1977) Synthesis of phospholipids and membranes in prebiotic conditions. Nature 266:355–357

Rao M, Eichberg J, Oro J (1982) Synthesis of phosphatidylcholine under possible primitive earth conditions. J Mol Evol (in press.)

Stoeckenius W (1978) Speculations about the evolution of halobacteria and of chemiosmotic mechanisms. In: Deamer DW (ed) Light transducing membranes. Academic Press, New York

Usher DA (1977) Early chemical evolution of nucleic acids: a theoretical model. Science 196:311–313

Received December 1, 1981/Revised December 14, 1981

THE CHEMICAL LOGIC OF A MINIMUM PROTOCELL

HAROLD J. MOROWITZ

Department of Molecular Biophysics and Biochemistry, Yale University, New Haven, CT 06511, U.S.A.

and

BETTINA HEINZ

Institute of Organic Chemistry, J.W. Goethe University, Frankfurt am Main, F.R.G.

and

DAVID W. DEAMER

Department of Zoology, University of California, Davis, CA 95616, U.S.A.

(Received 2 March, 1987)

Abstract. Traditional schemes for the origin of cellular life on earth generally suppose that the chance assembly of polymer synthesis systems was the initial event, followed by incorporation into a membrane-enclosed volume to form the earliest cells. Here we discuss an alternative system consisting of replicating membrane vesicles, which we define as minimum protocells. These consist of vesicular bilayer membranes that self-assemble from relatively rare organic amphiphiles present in the prebiotic environment. If some of the amphiphiles are primitive pigment molecules asymmetrically oriented in the bilayer, light energy can be captured in the form of electrochemical ion gradients. This energy could then be used to convert relatively common precursor molecules into membrane amphiphiles, thereby providing an initial photosynthetic growth process, as well as an appropriate microenvironment for incorporation and evolution of polymer synthesis systems.

We define biogenesis as that continuum of chemical processes which, acting under the constraints of thermodynamics, led from an essentially random mixture of inorganic and organic substances in the prebiotic environment to the first assemblies recognizable as living cells. Along this continuum there must have been a minimum protocell, which we define as an entity thermodynamically separated from the environment and able to replicate using available nutrient molecules and energy sources. Replication is defined as any energy-requiring growth process in which an organized assembly of molecules produces similar assemblies over time. We do not require sequence-mediated information transfer, nor a precise doubling of the assemblies. Finally, we limit our considerations to evolutionary processes involving relatively small molecules, as distinguished from macromolecular assemblies whose size would require a degree of structural specificity improbable in the earliest cells.

There is a certain generality in the preceding reasoning. In essence there are two major structural assembly processes employed by contemporary cells: the growth of specific polymers from monomers, and the growth of membrane bilayers from amphiphiles. It follows that biogenesis could have occurred either through a polymer-synthesizing replicating system that later became packaged in a boundary membrane, or a replicating membrane system that later acquired the ability to synthesize polymers. These two schemes can be summarized as follows:

(I) Small molecules → macromolecules → directed synthetic system → protocell → prokaryote

(II) Amphiphiles → protocell → directed synthetic system → prokaryote

Much of the literature on biogenesis has dealt with the former scheme; we suggest here that the latter scheme offers useful insights and new experimental approaches. In particular, we note that there are fewer specificity requirements at the early stages of the second scheme, because thermodynamic phases are much less restricted in this regard than stereospecific macromolecular structures. Even today cell membranes are the least specific of all organelle components in terms of the amphiphiles that can participate in their formation.

In principle the logic of replication could be separated from the chemical constraints, but we are interested only in those logical steps for which a possible chemical embodiment exists. In order to constitute biogenesis as a physical, in contrast to a historical science (see Smith and Morowitz, 1982) we will limit the logical steps to those which can take place using chemical and physical processes that are accessible in the laboratory. The logical minimum protocell that we construct will therefore be subject to experimental demonstration or falsification.

We shall place one further constraint on a theoretical model, a restricted statement of the principle of continuity: *from any stage in biogenesis, a continuous series of plausible transitions must lead backward in time to the non-living geochemistry of the planet, and forward in time to biochemically conventional cellular life.* This is designed to deal with the problems of biogenesis using the same epistemological considerations that have been so successful in normative science. It does not disallow discontinuities, but puts the burden of proof on any theory that requires discontinuities. The principle of continuity is for biogenesis a statement of the metaphysical requirement for simplicity (Margenau 1977) which was early recognized as a guiding motif in research. It appeared first as Occam's razor: 'Non sunt entia multiplicanda practer necessitate,' Occam's words appear to have been ignored in many treatises on the origin of life.

Within this framework, early replication may be represented in an overall way by the following:

Protocell + nutrients + energy → protocells + waste products + heat.

The existence of a protocell requires a phase separation from the environment. A heterogeneous coacervate droplet of the type suggested by Oparin (1928) was an early attempt to define such a phase separation. However, the coacervate concept developed within a paradigm of colloidal protoplasm, while we now know that contemporary cells are primarily organized by membranous compartments. The coacervate is therefore rejected by the principle of contuity. Similarly, we exclude proteinoid microspheres as protocells (Fox, 1965) in that they lack the universal barrier of a liquid crystalline, non-polar phase present in all contemporary cells.

The simplest protocell which fulfulls the principle of continuity is a bilayer vesicle made from a single species or mixture of small amphiphiles. If the membrane is relatively impermeable to certain solutes, it gives rise to a three phase system which is defined by an aqueous internal volume, the membrane phase and the aqueous external environment. If one phase represents a restricted microenvironment with respect to the total system, some of the reactions occurring therein can be thermodynamically improbable in an equilibrium system, but given an appropriate pathway can be driven energetically uphill by free energy available in the total system. Consider a protocell composed of a vesicular membrane. The external environment is a source of free energy and nutrients, while the internal volume provides a closed microenvironment in which directed chemical reactions can occur. Our primary aim here is to argue that this arrangement represents a plausible protocell which arises spontaneously from known chemical and physical principles, and is accessible in the laboratory. (See also Koch, 1985.)

We can now be more explicit in outlining the primary assumptions about a protocell whose logic arises from the principle of continuity. They are as follows:

(1) The most plausible structural components are amphiphilic compounds, and growth occurs by chemical transformation of nutrient precursors to amphiphiles.

(2) Light transduction by a primitive pigment system provides energy to drive the chemical transitions involved in growth of the protocell, and chemiosmotic proton gradients are a likely form of primary energy storage across the membrane (see Deamer, 1978, and Baltscheffsky, 1981, for reviews).

(3) Phosphate is the most probable chemical group to mediate the transition between light trapping and chemical changes in nutrient molecules that lead to growth (see Westheimer, 1987 for review). In order to activate phosphate so that it can undergo transfer reactions, we will assume that pyrophosphate is an intermediate.

One further feature is that the protocell interior is a chemical reactor where energy is constantly supplied as high energy phosphate bonds and presumably withdrawn as thermal energy. The internal chemical networks are organized both by the energy flow and by the fact that there is a single chemical form to the energy input, represented by the difference between the Gibbs free energy of pyrophosphate compared to two orthophosphates. Since metabolism universally involves a chemistry driven by phosphate group transfer, a continuous series of processes is provided from primordial chemistry to contemporary biochemistry.

We can now elaborate on these basic assumptions. The reaction product necessary for the growth of a system described in equation 1 is a bilayer-forming amphiphile. The chemical logic of equation 1 can be summarized as follows:

> Protocell + amphiphile precursors + chromophore + converter 1 + converter 2 + energy → protocells + waste products + heat.

The process described above can be decomposed into the following steps for a light-driven system:

(A) Energy input (photons) + chromophore → excited chromophore
(B) Excited chromophore + converter → electrochemically stored energy
(C) Electrochemically stored energy + converter 2 + low energy compound → high energy compounds
(D) High energy compounds + membrane precursors → membrane amphiphiles + reaction products.

The 'nutrients' for the above reactions consist of membrane precursers and low energy compounds. It is assumed that a chromophore is available in the environment, and partitions into the membrane phase. Otherwise the growth of the system would rapidly dilute the available chromophore. In a more complex system, one may imagine that the chromophore is synthesized together with the amphiphile from precursors (Heinz and Ried, 1981).

The initial protocell is a vesicle self-assembled as a bimolecular leaflet of amphiphiles and chromophores. The presence of a chromophore permits energy conversion within the lipid phase of the membrane. The low energy compounds are derived from the external environment by selective transport across the membrane, and high energy compounds are produced through photochemically coupled reactions in the internal volume. The reaction products are amphiphilic molecules that would normally be rare in the environment, but are concentrated in the membrane microenvironment through energy conversion processes. As the amphiphile is synthesized, it becomes incorporated into the existing membrane, and the area of membrane increases. At some point, the size of the membranous vesicle forming the protocell increases to the point that stabilizing forces are no longer able to maintain integrity, and the vesicle breaks down into two or more smaller vesicles (Rashevsky, 1938). This step represents replication of the protocell, and requires only that the smaller vesicles have a lower Gibbs free energy than the initial larger vesicle.

The logic outlined above is general, and follows from the principle of continuity. Attempts to impose specific chemical embodiments on the logic must necessarily be more speculative, but are essential in directing experimentation. We will first assume that randomly synthesized chromophores dissolve in the membrane in a radially asymmetric way. Although at first examination an asymmetry would seem thermodynamically implausible, in fact the difference in radius of curvature between the inner and outer halves of the bilayer imposes a steric asymmetry on vesicles of sufficiently small size. Any difference in the internal and external concentrations of ionic solutes would also impose an asymmetry by inducing differences in the electrical properties of the membrane surfaces, even though the inner and outer lipid leaflets are chemically identical.

An asymmetric distribution of chromophores permits vectorial translocation of charge, which in turn can be used as an energy source to drive thermodynamically unfavorable reactions that store the energy in chemical form. The details of how such a membrane can absorb energy and generate a transmembrane potential have been described earlier (Morowitz, 1981). We assume that chemiosmotic energy in the form of a transmembrane proton gradient represents a plausible intermediate between light energy and chemical energy (Koch, 1985). This energy source, coupled to formation of pyrophosphate bonds, would provide a primitive converter system. The thermodynamic plausibility of generating high energy phosphate bonds has also been discussed (Morowitz, 1978).

Proton gradients are plausible energetic intermediates for two reasons. First, protons are ubiquitous in aqueous solutions, and can be produced by numerous chemical reactions, including some driven by light energy. Second, protons have been demonstrated to have specialized conductive pathways in bilayer membranes which permit a very high permeability even in the absence of specific channels (Nichols and Deamer, 1980). Conductive specificity for protons can be provided by continuous chains of hydrogen bonds existing in the membrane as a peptide strand (Nagle and Morowitz, 1978) or associated water (Nichols and Deamer, 1980; Deamer and Bramhall, 1986). As a result, proton gradients are able to generate membrane potentials (Biegel and Gould, 1981; Deamer and Nichols, 1983; Cafiso and Hubbell, 1983) and it is likely that such potentials would be available in the postulated protocell. Step B above can thus be considered to result from the interaction between an excited chromophore and a proton conductance which together produce a transmembrane chemiosmotic potential of protons.

To introduce a chemical embodiment of the vesicle, consider a mixture of alkane derivatives such as the mixtures of long chain alcohols and alkyl phosphates described by Hargreaves and Deamer (1978). This vesicle population will grow if a reaction is available that produces a membranogenic species (alkyl phosphate) from a membrane precursor (the long chain alcohol) which is available in the environment. The alcohol would then be a nutrient which partitions into the membrane phase, where some of it is converted to the phosphate ester:

$$ROH + pyrophosphate \rightarrow R\text{-}PO4 + Pi.$$

The newly synthesized amphiphile would generate increased membrane area, the overall process being summarized as:

$$Vesicle + fatty\ alcohol + orthophosphate + converter + chromophores$$
$$\rightarrow alkyl\ phosphate\ (larger\ vesicle).$$

In summary, we suggest the following logic for a minimum protocell:

(1) The initial protocell consisted of a membranous vesicle formed from a relatively rare amphiphilic material available in the prebiotic environment, together with an ubiquitous chromophore that could partition into the membrane phase.

(2) A reaction exists by which a potential amphiphile (nutrient) common in the environment could be converted into amphiphile.

(3) That reaction can be driven by a photochemical process involving the chromophore and phosphate.

(4) As a result, increased amounts of amphiphile are synthesized, and these cause growth of the membrane vesicle through addition to the existing membrane structure.

(5) At some point the vesicle reaches an unstable size, and breaks up into two or more smaller vesicles which continue the process by taking in additional nutrients and chromophore from the environment.

The logic outlined here defines several important questions, and suggests useful experimental approaches:

(1) What mechanisms are available for primitive chromophores to generate a transmembrane proton gradient? Chemiosmotic transmembrane proton gradients are plausible intermediates between light energy and chemical energy. There are several possible chromophores which can be tested for their ability to accept light energy and transduce it into a form capable of activating a converter molecule. These include the fluorescent pteridine-like and flavin-like pigments produced thermochemically from certain amino acids (Heinz *et al.*, 1979; Heinz and Ried, 1981) the phorphyrins studied by Mercer-Smith and Mauzerall (1984) and the porphyrin — carotene system reported by Seta *et al.* (1985). The fluorescent polycyclic aromatic compounds that are major constituents of the organic components of carbonaceous chondrites represent another possibility (Basile *et al.*, 1984). Meteoritic organic amphiphiles also have the capacity to self-assemble into membranous structures, and are candidates for the initial self-assembling membrane system described here (Deamer, 1985).

(2) What reactions are available for pyrophosphate synthesis under prebiotic conditions? Heat energy can be stored as chemical energy in pyrophosphate, which can be produced simply by heating/dehydration of phosphate, but it would be desirable to demonstrate a reaction occurring in soution that involves light energy and membrane-bound chromophores. An alternative to pyrophosphate for such a system comes from recent work of Saygin (1981) on nonenzymatic photophosphorylation with visible light, in which chemical energy is stored as carbamylphosphate.

(3) What reaction mechanisms are possible for phosphorylation of amphiphiles? Plausible nutrients (amphiphile precursors) include long chain alcohols. Phosphate and sulfate derivatives of such alcohols have been shown to form stable membranes, and further experimentation should be directed toward group transfer processes which would permit the formation of amphiphiles from precursor molecules.

(4) What transport processes are available for nutrient translocation across bilayer membranes? Amphiphiles readily penetrate membrane barriers, and could enter a protocell simply by partitioning into the membrane phase. However, in the scheme proposed here, phosphate also participates in amphiphile formation, and bilayer membranes are relatively impermeable to small ionic compounds like phosphate. Therefore it would be desirable to establish a transport mechanism by which phosphate could be concentrated in the protocell. It is interesting to note in this regard that phosphate is a weak acid, and like other weak acids presumably will accumulate in interior volumes which are basic with respect to the external environment. If in fact transmembrane proton gradients were established in protocells, as postulated above, and had the same vector as in contemporary

prokaryotes (active transport outward, producing an alkaline interior) phosphate would be concentrated in the interiors with some degree of specificity (see also Deamer and Oro, 1980). The efficacy of this process can be tested in a model membrane system such as liposomes.

In conclusion, we have shown here how bioenergetic principles can be used to define a minimum protocell that could have embodied energy transduction, growth and replication. Given such a microenvironment, it is less difficult to imagine how conventional modes of information transfer, enzymatic catalysis and metabolism might be incorporated at a later evolutionary stage. It is clear that bioenergetic analysis of biogenesis substantially reorders the timing of various processes leading to the origin of cellular life forms, and we consider this changed temporal perspective to be an important heuristic feature of our model.

References

Baltscheffsky, H. (ed.): 1981, 'Molecular Origins and Evolution of Photosynthesis', *Biosystems* **14**, 1–147.

Basile, B. P., Middleditch, B. S., and Oro, J.: 1984, *Org. Geochem.* **5**, 211–216.

Biegel, C. M. and Gould, J. M.: 1981, *Biochemistry* **20**, 3474–3479.

Cafiso, J. D. and Hubbell, W. L.: 1983, *Biophys. J.* **44**, 49–57.

Deamer, D. W. (ed.): 1978, *Light Transducing Membranes: Structure, Function and Evolution*, Academic Press, New York.

Deamer, D. W.: 1985, *Nature* **317**, 792–794.

Deamer, D. W. and Bramhall, J.: 1986, *Chem. Phys. Lipids* **40**, 167–188.

Deamer, D. W. and Nichols, J. W.: 1983, *Proc. Natl. Acad. Sci. USA* **80**, 165–168.

Deamer, D. W. and Oró, J.: 1980, *BioSystems* **12**, 167–175.

Fox, S. W.: 1965, *Nature* **205**, 328–340.

Hargreaves, W. R., Mulvihill, S. J., and Deamer, D. W.: 1977, *Nature* **266**, 78–80.

Hargreaves, W. R. and Deamer, D. W.: 1978, *Biochemistry* **17**, 3759–68.

Heinz, B., Ried, W., and Dose, K.: 1979, *Agnew Chem. Engl. Ed.* **18**, 478.

Heinz, B. and Ried, W.: 1981, *Biosystems* **14**, 33.

Koch, A. L.: 1985, *J. Mol. Evol.* **21**, 270–277.

Margenau, H.: 1977, *The Nature of Physical Reality*, Ox Bow Press, Woodbridge, Connecticut.

Mercer-Smith, J. A. and Mauzerall, D. C.: 1984, *Photochem. Photobiol.* **39**, 397–405.

Morowitz, H. J.: 1978, *Am. J. Physiol.* **235**, R99–R114.

Morowitz, H. J.: 1981, *Biosystems* **14**, 41–48.

Nagle, J. F. and Morowitz, H. J.: 1978, *Proc. Natl. Acad. Sci. USA* **75**, 298–302.

Nichols, J. W. and Deamer, D. W.: 1980, *Proc. Natl. Acad. Sci. USA* **77**, 2038–2042.

Oparin, A. I.: 1957, *The Origin of Life on the Earth*, Oliver and Boyd, Edinburgh.

Rashevsky, N.: 1938, *Mathematical Biophysics*, The University of Chicago Press.

Seta, P., Bienvenue, E., Moore, A. L., Mathis, P., Bansasson, R. V., Lidell, P., Pessiki, P. J., Joy, A., Moore, T. A., and Gust. D.: 1985, *Nature* **316**, 653–655.

Smith, T. F. and Morowitz, H. J.: 1982, *J. Mol. Evol.* **18**, 265–282.

Saygin, O.: 1981, *Naturwissenschaften* **67**, 617.

Westheimer, F. H.: 1987, *Science* **235**, 1173–1178.

AMPHIPHILIC COMPONENTS OF THE MURCHISON CARBONACEOUS CHONDRITE: SURFACE PROPERTIES AND MEMBRANE FORMATION

D. W. DEAMER

Department of Zoology, University of California, Davis, CA 95616, U.S.A.

and

R. M. PASHLEY

Department of Chemistry, The Faculties, The Australian National University, Canberra ACT, Australia

(Received 10 May, 1988)

Abstract. We have investigated physicochemical properties of amphiphilic compounds in carbonaceous meteorites. The primary aim was to determine whether such materials represent plausible sources of lipid-like compounds that could have been involved as membrane components in primitive cells. Samples of the Murchison CM2 chondrite were extracted with chloroform-methanol, and the chloroform-soluble material was separated by two-dimensional thin layer chromatography. Fluorescence, iodine stains and charring were used to identify major components on the plates. These were then scraped and eluted as specific fractions which were investigated by fluorescence and absorption spectra, surface chemical methods, gas chromatography-mass spectrometry, and electron microscopy. Fraction 5 was strongly fluorescent, and contained pyrene and fluoranthene, the major polycyclic aromatic hydrocarbons of the Murchison chondrite. This fraction was also present in extracts from the Murray and Mighei CM2 chondrites. Fraction 3 was surface active, forming apparent monomolecular films at air-water interfaces. Surface force measurements suggested that fraction 3 contained acidic groups. Fraction 1 was also surface active, and certain components could self-assemble into membranous vesicles which encapsulated polar solutes. The observations reported here demonstrate that organic compounds plausibly available on the primitive Earth through meteoritic infall are surface active, and have the ability to self-assemble into membranes.

1. Introduction

Investigations of chemical reactions occurring in conditions simulating those of interstellar molecular clouds suggest that organic molecules would have been ubiquitous products (Hagen *et al.*, 1979; d'Hendecourt *et al.*, 1986). It is likely that at least some of these formed grain mantles on interstellar dust particles which in turn became incorporated into comets and planetesimals in solar accretion disks (Kerridge, 1983; Delsemme, 1984; Kerridge and Chang, 1985; Kerridge, 1986). Carbonaceous meteorites, which represent samples of the chemical components of the early solar system, contain amino acids and various hydrocarbon derivatives (Mueller, 1953; Anders *et al.*, 1973; Allamandola *et al.*, 1987) whose isotope composition strongly supports this contention (Kerridge, 1983; Yang and Epstein, 1984). It follows that meteoritic organics are likely to have contributed

Origins of Life & Evolution of the Biosphere, Volume 19 (1989), pp. 21-38.

to the carbonaceous material available on the prebiotic earth, particularly in the late accretion phases following crust formation and the condensation of oceans (Delsemme, 1984; Anders, 1973). Because of the possible role of meteoritic organics in prebiotic chemical evolution, their chemical and physical characteristics are of considerable interest.

The content of amino acids, monocarboxylic acids and hydrocarbon derivatives in carbonaceous meteorites has been the subject of several prior investigations (Kvenholden et al., 1970; Cronin and Moore, 1971; Oró et al., 1971; Studier et al., 1972; Lawless and Yuen, 1979; Anders and Hayatsu, 1980; Deamer, 1985; Basile et al., 1984). However, the question of possible amphiphilic molecules and pigments has not been thoroughly addressed. In earlier work (Deamer, 1985) samples of the Murchison carbonaceous chondrite were extracted under conditions expected to solubilize lipid-like compounds, and amphiphilic substances were found. In the present study we observed that the chloroform-soluble components are readily separated by two-dimensional thin layer chromatography. Investigations of the separated fractions with surface chemical techniques, GC-MS and electron microscopy enabled us to address the following questions: (1) What is the chemical composition of the material soluble in chloroform, and which components are surface active? (2) What are the physical properties of the surface active material? (3) Do any of the components form true membranes with the ability to encapsulate solutes?

2. Methods

2.1. METEORITE SAMPLES

Samples of the Murchison CM2 meteorite were obtained from two sources. The first was a gift of Dr. W. Compston of the Department of Earth Sciences, the Australian National University. This represented partially crushed material that had been separated by a density gradient on ethylene tetrabromide. The heavier fraction was taken for mineralogical studies, and the lighter fraction was used for some of the experiments reported here. The second sample was a pristine 90 g stone obtained from Dr. Edward Olsen, curator of the Field Museum Mineralogical Collection, Chicago IL. The stone had its fusion crust intact, except for one chip about two cm across. After reaching our laboratory at UC Davis, it was handled only with cleaned metal forceps and stored under argon. Samples were taken from interior regions after fracturing the stone, so that the content of fusion crust in the material to be extracted was minimized.

Three controls ruled out major contamination introduced by preparation and extraction procedures. The first control procedure involved parallel extractions of the Allende C3 chondrite, and the Murray and Mighei CM2 chondrites. The Allende meteorite is relatively low in organic content, while the Murray and Mighei extracts should resemble the Murchison in organic content. Any components appearing in all four sources in similar amounts would be suspect. Gas chromatography/mass spectrometry (GC-MS) was carried out on several of the fractions, and the procedural blanks, which included the entire procedure minus the meteoritic extracts, represented a second level of control.

A third control involved potential of contamination related to the site of the fall, an area of 60 km^2 near Murchison, Australia which included farm paddocks. Because we did not have information regarding the collection conditions of our Murchison sample, in the worst case some of the fluorescent material could have been soil contamination. We therefore performed our standard extraction and chromatography procedure on gram quantities of paddock soil, but no fluorescent material was found. Similar extracts of cattle dung showed several red fluorescent spots near the origin on TLC plates, together with one major fluorescent spot having Rf values near 0.5 in both TLC solvent systems. The red fluorescence is a marker for chlorophyll and its degradation products, and its absence in chromatograms of the Murchison material suggests that surface contamination did not contribute to the fluorescent compounds extracted from the meteorite.

Other possible sources of contamination include human skin oils, plasticizers from plastic containers, and absorbed smoke and hydrocarbon vapors, all of which could be added to the specimens during collection, transport and storage. Each of these possible contaminants must be addressed and excluded where possible. To this end, Anders and co-workers compared hydrocarbon content of hexane surface rinses and interior extracts of the Murchison chondrite. The surface rinses showed several isoprenoid derivatives while the interior was free of such probable terrestrial contaminants. (See Anders and Hayatsu, 1980 for review). For this reason, we used interior samples of the pristine stone for most of the work reported here. The previously crushed samples were used to establish chromatographic separations, surface chemical techniques, and for the surface force measurements of fraction 3.

2.2. EXTRACTION METHODS

Weighed samples of the meteorite (~1–2 g) were extracted at three pH ranges by a modification of a previously described method (Deamer, 1985) in which soluble organic components were permitted to partition between organic and aqueous phases. The rationale was to extract at widely different pH ranges so that ionized amphiphilic substances would have three opportunities to partition into the chloroform phase. All glassware was cleaned by sulfuric acid-potassium dichromate solutions, followed by six or more distilled-deionized water rinses. The sample was first ground in 1.0 ml of 0.1 M KOH, followed by addition of 3.0 ml chloroform/methanol 2:1 with continued grinding for about 1 minute. The slurry was centrifuged, the upper methanol-aqueous layer was removed by a pipette, and the chloroform layer decanted and saved. The pellet was returned to the morter together with 1.0 ml of 0.1 M sodium bicarbonate, pH 6, then re-extracted as described above. The pellet was finally acidified with 1.0 ml of 0.1 N H$_2$SO$_4$ and extracted a third time. All solvents were redistilled from Mallinkrodt reagent grade stocks. In typical extractions, a total of 1.5 mg per gram meteorite was solubilized in the combined chloroform fractions. The methanol/water fractions contain soluble polar components, including other fluorescent compounds. They were frozen under nitrogen, and will be made available to other investigators upon request. The ground mineral residue is also available upon request.

We found that the extracted material underwent subtle changes in physical properties over a period of several days' storage. For instance, aged samples tended to produce streaking, rather than individual spots on thin layer chromatograms. It seemed likely that the changes were caused by oxidation, and some experiments were therefore carried out entirely under nitrogen. This procedure did not significantly alter chromatographic patterns, and we concluded that the mixture was not being degraded by oxidation during extraction. However, we cannot rule out the possibility that some Murchison components were altered by atmospheric oxidation during storage in the interim period since it fell in 1969, and this must be taken into account in any investigation of meteoritic organic compounds. Another probable change is volatilization of smaller non-polar molecules such as short chain alcohols and hydrocarbons. These could be lost by evaporation over years of storage, or during extraction and chromatographic procedures.

2.3. THIN LAYER CHROMATOGRAPHY

Aliquots of the chloroform extract were dried under nitrogen, redissolved in smaller volumes, and applied to silicic acid TLC plates (10 × 10 cm, 0.2 mm thick, E. Merck, FRG), followed by development in n-hexane:diethyl ether 8:2 in the first dimension and chloroform in the second dimension. Five regions were defined by their fluorescent components under ultraviolet illumination. Two other regions were defined by components which produced UV-absorbant streaks on the TLC plate. To monitor non-fluorescent components, the plate was divided into nine equal squares. The fluorescent spots and non-fluorescent streaks were scraped individually from the plate, and afterwards each square was scraped. The scraped material was eluted with chloroform:methanol in small funnels produced by drawing out Pasteur pipettes in a flame and plugging them with glass wool. The eluates will be referred to as fractions 1 through 5 (fluorescent spots) streaks 1 and 2, and squares 1 through 9.

2.4. SPECTRAL MEASUREMENTS

Individual fractions were eluted as described above and the entire amounts diluted into 2.0 ml of chloroform or methanol. Ratio-corrected fluorescence excitation-emission spectra were obtained with an SLM Model 2000 Fluorometer (Urbana IL). Spectra of twelve standard polycyclic aromatic compounds were obtained under the same conditions for comparison with the meteoritic extracts.

Fourier-transform infrared spectra were obtained with a Perkin Elmer FTIR spectrometer (Model 1750) and UV-VIS spectra with a Perkin Elmer Lambda 3 spectrophotometer.

2.5. MASS SPECTROMETRY

Only fraction 5 and square 9 contained components sufficiently volatile for direct GC-MS analysis. Aliquots of the chloroform eluates of material scraped from the TLC plates were injected into the GC port of the VG ZAB-HS mass spectrometer at the Facility for

Advanced Instrumentation, UC Davis. Compounds were identified from their molecular ions and pattern of breakdown products, with the help of comparison mass spectra library searches, and related to previous analyses carried out with similar non-polar extracts (Smith and Kaplan, 1970; Studier *et al.*, 1972; Basile *et al.*, 1984).

2.6. ENCAPSULATION OF POLAR SOLUTES

Fraction 1 was eluted in chloroform from the scraped spot and dried. A typical yield was 30–50 μg per mg meteorite extract. Alkaline buffer was then added, together with 0.5 mM pyranine, an impermeant fluorescent dye, and the mixture was subjected to a freeze-thaw procedure which is known to cause encapsulation of solutes by lipids (Pick, 1981). The faintly turbid dispersion was then filtered through a Sephadex G-50 column (0.3 cm × 5.0 cm) to remove external dye, and the encapsulated dye was measured fluorometrically (excitation 430 nm, emission 510 nm). As a control, similar weights of egg phosphatidylcholine were carried through the same procedure in parallel so that encapsulation efficiencies could be directly compared.

2.7. ELECTRON MICROSCOPY

Electron microscopy was carried out on the total extract and fraction 1 from the TLC plate using osmium fixed material and freeze-fracture electron microscopy. Samples were dispersed in 10 mM sodium hydroxide, pH 9.5, and a 1% solution of osmium tetroxide and potassium ferrocyanide in water was added to fix the structures. Following centrifugation, the pellet was washed once in deionized water, followed by embedding in 2% agar to support the fragile membranes during dehydration and centrifugation. The agar block containing the pellet was post-fixed in 2% glutaraldehyde, dehydrated in acetone, embedded in Araldite and sectioned with a diamond knife. The sections were post-stained in 2% uranyl acetate and lead citrate. In other experiments, aliquots of the total extract were dried from the chloroform solution onto freeze-fracture planchettes, and alkaline buffer (10 mM disodium phosphate) containing 25% glycerol by volume was added. The samples were then frozen in liquid nitrogen-cooled Freon, followed by fracturing, shadowing and replication in a Balzers 410 microtome. Electron micrographs were obtained with a Phillips 400 instrument, using a operating voltage of 80 kV.

3. Results

3.1. CHROMATOGRAPHIC PATTERNS

A typical thin layer chromatographic pattern for the Murchison extract is shown in Figure 1A. Five major fluorescent regions were defined by this procedure, and two non-fluorescent streaks were also observed. A significant amount of yellow-brown pigment remained at the origin, which produced a highly fluorescent unresolved streak if developed afterwards in chloroform:methanol 2:1. The pattern of spots was similar (but not

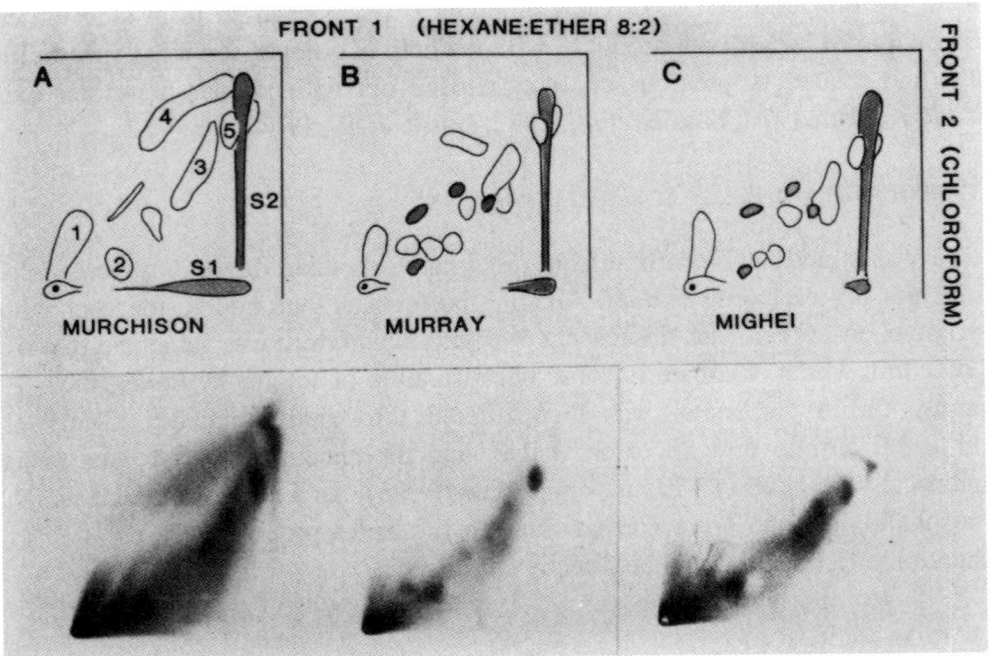

Fig. 1. Chromatographic patterns of meteoritic non-polar extracts. Approximately 1 mg of material was spotted at the origin and chromatograms were developed as described in the text, with 8:2 n-hexane:diethyl ether in the first dimension and chloroform in the second. Figure 1A shows a typical pattern for the Murchison meteorite extract, with five major fluorescent regions (numbered) and two non-fluorescent streaks which have been identified as elemental sulfur (see text). For comparison, figures 1 B and C show chromatographic patterns from the Murray and Mighei chondrites. Fluorescent material is indicated by open areas, and iodine-staining material by shaded areas. The fluorescent patterns of all three chondrites were similar, but not identical. Negative prints made from colored slides are shown below the illustrations. (The slides were made from different chromatograms of the three meteoritic extracts than those illustrated above.) Extracts of the Allende C3 chondrite, which is relatively low in organic carbon content, contained essentially no fluorescent material (not shown).

identical) when pristine samples of the Murchison were compared with previously crushed material, indicating that the treatment with ethylene tetrabromide had not markedly affected the organic content. In parallel extracts of the Allende meteorite sample, the amount of extracted organic material was below the sensitivity of our microbalance. Furthermore, only two very faint fluorescent spots were observed on TLC plates, and neither matched the pattern of the Murchison material. TLC patterns of extracts from the Murray and Mighei meteorites resembled the Murchison patterns (Figure 1B and C). The similarities in TLC patterns suggest that the fluorescent organic content of the three chondrites was produced by similar chemical processes.

Iodine staining and sulfuric acid charring were employed to determine whether components other than the fluorescent substance were present. The two streaks stained strongly with iodine, but gave no char with sulfuric acid. Because inorganic sulfur was associated with a scraped region of streak 2 in GC/MS data, sulfur was tested in a

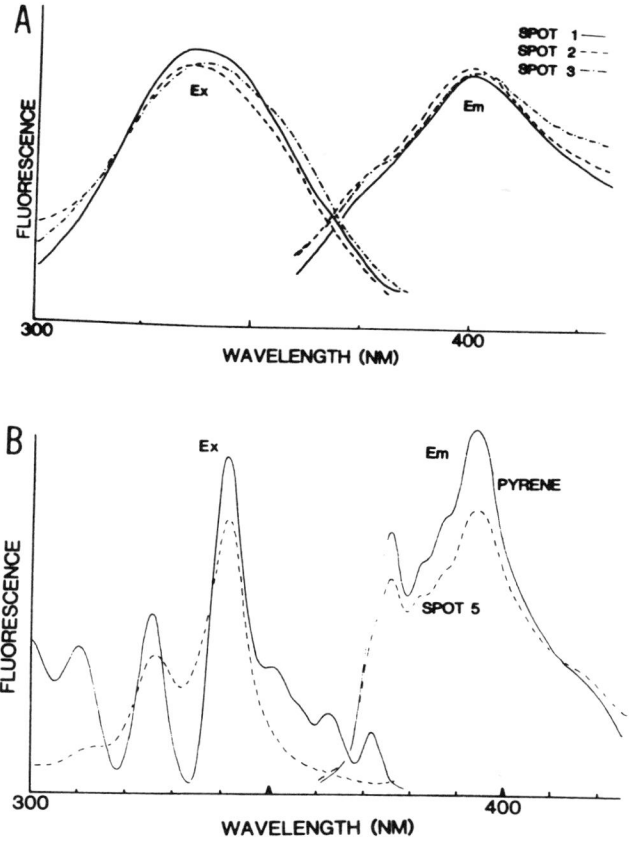

Fig. 2. Fluorescent regions were scraped and eluted for spectral analysis. Figure 2A shows fluorescence excitation-emission spectra obtained after eluting spots 1, 2, and 3 into chloroform. Figure 2B compares the spectrum of spot 5 (dashed line) with that of pyrene (solid line). Fluorescence given in arbitrary units.

separate chromatogram, and was found to produce identical streaks and iodine staining patterns. We conclude that streaks 1 and 2 are elemental sulfur.

3.2. SPECTRAL PROPERTIES

Scraped spots were eluted twice with 0.1 ml aliquots of chloroform/methanol 2:1 for investigations of spectral and surface properties. Excitation/emission fluorescence spectra of fractions 1, 2, and 3 diluted into 2 ml methanol are shown in Figure 2A. The spectra were similar, with excitation peaks near 340 nm and emission peaks near 400 nm, suggesting that they may be derivatives of the same fluorescent chromophore. The spectrum of fraction 5, in contrast, was relatively complex (Figure 2B). Basile *et al.* (1984) have shown that fluoranthene and pyrene compose most of the polycyclic aromatic hydrocarbons of the Murchison meteorite. We have superimposed a comparison spectrum of pyrene with this figure, and it is clear that its spectrum closely resembles that of fraction 5. Furthermore, pyrene and fluoranthene standards migrated with similar 2-

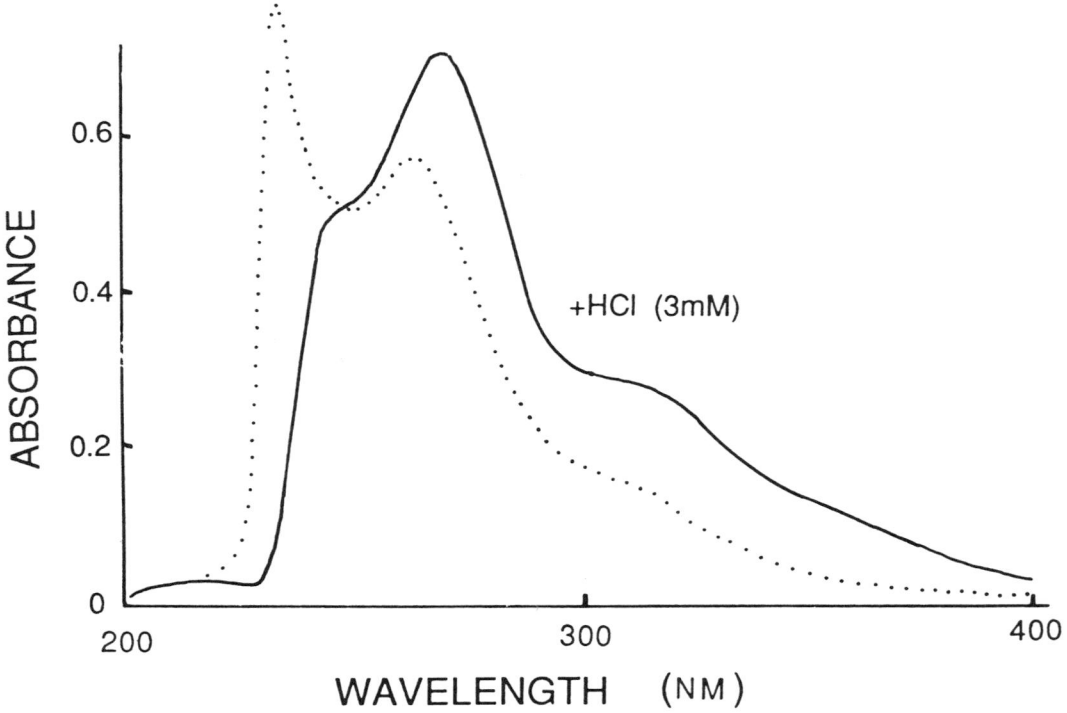

Fig. 3. Ultraviolet absorption spectrum of fraction 1. Approximately 50 μg of fraction 1 material was dissolved in 2 ml methanol, and absorption peaks were observed at 230 and 260 nm. Acidification by HCl addition reduced the 230 peak and enhanced the 260 peak, suggesting the presence of a titratable chromophore.

dimensional Rf values on TLC plates developed by the same solvent systems. GC-MS results also indicate the presence of pyrene and fluoranthene in this spot, as will described later.

The Murray and Mighei chondrites produced TLC patterns containing spot 5, and the fluorescence spectra were those expected for pyrene and fluoranthene mixture. The pyrene-fluoranthene spot may be common to CM2 chondrites, a possibility which should be investigated in a broader sampling of meteoritic material. Fraction 1 had interesting physical properties which will be described later, and was therefore investigated in more detail with other spectral techniques. Its UV spectrum in methanol under neutral and acidic conditions (3mM HCl) is shown in Figure 3. Two major peaks were apparent at 235 and 260 nm, and a shoulder at 310 nm. The 235 peak and 310 shoulder were markedly reduced at acidic pH, indicating the presence of a titratable group on the UV chromophore.

An infrared spectrum of fraction 1 is shown in Figure 4. The major bands at 2950, 2917, and 2861 are assigned to stretching and bending of $-CH_3$ and $-CH_2$ groups. The peak ratios are indicative of shorter or branched aliphatic chains. The 1724, 1700 and 1650 bands are assigned to carbonyl oxygen, and the 3400 band to hydroxyl groups. Surprisingly, no aromatic bands are apparent, even though the UV absorbance and fluorescence of fraction 1 suggests that polycyclic aromatic compounds must be present.

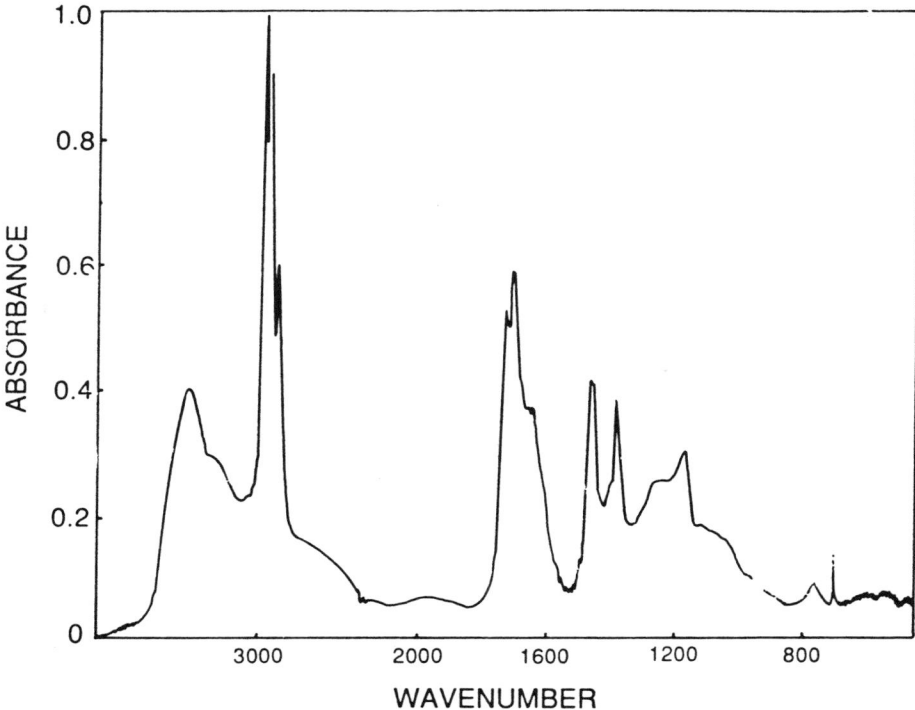

Fig. 4. FTIR spectrum of fraction 1. (See text for details.) Approximately 50 μm were dried from chloroform and dispersed in a KBr pellet.

3.3. MASS DETERMINATIONS

Table I shows the fractional masses of the major fluorescent spots and the eluted squares from the TLC plates. Nearly a fourth of the total mass remained at the origin, while spots 1, 2, 3 and 5 contained 5–8% each. (Spot 4 is typically quite diffuse, and was not analysed gravimetrically). The elemental sulfur in streaks 1 and 2 represented another 20% of the mass in the extract, and the remainder was distributed largely in squares 5 and 9 (7–8% each) the latter square containing the hydrocarbons of the sample. The fluorescent substances therefore represent about half of the material extractable in chloroform, with the remainder largely composed of sulfur and aliphatic hydrocarbons.

3.4. GC-MS OBSERVATIONS

Gas chromatograms of fraction 5 and square 9 are shown in Figure 5A and B. Major peaks were identified by mass spectrometry. Square 9 contained normal alkanes ranging from 13 to 23 carbons, with distribution patterns resembling those reported for other CM2 chondrites by Smith and Kaplan (1970) and for the Murchison by Anders et al. (1973). Fraction 5 was strongly fluorescent, and contained two prominent peaks at 202 daltons which were identified as pyrene and fluoranthene. Peaks of anthracene and elemental sulfur were also observed in this fraction.

TABLE I

Relative masses of components in Murchison non-polar extract following
separation by 2-dimensional thin layer chromatography.

Chromatographic separations were carried out as described in Figure 1. Spots 1–5 and streaks 1 and 2 were scraped from the plate. The remainder of the plate was then defined by a square which included the origin at lower left, and the furthest migration of non-polar iodine-staining material (hydrocarbons) at upper right. The enclosed area was divided into nine equal smaller squares with square 1 at lower left, square 9 at upper right, and each was scraped. Fractions were eluted from the scraped material with chloroform:methanol 2:1 as described in the methods section, followed by drying and weighing on a Cahn microbalance. In the experiment described here, 0.70 mg of extract was spotted on the plate, and 0.66 mg (94%) was recovered as the total weight of separated fractions. Similar recoveries and weights were obtained in two separate experiments.

Spots	μm	% of total mass	Squares	μg	% of total
Origin	174	26.0	1	4	0.6
1	52	7.9	2	22	3.3
2	44	6.7	3	10	1.5
3	42	6.4	4	23	3.5
4	not analysed		5	50	7.6
5	34	5.1	6	8	1.2
			7	16	2.4
			8	2	0.3
			9	48	7.3
Streak					
1	72	11			
2	58	9			

3.5. SURFACE ACTIVITY

Individual force/area curves were obtained for the isolated fluorescent spots, and these are shown in Figure 6A. The combined area of the films was about a tenth that of the area of the total extract, which was normalized as 1.0 on a logarithmic scale. When blanks were run with solvent alone, little surface activity was apparent. Control areas were scraped in regions neighboring the fluorescent spots, but their content of surface active material was minimal.

Spot 3 was well separated from other spots, and was chosen for more detailed investigations of surface properties. Following elution onto an air-water interface in a Langmuir trough apparatus (Adamson, 1982) the isolated material was compressed to a film pressure of about 18 mN per meter, at which the film was presumed to be tightly packed, and freshly cleaved mica surfaces were slowly drawn up through the interface to produce Langmuir-Blodgett deposition (Chan *et al.*, 1980). We found that when the subphase in the trough was distilled water, little disposition occurred, indicating that the amphiphilic substance was negatively charged in water, as is mica. This was further substantiated by

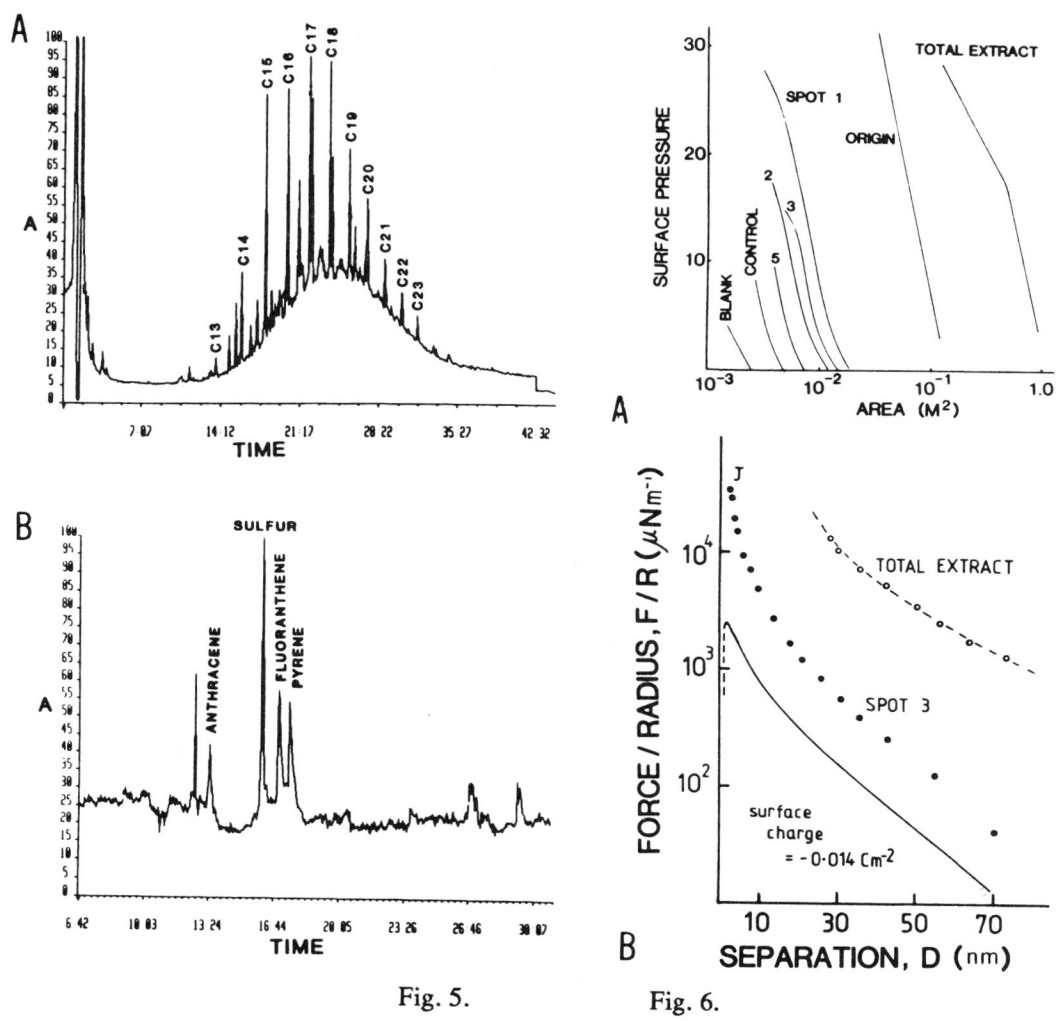

Fig. 5.   Fig. 6.

Fig. 5. Gas chromatograms of Murchison hydrocarbons. Alkanes ranging from 13 to 23 carbons were found in eluates of square 9 (Figure 5A) confirming earlier reports by Studier *et al.* (15). Two major polycyclic aromatic hydrocarbons were found in spot 5 and identified as pyrene and fluoranthene (Figure 5B) confirming earlier reports by Basile *et al.* (19). Peaks were identified by mass spectral analysis of molecular ions and fragment patterns. Column: 30 m DB-1, 0.25 mm I.D., 025 μm film. Temperature programs: 5 A: 40–260°C at 6° min^{-1}; 5 B: 70–250°C at 8° min^{-1}. Helium carrier gas, 35 cm/s.

Fig. 6. Properties of films of meteoritic amphiphiles spread on air-water interfaces. Figure 6A shows force-area isotherms plotted on a logarithmic scale. The isotherm for the total extract was calculated to show a normalized area of 1.0 at low surface pressures (given as mN m^{-1}) and other isotherms were plotted as fractional areas relative to that of the total extract so that the amount of surface active material in each spot could be estimated. The largest fraction was found at the origin and in spot 1, while other fluorescent components represented less than one percent of the area occupied by the total extract. Figure 6B shows interaction forces measured between two mica substrate surfaces coated with the total organic extract or spot 3. The solid line is the theoretical curve for negatively charged surfaces interacting in 10^{-4} M CaCl$_2$ (of Debye length 17.5 nm) where the charge is assumed to originate at the mica surface (i.e. at D = 0,). When the coated mica surfaces were caused to touch, the layer was displaced and a force maximum was observed at J, followed by adhesive contact at D = 0.

the fact that addition of 10^{-2} M CaCl$_2$ had the effect of inducing complete desposition, similar to the effect of calcium ion on monolayers of longer chain organic acids. Once the mica surfaces were successfully coated with the film from the calcium chloride solution, the interaction forces between adjacent layers could be accurately measured by the surface force technique developed by Israelachvili and Adams (1978). This method permits measurement of separation distances to an accuracy better than 0.2 nm and forces from about 0.1 uN to 1 mN between curved surfaces of approximately 1 cm radius. The coated surfaces were immersed in 10^{-4} M CaCl$_2$ to prevent desorption of the film.

Interaction forces were measured between films prepared from spot 3 deposited onto two mica substrates (Figure 6B). The parameter F/R is the total repulsive force between the surfaces divided by the mean radii, and D is the bare mica-mica separation distance. The surface films on the mica are electrostatically charged and the force decays with precisely the calculated Debye length for this electrolyte solution. The full curve is that calculated for charged surfaces in this solution using the Poisson-Boltzmann equation with a short range van der Waals attractive component (Pashley *et al.*, 1985). This theoretical line was calculated assuming that the charge originated at the mica surface. It is clear from comparison with the experimental results that this is an incorrect assumption. However, the theoretical line would produce the observed curve if it is assumed that the surface-active material formed a 10 nm layer on each mica surface, so that D is displaced 20 nm.

In an asymmetrical electrolyte solution such as CaCl$_2$ the sign of the surface charge is defined by the shape of the curve, and in this case the film was negatively charged with about one charge every 12 nm^2. At mica separations of less than 29 nm the layers were displaced, and eventually mica-mica adhesive contact was obtained. Separation and repeat measurements, although showing some hysteresis, gave similar force curves, indicating that the amphiphiles were readily re-adsorbed. For comparison, forces were also measured between deposited layers of the complete organic extract from the meteorite (Figure 6B). The forces in this case suggest deposition of a significantly thicker layer of about 30–40 nm and were markedly more hysteretic.

3.6. MEMBRANE FORMATION AND SOLUTE ENCAPSULATION

Many surface active amphiphiles that form monolayers also form bilayer membranes under appropriate conditions. In past work, we found that total extracts of the Murchison meteorite could produce membranous boundary structures (Deamer, 1985) and in the present study we investigated each fraction of the TLC plate for membranogenic activity. Most of the fractions simply produced droplets or viscous globules which adhered to the glass. However, certain components of spot 1 reliably produced vesicles which could be visualized by phase microscopy. We used electron microscopic methods to confirm the presence of membranous structures, and simultaneously tested their ability to encapsulate polar solutes. Following encapsulation of dye, dispersions of spot 1 material were examined by fluorescence microscopy. The majority of the particles were quite small, in the micron size range, and showed only the blue fluorescence typical of the meteoritic

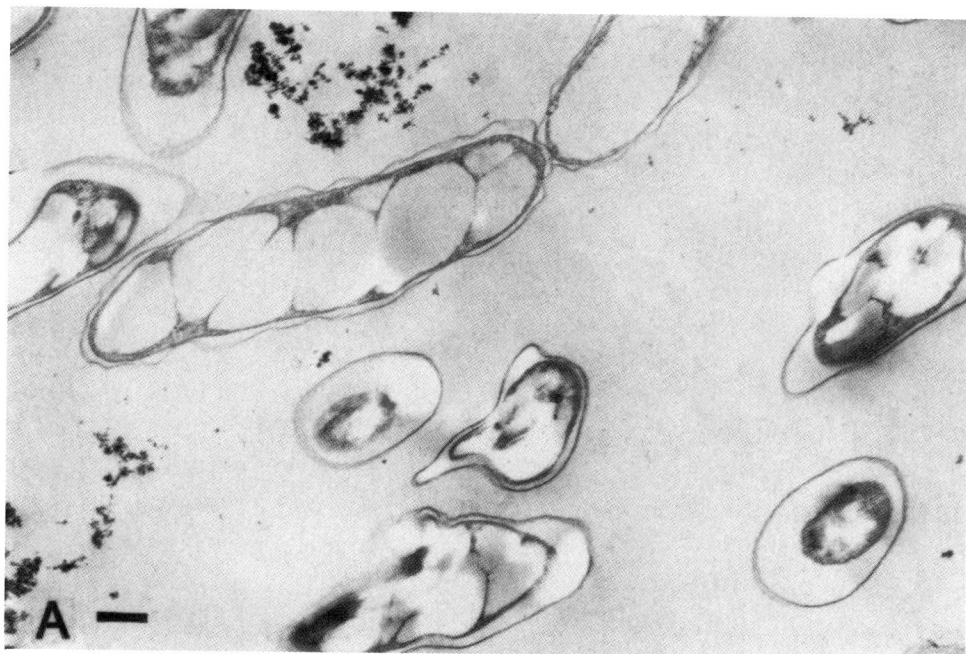

Fig. 7a.

extract. However, some of the vesicles clearly contained dye, which was characterized by a bright green fluorescence followed by photobleaching. The presence of encapsulated dye was confirmed quantitatively by the spectral measurements. When compared with the phospholipid, the meteoritic material was able to encapsulate with an efficiency about 0.1% that of the lipid, on a volume/mass basis.

Electron microscopic examination of thin sectioned material showed obvious membranes surrounding apparent droplets of non-membranous material (Figure 7A). Under high magnification, the membranes often revealed bilaminar structures typical of lipid bilayer membranes (Figure 7B). Freeze-fracture images revealed fracture planes in the fraction 1 dispersions as well (Figure 8A) confirming the structures observed in thin sections. Similar fracture planes were observed throughout dispersed samples of the total extract (Figure 8B) demontrating that apparent bilayer structures could self-assemble within the extracted mixture of meteoritic components.

Discussion

This report describes surface properties and membrane formation by amphiphilic molecules present in a carbonaceous chondrite. We investigated three fractions isolated by two-dimensional thin layer chromatography. The first represents relatively non-polar components which migrate farthest in non-polar solvent systems. Mass spectrometry and fluorescence spectra showed that this fraction is largely composed of a series of normal and branched chain alkanes ranging from 13 to 23 carbons in length, together with aromatic hydrocarbons – pyrene, fluoranthene and anthracene – which account for the

Fig. 7b.

Fig. 7a–b. When eluates of fraction 1 were dispersed in alkaline solutions, membranous vesicles formed (Figure 7A) which were composed of outer membranes enclosing non-membranous material in the interiors. At higher magnifications, trilaminar structures could be seen in the membranes (Figure 7B, inset). Bars show 0.2 μm.

fluorescence in spot 5. The presence of aliphatic and aromatic hydrocarbons confirms earlier reports (Kvenholden *et al.*, 1970; Smith and Kaplan, 1970); Lawless and Yuen, 1979; Basile *et al.*, 1984).

A second fluorescent component – fraction 3 – formed films sufficiently stable to be picked up by the Langmuir-Blodgett method. The films were negatively charged, about 10 nm thick, and are probably composed of polymeric organic acids with incorporated aromatic groups which are fluorescent chromophores. The compounds are polyfunctional, perhaps with several hydroxyl or carboxyl groups, because deposition did not

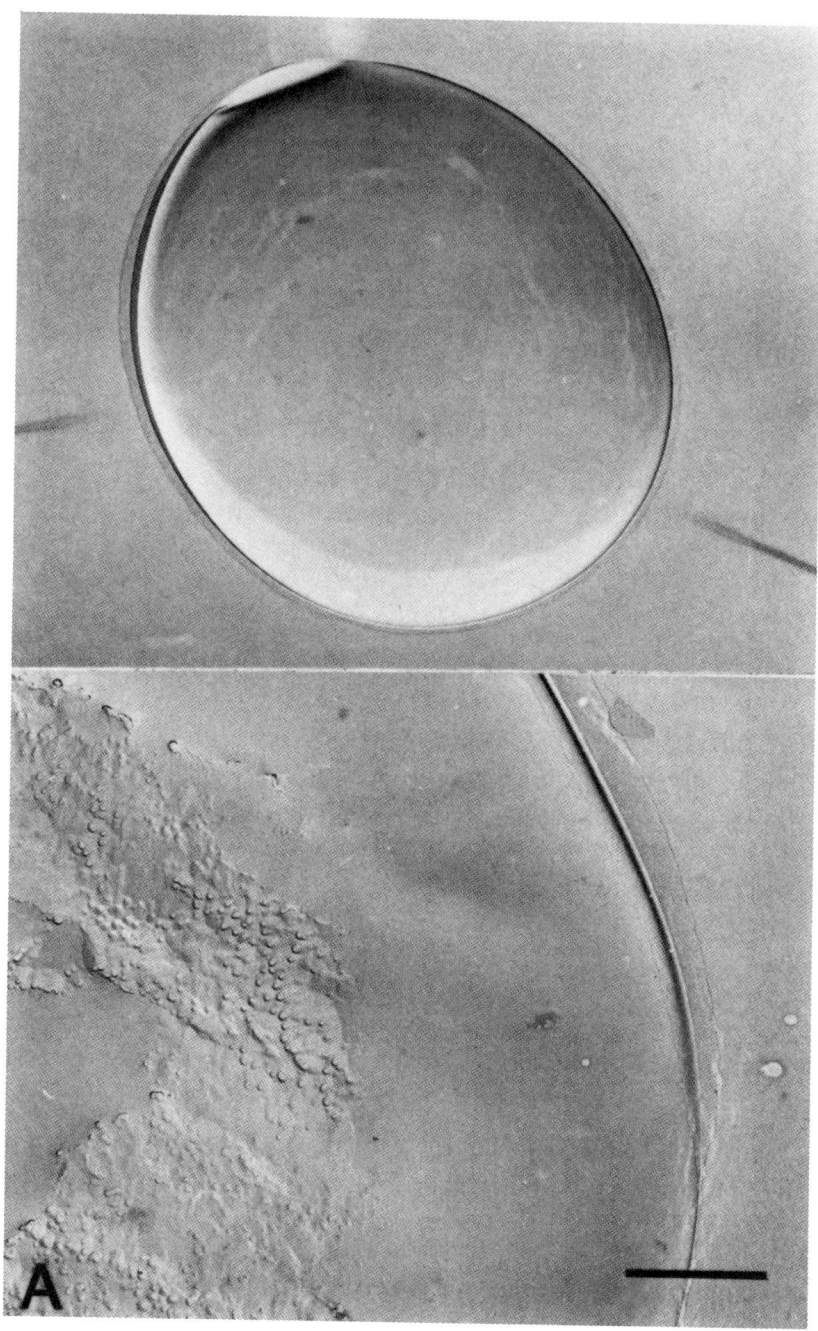

Fig. 8a.

produce a hydrophobic film with the properties expected for a typical monofunctional amphiphile. The latter has been well studied (Pashley *et al.*, 1985) and gives rise to a strong hydrophobic attractive force which was not observed here. Adsorbed films formed by this relatively large amphiphile (or mixture of similar amphiphiles) may therefore be

Fig. 8b.

Fig. 8a–b. Freeze-fracture images of vesicles. Fraction 1 produced vesicular structures which displayed clear fracture planes (Figure 8A, above) as well as internal structure in cross fractures (Figure 8A, lower micrograph). Fracture planes which occurred along surfaces resembled typical fracture planes in lipid membrane systems. Bar indicates 1.0 μm. Dispersions of the total extract also revealed fracture planes (Figure 8B) and internal structures in cross fractured particles (lower micrograph). Bars show 0.1 μm.

composed of a single molecular layer, polar at both surfaces. It is likely that the surface active material represents a sub-fraction of the 'organic polymer' which is soluble in chloroform under the extraction conditions described here. This fraction produces viscous globules under the microscope, and does not assemble into membranes.

A third fraction – spot 1 – formed membranes in dilute alkaline solutions. These were visualized by light microscopy, by thin sectioning of osmium-fixed specimens for electron microscopy, and by freeze-fracture methods. The membranes are approximately 10 nm thick, and show a trilaminar staining pattern with a 5.5 nm spacing between the densely stained lines, as expected for a bilayer of some amphiphilic material. The fact that they manifest fracture planes by the freeze-fracture technique is also consistent with bilayer organization. We conclude that a small, undefined fraction of Murchison components have amphiphilic properties which permit them to assemble into boundary membranes. This suggests that they contain sufficient hydrophobic character to stabilize a bilayer structure. The FTIR spectral patterns of fraction 1 are consistent with presence of significant amounts of hydrocarbon, as well as polar groups such as carbonyl oxygen and hydroxyls. Some of the polar groups are acidic, as indicated by UV spectral changes upon acidification described here, and the pK values established in our earlier investigation (Deamer, 1985).

In contemporary membrane lipids, hydrophobicity is provided by hydrocarbon chains of fatty acids with chain lengths ranging from 12 to 20 or more carbons. A plausible source of such chains in the prebiotic environment is not readily apparent, and it is significant that such chains are relatively abundant in carbonaceous chondrites. Smith and Kaplan (1970) demonstrated chain lengths ranging from 10 to 23 carbons in seven different CM2 chondrites, and a similar distribution was observed in the Murchison (Lawless and Yuen, 1979). It is interesting that Seleznev et al. (1977) showed that ultraviolet light produces amphiphilic compounds from alkanes illuminated over sea water under aerobic conditions. In preliminary studies we have confirmed that these reactions occur in carbon dioxide and argon atmospheres as well.

If amphiphilic substances derived from meteoritic infall and chemical evolution were available on the prebiotic earth following condensation of oceans, it follows that surface films would have been present at air-water interfaces. Similar films, of biological origin, occur today (Huehnerfuss et al., 1982) and contribute to the foam that accumulates in intertidal zones. Given the presence of surfactant films, it is likely that wind and wave action would have caused amphiphilic substances to accumulate on ancient beaches and tide pools. This material would thereby be concentrated for self-assembly into boundary structures with barrier properties relevant to function as early membranes.

The presence of fluorescent aromatic substances in carbonaceous meteorites is also significant. If such compounds contributed to the inventory of organic material in the surface films described above, they would provide pigment systems which could participate in the formation of early membranes and perhaps undergo photochemical reactions to capture light energy.

Acknowledgements

This study was supported in part by NASA Grant NAWG-1119. The VG ZAB-HS mass spectrometer was purchased in part with NIH grant RR01460-01. The authors wish to express their gratitude to the Guggenheim Foundation for supporting the travel related to the research (D.W.D.). We thank the Department of Applied Mathematics at the Australian National University for hosting a research visit, and the Department of Chemistry for providing space and facilities. We also thank John Mais for expert technical assistance, and Cynthia Lean, Helen McDougall, Michael Dunn and Mal Rasmussen for their valuable contributions. Daniel Jones and Kei Miyano were consultants on mass spectrometric techniques and interpretations.

References

Adamson, A. W.: 1982, *Physical Chemistry of Surfaces, 4th Ed.* J. Wiley, New York.

Allamandola, L. J., Sandford, S. A., and Wopenka, B.: 1987, *Science* **237**, 56–59.

Anders, E. Hayatsu, R., and Studier, M. H.: 1973, *Science* **182**, 781–790.

Anders, E.: 1973, in M. A. Gordon and L. E. Snyder (eds.), *Molecules in the Galactic Environment*, Wiley, New York, p. 429.

Anders, E. and Hayatsu, R.: 1980, *Topics in Current Chemistry* **99**, 1–37.

Basile, B. P., Middleditch, B. S., and Oró, J.: 1984, *Org. Geochem.* **5**, 211–216.

Chan, D. Y. C., Pashley, R. M., and White, L. R.: 1980, *J. Colloid Interface Sci.* **77**, 283–285.

Cronin, J. R. and Moore, C. B.: 1971, *Science* **172**, 1327–1329.

Deamer, D. W.: 1985, *Nature* **317**, 792–794.

Delsemme, A. H.: 1984, *Origins of Life* **14**, 51–60.

Hagen, W., Allamandola, L. J., and Greenberg, J. M.: 1979, *Astrophys. Space Sci.* **65**, 215–225.

d'Hendecourt, L. B., Allamandola, L. J., and Greenberg, J. M.: 1986, *Astron. Astrophys.* **158**, 119–134.

Huehnerfuss, H., Lange, P., and Walter, W.: 1982, *J. Marine Research* **40**, 209–216.

Israelachvili, J. N. and Adams, G. E.: 1978, *Faraday Transactions* **74**, 975–982.

Kerridge, J. F.: 1983, *Earth Planetary Science Letters* **64**, 186–200.

Kerridge, J. F. and Chang, S.: 1985, in D. C. Black and M. S. Matthews (eds.), *Protostars and Planets II*, University of Arizona Press, Tucson, p. 738.

Kerridge, J. F.: 1986, in J. Nuth and R. E. Stencel (eds.), *Interrelationships among Circumstellar, Interstellar and Interplanetary Dust*, NASA CP 2403, p. 37.

Kvenholden, K. A., Lawless, J., Pering, K., Peterson, E., Flores, J., Ponnamperuma, C., Kaplan, I. R., and Moore, C. B.: 1970, *Nature* **228**, 923–926.

Lawless, J. and Yuen, G.: 1979, *Nature* **282**, 396–398.

Mueller, G.: 1953, *Geochim. Cosmochim. Acta* **4** 1–10.

Oró, J., Gilbert, J. Lichtenstein, H., and Flory, D. A.: 1971, *Nature* **230**, 105–106.

Pashley, R. M., McGuiggan, P. M., Ninham, B. W., and Evans, D. F.: 1985, *Science* **229**, 1088–1089.

Pick, U.: 1981, *Arch. Biochem. Biophys.* **212**, 186–194.

Seleznev, S. A., Fedorov, L. M., Kuzina, S. I., and Mikhailov, A. I.: 1977, *Revue Française des Corps Gras* **24**, 191–193.

Smith, J. W. and Kaplan, I. R.: 1970, *Science* **167**, 1367–1370.

Studier, M. H., Hayatsu, R., and Anders, E.: 1972, *Geochim. Cosmochim. Acta* **36**, 189–215.

Yang, J. and Epstein, S.: 1984, *Nature* **311**, 544–547.

Section IV

ENERGETICS OF LIFE'S ORIGIN

ENERGETICS OF LIFE'S ORIGIN

Cellular organization, far from being an after-thought, must have been from the beginning part and parcel of the origin of life. The vital force, that vis vitae which will not be exorcised without proper explanation, has its roots in the astonishing degree of organization that pervades the living world from the molecular level to the organismic and societal. Biological order must be maintained by a continuous flux of energy. Therefore a believable biopoietic scheme is one that creates mounting levels of biological order naturally, by providing the means to convert the flux of energy into the organization of matter.

F. M. Harold, 1986

Energy sources

As we look at life processes going on around us, it is abundantly clear that energy is required at multiple levels and that processes of the living have evolved to capture energy and then to partition it with high efficiency to various energy-requiring reactions. The capture and utilization of energy by contemporary cells is a highly evolved and intricate process, the details of which are still being worked out by researchers. But there must have been much simpler energetic processes available on the early Earth, driving chemical reactions uphill toward increasingly complex organization.

Energy sources on the prebiotic Earth can be divided roughly into relatively high- and relatively low-energy inputs. Examples of high-energy sources include ionizing radiation, light energy, and electrical discharges, while low-energy sources include heat and dehydration. High-energy inputs are capable of activating small molecules —water, methane, ammonia—to form reactive molecules such as formaldehyde (HCHO) and hydrogen cyanide (HCN). These, in turn, can undergo secondary reactions to form stable molecules like amino acids and purines, which are significant because of their role as monomers in contemporary life processes.

An important principle constraining high-energy monomer synthesis on the early Earth was established by Stribling and Miller (1987): while there may be abundant sources of free energy capable of producing small reactive molecules, it is the actual energy *yield* from a given source that is significant. It is important, therefore, that investigators not only find potential pathways for prebiotic syntheses but estimate how much energy is captured in any given pathway as well.

While the formation of monomers requires high-energy inputs, the next step in chemical synthesis (formation of polymers) generally requires low-energy inputs. The actual links in the polymeric chain—ester, peptide, and glycoside bonds—are formed typically by condensation reactions in which the elements of water are lost from neighboring monomeric reactants. These anhydride bonds are labile and easily broken.

As noted earlier, both heat and dehydration are low-energy sources. They have the advantage of being ubiquitous in the environment, and, together, they are able to concentrate potential reactants and then drive condensation reactions. For example, Usher and McHale (1976) have shown that short RNA oligonucleotides, when dried and heated to temperatures typically produced by sunlight, can line up along a complementary template and then form phosphodiester linkages.

Yet heat and dehydration have the disadvantage of being relatively disorderly processes. Furthermore, because it is unlikely that monomers were present in pure form on the primitive Earth, prebiotic chemistry must have occurred in highly complex mixtures. Any components capable of reacting would have done so under these conditions, producing cross-linked polymers that would not have been particularly useful to living cells.

An important problem, then, involves the sorting of conditions to allow the orderly processes required for the synthesis of particular molecules. Two possible ways of sorting are observed in contemporary cell metabolism: the chemical activation of specific monomeric reactants and the catalysis of specific reaction pathways.

Chemical activation and catalyzed reaction pathways

Contemporary organisms employ chemical activation and catalyzed reaction pathways rather than heat energy to drive metabolism and synthesis of macromolecules. Plausible prebiotic mechanisms for selective activation of monomers are not readily apparent, however. This important research topic deserves more attention. One approach involves phosphate chemistry (Baltscheffsky, 1971). It has been known for many years that drying and heating phosphate solutions causes anhydride bonds to form between phosphate pairs. In the resulting molecule (*pyrophosphate*), the anhydride bonds store the energy made available by the heating-and-drying process. Pyrophosphate formation by heating is

291

among the simplest examples of energy trapping by a chemical bond. Some microorganisms use pyrophosphate as an energy source; ATP (adenosine triphosphate), the most common form of stored chemical energy used by contemporary organisms, contains pyrophosphate bonds in its molecular structure. So, a possible answer to the sorting problem posed above may be that order is imposed on a complex mixture of reactants because a few reactions—those incorporating phosphate, for instance—are highly probable and fast, while others are improbable and relatively slow.

Another important approach to chemical activation and polymerization is the demonstration by Inoue and Orgel (1983) that certain activated monomers of RNA are able to line up along template molecules and polymerize in a one-step replication cycle. The template acts as a specific catalyst for the reaction, and the driving energy is present in the chemical bonds linking imidazole groups to the ribonucleotides. The bond energy is released when phosphodiester linkages form between nucleotide monomers as the imidazole groups leave. The result is a non-enzymatic polymerization of RNA molecules as many as 30 nucleotides in length.

Although these observations are highly significant as demonstrations of catalyzed polymerization, the activated compounds and templates do not represent plausible prebiotic reactions. Schwartz and Orgel (1985) discuss this problem and propose possible ways to simplify the reaction pathways.

Was light a primary energy source?

Sunlight is the most abundant energy source available on the present Earth and was presumably equally abundant at the time of life's origin. Light energy today is captured by the pigment systems of photosynthetic bacteria, algae, and plants. When light energy is captured by the thylakoid membranes of chloroplasts, it causes an electron to be lost from a chlorophyll molecule and donated to an acceptor at a relatively high energy level. The electron then falls energetically "downhill" to a second photosystem, where it is again driven "uphill" by a second photon to a final acceptor molecule called NADP. The reduced form of this compound is a source of reducing power for reactions not requiring light energy, in which carbon dioxide is reduced to carbohydrates. Electrons are replaced on the chlorophyll by stripping them from water molecules with a concomitant release of molecular oxygen.

Was light the original energy source that drove prebiotic evolution toward the first living systems? If so, there must have been some form of pigment molecule available in the prebiotic environment. Plausible pigment molecules are not at all obvious, however. Chlorophyll itself is a highly evolved porphyrin-magnesium complex, and, while simpler porphyrins may have been one of its evolutionary precursors, there is no plausible prebiotic reaction pathway to porphyrins. An ideal pigment system could capture light energy and then use the energy to strip electrons from water and donate them to carbon dioxide, producing a reduced form of carbon. This has proven to be very difficult to achieve in the laboratory. One alternative is a simple proton gradient. Because pigment molecules spontaneously partition into lipid environments, early membranes may well have embedded a pigment molecule that, upon illumination, would have produced a proton gradient across the membrane. Two laboratory models of such a system have been described (Deamer and Harang, 1990; Deamer, 1992).

Electron transport and proton gradients are sources of free energy

All forms of life today use some form of electron transport as an internal source of free energy. As noted earlier, electron transport is directly involved in photosynthesis after the initial trapping of light energy by chlorophyll, and, in the mitochondria of both plant and animal cells, electrons are transported from nutrients to oxygen along specialized systems of enzymes. A variety of electron transport processes were potentially available on the early Earth. Wächtershäuser (1988) has suggested the formation of iron-sulfur minerals as one source; Corliss et al. (1981) have proposed that deep thermal vents might have provided a source of electrons in the form of reduced sulfur compounds. Although these ideas are more than interesting, there has been little supporting evidence for them from laboratory studies thus far.

Transport of protons is typically coupled to electron transport processes in membranes. When the chemical energy of nutrients such as glucose and fatty acid is captured by mitochondrial and bacterial membranes, electrons are taken from the nutrients and donated to acceptor molecules embedded in the membrane. The electrons then fall "downhill" to molecular oxygen, which is the final acceptor in an aerobic electron transport chain. During electron transport, protons are transported across the membrane to produce a chemiosmotic proton gradient that drives ATP synthesis (Mitchell, 1961, 1981).

Could there have been a primitive form of chemiosmotic energy available for the earliest living organisms? Given that first life was cellular, boundary membranes must have been present, and, because any membrane is a permeability barrier to solutes, concentration gradients across the membrane would have been inevitable. One gradient that is simple to generate involves proton uptake and release (as in most organic oxidation-reduction reactions). As noted by Deamer and Oró (1980), contemporary cells use the energy of proton gradients for substrate transport across membranes as well as for ATP synthesis. Because substrate transport is a relatively simple process, it may be that proton gradients were first used by primitive cells to accumulate nutrients and that coupling to the synthesis of high-energy bonds (pyrophosphate, ATP) was a later evolutionary step.

Reprinted papers on energy sources and utilization

The concept of energy yield is significant and must be taken into account in considering any potential prebiotic reaction leading to the synthesis of organic monomers such as amino acids. The first paper included in this section, then, is by Stribling and Miller (1987) and considers energy yields both as a general topic and specifically with respect to the production of hydrogen cyanide and amino acids on the early Earth.

The second paper, by Peter Mitchell (1981), discusses chemiosmosis—the coupling mechanism by which energy from light or electron transport is converted into an electrochemical proton gradient, which in turn is used in the synthesis of ATP. Although Mitchell does not directly address the origin of life, the basic concepts of chemiosmosis must be included in our understanding of early life processes. Were chemiosmotic proton gradients necessary for initiating life processes, or did they become involved in energy transduction at a later time? The third paper (Morowitz, 1981) explores the possible relationships between chemiosmotic principles and the origin of life.

The fourth paper, by Seta et al. (1985), demonstrates that a relatively simple membrane-pigment system is capable of producing charge separation and electron transfer with light as an energy source. The fifth paper, by Wilson and Maloney (1976), carries this topic further, focusing on the origin and evolution of chemiosmotic coupling and ion transport in cell membranes.

References

Baltscheffsky, H. (1971) Inorganic pyrophosphate and the origin and evolution of biological energy transformation. In: *Chemical Evolution and the Origin of Life*, R. Buvet and C. Ponnamperuma, eds., Amsterdam: North-Holland, pp. 466-474.

Corliss, J. B., Baross, J. A. and Hoffmann, S. E. (1981) An hypothesis concerning the relationship between submarine hot springs and the origin of life on Earth. *Oceanol. Acta* **4** (suppl.):59-69.

Deamer, D. W. (1992) Polycyclic aromatic hydrocarbons: Primitive pigment systems in the prebiotic environment. *Advances in Space Research* **12**:183-189.

Deamer, D. W. and Harang, E. (1990) Light-dependent pH gradients are generated in liposomes containing ferrocyanide. *BioSystems* **24**:1-4.

Deamer, D. W. and Oró, J. (1980) Role of lipids in prebiotic structures. *BioSystems* **12**:167-175.

Harold, F. M. (1986) *The Vital Force: A Study in Bioenergetics*. New York: W. H. Freeman.

Inoue, T. and Orgel, L. E. (1983) A nonenzymatic RNA polymerase model. *Science* **219**:859-862.

Mitchell, P. (1961) Coupling of phosphorylation to electron and hydrogen transfer by a chemiosmotic type of mechanism. *Nature* **191**:144-148.

Mitchell, P. (1981) Davy's electrochemistry: Nature's protochemistry. *Chemistry in Britain* **17**:14-23.

Morowitz, H. J. (1981) Phase separation, charge separation and biogenesis. *BioSystems* **14**:41-47.

Schwartz, A. W. and Orgel, L. E. (1985) Template-directed synthesis of novel, nucleic acid-like structures. *Science* **228**:585-587.

Seta, P., Bienvenue, E., Moore, A. L., Mathis, P., Bensasson, R. V., Liddell, P., Pessiki, P. J., Joy, A., Moore, T. A. and Gust, D. (1985) Photodriven transmembrane charge separation and electron transfer by a carotenoporphyrin-quinone triad. *Nature* **316**:653-655.

Stribling, R. and Miller, S. L. (1987) Energy yields for hydrogen cyanide and formaldehyde syntheses: The HCN and amino acid concentrations in the primitive ocean. *Origins of Life* **17**:261-273.

Usher, D. A. and McHale, A. H. (1976) Nonenzymic joining of oligoadenylates on a polyuridylic acid template. *Science* **192**:53-54.

Wächtershäuser, G. (1988) Pyrite formation, the first energy source for life: A hypothesis. *Syst. Appl. Microbiol.* **10**:207-210.

Wilson, T. H. and Maloney, P. C. (1976) Speculations on the evolution of ion transport mechanisms. *Fed. Amer. Soc. Exp. Biol. Proc.* **35**:2174-2179.

ENERGY YIELDS FOR HYDROGEN CYANIDE AND FORMALDEHYDE SYNTHESES: THE HCN AND AMINO ACID CONCENTRATIONS IN THE PRIMITIVE OCEAN

ROSCOE STRIBLING and STANLEY L. MILLER

Department of Chemistry B-017, University of California, San Diego, La Jolla, CA 92093, U.S.A.

(Received 20 November, 1986)

Abstract. Prebiotic electric discharge and ultraviolet light experiments are usually reported in terms of carbon yields and involve a large input of energy to maximize yields. Experiments using lower energy inputs are more realistic prebiotic models and give energy yields which can be used to estimate the relative importance of the different energy sources on the primitive earth.

Simulated prebiotic atmospheres containing either CH_4, CO or CO_2 with N_2, H_2O and variable amounts of H_2 were subjected to the spark from a high frequency Tesla coil. The energy yields for the synthesis of HCN and H_2CO were estimated. CH_4 mixtures give the highest yields of HCN while H_2CO is most efficiently produced with the CO mixtures. These results are a model for atmospheric corona discharges, which are more abundant than lightning and different in character. Preliminary experiments using artificial lightning are also reported.

The energy yields from these experiments combined with the corona discharge available on the earth, allows a yearly production rate to be estimated. These are compared with other experiments and model calculations. From these production rates of HCN (e.g. 100 nmoles cm^{-2} yr^{-1}) and the experimental hydrolysis rates, the steady state concentration in the primitive ocean can be calculated (e.g., 4×10^{-6} M at pH 8 and 0°). A steady state amino acid concentration of 3×10^{-4} M is estimated from the HCN production rate and the rate of decomposition of the amino acids by passage through the submarine vents.

1. Introduction

One of the goals of origin of life studies is to determine the composition of the primitive atmosphere and the organic compounds that could have been present in the primitive oceans. Without some knowledge of the organic compounds available, it is difficult to conduct realistic experiments in the further polymerization steps and the organization of these polymers into the first living organisms.

The types of organic compounds which could have been available in the primitive oceans will be determined by understanding such factors as the composition of the atmosphere, the sources of energy available to activate these components, the further reactions which occur in the oceans, and the rate of decomposition of organic compounds in the ocean. The usual prebiotic synthesis experiment takes a mixture of gases and applies a source of energy for a long period of time to obtain the maximum yield of organic compounds. This would be a good representation of synthesis on the primitive earth if there were only a single kind of energy available and the synthesis of organic compounds were not limited by this energy. However, there were a number of sources of energy on the primitive earth, as has been previously discussed (Miller *et al.*, 1976), and prebiotic synthesis was probably limited by the total quantity of carbon available in the atmosphere rather than by the energy available from the various sources (Miller and Schlesinger, 1984).

There is considerable controversy over the composition of the primitive atmosphere. (Chang *et al.,* 1983; Holland, 1984; Levine, 1985; Lewis and Prinn, 1984; Walker, 1977). Previous experiments in this laboratory have examined yields of amino acids (Schlesinger and Miller, 1983a) as well as hydrogen cyanide, formaldehyde, and ammonia (Schlesinger and Miller, 1983b) from various mixtures of H_2, CH_4, CO, CO_2, N_2, H_2O, and NH_3 using a Tesla coil as an energy source. These are carbon yields, which are product yields based on the initial carbon placed in the system. All of the mixtures were sparked for 48 hr. Aside from the likelihood of multiple activations of reactants and products, this rate of energy input is quite high. Thus, a 2 W spark for 48 hr is 173 000 J, or the present atmospheric corona discharge energy (12.5 J cm^{-2} yr^{-2}) available per cm^{-2} in 14 000 yr. But the atmosphere above each cm^2 of the earth's surface contains about 875 l of gas, so it would take 4×10^6 years for a given three liter volume (the volume of our reaction vessel) to receive 173 000 J of corona discharge.

A more realistic prebiotic experiment is to remove the products of the activation step so that they are not exposed a second time to the energy source. This is equivalent to transporting the products synthesized in the primitive atmosphere to the ocean, and thereby protecting them from destruction by atmospheric energy sources. Experiments of this type, using lower inputs of energy, give energy yields (moles of product J^{-1}) which can be used to estimate the relative importance of the different energy sources on the primitive earth. This paper will show that energy yields give a considerably different picture than carbon yields when considering various CH_4, CO, and CO_2 atmospheres.

2. Experimental

The spark discharge flask previously used to determine carbon yields (Schlesinger and Miller, 1983a, b) was modified by the addition of a thermometer well. The spark was provided by a continuous wave spark generator, Cenco Model 80721. This spark generator was modified for continuous operation by removing part of its Bakelite covering to allow for air circulation and cooling. Several different spark coils were examined with a frequency analyzer which indicated that the differences in power among the different spark coils were due to differences in the magnitude and not the range of frequencies. A spectrum analyzer and broad band oscilloscope showed a peak around 3.5 MHz with overtones detectable to 1000 MHz. The tungsten electrodes were cemented in place with Apiezon Type W sealing compound to minimize the variability of energy input due to differences in the length of the spark gap. The discharge and calibration flasks were insulated with polyurethane. In order to determine the total energy input into a discharge flask, the rise in temperature during the initial stage of a sparking run was compared to that produced by a known amount of power from a Hewlett- Packard 25 power supply passing through a Nichrome wire connecting the two electrodes in the calibration flask. The power generated by the spark coils ranged from 2 to 4 W.

Either 25 or 50 ml of water were added to the discharge flask and the air removed via a vacuum manifold system. After allowing sufficient time for the water to degas, the desired gases were introduced into the discharge flask. All reaction mixtures contained 100 torr of N_2 and 100 torr of either CH_4, CO, or CO_2. The amount of H_2 varied from 0 to 400 torr. All gases were C.P. grade and were used without further purification except for CH_4 and CO_2 which were condensed and distilled.

The discharge flask was insulated after filling and the temperature rise was recorded over a period of 20 min after turning on the spark. At predetermined times, ranging from 1 hr to 11 days, the sparking was stopped and the aqueous phase in the flask stirred for at least two hours to allow equilibration with the gas phase. A small portion of the solution in the flask was then withdrawn with care being taken to prevent air from entering the vessel. The sample was then stored in the refrigerator until analysis. The sparking was continued after the first and succeeding samples were withdrawn without changing the gas phase. Corrections were made for the amounts of HCN and H_2CO removed with the samples. The rise in temperature and the watts were measured with each restart.

The HCN concentrations in the samples were measured with an Orion cyanide electrode (Model 94-06). Prior to measurement, the pH of the samples were adjusted to 12.7 and dimedone added to complex the formaldehyde (Schlesinger and Miller, 1983b). This technique measures the cyanide present as free HCN and any which is bound in the hydroxynitriles. Only the cyanide occurring as aminonitriles is not measured since the dissociation is slow (Van Trump and Miller, 1981).

The H_2CO concentrations in the samples were measured with chromotropic acid using the procedure of West and Sen (1956). In order to measure the H_2CO bound in the glycolonitrile it is necessary to dissociate the glycolonitrile in basic solution, since the nitrile gives a very small color yield with chromotropic acid. At pH 10 and 1M SO_3^{-2}, the reaction

$$HOCH_2CN + SO_3^{-2} = CN^- + HOCH_2SO_3^-$$

converts more than 90% of the glycolonitrile to the formaldehyde-sulfite adduct. This adduct decomposes under the acid conditions of the chromotropic acid reaction, giving a full color yield for the formaldehyde. This technique is superior to the method previously used (Schlesinger and Miller, 1983b) of estimating the formaldehyde by the difference in the total HCN and the free HCN.

The major uncertainty in the energy yield determination is the energy of the spark. We estimate the energy yields to be accurate only within a factor of two. There is also an uncertainty of whether different corona discharges give the same energy yields since there is a limited understanding of the nature of the discharges and of the mechanisms of HCN and H_2CO synthesis.

3. Results

The first set of experiments investigated the time course of the production of HCN and H_2CO in three different atmospheres, $H_2/CH_4 = 1$, $H_2/CO = 2$, and $H_2/CO_2 = 3$. Figure 1 shows the yields of HCN carbon yields vs spark time. Figure 2 shows the H_2CO carbon yields vs spark time.

It is apparent that the relative carbon yields in these three atmospheres shown are dependent on the time taken for comparison, particularly for HCN. The decrease in carbon yields with time, when they might be expected to either increase or level off, could be due to the decomposition of HCN and H_2CO previously synthesized in the spark or by formation of glycine nitrile and other aminonitriles. The aminonitrile concentration was estimated by hydrolyzing the last day's sample and analyzing for amino acids. The measured carbon yields of glycine nitrile (10−20%) suggest that the decline in HCN and H_2CO yields after several days (Figures 1−2) could be accounted for by assuming that they reacted to form glycine nitrile.

In order to avoid the complications caused by long spark times, we estimated the energy yields by extrapolating the yields measured at 1, 3, and 9 hr back to zero time. Figure 3 shows typical extrapolations for three different gas mixtures. The extrapolations with CH_4 and CO are straight-forward. The CO_2 mixture shows a substantial increase in yield with time, suggesting that there may be an intermediate between CO_2 and HCN. Although there may be other possibilities, CH_4 and CO are reasonable candidates for this intermediate.

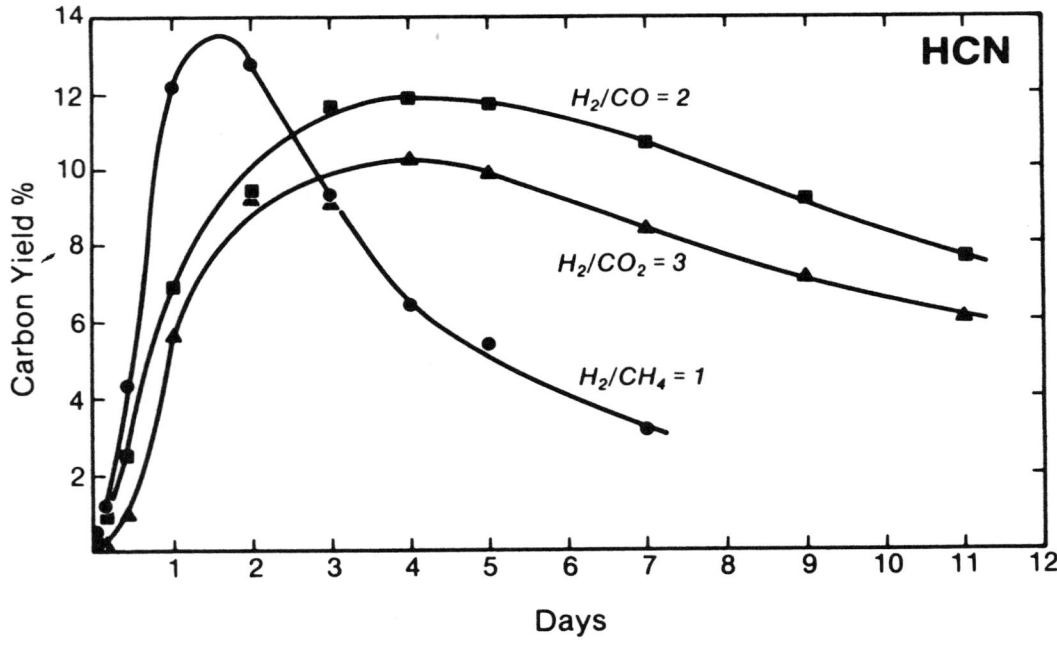

Fig. 1.   Time course for the production of HCN for three gas mixtures.

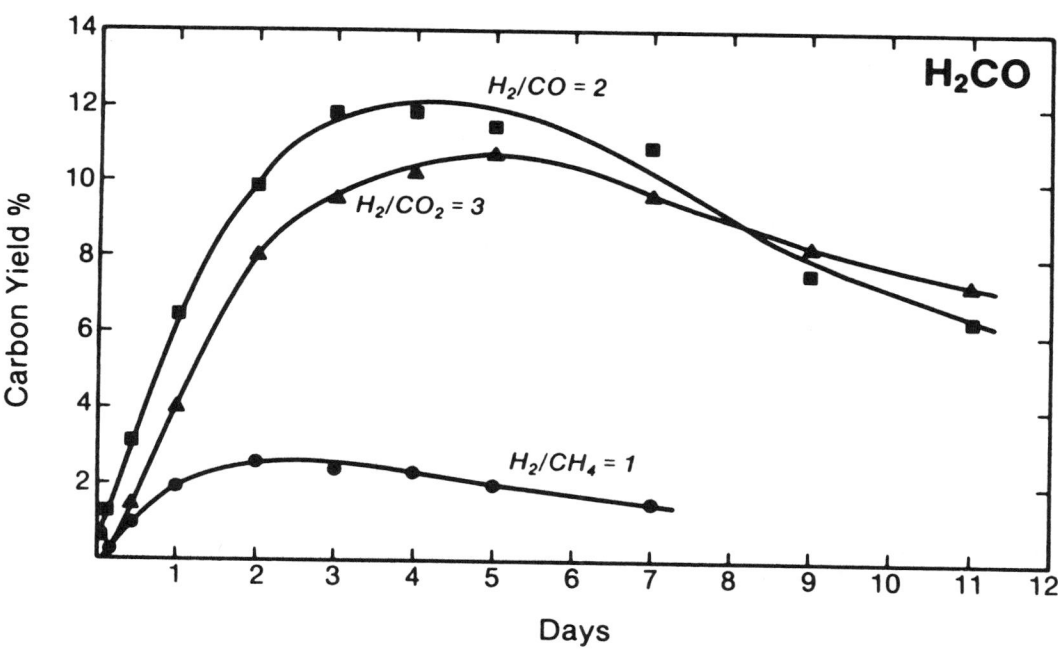

Fig. 2. Time course for the production of H_2CO for three gas mixtures.

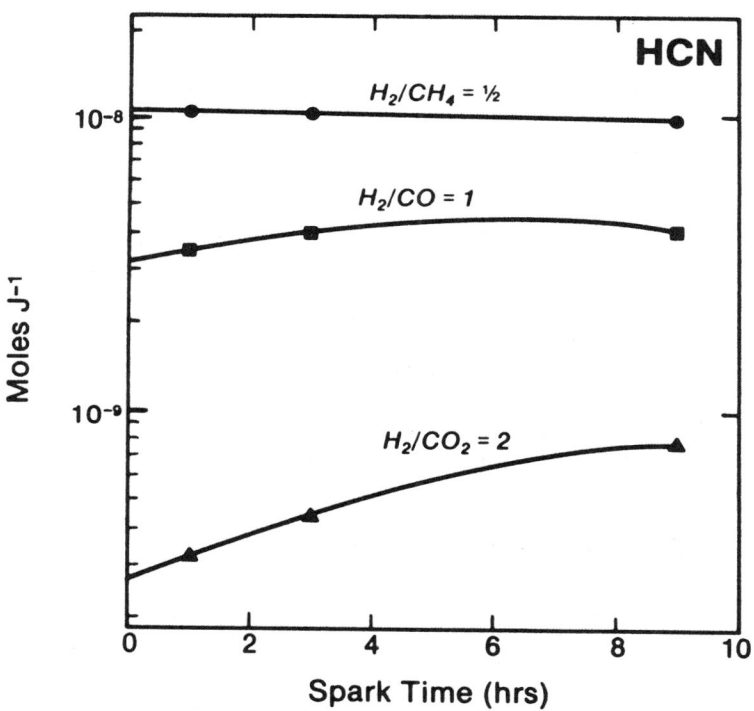

Fig. 3. Extrapolation of energy yields to obtain zero time value.

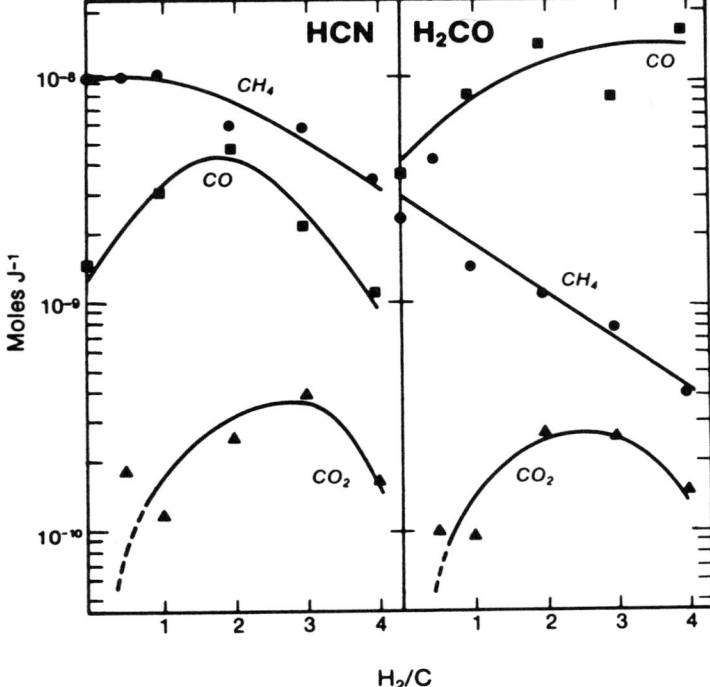

Fig. 4.   Zero time energy yields for different H_2/C ratios.

This study was extended to a range of H_2/CH_4, H_2/CO, and H_2/CO_2 mixtures. Figure 4 shows the energy yields of HCN and H_2CO for those atmospheres. The energy yields for the production of HCN ranged from 4 to 10 nmoles J^{-1} in the CH_4 atmospheres, from 1 to 5 nmoles J^{-1} in the CO atmospheres, and from 0.1 to 0.4 nmoles J^{-1} in the CO_2 atmospheres. For the production of H_2CO, the values ranged from 4 to 14 nmoles J^{-1} in the CO atmospheres, from 0.4 to 4 nmoles J^{-1} in the CH_4 atmospheres, and from 0.1 to 0.3 nmoles J^{-1} in the CO_2 atmospheres.

4. Discussion

It is apparent from the results of this study that using carbon yields to compare the relative efficiencies of different atmospheres for the prebiotic production of small molecules can be misleading. The problem is best illustrated in Figure 1, which shows that the carbon yields with different gas mixtures vary widely depending on the sparking time. After sparking for one day, it would appear that the CH_4 atmosphere would be much more efficient that either the CO or CO_2 mixtures. If the samples were taken after seven days of sparking, however, the efficiency of the CH_4 mixture would appear to be much less than the CO and CO_2 mixtures. The previous study of carbon and energy yields (Schlesinger and Miller, 1983 a, b; Miller and Schlesinger, 1984) used a 48 hour sparking period and in those cases the carbon yields of the CH_4, CO, and CO_2 were about the same except for the fall off at low H_2/CO and H_2/CO_2 ratios.

Energy yield data (Figure 4) to suggest quite a different interpretation. While there is an initial increase in the efficiency of HCN and H_2CO production with increasing H_2/C in several cases, the general trend is one of decreasing production at high H_2/C ratios. Also, at no H_2/C ratio do the yields generated from the CO_2 atmospheres reach the yields produced by the CH_4 and CO atmospheres. It is of interest to note that CH_4 is more efficient than CO in HCN production while the reverse is true for the synthesis of H_2CO.

The principle variable examined in these experiments was the H_2/C ratio. Other variables such as temperature, pressure and the mole fractions of the gases were not examined and might affect the energy yields, with the mole fractions probably being the most important.

5. Comparisons with Previous Discharge Experiments

There are only a few previous reports of HCN and H_2CO syntheses in which energy yields can be estimated. Experiments using a silent electric discharge and a flow system gave HCN carbon yields from $CH_4 + N_2$ as high as 36% and estimated energy yields as high as 27 nmoles J^{-1} (Toupance et al., 1975). The corresponding carbon and energy yields from $CH_4 + NH_3$ were 30% and 27 nmoles J^{-1}. Similar yields (26 nmoles J^{-1}) were obtained using a radio frequency-induced plasma (Capezzuto et al., 1973). These are in apparent agreement with our yields of about 10 nmoles J^{-1}.

Briner and co-workers conducted an extensive investigation of HCN synthesis using $CH_4 + NH_3$, $CH_4 + N_2$ and $CO + N_2 + H_2$ mixtures (Briner and Baerfuss, 1919; Briner et al., 1938; Briner and Hoefer, 1940). Most of their yields are in the range of 4.7 to 96 nmoles J^{-1}, although a few high frequency experiments gave yields as high as 382 nmoles J^{-1}. Briner's experiments were done to optimize yields for industrial synthesis, so it would be surprising if such high yields were possible under geological conditions.

Thornton and Sergio (1967) used a high frequency discharge on mixtures of CH_4 and H_2O. The yield of H_2CO was about 13 nmoles J^{-1} when the product was protected by absorption in liquid water after synthesis. The yield was about 1/3 of this without absorption. Koenig and Weinig (1927) used a flow discharge on mixtures of CO and H_2 and obtained a yield of 18 nmoles J^{-1}. Mourey et al., (1981) and Bossard and Toupance (1981) used a spark discharge on a CH_4 and water mixture and obtained a yield of only 0.29 nmoles J^{-1}. This yield was increased to 3 nmoles J^{-1} when liquid water was present to absorb the H_2CO (F. Raulin, personal communication; Mourey, 1982).

6. A Balance Sheet for HCN Synthesis on the Primitive Earth

A balance sheet for the production of small molecules on the primitive Earth can be prepared by combining the energy yields from different sources with the energy available from those sources. The generally accepted values for energy available from lightning and corona discharge are 4.2 and 12.6 $J\ cm^{-2}\ yr^{-1}$, respectively (Miller *et al.*, 1976).

There are few data available for HCN production by lightning. A calculation by Chameides and Walker (1981) gives energy yields for the production of HCN and NO from lightning bolts for various mixtures of N_2, CH_4, CO, CO_2, and H_2O. The calculations were based on the effects of the post-flash shock wave, which raises the region near the lightning bolt to a high temperature, followed by cooling and quenching of the high temperature equilibrium mixture of products at the freeze-out temperature (2000 to 2500 K for HCN). Their calculations show that the yield of HCN depends mainly on the O/C ratio in the presence of excess N_2, where O is the total oxygen ($CO + 2CO_2 + H_2O$, etc.) and C is the total carbon ($CH_4 + CO + CO_2$). The calculated yields do not depend strongly on the H/C ratio. The yields of HCN in CH_4 atmospheres are about 530 nmoles J^{-1}. In CO atmospheres the yield of HCN falls off to about 0.153 nmoles J^{-1}, and in CO_2 atmospheres the yield is calculated to be only about 0.00042 nmoles J^{-1}.

The spark discharge energy yields fall off much less in the sequence CH_4, CO and CO_2 as shown in Figure 4. These experimental results are not in contradiction to the lightning calculations since sparks and other electrical discharges are different in character from lightning bolts and their associated shock waves. These differences include production of HCN in electric discharges at O/C ratios where little HCN production is calculated in lightning bolts, synthesis of substantial amounts of NH_3, and a major role for H_2 in the HCN yields in electric discharges. In addition, formaldehyde would not be synthesized at all in the high temperature region of a lightning bolt. The role of shock waves is apparently minimal in the spark, the effective temperature of the spark being much lower than lightning, and ion-molecule and radical reactions appear to play a more important role in the products produced by the spark than in the frozen high temperature equilibrium of lightning. It is possible that additional products with lower freeze-out temperatures can be made in lightning bolts by reactions not considered by Chameides and Walker.

A preliminary measurement of the HCN energy yield from artificial lightning was obtained at Ames Research Center in collaboration with C. McKay and T. Scattergood. In a CH_4 atmosphere, we measured an HCN energy yield of 31 nmoles J^{-1}. This is an order of magnitude lower than the calculated yield. The energy yields in four CO atmospheres ranged from about 0.002 to 0.05 nmoles J^{-1} and the yields in four CO_2 atmospheres ranged from about 0.003 to 0.006 nmoles J^{-1}. The yields with CO and CO_2 are quite uncertain because the HCN measurements were near the limit of sensitivity of the cyanide electrode.

There has also been some work done using energy generated by expanding gases in shock wave tubes. Rao *et al.* (1967) investigated HCN synthesis from various hydrocarbons and N_2 in high temperature shock experiments. HCN carbon yields were about 15% for temperatures greater than 3000 K. The energy yields from these experiments [calculated by Bar-Nun and Shaviv (1975)] were 332 nmoles J^{-1}.

A photochemical source of HCN has been proposed by Zahnle (1986). Nitrogen atoms are produced by the Lyman continuum (796 to 912 A) and diffuse lower into the atmosphere to react with CH_2 and CH_3, producing HCN. The yields of HCN range between a maximum of 520 nmoles cm^{-2} yr^{-1} when the production of N atoms is limiting and 5 to 50 nmoles cm^{-2} yr^{-1} at lower CH_4 fluxes.

Inspection of the estimates discussed in this section, as well as the results of our experiments, shows that the HCN production rates from the various sources are comparable within an order of magnitude. This suggests that errors in any particular estimate will not affect the value chosen for model calculations.

7. A Balance Sheet for H_2CO Synthesis on the Primitive Earth

The balance sheet for the production of formaldehyde is limited to that which may be produced from either spark discharge or UV light. No H_2CO is generated in lightning bolts either in calculations (Chameides and Walker, 1981) or in our limited measurements of the samples generated from artificial lightning.

The data of Ferris and Chen (1975) can be used to estimate energy yields for the production of H_2CO from the action of UV light on a CH_4 atmosphere of about 3 nmoles J^{-1}. Experimental values for CO atmospheres range from about 4 nmoles J^{-1} (Bar-Nun and Hartman, 1978) to 13 nmoles/J (Bar-Nun and Chang, 1983).

The production rate of H_2CO in a model CO_2 atmosphere has been calculated by Pinto et al. (1980). The rain out rate is about 15 nmoles cm^{-2} yr^{-1}. The assumed atmosphere is relatively reducing since $H_2/CO_2 = 3$. Another calculation by Kasting et al. (1984) gives 1.6 to 45 nmoles cm^{-2} yr^{-1}. An important part of these calculations is the rain out rate of the H_2CO, since H_2CO absorbs U.V. at 3200 A. It is not clear whether this factor has been adequately treated. As in the case with HCN, the H_2CO production rates from the various sources, including the results of our experiments, are generally comparable.

8. The Concentration of HCN in the Primitive Ocean

These production rates of HCN allow a calculation of the steady state concentration in the primitive ocean. We assume that all the HCN produced in the atmosphere enters the ocean and remains there and that the only pathway for the loss of HCN is hydrolysis to formamide and then to formic acid. At steady state, the production rate (S_{HCN} in moles cm^{-2} yr^{-1}) equals the rate of hydrolysis

$$S_{HCN} = \frac{d(HCN)}{dt} V_0,$$

where (HCN) is the molar concentration, and V_0 is the volume of the ocean in liters cm^{-2}, now 300 1 cm^{-2}. Writing the hydrolysis rate of HCN as a pseudo first-order reaction

$$-\frac{d(HCN)}{dt} = k_1 (\Sigma\ HCN),$$

where k_1 is the pseudo first-order rate constant in yr^{-1} and depends on temperature and pH, and $\Sigma\ HCN = HCN + CN^-$.

The hydrolysis of HCN is both acid and base catalyzed. The acid catalyzed reaction has been investigated mostly in 2 to 10 M HCl and H_2SO_4 (Krieble and Noll, 1939; Rabinovitch and Winkler, 1942) and at 0.1 to 1M HNO_3 (Miller, unpublished). The rate is

$$-\frac{d(HCN)}{dt} = k_2^{H^+} (H^+) (HCN) \tag{1}$$

with an approximate value of $k_2^{H^+}$ given by

$$\log k_2^{H^+}\ (yr^{-1}) = 12.76 - 4720/T.$$

The rate of the base catalyzed hydrolysis is

$$-\frac{d(HCN)}{dt} = k_2^{OH^-} (OH^-) (HCN)$$

$$= \frac{k_2^{OH^-} K_w (\Sigma\ HCN)}{(H^+) + K_{HCN}}, \tag{2}$$

where $k_2^{OH^-}$ is the second order rate constant of OH^- hydrolysis,

$$K_{HCN} = (H^+)(CN^-)/(HCN),\ K_w = (H^+)(OH^-).$$

We use pK_w from Robinson and Stokes (1959) and pK_{HCN} from Schlesinger and Miller (1973).

$$p\dot{K}_w = -6.0846 + 4471.33/T + 0.017053T,$$

$$pK_{HCN} = -8.85 + 3802/T + 0.01786T.$$

The value of $k_2^{OH^-}$ can be obtained from the data reported by Sanchez *et al.* (1967) and Miller (unpublished)

$$\log k_2^{OH^-}\ (day^{-1}) = 16.62 - 4440/T.$$

In the pH region 6–9, Equation (2) is applicable with the acid catalyzed pathway making a negligible contribution. Equating the production rate and hydrolysis of HCN gives

$$\Sigma\ HCN = \frac{S}{V_0} \frac{[(H^+) + K_{HCN}]}{k_2^{OH^-} K_w} = \frac{S}{V_0 k_1}. \tag{3}$$

Taking $S = 100$ nmoles cm^{-2} yr^{-1} and $V_0 = 300$ l cm^{-2} we get the half lives and concentrations in Table I.

TABLE I

Half-lives and concentrations of HCN in the primitive ocean based on a HCN production rate of 100 nmoles cm^{-2} yr^{-1} and an ocean volume of 300 cm^{-2}

		0 °C	25 °C	50 °C
$t_{1/2}$ (yr)	pH8	7000	40	0.5
$t_{1/2}$ (yr)	pH7	70000	400	5
M_{HCN}	pH8	4×10^{-6}	2×10^{-8}	3×10^{-10}
M_{HCN}	pH7	4×10^{-5}	2×10^{-7}	3×10^{-9}

Equation (3) shows that the concentration of HCN will be increased proportionally to the production rate and to the (H$^+$), an inversely to the volume of the ocean. The results of Table I make it clear that low temperatures and pH favor higher concentrations of HCN, but even at pH7 and 0° a concentration of 3.5×10^{-5} M is much lower than the usual prebiotic experiment. Concentrations of HCN of 10^{-6} M are sufficient to make amino acids by the Strecker synthesis (Miller and Van Trump, 1981) but adenine synthesis would require a concentration mechanism, with freezing being the most likely (Sanchez *et al.*, 1967).

The spark discharge synthesis of HCN would have been only 1 nmole cm^{-2} yr^{-1} for H$_2$/CO$_2$ = 1 and 4 nmoles cm^{-2} yr^{-1} for H$_2$/CO$_2$ = 4, which is considerably lower than the 100 nmoles cm^{-2} yr^{-1} assumed in this calculation. This would lead to HCN concentrations 25 to 100 times lower than given in Table I, which might raise problems for amino acid and purine synthesis. If the Zahnle N atom synthesis were as efficient on the primitive earth as calculated, then this problem would not arise.

9. The Concentration of Amino Acids in the Primitive Ocean

The production rates of HCN and H$_2$CO estimated above permit us to calculate the rate of build up of amino acids in the primitive ocean and their steady state concentration. Provided the concentrations of HCN and aldehydes do not drop too low, the Strecker synthesis will be effective in the primitive ocean (Miller and Van Trump, 1981). We will assume an amino acid yield of 10% based on the HCN production. This is the approximate yield reported with CH$_4$, CO and CO$_2$ atmospheres (Schlesinger and Miller, 1983 a, b). The yield is much higher (~ 90%) with the Strecker synthesis, while the cyanide polymerization gives only about 1% (Lowe *et al.*, 1963; Ferris *et al.*, 1978).

Taking a combined HCN production rate of 100 nmoles cm^{-2} yr^{-1}, the steady state amino acid production rate would then have been 10 nmoles cm^{-2} yr^{-1}. Taking the volume of the ocean as 300 cm^{-2}, the increase in amino acid concentration would have been 3.3×10^{-11} moles yr^{-1}. Assuming no losses, this gives 3.3×10^{-4} M in 10 million years. In a low temperature ocean the losses from thermal decomposition should be low for amino acids such as glycine and alanine but considerably greater for less stable amino acids such as serine. The losses from adsorbtion on clays, ionizing radiation and ultraviolet light are much more difficult to estimate. The most important loss mechanism was probably the submarine vents, in which the sea water is heated to at least 350° thereby decomposing all the animo acids. Since the entire ocean on the average passes through the vents in 10 million years (Edmond *et al.*, 1982), the amino acid concentration could not have risen higher than 3×10^{-4} M with the above assumptions. Increasing the production rate of HCN or decreasing the size of the ocean would increase proportionally the steady state concentration of amino acids.

As discussed for the HCN production, CO_2 atmospheres give considerably lower HCN yields from spark discharges. Without the HCN production from the Zahnle N atom mechanism, the amino acid concentrations would be 25 to 100 times lower than the 3×10^{-4} M. There is no way at the present time to estimate the amino acid concentration needed for life to arise. A molarity of 3×10^{-4} seems substantial and a factor of 100 or 1000 lower seems almost too small. A detailed mechanism of the process leading to the origin of life is be needed to place constraints on the amino acid concentration as well as the atmosphere conditions and composition required for the synthesis of necessary amino acids.

Acknowledgement

This work was supported by NASA Grant NAGW-20. We thank Chris McKay and Tom Scattergood at Ames Research Center for permission to use the HCN energy yields from artificial lightning.

References

Bar-Nun, A. and Chang, S.: 1983, *J. Geophys. Res.* **88**, 6662.
Bar-Nun, A. and Hartman, H.: 1978, *Origins of Life* **9**, 93.
Bar-Nun, A. and Shaviv, A.: 1975, *Icarus* **24**, 197.
Bossard, A., Raulin, F., Mourey, D., and Toupance, G.: 1981, in Y. Wolman (ed.), *Origins of Life*, D. Reidel Publ. Co., Dordrecht, Holland, p. 83.
Bossard, A. and Toupance, G.: 1981, in Y. Wolman (ed.), *Origins of Life*, D. Reidel Publ. Co., Dordrecht, Holland, p. 93.
Briner, E. and Baerfuss, A.: 1919, *Helv. Chim. Acta.* **2**, 663.
Briner, E., Desbaillets, J., and Paillard, H.: *Helv. Chim. Acta.* **21**, 115.
Briner, E. and Hoefer, H.: 1940, *Helv. Chim. Acta.* **23**, 826.
Capezzuto, P., Cramarossa, F., Ferraro, G., Maione, P., and Molinari, E.: 1973, *Gazz. Chim. Ital.* **103**, 1153.
Chameides, W. L. and Walker, J. C. G.: 1981, *Origins of Life* **11**, 291.
Chang, S., DesMarais, D., Mack, R., Miller, S. L., and Strathearn, G. E.: 1983, in J. W. Schopf (ed.), *Earth's Earliest Biosphere Its Origin and Evolution,* Princeton University Press, Princeton, New Jersey, p. 53.

Edmond, J. M., Von Danm, K. L., McDuff, R. E., and Measures, C. I.: 1982, *Nature* **297**, 187.

Ferris, J. P. and Chen, C. T.: 1975, *J. Am. Chem. Soc.* **97**, 2962.

Ferris, J. P., Joshi, P., Edelson, E. H., and Lawless, J. G.: 1978, *J. Mol. Evol.* **11**, 293.

Holland, H. D.: 1984, *The Chemical Evolution of the Atmosphere and Oceans,* Princeton University Press, Princeton New Jersey.

Kasting, J. F., Pollack, J. B., and Crisp, D.: 1984, *J. Atmos. Chem.* **1**, 403.

Koenig, A. and Weinig, R.: 1927, *Chem. Abstr.* **21**, 3834.

Krieble, V. K. and Noll, C. I.: 1939, *J. Am. Chem. Soc.* **61**, 560.

Levine, J. S.: 1985, *The Photochemistry of Atmospheres,* Academic Press, Orlando, Florida.

Lewis, J. S. and Prinn, R. G.: 1984, *Planets and Their Atmospheres,* Academic Press, Orlando, Florida.

Lowe, C. U., Rees, M. W., and Markham, R.: 1963, *Nature* **199**, 219.

Miller, S. L. and Schlesinger, G.: 1984, *Origins of Life* **14**, 83.

Miller, S. L., Urey, H. C., and Oro, J.: 1976, *J. Mol. Evol.* **9**, 59.

Miller, S. L. and Van Trump, J. E.: in Y. Wolman (ed.), *Origins of Life,* D. Reidel Publ. Co., Reidel, Dordrecht, Holland, p. 135.

Mourey, D., Thesis, Univ. Paris VI, 1982.

Mourey, D., Raulin, F., and Toupance, G.: 1981, in Y. Wolman (ed.), *Origins of Life,* D. Reidel Publ. Co., Dordrecht, Holland, p. 73.

Pinto, J. P., Gladstone, G. R., and Yung, Y. L.: 1980, *Science* **210**, 183.

Rabinovitch, B. S. and Winkler, C.A.: 1942, *Canad. J. Res.* **20B**, 221.

Rao, V. V., MacKay, D., and Trass, O.: 1967, *Canad. J. Chem. Eng.* **45**, 61.

Robinson, R. A. and Stokes, R. H.: 1959, *Electrolyte Solutions,* Buttersworth, London, p. 363.

Sanchez, R. A. and Ferris, J. P., and Orgel, L. E.: 1966, *Science* **153**, 72.

Sanchez, P. A., Ferris, J. P., and Orgel, L. E.: 1967, *J. Mol. Biol.* **30**, 223.

Schlesinger, G. and Miller, S. L.: 1973, *J. Am. Chem. Soc.* **95**, 3729.

Schlesinger, G. and Miller, S. L.: 1983a, *J. Mol. Evol.* **19**, 376.

Schlesinger, G. and Miller, S. L.: 1983b, *J. Mol. Evol.* **19**, 383.

Thornton, J. D. and Sergio, R.: 1967, *Nature* **213**, 590.

Toupance, G., Raulin, F., and Buvet, R.: 1975, *Origins of Life* **6**, 83.

Van Trump, J. E. and Miller, S. L.: 1973, *Earth Planetary Sci. Lett.* **20**, 145.

Walker, J. C. G.: 1977, *Evolution of the Atmosphere,* Macmillan, New York.

West, P. W. and Sen, B.: 1956, *Z. Anal. Chem.* **153**, 177.

Zahnle, K. J.: 1986, *J. Geophys. Res.* **91**, 2819.

Davy's electrochemistry: Nature's protochemistry

Peter Mitchell

The growth of understanding and knowledge in chemistry—or chemical philosophy, as Humphry Davy properly called it[1]— occurs by a largely evolutionary process, as in all branches of natural philosophy.[2] Therefore, it is a special pleasure for me to have this opportunity to trace back some of the recent knowledge and ideas about proticity and protonmotive reactions in biology[3] to their identifiable intellectual evolutionary origins.

One of the chief amongst these origins was Davy's marvellously imaginative recognition and practical exploitation of what he described as the electromotive power of the giant voltaic pile used in his laboratory to overcome chemical affinity and drive chemical reactions—as in his famous discovery and isolation of potassium by electrolysis of potash in 1807. We all know that it was Davy's former assistant, Michael Faraday, who began to lay the quantitative foundations of electrochemistry around 1835—a few years after the end of Davy's life. But it was Davy himself who recognised the essentially motive chemical property of the electricity produced by the voltaic pile, and who, with the practical aid or conceptual inspiration of his electrochemical work, discovered not only potassium but also sodium, calcium, magnesium, barium, strontium and chlorine.

Following the discovery of the electrolysis of water by Nicholson and Carlisle, Davy showed, in 1806, that oxygen and hydrogen gas are the sole products of electrolytic decomposition of pure water. That tremendously important observation subsequently led W. R. Grove to observe the reverse process, in which the combination of oxygen and hydrogen *via* catalytically active platinum electrodes was used to generate electricity. Thus, in a remarkable letter to the *Philosophical Magazine*,[4] entitled 'On Voltaic Series and the Combination of Gases by Platinum', the far-sighted Grove, standing on Davy's shoulders, invented the hydrogen-burning fuel cell in 1839.

Grove recognised that the great promise of the fuel cell lay in the relatively high degree of reversibility with which it could interconvert chemical and electrical forces—or chemical and electrical energy, as most theoreticians, from Helmholtz onwards, have so far preferred to say. Thus, using the Newtonian principle that the forces of action and reaction are equal and opposite, Grove described a law of the conservation of forces—which we might appropriately describe as the first law of molecular mechanics—before Helmholtz formulated the law of conservation of energy known as the first law of thermodynamics.

Figure 1a shows a modern diagram of a hydrogen-burning fuel cell.[5] It is, perhaps, not self evident that such a fuel cell for generating electricity is also, potentially, a generator of *proticity*. It simply depends where one opens the circuit to conduct away the power for external use. In *Fig.1a*, the circuit is opened in the electron conductor to

Fig. 1. Hydrogen-burning fuel cell: (a) generating electricity; (b) generating proticity.

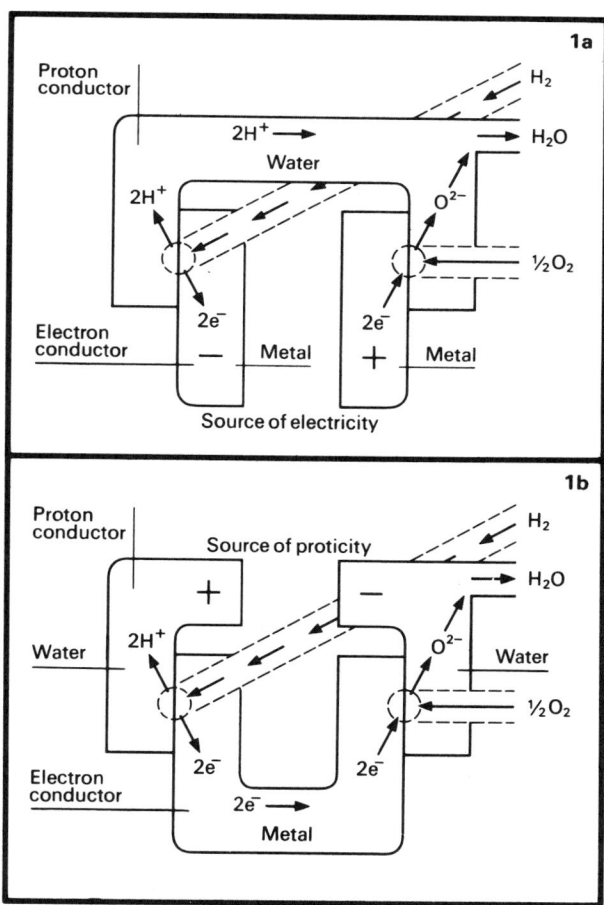

Peter Mitchell, who received the Nobel Prize for Chemistry in 1978, is director of research at the Glynn Research Institute, Bodmin, Cornwall, PL30 4AU. His work on chemiosmotic coupling in oxidative phosphorylation and energy transduction is aimed at resolving the question of how light energy and respiratory energy are conserved and transferred in cells. This article is a shortened version of the Humphry Davy Memorial Lecture delivered to the Chilterns and Middlesex Section of the CS and RIC 20 May 1980; the original title of the lecture was *From Davy's electronmotive power in chemistry to Nature's protonmotive power in biology*.

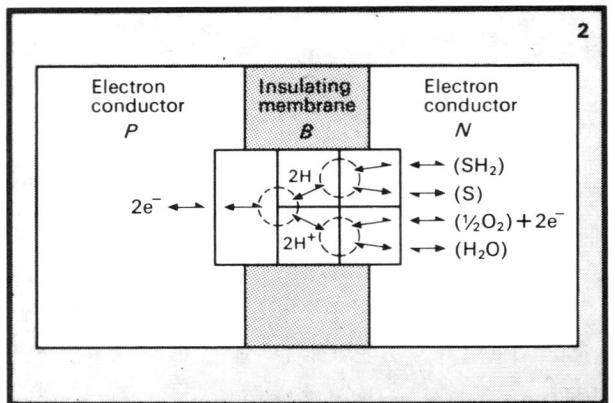

Fig. 2. Ligand conduction diagram of electronmotive oxidoreduction.

give electricity at the metallic terminals. In *Fig. 1b*, the circuit is opened in the proton conductor to give proticity at the aqueous terminals. I introduced the word *proticity* to describe the force and flow of the proton current or its equivalent,[6] just as one uses the word electricity to describe the force and flow of the electron current or its equivalent.

The fuel cell is a beautiful example of the truth of the principle, enunciated by Pierre Curie at the end of the last century,[7] that effects cannot be less symmetric than their causes. The phenomena of transport in the fuel cell arise from the intrinsically vectorial disposition of the chemical reactions (heterolytic half reactions) at the anisotropic metal/aqueous catalytic electrode interfaces.[8] Thus, the scalar group-potential differences of the chemical reactions are projected in space as vectorial chemical fields of force corresponding to the chemical group-potential gradients directed across the electrode interfaces.

It is an interesting fact in the evolution of chemical theory that Davy and Grove sought to describe chemical processes in terms of forces of chemical affinity, which implicitly retained their Newtonian vectorial (spatially-directed) property. But with the development of more exact quantitative thermodynamic and kinetic chemical theories, describing chemical transformation in terms of scalar energy changes (given by the abstract scalar products of the real co-linear vectorial forces and displacements of the interacting chemical particles), the vectorial transport aspect of chemistry tended to be eliminated from classical chemical theory. So much so that velocities in chemistry became scalar—as in reaction velocities of chemical kinetics—and certain theoreticians were even led to regard the forces of chemical affinity as being intrinsically scalar. But now that we are seeking to describe the interrelationship between chemical transformation and chemical translocation, especially in biochemistry, the time has come to put the vectorial forces and displacements of the chemical particles back into chemical theory.[9] That is, perhaps, one reason why the views of Davy and Grove seem to be so modern today.

The vectorially organised chemical reactions catalysed in the chemicomotive type of system, represented by the fuel cell, differ from the transformation reactions of classical chemistry in that they involve an obligatory process of transport of chemical species through specifically conducting membranes, plugs or interfaces. The

transport of a given chemical species is attributable to a net force (Greek ὠσμό ζ = push) through the specifically permeable region. Guggenheim described chemicomotive reactions under the general heading of membrane equilibria.[10] Therefore I introduced the term *chemiosmotic* to describe the type of chemical transformation that is catalysed by an anisotropic system (membrane, plug or interface), so that the process of chemical group (or electron) transfer obligatorily involves the osmotic translocation of one or more chemical species from the medium on one side of the system to that on the other. The special property of chemiosmotic systems is to interconvert chemical and osmotic forces and energy.[11]

The ligand-conduction diagram in *Fig. 2* represents an electronmotive oxidoreduction fuel cell similar to that of *Fig. 1a*, but using a hydrogenated solute SH_2 as the fuel instead of hydrogen gas. Components of the oxidant and reductant couples, shown at the *N* terminal of the cell for formal topological reasons, are bracketed to denote that there are pathways of access for these components passing through the *N* terminal and communicating with external reservoirs. The feasibility of such a system for the reversible and efficient interconversion of chemical and osmotic (in this case electronic) energy depends on two main requirements: (*i*) the chemical transformations (*ie* chemical bond exchanges) must be catalysed with high specificity and with low resistance only at the catalytic centres or interfaces marked by the dashed circles; (*ii*) the various chemical species must be conducted with high specificity and low resistance only through the ligand-conducting pathways marked by the brackets or by the arrows, the double barbs of which represent the (purely formal) forward direction of the translocations. In other words, the great problem for the designer of chemicomotive cells is to achieve the rapid conduction of specific ligands through the continuous pathways, including those of chemical transformation at the electrode interfaces, in the chemicomotive cell.

Electronmotive systems happen to be peculiarly favoured, relative to most other chemicomotive systems, by the fact that metals are very good and very specific conductors of electrons, whereas aqueous media containing acids, bases, pH buffers, or ion-exchange polymers, and certain other liquid or solid media, are moderately good conductors of protons (or their equivalent: OH^- or

Fig. 3. Ligand conduction diagram of oxideionmotive hydrodehydration.

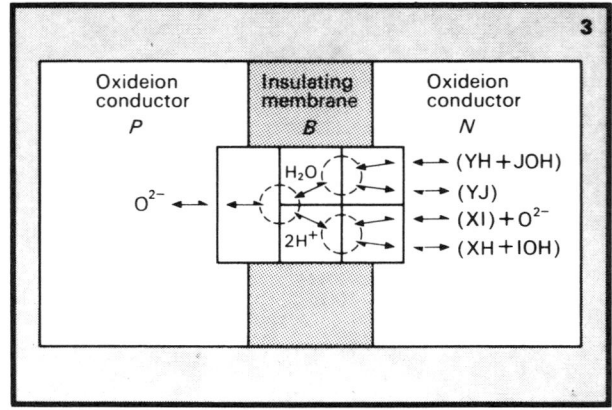

O^{2-} ions), but not of electrons. However, it is well known to the designers of electronmotive fuel cells, and to students of modern electrochemistry, that the ligand-conducting requirements for quantitatively good reversibility and high efficiency have been extremely difficult to fulfil in artificial electrochemical systems, especially when operating at the high chemical current and electric current densities that have to be achieved at the electrode interfaces in a fuel cell of compact and lightweight design.[8] Indeed, electrode reactions have come to be regarded as generally irreversible by some modern electrochemists, who have made a special study of what they call electrodics.[12]

For reasons which will become apparent later, it is interesting to consider an unconventional type of fuel cell that is driven, not by *oxidoreduction* (o/r), but by *hydrodehydration* (h/d), as illustrated in *Fig. 3*. This hypothetical system is driven by the transfer of the elements of H_2O from the h/d couple (YH + JOH)/YJ to the h/d couple (XH + IOH)/XI; and it is oxideionmotive but not electronmotive. The oxideionmotive h/d system of *Fig. 3* is similar in general principle to the electronmotive o/r system of *Fig. 2*, but it differs in chemical detail by involving the transfer of H_2O in place of 2H from the donor to the acceptor couples, and it requires ligand-conducting components of different specificity to catalyse the overall chemiosmotic reaction. The internal circuit components include specific conductors of H_2O and H^+, which could be appropriate aqueous media, and the terminals (and external circuit connection, not shown) must be conductors of oxide ions.

Our knowledge of specific oxideion conductors has its origin in the electric light source made of a bar of zirconia doped with a rare earth such as yttria or ceria, developed by Nernst at the end of the 19th century, and known as the Nernst glower.[13] Nernst—whose imaginative genius and practical enterprise resembled that of Davy—was aware that the temperature dependent electric conductivity of the glower was not attributable to free electrons, but depended on an ionic 'dissolved conducting material' that was replenished by external oxygen gas at the cathode when the glower was run on direct current. However, we owe to Wagner[14] the important explanation that the 'dissolved conducting material' of Nernst is mainly lattice vacancies that make possible the virtually pure conduction of O^{2-} ions through the zirconium oxide lattice.

So far in this tribute to Humphry Davy, I have been considering some aspects of the conscious evolution of ideas and knowledge about electrochemistry, and about artificial chemicomotive devices, that can be traced back to Davy's pioneering work at the beginning of the 19th century—through an intellectual evolutionary period of rather less than 200 years. But the unconscious evolutionary process of natural selection has had more than 10 million times as long as that to hit on ways of exploiting chemicomotive techniques for power transmission, nutrient transfer and other spatially orientated processes of movement in living organisms. Therefore, I think that, if Davy could have been consulted by biochemists in the 1950s about the problematical relationship between the scalar theory of 'bag-of-enzymes' metabolism and the vectorial theory of membrane transport, he would have agreed that, from the viewpoint of chemical philosophy, the vectorial development and action of chemical fields of force through chemiosmotic reaction systems might be

very important in biochemistry; and I think he would have advised us to investigate that possibility both theoretically and experimentally.

One would not have expected electronmotive systems coupled by electricity to play a part in biochemical power transmission because living cells lack the spatially extensive electron conductors—corresponding to metallic power lines of artificial systems—that would be required. However, the spatially extensive aqueous media, separated by insulating bimolecular lipid membranes, are moderately good conductors of proticity. Moreover, it happens to be very significant from the chemical point of view that the oxidoreduction and hydrodehydration fuel cell systems, shown in *Fig. 2* and *Fig. 3*, share the proton, but not the electron, as a common reactant. Therefore it is possible to conceive both o/r and h/d systems that are protonmotive, as shown in *Fig. 4*. The h/d system of *Fig. 4* differs from that of *Fig. 3* in that H_2O of the aqueous media is shown in place of the (YH + JOH)/YJ couple, but the general principle is the same. It is convenient to use the terms protonmotive redox loop and protonmotive hydrodehydration loop to describe systems of types *a* and *b* respectively in *Fig. 4*.

An obvious inference from *Fig. 4* is that if the protonmotive hydrolytic h/d loop system were reversible, and the membrane were topologically closed so as to isolate the conductors *P* and *N* except for their specific connection by the protonmotive plug-through or translaid components, the proticity-producing o/r loop system could drive the reversible hydrolytic h/d loop system backwards, and oxidative esterification would ensue. Twenty years ago, this type of system provided the elementary vectorial chemical basis of a chemiosmotic hypothesis of oxidative and photosynthetic phosphorylation that has recently come to be known as the chemiosmotic theory.[3]

Oxidative and photosynthetic phosphorylation are biochemical processes in which redox reactions drive a hydrodehydration reaction, so that adenosine diphosphate

Fig. 4. Ligand conduction diagrams of: (a) protonmotive oxidation; (b) protonmotive hydrolysis. The sum reaction could be oxidative esterification.

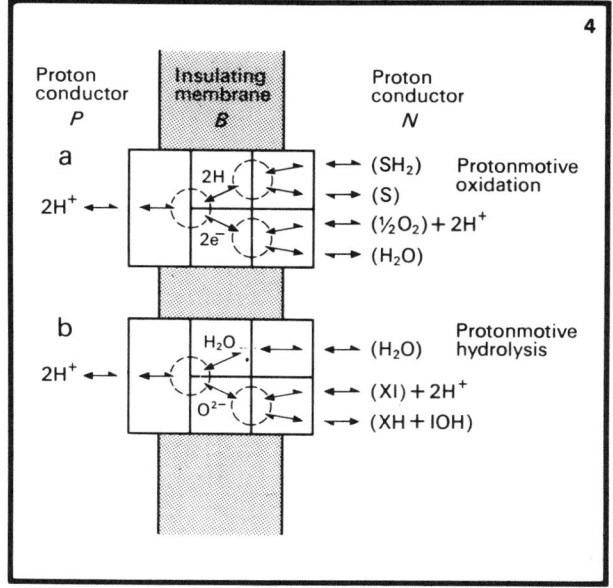

(ADP or ADPOH) is phosphorylated by inorganic phosphate (P or POH) to give adenosine triphosphate (ATP or ADPOP). More than 95 per cent of the energy of most animal cells is obtained by the process of oxidative phosphorylation, catalysed by small intracellular organelles known as mitochondria; and much of the energy harvested from sunlight for the synthesis of plant products is conserved in the corresponding process of photosynthetic phosphorylation, catalysed by light-absorbing green intracellular organelles known as chloroplasts. A wide variety of bacteria also obtain most of their energy through the processes of oxidative or photosynthetic phosphorylation, or through both.[3]

During the past two decades, the evolution of the chemiosmotic theory has depended on the painstaking experimental elucidation of the structure and function of the energy-conserving systems by a great number of biochemists, using a wide range of techniques. I am indebted to these colleagues for much of the information contained in the summary of knowledge that follows.[3]

Mitochondria, chloroplasts and bacteria contain inner bimolecular lipid-membrane systems that are topologically closed. These membrane systems (which have an area of around 50 to 100 m^2 per g of protein in the organelles) have an effective proton conductance (meaning $H^+ + OH^- + O^{2-}$ conductance) of only about 0.5 μmho cm^{-2}, whereas the effective specific proton conductance of the aqueous media containing pH buffers on either side of the membrane is millions of times greater than that of the hydrocarbon osmotic barrier domain B of the lipid membrane. Thus mitochondria, chloroplasts and bacteria contain a system of two spatially extensive proton conductors P and N, insulated from each other by the B domain of the lipid membrane.

Fig. 5. Direct chemiosmotic ligand conduction systems for: (a) mitochondrial oxidative phosphorylation; (b) chloroplast noncyclic photosynthetic phosphorylation. (Information from many sources.[3]) Each system consists essentially of four osmoenzymes translaid in the topologically closed membrane. In 5a the P side is the outside. In 5b the P side is the inside. Ancillary solute porters, antenna pigments and other systems are omitted for simplicity. 5a shows: NADH dehydrogenase (NADH deH), cytochrome c reductase (c reductase), cytochrome c oxidase (c oxidase) and reversible ATPase (F_0F_1). 5b shows: photosystem II (PS II) containing chlorophyll a_{II} (Chl a_{II}), plastocyanin reductase (PC reductase), photosystem I (PS I) containing chlorophyll a_I (Chl a_I) and reversible ATPase (CF_0CF_1). Hydrogen conductors are represented by flavin mononucleotide (FMN), by bound ubiquinone and plastoquinone (Q and PQ), by ubiquinone and plastoquinone pools in the membrane hydrocarbon B domain ([Q] and [PQ]), and by flavoprotein, Fp. Electron conductors are represented by ferredoxin (Fd) and other ironsulphur centres (FeS), by cytochromes (a, a_3, b, c, c_1), by copper centres (Cu) and manganese centres (Mn), and by electron donor pigments P680 and P700 with absorbance maxima at 680 nm and 700 nm respectively. Oxideion conductors are represented by F_1 and CF_1 containing magnesium centres (Mg). Proton conductors are represented by the aqueous domains P and N, by the acid-base relay system (R^-/RH) passing into F_1 and CF_1 through F_0 and CF_0, and possibly by unnamed components in the redox osmoenzymes. Photons are represented by $h\nu$, and the open arrows represent photon driven electron translocation. Broken arrows in PC reductase represent reactions that require further confirmation.

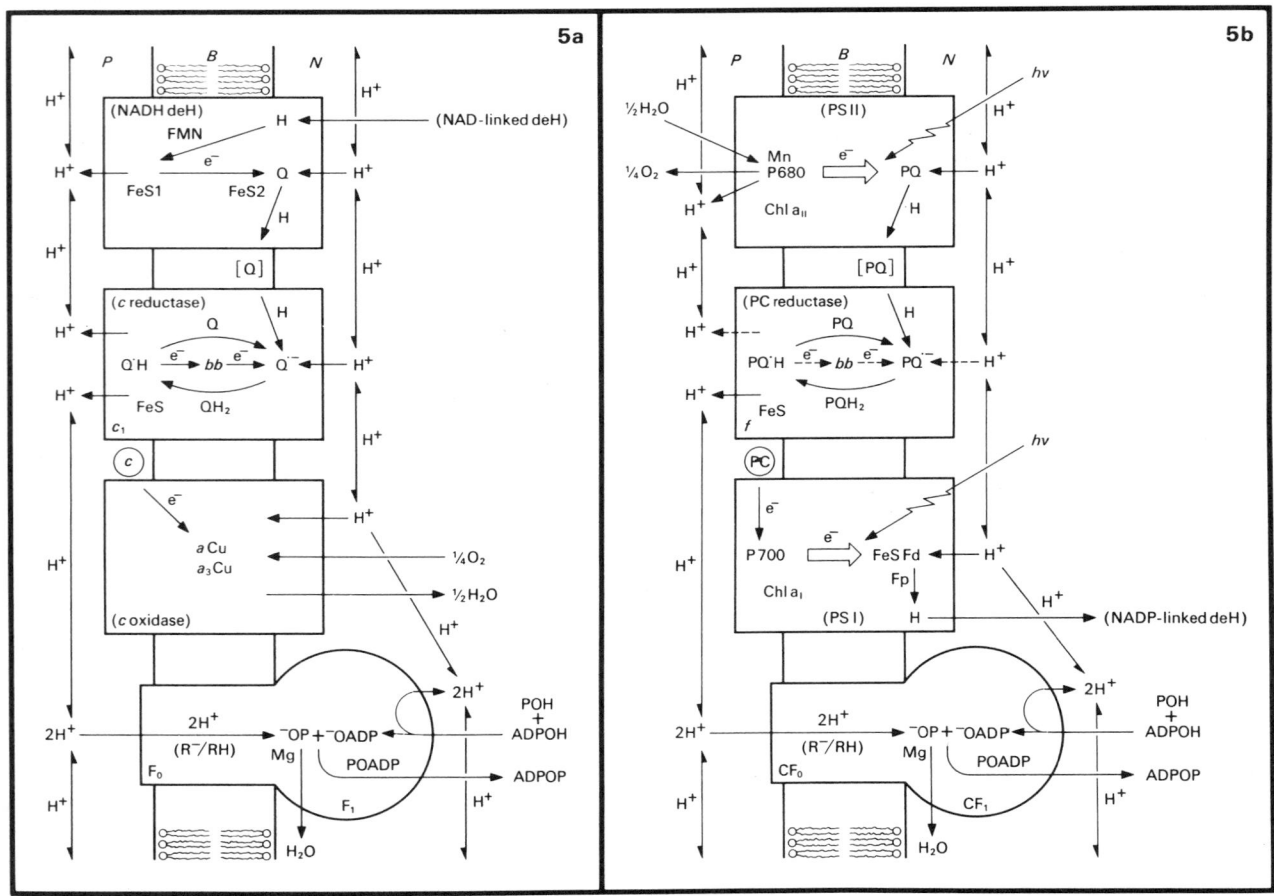

As anticipated by the chemiosmotic hypothesis, the two proton conductors P and N are brought to relatively positive and negative protonic potentials by osmoenzyme molecules of the redox or photoredox chain, which are plugged through the insulating 4 nm thick barrier domain B of the membrane. The proticity thus generated at a potential difference Δp of 200 to 300 mV can be drawn upon for the phosphorylation of ADP by molecules of a reversible protonmotive ATPase, which are also plugged through the B domain of the membrane.

The diagrams in *Fig. 5* interpret and summarise recent experimental information about the mitochondrial respiratory chain (*Fig. 5a*), the chloroplast non-cyclic photoredox chain (*Fig. 5b*), and the reversible ATPases or ATP synthases (F_0F_1 and CF_0CF_1), driven by these systems. Some details in these schemes are still subject to revision and extension, but I think that the general chemical and topological principles are now well established experimentally. The two schemes are remarkably similar.

The protonmotive respiratory chain that oxidises reduced nicotinamide adenine dinucleotide (NADH) in mitochondria (*Fig. 5a*) consists of three osmoenzymes, cytochrome c oxidase (c oxidase), cytochrome c reductase (c reductase) and NADH dehydrogenase (NADH deH), the monomeric forms of which have molecular weights in the range 100 to 600 kilodaltons. Each of these osmoenzymes is made up of a number of polypeptide subunits, and they contain several redox-functional centres that carry electrons or hydrogen atoms or both. They have a dynamically stable translaid orientation through the hydrocarbon B domain of the membrane because the surfaces of their P and N poles are hydrophilic, and the intermediate region of their surfaces is lipophilic. Thus, although they are virtually prevented from turning end-over-end, they are very mobile laterally in the fluid lipid membrane, and they undergo rapid rotational diffusion about axes normal to the surfaces of the B domain. The molecules of c oxidase and c reductase are connected electronically by cytochrome c (c)—a small basic haemoprotein with a large dipole moment—which is present at the P surface of the membrane, and may diffuse rapidly between c oxidase and c reductase by a surface conduction mechanism. The molecules of NADH dehydrogenase and c reductase are electronically or hydrogenically connected by ubiquinone (Q)—a very hydrophobic substituted benzoquinone with a long polyisoprenoid side chain, which is retained by, but is very mobile in, the hydrocarbon B domain. There are about equal numbers of c oxidase and c reductase molecules in the membrane, about 10 times as many Q molecules, but only about a quarter as many NADH dehydrogenase molecules.

Besides NADH dehydrogenase, there are other Q-reducing enzymes in the membrane (not shown in *Fig. 5a*)—notably succinate dehydrogenase, α-glycerophosphate dehydrogenase, choline dehydrogenase and electron-transfer flavoprotein (ETF) dehydrogenase. Unlike NADH dehydrogenase, the latter are not protonmotive themselves—in keeping with the fact that the $(NADH + H^+)/NAD^+$ couple which oxidises moderately reducing substrates of NAD-linked dehydrogenases (NAD-linked deH) in the mitochondria, operates around a redox potential of -300 mV, whereas the substrate couples of the other Q-linked dehydrogenases produce redox potentials only slightly below 0 mV, near the operating

redox potential of the QH_2/Q couple.

The F_0F_1 and CF_0CF_1 ATPase molecules have a molecular weight of about 450 kilodaltons; and they consist of a number of polypeptide subunits. They are more physically asymmetric than the other osmoenzyme molecules, as indicated crudely in *Fig. 5*; and the lipophilic F_0 or CF_0 and hydrophilic F_1 or CF_1 parts have been shown to have separate functions. F_0 or CF_0 is a specific proton-conducting component, and F_1 or CF_1 catalyses reversible oxideionmotive or effectively protonmotive ATP hydrolysis.

In the chloroplast photoredox chain system of *Fig. 5b*, photosystem I (PS I), plastocyanin reductase (PC reductase) and photosystem II (PS II) are respectively somewhat analogous to c oxidase, c reductase and NADH dehydrogenase of the mitochondrial respiratory chain (*Fig. 5a*); and the copper protein plastocyanin (PC), and plastoquinone (PQ) have functions and distributions in the chloroplast photoredox chain system that correspond to those of c and Q respectively in the mitochondrial respiratory chain system. The chloroplast photoredox chain system (*Fig. 5b*) is distinguished from the mitochondrial respiratory chain system in that the electron translocation reactions are driven by photosystems I and II, containing chlorophylls I and II (Chl I and II) and corresponding electron-donor pigments with absorbances at 700 nm and 680 nm (P700 and P680) respectively.

The essential point that I am endeavouring to illustrate in *Fig. 5* is that the observed protonmotive function of the biochemical systems can be explained by the same elementary chemical principle of specific vectorial ligand conduction that is applicable to artificial chemicomotive fuel cell systems. The alternation of hydrogen-conducting and electron-conducting components down the respiratory chain system (*a*), and down the photoredox chain system (*b*), in the looped configurations shown in *Fig. 5* is sufficient to explain, not only what may be the correct protonmotive stoicheiometry, but also many kinetic and thermodynamic characteristics of these interesting systems.[3]

The specific ligand conduction mechanism of the protonmotivated F_0F_1 and CF_0CF_1 ATP synthases outlined in *Fig. 5* depends on the notion of oxideion conduction, as in the artificial protonmotive hydrolase of *Fig. 4b*. No chemical intermediates, other than those directly involving inorganic phosphate and ADP appear to participate in the ATPase or ATP synthase reactions; but catalytic activity is absolutely dependent on magnesium or certain other divalent cations. Therefore, two mechanisms of oxideion conduction (or its equivalent) suggest themselves.[15, 16] First, as indicated in *Fig. 5*, the conduction pathways for phosphate and adenine nucleotides between the N domain and the P-O bond-unlatching active site of F_1 and CF_1 may be specific for given protonation states represented by $PO^- + ADPO^-$ and ADPOP—giving net conduction of O^{2-} in the $N \rightarrow P$ direction during ADP phosphorylation. Second, the magnesium complexed at the active site may facilitate the conduction of the anionic O^{2-} group between the phosphate/phosphorylium (PO^-/P^+) couple and the water/proton ($H_2O/2H^+$) couple at the P side of the active site of F_1 and CF_1. The O^{2-} group might thus be conducted at the P side of the active site as complexes of magnesium (oxide/hydroxide/hydrate) in a Lewis acid and general acid–base type of catalytic mechanism[3, 15, 16] involving proton-conducting, dipole-inducing and other functional groups in the active site domain of the ATP synthase.

Essentially, the chemistry of the Lewis acid and general acid–base catalysis mechanism of ADP phosphorylation outlined in *Fig. 5* involves conduction of protons through F_0 and into F_1 *via* a general acid–base relay system represented by (R^-/RH), successive nucleophilic attacks by two of these ligand-conducted protons on the O^{2-} group that leaves the inorganic phosphate phosphorus centre with both valency electrons, and nucleophilic attack on the remaining positive phosphorus centre by the anionic oxygen of $ADPO^-$.[16] The pulling effect of the ligand-conducted protons, inducing the O^{2-} group to leave the phosphorus centre, may be represented simply as the incipient formation of an oxonium group $(-OH_2^+)$ that would accept an electron from the phosphorus centre and leave as H_2O. But this is not an essential feature of the chemistry of the proposed direct chemiosmotic mechanism.[15] The O^{2-} group may, in fact, leave the phosphorus centre as a complexed OH^- ion (involving magnesium and other ligands in the active site) before attack and conversion to H_2O by the second proton.[16, 17]

I must interject here that many of my biochemical colleagues do not accept this chemically orientated direct chemiosmotic view of the protonmotive mechanisms, but prefer indirect conformationally coupled mechanisms.[18]

It is especially noteworthy in the context of Davy's view of motive power in chemistry that we regard the specific ligand *conduction* property of chemiosmotic systems––both conceptually and in reality––as a spatial extension and a mechanistic evolution of the property of specific ligand *binding* in classical chemistry and biochemistry. As examples of specific ligand *binding*, we have the electron accepting action of haem groups, iron–sulphur centres and copper centres, the hydrogen accepting action of flavin centres and quinone centres, the proton accepting action of acid/base centres, and oxideion accepting action of the $ADPOP/(ADPO^- + PO^-)$ centres and possibly of magnesium centres. But the action of specific ligand *conduction* through osmoenzyme molecules requires additional dynamic topological, physical and chemical specifications that facilitate the diffusion of the ligands through and between appropriately orientated chemical centres, and along uniquely articulated conformationally mobile pathways. Although we are not yet able to characterise these ligand conduction mechanisms in detail, even in the case of electron conduction and proton conduction, we know that, as in enzymic catalysis generally, they are much more subtle and minutely organised, and much more specific and potentially reversible, than the electrode reaction mechanisms and conduction mechanisms in artificial fuel cells.[3]

The minute scale of the natural systems allows the use of relatively short circuit elements, and relatively great contact areas. More than a decade ago it was estimated that the average distance between the respiratory chain and ATPase molecules in the mitochondrial inner membrane would be only about 20 nm, so that, as indicated in *Fig. 5*, the functional proton current in what was called the proton circuit network would flow tangentially to the membrane surface round very short circuit elements between neighbouring proticity producing and proticity consuming molecules.[19]

The protonic conductances of the aqueous P and N domains may well be higher than their bulk conductances,

but the functional proton current would tend to flow by lines of least resistance between the poles of nearest-neighbour proticity producing and consuming molecules, close to the surfaces of the B domain, even if the proton conductance of the P and N domains were isotropic. I must, however, point out that, contrary to some recent suggestions,[20] this localisation of the functional proton current, which is characteristic of the classical chemiosmotic systems, is not compatible with the type of localised protonic anhydride coupling mechanism described by Williams.[17] Williams's mechanism would require the transfer of anhydrous or only partially hydrated protons *via* a non-aqueous 'dislocated phase' situated *in the membrane* to protect it from attack and de-energisation by water.[21] By contrast, the high protonic potential P domain of the chemiosmotic systems is known to be at the same total water potential as the N domain, because the hydrocarbon medium of the B domain is highly permeable to water. Further, the equilibration of the total potential of hydrated protons throughout the aqueous P domain has been shown to be much too fast, relative to the rate of onset of ATP synthesis, to be compatible with the Williams type of localised energy-rich protonic anhydride mechanism.[22] It is relevant that the rapid lateral diffusional motion of the translaid osmoenzyme molecules would tend to smear out local potential differences near the poles of the proticity producing and consuming molecules, and help to achieve reversible and efficient osmotic coupling and energy transfer between the chemiosmotic reactions. For this reason, the close identification of natural chemiosmotic reactions with artificial electrodic ones, in which the current flow is normal to a troublesome macroscopic electrode/electrolyte interface, may be more misleading than helpful to the development of chemiosmotic theory.[20] On the other hand, I think that much may be learned from the natural protonmotive chemiosmotic systems that could be useful for improving the design of the artificial electronmotive electrodic systems.[5, 16] Indeed, Nature's protonmotive power in biology seems to dominate in more ways than one.

Davy's brain, with which he opened the chapter of our understanding of electrochemistry, was powered by proticity. That proticity produced ATP in his nerve cell mitochondria; and that ATP was used by the Na^+/K^+-motive ATPase molecules translaid through his nerve cell membranes; and the transmembrane potential differences of Na^+ and K^+, thus produced, were used to power the nerve impulses required for Davy's thought processes.

Conclusions

In this article, I have concentrated attention on the use of proticity for ADP phosphorylation in mitochondria and chloroplasts. But, for the sake of simplicity and brevity, I have omitted to mention important protonically coupled ancillary systems which, especially in mitochondria, are involved in the conduction of inorganic ions, organic substrates, and ATP and ADP, *via* solute-specific porter molecules translaid through the B domain of the membrane.[3] To indicate the far-reaching scope of Nature's protonmotive power in biology, I must just mention that there are particularly interesting bacterial redox chain, photoredox chain, pyrophosphatase, solute porter and other protonically coupled systems. Moreover, proticity is used as a very widespread energy currency, providing power for a variety of systems, not only in bacteria, mitochondria and chloroplasts, but also in a number of other organelles that are present in modern eukaryotic cells, and which, like mitochondria and chloroplasts, appear to be the distant descendants of ancient bacteria that have become domesticated and enslaved within their modern eukaryotic masters. Perhaps the two most fascinating adaptations of the use of proticity are: for protic heating, by a sensitively controlled short-circuiting system in the mitochondria of brown fat cells in hibernating animals; and for protic locomotion, in which a reversible rotatory protic motor is used to propel certain bacteria by rotating a helical flagellum like the screw of a ship.[3,16]

One wonders who has been the more imaginative—Davy or Nature? Perhaps it is all part of the same partially accidental inventive evolutionary process.

Just over a century ago, the great Dutch chemist van't Hoff gave a delightful lecture on the subject of imagination in science.[23] To end my tribute to Davy, I cannot do better than quote van't Hoff's words in praise of Davy's genius. 'His discoveries were the fruits of that great gift which Buckle describes: "There is a spiritual, a poetic, and for aught we know a spontaneous and uncaused element in the human mind, which ever and anon, suddenly and without warning, gives us a glimpse and a forecast of the future, and urges us to seize the truth as it were by anticipation."'

References

1. H. Hartley, *Humphry Davy*. London: Nelson & Sons, 1966.
2. K. R. Popper, *Objective knowledge: an evolutionary approach.* Oxford: Clarendon, 1972.
3. P. C. Hinkle and R. E. McCarty, *Sci. Am.*, 1978, **238**, 104; P. Mitchell, *Eur. J. Biochem.*, 1979, **95**, 1; P. Mitchell, *Science*, 1979, **206**, 1148; P. Mitchell, in *Membrane bioenergetics* (C. P. Lee and G. Schatz eds).), p 361. Reading, Mass: Addison-Wesley, 1979; P. Mitchell, *Ann. N.Y. Acad. Sci.*, 1980, **341**, 564.
4. W. R. Grove, *Phil. Mag. Ser. 3*, 1839, **14**, 127.
5. P. Mitchell, *Fed. Proc.*, 1967, **26**, 1335.
6. P. Mitchell, *FEBS Symp.*, 1972, **28**, 353.
7. P. Curie, *J. Phys. 3ème Ser.*, 1894, 393.
8. H. A. Liebhafsky and E. J. Cairns, *Fuel cells and fuel batteries.* New York: Wiley, 1968.
9. P. Mitchell, *Symp. Soc. Gen. Microbiol.*, 1977, **27**, 383.
10. E. A. Guggenheim, *Modern thermodynamics by the methods of Willard Gibbs.* London: Methuen, 1933.
11. P. Mitchell, *Chemiosmotic coupling and energy transduction.* Bodmin, England: Glynn Research, 1968; P. Mitchell, *Theor. Exp. Biophys.*, 1969, **2**, 159.
12. J. O'M. Bockris and A. K. N. Reddy, *Modern electrochemistry, Vols. I and II.* Cambridge: C U P, 1970.
13. W. Nernst, *Z. Electrochem.*, 1899, **6**, 41; W. Nernst and W. Wald, *Z. Electrochem.*, 1900, 7, 373.
14. C. Wagner, *Naturwiss.*, 1943, **31**, 265.
15. P. Mitchell, *FEBS Lett.*, 1974, **43**, 189; P. Mitchell, *FEBS Lett.*, 1975, **50**, 95.
16. P. Mitchell, *Biochem. Soc. Trans.*, 1976, 4, 399.
17. P. Mitchell, *FEBS Lett.*, 1977, **78**, 1.
18. For further discussion and references, see (3, 16, 17).
19. P. Mitchell, *Biol. Rev. Cambridge Philos. Soc.*, 1966, **41**, 445; P. Mitchell, *Nature (London)*, 1967, **214**, 1327.
20. D. B. Kell, *Biochem. Biophys. Acta*, 1979, **549**, 55; E. C. Slater, *Trends in Biological Sciences*, 1980, **5**, No. 5.
21. R. J. P. Williams, *Curr. Top. Bioenerg.*, 1969, **3**, 79; R. J. P. Williams, *FEBS Lett.*, 1975, **53**, 123; R. J. P. Williams, *FEBS Lett.*, 1979, **102**, 126.
22. W. Junge, W. Ausländer, A. J. McGeer and T. Runge, *Biochim. Biophys. Acta*, 1979, **546**, 121; J. W. Davenport and R. E. McCarty, *Biochim. Biophys. Acta*, 1980, **589**, 353; C. Vinkler, M. Avron and P. D. Boyer, *J. Biol. Chem.*, 1980, **255**, 2263.
23. J. H. van't Hoff, *Imagination in science*. An inaugural lecture given as Professor of Chemistry, Mineralogy and Geology at the State University of Amsterdam in 1878. Translated by G. F. Springer. *Mol. Biol. Biochem. Biophys.*, 1967, **I**.

PHASE SEPARATION, CHARGE SEPARATION AND BIOGENESIS

HAROLD J. MOROWITZ

Yale University New Haven, Connecticut, U.S.A.

(Received July 15th, 1980)
(Revision received January 2nd, 1981)

The current view of bioenergetics postulates transmembrane charge separation as a primary mechanism of energy storage and transformation. Using that bioenergetic view we examine possible methods of photon driven transduction in primordial vesicles. Two possible types of proton pumps are analyzed and a method of anaerobic photophosphorylation is discussed. Using these principles we theorize about the formation of prebiotic photochemical vesicles utilizing the same transmembrane energy conversions characteristic of contemporary cellular systems.

The strong evidence for a transmembrane difference of electrochemical potential of hydrogen as a necessary intermediate in the biological utilization of redox energy (Boyer et al., 1977) allows us to reexamine the origins of photosynthesis and life from new perspectives. This paper will deal with certain features of the physics of the transformation of electromagnetic energy into the chemical potential energy of reasonably stable compounds. These ideas will be applied to certain aspects of biogenesis to examine that subject in light of our newer knowledge of bioenergetics.

Photochemical energy transduction may be divided into homogeneous phase and heterogeneous phase processes. In the former the absorption of photons is followed by a sequence of reactions leading to product molecules which may store some of the original electromagnetic energy as chemical potential. This is classical photochemistry and accounts for much of the literature of that subject. Biological transductions appear not to utilize this type of mechanism, but employ heterogeneous phase processes. We must either assume that there has been an evolutionary change in the type of photochemistry carried out or the earliest system used the heterogenous mode. On the principle of organic and evolutionary continuity, in the absence of evidence to the contrary, we assume the second method.

Heterogeneous phase photochemical reactions allow for interfacial phenomena, charge separation and chemical species separation. These phenomena require a minimum of two phases and may involve a third phase separating two others. The model system of particular interest is the bimolecular leaflet membrane vesicle separating two aqueous phases. This embodies most general features of heterogeneous phase reactions and relates directly to the type of processes known to occur in cells and organelles.

There have been a number of suggestions in recent years that the early Earth's oceans were covered with a hydrocarbon layer (Lasaga et al., 1971). Current views favor a less reducing atmosphere but still allow for the presence of some primordial hydrocarbon. The irradiation of an oil-water system by short wavelength ultraviolet can lead to amphiphilic molecules and bilayer formation (Folsome and Morowitz, 1969). Onsager (1974) has discussed the formation of vesicles as probable local energy minima in an aqueous amphilphilic system. Thus, primordial protocells could have formed spontaneously, providing a milieu for heterogeneous phase, photochemical reactions. The vesicle provides aqueous phases separated by a barrier of low dielectric constant.

For purposes of examining the heterogeneous case, consider a bimolecular leaflet of width L between two aqueous phases. Electrochemical processes can occur if charged species are carried across the membrane or the membrane itself becomes polarized. If the membrane occurs as a closed vesicle we can distinguish an inner and outer phase. Asymmetries occur in the membrane since the inner and outer radii of curvature vary by the distance L. Therefore any asymmetric molecule dissolved in the bilayer would be expected to have a statistically preferred orientation.

In the model system under discussion we assume that the heterogeneous phase process will lead to macroscopic potentials by transport across the membrane from one phase to another of one of the following entities: (1) electrons; (2) protons; and (3) large ions.

The migration of electrons can take place by metallic conduction and electron hopping from donor to receptor. In addition, electrical polarization can follow from the redistribution of electronic charge within a covalently bonded molecule. Metallic conduction in general requires a regular three dimensional array, that is, it would require a metal or metal-like phase embedded in the membrane. Since such conductors have not been reported, we can at this stage shelve our discussion of normal metallic conductance. The possibility of electronic conductance by conjugated carbon chains such as carotenoids must also be considered. The conductive properties of interest are likely to be those of individual molecules or small groups of molecules embedded in a membrane.

The hopping of an electron from a donor to a receptor is normally described as an oxidation reduction process. In homogeneous phase chemical reactions the transfer occurs on collision and there are no macroscopic electrical asymmetries associated with the electron transfer. If the donor and receptor are fixed in space then an electrical asymmetry must be associated with the oxidation reduction reaction. The migration of electrons is either by tunneling or mobile carriers between the centers. In fixed solid state systems the dominant process is tunneling.

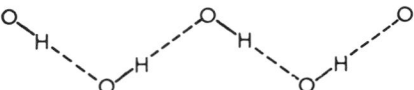

Fig. 1. A proton channel consists of a contiguous chain of hydrogen bonds. The conduction of protons along such a chain is discussed in detail in Nagle and Morowitz (1978).

While electron transfer has been studied in electrochemical systems for a long time, the chemiosmotic hypothesis has led to considerations of mechanisms of proton transfer (Morowitz, 1978). The general mechanisms in hydrogen bonded crystals such as ice consists of a two-step process of an ion migration followed by a fault migration (Nagle and Morowitz, 1978). This transfers a proton and returns the system to its original configuration. The process, shown in Fig. 1, was first discovered in the anomalous conductance of ice. The minimum requirement for this type of process is a continuous chain of hydrogen bonds in some kind of regular array. The occurrence of hydrogen bonds is a major feature of biological macromolecules, examples being base pairing in DNA and the role of hydrogen bonds in stabilizing the α-helical configurations in polypeptides. A number of models have been introduced to demonstrate how regular chains of hydrogen bonds might exist in protein molecules of the appropriate amino acid sequence.

Proton mobility over an appreciable spatial domain occurs in microscopic proton semiconductors which have been referred to as proton wires. Such devices make possible protochemical processes which are analogous to electrochemical processes in the following way:

	Proto-chemistry	Electro-chemistry
Mobile element	Proton	Electron
Chemical reaction	Acid-base dissociation	Oxidation-reduction

What emerges in bioenergetics is the continuous coupling of electrochemical and protochemical processes, a phenomenon to which we shall later return.

In addition to proton conductors, there have been a number of reported cases of inorganic solids with high ionic conductance (Farrington and Briant, 1979). Yao and Kummer (1967), for example, report a very high sodium ion conductivity in crystalline sodium beta aluminum. There are three types of processes which are normally considered to describe ion migration. In the vacancy model, a hole (missing ion in a crystal) migrates through successive ions hopping in a direction opposite to the vacancy motion. The interstitial model involves an excess ion which migrates from one interstitial site to another. The interstitialcy process involves an interstitial ion moving into a lattice site in a cooperative motion with a lattice ion which moves into an interstitial site. While ionic conductances of this type have not been demonstrated in biological systems, it remains an interesting subject for investigation.

With the preceding as background we shall make a series of assumptions about biogenesis and early photosynthetic systems and then work out some consequences of those assumptions.

I. Active living systems, indeed any unstable far from equilibrium system, require a continuous energy flux.

II. At present that energy is supplied to the global ecosystem almost entirely by solar radiation.

III. That energy source has been continuous and supplied the energy for the earliest cellular system.

IV. Transmembrane charge separation is a primitive mode of energy storage.

V. The primary physical problem in prebiotic energy transduction is the coupling of photon absorption with charge transfer. Upon absorption of a photon many chromophores are in excited states with high dipole moments (Barltrop and

Coyle, 1975; Shablya, 1968; Matthies and Stryer, 1976). This charge asymmetry may interact with chains of hydrogen bonds in the following ways:

1. The shift of electrons away from an atom bonded to a dissociable hydrogen will lower the pK. The group may become sufficiently acidic to dissociate and inject the proton into a hydrogen conducting channel if one is present in the immediate neighborhood.

2. The electrostatic field of the excited chromophore might pump protons if the field has a component parallel to the axis of the proton channel.

The first method has been discussed by Warshel (1979) in postulating a model of *Halobacterium halobium* proton pumping. The conducting channel is not specified. In our treatment we shall assume that there exist in the membranes chains of contiguous hydrogen bonds capable of ice-like proton conductance.

Since our purpose is to show how photon absorption can be coupled to proton pumping, we consider two abstract cases shown in Figs. 2 and 3.

In each case a chromophore of length P absorbs a photon and experiences a change of dipole moment of μ. This is equivalent to a formal charge of μ/P at one end of the molecule and $-\mu/P$ at the other end. In the first case there is a proton with a high pK, located a distance p from the positive end of the dipole and at a terminus of a proton

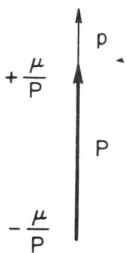

Fig. 2. A photon induced dipole moment μ is shown a distance p from a dissociable group. Upon induction of the dipole there is a pK drop.

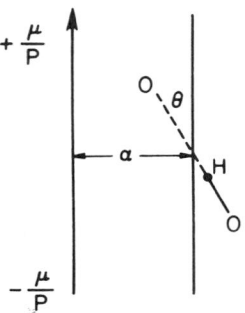

Fig. 3. A photon induced dipole lying parallel to a proton conducting channel. The picture shows a single H—O- - - -H linkage.

TABLE 1

ΔpK calculated as a function of the dielectric constant, κ, and the separation distance p

κ/p	1	1.5	2	2.5	3	3.5	4.0
2	23.6	15.3	11.1	8.6	7.0	5.8	5.0
3	15.8	10.2	7.4	5.8	4.7	3.9	3.3
4	11.8	7.6	5.6	4.3	3.5	2.9	2.5
5	9.5	6.1	4.4	3.5	2.8	2.3	2.0
6	7.8	5.1	3.7	2.1	2.3	1.9	1.7
7	6.8	4.4	3.2	2.5	2.0	1.7	1.4
8	5.9	3.8	2.8	2.2	1.8	1.5	1.2
9	5.2	3.4	2.5	1.9	1.6	1.3	1.1
10	4.7	3.1	2.2	1.7	1.4	1.2	1.0

conducting chain. We shall calculate the change in pK of that proton which is a measure of the tendency to inject it into the chain and thus pump protons. The electrostatic potential experienced by the dissociable proton is

$$V = \frac{1}{4\pi\epsilon_0} \frac{\mu}{\kappa P} \left(\frac{1}{p} - \frac{1}{P+p} \right)\epsilon$$

where $1/4\pi\epsilon_0$ is the electrical permittivity of vacuum, κ is the dielectric constant, and ϵ the elementary charge. Since

$$pK = \frac{\Delta G}{2.3RT}$$

and NV, the product of Avagodro's number and the voltage, measures the change in ΔG, then

$$pK = \frac{-V}{2.3RT}$$

$$= \frac{1}{2.3kT} \left(\frac{1}{4\pi\epsilon_0} \right) \left(\frac{\mu\epsilon}{\kappa P} \right) \left(\frac{1}{p} - \frac{1}{P+p} \right)$$

$1/4\pi\epsilon_0$ has a value of 9.0×10^9 Nm²/C²; ϵ has a value of 1.60×10^{-19} C. We will choose P as $15A$ and μ as 15 Debye units. These are reasonable values based on experimental evidence for a molecule like retinal. The value of μ/P is then 0.33×10^{-19} C. Boltzmann's

constant, k, is 1.381×10^{-23} J/K, and T will be taken as 300 K. We then get

$$\Delta pK = \frac{-50.4}{\kappa} \left(\frac{1}{p} - \frac{1}{15+p} \right)$$

where p is given in Angstrom units. It is difficult to estimate κ or p with precision. In non-aqueous membrane interiors the dielectric constants will range from 2 to 10. We calculate ΔpK-values for a range of parameters. The results are given in Table 1.

The large decreases in pK shown in the table could make adjacent groups considerably more acidic and could act as a mechanism of proton injection if a conducting channel is available.

In the second model we assume that the conducting element consists of a continuous set of hydrogen bonds shown in Fig. 3.

The O—H bond distance is assumed to be $0.97A$, the O—O distance is assumed to be $2.64A$ and the H—O—H angle is taken as $111°28'$. The potential energy diagram of a proton is shown in Fig. 4.

Consider the hydrogen bond in the center of the chain. We are interested in calculating the potential difference between the possible proton loci due to the dipolar field shown in the previous problem. We establish the origin of the coordinate system at a point midway between the potential minima shown in Fig. 4.

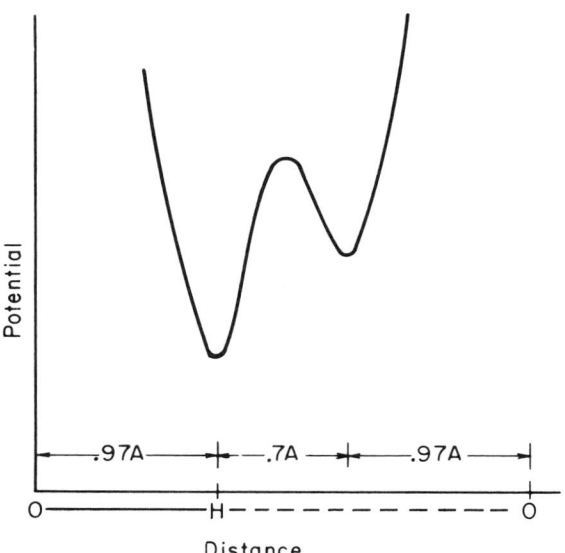

Fig. 4. The potential of a proton lying between two oxygens. The proton located in the left well is covalently bonded to the left oxygen and hydrogen bonded to the right oxygen. When it is in the right well an ion pair is formed with a negative charge associated with the left oxygen and a positive charge with the right.

The important coordinates then are

	x	y
Positive charge	$-\alpha$	$P/2$
Negative charge	$-\alpha$	$-P/2$
Primary proton position (I)	0.20	0.29
Secondary proton position (II)	-0.20	-0.29

The difference in potential between the two positions is

$$\Delta V = \frac{1}{\kappa}\left(\frac{1}{4\pi\epsilon_0}\right)\left(\frac{\mu\epsilon}{P}\right)$$

$$\times\left[\left(\frac{1}{\sqrt{(\alpha+0.20)^2+\left(\frac{P}{2}+0.29\right)^2}}\right)\right.$$

$$-\frac{1}{\sqrt{(\alpha+20)^2+\left(\frac{P}{2}-0.29\right)^2}}$$

$$-\frac{1}{\sqrt{(\alpha-0.20)^2+\left(\frac{P}{2}-0.29\right)^2}}$$

$$\left.+\frac{1}{\sqrt{(\alpha-0.20)^2+\left(\frac{P}{2}+0.29\right)^2}}\right]$$

If we assume an α of $2.5A$ and κ of 2 we get ΔV values of the order of kT. The results are quite sensitive to α and P and values of up to $10kT$ are possible. For each kT increment we bias the hydrogen bond to the ion pair formed by a factor of e. The formation of such ionic configurations is the beginning of proton pumping. The migration of the pair members is further biased by the dipole field, thus leading to further pumping. In any case at least two mechanisms exist which can convert the absorption of a photon into proton pumping.

The experimental model of an elementary photo-driven proton pump is the purple membrane of Halobacterium halobium. In this system the active element is bacterial rhodopsin, which consists of a relatively small protein (mol. wt 26 000) complexed to a molecule of retinal. Our previous analysis presents possible modes of operation. The further study of this system plus more detailed electrostatic calculations of charge distributions and associated energies should allow us to propose more realistic modes of prebiotic proton pumps.

The back flow of protons from a high potential to a low potential can be coupled to the protochemical synthesis of phosphate esters (Morowitz, 1978). Polyphosphates are ideal compounds for this type of coupling because of the number of dissociable groups

TABLE 2

pK values of ortho-, pyro- and tetraphosphates

	Ortho-phosphate	Pyro-phosphate	Tetra-phosphate
pK_1	2.14	1	Completely dissociated
pK_2	7.10	2	1
pK_3	12.32	6.57	2.30
pK_4		9.62	6.26
pK_5			8.90

with pK-values in the range between strong acids and strong bases. This may be seen in Table 2. The mechanism we are invoking is a generalization of those used in oxidative phosphorylation and photophosphorylation. It involves the phosphates being protonated at a binding site attached to a transmembrane proton conducting element. The phosphate bond is effectively formed at the high proton activity characteristic of the high potential side of the membrane. When the polyphosphate is released it dissociates, releasing protons. The negative free energy of this acid-base dissociation sums with the positive free energy of the phosphate ester formation, resulting in a zero or negative total free energy. This drives the polyphosphate synthesis reaction by the transport of protons from a high chemical potential to a low chemical potential.

We now set forth a scenario of prebiotic events.

(1) On the early planet there existed appreciable quantities of hydrocarbons of sufficient molecular weight to exist in the liquid phase. These were distributed so as to generate appreciable hydrocarbon-water interface.

(2) The flow of energy (solar UV, lightning, volcanoes, etc.) resulted in the synthesis of numerous compounds including amino acids, ring compounds and other compounds with π-electrons, amphiphilic molecules, and heterocyclic rings. Because of the presence of water it is certain that phosphorus existed in the form of phosphate.

(3) The amphiphilic molecules condensed in a wide variety of colloidal forms including closed bilayer vesicles which partitioned the interior and exterior phases.

(4) A number of molecules with π-electrons dissolved in the interior of the bilayers. Their distribution was asymmetric due to the spatial inhomogeneity of the curved vesicle elements.

(5) Due to this asymmetry the absorption of photons led to transmembrane electrical potentials and in favorable cases to ion transport.

(6) Some vesicles were spanned by structures lipophilic on the outside but stabilized by contiguous hydrogen bonds in the cores. Such molecular aggregates are hydrogen channels or proton wires and can convert the back flow of protons into the synthesis of phosphate esters.

(7) The system we have described is an anaerobic photo-phosphorylator. Under appropriate conditions it will lead to the continuous synthesis of pyrophosphate from orthophosphate within the interior of the vesicles. The fact that this principle energy input is pyrophosphate will structure the emerging chemical network with the selection of a subset of reactions dominated by phosphate transfers.

At this point our scenario comes to an end, hovering on the brink of biogenesis. It inverts the usual view and moves from energetics to genetics. This has the distinct advantage of removing vast randomness from the early chemistry and forcing the kind of biochemistry that developed by the specific form of the energy input. It is a view of biogenesis that takes due cognizance of what we now know about bioenergetics.

References

Barltrop, J.A. and J.D. Coyle, 1975, in Excited States in Organic Chemistry (John Wiley and Sons).

Boyer, P.D., B. Chance, L. Ernster, P. Mitchell, E. Racker and E.C. Slater, 1977, Annu. Rev. Biochem. 46, 955.

Farrington, G.C. and J.L. Briant, 1979, 204, 1371.

Folsome, C.E. and H.J. Morowitz, 1969, Space Life Sci. 1, 538.

Lasaga, A.C., H.D. Holland and M.D. Dwyer, 1971, Science 174, 53.

Matthies, R. and L. Stryer, 1976, PNAS 73, 2169.

Morowitz, H.J., 1978, Am. J. Physiol. 235, R99.

Nagle, J.F. and H.J. Morowitz, 1978, PNAS 75, 298.

Onsager, L., 1974, in Qunatum Statistical Mechanics in the Natural Sciences (Plenum Publ. Corp.).

Shablya, A.V., 1968, in Elementary Photo-processes in Molecules. B. Neporent (ed.) Consultants Bureau.

Warshel, A., 1979, Photochem. Photobiol. 30, 285.

Yao, Y.-F. and J.T. Kummer, 1967, J. Inorg. Nucl. Chem. 29, 2453.

Photodriven transmembrane charge separation and electron transfer by a carotenoporphyrin–quinone triad

Patrick Seta*, Elisabeth Bienvenue*, Ana L. Moore*†‡, Paul Mathis‡, Rene V. Bensasson§, Paul Liddell†, Peter J. Pessiki†, Anna Joy†, Thomas A. Moore*†‡ & Devens Gust†§

* Laboratoire de Physico-Chimie des Systèmes Polyphasés, Associé au CNRS (UA. 330), Route de Mende, BP 5051, 34033 Montpellier Cedex, France
† Department of Chemistry, Arizona State University, Tempe, Arizona 85287, USA
‡ Département de Biologie, CEN Saclay, 91191 Gif-sur-Yvette Cedex, France
§ Laboratoire de Biophysique, INSERM U.201, ERA 951 du CNRS, Muséum National d'Histoire Naturelle, 61 Rue Buffon, 75005 Paris, France

Photochemical transmembrane electron transfer processes are an integral part of natural photosynthetic solar energy conversion and are also central to the design of biomimetic energy conversion schemes[1-6]. Here we report the synthesis and membrane-associated photoelectrochemical properties of carotenoporphyrin–quinone triad (I), a compound containing a photochemically active porphyrin and electron donor and acceptor moieties, and with the molecular architecture necessary to span a phospholipid bilayer. On excitation of compound 1 by visible light, charge is separated across a planar phospholipid bilayer membrane (BLM) in an intramolecular step; in the presence of suitable electron donor and acceptor species in the aqueous phases, a steady-state photocurrent is observed in an external circuit bridging the BLM. Artificial membranes containing I thus mimic key features of the photodriven transmembrane electron transfer processes characteristic of photosynthetic organisms.

Photodriven charge separation in single-phase solution in compounds similar to I has been shown to be a two-step process involving electron transfer from the photoexcited porphyrin singlet state to the attached quinone, followed by electron transfer from the carotenoid to the porphyrin radical cation[7-10]. The resulting species, $C^{+\cdot}-P-Q^{-\cdot}$, is formed within ~100 ps of excitation and has a solvent-dependent lifetime on the microsecond timescale.

Triad I and its related forms II and III (see above) were synthesized by routes similar to those used previously[7,11]. Their structures were verified using 500-MHz ^{1}H-NMR spectroscopy[12].

The photoelectrochemical experiments were carried out in the apparatus described previously[13] (see Fig. 1a). The triad was dissolved in the appropriate membrane-forming solution (see Fig. 2 legend) and a small drop applied to the orifice in the Teflon divider. The phospholipid solution gradually thinned and formed a bilayer, as detected by coulostatic measurements[13]. The current flowing through the cell, and therefore through the BLM, was detected as a function of time and displayed on a digital oscilloscope. In a typical experiment with triad I in the membrane, 1×10^{-2} M $Fe(CN)_6K_3$ was dissolved in the aqueous phase on side 1 and 1×10^{-2} M ascorbic acid was placed on side 2. In the dark, no significant current was observed. However, irradiation of the BLM with 600-nm light from the laser produced an immediate photocurrent (Fig. 2). Typically (Fig. 2a), the photocurrent consisted of a transient component that rose with the opening time of the shutter (<1 ms) and quickly decayed to a steady-state value (typically 50–500 pA). The peak of the transient current and the steady-state value varied considerably between samples. Generally, there was a strong positive correlation between the extent of bilayer formation, as estimated from the coulostatic measurements, and the magnitude of the photocurrent. Moreover, on average, photocurrents obtained when the membrane-forming solution was prepared with hexadecane were larger than those when decane was used. This observation is consistent with the fact that solvents comprising hydrocarbon of longer chain length yield thinner membranes[14]. Figure 2b shows a photocurrent in which the transient component was not observed; such signals were less common than the first type, but occurred more frequently when the aqueous phases were at low pH and when the membranes were more than 10 min old and very thin (that is, when hexadecane was used as the solvent).

The currents discussed above are clearly the result of photoinduced electron transport across the phospholipid bilayer. We propose the following explanation for the results. The triad is a rod-like, amphipathic molecule that would be expected to lie in the bilayer as shown in Fig. 1b. The hydrocarbon-like carotenoid moiety (which extends ~39 Å from the porphyrin centre) should be well anchored in the low-dielectric membrane interior and should be aligned to some extent with the fatty-acid chains. The polar dipyridyl porphyrin moiety, on the other hand, should lie in the higher-dielectric region near the head groups. This is particularly the case at pH 3, where one would assume that the dipyridyl porphyrin was positively charged. Because of its short connecting link to the porphyrin, the quinone must also reside near the membrane surface. In addition, the membrane is asymmetric by virtue of the fact that even though the triad is added to the membrane-forming solution in the hydroquinone form II (because it is more soluble), the addition of $Fe(CN)_6K_3$ to the solution on side 1 (Fig. 1a) converts the molecules having their hydroquinone moieties at the side 1 aqueous interface to the quinone form. Those molecules having their hydroquinone moieties at the side 2 interface remain in that oxidation state. (This interfacial oxidation was demonstrated in bulk two-phase mixtures of the membrane-forming solution and aqueous ferricyanide, using chromatographic detection. No detectable decomposition of the pigments in such bulk mixtures was noted even after several hours of contact.)

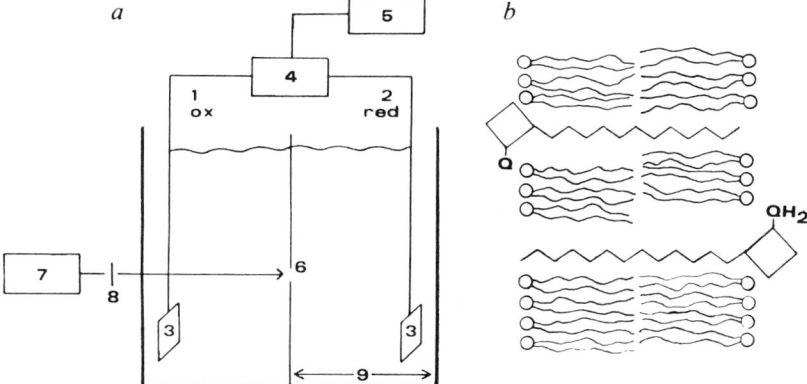

Fig. 1 a, Schematic diagram of a BLM cell for photoelectrical measurements: 1, side 1 containing oxidant ($Fe(CN)_6K_3$), phosphate buffer and NaCl; 2, side 2 containing reductant (sodium ascorbate), phosphate buffer and NaCl; 3, Ag/AgCl electrode; 4, Keithley model 427 current-to-voltage converter; 5, Tektronix digital storage oscilloscope; 6, 1-mm-diameter orifice in thin Teflon divider; 7, laser; 8, shutter; 9, cell machined from block of Teflon. b, Schematic diagram showing the postulated arrangement of triads in the phospholipid bilayer. Note that triad I (labelled Q) is shown at the aqueous interface on side 1 and triad II (QH₂) at the interface on side 2.

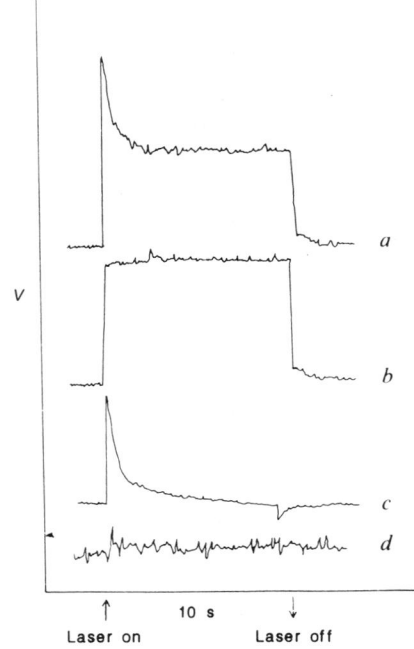

Irradiation of a triad quinone I in such a bilayer is postulated to produce the $C—P^{+\cdot}—Q^{-\cdot}$ charge-separated state, and ultimately the $C^{+\cdot}—P—Q^{-\cdot}$ state, as observed in similar monophasic systems[7-10]; this would lead to a quinone radical anion near the aqueous interface on side 1, and a carotenoid radical cation in the interior of the bilayer. The quinone radical anion then donates an electron to ferricyanide to regenerate the neutral quinone, and the carotenoid radical cation, which can nearly span the bilayer, can accept an electron from the ascorbate on side 2 to regenerate the neutral carotenoid. The net result is the vectorial transfer of an electron from side 2 to side 1 and the regeneration of the triad photocatalyst. The charge separation across the membrane is countered by electron flow through the electrodes and detected as current in the external circuit.

There are several lines of evidence supporting the above interpretation of the experimental results. First, laser-flash photolysis of phosphatidylserine (or ethanolamine) liposomes containing compound I with 590-nm light resulted in a transient species having a maximum wavelength (λ_{max}) of 970 nm; the spectrum of this transient confirmed that it was indeed the carotenoid radical cation. Thus, the postulated charge separation does occur in the phospholipid bilayer. Furthermore, when the electron donor (ascorbate) was omitted from side 2, large transient photocurrents were observed on illumination, but a steady-state photocurrent was never detected (Fig. 2c). Similarly, no steady-state current was observed in the absence of ferricyanide. In addition, no transmembrane electrical response was detected for triad III in the conditions used for I (Fig. 2d). In III, the hydroquinone moiety is protected as the dimethyl ether, and formation of the intramolecular charge-separated state is precluded. These three control experiments confirm the involvement of the transmembrane charge-separated state of triad I and the aqueous-phase redox couples in the observed steady-state photocurrents.

A final piece of evidence concerns the turnover number of the triad in the BLM. If the proposed scheme is indeed correct, then each triad I molecule should be capable of transporting many electrons across the bilayer. At the excitation wavelength of 600 nm, the porphyrin absorption cross-section is $\sim 1.7 \times 10^{-17} cm^2$ per molecule and the laser intensity was $\sim 3 \times 10^{18}$ quanta per s cm^2, yielding ~ 50 excitations per s molecule. From the composition of the membrane-forming solution, there are an estimated 10^{10} molecules per BLM and therefore 5×10^{11} excitations per s in the BLM. At a quantum yield of $\sim 10\%$ (refs 7-10), this corresponds to a photocurrent of ~ 4 nA, a value recorded in some of the BLM experiments. Although very approximate, this result is certainly consistent with a steady-state process having a high turnover number and capable of operating for many seconds.

Note that this example of transmembrane electron transfer is essentially the light-driven catalysis of the thermodynamically spontaneous oxidation of ascorbate by ferricyanide ion. None of the intramolecular redox potential of the charge-separated

state $C^{+\cdot}—P—Q^{-\cdot}$ is conserved as transmembrane chemical potential. However, photocurrents opposing the applied electrical potential were observed even when side 1 was 100 mV negative to side 2. The judicious choice of redox couples and a properly asymmetrized membrane should ultimately make it possible to achieve net energy conservation in this system.

In photosynthetic membranes, carotenoids prevent the sensitization of oxygen by chlorophyll triplet-state species and thereby perform a photoprotective function[15,16]. These carotenoporphyrin triads have sufficient electronic coupling between the carotene and porphyrin moieties[11], via the amide linkage, to mediate extremely rapid triplet–triplet energy transfer $(C—^3P—Q \rightarrow {}^3C—P—Q)^9$. Therefore, even though excitation of the hydroquinone triads results in a high yield of porphyrin triplet, energy transfer to the attached carotenoid occurs on the nanosecond timescale and precludes singlet oxygen formation. These artificial membranes are thus photoprotected in precisely the same way as native photosynthetic membranes.

The photodriven charge separation by carotenoporphyrin-quinone triads across artificial membranes, as demonstrated here, serves as a paradigm for many transmembrane electrochemical events in energy-transducing biological membranes, and will allow detailed investigations of photodriven charge separation across both chemical and electrical potentials, which should in turn lead to a better understanding of photosynthetic energy conversion.

Fig. 2 Transmembrane photocurrents observed on irradiation of triads in BLM systems. *a*, pH 6.5, peak current 200 pA, 1×10^{-2} M ascorbate in side 2 and 1×10^{-2} M $Fe(CN)_6K_3$ in side 1; *b*, pH 2.7, current 150 pA, same redox couples as in *a*; *c*, pH 5, peak current 29 pA, oxidizing couple as in *a*, no reductant in side 2, hexadecane as the solvent; *d*, triad III, noise peaks are ~ 5 pA, conditions as in *a*. The membrane-forming solutions used phosphatidylserine or phosphatidylethanolamine (3 mg), decane (100 μl) and ~ 0.05 mg of triad. In each case the membrane showed substantial thinning, as measured by coulostatic response, greater than 5 GΩ d.c. resistance, and a dark current of <5 pA. Each aqueous phase contained 0.1 M NaCl, 1×10^{-3} M phosphate buffer and a freshly prepared Ag/AgCl electrode. The sign of the observed currents is consistent with a negative charge carrier flowing from side 2 to side 1 through the bilayer. In these experiments there was no pH gradient (ΔpH) across the membrane. For *a*, *b* and *c*, triad II was used in the membrane-forming solution as described in the text.

This material is based on work supported by the NSF under grants CHE 8209348 and INT 8212583 (CNRS–NSF Exchange Program), CNRS grant 3064 and NATO grant RG.083.81. We thank Hoffman-La Roche, Inc. for a gift of β-apo-8'-carotenal.

Received 26 April; accepted 18 June 1985.

1. Dutton, P. L., Prince, R. C. & Tiede, D. M. *Photochem. Photobiol.* **28**, 939–949 (1978).
2. Sauer, K. *Acc. chem. Res.* **11**, 257–264 (1978).
3. Mathis, P. & Paillotin, G. in *The Biochemistry of Plants* Vol. 8 (eds Hatch M. D. & Boardman, N. K.) 97–161 (Academic, New York, 1981).
4. Calvin, M. in *Photochemical Conversion and Storage of Solar Energy* (ed. Connolly, J. S.) 1–26 (Academic, New York, 1981).
5. Porter, G. *Proc. R. Soc.* A**362**, 281–303 (1978).
6. Barber, J. *Photosynthesis in Relation to Model Systems* (Elsevier, New York, 1979).
7. Moore, T. A. *et al. Nature* **307**, 630–632 (1984).
8. Gust, D. *et al. Photochem. Photobiol.* **37S**, S46 (1983).
9. Moore, T. A. *et al. Adv. Photosynthesis Res.* **1**, 729–732 (1984).
10. Gust, D. & Moore, T. A. *J. Photochem.* (in the press).
11. Gust, D. *et al. J. Am. chem. Soc.* (in the press).
12. Chachaty, C. *et al. Org. magn. Reson.* **22**, 39–46 (1984).
13. Bienvenue, E., Seta, P., Hofmanova, A., Gavach, C. & Momenteau, M. *J. electroanalyt. Chem.* **162**, 275–284 (1984).
14. Fettiplace, R., Andrews, D. M. & Hayden. D. A. *J. Membrane Biol.* **5**, 277–296 (1971).
15. Mathis, P., Butler, W. L. & Sato, K. *Photochem. Photobiol.* **30**, 603–614 (1979).
16. Bensasson, R. V. *et al. Nature* **290**, 329–332 (1981).

Speculations on the evolution of ion transport mechanisms

T. HASTINGS WILSON AND PETER C. MALONEY

Department of Physiology, Harvard Medical School, Boston, Massachusetts 02115

Even during the very earliest stages of cellular evolution, a plasma membrane was required to enclose those macromolecules and metabolic intermediates that were essential for growth and cell division. While this structure was successful in limiting the loss of biologically important substances, it posed problems of its own, by defining a compartment that was not in osmotic equilibrium with the environment. If a cell contains macromolecules that cannot penetrate the membrane there will be an inward diffusion of salt and water, causing swelling and eventual lysis. No osmotic problem would have arisen had the primitive membrane been impermeable to salt and water. However, all biological membranes show considerable permeability to water and a low but definite permeability to small molecular weight cations and anions. Evolution has provided two solutions that give long-term stability in the face of such an osmotic crisis. One has been the development of a rigid cell wall; the other has been the development of specific ion pumps that utilize metabolic energy to actively extrude those salts which diffuse into the cell (19, 34, 42).

It is reasonable to suppose that simple cell membranes, unprotected by rigid cell walls, existed before the development of the complex apparatus required for cell wall biosynthesis (3). Under these conditions, the alternative solution for the osmotic prob-

ABSTRACT

Primitive cells evolved a plasma membrane to restrict the loss of important molecules. The osmotic problems that then arose were solved in one of several ways. Of major importance was the evolution of specific ion pumps, to actively extrude those salts whose inward diffusion would have led to swelling and lysis. In addition, these pumps allowed the cell to store energy in the form of ion gradients across the membrane. Thus, even in the earliest stages, the evolution of ion transport systems coincided with the development of mechanisms which catalyzed energy transformations. It is postulated that an "ATP"-driven proton pump was one of the first ion transport systems. Such a proton pump would extrude hydrogen ions from the cell, establishing both a transmembrane pH gradient (alkaline inside) and a membrane potential (negative inside). This difference in electrochemical potential for protons (the protonmotive force) could then drive a variety of essential membrane functions, such as the active transport of ions and nutrients. A second major advance was the evolution of an ion transport system that converted light energy into a form which could be used by the cell. The modern model for this is the "purple membrane" of *Halobacterium halobium*, which catalyzes the extrusion of protons after the capture of light. The protonmotive force generated by such a light-driven proton pump could then power net synthesis of ATP by a reversal of the ATP-driven proton pump. A third important evolutionary step associated with ion transport was the development of a system to harness energy released by biological oxidations. Again, the solution to this problem was to conserve energy as a protonmotive force by coupling the activity of a respiratory chain to the extrusion of protons. Finally, with the development of animal cells a more careful regulation of internal and external pH was required. Thus, an ATP-driven Na$^+$–K$^+$ pump replaced the proton-translocating ATPase as the major ion pump found in plasma membranes. — **Wilson, T. H., and P. C. Maloney.** Speculations on the evolution of ion transport mechanisms. *Federation Proc.* 35: 2174–2179, 1976.

lem must have been utilized. Such primitive cells must have possessed the machinery necessary for ion pumping. What was the nature of this primitive ion pump?

In attempting to answer this question, we first assume that one or more of the ion transport systems found in modern cells provides a useful model for ion transport in the hypothetical "first cell." The predominant ion pump in the plasma membrane of animal cells is an ATP-driven Na$^+$–K$^+$ transport system. However, it seems unlikely that this is representative of the ion pump in primitive cells, for the primary transfer of sodium and potassium is not usually found in bacteria or blue-green algae, whose evolution preceded that of animal cells. Instead, it is our speculation that the ion transport in present day bacteria retains the same general properties that were associated with ion transport in the first

* EVOLUTION OF TRANSPORT THROUGH MEMBRANES—Session V of the FASEB Conference, **Genetics and Biological Evolution,** presented at the 59th Annual Meeting of the Federation of American Societies for Experimental Biology, Atlantic City, New Jersey, April 17, 1975. Chairman: T. H. Wilson.

living cells. This speculation is made despite the fact that bacteria have solved their major osmotic problem by using a rigid cell wall. Ion pumps have evolved not only to accommodate osmotic stress, but also to enable the cell to both generate and utilize the potential energy represented by ion gradients across the membrane. Thus, one might expect that even after the adoption of cell walls, primitive cells and their descendants would continue to exploit the ion transport mechanisms developed previously. With this assumption, it is of interest to look among the ion transport systems of modern microorganisms, especially the anaerobes, for possible insights into the nature of primitive ion pumps.

PROTON ATPase

One of the fundamental energy transducing devices in microorganisms is a reversible proton-translocating ATPase (11, 28, 39). In the anaerobe, which uses only fermentations to generate metabolic energy, this enzyme couples the hydrolysis of ATP to the extrusion of protons from the cell (Fig. 1). This establishes an electrochemical potential difference for protons that, in turn, provides the driving force for a variety of membrane functions. However, in aerobic or photosynthetic bacteria, other reactions establish an electrochemical gradient for protons, and net synthesis of ATP occurs by reversal of this ATPase. It is the primary thesis of this article that such

Figure 1. Proton-translocating ATPase. Hydrolysis of ATP by the membrane-bound ATPase is coupled to electrogenic proton transport out of the cell. Inward movement of protons drives sodium extrusion.

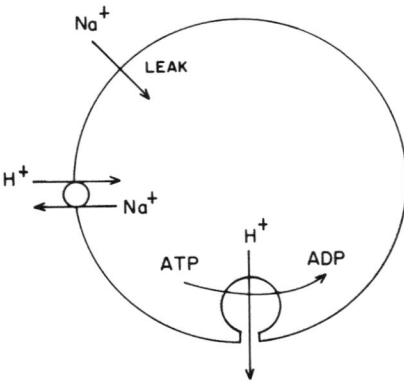

a proton-translocating machine is representative of the most primitive form of ion transport. We hope to show how such a proton pump could carry out the osmoregulation needed by early cells, and to point out how subsequent evolution could have taken advantage of a preexisting proton pump in the design of more complex systems to catalyze oxidative and photosynthetic phosphorylations.

The experiment outlined in Fig. 2 illustrates the transport of hydrogen ions driven by the hydrolysis of ATP. Addition of ATP to a suspension of everted bacterial membrane vesicles led to a prompt rise in the pH of the medium, indicating an inward movement of protons.

OSMOREGULATION AND A PROTON PUMP

The most primitive cell probably powered its proton pump with ATP or some other high energy compound derived from intermediary metabolism. The diagram given in Fig. 1 illustrates the mechanism utilized by present day microorganisms. The membrane-bound Mg^{2+}, Ca^{2+}-dependent ATPase hydrolyzes ATP on the inner surface of the membrane, driving the active transport of protons out of the cell, against their electrochemical gradient. This outward movement of hydrogen ions generates a transmembrane pH gradient (inside alkaline) as well as a membrane potential (inside negative). In the terminology of Mitchell (25–27), who first described such a proton translocating ATPase, chemical energy dissipated by the hydrolysis of ATP is thus conserved as a "protonmotive force" (an electrochemical gradient for protons). For osmoregulation, the energy stored in this protonmotive force must be utilized to bring about the net export of, for example, sodium and chloride. To extrude sodium ions, one might postulate the existence of specific membrane carriers that catalyze the exchange of Na^+ for H^+ (Fig. 1). This mechanism would drive sodium out of the cell at the expense of the transmembrane pH gradient, but would have no direct effect on the membrane potential. Instead, the electrical component of the protonmotive force would be utilized to extrude chloride. Chloride

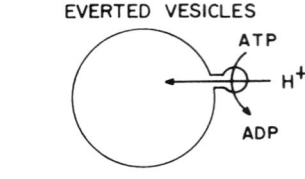

EVERTED VESICLES

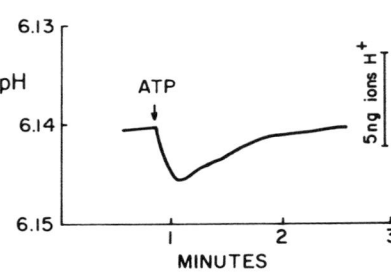

Figure 2. ATP-driven proton transport in *E. coli*. ATP was added to everted vesicles of *E. coli* suspended in a lightly buffered medium. Hydrolysis of ATP led to an inward movement of protons that was reflected by an alkaline shift of the external pH. Redrawn from Hertzberg and Hinkle (11).

ions, which have entered the cell along with sodium ions (when the pump is inactive), would be driven out when a membrane potential (negative inside) is established by the proton pump. Stated in a different way, there is a primary extrusion of positively charged hydrogen ions, followed passively by the outward movement of chloride. Hydrogen ions would then reenter the cell in an electrically neutral exchange for sodium ions. Thus, by means of a circulation of protons, the cell may carry out osmoregulation in the absence of a cell wall.

There is a good deal of evidence that supports this general view. Clearly, the cell membrane must have a low permeability to protons; Scholes and Mitchell (30) have shown this directly for *Micrococcus denitrificans*. In addition, a $Na^+–H^+$ exchange diffusion system may be demonstrated in both the anaerobe *Streptococcus faecalis* (8) and the facultative anaerobe *Escherichia coli* (39). Substantial electrical gradients, generally of –100 mV or greater, are found across the plasma membranes of *S. faecalis* (7), *S. lactis* (15, 17), *E. coli* (6), as well as *Neurospora crassa* (32). Finally, Harold and his collaborators (10) have demonstrated that metabolizing cells of *S. faecalis* can maintain a pH gradient (inside alkaline) of about 1 pH unit.

We have postulated that primitive cells utilized the energy stored in a protonmotive force to achieve osmoregulation. It is also probable that such cells took advantage of the protonmotive force to power other membrane events. Certainly this is true in present day microorganisms. For example, it has been suggested that in *S. faecalis* the intracellular accumulation of potassium, characteristic of most living cells, is driven by the membrane potential (9). In microorganisms additional functions that utilize the protonmotive force include the membrane-bound transhydrogenase reaction (4), flagellar motion (18, 33), and the active transport of nutrients, such as sugars and amino acids (12, 14, 36). In fact, osmoregulation and nutrient transport may have been the two most important functions linked to proton movements in the primitive cell.

COTRANSPORT OF H⁺ AND NUTRIENTS

A strong selective advantage must have accompanied the development of systems that actively transported nutrients into early cells (16). Mitchell (26) was the first to suggest that certain active transport systems in microorganisms catalyze the cotransport of substrate molecules and hydrogen ions. Thus, the input of metabolic energy necessary for the accumulation of substrate comes from a dissipation of the protonmotive force as protons reenter the cell along with substrate. It does not seem unlikely that such systems existed in primitive cells, for if a proton pump was already present, then a ready source of power was at hand.

Recent work has provided strong evidence in support of the idea that bacteria couple the movement of protons and certain substrates. One prediction of this model is that in energy-depleted cells (in which the protonmotive force is close to zero) the entry of substrate (such as lactose), down its concentration gradient, would be accompanied by the entry of protons. This was first shown by West (36–38), who studied the lactose transport system of *E. coli*. Under anaerobic conditions, succinate-grown cells were incubated in a lightly buffered medium containing iodoacetate. Anaerobiosis prevented respiration-

driven proton extrusion; iodoacetate inhibited ATP production from glycolysis, thus preventing ATP-driven proton extrusion. Measurements were then made of the extracellular pH. Upon addition of lactose to transport-positive, β-galactosidase-negative cells, there was a prompt alkalinization of the medium, indicating that protons had been taken up by the cell (Fig. 3). One proton was carried into the cell for each lactose molecule that entered. Moreover, mutants in which the lactose membrane carrier catalyzed the entry (and exit), but not the accumulation of substrate, failed to show this coupled movement of sugar and protons (41). Thus, one may correlate *active* transport with the capacity of the membrane carrier to participate in the circulation of protons.

The second prediction made by the H⁺–nutrient cotransport model is that accumulation of substrate will occur when protons can flow into the cell, down their own concentration gradient. Experimental support (14) for this is shown in Fig. 4. Washed cells of the anaerobe *S. lactis* were suspended at pH 8 in a medium without an energy source. Under these conditions there was no accumulation of the lactose analogue thiomethylgalactoside (TMG). However, marked accumulation of TMG occurred when a pH gradient was imposed across the membrane by suspending cells at pH 6. The subsequent fall in intracellular sugar paralleled a reduction in the protonmotive force; the inward movement of protons (via the trans-

Figure 3. Proton influx driven by sugar entry in *E. coli*. Lactose was added to cells of *E. coli* (β-galactosidase negative) suspended in a lightly buffered medium under anaerobic conditions and in the presence of 1 mM iodoacetate. Redrawn from West (36).

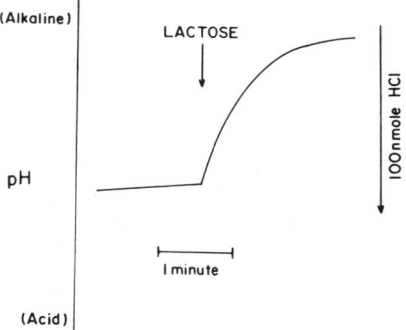

port carrier as well as nonspecific pathways) acidified the interior and may also have developed a membrane potential, positive inside.

At this point we have described three kinds of transport functions that may be energized by the primary extrusion of protons. Cation exchange (e.g., Na⁺ for H⁺) would draw on the transmembrane pH gradient, whereas the extrusion of anions or the accumulation of specific cations would be driven by the membrane potential. Finally, both the pH gradient and the membrane potential provide energy for substrate accumulation when a proton-substrate cotransport system is present.

LIGHT-DRIVEN PHOSPHORYLATION

One of the major events in evolution was the development of a mechanism that converted the energy of light into a form that could be used by the cell. Clearly, it is significant that present-day plant and bacterial cells conserve light energy in the same way that they conserve the energy derived from ATP hydrolysis —as a protonmotive force. Marine microorganisms such as *Halobacterium halobium* contain a light-sensitive transport protein, the so-called "purple membrane," which may be used as a modern model of the relatively primitive device that early cells used to trap light. As shown in Fig. 5, energy made available by the capture

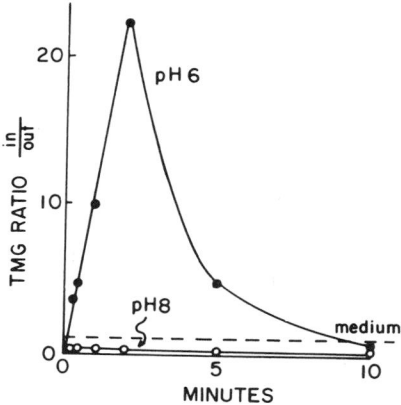

Figure 4. Accumulation of sugar driven by inward movement of protons. Cells of *S. lactis*, previously grown at pH 7, were suddenly exposed to thiomethylgalactoside (TMG) in a medium at pH 6 or pH 8. Redrawn from Kashket and Wilson (14).

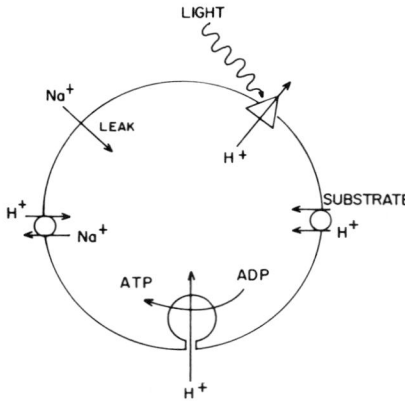

Figure 5. Light-driven proton transport. Light-driven proton extrusion is catalyzed by a specific membrane protein (triangle). This establishes a protonmotive force that is utilized for ATP synthesis, substrate accumulation, and sodium export.

of photons is used to extrude hydrogen ions from the cell. This, of course, gives rise to a protonmotive force that can power various other functions, as discussed above. The appearance of such a light-driven proton pump must have been of great value to early cells, for storage of light energy as a protonmotive force offered an alternative way of making ATP. The proton-translocating ATPase could now be driven in reverse, catalyzing net ATP synthesis rather than hydrolysis.

The experiment given in Fig. 6 was taken from the work of Racker and Stoeckenius (29) and gives an example of light-driven proton transport. Purified membrane protein from *H. halobium* was incorporated into artificial lipid vesicles. Subsequent illumination caused net movement of hydrogen ions across the vesicle membrane. In this artificial system, the orientation of the proton pump is opposite to that of the intact cell. Using intact cells, Danon and Stoeckenius (5) have demonstrated ATP synthesis driven by such a light-induced protonmotive force.

ATP SYNTHESIS
VIA THE PROTON ATPase

In the previous section, we have assumed that the primitive proton ATPase, which had originally catalyzed only ATP hydrolysis, was suitable for the purpose of net ATP synthesis. Recent studies support this

idea. The membrane-bound ATPase of the anaerobe *S. lactis* normally acts to split ATP, but this reaction can be readily reversed in the laboratory when the appropriate protonmotive force is artifically imposed. As expected, this reversal of ATPase activity may be driven by either a pH gradient, a membrane potential, or the appropriate combination of these two (21, 22). An example of this is given in Fig. 7, which shows ATP synthesis driven by a pH gradient. Washed cells were incubated in buffer containing high potassium, so that after the addition of valinomycin (a potassium ionophore) the membrane potential would be set at a low value. Acid was then added to lower the external pH from pH 7.9 to pH 3.2. After the imposition of this protonmotive force a marked synthesis of ATP was observed. ATP synthesis did not occur in the presence of the ATPase inhibitor dicyclohexylcarbodiimide (DCCD), nor was it found when cells were treated with the proton conductor carbonylcyanide-*p*-trifluoromethoxyphenylhydrozone (CCFP). Similar experiments have also been performed in *E. coli* (21, 43).

EVOLUTION OF A
RESPIRATORY PROTON PUMP

At the time of the formation of the first cell, about 3 billion years ago, photolysis of water had provided an atmospheric oxygen level of about 0.002% (Fig. 8). Since this low level of oxygen could not support respiration (as found in modern cells), primitive organisms probably relied on fermentations to supply the greater part of their metabolic energy. Furthermore, these cells must have lived below the surface of the sea, for the low level of oxygen could not supply the ozone needed to protect against ultraviolet light. With the appearance of the first algae, about 2 billion years ago, oxygen from photosynthesis began to appear, elevating oxygen in the atmosphere to about 0.02%. At this point life was still present only at the bottom of fresh water ponds or seas because of the high surface density of UV light. By about 600 million years ago, oxygen had risen to 0.2%. Two factors contributed to the enormous proliferation of marine life

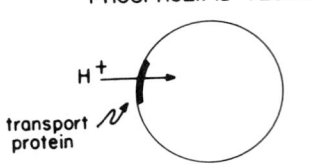

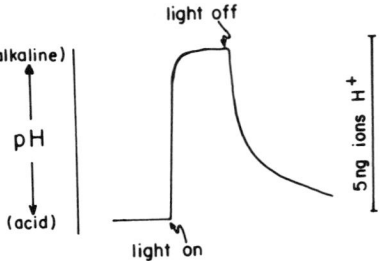

Figure 6. Light-driven proton uptake. Purified membrane protein (purple membrane) from *Halobacterium halobium* was incorporated into artificial phospholipid vesicles. When the light was turned on this protein catalyzed proton movement into the vesicles which was reflected in an alkaline shift in the pH of the medium. Redrawn from Racker and Stoeckenius (29).

that occurred at this time. The ozone shield was able to block enough UV light so that survival close to the warm surface of the water was possible. In addition, the increased level of oxygen allowed the development of efficient respiratory pathways. Thus, at this "Pasteur point" fermentation gave way to respiration as a major

Figure 7. ATP synthesis by reversal of the proton-translocating ATPase. Washed cells of the anaerobe *S. lactis* were suspended in phosphate buffer (pH 7.9) containing 0.3 M potassium. Because internal potassium was about 0.6 M, after the addition of valinomycin, only a small membrane potential was present. HCl was then added to lower the external pH to 3.2. See text for abbreviations. From Maloney and Wilson (22).

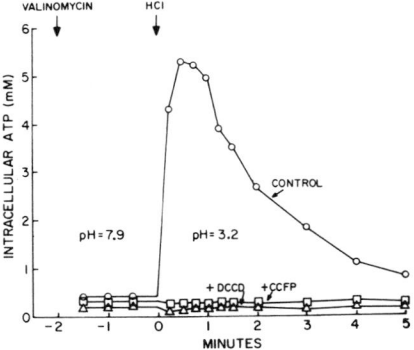

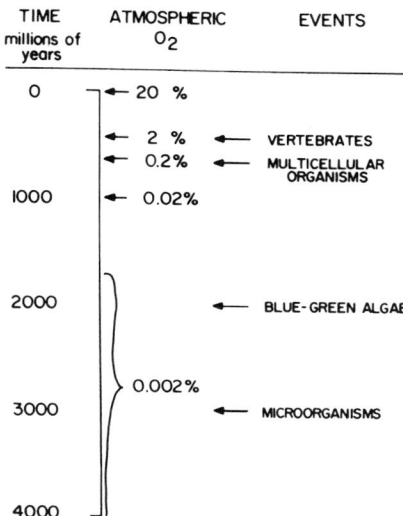

Figure 8. Oxygen content of the atmosphere and biological evolution. Data taken from Berkner and Marshall (2).

source of metabolic energy. During the next 150 million years, the oxygen content rose to about 2%, further increasing the ozone shield so that life was possible on dry land.

Primitive oxidative pathways certainly preceded the availability of oxygen and early cells may have used a wide variety of inorganic and organic substances as terminal electron acceptors. We are speculating that the appearance of a proton-translocating ATPase preceded the evolution of such oxidative pathways. Thus, to efficiently utilize the energy released from oxidative reactions, it was only necessary to develop membrane-bound systems that coupled the oxidation of substrates to the extrusion

Figure 9. Respiration-driven proton pump. Oxidation of substrates by the respiratory chain is coupled to the extrusion of protons. This establishes a protonmotive force that is utilized for ATP synthesis, substrate accumulation, and sodium export.

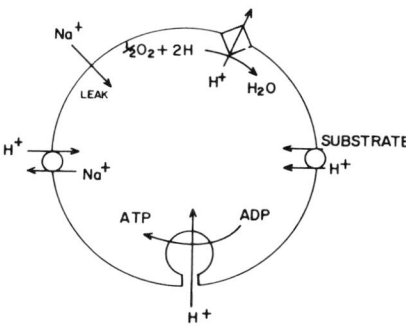

of protons. Net synthesis of ATP could then occur by reversal of the ATPase reaction. Just as with photophosphorylation, oxidative phosphorylation was possible as soon as these two independently developed proton transport systems found themselves within the same membrane. The primitive systems that catalyzed oxidative phosphorylation may have had the equivalent of only one "coupling site" for ATP synthesis, instead of the three sites found in modern cells. Such coupling sites, of course, correspond to those oxidative reactions that extrude protons. Subsequently, the continued use of oxygen tension in the environment provided selective pressure for the evolution of the more complex respiratory chains found today (Fig. 9).

The experiment outlined in Fig. 10 illustrates proton pumping by the respiratory chain of *Micrococcus denitrificans* (31). Cells were incubated under anaerobic conditions in a medium of low buffering power. When a pulse of oxygen was introduced, the oxidation of endogenous substrates led to a rapid acidification of the external medium. As discussed previously, this vectorial translocation of hydrogen ions would generate the protonmotive force needed to power other membrane events.

In microorganisms the respiratory chain has remained embedded in the plasma membrane, while in animal cells this apparatus is sequestered in the inner mitochondrial membrane. Although we do not wish to become involved in the controversy over the origins of mitochondria (23, 35), it is noteworthy that there is a remarkable similarity between the mechanisms of oxidative phosphorylation (with all their proton-translocating devices) in bacteria and mitochondria.

Na⁺, K⁺ DEPENDENT ATPase

At the present time the best studied ion pump is the sodium–potassium transport system of animal cells (Fig. 11) (13). In this case, the hydrolysis of ATP drives the extrusion of three sodium ions in exchange for two potassium ions. Since this system is not found in most microbial or simple plant cells, we believe it evolved much later than the proton pump. In effect, however, the animal cell sodium

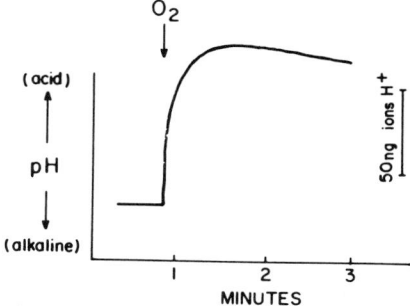

Figure 10. Respiration-drive proton extrusion. Cells of *Micrococcus denitrificans* were suspended in 150 mM KCl–3 mM glycylglycine buffer at pH 7 under anaerobic conditions. When a pulse of oxygen was introduced, acidification of the external medium occurred. Redrawn from Scholes and Mitchell (31).

pump plays essentially the same role as its predecessor. This Na⁺,K⁺ ATPase *1*) provides osmoregulation by allowing net extrusion of salts and water; *2*) provides an internal ionic environment (high K⁺, low Na⁺) faborable to many enzymatic reactions and essential for maintenance of the electrical properties of muscle and nerve; and *3*) establishes an electrochemical gradient for the sodium ion that may be used to drive the active transport of sugars, amino acids, and other solutes. Whereas in microorganisms (and primitive cells) a flow of hydrogen ions connects the various energy transducing systems, in animal cells it is the circulation of sodium ions that provides the currency for such energy transformation at the plasma membrane. This seems to be an important distinction, and may have arisen because of the advantage of a more careful regulation of intracellular and extracellular pH.

Figure 11. Na⁺, K⁺-translocating ATPase of animal cells.

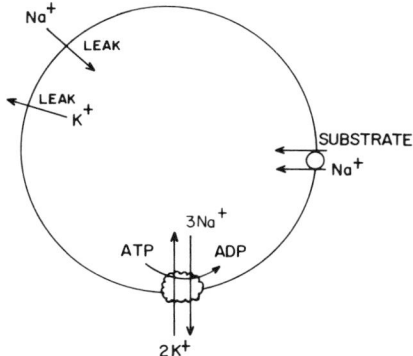

CONCLUSIONS

Our hypothesis on the evolution of ion translocating systems focuses attention on the possible importance of proton pumps in primitive cells. Because of the widespread distribution of proton pumps among modern bacterial, plant, and animal cells, we have speculated that the most primitive form of ion transport was associated with the movement of hydrogen ions. As a model for this first proton pump we chose the proton-translocating ATPase found in present-day microorganisms. The primary transfer of protons endowed primitive cells with a source of power that could drive a variety of secondary reactions, such as osmoregulation and the accumulation of essential nutrients. In the development of the more complex systems catalyzing oxidative and photophosphorylation, subsequent evolution took advantage of the preexisting proton-translocating ATPase.

In animal cells it became of selective advantage to sequester proton pumps within various organelles. In this case, evolution of a Na^+,K^+-ATPase allowed the active transport of ions across the plasma membrane, providing energy for osmoregulation, ion gradients, and nutrient accumulation. The wide variety of other ion transport mechanisms evolved when selective pressures led to specialization of cell function. If it is true that active transport of hydrogen ions represents the primitive form of ion transport, one might expect to find that certain specialized functions in higher forms have retained some aspect of the original proton pump. In this connection, it is of interest that ATP-driven hydrogen ion transport has been postulated to occur in the gastric mucosa (20), lysosomes (24) and chromaffin granules (1). 📭

REFERENCES

1. **Bashford, C. L., G. K. Radda and G. A. Ritchie.** *FEBS Lett.* 50: 21, 1975.
2. **Berkner, L. V., and L. C. Marshall.** *Discuss. Faraday Soc.* 37: 122, 1964.
3. **Calvin, M.** *Chemical Evolution.* New York: Oxford Univ. Press, 1969.
4. **Cox, G. B., N. A. Newton, J. D. Butlin and F. Gibson.** *Biochem. J.* 125: 489, 1971.
5. **Danon, A., and W. Stoeckenius.** *Proc. Natl. Acad. Sci. USA* 71: 1234, 1974.
6. **Griniuvienè, B., V. Chmieliauskaitè and L. Grinius.** *Biochem. Biophys. Res. Commun.* 56: 206, 1974.
7. **Harold, F. M., and D. Papineau.** *J. Membr. Biol.* 8: 27, 1972.
8. **Harold, F. M., and D. Papineau.** *J. Membr. Biol.* 8: 45, 1972.
9. **Harold, F. M., J. R. Baarda and E. Pavlasova.** *J. Bacteriol.* 101: 152, 1970.
10. **Harold, F. M., E. Pavlasova, and J. R. Baarda.** *Biochim. Biophys. Acta* 196: 235, 1970.
11. **Hertzberg, E. L., and P. C. Hinkle.** *Biochem. Biophys. Res. Commun.* 58: 178, 1974.
12. **Hirata, H., K. Altendorf and F. M. Harold.** *J. Biol. Chem.* 249: 2939, 1973.
13. **Jacob, S. W.,** editor. Properties and functions of (Na^+-K^+)-activated adenosine triphosphatase. *Ann. N.Y. Acad. Sci.* 1974, vol. 242.
14. **Kashket, E. R., and T. H. Wilson.** *Proc. Natl. Acad. Sci. USA* 70: 2866, 1973.
15. **Kashket, E. R., and T. H. Wilson.** *Biochem. Biophys. Res. Commun.* 59: 879, 1974.
16. **Kusch, M., and T. H. Wilson.** *Biochim. Biophys. Acta* 311: 109, 1973.
17. **Laris, P. C., and H. A. Pershadsingh.** *Biochem. Biophys. Res. Commun.* 57: 620, 1974.
18. **Larsen, S. H., J. Adler, J. J. Gargus and R. W. Hogg.** *Proc. Natl. Acad. Sci. USA* 71:1239, 1974.
19. **Leaf, A.** *Ann. N.Y. Acad. Sci.* 72: 398, 1959.
20. **Lee, J., G. Simpson and P. Scholes.** *Biochem. Biophys. Res. Commun.* 60: 825, 1974.
21. **Maloney, P. C., E. R. Kashket and T. H. Wilson.** *Proc. Natl. Acad. Sci. USA* 71: 3896, 1974.
22. **Maloney, P. C., and T. H. Wilson.** *J. Membr. Biol.* 25: 285, 1975.
23. **Margulis, L.** *Origin of Eukaryote Cells.* New Haven: Yale Univ. Press, 1970.
24. **Mego, J. L., R. M. Farb and J. Barnes.** *Biochem. J.* 128: 763, 1972.
25. **Mitchell, P.** *Nature (London)* 191: 144, 1961.
26. **Mitchell, P.** *Biochem. Soc. Symp.* 22: 142, 1963.
27. **Mitchell, P.** *Biol. Rev. Cambridge Philos. Soc.* 41: 445, 1966.
28. **Mitchell, P., and J. Moyle.** In: *Membrane Adenosine Triphosphatases and Transport Processes,* edited by J. R. Bronk. London: Biochem. Soc. Spec. Publ. no. 4, 1974, p. 91.
29. **Racker, E., and W. Stoeckenius.** *J. Biol. Chem.* 249: 662, 1974.
30. **Scholes, P., and P. Mitchell.** *J. Bioenerg.* 1: 61, 1970.
31. **Scholes, P., and P. Mitchell.** *J. Bioenerg.* 1: 309, 1970.
32. **Slayman, C. L.** *J. Gen. Physiol.* 49: 93, 1965.
33. **Thipayathasana, P., and R. C. Valentine.** *Biochim. Biophys. Acta* 347: 464, 1974.
34. **Tosteson, D. C.** In: *The Cellular Functions of Membrane Transport,* edited by J. F. Hoffman. Englewood Cliffs, NJ: Prentice-Hall, 1964, p. 3.
35. **Uzzell, T., and C. Spolsky.** *Am. Sci.* 62: 334, 1974.
36. **West, I. C.** *Biochem. Biophys. Res. Commun.* 41: 655, 1970.
37. **West, I. C., and P. Mitchell.** *Biochem. J.* 132: 587, 1973.
38. **West, I. C., and P. Mitchell.** *J. Bioenerg.* 3: 445, 1972.
39. **West, I. C., and P. Mitchell.** *Biochem. J.* 144: 87, 1974.
40. **West, I. C., and P. Mitchell.** *FEBS Lett.* 40: 1, 1974.
41. **West, I. C., and T. H. Wilson.** *Biochem. Biophys. Res. Commun.* 50: 551, 1973.
42. **Wilson, T. H.** *Science* 120: 104, 1954.
43. **Wilson, D. M., J. F. Alderete, P. C. Maloney and T. H. Wilson.** *J. Bacteriol.* 126: 327, 1976.

Section V

BIOINFORMATIONAL MOLECULES

BIOINFORMATIONAL MOLECULES

It has not escaped our notice that the specific pairing we have postulated immediately suggests a possible copying mechanism for the genetic material.

J. D. Watson and F. H. C. Crick, 1953

When the structure of DNA was proposed by Watson and Crick in April, 1953, their brief paper in *Nature* closed with the statement quoted above. It was evident from the double helix structure that one strand of the molecule could pass on an information-containing sequence if it were used as a template for the synthesis of a second strand of DNA (*replication*). Furthermore, it soon became clear that the same information could be passed on to a second kind of molecule (*transcription*) and then delivered to another part of the cell in the form of messenger RNA. Last, the information could contain a code that would guide the synthesis of yet a third kind of molecule—protein—if it were fed through machinery that could use the code to synthesize another sequenced molecule (*translation*). Thus, in one inspired stroke, several essential aspects of contemporary life processes were revealed.

Contemporary forms of life use a closed loop of information and catalysis

We generally understand that information can be contained in orderly arrays of symbols, which can be as simple as the binary bits of computational languages (0 and 1) or as complex as Balinese dances. Information content implies order, so that information is inevitably lost when disorder (randomness) is imposed on such an array. Information also implies the possibility of communication between a sender and a receiver. That is, we might not know whether a coded sequence of symbols contains information or is random until it is received and decoded. This latter point leads to the concept of codes: information need not be communicated directly by a sequence of symbols but can be encoded in an alternative set of symbols.

We will focus here on the specific example of biological information. When we think about contemporary living cells, we understand them to have a cycle of information replication in which the stored information of DNA is transcribed to messenger RNA, which then associates with the ribosome where the information is translated into protein sequences by polymerization of amino acids. But the simplest replicating system in the RNA-world model was the polymerase capable of catalyzing an RNA-mediated polymerization of nucleic acids. If protein synthesis were already present, replication would necessarily be followed by a crude form of transcription-translation, thereby closing the loop. Evolution could take place in a lineage of living cells that contained such replication systems, provided there was a mechanism that introduced changes into nucleic acid sequences (thereby altering the protein encoded by any specific sequence).

How did the first bioinformational molecules arise?

Given that the process we see today has evolved over 3 billion years, we can ask how the first such molecules were produced on the prebiotic Earth. With this question, we have reached the end of our present knowledge, and we can mention only a few ways in which it may have occurred. The possibilities were summarized by Orgel (1987):

1. Early functional proteins replicated directly. They 'invented' nucleic acids and were ultimately enslaved by them.

2. Early nucleic acids or related molecules replicated directly. They 'invented' protein synthesis. Uncoded polypeptides may or may not have been involved in the earliest precoding replication mechanism.

3. Nucleic acid replication and genetic coding of proteins coevolved.

4. The first form of life on the earth was based on some inorganic or organic system unrelated to proteins or nucleic acids.

The first possibility has been championed by Sidney Fox, who has synthesized proteinlike polymers ("proteinoids") by heating dry mixtures of amino acids (see Section II). Fox has argued that the amino acid composition of the polymers differs from that of the original mixture, from which he concludes that such polymers contain intrinsic information.

The second possibility has received considerable recent interest, arising from discoveries by Thomas Cech and by Sidney Altman that certain RNA sequences are biologically active catalysts, now called ribozymes.

The third possibility was formalized by Manfred Eigen in his model of *hypercycles*, a mathematical network that introduces higher-order catalytic action in the form of positive feedback loops among members of different classes of self-replicating macromolecules (Eigen and Schuster, 1979).

The fourth possibility has been expanded

independently by Graham Cairns-Smith (1982), who proposed that certain structures of the clay mineral surface could act as primitive "genes" by assembling organic compounds into specific configurations and aiding their self-assembly. The resulting macromolecules would later be able to code for their own self-assembly in a more efficient process, thus "taking over" the genetic function.

Although each of the above concepts has merit, each also has significant limitation. While condensation of amino acids into polymers occurs readily under laboratory conditions, it seems less plausible under prebiotic conditions involving complex, highly dilute mixtures of amino acids and other organic solutes in sea water. Furthermore, mechanisms for information transfer between two generations of polymerized amino acids are limited. Nucleic acids, on the other hand, readily perform information transmission. They face even more severe difficulties in their original synthesis, however, in that three different components (purine and pyrimidine bases, ribose and deoxyribose sugars, and phosphate) must be assembled into monomeric nucleotides that, in turn, must be precisely linked through covalent bonds to form a molecule capable of replication.

The last possibility—that life began on a mineral substrate—has the advantage of incorporating a ubiquitous surface, such as clay minerals, as an organizing template. Clay surfaces may also have catalytic properties, as demonstrated by Ferris et al. (1988) on montmorillonite clay. However, there is as yet no convincing experimental evidence that mineral surfaces are required to provide a transition between inorganic and organic structures.

Replication may occur in the absence of enzyme catalysis

If we simply mix together the monomers of a nucleic acid such as RNA, nothing much will happen. The reason has to do with a fundamental law of chemistry: polymerization of monomers requires energy. But what if we could somehow provide energy to each monomeric molecule and then set off a chain reaction in which the energy of each monomer were sequentially activated? What follows is an example of such a specific monomeric activation. Solutions of guanosine monophosphate (GMP) can be prepared in which each GMP has a methyl-imidazole group attached to its phosphate through an ester bond (2-MeImpG). The 2-MeImpG should be able to lose the MeImp group with the simultaneous formation of chemical bonds between phosphate and ribose groups, thereby producing a long chain of polyguanosine (poly-G). Over a period of days, however, very little reaction takes place in such a solution. But if a small amount of an RNA polymer consisting of polycytidine (poly-C) is then added and the solution examined a few days later, most of the monomers have been linked into poly-G polymers 30 or more monomers long.

This reaction mechanism, described by Inoue and Orgel (1983), is clearly related to possible prebiotic replication mechanisms. That is, the poly-C RNA polymer can act as a template if chemically activated monomers are present. The 2-MeImpG monomers line up along the poly-C polymer through Watson-Crick base pairing and are then able to form the phosphodiester linkages that hold all nucleic acid polymers together. Here is a catalyzed polymerization reaction in which the poly-C RNA is acting as a template and the 2-MeImp group attached to the G is acting as a polymerase.

There may have been an RNA world

If the RNA enzyme could use another copy of itself as a template, RNA self-replication could be achieved. Thus, it seems possible that RNA catalysts might have played a part in prebiotic nucleic acid replication, prior to the availability of useful proteins.

T. R. Cech, 1986

One of the most exciting developments concerning bioinformational molecules in the past decade was the discovery that certain strands of RNA are catalysts (Kruger et al., 1982; Guerrier-Takada et al., 1983; Zaug and Cech, 1986). The catalytic self-splicing of RNA discovered in the Cech laboratory involves a ribosomal RNA from the ciliate *Tetrahymena*, in which an intron sequence was found to remove itself from a strand of precursor RNA while joining the two ends to form the functional RNA molecule. Cech (1986) also showed how RNA could act as its own polymerase, catalyzing the reaction in which cytidylic acid is polymerized on a self-contained template to form polycytidylic acid, a polymer of RNA.

The discovery of ribozymes potentially solves the chicken-egg problem of whether nucleic acids or proteins came first in the origin of life. That is, if both catalytic and informational capacity were present in the same molecule, the problem is greatly simplified: that molecule could act as a catalyst for its own replication.

Following earlier work by Woese (1967), Crick (1968), and Orgel (1968), three different investigators suggested in 1986 that life had appeared with the emergence of RNA-based cells, prior both to proteins and to DNA (Alberts, 1986; Gilbert, 1986; Lazcano, 1986). Walter Gilbert proposed an "RNA world" scenario as follows:

> The first stage of evolution proceeds... by RNA molecules performing the catalytic activities necessary to assemble themselves from a nucleotide soup.... At the next stage, RNA molecules began to synthesize proteins, first by developing RNA adapter molecules that can bind activated amino acids and then by arranging them according to an RNA template using other RNA molecules such as the RNA core of the ribosome. This process would make the first proteins, which would simply be better enzymes than their RNA counterparts.... Finally, DNA appeared on the scene, the ultimate holder of information copied from the genetic RNA molecules by reverse transcription.... RNA is then relegated to the intermediate role that it has today—no longer the center of the stage, displaced by DNA and the more effective protein enzymes.

This is a persuasive argument, in large part because it seems likely that there must have been a simpler version of catalysis and information transfer in the earliest forms of life. Although this is a satisfying step forward, many difficult questions remain, not the least of which concerns the pathways by which long molecules of RNA were synthesized and maintained in a stable condition long enough to participate in early evolution. Some of these concerns have been clearly stated by Joyce (1989).

Reprinted papers on bioformational molecules

The first paper included in this section is by Horowitz (1945), written in the mode of "one gene, one enzyme," before the structure and function of DNA were understood. Although much of the paper deals with Mendelian inheritance, mutations, and natural selection, in the last paragraphs Horowitz turns his attention to Oparin and the origin of life, relating these concepts to the evolution of biosynthesis. While Oparin was the first to discuss selection of macromolecular systems in the prebiotic environment, here Horowitz is the earliest to discuss the origin of metabolic pathways.

The next paper, by Usher and McHale (1976), shows how RNA templates can catalyze the formation of phosphodiester bonds and then select the natural 3', 5'-bond. The following paper (Usher, 1977) relates this observation to the prebiotic environment.

The fourth paper, by Inoue and Orgel (1983), extends nonenzymatic polymerization of RNA to the use of activated monomers. The fifth paper, by Schwartz and Orgel (1985), discusses a series of simpler, nucleic-acid-like structures that may be more plausible as prebiotic macromolecules. We have included here a short note by Tjivikua et al. (1990) that describes another version of chemical self-replication.

The following series of papers deals with catalytic RNA molecules, or ribozymes. The paper by Zaug and Cech (1986) describes the reaction mechanism of the self-splicing ribozyme from *Tetrahymena*. In the next two papers, Cech (1986) shows how RNA catalysis could have preceded protein enzymes on the early Earth, and Gilbert (1986) coins the phrase "the RNA world" to describe this point in evolution. A paper by Doudna and Szostak (1989) describes an attempt to establish an evolving system of ribozymes.

The papers by Noller et al. (1992) and by Piccirilli et al. (1992) considerably expand our understanding of RNA capabilities. Earlier work in Cech's and in Altman's research labs had already shown that, in its catalytic activity, RNA can function as an enzyme. In providing strong evidence that RNA can catalyze the formation of peptide bonds, Noller and his co-workers show that RNA can make protein. Piccirilli, with the Cech laboratory, reports an RNA-catalyzed hydrolysis of amino acid-RNA bonds, a discovery that suggests that the opposite reaction—the ribozyme-mediated amino-acid charging of an RNA molecule—can also take place.

We close with a broad overview by Joyce (1991) and a discussion of the constraints placed on an RNA world by the prebiotic environment.

References

Alberts, B. M. (1986) The function of the hereditary materials: Biological catalyses reflect the cell's evolutionary history. *American Zoologist* **26**:781-796.

Cairns-Smith, A. G. (1982) *Genetic Takeover and the Mineral Origins of Life.* Cambridge, U.K.: The Cambridge University Press.

Cech, T. R. (1986) A model for the RNA-catalyzed replication of RNA. *Proc. Natl. Acad. Sci. USA* **83**:4360-4363.

Crick, F. H. C. (1968) The origin of the genetic code. *J. Mol. Biol.* **38**:367-379.

Doudna, J. A. and Szostak, J. W. (1989) RNA-catalysed synthesis of complementary-strand RNA. *Nature* **339**:519-522.

Eigen, M. and Schuster, P. (1979) *The Hypercycle: A Principle of Natural Self-Organization.* Berlin: Springer-Verlag.

Ferris, J. P., Huang, C.-H. and Hagan, W. J. (1988) Montmorillonite: A multifunctional mineral catalyst for the prebiological formation of phosphate esters. *Orig. Life Evol. Biosphere* **18**:121-133.

Fox, S. W. and Harada, K. (1958) Thermal copolymerization of amino acids to a product resembling protein. *Science* **128**:1214.

Gilbert, W. (1986) The RNA world. *Nature* **319**:618.

Guerrier-Takada, C., Gardiner, K., Marsh, T., Pace, N. and Altman, S. (1983) The RNA moiety of ribonuclease P is the catalytic subunit of the enzyme. *Cell* **35**:849-857.

Horowitz, N. H. (1945) On the evolution of biochemical syntheses. *Proc. Natl. Acad. Sci. USA* **31**:153-157.

Inoue, T. and Orgel, L. E. (1983) A nonenzymatic RNA polymerase model. *Science* **219**:859-862.

Joyce, G. F. (1989) RNA evolution and the origins of life. *Nature* **338**:217-224.

Joyce, G. F. (1991) The rise and fall of the RNA world. *The New Biologist* **3**:399-407.

Kruger, K., Grabowski, P. J., Zaug, A. J., Sands, J., Gottschling, D. E. and Cech, T. R. (1982) Self-splicing RNA: Autoexcision and autocyclization of the ribosomal RNA intervening sequence of *Tetrahymena*. *Cell* **31**:147-157.

Lazcano, A. (1986) Prebiotic evolution and the origin of cells. *Treb. Soc. Cat. Biol.* **39**:73-103.

Noller, H. F., Hoffarth, V. and Zimniak, L. (1992) Unusual resistance to protein extraction procedures. *Science* **256**:1416-1419.

Orgel, L. E. (1968) Evolution of the genetic apparatus. *J. Mol. Biol.* **38**:381-393.

Orgel, L. E. (1987) Evolution of the genetic apparatus: A review. *Cold Spring Harbor Symp. Quant. Biol.* **52**:9-16.

Piccirilli, J. A., McConnell, T. S., Zaug, A. J., Noller, H. F. and Cech, T. R. (1992) Aminoacyl esterase activity of the *Tetrahymena* ribozyme. *Science* **256**:1420-1424.

Schwartz, A. W. and Orgel, L. E. (1985) Template-directed synthesis of novel, nucleic acid-like structures. *Science* **228**:585-587.

Tjivikua, T., Ballester, P. and Rebek, J. (1990) A self-replicating system. *J. Am. Chem. Soc.* **112**:1249-1250.

Usher, D. A. (1977) Early chemical evolution of nucleic acids: A theoretical model. *Science* **196**:311-313.

Usher, D. A. and McHale, A. H. (1976) Hydrolytic stability of helical RNA: A selective advantage for the natural 3',5'-bond. *Proc. Natl. Acad. Sci. USA* **73**:1149-1153.

Watson, J. D. and Crick, F. H. C. (1953) Structure of deoxyribose nucleic acid. *Nature* **171**:737-738.

Woese, C. R. (1967) *The Genetic Code – The Molecular Basis for Genetic Expression.* New York: Harper and Row.

Zaug, A. J. and Cech, T. R. (1986) The intervening sequence RNA of *Tetrahymena* is an enzyme. *Science* **231**:470-475.

ON THE EVOLUTION OF BIOCHEMICAL SYNTHESES

By N. H. Horowitz

SCHOOL OF BIOLOGICAL SCIENCES, STANFORD UNIVERSITY, CALIF.

Communicated April 23, 1945

Although it has been recognized for a long time that the biochemistry of the organism is conditioned by its genetic constitution, a more precise definition of this dependence has not been possible until recently. A considerable amount of evidence now exists for the view that there is a one-to-one correspondence between genes and biochemical reactions. This concept, foreshadowed in the work of Garrod[1] on human alcaptonuria, accounts in a satisfactory way for the inheritance of pigment formation in guinea pigs,[2] insects[3] and flowers,[4] and the synthesis of essential growth factors in *Neurospora*.[5] It appears from these studies that each synthesis is controlled by a set of non-allelic genes, each gene governing a different step in the synthesis. As to the nature of this control, it is probable that the primary action of the gene is concerned with enzyme production. That genes can direct the specificities of proteins has been shown in the case of many antigens,[6] while several mutations demonstrably affecting the production of enzymes have been reported.[6] Evidence on the postulated gene-enzyme relationship is in most cases, however, still circumstantial; this is partly because of technical difficulties involved in the study of synthetic, or free-energy consuming reactions *in vitro*, and partly because of the insufficiency of biochemical information on those reactions which happen to be susceptible of genetic analysis.

As a corollary of the above hypothesis, each biosynthesis depends on the direct participation of a number of genes equal to the number of different, enzymatically catalyzed steps in the reaction chain. In attempting to account for the evolutionary development of such a reaction chain one meets in a clear form the problem of explaining macroevolutionary changes in terms of microevolutionary steps. The individual reactions making up the chain are of value to the organism only when considered collectively and in view of the ultimate product. Regarded individually, intermediate substances cannot, in general, be assumed to have physio-

Reprinted with permission of the author from *Proc Nat'l Acad Sci USA*, Volume 31, pp. 153-157. © 1945 by the National Academy of Sciences.

logical significance, and the ability to produce them does not of itself confer a selective advantage. An example from *Neurospora* genetics will serve to illustrate this point. At the present time seven different genes are known to be concerned in the synthesis of arginine by the mold.[7] The inactivation of any one prevents the synthesis from taking place. On the basis of the above hypothesis, at least seven different catalyzed steps must occur in the synthesis. Several of the steps have been identified and controlling genes assigned to each. Two of the intermediates in the chain have been shown to be the amino acids ornithine and citrulline. Unlike arginine, neither of these substances is a general constituent of proteins. Aside from their function as precursors, they are apparently of no further use to the organism.

While the above example probably represents the general case, there are also well-known instances in which precursors serve independent functions. Thus, arginine, glycine and methionine are precursors of creatine in the rat,[8] but the synthesis goes through the non-functional intermediate, glycocyamine. On the other hand, acetylcholine may be synthesized from choline in one step.[9] In cases such as these, the problem is that of accounting for the synthesis of the precursors.

Since natural selection cannot preserve non-functional characters, the most obvious implication of the facts would seem to be that a stepwise evolution of biosyntheses, by the selection of a single gene mutation at a time, is impossible. It will be shown below that this is not a necessary conclusion, but that under special conditions the stepwise evolution of long-chain syntheses may occur. First, however, an alternative to stepwise evolution will be considered; that is, the origin of a new reaction chain through the chance combination of the necessary genes.

Although the probability of the origin of a useful character through the chance association of many genes may be small, it is never zero. Indeed, a consideration of the statistical consequences of the interaction of mutation, Mendelian inheritance, and natural selection has led Wright[10] to the conclusion that such chance associations may be of major importance in evolution. He has analyzed the evolutionary possibilities of various types of breeding structures and has shown that under certain conditions an extensive trial and error mechanism exists, whereby the species can test numerous combinations of non-adaptive genes. The breeding structure which most favors this type of evolution is that of a large population divided into many small, partially isolated groups. Within each group the cumulative effects of the accidents of sampling among the gametes are of major significance in determining gene frequencies, but the penalty of fixation of deleterious genes, ordinarily incurred under inbreeding, is avoided by exchange of migrants with other groups. The pressures of forward and reverse mutations, which between them determine an equilibrium frequency for non-adaptive genes in large, random-breeding populations, become of minor importance. As a consequence, a random drift of gene frequencies occurs. If, by chance, one group finds a particularly favorable combination of genes, a process of intergroup selection comes into play, whereby the favorable combination is spread to the population at large.

This model provides a means for the evolution of a new gene combination in spite of unfavorable mutation rates to active alleles and in the absence of selection of individual genes. It is thus favorable for the evolution of systems of individually non-adaptive, but collectively adaptive, genes. The effectiveness of the process would seem to be strongly dependent on the size of the gene combination required, however, decreasing approximately exponentially with increasing numbers of genes, other factors remaining constant. There would result a tendency toward the evolution of short reaction chains involving the recombination of molecular units already available. There is no doubt that a conservative tendency of this sort actually exists in nature. The wide variety of biologically important compounds built up on the pyrrole nucleus, to mention but one example, is a case in point.

The application of Wright's theory to the particular problem under consideration is limited by the fact that it operates only under biparental reproduction. It is probable that a large number of basic syntheses evolved prior to sexual reproduction. The universal distribution among living forms of certain classes of compounds—viz., the amino acids, nucleotides and probably the B vitamins—identifies them as essential ingredients of living matter. The synthesis of these substances must have evolved very early in geologic time, as a necessary condition for further progress, although loss of certain syntheses may have occurred in the later differentiation of some forms. It is therefore desirable to search for another solution of the problem applicable to compounds of this type, preferably one in which a minimum burden is placed on chance and a maximum one on directed evolutionary forces. It is thought that the following suggestion, while definitely a speculation, offers a possible solution along these lines.

In essence, the proposed hypothesis states that the evolution of the basic syntheses proceeded in a stepwise manner, involving one mutation at a time, but that the order of attainment of individual steps has been in the reverse direction from that in which the synthesis proceeds—i.e., the last step in the chain was the first to be acquired in the course of evolution, the penultimate step next, and so on. This process requires for its operation a special kind of chemical environment; namely, one in which end-products and potential intermediates are available. Postponing for the moment the question of how such an environment originated, consider the operation of the proposed mechanism. The species is at the outset assumed to be heterotrophic for an essential organic molecule, A. It obtains the substance from an environment which contains, in addition to A, the substances B and C, capable of reacting in the presence of a catalyst (enzyme) to give a molecule of A. As a result of biological activity, the amount of available A is depleted to a point where it limits the further growth of the species. At this point, a marked selective advantage will be enjoyed by mutants which are able to carry out the reaction $B + C = A$. As the external supplies of A are further reduced, the mutant strain will gain a still greater selective advantage, until it eventually displaces the parent strain from the population. In the A-free environment a back mutation to the original stock will be lethal, so we have at the same time a theory of lethal genes. The majority of biochemical mutations in *Neurospora* are lethals of this type.

In time, B may become limiting for the species, necessitating its synthesis from other substances, D and E; the population will then shift to one characterized by the genotype $(D + E = B, B + C = A)$. Given a sufficiently complex environment and a proportionately variable germ plasm, long reaction chains can be built up in this way. In the event that B and C become limiting more or less simultaneously, another possibility is opened. Under these circumstances symbiotic associations of the type $(F + G \neq C, D + E = B)(F + G = C, D + E \neq B)$ will have adaptive value.

This model is thus seen to have potentialities for the rapid evolution of long chain syntheses in response to changes in the environment. As has been pointed out by Oparin[11] the hypothesis of a complex chemical environment is a necessary corollary of the concept of the origin of life through chemical means. The essential point of the argument is that it is inconceivable that a self-reproducing unit of the order of complexity of a nucleoprotein could have originated by the chance combination of inorganic molecules. Rather, a period of evolution of organic substances of ever-increasing degree of complexity must have intervened before such an event became a practical, as distinguished from a mathematical, probability. Or, put in another way, any random process which can have produced a nucleoprotein must at the same time have led to the production of a profusion of simpler structures. Oparin has considered in some detail the possible modes of origin of organic compounds from inorganic material and cites a number of known reactions of this type, together with evidences of their large-scale occurrence on the earth in past geologic ages. He concludes that in the absence of living organisms to destroy them highly complex organic systems can have developed. The first self-duplicating nucleoprotein originated as a step in this process of chemical evolution. The origin of living matter by physicochemical means thus presupposes the existence of a highly complex chemical environment.

To summarize, the hypothesis presented here suggests that the first living entity was a completely heterotropic unit, reproducing itself at the expense of prefabricated organic molecules in its environment. A depletion of the environment resulted until a point was reached where the supply of specific substrates limited further multiplication. By a process of mutation a means was eventually discovered for utilizing other available substances. With this event the evolution of biosyntheses began. The conditions necessary for the operation of the mechanism ceased to exist with the ultimate destruction of the organic environment. Further evolution was probably based on the chance combination of genes, resulting to a large extent in the development of short reaction chains utilizing substances whose synthesis had been previously acquired.

[1] Garrod, A. E., *Inborn Errors of Metabolism*, Oxford University Press (1923).

[2] Wright, S., *Biol. Symposia*, **6**, 337–355 (1942).

[3] Ephrussi, B., *Quart. Rev. Biol.*, **17**, 327–338 (1942).

[4] Lawrence, W. J. C., and Price, J. R., *Biol. Rev.*, **15**, 35–58 (1940).

[5] Horowitz, N. H., Bonner, David, Mitchell, H. K., Tatum, E. L., and Beadle, G. W., *Am. Nat.*, in press (1945).

[6] Summarized in Wright, S., *Physiol. Rev.*, **21**, 487–527 (1941).

[7] Srb, A., and Horowitz, N. H., *Jour. Biol. Chem.*, **154**, 129–139 (1944).

[8] Summarized in Schoenheimer, R., *The Dynamic State of Body Constituents*, Harvard University Press (1942).

[9] Lipmann, F., *Advances in Enzymology*, **1**, 99–162 (1941).

[10] Wright, S., *Bull. Am. Math. Soc.*, **48**, 223–246 (1942). Contains summary of earlier papers.

[11] Oparin, A. I., *The Origin of Life*, trans. by S. Morgulis, Macmillan, New York (1938).

Hydrolytic stability of helical RNA: A selective advantage for the natural 3',5'- bond

(oligoadenylates/RPC-5 chromatography/2',5'-link/kinetics/prebiotic)

D. A. USHER* AND A. H. McHALE

Spencer T. Olin Chemistry Laboratory, Cornell University, Ithaca, New York 14853

Communicated by Jerrold Meinwald, January 30, 1976

ABSTRACT Dodecaadenylic acid containing a single 2',5'-linkage at a defined position was formed by the coupling of two hexamers on a poly(U) template at 2°. The rate of hydrolysis of this dodecamer was compared with that of a dodecamer that contained only the natural 3'-5'-linkages. At 40°, in 1 M aqueous ethylenediamine at pH 8 in the absence of poly(U), both dodecamers hydrolyzed at comparable rates, but the addition of two equivalents of poly(U) caused a 7-fold *increase* in the initial rate of hydrolysis of the oligomer containing the 2',5'-bond, and a 5-fold *decrease* in the initial rate of hydrolysis of the natural oligomer. When the oligomers are fully constrained in helical form, the ratio of the rates of cleavage of one 2',5'-bond to one 3',5'-bond under these conditions is probably about 900:1. The use of the 3',5'-bond, in combination with a right-handed helix, appears to have had a large selective advantage over the use of the 2',5'-bond for the storage of genetic information.

Naturally occurring RNA appears to contain solely the 3',5'-internucleotide linkage. By contrast, successful attempts to demonstrate the template-directed non-enzymatic polymerization of ribonucleotides, which are of interest as models of prebiotic synthesis, have almost invariably resulted in the production of a large excess of the unnatural 2',5'-isomer (1). Thus, coupling of adenosine 2',5'-cyclic phosphate (A>p) on a poly(U) template is catalyzed by aqueous ethylenediamine, but the resulting dimer contains about 97% of the 2',5'-linkage (2). It has been suggested (3) that these two observations may be related: that the constraint of the helix forces the formation of the 2',5'-bond from the cyclic phosphate, and also stabilizes a 3',5'-bond against hydrolysis in mildly basic solution. As a corollary, the 2',5'-bond is still labile when the constituent nucleotides are part of a right-handed helix, and thus storage of "genetic" information is better accomplished by use of the 3',5'-bond. Previous evidence for the stability of the 3',5'-bond in helical conformation comes from the isolation of oligo(purines) from the base-catalyzed partial hydrolysis of single-stranded RNA. Purine-rich regions have a higher helical content than pyrimidine-rich regions, and better resist alkaline hydrolysis (4, 5).

We have made an unambiguous test of the ideas expressed above by measuring the rate of hydrolysis at pH 8 of two dodecaadenylates, one (I) that is entirely 3',5'-linked, the other (II) that contains a single 2',5'-bond at the central position. The measurements have been made both in the presence and in the absence of poly(U). The preparation of the synthetic dodecamer (II) could not be achieved by the

use of any known enzyme, and the equilibrium calculations of Renz *et al.* (2), appeared to suggest that a cyclic-phosphate coupling reaction, while giving the required 2',5'-bond, would result in far too small a yield of a dodecamer for this method to be of use. However, we hoped that if the above arguments (3) were indeed correct, it would prove possible to prepare dodecamer II from two 3',5'-linked hexamers by allowing the 5'-hydroxyl of one to react with the 2',3'-cyclic phosphate terminus of the other. The presence of a poly(U) template would provide protection for the 3',5'-bonds by the resulting helical conformation, and at the same time would facilitate the coupling reaction.

EXPERIMENTAL

Spectra were measured on a Cary 15 spectrophotometer using a specially constructed thermostated cell block (6). The temperature, held to ±0.1°, was measured with a calibrated YSI model 421 thermistor as one arm of an unbalanced Wheatstone bridge. Sample weights were measured with a Cahn model G-2 Electrobalance. High pressure liquid chromatography (HPLC) on RPC-5 (7, 8) (0.4 × 30 cm, stainless steel column), at room temperature was used for quantitative analysis of reaction products. A Waters Associates ALC 202 chromatograph with a UV detector and model 660 solvent programmer was used: the output at 254 nm was recorded on a Hewlett-Packard model 7127A recorder. Baseline separation of $(A)_5Ap$ from $(A)_5A>p$, and $(A)_{11}Ap$ from $(A)_{11}A>p$ was accomplished using a nonlinear gradient (no. 5, 2 hr) 0.2–0.5 M KCl (0.01 M Tris/Tris·HCl), pH 8.1) at a rate of 1.0 ml/min. With both the hexamer and dodecamer, some separation between the isomeric 2'- and 3'-terminal monophosphates was also seen. In both cases the terminal monoesters elute before the related terminal cyclic diesters. The same column was used with a linear gradient (1 hr) of 0.0025–0.04 M $KClO_4$ (0.01 M Tris/Tris·HCl, pH 8.1) at a rate of 1.0 ml/min to separate oligoadenylates from each other, and from poly(U). In this buffer, $(A)_5Ap$ eluted in about the same volume as $(A)_5A>p$. Poly(U) was periodically stripped from the column with 0.11 M $KClO_4$. Individual peaks to beyond a degree of polymerization of 60 (DP60) were countable when the gradient was extended to 0.06 M $KClO_4$.

Mono- and dinucleotides and adenosine were determined by high pressure chromatography (9) on either AS-Pellionex-SAX (Reeve-Angel, lot 152010) (1 mm × 100 cm; 0.02 M potassium phosphate, pH 5.3) or AL-Pellionex-WAX (Reeve-Angel, lot 100020) (1 mm × 53 cm; 0.001 M ammonium sulfate, pH 5.4 with acetic acid). Quantitative estimation of peak areas was done by cutting-and-weighing, after retracing the curves on Albanene paper.

Desalting of oligomer solutions was carried out with a Biogel P-2 column (Biorad, 200–400 mesh, 1.1 × 41 cm), using

---

Abbreviations: BAP, bacterial alkaline phosphatase; HPLC, high pressure liquid chromatography; A>p, adenosine 2',3'-cyclic phosphate; DP, degree of polymerization; percentages are given as % of total nucleoside units present; molar absorptivities are given per nucleoside unit.

* To whom correspondence should be addressed.

0.02 M triethylammonium bicarbonate (pH 8) as an eluant. Distilled, deionized water (specific resistance 2-3 MΩ) was used in all work; glassware was cleaned in acid chromate solution to remove ribonucleases, and gloves were used in critical work. Poly(U) (Schwarz/Mann, lot W-2069, K-salt, average molecular weight of 1.56×10^6) was dialyzed twice against 0.5 M NH$_4$Cl containing 25 mM Na$_2$EDTA (pH 7), 1 day each, and twice against water, 1 day each (2, 10, 11). Poly(A) (Sigma) was used as received. RNase-T$_2$ (Calbiochem, lot 300248, EC 2.7.7.17) was used as an aqueous solution of 8.76 units/ml. Bacterial alkaline phosphatase (BAP) (11.2 mg/ml, lot 3BA, EC 3.1.3.1), and RNase-A (3070 units/mg, lot 9DA) were obtained from Worthington. RPC-5 was prepared from 2 ml of Adogen 464 (Aldrich) and 50 g of Kel F powder, (Kel F 82, KF6301, 3M Co.): the components in chloroform (75 ml) were combined in a Waring Blendor and then air dried (7). Triethylammonium bicarbonate buffer at pH 8 was prepared from an aqueous solution of freshly distilled triethylamine (Eastman, bp 87.0–87.8°) and gaseous CO$_2$ at 0°. Ethylenediamine (Aldrich) was distilled under nitrogen (bp 99–105°).

Preparation of Oligoadenylates. Oligoadenylates with a 2′,3′-cyclic phosphate terminus were produced by controlled hydrolysis of poly(A) with RNase-A (12). Poly(U) (292 mg) was incubated with RNase-A (23 mg) in 0.1 M Tris-acetate buffer, pH 8.02 (40 ml), at 37° for 90 min, with constant stirring. Bentonite (B.D.H., 117 mg) was added (13), the mixture stirred for 1 hr at 0°, and then filtered through a pad of clean celite, which was rinsed two times with 40 ml of water. The filtrate, diluted with an additional 80 ml of water was run onto a column (2.2 × 38 cm) of DEAE-Sephadex A-25 (Pharmacia) in the bicarbonate form, and oligomers (DP 2 through 13) were collected by elution with a gradient of 0.01–1.0 M triethylammonium bicarbonate. The yield was about 37 μmol of nucleotide phosphate as the hexamer and 26 μmol as the dodecamer. The identity of the various oligomers was checked by chromatography on RPC-5 against authentic samples (Miles Laboratories, up to DP 6) and by peak counting. The pooled fractions were evaporated under reduced pressure, at a temperature below 30°

Opening of the *terminal* cyclic phosphate without cleavage or isomerization of the internucleotide bonds was achieved (14) by treatment of an individual oligomer with 0.1 M HCl at 25° for 4 hr. The reaction was stopped by addition of ethylenediamine to pH 7, and the mixture desalted on Biogel P-2. The ring-opening reaction was checked by treatment of the product with BAP: removal of the terminal 2′ or 3′ phosphate could be followed by RPC-5 chromatography, while treatment of the cyclic phosphate with BAP gave no reaction.

Formation of Dodecaadenylate (II): (A)$_5$A(2′-5′)(A)$_5$Ap. The hexamer as used in this reaction was resistant to BAP and could be completely digested to Ap by RNase-T$_2$ (15, 16); at the start of the reaction it contained 5.2% ring-opened material. The lithium salt of (A)$_5$A>p (19 μmol of A) with poly(U) (38 μmol of U) in 1 M aqueous ethylenediamine hydrochloride (0.75 ml), at pH 8, was kept at 2° for 5 days. The yield of dodecamer at this time was about 24%. Aqueous methanol was added so that the final volume was 10 ml, 10% in methanol (vol), and this solution was run onto a column of DEAE-Sephadex, (1.1 × 15 cm, bicarbonate form). A linear gradient of 0.01 M to 1.5 M triethylammonium bicarbonate was used (total volume 1200 ml), and the eluate was passed via a flow-through cell (monitored at 260

nm by a Zeiss PMQ-II spectrophotometer) to a Gibson Microfractionator, model FC-80E, and 7 ml fractions were collected. (A)$_5$A>p peaked at fraction 75, (A)$_5$Ap at fraction 85, and the dodecamer at fraction 110 (the forms with the different termini were not resolved). Fractions containing the dodecamer were pooled, and rotary-evaporated at a temperature below 30° with the use of an oil-diffusion pump and liquid nitrogen traps. A portion of the product was saved for the determination of the percentage of 2′,5′-bonds (see below), and the remainder was treated with HCl, as described above, to open any remaining cyclic phosphate termini. Chromatography on RPC-5 showed that the product consisted of 91% dodecamer, 4% hexamer, and less than 1% each of the other oligomers, monomer through undecamer.

Percentage of 2′,5′-Bond in Dodecamer II. Dodecamer II (0.01 μmol) was incubated at 37° for 85 min with RNase-T$_2$ (2 μl; 8.76 units/ml) in 10 μl of 0.05 M ammonium acetate, pH 4.5, containing 20% vol/vol methanol, and 0.03 mM Na$_2$EDTA (under these conditions the hexamer that was used to form the dodecamer was degraded completely to Ap). The products of the digestion had the same retention volumes on HPLC as authentic Ap and either 2′,5′- or 3′,5′-linked AAp. (The isomers were not resolved at this stage.) Treatment of the AAp with BAP gave AA which had the same retention volume as authentic A(2′-5′)A and different from that of A(3′-5′)A. The amount of A(2′-5′)Ap formed (15, 16) in the RNase-T$_2$ digestion of a known amount of dodecamer II was calculated by comparing the size of the peak with that given by injection on HPLC of a known amount of A(3′-5′)Ap, assuming that the molar absorptivities of A(2′-5′)Ap and A(3′-5′)Ap at 254 nm are in the same ratio as those of A(2′-5′)A and A(3′-5′)A at 254 nm. The concentrations of standard solutions of the A(2′-5′)A and A(3′-5′)A isomers were calculated from the absorbances of the solutions at 258 nm and published values of the molar absorptivities at 258 nm (17). The percentage of the central bond of the dodecamer II that is 2′,5′-linked is then six times the percentage of A(2′-5′)Ap produced in the digestion. Treatment of several different aliquot samples from the same preparation of dodecamer II gave from 76% to over 100% central 2′-5′-linkage: this large spread may in part reflect the difficulty of injecting precise volumes in the μl range. However, from the rate and products of hydrolysis of dodecamer II it can be calculated that there is about 95% of a labile bond at the central position; this figure is consistent with the 97% found by Renz *et al.* (2) for dimerization of A>p on a poly(U) template under the same conditions.

Quantitation of Oligoadenylates and Poly(U). The concentration of stock solutions of hexamer, dodecamer I, dodecamer II, and poly(U), were determined from the absorbance spectrum of aliquot samples after treatment with RNase-T$_2$ in 0.05 M ammonium acetate, pH 4.5. The spectrum (240–290 nm) was recorded at 25° before addition of the enzyme, the digestion was then performed at 37° until the absorbance at 260 nm was constant, and the spectrum was again recorded at 25°. Enzyme was also added to the acetate buffer in the reference cell. Literature values (18) of molar absorptivities of Ap (15.0 × 10^3 M^{-1} cm^{-1}) and Up (9.90 × 10^3 M^{-1} cm^{-1}) at 260 nm in 0.01 M potassium phosphate buffer, pH 7 at 25°, were converted to 0.05 M ammonium acetate, pH 4.5 (15.1 × 10^3 M^{-1} cm^{-1} and 9.70 × 10^3 M^{-1} cm^{-1}, respectively). The formation of A(2′-5′)Ap from the dodecamer II was taken into account: it was assumed that 16% of A(2′-5′)Ap was formed, and that its molar absorptivity at 260 nm was the same as that of A(2′-5′)A [measured as 12.8×10^3 M^{-1} cm^{-1}, based on the literature value (17) of 12.9×10^3 M^{-1} cm^{-1} at 258 nm, pH 7]. The molar absorptivities of the other oligoadenylates were estimated from the equation

$$\epsilon_N = \epsilon_\infty + (\epsilon_0 - \epsilon_\infty)/N, \qquad [1]$$

where ϵ_N is the molar absorptivity of an oligomer of chain length N, and ϵ_0 and ϵ_∞ are those of a single residue and an infinite chain respectively (19). Values of ϵ_0 (17.19×10^3 M^{-1} cm^{-1}) and ϵ_∞ (10.32×10^3 M^{-1} cm^{-1}) at 257 nm were obtained from an unweighted linear least squares fit of this equation to the data compiled by Blake and Fresco (20). The values for $N = 1$, 7, and 8 were not included in the least squares fit, but the literature value for $N = 1$ was used in subsequent work. These values were used to weight the relative sizes of the peaks from RPC-5 HPLC. The detector of the HPLC operates at 254 nm, but the values predicted by the above equation were found to fit our experimentally determined values for $N = 6$ and $N = 12$ at 254 nm within 3%, and were therefore used for weighting the peaks without further correction (21).

Mixing Curves (22). The stoichiometry of the (A)$_{11}$Ap·poly(U) complex was determined using Job's method of continuous variations. The absorbance of the solutions at 40° for a fixed nucleotide base concentration of 30 mM was measured at 260 nm and 280 nm in a variable pathlength cell (Beckman) with a 25 μm teflon spacer. The solvent was 1.0 M ethylenediamine hydrochloride, at pH 8, 1 mM in Na$_2$EDTA. The curve was run in duplicate and was consistent with the formation of the triple helix A·2U. The mixing curve for the (A)$_5$Ap·poly(U) system was determined at 2° with a total nucleotide base concentration of 75 mM and also was consistent with the formation of the triple helix (21).

Melting Curves (22). The thermal denaturation of the complex (A)$_{11}$Ap·2poly(U) (at a ratio A:U of 0.3:0.7) was followed by measuring the absorbance at 260 nm and 280 nm as a function of temperature. The same cell, solvent, and total nucleotide base concentration was used as for the mixing curves: a single transition was observed at both wavelengths, with a t_m of 76°. The (A)$_5$Ap·2poly(U) complex, at a total nucleotide base concentration of 75 mM (0.3A:0.7U), also gave a single transition, with a t_m of 70° (21). When the curves were repeated using just the oligoadenylates, the expected noncooperative broad transitions were obtained (21, 22). No correction was made for thermal expansion of the solvent; the pathlength of the cell was determined solely by the teflon spacer, which changed less than 1% in thickness when going from 0° to 100° (23). The buffer absorbance was constant over the temperature range used. The curves were run in duplicate.

Hydrolysis of Dodecaadenylates I and II. Hydrolytic reactions were run at 40.8° ± 0.01° in 3 ml borosilicate-glass centrifuge tubes. The sample volume was 50 μl, and a teflon rod was inserted into the tube to within a few mm of the surface of the liquid to reduce the air space. The tube was then capped tightly with Parafilm. The solution was 10 mM in adenine bases, and 20 mM in uracil bases in those experiments in which poly(U) was present. The buffer was 1.0 M ethylenediamine hydrochloride, at pH 8; the addition of 10 mM Na$_2$EDTA did not affect the rates or products of hydrolysis. In one experiment, to test for the presence of nucleases, dodecamer I was hydrolyzed in 0.10 M Tris-acetate, at pH 7.8, containing 10 mM Na$_2$EDTA. The rate was very slow. At suitable times aliquot samples (5 μl) were withdrawn and mixed with 2.5 μl n-propanol in 20 μl of water. A

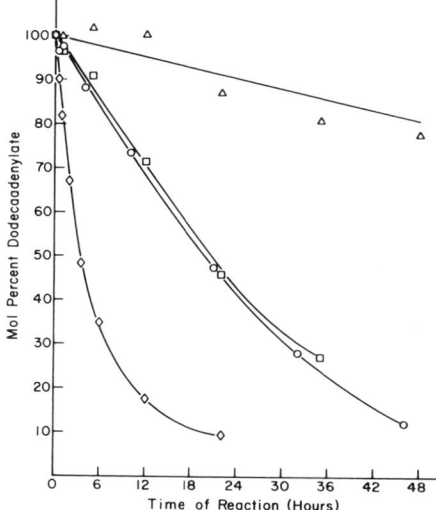

FIG. 1. Disappearance of dodecamer in 1.0 M ethylenediamine hydrochloride, pH 8, at 40°. A correction has been made for the small amount of degradation present in the starting material. △, dodecamer I with poly(U); □, dodecamer I without poly(U); ○, dodecamer II without poly(U); ◇, dodecamer II with poly(U).

further 22.5 μl of water was added, and 20 μl of the mixture analyzed on the RPC-5 column, using the KClO$_4$ buffer. This procedure allowed good recovery of oligoadenylates even in the presence of poly(U), and was found preferable to the use of a heated inlet system (to denature the helix), or a trichloroacetate buffer. A small amount of degradation had occurred during the previous manipulation of dodecamers I (15%) and II (10%). In the case of dodecamer I there was, in addition, about 5% of the tridecamer present at the start, arising from incomplete separation in the original DEAE-Sephadex chromatography. The disappearance of dodecamer for each case is shown in Fig. 1, and the appearance of hexamer in Fig. 2.

RESULTS AND DISCUSSION

The formation of dodecamer II was accomplished in 24% yield by incubating (A)$_5$A>p with two equivalents of poly(U) in ethylenediamine buffer at pH 8. When (A)$_5$Ap was employed in place of the cyclic phosphate, no dodecamer was formed, and when the poly(U) was omitted, the yield was less than 1%. The reaction appears to take place in a triple helix (A·2U), the geometry of which (24) is such that nucleophilic attack by the 5′-hydroxyl can occur most readily on the tetrahedral face of the phosphorus that is directly opposite the 3′-oxygen of the cyclic phosphate. An "in-line" displacement occurs (25), resulting in the cleavage of the 3′-O-P bond. Even if a pentacoordinate intermediate were formed during the reaction, it would be prevented from pseudorotation by the strong preference of the anionic oxygens to stay in the basal plane (26, 27). Thus the product has the 2′,5′-linkage. After 5 days of reaction time, the only peaks visible by DEAE-Sephadex chromatography were hexamer, dodecamer, and octadecamer. Our evidence for the structure of the dodecamer II includes the following: (1) dodecamer II with a cyclic phosphate terminus elutes at the same position as (A)$_{11}$A>p on RPC-5 chromatography in KCl/Tris buffer, and undergoes the same change in elution position on opening the ring with 0.1 M HCl. The same sep-

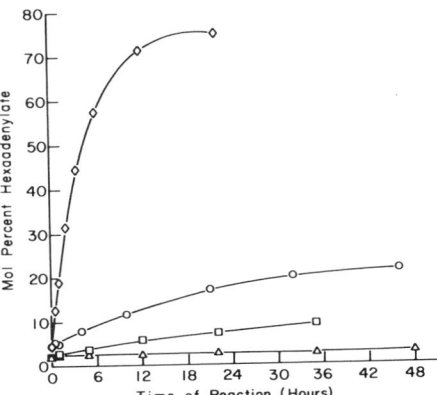

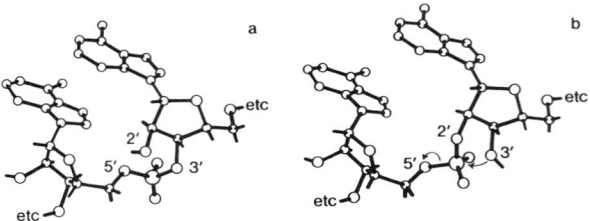

FIG. 3. (a) Part of a 3',5'-linked oligoadenylate as found in the triple helix poly(A)·2poly(U) (24). The poly(U) strands are not shown, and the phosphorus has been moved slightly towards the 2'-hydroxyl. If the 2'-hydroxyl were to attack phosphorus, a P-O⁻ would be forced to take up an apical position. Cleavage is slow. (b) The adenylate units are unchanged in position, but the bond is now the 2',5'-isomer. The 3'-hydroxyl is held in the correct position for cleavage to occur by an in-line mechanism. Both the 3'- and 5'-oxygens can take up apical positions in the trigonal bipyramidal intermediate or transition state. Cleavage is relatively rapid.

FIG. 2. Appearance of hexamer during the degradation of dodecamers I and II (see Fig. 1). Small amounts of hexamer were present at the start of the reaction: no correction has been made for this. Symbols are as in Fig. 1.

aration between the 2'- and 3'-phosphate termini can then be seen in both cases. (2) The rate of formation of dodecamer II is at a maximum at the start of the reaction (D. A. Usher and A. H. McHale, in preparation), when over 90% of the hexamer has a cyclic phosphate terminus. (3) Treatment of dodecamer II with RNase-T₂, and then with BAP gave a compound that had the same behavior on HPLC as authentic A(2'-5')A. (4) The yield of octadecamer is that expected from theory, based on the rates of formation and degradation of the dodecamer (D. A. Usher and A. H. McHale, in preparation). It would be difficult to reconcile points (1) and (2) with the existence of any significant amount of head to head coupling (3',3'-bonds, pyrophosphates, etc.).

Degradation Reactions. The cleavage reactions were run at 40°, a temperature chosen so that both dodecamers would be strongly helical in the presence of added poly(U), but would have a relatively low helical content in its absence. The cyclic phosphate termini were converted to the mixed 2'- and 3'-phosphates in order to prevent any further coupling during the hydrolysis reactions. A t_m of 76° was found for the (A)₁₁Ap·2poly(U) triple helix at the same total nucleotide and buffer concentrations as used in the degradation reaction. This figure is somewhat higher than one previously reported (28), presumably due to the higher nucleotide concentration employed here. There was no evidence for the formation of any other complex. The t_m for the triple helix of dodecamer II with 2poly(U) was not measured directly since the 2',5'-bond would have degraded appreciably during the time required to complete the measurement. However, it is likely that it would be similar to the figure of 76° found for the dodecamer I complex. The one 2',5'-bond is flanked on each side by five 3',5'-bonds, yet the t_m for a triple helix of poly(U) with the *all* 2',5'-linked octamer has previously been reported (28) to differ by only 6° from that of the related complex with the all 3',5'-octamer. It is also unlikely that the presence of this one 2',5'-bond will induce a major change in the structure of the helix, e.g., to a left-handed form (29, 30).

Fig. 1 shows the disappearance of dodecamers I and II; in the absence of poly(U) both dodecamers hydrolyze at about the same rate (about 0.03 hr⁻¹), but the addition of two equivalents of poly(U) causes a 5-fold decrease in the initial rate of loss of I, while II hydrolyzes about seven times more rapidly. It must be stressed that these figures refer to initial

rates; the kinetics become complex at later times as shorter oligomers accumulate, and the melting temperatures decrease. The potential reversibility of the cleavage reaction has also been taken into account. The distribution of products is striking; dodecamer II in the presence of poly(U) gives as an initial breakdown product over 90% of the hexamer, in spite of there being only one 2',5'-bond to ten 3',5'-bonds (Fig. 2). In dodecamer I there are eleven 3',5'-bonds. Thus, the specific rate for one 2',5'-bond in the triple helix is about 0.18 hr⁻¹, while that for one 3',5'-bond under the same constraint appears to be about ¹⁄₁₁ of 0.006 hr⁻¹, or 5.5×10^{-4} hr⁻¹. However, there is a further correction that must be made to these figures. The formation of undecamer plus monomer account for about 38% of the disappearance of dodecamer I in the presence of poly(U), and this appears to be due to an end, or fraying effect (28, 31, 32). This explanation can also account for an increasing percentage of what appears to be pentamer that is formed in the cleavage reaction of dodecamer II with poly(U); 0–4% of the products in the early stages which increases to 8% after 22 hr. It probably arises directly from the hexamer, which is present in a large concentration after a few hours. The two end residues, although 3',5'-linked, are not firmly held in the helix, and are therefore not as fully protected. If the above comparison is made by considering the central firmly-held bonds of helical dodecamer I, the rate of cleavage drops to 2×10^{-4} hr⁻¹, and one 2',5'-bond is seen to cleave about 900 times faster than one fully constrained 3',5'-bond.

Hydroxide-catalyzed cleavage of RNA is thought to involve intramolecular nucleophilic attack on phosphorus by the anion of the neighboring hydroxyl group (33). We feel that the most likely explanation for the present behavior is that in the helix, attack on phosphorus by the ionized 2'-hydroxyl of I would initially force an anionic phosphoryl oxygen into an apical position. This is energetically unfavorable (26, 27), and thus this adjacent displacement does not readily occur (Fig. 3).† In contrast to this, the addition of poly(U) to dodecamer II causes the 2',5'-linkage to be held

---

† As an alternative viewpoint, the diester conformation in the 2',5' or "in-line" case may correspond to a higher ground state energy than in the 3',5' or "adjacent" case (34, 35). If the difference were reduced in the transition state, this could account for the faster rate in the case of the 2',5'-linkage. This is consistent with the observation that helices formed from poly(U) and 2',5'-linked oligoadenylates always have lower values of t_m than those formed from the 3',5'-linked oligomers (28).

in such a way as to favor an in-line displacement by the ionized 3'-hydroxyl. In this sense poly(U) could be considered a catalyst for the cleavage of this 2',5'-bond, increasing the rate by a modest factor of 18, very largely through orientation effects. The catalytic function of the ethylenediamine is not yet known in detail, although it appears to act kinetically as a general base (2). In this reaction then, the acid or base catalysis is intermolecular, unlike the case of an enzyme such as ribonuclease-A, which supplies the acid-base groups as well as ensuring the correct orientation of binding (36).

In the absence of poly(U), the average initial rate of hydrolysis of one 3',5'-bond is about 0.0023 hr^{-1} (in dodecamer II) or 0.003 hr^{-1} (in dodecamer I), compared to about 0.01 hr^{-1} for the 2',5'-bond. The faster rate of hydrolysis for the 2',5'-bond may merely be due to the known residual helical character of an oligoadenylate when no complementary polynucleotide is present (22). If the rates of cleavage of the 2',5'- and 3',5'-bonds were identical in the absence of any helix, then a residual 4% helix would account for this rate difference. Though it is highly probable that in the absence of poly(U) the conformational restriction would be less stringent than in the triple helix, and the figure of 4% would then be low. The rate of cleavage of the central 3',5'-bonds of dodecamer I is decreased by a factor of 15 on the addition of poly(U), and this figure is consistent with earlier work (4, 5) on the stability towards alkaline hydrolysis shown by purine-rich regions of single-stranded RNA. The purine bases show a greater tendency than the pyrimidines to stack, and it was thought that this ordering protected the diester bonds against cleavage, although the reason for this effect was not understood. It appears probable that these arguments apply also to the double helix (24), and to bases other than adenine. We conclude that if the original polynucleotides were partly or wholly in the ribose series (1), the use of the 3',5'-bond, in combination with a right-handed helix, had a large selective advantage over the use of the 2',5'-bond (helical or not) for the storage of genetic information.

We thank Drs. Elson, Widom, Loudon, and Uhlenbeck for helpful discussions, the 3M Company and Miles Laboratories for gifts of material, and the National Science Foundation (Grant BMS 72-02370) for generous support of this research. A.H.M. was supported in part by a Training Grant from the National Institutes of Health (GM 00834).

1. Orgel, L. E. & Lohrmann, R. (1974) Acc. Chem. Res. 7, 368–377.
2. Renz, M., Lohrmann, R. & Orgel, L. E. (1971) Biochim. Biophys. Acta 240, 463–471.
3. Usher, D. A. (1972) Nature New Biol. 235, 207–208.
4. Lane, B. G. & Butler, G. C. (1959) Biochim. Biophys. Acta 33, 281–283.
5. Bock, R. M. (1967) in Methods in Enzymology, eds. Grossmann, L. & Moldave, K. (Academic Press, New York), Vol. 12A, pp. 218–221.
6. Usher, D. A., Richardson, D. I., Jr. & Oakenfull, D. G. (1970) J. Am. Chem. Soc. 92, 4699–4712.
7. Pearson, R. L., Weiss, J. F. & Kelmers, A. D. (1971) Biochim. Biophys. Acta 228, 770–774.
8. Walker, G. C., Uhlenbeck, O. C., Bedows, E. & Gumport, R. I. (1975) Proc. Nat. Acad. Sci. USA 72, 122–126.
9. Usher, D. A., McHale, A. H. & Yee, D. (1975) Anal. Chem. 47, 783–784.
10. Westhead, E. W. & McLain, G. (1966) in Biochemical Preparations, ed. Maehly, A. C. (John Wiley and Sons, Inc., New York), Vol. 11, p. 40, footnote 10.
11. Craig, L. C. & King, T. P. (1962) in Methods of Biochemical Analysis, ed. Glick, D. (Interscience, New York), Vol. 10, pp. 175–199.
12. Beers, R. F., Jr. (1960) J. Biol. Chem. 235, 2393–2398.
13. Singer, B. & Fraenkel-Conrat, H. (1961) Virology 14, 59–65.
14. Markham, R. (1957) in Methods in Enzymology, eds. Colowick, S. P. & Kaplan, N. O. (Academic Press, New York), Vol. 3, pp. 805–810.
15. Rushizky, G. W. & Sober, H. A. (1963) J. Biol. Chem. 238, 371–376.
16. Egami, F., Takahashi, K. & Uchida, T. (1964) in Progress in Nucleic Acid Research and Molecular Biology, eds. Davidson, J. N. & Cohn, W. E. (Academic Press, New York), Vol. 3, pp. 59–101.
17. Kondo, N. S., Holmes, H. M., Stempel, L. M. & Ts'o, P. O. P. (1970) Biochemistry 9, 3479–3498.
18. Volkin, E. & Cohn, W. E. (1954) in Methods of Biochemical Analysis, ed. Glick, D. (Interscience, New York), Vol. 1, pp. 287–305.
19. Bloomfield, V. A., Crothers, D. M. & Tinoco, I., Jr. (1974) in Physical Chemistry of Nucleic Acids (Harper and Row, New York), p. 299.
20. Blake, R. D. & Fresco, J. R. (1973) Biopolymers 12, 775–789.
21. McHale, A. H. (1976) Ph.D. Dissertation, Cornell University, Ithaca.
22. Bloomfield, V. A., Crothers, D. M. & Tinoco, I., Jr. (1974) in Physical Chemistry of Nucleic Acids (Harper and Row, New York), chap. 6.
23. Hodgman, C. D., ed. (1961–1962) Handbook of Chemistry and Physics (Chemical Rubber Publishing Co., Cleveland, Ohio), p. 1557.
24. Arnott, S. & Bond, P. J. (1973) Nature New Biol. 244, 99–101.
25. Usher, D. A. (1969) Proc. Nat. Acad. Sci. USA 62, 661–667.
26. Westheimer, F. H. (1968) Acc. Chem. Res. 1, 70–78.
27. Frank, D. S. & Usher, D. A. (1967) J. Am. Chem. Soc. 89, 6360–6361.
28. Michelson, A. M. & Monny, C. (1967) Biochim. Biophys. Acta 149, 107–126.
29. Lipsett, M. N., Heppel, L. A. & Bradley, D. F. (1961) J. Biol. Chem. 236, 857–863.
30. Tazawa, I., Tazawa, S., Stempel, L. M. & Ts'o, P. O. P. (1970) Biochemistry 9, 3499–3514.
31. Patel, D. J. (1975) Biochemistry 14, 3984–3989, and references therein.
32. Borer, P. N., Kan, L. S. & Ts'o, P. O. P. (1975) Biochemistry 14, 4847–4863.
33. Bruice, T. C. & Benkovic, S. J. (1966) in Bioorganic Mechanisms (W. A. Benjamin, Inc., New York), Vol. II, pp. 37–48.
34. Perahia, D., Pullman, B. & Saran, A. (1974) Biochim. Biophys. Acta 340, 299–313.
35. Newton, M. D. (1973) J. Am. Chem. Soc. 95, 256–258.
36. Richards, F. M. & Wyckoff, H. W. (1971) "Bovine Pancreatic Ribonuclease," in The Enzymes, ed. Boyer, P. D. (Academic Press, New York), 3rd ed., Vol. IV, pp. 647–806.

Early Chemical Evolution of Nucleic Acids:
A Theoretical Model

Abstract. *Recent experimental work suggests a possible cyclical pathway for early prebiotic oligonucleotide formation that involves (i) dry-state (nontemplate) synthesis of random copolymers with mixed 2',5' and 3',5' bonds, (ii) passage of these oligomers into solution at low temperatures, and (iii) a preferential hydrolysis of the 2',5' bond in any short helices that have formed. This early system could have selected for complementary sequences that were largely 3',5'-linked, but may not have selected efficiently for a single enantiomer of ribose.*

In a recent review (*1*) of theories concerning the origin of life, the assumption was made that the oligonucleotides formed in prebiotic times would have been better able to resist hydrolytic degradation if they could have adopted a helical conformation. A further assumption was made that there existed a nonenzymic mechanism by which a single-stranded RNA molecule could be replicated with a fidelity such that one error was made per 100 bases. Unfortunately, actual attempts to demonstrate template-directed formation of oligoribonucleotides under simulated prebiotic conditions have invariably led (*2, 3*) to the production of a large excess of the unnatural or 2',5' internucleotide bond (*4*). Further, the rate of hydrolysis of this 2',5' bond actually increases when the oligonucleotide of which it is a part forms a right-handed helix (*5*). By contrast, a modest excess of the natural 3',5' bond is formed (*6*) in dry-state polymerization of adenosine 2',3'-phosphate catalyzed by ethylenediamine, but this reaction occurs in the absence of a template. We now report that these recent experimental observations combined with the earlier theory (*1*) show how, under the influence of a daily heating and cooling cycle, a system of oligonucleotide formation could have arisen that selected for (i) the natural 3',5' bond, (ii) longer oligomers, and (iii) complementary sequences. The possible importance in prebiotic chemistry of the natural cycles in temperature and humidity (day and night, tides, seasons) has been explored by others, both in theory (*1*) and through experiment (*7*). In addition, it has been suggested that simulation experiments are more convincing if they are carried out at temperatures that resemble those found on the earth today (*2*), and a recent calculation of the average temperature on the prebiotic earth (about 3.5 to 4 billion years ago) lends support to this suggestion (*8*). The conditions given in Fig. 1 may be taken to represent a semidesert locale that has a reasonably large swing in surface temperature and humidity between day and night. The actual temperatures shown were measured in the Namib Desert (South-West Africa) by Buskirk over a 26-hour period in August 1973 (*9*).

The cyclical scheme is shown in Fig. 1, and assumes the existence of ribonucleoside 2'- and 3'-monophosphates (*2*), which, in turn, are converted into the nucleoside 2',3'-phosphates—for example, by reaction with a condensing agent (*10*), or by heating with a catalyst. The cycle starts with the dry-state synthesis of random copolymer oligomers. This occurs at slightly elevated temperatures in the presence of a catalyst (such as imidazole) (*6*) but is not directed by a template, and the ratio of natural to unnatural bonds formed is about 2 : 1. As the sun sets, the surface temperature decreases, and the oligomers go into solution in the dew which forms. At these low temperatures, some of the oligomers will find complementary partners with which they can form short helices (*11*). As the sun rises on the following morning, the solution warms up, and a preferential degradation of the helical 2',5' bonds will occur (*5*). Helical 3',5' bonds will be protected, while nonhelical 2',5' or 3',5' bonds will degrade at an intermediate rate. The slow rise in temperature is important in that a sudden increase would merely melt the helices and not allow the selective degradation of the "unnatural" bonds to occur. The most stable species under these circumstances are long, complementary oligomers that are entirely 3',5'-linked (higher melting temperature, T_m). Helices that contain a number of gaps or nicks from hydrolysis of 2',5' bonds will have lowered thermal stability and a greater tendency to denature as the temperature increases. Eventually, the solution will dry out, and the residue will consist of a mixture of random copolymers of shorter chain length but with an increased fraction of 3',5' bonds than previously. In some of these oligomers, the terminal phosphate will be in the form of the 2' or 3' monoester, and in others as the 2',3'-phosphate (*12*). In order to ensure further coupling between these oligomers, the terminal phosphates should be in the 2',3' form. Reactivation of the 2' and 3' monoesters to give the cyclic phosphate without causing simultaneous degradation of the internucleotide diester linkages is not a serious problem in principle. The 2' (or 3') terminal monoesters exist as the dianion at mildly alkaline pH, whereas the internal diester bonds are monoanions and are less nucleophilic toward a potential activating reagent. Provided that the rate of renewed dry-state synthesis from these remaining oligomers exceeds the rate of their breakdown, the cycle is completed and conceivably could go on to produce longer complementary oligomers that are largely 3',5'-linked. An upper limit to the length of the chains appears to be set by (i) the nonzero rate of hydrolysis of the helical 3',5' bonds, (ii) the increasing probability, as the chain length increases, that the rate of chain cleavage will exceed the rate of coupling during dry-state synthesis, and (iii) the decreasing probability, as the chain length increases, that random synthesis or coupling will generate complementary sequences. At present, it does not appear possible to calculate the upper limit from the experimental data that are now available.

The above scheme is not a template-directed synthesis, and therefore by-

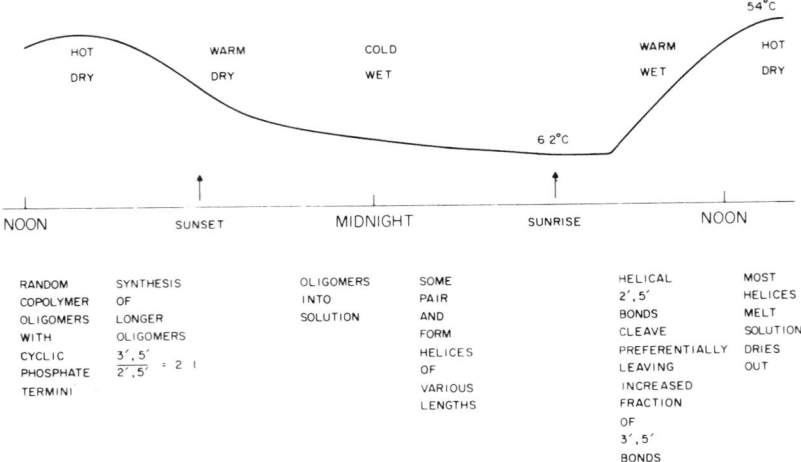

Fig. 1. One possible scheme for the synthesis of oligonucleotides under prebiotic conditions. The surface temperatures shown (continuous line) were measured on a west-facing slope in the Namib desert in August 1973 (9).

passes the known difficulty of including pyrimidine monomers in such a process (2). However, a system capable of direct template synthesis eventually must have arisen, possibly via the aggregation mechanism suggested by Kuhn (1). Here the problem is again not insuperable, as Orgel and Lohrman (2) have shown that 5'-activated pyrimidine nucleotides can be incorporated in a template-directed synthesis if they are carried to the template as purine-containing oligomers such as pUpG (U, uridine; G, guanosine).

Information relating to the potential enantiomeric specificity of this reaction is sparse. Both the 2',5' and 3',5' isomers of (L,L)-ApA (A, adenosine) have been found to form a right-handed triple helix with two strands of (D)-polyuridylic acid (13). The 2',5' isomer seems to resist inclusion in this right-handed helix less strongly than does the 3',5' isomer. Indeed the T_m for the helix with (L,L)-2',5' ApA is almost equal to that for the similar helix of (D,D)-3',5'-ApA, while that for (L,L)-3',5'-ApA is several degrees lower (13). I concur with Tazawa *et al.* (13) that, as shown by CPK (Corey, Pauling, Koltun) molecular models, there is no obvious stereochemical hindrance to the inclusion of (L)-nucleoside units in a right-handed helix [see (1)]. Further, it appears from the models that even when the nucleotides involved are of the L configuration, and are constrained in a right-handed helix, the 2',5' bond should again be more labile than the 3',5' bond. This prediction is independent of whether the (L)-nucleotide is the 5' or the 2'(3') partner to the bond,

or, indeed, whether both partners are (L)-nucleotides. The 2',5' isomer has the correct geometry for an in-line displacement, the 3',5' for an adjacent displacement (5). The same argument must, of course, apply to the (D)-nucleotides in a left-handed helix, but at present the only experimental proof of such a difference in hydrolytic stability relates to the all-(D) oligomers in a right-handed helix (5). It is not clear to what extent the results of Tazawa *et al.* (13) may be applied to random or block copolymers of (L)- and (D)-nucleotides, and more experimental work on these systems is required. However, it seems possible that the inclusion of, for example, (L)-nucleotides in a right-handed helix that is composed predominantly of (D)-nucleotides may tend to lower T_m if the L units are 3',5'-linked, but have relatively little effect on T_m if they are 2',5'-linked. In the first case, the helix will melt at a somewhat lower temperature and thus become more susceptible to hydrolysis; in the second case, the helix will be relatively resistant to melting initially, but the helical form will cause a relatively rapid hydrolysis of the 2',5' bonds to occur. If this conjecture has any validity, it would mean that a selection for one enantiomer could take place, but more weakly than the selection for the 3',5' bond. A stronger selection for one enantiomer may have occurred later with the use of 5'-activated nucleotides (14).

D. A. USHER

Spencer T. Olin Chemistry Laboratory,
Cornell University,
Ithaca, New York 14853

References and Notes

1. H. Kuhn, *Angew. Chem. Int. Ed. Engl.* **11**, 798 (1972).
2. L. E. Orgel and R. Lohrmann, *Acc. Chem. Res.* **7**, 368 (1974).
3. D. A. Usher and A. H. McHale, *Science* **192**, 53 (1976).
4. From an analysis of the geometry of the reaction [D. A. Usher, *Nature (London) New Biol.* **235**, 207 (1972)], it appears unlikely that an efficient template-directed system will be found that can yield the 3',5' bond by attack of a 5'-hydroxyl on the phosphorus of a nucleoside 2',3'-phosphate. This is significant, for the degradation of RNA by mild alkaline hydrolysis (nonenzymic) proceeds via the nucleoside 2',3'-phosphates and yields ultimately a mixture of the 2'- and 3'-monophosphates [D. M. Brown and A. R. Todd, *J. Chem. Soc.* (1952), p. 52]. These, upon activation, give again the 2',3'-phosphates. Thus in a nonenzymic system undergoing both synthesis and degradation of RNA, the 2',3'-phosphates may be expected to be the most abundant activated nucleotides present. There is no evidence that present-day enzymes make use of the 2',3'-phosphates in any template copying reaction that gives rise to the 3',5' bond. Instead, nucleotides activated at the 5' position are employed, and in this manner the geometrical restriction referred to above is obviated. Thus a reasonable evolutionary sequence may have been: early formation of oligomers as outlined in this report, followed by the evolution of "enzymes" that selectively gave the 3',5' linkage from 5'-activated nucleotides. A relatively small change in these same enzymes would then enable them to synthesize 3',5'-linked oligomers from 5'-activated deoxynucleotides, with a further gain in hydrolytic stability of the product.
5. D. A. Usher and A. H. McHale, *Proc. Natl. Acad. Sci. U.S.A.* **73**, 1149 (1976). The relative rates at 40°C, pH 8, 1M aqueous ethylenediamine were, 900 : 50 : 15 : 1 for 2',5' helical : 2',5' nonhelical : 3',5' nonhelical : 3',5' helical. The "nonhelical" figures probably reflect the presence of some residual helix in the oligoadenylates used.
6. M. S. Verlander, R. Lohrmann, L. E. Orgel, *J. Mol. Evol.* **2**, 303 (1973).
7. M. W. Neuman, W. F. Neuman, K. Lane, *Curr. Mod. Biol.* **3**, 277 (1970); M. J. Bishop, R. Lohrmann, L. E. Orgel, *Nature (London)* **237**, 162 (1972).
8. M. H. Hart, "The evolution of the atmosphere of the earth", Third College Park Colloquia on Chemical Evolution, College Park, Md., 1 October 1976. See also C. Sagan and G. Mullen, *Science* **177**, 52 (1972). Calculations of this type have not yet been extended to provide information on the expected variations in surface temperature. Many additional variables would need to be considered [for example, the probable faster rate of rotation of the earth; W. Munk, *Q. J. R. Astron. Soc.* **9**, 352 (1968)] in order to obtain even a rough estimate of the daily temperature extremes.
9. R. Buskirk, in preparation.
10. J. P. Ferris, G. Goldstein, D. J. Beaulieu, *J. Am. Chem. Soc.* **92**, 6598 (1970); see J. Oró and E. Stephen-Sherwood, *Origins Life* **5**, 159 (1974).
11. P. N. Borer, B. Dengler, I. Tinoco Jr., O. C. Uhlenbeck, *J. Mol. Biol.* **86**, 843 (1974).
12. D. A. Usher and A. H. McHale, in preparation.
13. I. Tazawa *et al.*, *Biochemistry* **9**, 3499 (1970).
14. H. Schneider-Bernloehr, R. Lohrmann, J. Sulston, L. E. Orgel, H. T. Miles, *J. Mol. Biol.* **47**, 257 (1970).
15. Supported by NSF grant PCM76-11944. I thank Dr. R. Buskirk for permission to reproduce the temperatures shown in Fig. 1.

22 October 1976; revised 17 December 1976

A Nonenzymatic RNA Polymerase Model

Abstract. *Polynucleotide templates containing C (cytidine) as the major component facilitate the synthesis of oligonucleotides from mixtures of the activated mononucleotide derivatives (as indicated by structure 1 in the text). A nucleotide is incorporated into oligomeric products if and only if its complement is present in the template. The reaction has a high fidelity and produces products with mean chain lengths of six to ten nucleotides. Bases other than guanosine are incorporated within oligomers or at their 3' termini, but rarely at their 5' termini.*

The activated nucleotide 2-MeImpG (*1*) undergoes a template-directed condensation on poly(C) to give predominantly 3',5'-linked oligo(G)'s in excellent yield (*2, 3*). Synthesis proceeds in the 5' to 3' direction. The mean chain length of the product can be as high as 15, and oligomers with chain length substantially in excess of 30 can be detected. The fidelity of the reaction is high—U, C, and A are discriminated against by factors of 100 to 500, depending on the conditions of the reaction. In collaboration with Frazier and Miles (*4*), we showed that this reaction occurs in a Watson-Crick double helix and yields double-stranded poly(C)·oligo(G) melting above 100°C as the initial product.

The corresponding reactions involving other monomer-homopolymer double helices cannot be carried out for various reasons. Poly(U) forms triple helices rather than double helices with monomeric adenosine derivatives, while poly(G) forms a very stable self-structure

1

(N = U, C, A or G)

ture that prevents G-C interaction (*5*). Poly(A) does not interact with U derivatives strongly enough to overcome the poor U-U stacking and so does not form helical complexes poly(A)·U of any kind (*5*). It is possible to circumvent these difficulties by using copolymers of C with one or more additional bases as templates. A poly(CU) template cannot form triple-stranded segments if C is present in sufficient amount. Similarly, poly(CG) copolymers cannot form long-interchain G-C self-structure, although they may contain short-intrachain self-structure based on C·G pairing if the G content is high enough. Finally, poly(CA) does interact with U derivatives, since U can stack next to G's even though it will not stack between other U's. In this report we present our findings on the efficiency and fidelity of incorporation of complementary bases on random copolymer templates containing C as a major constituent.

All reactions were carried out at 0°C and pH 8 and were analyzed after 7 days and again after 21 days (*6*). The reaction mixtures contained 1.2*M* NaCl, 0.22*M* MgCl₂, 0.4*M* 2,6-lutidine·HCl buffer, 0.05*M* 2-MeImpU, 0.05*M* 2-MeImpC, 0.05*M* 2-MeImpA, 0.05*M* 2-MeImpG, and an amount of template equivalent to

a 0.01*M* solution of total mononucleotides. Tubes were prepared in sets of four, and a different ^{14}C-labeled substrate (0.25 μCi) was added to each of them; for example, ^{14}C-labeled 2-MeImpG was added to the first tube of each set. The final volume in each tube was 20 μl.

We determined the percentage of total radioactive material incorporated into longer oligomers by carrying out paper chromatography in a mixture of *n*-propanol, ammonia, and water (55:10:35) and measuring, with a scintillation counter, the radioactivity remaining at the origin of the paper (*6*) (Table 1). The yield expected for a completely efficient incorporation of a nucleotide N is $0.2\,p$, where p is the molecular proportion of N in the template; for example, 5 percent incorporation of U on a poly(C₃A) template. We have also calculated the efficiency with which a nucleotide in the template directs the incorporation of its complement into oligomeric products long enough to remain at the origin of our chromatograms. These derived data are also included in Table 1.

To determine the way in which a nucleotide is distributed between 5'-terminal, internal, and 3'-terminal positions, we degraded selected oligomeric products with alkali and analyzed them by electrophoresis (*6*). The results of this analysis are presented in Table 2, together with estimates of the mean chain lengths derived from them.

In our chromatographic system, oligomers four or more units long and rich in G. and other oligomers more than five units long, remain very close to the origin of the chromatograms. The first row

Reprinted with permission from *Science*,
Volume 219, pp. 859-862. © 1983 by the American Association for the Advancement of Science.

of Table 1 confirms that negligible amounts of material as long as this are formed from a mixture of activated nucleotides in the absence of a template.

The data in the second row confirm that G is incorporated in almost quantitative yield on poly(C) and that the fidelity of this reaction is high. The next three rows confirm that, under the conditions of our experiment, the other homopolymers are not efficient templates. The very small amounts of material incorporated on these homopolymers are possibly attached to the ends of the template strands. Previously reported experiments have shown that poly(U) does facilitate the formation of oligomers of A up to the octamer from ImpA, but the reaction conditions were more favorable (7). We can, in fact, detect short oligomers of A, up to at least the tetramer on our paper chromatograms of reaction mixtures containing poly(U), while no comparable amounts of such oligomers are formed on other homopolymers or in the absence of a template.

The data presented in the remainder of Table 1 substantially extend the results of Ninio and Orgel (8). They show that random copolymers containing C as the major component facilitate the incorporation of substantial amounts of a base other than G into oligomeric products if and only if the complementary base is present in the template. In this restricted sense, our reaction models an RNA polymerase. We do not know how far the resemblance will extend.

The fidelity of synthesis on random copolymers cannot be determined unambiguously from incorporation data. First, it is not possible to determine whether a base complementary to a component of the template has been incorporated correctly on its complement or erroneously on some other base. This is unfortunate, since the erroneous incorporation of small amounts of G on U by wobble-pairing could never be detected on a polymer rich in C. The absence of U incorporation on poly(CG) templates, however, suggests that wobble-pairing may not be an important source of misincorporation. Second, it is not possible to partition misincorporation between different components of the template. Thus, C misincorporated on a poly-(CUA) template could have paired with C, U, or A (or could have attached to chain termini independently of base pairing).

In view of these ambiguities, we can draw only general conclusions from the data in Table 1. The main result is clear—the fidelities of all the reactions, insofar as they can be determined from the incorporation data, are high. The extent of misincorporation of C and A on mixed templates is not significantly different from the extent of misincorporation on poly(C). The misincorporation of U on poly(CU) may be greater than on poly(C), but if it is the effect is very small. Considering the simplicity of the system, the overall fidelity is surprisingly high.

The final proof that some mixed template sequences can be copied accurately must come from data obtained with oligonucleotide templates of known sequence. In the meantime, the incorporation data in Table 1 make it virtually

Table 1. Efficiency of incorporation of 2-MeImpU, 2-MeImpC, 2-MeImpA, and 2-MeImpG on polynucleotide templates. The numbers that are not underlined represent the fraction of the total radioactivity remaining at the origin of the paper chromatograms. Underlined numbers are efficiencies, defined as the ratio of the amount of base incorporated into products to the amount of its complement in the template, expressed as a percentage. To obtain these numbers the corrected incorporation of U, C, or A on a copolymer template was estimated by subtraction of the average incorporation of the base on poly(U), poly(C), poly(A), and poly(G) templates, from the observed incorporation on the relevant copolymer. In the case of G, we subtracted the average incorporation on poly(U), poly(A), and poly(G). The corrected amount of the base incorporated was then compared with the amount of its complement in the template, to obtain the efficiency. Yields shown in parentheses are uncertain because the amount of incorporation on the template is comparable to that in some of the controls.

Compositions: (a) $C_4(G_{0.42}U_{0.42}A_{0.15})$; (b) $C_2(G_{0.49}U_{0.30}A_{0.21})$; (c) $C_{7.4}(U_{0.67}A_{0.33})$; (d) $C_{3.6}(U_{0.69}A_{0.31})$; (e) $C_{2.8}(U_{0.57}A_{0.43})$; (f) $C_2(U_{0.6}A_{0.4})$; and (g) $C_{0.9}(U_{0.55}A_{0.45})$.

| Template | | 2-MeImpN | | | | | | |
	G 7d	G 21d	C 7d	C 21d	A 7d	A 21d	U 7d	U 21d
–	0.08	0.21	0.15	0.14	0.03	0.03	0.03	0.03
C	19.2	19.4	0.18	0.25	0.10	0.15	0.08	0.13
(C efficiency)	_96_	_97_						
G	0.15	0.23	0.15	0.14	0.03	0.03	0.03	0.03
U	0.07	0.50	0.13	0.23	0.09	0.11	0.03	0.09
A	0.07	0.18	0.13	0.16	0.03	0.14	0.02	0.04
CG 8:1	15.6	15.6	1.74	1.86	0.08	0.17	0.09	0.14
(eff)	_87_	_85_	_72_	_75_				
CG 5.4:1	13.2	16.8	2.32	2.51	0.09	0.15	0.11	0.17
(eff)	_78_	_97_	_69_	_73_				
CG 1.9:1	0.34	1.64	0.20	0.37	0.04	0.08	0.03	0.05
(eff)	_(2)_	_11_	_(0.7)_	_(2.4)_				
CU 11.3:1	19.1	19.9	0.18	0.24	0.95	1.06	0.11	0.16
(eff)	_104_	_107_			_54_	_59_		
CU 6.9:1	17.2	17.1	0.16	0.22	1.44	1.52	0.11	0.17
(eff)	_98_	_96_			_55_	_56_		
CU 2.9:1	14.5	14.1	0.18	0.22	2.58	2.76	0.14	0.20
(eff)	_97_	_96_			_49_	_53_		
CU 1.3:1	9.41	10.4	0.18	0.22	2.29	2.94	0.13	0.21
(eff)	_83_	_88_			_26_	_32_		
CA 9.5:1	17.3	17.9	0.22	0.16	0.11	0.19	1.18	1.31
(eff)	_95_	_97_					_61_	_67_
CA 5.1:1	15.0	16.2	0.25	0.36	0.13	0.22	1.57	1.79
(eff)	_90_	_96_					_48_	_53_
CA 2.9:1	9.46	10.8	0.21	0.31	0.10	0.18	1.07	1.42
(eff)	_63_	_70_					_21_	_27_
CA 1.9:1	7.07	8.46	0.16	0.28	0.10	0.18	0.98	1.40
(eff)	_53_	_62_					_14_	_19_
CA 0.89:1	1.28	2.50	0.14	0.20	0.04	0.11	0.18	0.33
(eff)	_13_	_24_					_(1.6)_	_(2.9)_
CGUA[a]	13.4	13.6	0.97	1.17	0.75	0.93	0.46	0.62
(eff)	_83_	_83_	_51_	_60_	_42_	_51_	_72_	_95_
CGUA[b]	5.47	9.16	0.51	0.94	0.30	0.68	0.24	0.49
(eff)	_40_	_66_	_12_	_25_	_12_	_30_	_15_	_32_
CUA[c]	16.7	17.1	0.19	0.26	0.82	0.88	0.51	0.60
(eff)	_94_	_95_			_48_	_50_	_60_	_68_
CUA[d]	14.5	14.5	0.19	0.25	1.23	1.30	0.64	0.80
(eff)	_92_	_91_			_39_	_40_	_40_	_50_
CUA[e]	11.8	12.6	0.19	0.25	1.38	1.49	0.59	0.77
(eff)	_80_	_84_			_44_	_47_	_25_	_31_
CUA[f]	6.60	7.60	0.20	0.28	1.35	1.68	0.47	0.69
(eff)	_49_	_55_			_32_	_39_	_17_	_24_
CUA[g]	1.55	5.28	0.13	0.25	0.30	0.79	0.07	0.28
(eff)	_16_	_54_			_(4.0)_	_13_	_(0.7)_	_(5.0)_

certain that limited regions of each template are copied accurately. They do not establish that extended processive synthesis of the kind that is the rule with DNA and RNA polymerase occurs. However, they strongly suggest that a primitive transcription of informational macromolecules has been achieved for the first time in a nonenzymatic system.

The mean chain lengths of the products of all of these reactions lie in the range 6 to 10 (Table 2). In every case, a nucleotide other than G occupies internal positions in oligomeric products 60 to 80 percent of the time, and occurs at the 3'-terminus 20 to 40 percent of the time. Bases other than G rarely occur at the 5'-terminus. This suggests, but does not prove, that oligomer synthesis on random copolymers, like the synthesis of oligo(G)'s on poly(C) (9), is initiated by the synthesis of an ImpGpG dimer, which then extends in the $5' \rightarrow 3'$ direction by the addition of activated monomers. It has been shown (10) that in the reactions on each of the random poly(UC) templates, G and A are distributed in more or less constant ratio among oligomers from 5 to more than 20 nucleotides in length.

It has not proved possible to determine accurately the proportion of 2',5' and 3',5' linkages formed when bases other than G are incorporated into oligomers. Our preliminary results indicate that 3',5' linkages are substantially in excess of 50 percent whenever U, C, or A is incorporated at the 3'-terminus of a chain. When U, C, or A is incorporated internally, we find that the sequence $pG^{3'}p^{5'}X^{3'}p^{5'}G$ is present in substantially more than 40 percent of the product, suggesting again that 3',5' linkages are present substantially more than 60 percent of the time. We will need to carry out experiments with much higher levels of radioactivity in order to obtain more accurate data on the regiospecificity of these reactions.

On templates rich in C there is little further change in the amounts of monomers incorporated into oligomers after 7 days (Table 1). However, on templates containing substantial amounts of bases other than C, incorporation increases significantly between 7 and 21 days. Further experiments are needed to determine whether the slower reaction rate on the latter templates is due to the difficulty of forming ImpGpG initiators or due to the slowness of propagation past sequences of "non-C" bases.

Much further experimental work is needed to provide the basis for a detailed description of oligomer synthesis on heteropolymer templates. The following discussion of the template properties characteristic of individual copolymers is therefore tentative.

Poly(CA) templates should provide the simplest picture since they cannot form internal self-structures or wobble-pairs with the monomers. The efficiency of incorporation of both G and U on poly(CA) templates falls steadily as the A content of the template increases. This trend is to be anticipated since the weakness of the stacking forces between U and other bases will lower the stability of the reactive template-oligomer-monomer complex. However, the extent of incorporation of U is surprising, since efforts to incorporate U or T on A-containing templates in other systems have failed (8).

We believe that the most likely explanation of the unexpectedly high G:U ratio in the products is "looping-out." It has been demonstrated (11) that when a copolymer such as poly(A_8U) is mixed with poly(U), an organized helix is formed in which U pairs with A almost exactly as in a homopolymer poly(A), poly(U) structure, but with intervening nonpairing U bases looped out. Looping-out would certainly occur with poly(CA) if mixed with poly(G), or with long enough oligo(G)'s.

In our system, G is incorporated into oligomers more rapidly than U, because G stacks better than U and occupies the template more effectively. Once long enough oligo(G)'s are formed, they are liable to migrate to new positions, even if a loop-out is involved, since the free-energy of looping-out is not excessive in relation to the thermal energy. Looping out of this kind will slow down the incorporation of U, and facilitate extension of G sequences beyond the length of the runs of C present in the template. We have evidence for unexpectedly long runs of G in our products. At present we are unable to suggest any other equally plausible explanation of the composition of the oligomeric products.

Poly(CU) templates support fairly efficient synthesis over a wide range of polymer compositions. The efficiency of incorporation of G is close to 100 percent for polymers with ratios of C to U from 11.3 to 2.9. The incorporation of A falls only slightly, from 63 to 54 percent over the same range of composition. In this case we can suggest two mechanisms that lead to an incorporation of A which is substantially less than the incorporation of G. Wobble-pairing between G and U could be important if U's in the template are masked by G's in the product. Looping-out is also a possible complication with this as with all other mixed templates. The efficiency of incorporation of A is substantially lower on the 1.3:1 template than on templates with a greater excess of C. Presumably, this template contains substantial runs of U, on which synthesis is relatively slow or inefficient.

The reaction on random poly(CG)

Table 2. Mean chain lengths and positions of incorporation in mixed template reactions. Yields of pNp, Np, and N correspond to the nucleoside N in 5'-terminal, internal, and 3'-terminal positions, respectively. The sums of these yields, weighted by the nucleotide abundances, are given as pXp, Xp, and X, respectively.

Compositions: (a) $C_4(G_{0.42}U_{0.42}A_{0.15})$; (b) $C_2(G_{0.49}U_{0.30}A_{0.21})$; and (c) $C_{0.9}(U_{0.55}A_{0.45})$.

Template		Time (day)	pGp:Gp:G	pCp:Cp:C	pAp:Ap:A	pUp:Up:U	pXp:Xp:X	Mean Chain Length
CG	5.4:1	7	17:70:13	4:72:25			15:70:15	6.7
	1.9:1	21	11:74:15	7:61:32			10:73:17	6.3
CU	6.9:1	7	12:79:9		1:70:29		11:78:11	9.1
	2.9:1	7	14:79:8		2:77:20		12:78:10	9.8
	1.3:1	21	10:83:8		2:80:18		8:82:10	10.2
CA	5.1:1	7	12:78:10			3:57:41	11.76:13	7.8
	2.9:1	7	13:76:11			2:65:33	13:76:11	8.9
	1.9:1	7	18:70:12			4:66:30	17:69:14	6.9
	0.89:1	21	13:73:14			4:80:16	12:74:14	7.3
CGUA[a]		21	9:84:7	3:71:26	4:71:25	4:58:29	8:81:11	9.3
CGUA[b]		21	13:79:8	4:71:25	9:67:24	3:63:34	12:76:12	8.3
CUAC[c]		21	12:77:11		5:79:16	10:72:19	11:79:10	9.9

templates leads to an efficient incorporation of both G and C for a 5.4:1 template, but a very low efficiency for a 1.9:1 template. This reduced efficiency is probably due to the extensive and stable self-structure of poly(CG) templates containing a substantial proportion of G.

The few experiments with more complex templates indicate that a wide variety of behaviors is to be anticipated. It is encouraging that only templates containing all four bases facilitate the incorporation of all four bases into product with reasonable efficiency.

We anticipate that even relatively short oligonucleotide sequences will display an enormous diversity in their nonenzymatic template chemistry. Templates will differ in the efficiencies with which they initiate, the rates at which they propagate, the probabilities that they terminate prematurely, and so on. This diversity corresponds, at the level of molecular structure, to the phenotypic differences in fecundity that distinguish organisms, and must make a substantial contribution to the "variation" on which natural selection at the level of molecular replication acts. It seems possible that an extension of our scheme, or of a scheme similar to it, might make it possible to study natural selection in a nonenzymatic system.

The relevance of our results to discussion of the origins of life is necessarily indirect, since it is unlikely that 2-methylimidazole was abundant on the primitive Earth. Nonetheless, the demonstration of a relatively simple reaction in which preformed copolymer templates containing all four of the natural bases direct the synthesis of their complements is significant. It makes it more plausible that the replication of polynucleotides was important at a very early stage in the evolution of life.

T. Inoue
L. E. Orgel
*Salk Institute for Biological Studies,
San Diego, California 92138*

References and Notes

1. Abbreviations: A, adenosine; C, cytidine; G, guanosine; U, uridine; pN(N is A, C, G, or U), nucleoside 5'-phosphate; pNp, 5'-phosphonucleoside 2'(3')-phosphate; 2-MeImpN, (nucleoside 5'-phosphor)-2-methylimidazolide.
2. T. Inoue and L. E. Orgel, *J. Am. Chem. Soc.* **103**, 7666 (1981).
3. T. Inoue and L. E. Orgel, *J. Mol. Biol.* **162**, 201 (1982).
4. J. Frazier and T. Miles, *ibid.*, p. 208.
5. P. O. P. Ts'o, *Basic Principles in Nucleic Acid Chemistry* (Academic Press, New York, 1974), chap. 6.
6. R. Lohrmann and L. E. Orgel, *J. Mol. Biol.* **142**, 555 (1980); P. K. Bridson and L. E. Orgel, *ibid.* **144**, 567 (1980).
7. R. Lohrmann and L. E. Orgel, *J. Mol. Evol.* **12**, 237 (1979).
8. J. Ninio and L. E. Orgel, *ibid.*, p. 91.
9. H. Fakhrai, unpublished results.
10. C. Taylor, unpublished results.
11. A. J. Lamont and J. R. Fresco, *Progr. Nucleic Acid Res. Mol. Biol.* **15**, 185 (1975).
12. This work was supported by grant GM-13435 from the National Institutes of Health and by NASA grant NGR 05-067. We thank A. R. Hill for technical assistance and Dr. H. Fakhrai for a gift of copolymer templates.

7 June 1982; revised 1 October 1982

Template-Directed Synthesis of Novel, Nucleic Acid–Like Structures

Abstract. *In studying the origins of life, it is important to examine reactions of substrate mixtures that could plausibly have accumulated on the primitive earth. Nucleoside diphosphates would probably have been synthesized along with the standard nucleotides under prebiotic conditions. For these reasons, the template-directed reactions of activated derivatives of these diphosphates, alone or mixed with activated nucleotides, were investigated. An activated derivative of deoxyguanosine 3',5'-diphosphate condensed efficiently on a polycytidylate template to give oligonucleotide analogues in which each 3',5'-phosphodiester bond was replaced by a pyrophosphate linkage. Oligomers were formed even in the absence of a template, but much more slowly. Template-directed condensation occurred also with an analogous deoxyadenosine derivative on polyuridylic acid and with an analogous acycloguanosine derivative on polycytidylic acid.*

All naturally occurring nucleic acids have in common a backbone formed by joining the 3'- and 5'-OH groups of successive nucleosides via a phosphodiester bond. However, a double-stranded structure analogous to the Watson-Crick helix is sometimes stable even when the covalently linked backbone of one strand is interrupted or modified. Thus polycytidylic acid [poly(C)] (*1*) will form stable, double-helical complexes with a variety of monomeric derivatives of guanine, including guanosine and guanosine 5'-phosphate (*2*), while long, 2',5'–linked oligoguanylic acids [oligo(G)'s] can be formed on poly(C) and form a double helix with the template (*3*).

These observations raise two general questions. Is the energy of base-pairing between G and C (or A and U), plus energy obtained by stacking base pairs on top of each other, sufficient to guarantee the stability of a variety of DNA-like double helices with one or both backbones very different from those familiar in biology? If so, can such oligonucleotide analogues be formed by template-directed synthesis?

As a first step toward answering these questions we have prepared and studied the template-directed reactions of the activated nucleotide analogues **1**, **2**, and **3** (Fig. 1) that can condense to form oligomers in which the phosphate group of the oligonucleotides is replaced by a pyrophosphate group. We chose these compounds because we have already shown that the formation of a pyrophosphate from a nucleotide and a nucleoside 5'-phosphoimidazolide in aqueous solution is a general and efficient reaction (*4*).

Phosphodiesterase I (snake venom phosphodiesterase), 2'-deoxyadenosine 3',5'-diphosphate and 2'-deoxyguanosine 3',5'-diphosphate were purchased from Pharmacia P-L. The 9-(1,3-dihydroxy-2-propoxy)methylguanine(Ḡ) (*5*) was converted to the diphosphate with phosphoryl chloride by a modification of a published procedure (*6*), and its identity was confirmed by comparison in several chromatographic systems with an authentic sample (*5*). The diimidazolides (**1**, **2**, and **3**) were prepared by a standard procedure that has been used for the synthesis of 5'-phosphoimidazolides of nucleotides (*7*).

Prior to analysis by high-performance liquid chromatography (HPLC) on RPC-5 (*7*), samples of the reaction mixtures were incubated at pH 4 (room temperature) overnight to hydrolyze phosphoimidazolides. If the overnight hydrolysis with acid was omitted, a complex family of products, many of which contained one or more phosphoimidazolide groups, was obtained. Individual oligomers were collected from the RPC-5 column, desalted by dialysis, and degraded at room temperature with phosphodiesterase I (Pharmacia; 0.01 unit per 0.01 optical density unit of nucleotide, in 0.1M tris-HCl containing 0.01M CaCl$_2$, pH 8.5). The pdGp (*1*) obtained by phosphodiesterase digestion of isolated oligomers was identified by co-chromatography with authentic material on RPC-5.

Compound **1** condensed rapidly on a poly(C) template to form a homologous series of product oligomers with chain lengths up to about 20 (Fig. 2A). The ultraviolet absorption of guanine in the

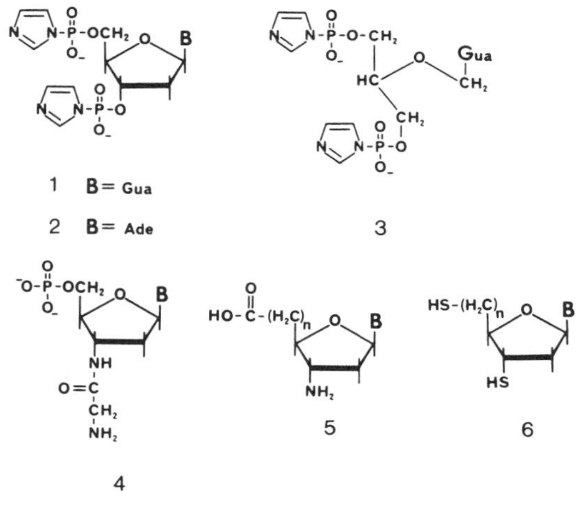

Fig. 1. The chemical names of the derivatives used are the following: **1**, 2'-deoxyguanosine-3',5'-diphosphoimidazolide; **2**, 2'-deoxyadenosine-3',5'-diphosphoimidazolide; **3**, 9-(1,3-dihydroxy-2-propoxy)-methylguanine-1,3-diphosphoimidazolide; **4**, 3'-glycylamido-2',3'-dideoxynucleoside 5'-phosphate; **5**, homologues of 3'-amino-2',3'-dideoxyuronic acids; **6**, homologues of 3',5'-dithio-2',3',5'-trideoxypentafuranosylnucleosides. For derivatives **4**, **5**, and **6**, B is a nucleoside base.

4

products from preparative scale reactions was equivalent to about 70 percent of the cytosine in the template. The efficiency of the reaction (the ratio of the amount of G incorporated into products to the amount of C in the template) approached 100 percent after longer times. When the composition of the reaction mixture was varied, high yields of products were obtained with a salt concentration as high as $0.8M$ $MgCl_2$ and $1.0M$ NaCl or as low as $0.1M$ $MgCl_2$ and $0.1M$ NaCl. The yields of products decreased significantly in the absence of $MgCl_2$. However, if sufficient NaCl (1.0 to $5.0M$) was present, the yields were still substantial.

The optimal pH for the oligomerization of ImpdGpIm (*1*) on a poly(C) template was about 6.5. At this pH, a triple-stranded rather than a double-stranded helix may be present. However, a large template effect was also observed at pH 8.1, a pH at which only double-stranded

complexes of poly(C) and monomeric G derivatives are stable (*8*).

The time course of the non–template-directed oligomerization of *1* was very different (Fig. 2B). The reaction produced only traces of oligomers containing eight or more nucleotides after 3 days. At later times, longer oligomers accumulated until, after 2 weeks, the overall pattern of products was not very different from that obtained in the template-directed reaction. However, the yield of longer oligomers from the non-template reaction never exceeded 50 percent of the yield obtained within 3 days in the template-directed reaction. The slow, non–template-directed reaction apparently stopped because of nearly complete hydrolysis of the starting imidazolides (half-life about 2 days at 0°C).

Enzymatic digestion of individual product oligomers isolated from the reaction of *1* on poly(C) suggested that the mode of linkage was the expected inter-

nucleoside pyrophosphate. Incubation of the hexamer with phosphodiesterase I, which is known to cleave pyrophosphate bonds (*9*), produced a series of lower oligomers at early times. The average chain length decreased with increasing time. After long times the final product was the monomer pdGp. The length of time it takes to achieve complete degradation depends on the amount of substrate and the amount of enzyme; 0.01 absorbance unit of oligomer was degraded completely by 0.1 unit of enzyme in 1 hour (pH 8.5, 37°C, $0.01M$ Ca^{2+}).

The product oligomers obtained after 3 days from the self-condensation of the adenylic acid analogue, *2*, in the presence of poly(U) were comparable in length and yield (Fig. 2C) with those obtained in the poly(C)-G system. In the absence of template, very little oligomeric product was observed (Fig. 2D). The acycloguanylic acid analogue *3* also underwent a template-directed condensation (Fig. 2E), but less product was formed than in reactions of the deoxynucleotide analogues. As in the other systems studied, the non–template-directed reaction (Fig. 2F) produced only small amounts of oligomers within 3 days.

An initial impression that the distribution of product peaks was simpler in the presence of template, indicating a greater regiospecificity of reaction, needs to be substantiated by detailed structural studies. We believe that the first stage in the formation of pyrophosphate bonds is the hydrolysis of a proportion of the phosphoimidazole groups to free phosphate groups. In the second step, these free phosphates attack unhydrolyzed phosphoimidazolide groups to form pyrophosphate bonds (*4*). However, we cannot exclude the possibility that the direct condensation of two phosphoimidazole groups also contributes to the products.

In previous experiments on template-directed synthesis, specific, activated nucleoside 5'-phosphates were usually used as substrates, and 3',5'-linked oligonucleotides were sought as products. However, it is unlikely that chemically pure substrates of this kind, for example 2-MeImpG (*1*), could have accumulated on the primitive earth. It is important to use the experience gained in experiments with specific substrates to design reactions that proceed efficiently with more plausibly prebiotic substrate mixtures.

The reaction of a nucleoside phosphate with a nucleoside phosphoimidazolide is general, and much more efficient than the corresponding reaction of a nucleoside. Non–template-directed syn-

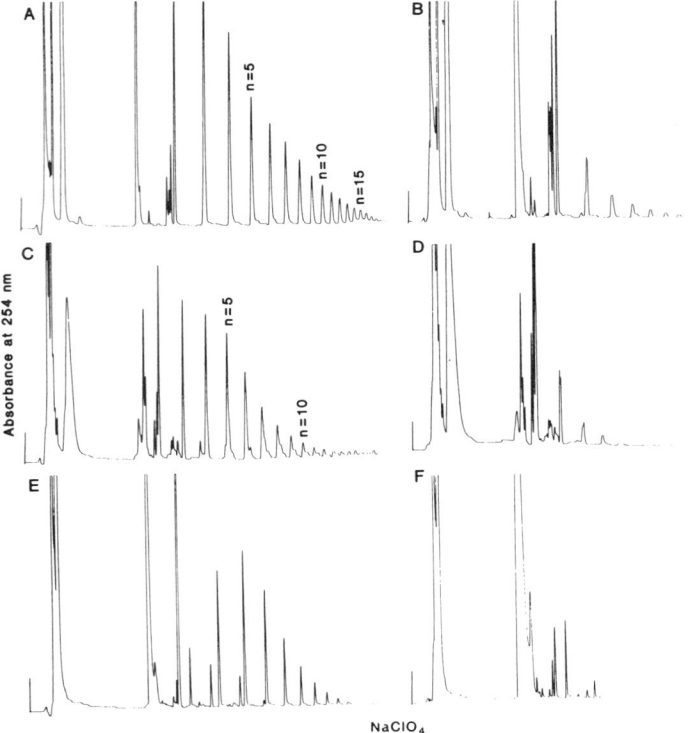

NaClO₄

Fig. 2. HPLC analyses of reaction products from the self-condensation of 2'-deoxynucleoside-3',5'-diphosphoimidazolides of G and A and of the corresponding derivative of an analogue of G in the presence and absence of polynucleotide templates. Reaction mixtures were prepared as previously described (*7*). (A) The reaction mixtures contained $0.1M$ ImpdGpIm, $0.1M$ poly(C), $0.2M$ $MgCl_2$, $0.1M$ NaCl, and $0.2M$ 2,6-lutidine-HCl (pH 6.5) and was kept at 0°C for 3 days. Chain lengths are given by n. (B) As (A), but without poly(C). (C) The reaction mixture contained $0.1M$ ImpdApIm, $0.1M$ poly(U), $0.2M$ $MgCl_2$, $0.1M$ NaCl, and $0.2M$ 2,6-lutidine-HCl (pH 6.5) and was kept at 0°C for 3 days. (D) As (C), but without poly(U). (E) The reaction mixture contained $0.1M$ ImpGpIm, $0.1M$ poly(C), $0.2M$ $MgCl_2$, $1.0M$ NaCl, and $0.2M$ 2,6-lutidine-HCl (pH 6.5) and was kept at 0°C for 3 days. (F) As (E), but without poly(C). Reactions were stopped by the addition of an excess of EDTA and, after overnight hydrolysis at pH 4, were analyzed on RPC-5 in $0.02M$ NaOH containing $0.002M$ tris-HClO₄ and with a linear gradient of NaClO₄ (0 to $0.04M$ over 60 minutes) at a flow rate of 1.0 ml/min.

thesis of short oligomers of deoxynucleosides linked by pyrophosphate bonds is, therefore, easier than the formation of equivalent oligonucleotides. Furthermore, template-directed reactions are, as we have shown, much less dependent on the choice of the activated substrate and the reaction conditions. Even the very simple and potentially prebiotic acyclonucleoside derivative ImpḠpIm (*1*) was an adequate substrate.

In other experiments we have shown that poly(C) will direct the efficient copolymerization of 2-MeImpG and ImpdGpIm (*1*) at all ratios of the two substrates (*10*). Thus it seems plausible that a poly(C) template could bring about a copolymerization of components selected from a complex mixture of partially and fully phosphorylated G derivatives. If, as we expect, a poly(C) analogue in which pyrophosphate groups have partially or fully replaced the normal phosphate groups will itself act as a template for the condensation of G derivatives, it will indicate that template-directed synthesis is much less demanding than has been demonstrated previously.

Our results also suggest, but certainly do not prove, that a great variety of template-directed reactions may be possible with unconventional substrates. Substrates **4** and **5**, for example, in the presence of poly(C), might react to form amide-linked polymers when the phosphate is activated in **4** (*11*) or the carboxyl in **5**, while substrate **6** might form disulfide-linked oligomers on oxidation (Fig. 1). While such reactions may not be interesting in the context of the origins of life on the earth, they would greatly extend the scope of template-directed organic synthesis.

ALAN W. SCHWARTZ*
LESLIE E. ORGEL
*Salk Institute for Biological Sciences,
San Diego, California 92138*

References and Notes

1. Abbreviatons: A, adenosine; C, cytidine; G, guanosine; U, uridine; Ḡ, 9-(1,3-dihydroxy-2-propoxy)methylguanine; dG, 2′-deoxyguanosine; pN (N is A, G, or dG), the 5′-phosphate of N; 2-MeImpN, the 2-methylimidazolide of pN; Np, the 3′-phosphate of N; ImpNpIm, the 3′,5′-diphosphoimidazolide of N; ImpGpIm, the 1,3-diphosphoimidazolide of Ḡ (compound **3**).
2. P. O. P. Ts'o, *Basic Principles in Nucleic Acid Chemistry* (Academic Press, New York, 1974), vol. 1, pp. 453–584.
3. P. K. Bridson, H. Fakhrai, R. Lohrmann, L. E. Orgel, M. van Roode, in *Origin of Life*, Y. Wolman, Ed. (Reidel, Jerusalem, 1981), pp. 233–239.
4. B. C. F. Chu and L. E. Orgel, *Biochim. Biophys. Acta* **782**, 103 (1984).
5. The 9-(1,3-dihydroxy-2-propoxy)methylguanine was received from D. W. Barry, Burroughs Wellcome Company, Research Triangle Park, N.C. A sample of the diphosphate derivative was given to us by J. P. H. Verheyden, Syntex Research, Palo Alto, Calif.
6. L. A. Slotin, *Synthesis* **1977**, 737 (1977).
7. G. F. Joyce, T. Inoue, L. E. Orgel, *J. Mol. Biol.* **176**, 279 (1984).
8. F. B. Howard, J. Frazier, M. N. Lipsett, H. T. Miles, *Biochem. Biophys. Res. Commun.* **17**, 93 (1964).
9. H. Matsubara, S. Hasegawa, S. Fujimura, T. Shima, T. Sugimura, M. Fatai, *J. Biol. Chem.* **245**, 3606 (1970).
10. A. W. Schwartz, unpublished results.
11. For a related template-directed reaction, see J. L. Shim, R. Lohrmann, L. E. Orgel, *J. Am. Chem. Soc.* **96**, 5283 (1974).
12. This work was supported by National Aeronautics and Space Administration grant NGR 05067. We are indebted to A. Hill for technical assistance, R. Brown for manuscript preparation, and G. von Kjedrowskj for carrying out the preparation of pGp from G.
* Present address: Department of Exobiology, Faculty of Science, University of Nijmegen, Toernooiveld, 6525 ED Nijmegen, The Netherlands.

20 August 1984; accepted 27 November 1984

A Self-Replicating System

T. Tjivikua, P. Ballester, and J. Rebek, Jr.*

*Departments of Chemistry
Massachusetts Institute of Technology
Cambridge, Massachusetts 02139
University of Pittsburgh
Pittsburgh, Pennsylvania 15260*

Received September 25, 1989

The ability of nucleic acids to act as templates for self-replication has been unique. In living systems, single strands act as templates during phosphate transfer reactions. In nonenzymatic systems, autocatalysis[1] can be observed during these reactions. Here we show that base pairing can also enhance acyl transfer reactions.

The reaction involves the coupling of the amino adenosine **1** to the pentafluorophenyl ester **2a** (Scheme I), and its shows the following features:

1. *The reaction exhibits autocatalysis.* Addition of the product **3** the reaction mixture results in an increase in the initial coupling rate (Figure 1, entries c–e).

2. *The product is self-complementary.* Dimerization occurs (**3 → 4**, Scheme I), and the value measured by NMR dilution studies[2] for the dimerization constant is $K_d = 630$ M^{-1}. This value is consistent with expectations involving base pairing and aryl stacking, with some attenuation by steric effects in the middle of the structure.[3]

3. *The reactions of 2a proceed through the formation of base-paired complexes.* The imide **2a** reacts some 10 times faster than the *N*-methyl derivative **2b**, a factor that can hardly be attributed to classic steric effects at a site so remote from the reacting centers. Moreover, 2,6-bis(acylamino)pyridine acts as a competitive inhibitor by forming a nonproductive complex, **5** (Scheme II). Titration of **2a** with 5-O-acetyl-2',3'-O-iso-

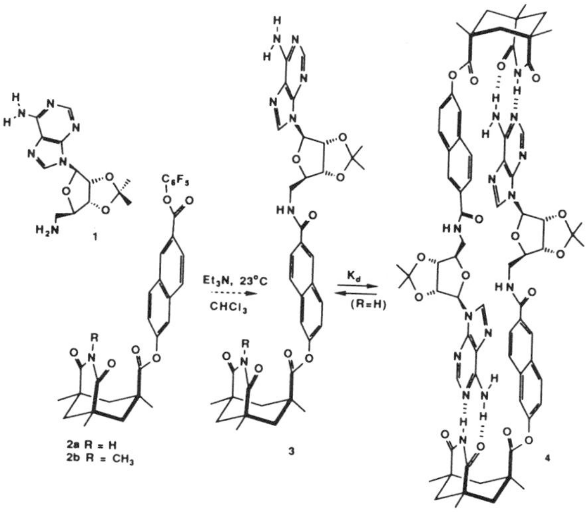

* Address correspondence to this author at the Massachusetts Institute of Technology.

(1) Pitha, P. M.; Pitha, J. *Nat. New Biol.* **1972**, *98*, 78–80. von Kiedrowski, G. *Angew. Chem., Int. Ed. Engl.* **1986**, *25*, 932–935. von Kiedrowski, G.; Woltzka, B.; Helbing, J. *Ibid.* **1989**, *28*, 1235–1237. Zielinski, W. S.; Orgel, L. E. *Nature* **1987**, *327*, 346–347. Visscher, J.; Bakker, C. G.; van der Woerd, R.; Schwartz, A. *Science* **1989**, *244*, 329–331. For double-stranded template effects in *triple-helix* formation, see: Luebke, K. J.; Dervan, P. B. *J. Am. Chem. Soc.* **1989**, *111*, 8733–8735.

(2) Wilcox, C. S.; Cowart, M. D. *Tetrahedron Lett.* **1986**, *27*, 5563–5566.

(3) Williams, K.; Askew, B.; Ballester, P.; Buhr, C.; jeong, K. S.; Jones, S.; Rebek, J., Jr. *J. Am. Chem. Soc.* **1989**, *111*, 1090–1094. Only Watson–Crick forms of base pairing are shown, but NOE experiments[4] indicate the presence of Hoogsteen forms as well.

(4) See, for example: Askew, B.; Ballester, P.; Buhr, C.; Jeong, K. S.; Jones, S.; Parris, K.; Williams, K.; Rebek, J., Jr. *J. Am. Chem. Soc.* **1989**, *111*, 1082–1090.

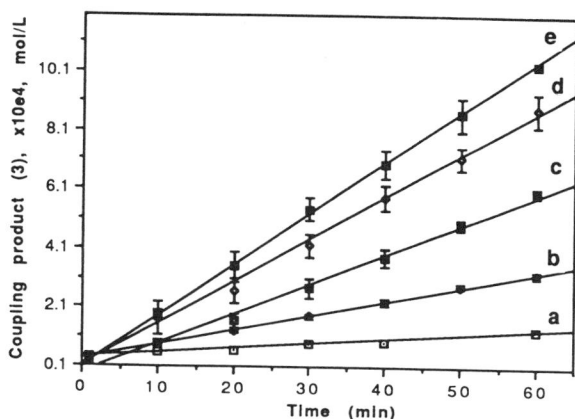

Figure 1. Initial rates of product formation (Scheme I). Plots of initial appearance of coupling products vs time as determined by HPLC. All reactions were performed with initial concentrations of [**1**] = [**2**] = 8.2 mM in CHCl$_3$ with 4 equiv of Et$_3$N added as a general base. Each run was independently performed in triplicate, and error bars represent standard deviations for each point. (a) Reaction of **1** and the *N*-methylated **2b**. (b) Reaction of **1** and **2a** with 1 equiv of 2,6-bis(acyl-amino)pyridine. (c) Reaction of **1** and **2a**. (d) Reaction of **1** and **2a** with 0.2 equiv of **3** added as autocatalyst. (e) As in d with 0.5 equiv of **3**.

Scheme I

Scheme II

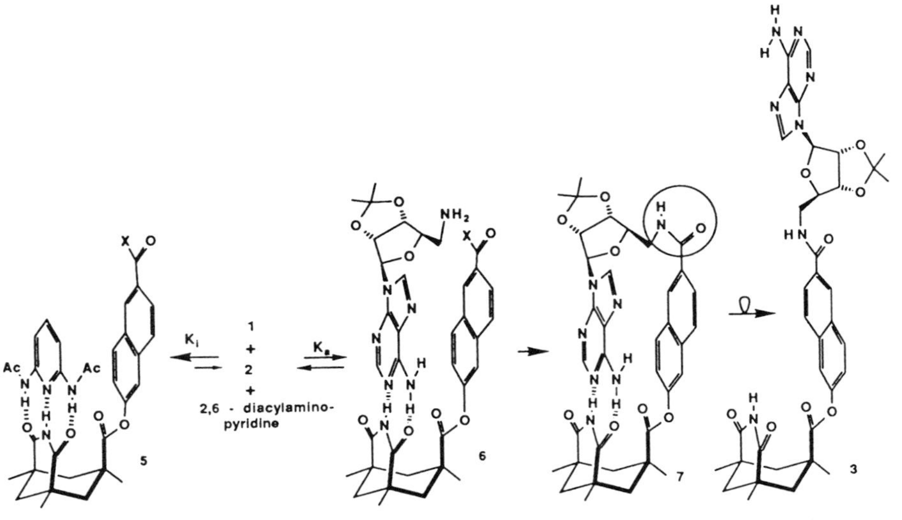

Scheme III

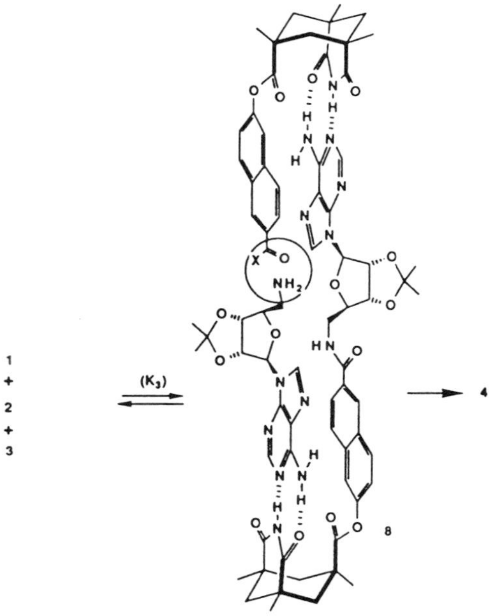

Scheme IV

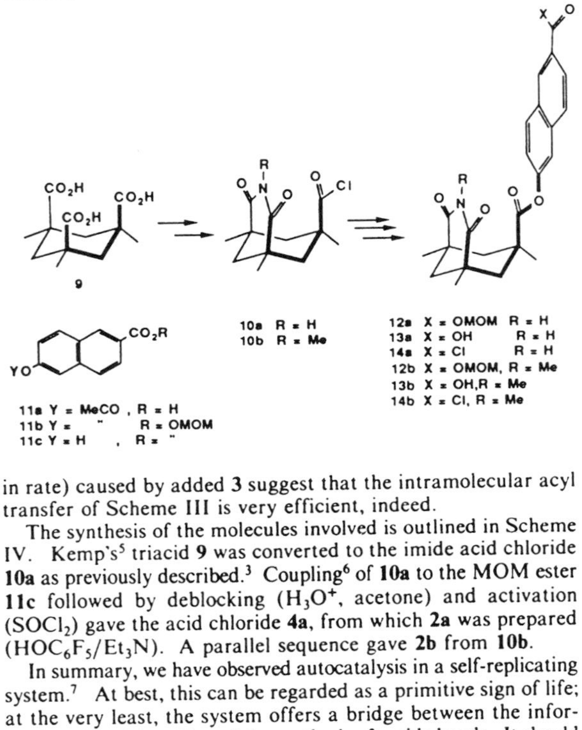

propylideneadenosine gave an association K_a of 60 M^{-1} under these conditions, a value that serves as a reasonable estimate of K_a for the formation of complex **6**.

4. *The initial product of the intramolecular acyl transfer is a cis amide.* Modeling shows that the distance between the hydrogen-bonding surface and the active ester can be spanned only by such a conformation of the amide. The product does not appear to remain in this conformation; rather, cis → trans isomerization occurs (**7** → **3**, Scheme II) and the hydrogen-bonding surfaces become exposed for dimerization or autocatalysis.

Three mechanisms are involved in the reaction of **1** with **2a**: the slow aminolysis of Scheme I, the more efficient base-paired version of Scheme II, and the autocatalytic component of Scheme III. An estimate can be made for the formation of the termolecular complex **8**. Assuming independent (noncooperative) binding sites on the template **3**, the value for K_3 should be K_a^2 or 3600 M^{-2}. Using this figure, one can calculate that only about 2% of **1** and **2a** are in the form of **8** under the autocatalytic conditions (runs d and e, Figure 1) while about 25% are in the form of the base-paired **6**. The observed enhancements (>40% in rate) caused by added **3** suggest that the intramolecular acyl transfer of Scheme III is very efficient, indeed.

The synthesis of the molecules involved is outlined in Scheme IV. Kemp's[5] triacid **9** was converted to the imide acid chloride **10a** as previously described.[3] Coupling[6] of **10a** to the MOM ester **11c** followed by deblocking (H$_3$O$^+$, acetone) and activation (SOCl$_2$) gave the acid chloride **4a**, from which **2a** was prepared (HOC$_6$F$_5$/Et$_3$N). A parallel sequence gave **2b** from **10b**.

In summary, we have observed autocatalysis in a self-replicating system.[7] At best, this can be regarded as a primitive sign of life; at the very least, the system offers a bridge between the information of nucleic acids and the synthesis of amide bonds. It should be possible to design systems capable of peptide synthesis on a nucleic acid backbone and thereby provide models for events that occurred some time ago.

Acknowledgment. We thank the National Science Foundation for support and David Chalfoun for help with computations.

(5) Kemp, D. S.; Petrakis, K. S. *J. Org. Chem.* **1989**, *46*, 5140–5143. Commercially available from the Aldrich Chemical Co.

(6) All new compounds were characterized by a full complement of high-resolution spectra; details will be published elsewhere. The amine **1** was obtained as described by Kolb et al.: Kolb, M.; Danzin, C.; Barth, J.; Claverie, N. *J. Med. Chem.* **1982**, *25*, 550–556.

(7) For a recent, beautiful example of a template-directed substitution reaction, see: Kelly, T. R.; Zhao, C.; Bridger, G. J. *J. Am. Chem. Soc.* **1989**, *111*, 3744–3745.

Research Articles

The Intervening Sequence RNA of *Tetrahymena* Is an Enzyme

ARTHUR J. ZAUG AND THOMAS R. CECH

A shortened form of the self-splicing ribosomal RNA (rRNA) intervening sequence of *Tetrahymena thermophila* acts as an enzyme in vitro. The enzyme catalyzes the cleavage and rejoining of oligonucleotide substrates in a sequence-dependent manner with $K_m = 42$ μM and $k_{cat} = 2$ min^{-1}. The reaction mechanism resembles that of rRNA precursor self-splicing. With pentacytidylic acid as the substrate, successive cleavage and rejoining reactions lead to the synthesis of polycytidylic acid. Thus, the RNA molecule can act as an RNA polymerase, differing from the protein enzyme in that it uses an internal rather than an external template. At *p*H 9, the same RNA enzyme has activity as a sequence-specific ribonuclease.

IN RNA SELF-SPLICING, THE FOLDED STRUCTURE OF AN RNA molecule mediates specific cleavage-ligation reactions (*1–5*). Self-splicing exemplifies intramolecular catalysis (*6*) in that the reactions are accelerated many orders of magnitude beyond the basal chemical rate (*7, 8*). The reactions are highly specific, as seen in the choice of a free guanosine nucleotide as a substrate in the self-splicing of the *Tetrahymena* ribsomal RNA precursor (pre-rRNA)

and other RNA's containing group I intervening sequences (*1–3, 7*). Furthermore, the cleavage-ligation activity mediates a series of splicing, cyclization, and reverse cyclization reactions, suggesting that the active site is preserved in each reaction (*9, 10*). However, the RNA is cleaved and rejoined during self-splicing; because the RNA is not regenerated in its original form at the end of the reaction, it is not an enzyme. The RNA moiety of ribonuclease P, the enzyme responsible for cleaving transfer RNA (tRNA) precursors to generate the mature 5' end of the tRNA, has been the only example of an RNA molecule that meets all criteria of an enzyme (*11–13*).

Following self-splicing of the *Tetrahymena* rRNA precursor, the excised IVS RNA (*14*) undergoes a series of RNA-mediated cyclization and site-specific hydrolysis reactions. The final product, the L − 19 IVS RNA, is a linear molecule that does not have the first 19 nucleotides of the original excised IVS RNA (*9*). We interpreted the lack of further reaction of the L − 19 species as an indication that all potential reaction sites on the molecule that could reach its active site (that is, intramolecular substrates) had been consumed; and we argued that the activity was probably unperturbed (*9*). We have now tested this by adding oligonucleotide substrates to the L − 19 IVS RNA. We find that each IVS RNA molecule can catalyze the cleavage and rejoining of many oligonucleotides. Thus, the L − 19 IVS RNA is a true enzyme. Although the enzyme can act on RNA molecules of large size and complex sequence, we have found that studies with simple oligoribonucleotides like pC5 (pentacytidylic acid) have been most valuable in

Thomas R. Cech is a professor and Arthur J. Zaug is a research associate in the Department of Chemistry and Biochemistry, University of Colorado, Boulder, 80309-0215. Send correspondence to T.R.C.

Reprinted with permission from *Science*,
Volume 231, pp. 470–475. © 1986 by the American Association for the Advancement of Science.

revealing the minimum substrate requirements and reaction mechanism of this enzyme. These studies are presented below.

The L − 19 IVS RNA catalyzes the cleavage and rejoining of oligonucleotides.

Unlabeled L − 19 IVS RNA was incubated with $5'$-^{32}P-labeled pC_5 in a solution containing 20 mM $MgCl_2$, 50 mM tris-HCl, pH 7.5. The pC_5 was progressively converted to oligocytidylic acid with both longer and shorter chain length than the starting material (Fig. 1A). The longer products extended to at least pC_{30}, as judged by a longer exposure of an autoradiogram such as that shown in Fig. 1A. The shorter products were exclusively pC_4 and pC_3. Incubation of pC_5 in the absence of the L − 19 IVS RNA gave no reaction (Fig. 1C).

Phosphatase treatment of a 60-minute reaction mixture resulted in the complete conversion of the ^{32}P radioactivity to inorganic phosphate, as judged by polyethyleneimine thin-layer chromatography (TLC) in 1M sodium formate, pH 3.5 (15). Thus, the $5'$-terminal phosphate of the substrate does not become internalized during the reaction, and the substrate is being extended on its $3'$ end to form the larger oligonucleotides. When C_5*C was used as the substrate and the products were treated with ribonuclease T_2 or ribonuclease A, the ^{32}P radioactivity was totally converted to C*p (15). Thus, the linkages being formed in the reaction are exclusively $3',5'$-phosphodiester bonds. The products of the C_5*C reaction were totally resistant to phosphatase treatment.

The reaction was specific for ribonucleotides, no reaction taking place with d-pC_5 (Fig. 1B) or d-pA_5 (15). Among the oligoribonucleotides, pU_6 was a much poorer substrate than pC_5 or pC_6 (Fig. 1D), and pA_6 gave no reaction (15).

No reaction occurred when magnesium chloride was omitted. The enzyme activity was approximately constant in the range 5 to 40 mM $MgCl_2$ (15). The 20 mM concentration was routinely used to circumvent the potential effect of chelation of Mg^{2+} by high concentrations of oligonucleotide substrates.

The L − 19 IVS RNA is regenerated after each reaction, such that each enzyme molecule can react with many substrate molecules. For example, quantitation of the data shown in Fig. 1G revealed that 16 pmol of enzyme converted 1080 pmol of pC_5 to products in 60 minutes. Such numbers underestimate the turnover number of the enzyme; because the initial products are predominantly C_6 and C_4, it is likely that the production of chains of length greater than six or less than four involves two or more catalytic cycles. Quantitation of the amount of radioactivity in each product also provides some indication of the reaction mechanism. At early reaction times, the amount of radioactivity (a measure of numbers of chains) in products larger than pC_5 is approximately equal to that found in pC_4 plus pC_3, consistent with a mechanism in which the total number of phosphodiester bonds is conserved in each reaction. As the reaction proceeds, however, the radioactivity distribution shifts toward the smaller products. This is most likely due to a competing hydrolysis reaction also catalyzed by the L − 19 IVS RNA, as described below.

The rate of conversion of 30 μM pC_5 to products increases linearly with L − 19 IVS RNA enzyme concentration in the range 0.06 to 1.00 μM (15). At a fixed enzyme concentration (Fig. 1, E to G), there is a hyperbolic relation between the reaction rate and the concentration of pC_5. The data are fit by the Michaelis-Menten rate law in Fig. 2. The resulting kinetic parameters are K_m = 42 μM and k_{cat} = 1.7 min^{-1}

The stability of the enzyme was determined by preliminary incubation at 42°C for 1 hour in the presence of Mg^{2+} (standard reaction conditions) or for 18 hours under the same conditions but without Mg^{2+}. In both cases, the incubated enzyme had activity indistinguishable from that of untreated enzyme tested in parallel, and no degradation of the enzyme was observed on polyacrylamide gel electrophoresis (15). Thus, the L − 19 IVS RNA is not a good

substrate. The enzyme is also stable during storage at −20°C for periods of months. The specific activity of the enzyme is consistent between preparations.

Covalent intermediate. When C_5*p was used as a substrate, radioactivity became covalently attached to the L − 19 IVS RNA (Fig. 3A) (16). This observation, combined with our previous knowledge of the mechanism of IVS RNA cyclization (9, 10, 17,

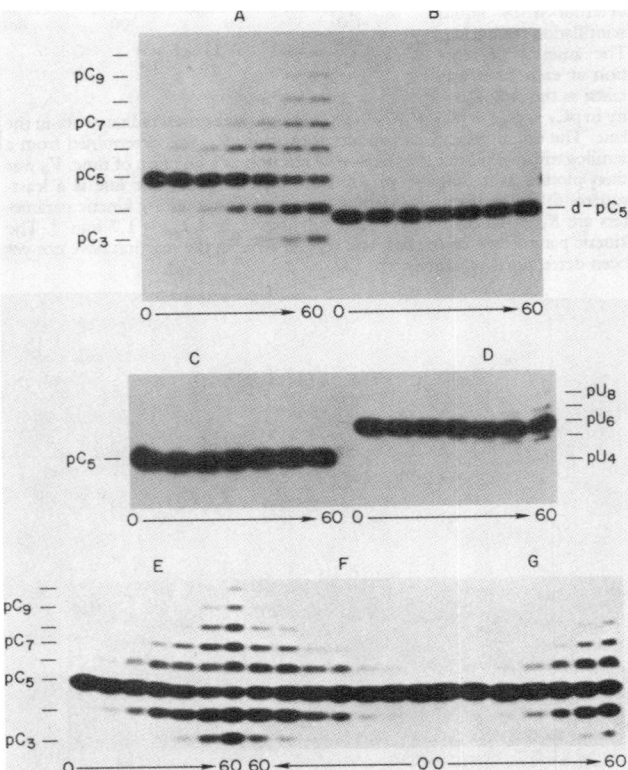

Fig. 1. The L − 19 IVS RNA catalyzes the cleavage and rejoining of oligoribonucleotide substrates; (A) 10 μM pC_5 and (B) 10 μM d-pC_5, both with 1.6 μM L − 19 IVS RNA; (C) 45 μM pC_5 in the absence of L − 19 IVS RNA; (D) 45 μM pU_6 with 1.6 μM L − 19 IVS RNA; (E) 10 μM pC_5, (F) 50 μM pC_5 and (G) 100 μM pC_5, all with 1.6 μM L − 19 IVS RNA. Oligonucleotides were $5'$-end labeled by treatment with [γ-^{32}P]ATP and polynucleotide kinase; they were diluted with unlabeled oligonucleotide of the same sequence to keep the amount of radioactivity per reaction constant. The L − 19 IVS RNA was synthesized by transcription and splicing in vitro. Supercoiled pSPTT1A3 DNA (30) was cut with Eco RI and then transcribed with SP6 RNA polymerase (31) for 2 hours at 37°C in a solution of nucleoside triphosphates (0.5 mM each), 6 mM $MgCl_2$, 4 mM spermidine, 10 mM dithiothreitol, 40 mM tris-HCl, pH 7.5, with 100 units of SP6 RNA polymerase per microgram of plasmid DNA. Then NaCl was added to a final concentration of 240 mM and incubation was continued at 37°C for 30 minutes to promote excision and cyclization of the IVS RNA. Nucleic acids were precipitated with three volumes of ethanol and redissolved in 50 mM CHES, pH 9.0; $MgCl_2$ was added to a final concentration of 20 mM, and the solution was incubated at 42°C for 1 hour to promote site-specific hydrolysis of the circular IVS RNA to give L − 19 IVS RNA (9). The reaction was stopped by the addition of EDTA to 25 mM. The L − 19 IVS RNA was purified by preparative gel electrophoresis and Sephadex G-50 chromatography. Labeled oligonucleotides were incubated with unlabeled L − 19 IVS RNA at 42°C in 20 mM $MgCl_2$, 50 mM tris, pH 7.5, for 0, 1, 2, 5, 10, 30, and 60 minutes. Reactions were stopped by the addition of EDTA to a final concentration of 25 mM. Products were analyzed by electrophoresis in a 20 percent polyacrylamide, 7M urea gel, autoradiograms of which are shown.

Fig. 2. Kinetics of conversion of pC$_5$ to larger and smaller oligonucleotides with 1.6 μM L – 19 IVS RNA. Products were separated by polyacrylamide gel electrophoresis. With the autoradiogram as a guide, the gel was cut into strips and the radioactivity in each RNA species was determined by liquid scintillation counting. The amount of reaction at each time was taken as the radioactivity in pC$_3$ + pC$_4$ + pC$_6$ + pC$_7$ + . . . divided by the total radioactivity in the lane. The initial velocity of product formation, V_o, was determined from a semilogarithmic plot of the fraction of reaction as a function of time. V_o was then plotted as a function of substrate concentration; the line is a least-squares fit to the Michaelis-Menten equation. The resulting kinetic parameters are K_m = 42 μM, V_{max} = 2.8 μM min^{-1}, and k_{cat} = 1.7 min^{-1}. The kinetic parameters for the first and second steps in the reaction have not yet been determined separately.

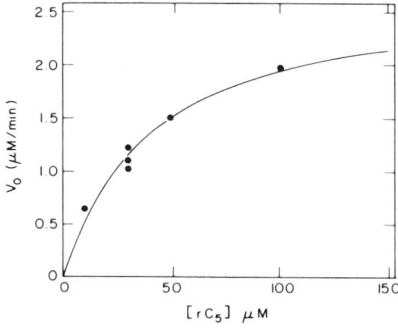

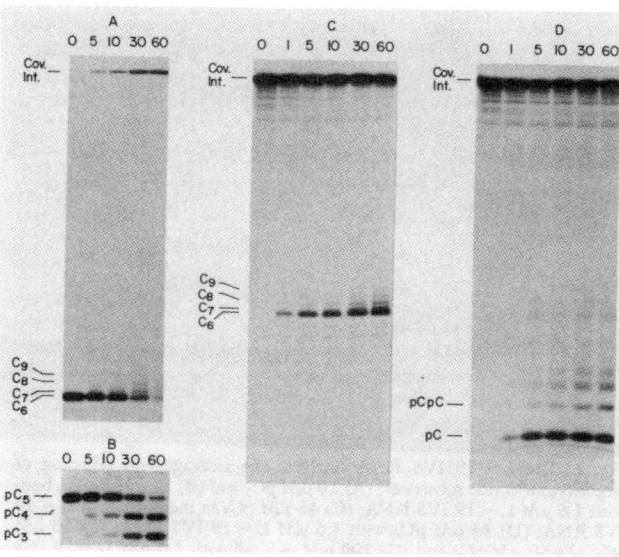

Fig. 3. Formation and resolution of the covalent enzyme-substrate intermediate. (A) To drive the formation of the covalent L – 19 IVS RNA-substrate intermediate, 8.5 nM C$_5$$\overset{*}{p}$C was treated with 0.16 μM L – 19 IVS RNA under standard reaction conditions for 0 to 60 minutes. (B) $\overset{*}{p}$C$_5$ (0.01 μM) was reacted with 0.16 μM L – 19 IVS RNA. Cleavage occurred normally, but there was very little rejoining. (C) Labeled covalent intermediate was prepared as in (A) (60 minutes) and purified by electrophoresis in a 4 percent polyacrylamide, 8M urea gel. It was then incubated with 10 μM unlabeled C$_5$ under standard reaction conditions for 0 to 60 minutes. The product designated C$_6$ comigrated with labeled C$_6$ marker (not shown). (D) Isolated covalent intermediate as in (C) was incubated under site-specific hydrolysis conditions (20 mM MgCl$_2$, 50 mM CHES, pH 9.0) at 42°C for 0 to 60 minutes. Positions of labeled mono- and dinucleotide markers are indicated. In the 10- and 30-minute lanes of (A) and the 10-, 30-, and 60-minute lanes of (C), band compression (reduced difference in electrophoretic mobility) is seen between C$_6$ and C$_7$ and to a lesser extent between C$_7$ and C$_8$. This is due to the absence of a 5' phosphate. Thus, the charge-to-mass ratio is increasing with chain length, whereas with 5'-phosphorylated oligonucleotides the charge-to-mass ratio is independent of chain length. When such products were phosphorylated by treatment with polynucleotide kinase and ATP, the distribution was converted to the normal spacing as in Fig. 1 (15).

18), led to a model for the reaction mechanism involving a covalent enzyme-substrate intermediate (Fig. 4).

This reaction pathway is supported by analysis of reactions in which a trace amount of $\overset{*}{p}$C$_5$ was incubated with a large molar excess of L – 19 IVS RNA. The cleavage reaction occurred with high efficiency, as judged by the production of $\overset{*}{p}$C$_4$ and $\overset{*}{p}$C$_3$, but there was very little synthesis of products larger than the starting material (Fig. 3B; compare to Fig. 1A). These data are easily interpreted in terms of the proposed reaction pathway. The first step, formation of the covalent intermediate with release of the 5'-terminal fragment of the oligonucleotide, is occurring normally. The first step consumes all the substrate, leaving insufficient C$_5$ to drive the second transesterification reaction.

The model shown in Fig. 4 was tested by isolating the covalent enzyme-substrate complex prepared by reaction with C$_5$$\overset{*}{p}$C and incubating it with unlabeled C$_5$. A portion of the radioactivity was converted to oligonucleotides with the electrophoretic mobility of C$_6$, C$_7$, C$_8$, and higher oligomers (Fig. 3C). In a confirmatory experiment, the covalent complex was prepared with unlabeled C$_5$ and reacted with $\overset{*}{p}$C$_5$. Radioactivity was again converted to a series of higher molecular weight oligonucleotides (15). In both types of experiments the data are readily explained if the covalent complex is a mixture of L – 19 IVS RNA's terminating in . . .GpC, . . .GpCpC, . . .GpCpCpC, and so on. Because they can react with C$_5$ to complete the catalytic cycle, these covalent enzyme-substrate complexes are presumptive intermediates in the reaction (Fig. 4). A more detailed analysis of the rate of their formation and resolution is needed to evaluate whether or not they are kinetically competent to be intermediates. We can make no firm conclusion about that portion of the enzyme-substrate complex that did not react with C$_5$. This unreactive RNA could be a covalent intermediate that was denatured during isolation such that it lost reactivity, or it could represent a small amount of a different enzyme-substrate complex that was nonproductive and therefore accumulated during the reaction.

The G^{414}-A^{16} linkage in the C IVS RNA, the G^{414}-U^{20} linkage in the C' IVS RNA, and the G^{414}-U^{415} linkage in the pre-rRNA are unusual phosphodiester bonds in that they are extremely labile to alkaline hydrolysis, leaving 5' phosphate and 3'-hydroxyl termini (9, 19). We therefore tested the lability of the G^{414}-C linkage in the covalent enzyme-substrate intermediate by incubation at pH 9.0 in a Mg^{2+}-containing buffer. This treatment resulted in the release of products that comigrated with pC and pCpC markers and larger products that were presumably higher oligomers of pC (Fig. 3D). Thin-layer chromatography was used to confirm the identity of the major products (15). In those molecules that released pC, the release was essentially complete in 5 minutes. Approximately half of the covalent complex was resistant to the pH 9.0 treatment. Once again, we can make no firm conclusion about the molecules that did not react. The lability of the G^{414}–C bond forms the basis for the L – 19 IVS RNA acting as a ribonuclease (Fig. 4).

A competitive inhibitor. Deoxy C$_5$, which is not a substrate for L – 19 IVS RNA–catalyzed cleavage, inhibits the cleavage of pC$_5$ (Fig. 5A). Analysis of the rate of the conversion of pC$_5$ to pC$_4$ and pC$_3$ as a function of d-C$_5$ concentration is summarized in Fig. 5, B and C. The data indicate that d-C$_5$ is a true competitive inhibitor with the inhibition constant K_i = 260 μM. At 500 μM, d-A$_5$ inhibits the reaction only 16 percent as much as d-C$_5$. Thus, inhibition by d-C$_5$ is not some general effect of introducing a deoxyoligonucleotide into the system but depends on sequence.

The formation of the covalent enzyme-substrate intermediate (EpC) can be represented as

$$E + C_5 \underset{k_{-1}}{\overset{k_1}{\rightleftharpoons}} E \cdot C_5 \overset{k_2}{\longrightarrow} EpC + C_4$$

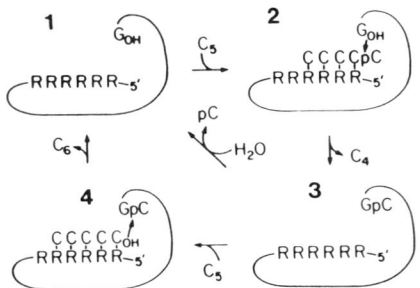

Fig. 4. Model for the enzymatic mechanism of the L − 19 IVS RNA. The RNA catalyzes cleavage and rejoining of oligo(C) by the pathway $1 \to 2 \to 3 \to 4 \to 1$. The L − 19 IVS RNA enzyme (1) is shown with the oligopyrimidine binding site (RRRRRR, six purines) near its 5′ end and G^{414} with a free 3′-hydroxyl group at its 3′ end. The complex folded core structure of the molecule (23, 24, 32) is simply represented by a curved line. The enzyme binds its substrate (C_5) by Watson-Crick base-pairing to form the noncovalent enzyme-substrate complex (2). Nucleophilic attack by G^{414} leads to formation of the covalent intermediate (3). With the pentanucleotide C_5 as substrate, the covalent intermediate is usually loaded with a single nucleotide, as shown; with substrates of longer chain length, an oligonucleotide can be attached to the 3′ end of G^{414}. If C_5 binds to the intermediate (3) in the manner shown in (4), transesterification can occur to give the new product C_6 and regenerate the enzyme (1). Note that all four reactions in this pathway are reversible. When acting as a ribonuclease, the L − 19 IVS RNA follows the pathway $1 \to 2 \to 3 \to 1$. The covalent intermediate (3) undergoes hydrolysis, releasing the nucleotide or oligonucleotide attached to its 3′ end (in this case pC) and regenerating the enzyme (1).

If $k_{-1} \gg k_2$, then $K_m = k_{-1}/k_1$, the dissociation constant for the noncovalent E · C_5 complex. The observation that the K_i for d-C_5 is within an order of magnitude of the K_m for C_5 can then be interpreted in terms of d-C_5 and C_5 having similar binding constants for interaction with the active site on the enzyme. This fits well with the idea that the substrate binds to an oligopurine (R_5) sequence in the active site primarily by Watson-Crick base-pairing, in which case the $C_5 \cdot R_5$ duplex and the d-$C_5 \cdot R_5$ duplex should have similar stability.

Enzyme mechanism and its relation to self-splicing. The stoichiometry of the reaction products (equimolar production of oligonucleotides smaller than and larger than the starting material), the lack of an ATP or GTP (adenosine triphosphate; guanosine triphosphate) energy requirement, the involvement of a covalent intermediate, the specificity for oligoC substrates, and the competi-

tive inhibition by d-C_5 lead to a model for the enzyme mechanism (Fig. 4). The L − 19 IVS RNA is proposed to bind the substrate noncovalently by hydrogen-bonded base-pairing interactions. A transesterification reaction between the 3′-terminal guanosine residue of the enzyme and a phosphate ester of the substrate then produces a covalent enzyme-substrate intermediate.

Transesterification is expected to be highly reversible. If the product C_4 rebinds to the enzyme, it can attack the covalent intermediate and reform the starting material, C_5. Early in the reaction, however, the concentration of C_5 is much greater than the concentration of C_4; if C_5 binds and attacks the covalent intermediate, C_6 is produced (Fig. 4). The net reaction is $2 C_5 \to C_6 + C_4$. The products are substrates for further reaction, for example, $C_6 + C_5 \to C_7 + C_4$ and $C_4 + C_5 \to C_3 + C_6$. The absence of products smaller then C_3 is explicable in terms of the loss of binding

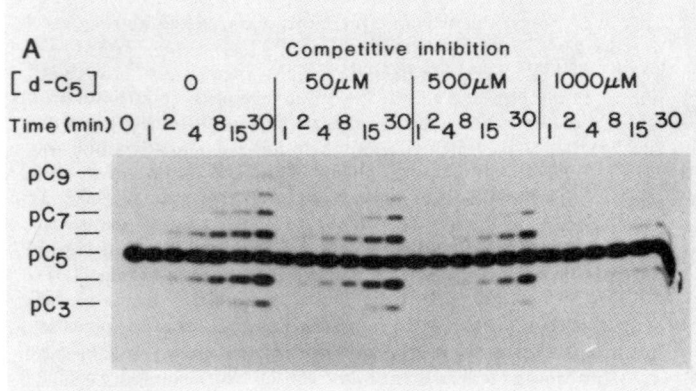

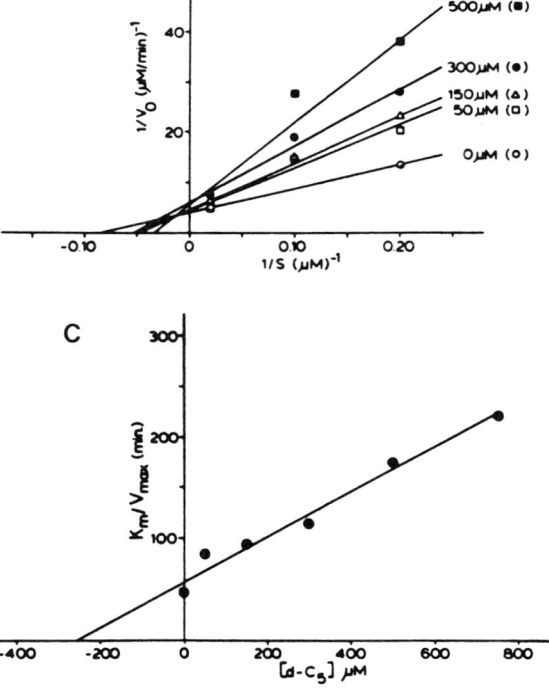

Fig. 5. Competitive inhibition of the pC_5 reaction by d-C_5. (A) 5 μM $\overset{*}{p}C_5$, shown unreacted in lane O, was incubated with 0.16 μM L − 19 IVS RNA under standard reaction conditions. Reactions were done in the absence of d-C_5 or in the presence of 50 μM, 500 μM, or 1000 μM d-C_5 as indicated. (B) Lineweaver-Burk plots of the rate of conversion of pC_5 to $pC_4 + pC_3$ in the presence of (○) 0 μM, (□) 50 μM, (△) 150 μM, (●) 300 μM, or (■) 500 μM unlabeled d-C_5. The analysis was limited to the smaller products because their production is affected only by the first transesterification reaction (Fig. 4). Although d-C_5 is inactive in the first transesterification reaction, it has some activity as a substrate in the second transesterification reaction (15) and therefore could affect the production of chains of length greater than 5. (C) K_m/V_{max}, determined from the slopes of the lines in (B), is plotted against the inhibitor concentration. The x-intercept gives the negative of K_i; $K_i = 260$ μM.

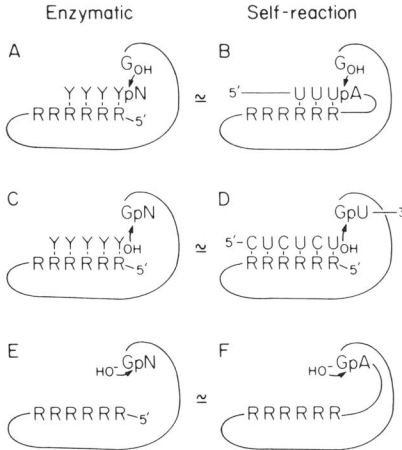

Enzymatic   Self-reaction

Fig. 6. Relation of reactions catalyzed by the L − 19 IVS RNA to self-splicing and the related IVS RNA-mediated reactions. Formation of the covalent enzyme-substrate intermediate (A) is analogous to IVS RNA autocyclization (B). Resolution of the enzyme-substrate intermediate (C) is analogous to exon ligation (D) or the reversal of cyclization (10). Hydrolysis of the enzyme-substrate intermediate (E) is analogous to site-specific hydrolysis of the circular IVS RNA (F) or of the pre-rRNA (19).

interactions relative to C_4 (C_3 could form only two base pairs in the binding mode that would be productive for cleavage).

The transesterification reactions are conservative with respect to the number of phosphodiester bonds in the system. Thus, RNA ligation can occur without an external energy source as is required by RNA or DNA ligase. Hydrolysis of the covalent intermediate competes with transesterification. The net reaction is $C_5 + H_2O \rightarrow C_4 + pC$, with the L − 19 IVS RNA acting as a ribonuclease.

On the basis of our current understanding of the reaction, the catalytic strategies of the L − 19 IVS RNA enzyme appear to be the same as those used by protein enzymes (20). First, the RNA enzyme, like protein enzymes, forms a specific noncovalent complex with its oligonucleotide substrate. This interaction is proposed to hold the oligonucleotide substrate at a distance and in an orientation such as to facilitate attack by the 3′-hydroxyl of the terminal guanosine of the enzyme. Second, a covalent enzyme-substrate complex is a presumptive intermediate in the L − 19 IVS RNA reaction. Covalent intermediates are prevalent in enzyme-catalyzed group transfer reactions. Third, the phosphodiester bond formed in the covalent intermediate is unusually susceptible to hydrolysis, suggesting that it may be strained or activated to facilitate formation of the pentavalent transition state upon nucleophilic attack (8, 9). Similarly, protein catalysts are thought to facilitate the formation of the transition state, for example, by providing active site groups that bind the transition state better than the unreacted substrate (6, 21). Thus far there is no evidence that another major category of enzyme catalysis, general acid-base catalysis, occurs in the L − 19 IVS RNA reactions, but we think it likely that it will be involved in facilitating the required proton transfers.

Each L − 19 IVS RNA-catalyzed transesterification and hydrolysis reaction is analogous to one of the steps in *Tetrahymena* pre-rRNA self-splicing or one of the related self-reactions (Fig. 6). Thus, the finding of enzymatic activity in a portion of the IVS RNA validates the view that the pre-rRNA carries its own splicing enzyme as an intrinsic part of its polynucleotide chain. It seems likely that the C_5 substrate binding site of the L − 19 IVS RNA is the

oligopyrimidine binding site that directs the choice of the 5′ splice site and the various IVS RNA cyclization sites (10, 18, 19, 22). Although the location of this site within the IVS RNA has not been proved, the best candidate is a portion of the "internal guide sequence" proposed by Davies and co-workers (23). Michel and Dujon (24) show a similar pairing interaction in their RNA structure model. The putative binding site, GGAGGG, is located at nucleotides 22 to 27 of the intact *Tetrahymena* IVS RNA and at positions 3 to 8 very near the 5′ end of the L − 19 IVS RNA. If this is the substrate binding site, site-specific mutation of the sequence should change the substrate specificity of the enzyme in a predictable manner.

RNA polymerase or RNA restriction endonuclease? With C_5 as a substrate, the L − 19 IVS RNA makes poly(C) with chain lengths of 30 nucleotides and longer. The number of P-O bonds is unchanged in the process. In the synthesis of poly(C) on a poly(dG) template by RNA polymerase, one CTP is cleaved for each residue polymerized. Thus, the RNA polymerase reaction is also conservative with respect to the number of P-O bonds in the system. The L − 19 IVS RNA can therefore be considered to be a poly(C) polymerase that uses C_4pC instead of pppC as a substrate. It incorporates pC units at the 3′ end of the growing chain and releases C_4; the C_4 is analogous to the pyrophosphate released by RNA polymerase. Synthesis is directed by a template, but the template is internal to the RNA enzyme. It may be possible to physically separate the template portion from the catalytic portion of the RNA enzyme with retention of activity. If so, the RNA enzyme could conceivably act as a primordial RNA replicase, catalyzing both its own replication and that of other RNA molecules (25).

In its ribonuclease mode, the L − 19 IVS RNA is expected to have specificity similar to that of the IVS RNA cyclization reaction (10, 18). That is, it recognizes three or more nucleotides in choosing a reaction site. Protein ribonucleases that are active on single-stranded RNA substrates have specificity only at the mononucleotide level (for example, ribonuclease T_1 cleaves after guanosine). Thus the L − 19 has more base-sequence specificity for single-stranded RNA than any known protein ribonuclease, and may approach the specificity of some of the DNA restriction endonucleases. An attractive feature of this new RNA ribonuclease is the possibility of completely and predictably changing its substrate specificity by altering the sequence of the internal binding site.

How good an enzyme? The L − 19 IVS RNA catalyzes the cleavage-ligation of pC_5 with $K_m = 42~\mu M$, $k_{cat} = 2~min^{-1}$, and $k_{cat}/K_m = 1 \times 10^3~sec^{-1}~M^{-1}$. The K_m is typical of that of protein enzymes. The k_{cat} and k_{cat}/K_m are lower than those of many protein enzymes. However, k_{cat} is well within the range of values for proteins that recognize specific nucleic acid sequences and catalyze chain cleavage or initiation of polymerization. For example, Eco RI restriction endonuclease cleaves its recognition sequence in various DNA substrates, including a specific 8-bp DNA fragment, with $k_{cat} = 1~min^{-1}$ to $18~min^{-1}$ (26). The k_{cat} is also similar to that of the RNA enzyme ribonuclease P, which cleaves the precursor to tRNA with $k_{cat} = 2~min^{-1}$ (11, 13).

Another way to gauge the catalytic effectiveness of the L − 19 IVS RNA is to compare the rate of the catalyzed reaction to the basal chemical rate. A transesterification reaction between two free oligonucleotides has never been observed, and hence the uncatalyzed rate is unknown. On the other hand, the rate of hydrolysis of simple phosphate diesters has been studied (27, 28). The second-order rate constant for alkaline hydrolysis of the labile phosphodiester bond in the circular IVS RNA (8) is 12 orders of magnitude higher than that of dimethyl phosphate (27) and ten orders of magnitude higher than that expected for a normal phosphodiester bond in RNA (29). On the basis of the data of Fig. 3D, the covalent enzyme-substrate complex undergoes hydrolysis at approximately the same rate as the equivalent bond in the circular IVS RNA. Thus, we estimate that the $L - 19$ IVS RNA in its ribonuclease mode enhances the rate of hydrolysis of its substrate about 10^{10} times.

REFERENCES AND NOTES

1. T. R. Cech, A. J. Zaug, P. J. Grabowski, *Cell* **27**, 487 (1981); K. Kruger *et al.*, *ibid.* **31**, 147 (1982).
2. G. Garriga and A. M. Lambowitz, *ibid.* **39**, 631 (1984).
3. G. Van der Horst and H. F. Tabak, *ibid.* **40**, 759 (1985).
4. F. K. Chu, G. F. Maley, M. Belfort, F. Maley, *J. Biol. Chem.* **260**, 10680 (1985).
5. C. L. Peebles *et al.*, *Cell*, in press; R. Van der Veen *et al.*, *ibid.*, in press.
6. A. Fersht, *Enzyme Structure and Mechanism* (Freeman, New York, ed. 2, 1985).
7. B. L. Bass and T. R. Cech, *Nature (London)* **308**, 820 (1984).
8. A. J. Zaug, J. R. Kent, T. R. Cech, *Biochemistry* **24**, 6211 (1985).
9. ——, *Science* **224**, 574 (1984).
10. F. X. Sullivan and T. R. Cech, *Cell* **42**, 639 (1985).
11. C. Guerrier-Takada *et al.*, *ibid.* **35**, 849 (1983).
12. C. Guerrier-Takada and S. Altman, *Science* **223**, 285 (1984).
13. T. L. Marsh, B. Pace, C. Reich, K. Gardiner, N. R. Pace, in *Sequence Specificity in Transcription and Translation*, R. Calendar and L. Gold Eds., *UCLA Symposium on Molecular and Cellular Biology* (Plenum, New York, in press); T. L. Marsh and N. R. Pace, *Science* **229**, 79 (1985).
14. Abbreviations: IVS, intervening sequence or intron; $L - 19$ IVS RNA (read "L minus 19"), a 395-nt RNA missing the first 19 nt of the L IVS RNA (the direct product of pre-ribosomal RNA splicing); p̊, ^{32}P within an oligonucleotide, that is, C_5pC is CpCpCpCp^{32}pC and pC_5 is ^{32}pCpCpCpCpC; d-C_5, deoxyC$_5$.
15. A. Zaug and T. Cech, unpublished data.
16. The radioactive phosphate was bonded covalently to the $L - 19$ IVS RNA as judged by the following criteria: it remained associated when the complex was isolated and subjected to a second round of denaturing gel electrophoresis; it was released in the form of a mononucleotide upon RNase T$_2$ treatment; and it was released in the form of a series of unidentified oligonucleotides upon RNase T$_1$ treatment (*15*). These results are consistent with a series of covalent enzyme-substrate complexes in which various portions of C_5pC are linked to the $L - 19$ IVS RNA via a normal $3',5'$-phosphodiester bond. A more complete structural analysis of the covalent complexes is in progress.
17. A. J. Zaug, P. J. Grabowski, T. R. Cech, *Nature (London)* **301**, 578 (1983).
18. M. Been and T. R. Cech, *Nucleic Acids Res.* **13**, 8389 (1985).
19. T. Inoue, F. X. Sullivan, T. R. Cech, *J. Mol. Biol.*, in press.
20. W. P. Jencks, *Catalysis in Chemistry and Enzymology* (McGraw-Hill, New York, 1969).
21. T. N. C. Wells and A. R. Fersht, *Nature (London)* **316**, 656 (1985).
22. T. Inoue, F. X. Sullivan, T. R. Cech, *Cell* **43**, 431 (1985).
23. R. W. Davies *et al.*, *Nature (London)* **300**, 719 (1982); R. B. Waring, C. Scazzocchio, T. A. Brown, R. W. Davies, *J. Mol. Biol.* **167**, 595 (1983).
24. F. Michel and B. Dujon, *EMBO J.* **2**, 33 (1983).
25. T. R. Cech, in preparation.
26. P. J. Greene *et al.*, *J. Mol. Biol.* **99**, 237 (1975); P. Modrich and D. Zabel, *J. Biol. Chem.* **251**, 5866 (1976); R. D. Wells, R. D. Klein, C. K. Singleton, *Enzymes* **14**, 157 (1981); C. A. Brennan, M. B. Van Cleve, R. I. Gumport, in preparation; B. Terry, W. Jack, P. Modrich, in preparation.
27. J. Kumamoto, J. R. Cox, Jr., F. H. Westheimer, *J. Am. Chem. Soc.* **78**, 4858 (1956); P. C. Haake and F. H. Westheimer, *ibid.* **83**, 1102 (1961).
28. A. J. Kirby and M. Younas, *J. Chem. Soc. Ser. B.* (1970), p. 1165; C. A. Bunton and S. J. Farber, *J. Org. Chem.* **34**, 767 (1969).
29. The rate of nucleophilic attack by hydroxide ion on phosphate esters is sensitive to the pK_a of the conjugate acid of the leaving group. A phosphate in RNA should be more reactive than dimethyl phosphate, because $pK_a = 12.5$ for a nucleoside ribose and $pK_a = 15.5$ for methanol [values at 25°C from P. O. P. Ts'o, *Basic Principles in Nucleic Acid Chemistry* (Academic Press, New York, 1974), vol. 1, pp. 462–463 and P. Ballinger and F. A. Long, *J. Am. Chem. Soc.* **82**, 795 (1960), respectively]. On the basis of the kinetic data available for the alkaline hydrolysis of phosphate diesters (*27, 28*), the slope of a graph of the logarithm of the rate constant for hydrolysis as a function of pK_a can be roughly estimated as 0.6. Thus, RNA is expected to be more reactive than dimethyl phosphate by a factor of $10^{0.6\ (15.5-12.5)} = 10^{1.8}$. The estimate for RNA pertains to direct attack by OH$^-$ on the phosphate, resulting in $3'$-hydroxyl and $5'$-phosphate termini. Cleavage of RNA by OH$^-$-catalyzed transphosphorylation, producing a $2',3'$-cyclic phosphate, is a much more rapid (intramolecular) reaction but is not relevant to the reactions of the $L - 19$ IVS RNA.
30. J. V. Price and T. R. Cech, *Science* **228**, 719 (1985).
31. E. T. Butler and M. J. Chamberlin, *J. Biol. Chem.* **257**, 5772 (1982); D. A. Melton *et al.*, *Nucleic Acids Res.* **12**, 7035 (1984).
32. T. R. Cech *et al.*, *Proc. Natl. Acad. Sci. U.S.A.* **80**, 3903 (1983); T. Inoue and T. R. Cech, *ibid.* **82**, 648 (1985).
33. We thank O. Uhlenbeck for gifts of oligoribonucleotides; J. Beltman, J.-Y. Tang, and M. Caruthers for oligodeoxyribonucleotides; and A. Sirimarco and M. Gaines for preparation of the manuscript and illustrations. Supported by American Cancer Society grant NP-374B, the National Foundation for Cancer Research, NIH grant GM28039, and an NIH Research Career Development Award (T.R.C.).

25 November 1985; accepted 27 December 1985

A model for the RNA-catalyzed replication of RNA[†]

(origin of life/ribozyme/RNA polymerase/intron/RNA splicing)

Thomas R. Cech

Department of Chemistry and Biochemistry, University of Colorado, Boulder, CO 80309-0215

Communicated by Norman H. Horowitz, February 27, 1986

ABSTRACT   A shortened form of the self-splicing ribosomal RNA intervening sequence of *Tetrahymena thermophila* has enzymatic activity as a poly(cytidylic acid) polymerase [Zaug, A. J. & Cech, T. R. (1986) *Science* **231**, 470–475]. Based on the known properties of this enzyme, a detailed model is developed for the template-dependent synthesis of RNA by an RNA polymerase itself made of RNA. The monomer units for RNA synthesis are tetra- and pentanucleotides of random base sequence. Polymerization occurs in a 5′-to-3′ direction, and elongation rates are expected to approach two residues per minute. If the RNA enzyme could use another copy of itself as a template, RNA self-replication could be achieved. Thus, it seems possible that RNA catalysts might have played a part in prebiotic nucleic acid replication, prior to the availability of useful proteins.

The question about the origin of life often appears as [the] question . . . *Which came first, the protein or the nucleic acid?*—a modern variant of the old chicken-and-the-egg problem. The term "first" is usually meant to define a causal rather than a temporal relationship, and the words "protein" and "nucleic acid" may be substituted by "function" and "information." The question in this form, when applied to the interplay of nucleic acids and proteins as presently encountered in the living cell, leads *ad absurdum*, because "function" cannot occur in an organized manner unless "information" is present and this "information" only acquires its meaning *via* the "function" for which it is coding. [M. Eigen (1)]

The finding of self-splicing RNA (2–8) and RNA with ribonuclease activity (9–11) has been widely interpreted to provide a possible resolution of the "chicken-and-the-egg problem" in favor of RNA (9, 12–17). That is, the presence of both "information" and "function" in the same RNA molecule might, in principle, allow it to catalyze its own replication. Yet neither self-splicing RNA nor RNase P has any obvious relationship to the process of nucleic acid replication as it occurs in contemporary cells. Thus, while it has been a logical extrapolation of known facts to envision RNA catalysis of prebiotic RNA recombination (2, 18, 19) or RNA processing (9, 13), it has been more difficult to envision RNA catalysis of prebiotic nucleic acid synthesis.

Arthur Zaug and I (20) recently described a system in which a 395-nucleotide form of the *Tetrahymena* ribosomal RNA intervening sequence (IVS) acts as an RNA cleavage–ligation enzyme. Using the same activity that it employs in the self-splicing and autocyclization reactions, the RNA enzyme converts pentacytidylic acid (pC$_5$) to poly(C) with multiple turnover. Thus, it acts as a poly(C) polymerase, synthesizing RNA in a 5′-to-3′ direction. The enzyme could also be designated terminal cytidylyltransferase or oligo(C) dismutase. The term "polymerase" is chosen to emphasize

the template dependence of the enzyme. The RNA enzyme differs from protein RNA polymerases in that it uses an internal rather than an external nucleic acid template.

I now extrapolate these findings to develop an entirely RNA-based model for prebiotic RNA replication. In so doing, I am in no way trying to present an historical account of the events that occurred early in evolution. Instead, I simply wish to establish the principle that an RNA polymerase, itself made of ribonucleic acid, might have played a key role in prebiotic nucleic acid replication independent of the availability of proteins. The model is intended to complement rather than compete with other models for prebiotic nucleic acid replication, such as those of Orgel, Usher, and colleagues (21–25).

The L − 19 IVS RNA Is a Poly(C) Polymerase. The L − 19 IVS RNA converts pC$_5$ (or any pC$_n$ with $n \geq 4$) to both larger and smaller oligomers of C, with K_m = 40 μM and a turnover number of 2 per min (20). Chain lengths up to 30 are produced after a 1-hr reaction, at which time the substrate is depleted. The reaction is specific for oligo(C) substrates, there being very little reaction with oligo(U) and none with oligo(A) or oligo(dC). The proposed mechanism is illustrated in Fig. 1. The individual steps in the mechanism are intermolecular versions of RNA self-splicing and IVS RNA cyclization.

At first glance, the activity of the L − 19 IVS RNA seems very different from that of RNA polymerase; the former catalyzes RNA recombination, giving no net change in the number of phosphodiester bonds, whereas the latter clearly gives net synthesis of nucleic acid. In fact, however, both reactions are strictly conservative with respect to the number of O—P bonds in the system. The L − 19 IVS RNA uses C$_5$ (or C$_4$) instead of CTP as a substrate. It incorporates pC units at the 3′ end of the growing chain and releases C$_4$ (or C$_3$), which is analogous to the pyrophosphate released by RNA polymerase (Fig. 2).

The reaction shown in Fig. 1 has further similarities to contemporary RNA polymerase reactions. Chain elongation occurs in a 5′→3′ polarity. All products have 3′-hydroxyl termini. The covalent linkages are exclusively 3′,5′-phosphodiester bonds (20).

The L − 19 IVS RNA is thought to recognize oligo(C) substrates by Watson–Crick base-pairing to an oligonucleotide binding site (internal template) with the sequence GGAGGGA (26–29). This binding site was originally predicted from RNA structure models to be the part of the "internal guide sequence" that pairs with the 5′ exon (ref. 30; see also

Abbreviations: IVS, intervening sequence (intron); L − 19 IVS RNA (read "L minus 19"), a 395-nucleotide RNA missing the first 19 nucleotides of the linear IVS RNA that is the direct product of *Tetrahymena* pre-ribosomal RNA splicing; ribozyme, an RNA molecule that shows intramolecular catalysis or acts as an enzyme; ribozyme*, a hypothetical ribozyme that has the known enzymatic activity of the L − 19 IVS RNA but is dependent on an external template and able to incorporate all four nucleotides.
[†]Presented at the Origin of Life Symposium held in honor of Norman Horowitz at the California Institute of Technology on October 15, 1985.

Reprinted with permission of the author from *Proc Nat'l Acad Sci USA*, Volume 83, pp.4360–4363. © 1986 by the National Academy of Sciences.

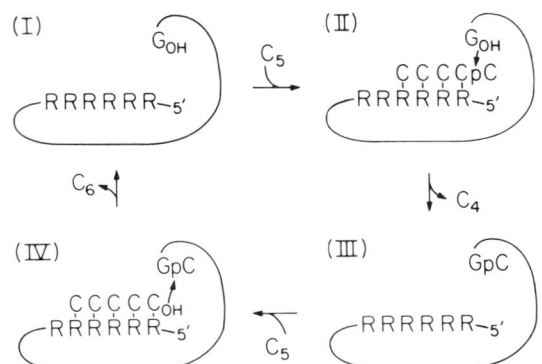

FIG. 1. Proposed mechanism of polymerization of cytidylic acid by a real RNA enzyme, the L − 19 IVS RNA (figure adapted from ref. 20, with permission of the copyright holder). The L − 19 IVS RNA enzyme (**I**) is shown with the pyrimidine oligonucleotide ("oligopyrimidine") binding site (RRRRRR, 6 purines) near its 5′ end and guanosine-414 (G^{414}) with a free 3′-hydroxyl group at its 3′ end. The complex folded core structure of the molecule is simply represented by a curved line. The enzyme binds its substrate (C_5) by Watson–Crick base-pairing to form the noncovalent enzyme–substrate complex (**II**). Nucleophilic attack by G^{414} leads to formation of the covalent intermediate (**III**), which contains a high-energy GpC bond (20, 26). If C_5 binds to the intermediate in the manner shown (**IV**), transesterification can occur to give the new product C_6 and regenerate the enzyme (**I**). As the concentration of the product C_6 increases, it can be used as a substrate to give C_7, and so on. With longer oligonucleotides as substrates, the enzyme is not restricted to attacking at the 3′ end of the substrate; thus, the covalent intermediate can have more than one C residue esterified to G^{414} (20).

ref. 31). The IVS can be altered to recognize a different 5′ exon sequence by changing its internal guide (M. Been and T.R.C., unpublished data). Other group-I introns recognize quite different 5′ exon sequences, presumably because they have different internal guide sequences (32, 33). Therefore, as RNA polymerases they are predicted to have different substrate specificity.

Although the 5′ exon sequences of different group-I introns vary widely, there is one conserved position. The 5′ splice is preceded by a conserved U residue that is thought to pair with a conserved G residue at the 5′ end of the exon binding site within the intron (31–33). This interaction may be obligatory for the first step of RNA self-splicing (ref. 29; L. Barfod and T.R.C., unpublished data), but recent results indicate that it may not be important for L − 19 IVS RNA catalysis (A. Zaug, R. Kierzek, M. Caruthers, and T.R.C., unpublished data). In any case, the ability of the L − 19 IVS RNA to polymerize RNA in a template-dependent manner with no restriction on the base sequence of its internal template is, at present, conjecture rather than established fact.

An RNA Enzyme as an RNA Polymerase. An effective RNA polymerase must not only be able to incorporate all four nucleotides into a growing chain, but it must also utilize an external rather than an internal template so that it can copy

Enzyme	Substrate
RNA Polymerase	pppC$_{OH}$
L - 19 IVS RNA	C$_4$pC$_{OH}$

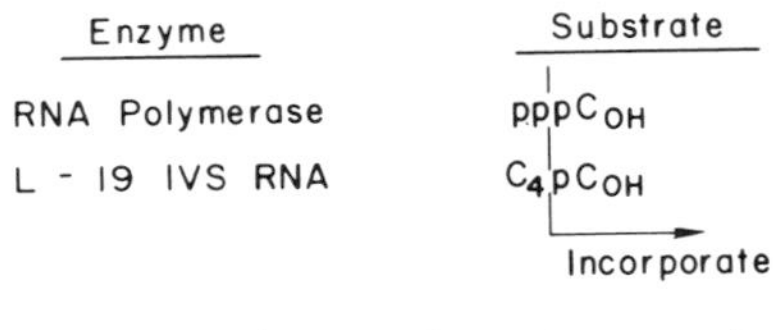

FIG. 2. Substrate requirements for contemporary (protein) RNA polymerases [using a poly(dG) template] and the L − 19 IVS RNA. Both polymerization reactions are conservative with respect to O—P bonds.

chains of any length and sequence. It seems possible that the L − 19 IVS RNA might retain activity if its template (internal guide sequence) were dissociated from its catalytic portion. The internal guide sequence is presumably oriented very precisely with respect to the critical conserved sequence elements (34–36) and the 3′-terminal G residue to allow self-splicing and L − 19 IVS RNA activity. It seems likely that this orientation is provided by multiple sequence-independent interactions, perhaps interactions with the phosphates or the 2′-hydroxyl groups of the ribose moieties. If its internal template were deleted, the molecule might assemble with an external template to give an active complex.

For the purpose of the RNA polymerase model, I define ribozyme* as an RNA enzyme with the known catalytic activity of the L − 19 IVS RNA but dependent on an external template and able to incorporate all four nucleotides. The proposed mechanism of RNA-catalyzed RNA polymerization is shown in Fig. 3A. The ribozyme* already has a nucleotide N loaded on its 3′-terminal guanosine (**I**). The ribozyme* is noncovalently bound to the template RNA, the template assuming the position occupied by the internal guide sequence in Fig. 1. The ribozyme* either slides along the template or transiently associates and dissociates. The solution also contains a collection of tetra- and pentanucleotides of random base sequence to provide monomer units. If one of these oligonucleotides binds to a complementary sequence on the template adjacent to the activated nucleotide N at the 3′ end of ribozyme*, transesterification can occur to transfer N from the enzyme to the oligonucleotide (**II**). This reaction, which is thermodynamically favorable (26), is equivalent to reaction **IV** → **I** of L − 19 IVS RNA catalysis (Fig. 1). The

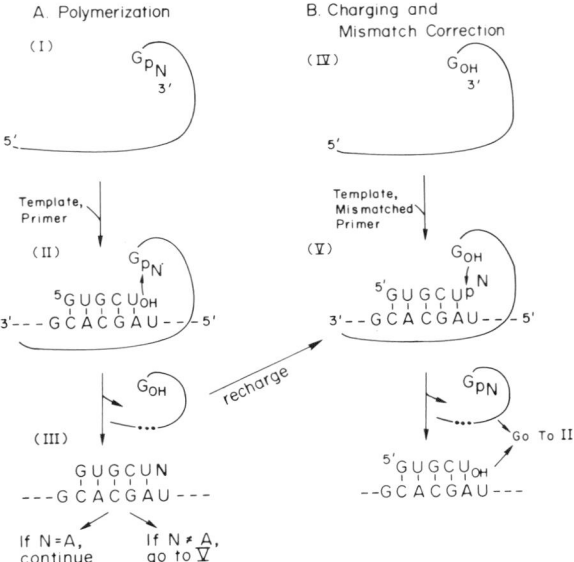

FIG. 3. Mechanism of template-dependent RNA polymerization by ribozyme*, a hypothetical RNA enzyme with enzymatic activity similar to that of the L − 19 IVS RNA. Ribozyme* is missing the oligopyrimidine binding site of the L − 19 IVS RNA (Fig. 1). (*A*) Charged ribozyme* (**I**) with a nucleotide N esterified to its 3′-terminal G residue, interacts with a template and primer to form complex **II**. Ternary complex **II** is envisioned to have the same structure as binary complex **IV** of Fig. 1. The enzyme facilitates the attack of the 3′-terminal hydroxyl group of the primer on the phosphate preceding N, transferring N to the growing chain (**III**) and releasing uncharged ribozyme*. (*B*) Uncharged ribozyme* (**IV**) interacts with a template-primer system containing a terminal mismatched nucleotide (**V**). The enzyme facilitates the attack of the 3′-terminal hydroxyl group of the ribozyme* on the phosphate preceding N, thereby recharging the enzyme.

transfer reaction is proposed to occur with highest efficiency if N can also base-pair with the template strand; i.e., if N = A in Fig. 3 (II). [The equivalent "rule" for RNA self-splicing would be for the first nucleotide of the IVS to be complementary to the nucleotide preceding the conserved G in the internal guide sequence. This rule is followed by most group-I introns (33), but the requirement for pairing of these bases has not yet been critically tested.]

If the newly added base N is properly paired with the template, another charged ribozyme* can bind and chain elongation can continue. On the other hand, if N is mispaired, the primer will be a very poor attacking group for further polymerization but a very good reactant for a charging reaction (Fig. 3B). Thus, the reversibility of the reaction assures attainment of equilibrium and gives an opportunity for mismatches to be corrected. Such a mechanism can give reasonable fidelity in copying ($<10^{-2}$ errors per step), provided that the rate of adding a matched nucleotide is at least 20-fold greater than the rate of adding a mismatched nucleotide, and that the rate of removing a mismatched nucleotide is at least 20-fold greater than the rate of removing a matched nucleotide. This is an energy-efficient correction process, because instead of removing the mispaired nucleotide by hydrolysis, it uses it to recharge a ribozyme*. However, it does not provide the high fidelity that can be obtained by proofreading mechanisms in which there is excess consumption of high-energy bonds.

The charging reaction (Fig. 3B) need not take place on the same template but could take place on a primer–template pair anywhere in the system. An uncharged ribozyme* with a free 3' hydroxyl (IV) associates with a template containing a primer that is not base-paired at its 3' end (V). The 3'-terminal G attacks the phosphate preceding nucleotide N, forming the covalent GpN bond and recharging the ribozyme*. The reaction is equivalent to reaction II → III of L − 19 IVS RNA catalysis (Fig. 1). It has an unfavorable equilibrium constant (26) but can be driven by an excess concentration of oligonucleotide.

Depending on the size distribution of oligonucleotide substrates, ribozyme* might be charged by more than a single nucleotide. This can be accommodated in the model. For example, if ribozyme* were charged with a dinucleotide, it would transfer two nucleotides to the primer in a single step. If both nucleotides could pair with the template, they would be retained. If neither paired, they would be removed by the correction mechanism. If the first nucleotide were matched but the 3'-terminal nucleotide were mismatched, the latter could be removed, giving a net extension of one nucleotide.

Early in the reaction, when the oligonucleotide primers are short, they would be expected to pair with the template only transiently, as in the reactions catalyzed by the L − 19 IVS RNA. However, as the primer is lengthened it would become stably paired with the template. Because each step in the polymerization reaction is reversible, pairing would be important to help drive polymerization to completion. It would also be important for the accuracy of the replication process. As the primer becomes longer it will be able to pair stably even if there is an internal mismatch. If it were continually dissociating and reassociating, it could pair to sequences other than the one that served as its template. If the match were imperfect, errors would accumulate. On the other hand, if the primer remained bound, the only opportunity for introducing new errors would be at the 3' end of the growing chain, where the mismatch correction mechanism is operative. At all stages of the reaction, the ribozyme* would operate in a distributive rather than a processive manner, since it must dissociate from the template–primer to be recharged. In this manner the reaction is dissimilar to contemporary transcription and replication, which are processive.

RNA Self-Replication. The model for RNA-catalyzed, template-dependent RNA polymerization provides the key element for a complete model for RNA self-replication (Fig. 4). The general scheme is similar to one published recently by Sharp (15). The replication cycle begins with a double-stranded RNA (I). (This double-stranded form is a useful mental construct but, as described below, need not exist as an intermediate.) The double-stranded RNA undergoes strand-separation—e.g., thermal denaturation under the influence of the heat of the sun (21). One of the single strands (the plus strand) folds to form the ribozyme* (II). The complementary (minus) strand serves as the template. The oligonucleotides that serve as the primers and as the source of monomer units have been previously synthesized—e.g., by the template-directed nonenzymatic polymerization processes described by Orgel and colleagues (22–24).

Polymerization and mismatch correction proceed according to the mechanism described in Fig. 3 (III). The end result can be re-formation of a double-stranded RNA (I). Alternatively, it is attractive to envision strand-displacement taking place during replication (17), so that the intermediate (III) is directly converted to the single-stranded product (II) without going through a form that is double-stranded throughout its length (I). This might be accomplished if local regions of the RNA were able to form transient intramolecular base-pairing that competed with the intermolecular base-pairing, as occurs during the replication of single-stranded bacteriophage RNA (37, 38).

In a subsequent round of replication, one of the catalyst strands can serve as the template for the production of another minus-strand RNA. Thus, the same ribozyme* molecule can serve both as a functional catalyst and as an informational entity.

In considering the origin-of-life implications of RNA-catalyzed RNA polymerization, it is important to realize that a ribozyme* would not be restricted to using itself and its complement as templates. Other RNA molecules in its environment would also be replicated. Some of these might have useful enzymatic activities, for example as specific RNA-processing enzymes like RNase P (9–11). Other RNA molecules might be able to bind an amino acid as well as a portion of an RNA template, thereby serving as primitive transfer RNAs (39, 40). Still other RNA molecules might facilitate the binding of two transfer RNAs at adjacent sites

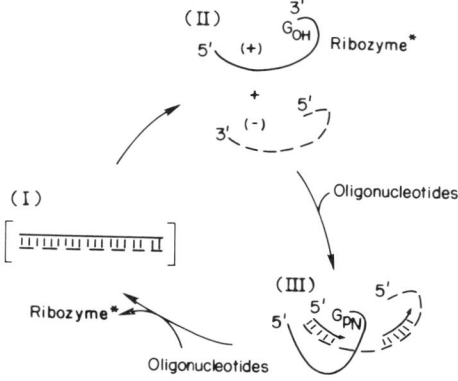

FIG. 4. Self-replication scheme. Double-stranded RNA (I) undergoes strand separation to give ribozyme* [(+)-strand] and the complementary (−)-strand (II). The ribozyme* catalyzes synthesis of a new (+)-strand, using the (−)-strand as a template (III). The detailed mechanism of replication is shown in Fig. 3. If more than one primer is elongated on the same template, the resulting fragments can be spliced together (15). Completion of replication gives a double-stranded RNA (I). Square brackets indicate that I is not an obligatory intermediate in the replication cycle (see text).

on an RNA template and catalyze peptide-bond formation, thereby serving as primitive ribosomal RNAs (41, 42). As peptides and proteins became available, some of them would interact with the RNA catalysts and enhance their activity or modulate it in useful ways. The ribozymes would begin to work as ribonucleoprotein particles. The major point of this paper, however, is that it is now reasonable to envision prebiotic nucleic acid replication in an entirely RNA-based system, prior to the advent of any translational machinery or other source of proteins.

Further Evaluation of the Model. The scheme for RNA-catalyzed RNA replication shown in Figs. 3 and 4 does not address some important details. To what extent would replication be blocked by structured regions in the RNA template? If replication proceeded simultaneously from two or more primers on the same template, could the fragments be ligated? [Such splicing could presumably be RNA-catalyzed; Sharp (15) has proposed a replication scheme based entirely on such RNA splicing.] How could the mismatch correction mechanism be restricted to working on the growing chain and prevented from catalyzing breakdown of the template? Although it is possible to invent solutions, there is not yet sufficient experimental basis for evaluating the gravity of these problems.

Further, it is not clear how small an RNA molecule could function as ribozyme*. Based on nucleotide-deletion studies with the *Tetrahymena* IVS RNA, it seems possible that efficient cleavage–ligation activity might require an RNA enzyme as large as 300 nucleotides (ref. 43; G. Dinter-Gottlieb, L. Dokken, and T.R.C., unpublished data). At the other extreme, the core structure of group-I introns (31–33) contains only about 100 nucleotides, so there is hope that smaller molecules might have substantial catalytic activity.

The scheme for RNA-catalyzed RNA replication appears to have several advantages over schemes for nonenzymatic polymerization of activated mononucleotides, in which the only catalysis involves the alignment of the monomer units by their interaction with the template (23). The most obvious is the rate acceleration. If ribozyme* could work at the rate of the L − 19 IVS RNA, it could achieve rates of chain elongation of two residues per minute (20), 1000 times the rate of polymerization of activated mononucleotides or oligonucleotides (23, 25). The higher rate might be necessary for the establishment and maintenance of a prebiotic replication system, because RNA is not infinitely stable and polymerization of reasonably long RNA molecules must occur fast enough to compete with random hydrolysis. Second, the L − 19 IVS RNA works efficiently in dilute solutions. It requires only micromolar concentrations of oligonucleotides, presumably because binding is facilitated by stacking and other interactions in the enzyme active site. The efficient nonenzymatic polymerization of activated mononucleotides or oligonucleotides, on the other hand, requires concentrations of 25–100 mM (23–25). Third, the enzymatic synthesis utilizes probable monomer units—oligonucleotides. It seems reasonable that oligonucleotide synthesis preceded polynucleotide synthesis, so that oligonucleotides were already present in the prebiotic environment where the first polynucleotides were synthesized (21). Finally, the enzymatic synthesis does not totally consume its monomer units but only reduces their length. Residual oligonucleotides that are too short to serve as monomers should be ideal reactants for extension by the nonenzymatic reactions, which in some cases have been shown to become more efficient once a dinucleotide or oligonucleotide has been formed (44). Thus, the enzymatic and nonenzymatic processes could be synergistic.

I thank L. Orgel, J. Hopfield, and O. Uhlenbeck, for critical comments on the manuscript, and A. Sirimarco and M. Neary for preparation of the manuscript and illustrations. This work was supported by Grant GM28039 from the National Institutes of Health and by a fellowship from the John Simon Guggenheim Memorial Foundation.

1. Eigen, M. (1971) *Naturwissenschaften* **58**, 465–523.
2. Cech, T. R., Zaug, A. J. & Grabowski, P. J. (1981) *Cell* **27**, 487–496.
3. Kruger, K., Grabowski, P. J., Zaug, A. J., Sands, J., Gottschling, D. E. & Cech, T. R. (1982) *Cell* **31**, 147–157.
4. Garriga, G. & Lambowitz, A. M. (1984) *Cell* **39**, 631–641.
5. Van der Horst, G. & Tabak, H. F. (1985) *Cell* **40**, 759–766.
6. Chu, F. K., Maley, G. F., West, D. K., Belfort, M. & Maley, F. (1986) *Cell* **45**, 157–166.
7. Peebles, C. L., Perlman, P. S., Mecklenburg, K. L., Petrillo, M. L., Tabor, J. H., Jarrell, K. A. & Cheng, H.-L. (1986) *Cell* **44**, 213–223.
8. Van der Veen, R., Arnberg, A. C., Van der Horst, G., Bonen, L., Tabak, H. F. & Grivell, L. A. (1986) *Cell* **44**, 225–234.
9. Guerrier-Takada, C., Gardiner, K., Marsh, T., Pace, N. & Altman, S. (1983) *Cell* **35**, 849–857.
10. Guerrier-Takada, C. & Altman, S. (1984) *Science* **223**, 285–286.
11. Marsh, T. L. & Pace, N. R. (1985) *Science* **229**, 79–81.
12. Lewin, R. (1982) *Science* **218**, 872–874.
13. Altman, S. (1984) *Cell* **36**, 237–239.
14. Cech, T. R. (1985) *Int. Rev. Cytol.* **93**, 3–22.
15. Sharp, P. A. (1985) *Cell* **42**, 397–400.
16. Darnell, J. E. & Doolittle, W. F. (1986) *Proc. Natl. Acad. Sci. USA* **83**, 1271–1275.
17. Pace, N. R. & Marsh, T. L. (1986) *Origins of Life*, in press.
18. Reanney, D. C. (1979) *Nature (London)* **277**, 597–600.
19. Zaug, A. J. & Cech, T. R. (1985) *Science* **229**, 1060–1064.
20. Zaug, A. J. & Cech, T. R. (1986) *Science* **231**, 470–475.
21. Orgel, L. E. & Lohrmann, R. (1974) *Acc. Chem. Res.* **7**, 368–377.
22. Bridson, P. K. & Orgel, L. E. (1980) *J. Mol. Biol.* **144**, 567–577.
23. Inoue, T. & Orgel, L. E. (1982) *J. Mol. Biol.* **162**, 201–217.
24. Inoue, T. & Orgel, L. E. (1983) *Science* **219**, 859–862.
25. Usher, D. A. & McHale, A. H. (1976) *Science* **192**, 53–54.
26. Sullivan, F. X. & Cech, T. R. (1985) *Cell* **42**, 639–648.
27. Inoue, T., Sullivan, F. X. & Cech, T. R. (1985) *Cell* **43**, 431–437.
28. Been, M. D. & Cech, T. R. (1985) *Nucleic Acids Res.* **13**, 8389–8408.
29. Inoue, T., Sullivan, F. X. & Cech, T. R. (1986) *J. Mol. Biol.* **188**, in press.
30. Waring, R. B., Scazzocchio, C., Brown, T. A. & Davies, R. W. (1983) *J. Mol. Biol.* **167**, 595–605.
31. Michel, F. & Dujon, B. (1983) *EMBO J.* **2**, 33–38.
32. Davies, R. W., Waring, R. B., Ray, J. A., Brown, T. A. & Scazzocchio, C. (1982) *Nature (London)* **300**, 719–724.
33. Waring, R. B. & Davies, R. W. (1984) *Gene* **28**, 277–291.
34. Cech, T. R. (1983) *Cell* **34**, 713–716.
35. Waring, R. B., Ray, J. A., Edwards, S. W., Scazzocchio, C. & Davies, R. W. (1985) *Cell* **40**, 371–380.
36. Burke, J. M., Irvine, K. D., Kaneko, K. J., Kerker, B. J., Oettgen, A. B., Tierney, W. M., Williamson, C. L., Zaug, A. J. & Cech, T. R. (1986) *Cell* **45**, 167–176.
37. Mills, D. R., Kramer, F. R. & Spiegelman, S. (1973) *Science* **180**, 916–927.
38. Kramer, F. R. & Mills, D. R. (1981) *Nucleic Acids Res.* **9**, 5109–5124.
39. Crick, F. H. C. (1968) *J. Mol. Biol.* **38**, 367–379.
40. Orgel, L. E. (1968) *J. Mol. Biol.* **38**, 381–393.
41. Woese, C. R. (1972) in *Exobiology*, ed. Ponnamperuma, C. (North-Holland, Amsterdam), pp. 301–341.
42. Noller, H. F. (1984) *Annu. Rev. Biochem.* **53**, 119–162.
43. Price, J. V., Kieft, G. L., Kent, J. R., Sievers, E. L. & Cech, T. R. (1985) *Nucleic Acids Res.* **13**, 1871–1889.
44. Inoue, T., Joyce, G. F., Grzeskowiak, K., Orgel, L. E., Brown, J. M. & Reese, C. B. (1984) *J. Mol. Biol.* **178**, 669–676.

Origin of life

The RNA world

from Walter Gilbert

UNTIL recently, when one thought of the varied molecular processes at the origin of life, one imagined that the first self-replicating systems consisted of both RNA and protein. RNA served to hold information, whereas protein molecules provided all the enzymic activities needed to make copies of RNA and to reproduce themselves. The cycle that developed a self-replicating system out of the primitive soup of amino acids and nucleotides had two radically different components[1].

Now it seems possible that the informational and catalytic properties of these two components may be combined in a single molecular species. Last week in these columns Frank Westheimer[2] described the discovery of enzymic activities in the RNA molecules of *Escherichia coli*, in which ribonuclease-P cuts phosphodiester bonds during the maturation of the transfer RNA molecule[3,4], and of *Tetrahymena*, whose ribosomal RNA contains a self-splicing exon[5-7]. If there are two enzymic activities associated with RNA, there may be more. And if there are activities among these RNA enzymes, or ribozymes, that can catalyse the synthesis of a new RNA molecule from precursors and an RNA template, then there is no need for protein enzymes at the beginning of evolution. One can contemplate an RNA world, containing only RNA molecules that serve to catalyse the synthesis of themselves[8].

The self-splicing intron is an RNA element that can splice itself out of an RNA molecule. This reaction should be reversible, and the intron could splice itself back into an appropriate nucleotide sequence. Thus, in the RNA world, such introns could both remove and insert themselves into the background of replicating RNA molecules. The significance of this is not the simple insertion and removal of introns, but the fact that two introns, separated from each other by another RNA element, an exon, can combine with each other so as to remove as a unit both themselves and the intervening exon from one RNA molecule and to insert into another. Thus, self-inserting introns can create transposons to move exons around. This property provides RNA with a major evolutionary facility that it otherwise lacks — recombination, the ability to produce new combinations of genes. Of course the self-replicating molecules would in any case have evolved slowly by miscopying, that is, by mutation. But transposons provide the equivalent of sex — the infectious transmission of genetic elements from one organism to another. Recombination and sex are powerful devices to permit a more useful exon to pass from one replicating structure to an unrelated one.

This picture of the RNA world is one of replicating molecules that reassort exons by transposable elements created by introns. This process builds and remakes RNA molecules by chunks and also permits the useful distinction between information and function. Information storage needs to be one-dimensional, for ease of copying, but molecules with enzymic functions tend to be tight three-dimensional structures, whose forms are unrelated to the demands of any copying mechanism. (This dichotomy is most obvious today between the linear order along DNA and the structure of proteins.) In the RNA world, the structure that would be replicated has the full complement of introns. Some of the daughters, by splicing out all their introns, would convert to functional molecules, the ribozymes. A remnant of this process may be the structure of transfer RNA, where a compact secondary structure is broken up by the insertion of an intron.

The first stage of evolution proceeds, then, by RNA molecules performing the catalytic activities necessary to assemble themselves from a nucleotide soup. The RNA molecules evolve in self-replicating patterns, using recombination and mutation to explore new functions and to adapt to new niches. By using RNA cofactors, such as nicotinamide adenine dinucleotide and flavin mononucleotide they then develop an entire range of enzymic activities[9]. At the next stage, RNA molecules began to synthesize proteins, first by developing RNA adapter molecules that can bind activated amino acids and then by arranging them according to an RNA template using other RNA molecules such as the RNA core of the ribosome. This process would make the first proteins, which would simply be better enzymes than their RNA counterparts. I suggest that protein molecules do not carry out enzymic reactions of a different nature from RNA molecules but are able to perform the same reactions more effectively and rapidly, and hence will eventually dominate. These protein enzymes are encoded by RNA exons, thus they, in turn, are built up of mini-elements of structure.

Finally, DNA appeared on the scene, the ultimate holder of information copied from the genetic RNA molecules by reverse transcription. After double-stranded DNA evolved there exists a stable linear information store, error-correcting because of its double-stranded structure but still capable of mutation and recombination. RNA is then relegated to the intermediate role that it has today — no longer the centre of the stage, displaced by DNA and the more effective protein enzymes. But a few RNA enzymic activities still exist, the two described recently[3-7], and possibly others in the role of ribosomal RNA or in the splicing of eukaryotic messenger RNA. The relic of this process is the intron/exon structure of genes, left imprinted on DNA from the RNA molecules that earlier encoded proteins, a residue of the basic mechanism of RNA recombination. □

1. Eigen, M., Gardiner, W., Schuster, P. & Winkler-Oswatitsch, R. *Sci. Am.* **244**, 88 (1981).
2. Westheimer, F.H. *Nature* **319**, 534 (1986).
3. Guerrier-Takada, C., Gardiner, K., Marsh, T., Pace, N. & Altman, S. *Cell* **35**, 849 (1983).
4. Guerrier-Takada, C., & Altman, S. *Science* **223**, 285 (1984).
5. Kruger, K. *et al. Cell* **31**, 147 (1982).
6. Cech, T.R. *Int. Rev. Cytol.* **93**, 3 (1985).
7. Zaug, A.J. & Cech, T.R. *Science* **231**, 470 (1986).
8. Sharp, P.A. *Cell* **42**, 397 (1985).
9. White, H.B. III *J. molec. Evol.* **7**, 101 (1976).

Walter Gilbert is at the Biological Laboratories, Harvard University, 16 Divinity Avenue, Cambridge, Massachusetts 02138, USA.

RNA-catalysed synthesis of complementary-strand RNA

Jennifer A. Doudna & Jack W. Szostak

Department of Molecular Biology, Massachusetts General Hospital, Boston, Massachusetts 02114, USA

The *Tetrahymena* ribozyme can splice together multiple oligonucleotides aligned on a template strand to yield a fully complementary product strand. This reaction demonstrates the feasibility of RNA-catalysed RNA replications.

THE recent identification of several catalytically active RNA molecules[1-6] has led to extensive speculation concerning the role of RNA in the origin of life[7-11] A self-replicating RNA or related polynucleotide is thought to be the key intermediate in the evolution of living systems from prebiotic chemicals. The identification or design of an RNA replicase would therefore be helpful in establishing the feasibility of this pathway, and might also allow the construction of simple proto-cells in the laboratory. The protozoan *Tetrahymena* self-splicing intron[1,2,6] is of particular interest in this respect because of the variety of phosphoester transfer reactions that it can catalyse. The activities of this ribozyme, demonstrated in a series of experiments by Cech and coworkers, include that of a ribonuclease, phosphotransferase, acid phosphatase and RNA restriction endonuclease[12-14]

An enzyme capable of catalysing transesterification reactions on RNA substrates is potentially capable of catalysing RNA polymerization. Indeed, Zaug and Cech[12] and Been and Cech[15] have shown that the intron will catalyse limited polymerization of ribonucleotides onto a short primer annealed to a sequence within the intron. But the primer could be extended only to a maximum of ~15 nucleotides, and the nucleotides were added beyond the end of the template[15]. These experiments, although demonstrating the ability of the *Tetrahymena* ribozyme to polymerize RNA, also pointed out the problems to be overcome before replicase activity could be approached: first, the template must be on a separate molecule from the replicase; second, the polymerase activity must be template directed; and third, the nucleotide specificity that ensures accurate splicing must be overcome to allow copying of an arbitrary template sequence. Here we show that these three problems can be overcome, and that a modified version of the *Tetrahymena* ribozyme can catalyse the formation of an RNA molecule complementary to a template strand.

Catalysis on an independent template

The initial step in self-splicing (Fig. 1) is the attack of the 3′ hydroxyl of a free guanosine on a specific phosphate in a stem-loop of the intron referred to as P1. A phosphoester transfer reaction occurs in which the exon-intron junction is cleaved and the G becomes attached to the 5′ end of the intron. P1 is the first of a series of base-paired stems in the intron; the 3′ strand of P1 (the internal guide sequence) has been used as a primer-binding site for primer elongation experiments[12,15]. To overcome the requirement that a replicase act on a template that is a separate molecule, we synthesized the isolated P1 stem-loop (Fig. 1a) by transcription of a synthetic oligonucleotide, and designed a modified version of the ribozyme extending from stem P2 to P9 (Fig. 1b). We found that this enzyme RNA catalyses the site-specific attack of guanosine on the isolated P1 stem (Fig. 2), but that the K_m for free P1 was very high (>0.1 mM). This weak interaction probably reflects the fact that

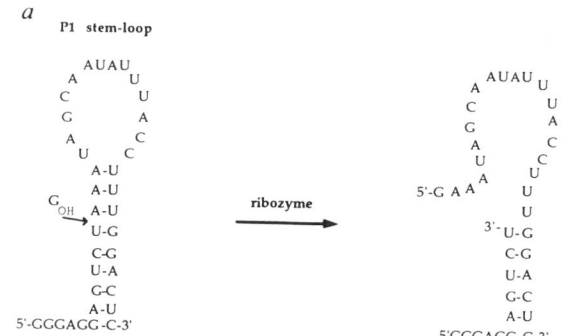

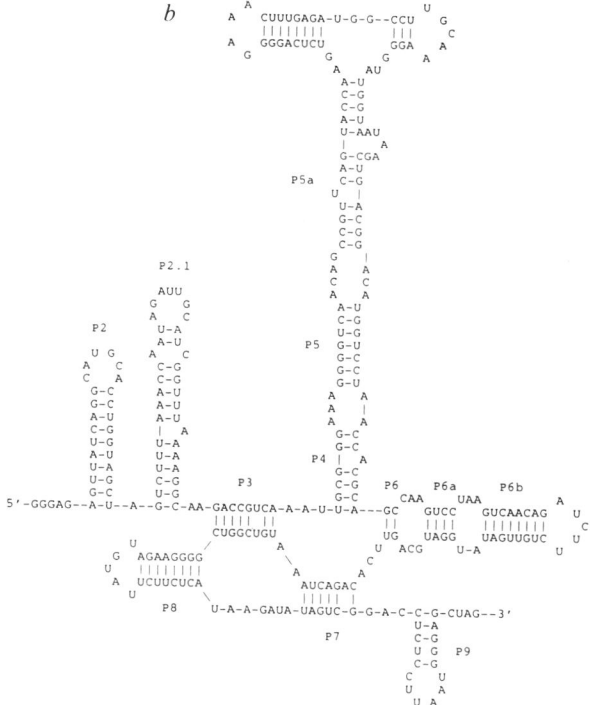

FIG. 1 *a*, Cleavage of an independent P1 substrate. The diagram illustrates the structure of the substrate used in this experiment and the guanosine cleavage reaction catalysed by the ribozyme. P1 is normally the first stem-loop of the intron, containing the 5′ exon-intron junction and the internal guide sequence; in our experiments it is a separate molecule. *b*, Sequence and secondary structure of the modified *Tetrahymena* intron used in this paper. This molecule is missing the P1 stem-loop, the 3′ stem-loops P9.1 and P9.2, and the 3′ intron-exon junction. RNA was synthesized by T7 RNA polymerase runoff transcription of plasmid pJD1100 digested with the restriction enzyme *Nhe*I, and purified either by electrophoresis on a 6% polyacrylamide, 7M urea gel, or by chromatography on a Sephadex G-75 spin column. In either case, the RNA was extracted with phenol, precipitated with ethanol, and stored in distilled water at −80 °C.

there are few sequence or size requirements for recognition of P1 by the core intron[16-18]

As a first step towards the assembly of oligonucleotides on an external template, we synthesized two RNA oligonucleotides that correspond to the products of guanosine attack on P1 (Fig. 3a). The sequences are different from that of wild-type P1, apart from the U·G base pair at the site of the guanosine cleavage reaction. The two oligonucleotides anneal to form a complex that consists of an RNA primer annealed to a partial hairpin RNA, with a guanosine residue extruded from the helix at the gap. Incubation of this complex with the modified ribozyme resulted in the extremely efficient regeneration of intact P1, with the release of free guanosine (Fig. 3b). This is the equivalent of the reverse of the reaction shown in Fig. 1a. The oligonucleotide ligation reaction was essentially complete after one hour, with about 250 turnovers of substrate per enzyme.

Spermidine overcomes sequence specificity

We have previously shown, in a different system, that only the wobble base pairs U·G and C·A in the intact P1 stem allow efficient guanosine attack[18]. To evaluate the base-pair requirements for the reverse reaction (oligonucleotide ligation), we synthesized four primers ending in either A, C, G or U, and four partial hairpins with either A, C, G or U opposite the last base of the primer (Fig. 4a). The 16 primer-hairpin combinations were then tested for ligation by the ribozyme (Fig. 4b). In contrast to the cleavage reaction, only the U·G and C·G base combinations allow efficient ligation. Ligation occurs to a lesser extent with the A·G combination.

The same set of primer-template combinations was tested for ligation by the intron under various reaction conditions. Because we have hypothesized that base-pair geometry at the reaction site is critical[18], it seemed possible that subtle changes in the structure of either P1 or the ribozyme might allow ligation to occur with other base-pair combinations. One of the conditions tested was the addition of 5 mM spermidine to the reaction (Fig. 4c). This led to the efficient ligation of all the substrate complexes with either Watson-Crick or wobble (U·G, C·A or A·G) base pairs at the ligation junction. A set of 10 related polyamines was tested for similar effects, and none was as effective as

spermidine, although putrescine and spermine work to a lesser extent. Spermidine overcomes the inherent sequence specificity of the ribozyme, and should, in principle, allow the production of a faithful (that is, entirely Watson-Crick) copy of a template.

Template-directed oligonucleotide ligation

Previous work had shown that the loop of the P1 stem did not seem to be important in the guanosine-attack reaction. This suggested that we could simply dispense with the P1 loop in the reverse reaction, and try to ligate together two short oligonucleotides aligned on a longer third oligonucleotide which would act as a template. Of the two shorter oligonucleotides, the 5′ one is referred to as the primer, and the 3′ one as the ligator. The ligator oligonucleotide begins with a guanosine residue that is not paired with the template, just as in the P1 regeneration experiments described above. Several different primer, ligator and template combinations were designed: three substrates contained a U·G at the ligation junction, whereas four others contained the Watson-Crick base pairs at the junction. In each case, the ribozyme catalysed the ligation of the primer and the ligator in a template-dependent manner (Fig. 5). The number of turnovers per enzyme molecule ranged from 100 in a 15-min reaction for the U·G substrates, to 30-100 in 60 min for the C·G, G·C and A·U substrates, to a low of 5 in 60 min for a U·A substrate. The ligated products were characterized by T1 digestion; in all cases the expected fragments were observed. These experiments show that the ribozyme can catalyse oligonucleotide ligation which is independent of sequence.

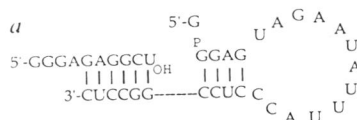

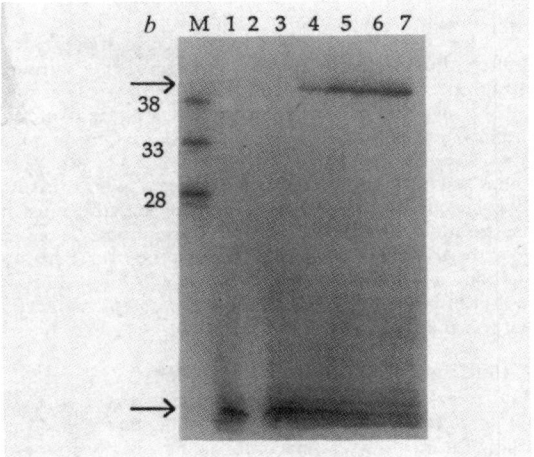

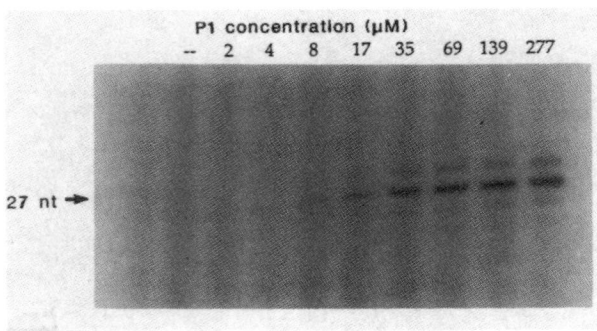

FIG. 2 Cleavage of a P1-like substrate by the modified *Tetrahymena* ribozyme. P1 RNA was prepared by T7 transcription of a synthetic oligodeoxynucleotide[19], followed by purification on a 20% acrylamide, 7 M urea gel. Enzyme reactions contained 10 mM NH₄Cl, 20 mM MgCl₂, 30 mM Tris-HC1 pH 7.4, 1 mM aurin trichloroacetic acid, 0.2 μM enzyme (prepared as in Fig. 1b), 200 μM [α-³²P]GTP, and the indicated concentration of P1 RNA, in a 5-μl reaction volume. Reactions were incubated at 58 °C for 20 min, then stopped by the addition of an equal volume of 90% formamide, 10 mM Tris-HC1 pH 8.0, 1 mM EDTA, 0.2% bromphenol blue and xylene cyanol. Aliquots were electrophoresed on a 15% acrylamide, 7 M urea gel; the gel was then dried and autoradiographed. Lanes (left to right): no enzyme, no substrate, increasing substrate concentration. The major band is the expected product of the reaction (see text and Fig. 1a). The minor bands are due to 3′ heterogeneity of the substrate, as determined by 5′-end-labelling with polynucleotide kinase.

FIG. 3 a, Substrate ribo-oligonucleotides for P1 regeneration. RNAs corresponding to the products of a P1 cleavage reaction were prepared and purified as described above, so that the enzymatic reversal of the cleavage reaction could be followed. The 5′ primer oligonucleotide was internally labelled by including 0.5 μCi per μl of [α-³²P]GTP in the transcription reaction b, Time course of P1 regeneration. Reactions contained 10 mM NH₄Cl, 20 mM MgCl₂, 30 mM Tris HC1 pH 7.4, 1 mM aurin trichloroacetic acid, 0.2 μM enzyme and 50 μM of each RNA substrate oligonucleotide in a total volume of 5 μl. Reactions were incubated at 58 °C, and stopped by the addition of formamide/dye solution as above. Aliquots were electrophoresed on a 20% acrylamide, 7 M urea gel; the gel was dried and autoradiographed. Lane M, DNA size markers; lane 1, no template oligonucleotide, 60 min reaction; lane 2, no primer oligonucleotide, 60 min reaction; lanes 3-7, complete reactions incubated for 0, 15, 30, 45 and 60 min, respectively. Bottom arrow, labelled primer; top arrow, ligated product.

Ligation of aligned multiple oligonucleotides

To see if the ligation reaction explored above could be used to assemble a longer complementary-strand RNA, three sets of templates and complementary oligonucleotides were made (Fig. 6a). In the first set, four identical oligonucleotides align on one long template, with U·G base pairs at each ligation junction. With equimolar template and ribozyme in the reaction, products corresponding to the ligation of two, three or all four oligonucleotides are visible within 5 min of incubation (Fig. 6b). The ratio of full-length to shorter products increases over time, with

concomitant release of free GMP. With smaller amounts of enzyme, the reaction proceeds more slowly.

We considered the possibility that the longer products in the above reaction were generated by a ligation of two oligonucleotides, dissociation from the template and subsequent reannealing to the template with the original ligator in the first position on the template. Longer oligonucleotides could then form by repeated cycles of a reaction very similar to two-oligonucleotide ligation. This model was eliminated by using a template on which four different oligonucleotides could anneal (Fig. 6a), so that each ligation reaction would have to take place at a different position on the template. The course of the reaction was similar to the first multiple ligation (Fig. 6c). The sequence of the full-length product was confirmed by dideoxy sequencing with reverse transcriptase.

In the two experiments described above, the product strand is not fully complementary to the template strand, because of the U·G base pairs at the ligation junctions. We therefore synthesized a template on which five small RNAs could anneal, such that each of the four ligation junctions consisted of a

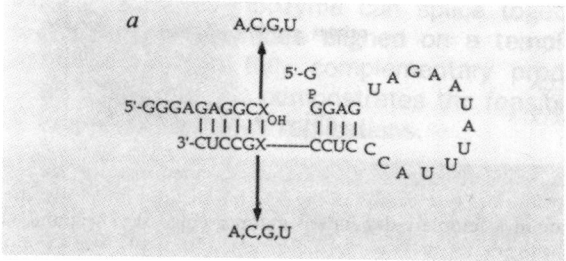

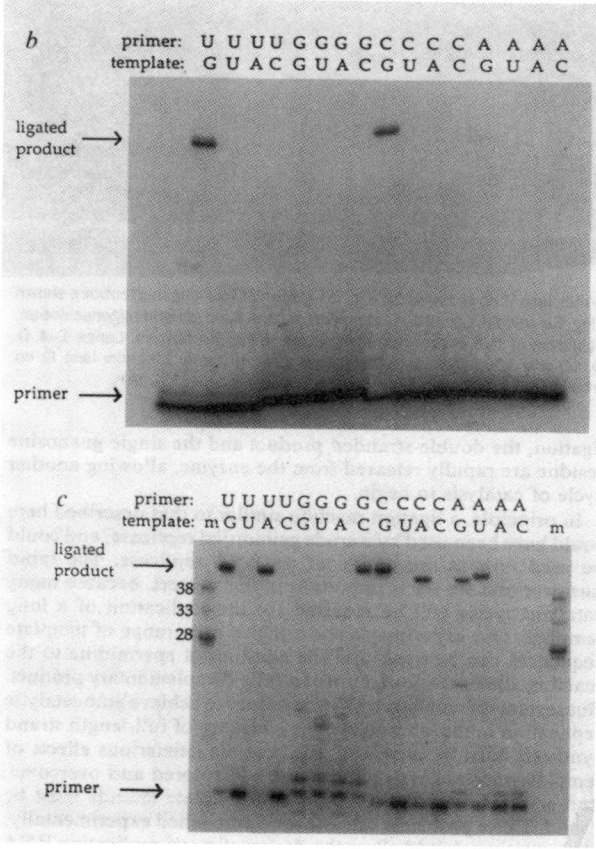

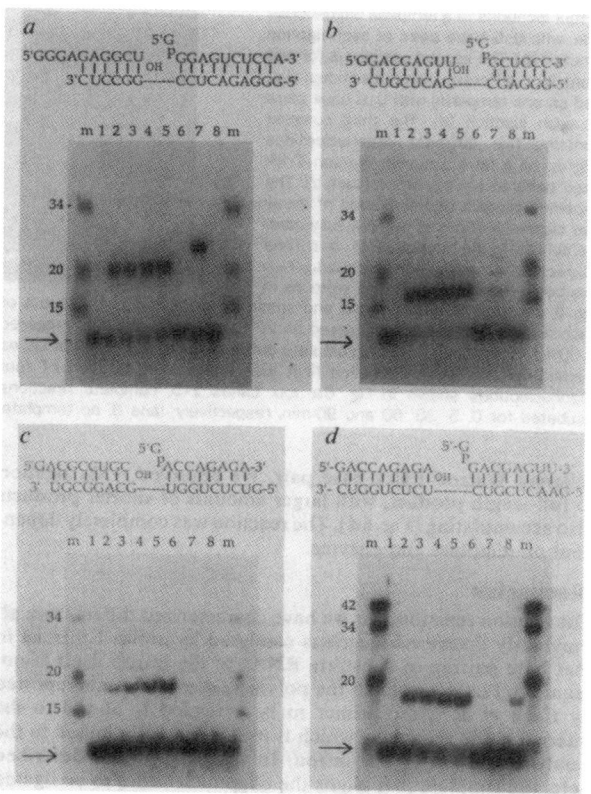

FIG. 4 Effect of the base pair at the ligation junction on P1 regeneration. a, We synthesized four primers and four templates, differing at the indicated position. These sets of oligonucleotides can be combined to yield P1 regeneration substrates with any base–base combination at the ligation junction. b, P1 regeneration reactions. Reactions were as described above, except that 20 μM oligonucleotides were used, and incubations were at 58 °C for 60 min. Each reaction contained the indicated base–base combination at the ligation junction. The reaction works well only with the U·G and C·G combinations. c, Effect of spermidine on P1 regeneration. Reactions were as above, except for the addition of 5 mM spermidine, which allows U·G, C·A, A·G and all Watson–Crick combinations to ligate efficiently. Ligated products with different sequences have slightly different mobilities.

FIG. 5 Template-directed oligonucleotide ligations. Four examples of the ligation of two oligonucleotides aligned by a template oligonucleotide are shown. RNAs were prepared and purified as described above, except that the primer RNAs were labelled with [γ-³²P]ATP and polynucleotide kinase. Reaction conditions were as in Fig. 4, with 0.2 μM enzyme, 20 μM oligonucleotides and 5 mM spermidine. Reactions were stopped by the addition of 3 vol 9.3 M urea, 33 mM EDTA. To denature product template complexes completely, reaction mixes were separated by electrophoresis on a 20% polyacrylamide, 90% formamide gel. Lanes M, RNA oligonucleoide size markers; lanes 1–5, time course, 0, 15, 30, 45 and 60 min, respectively; lanes 6, no template oligonucleotide, 60-min reaction; lanes 7, no ligator oligonucleotide, 60 min reaction; lanes 8, no enzyme (except d; lane 7, no enzyme; lane 8, no ligator), 60 min reaction. In each case, the arrow represents the position of the primer. a, b, Substrate complexes with a U·G base pair at the ligation junction; c, d, complexes with Watson–Crick base pairs at the junction.

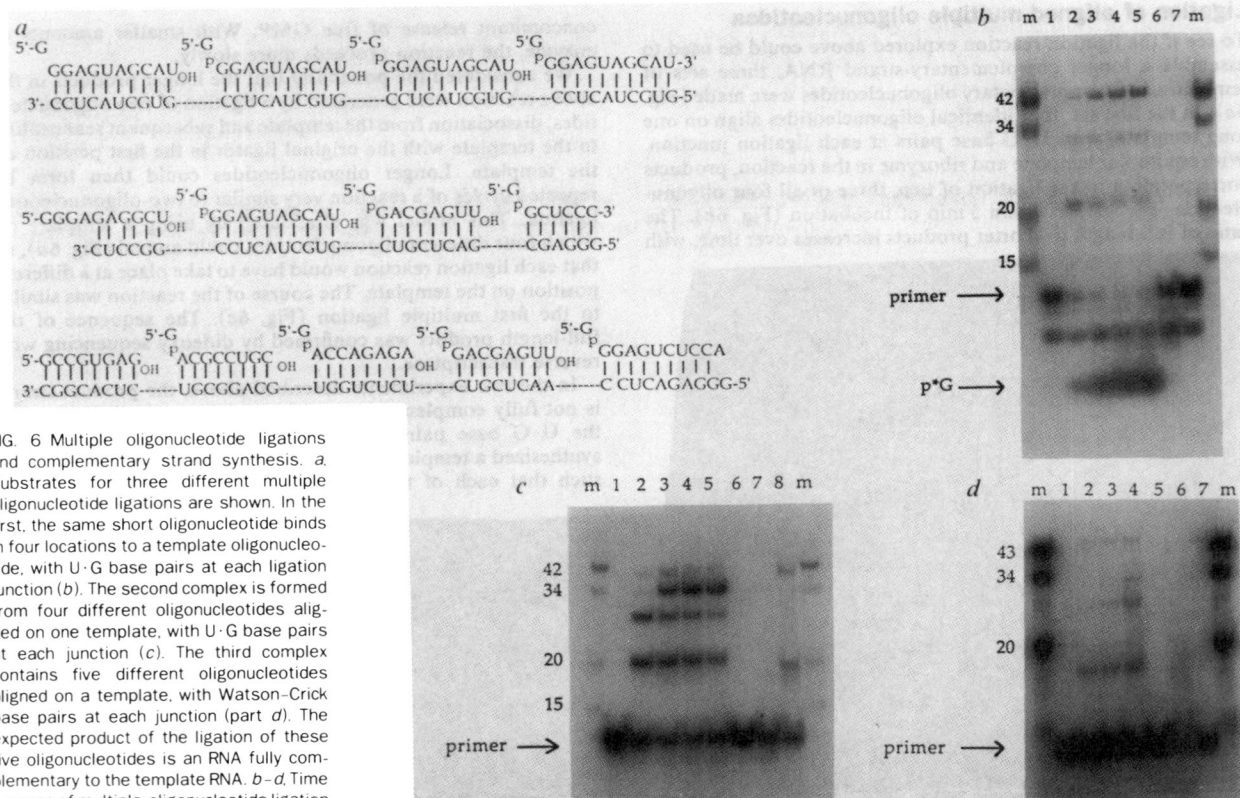

FIG. 6 Multiple oligonucleotide ligations and complementary strand synthesis. *a.* Substrates for three different multiple oligonucleotide ligations are shown. In the first, the same short oligonucleotide binds in four locations to a template oligonucleotide, with U·G base pairs at each ligation junction (*b*). The second complex is formed from four different oligonucleotides aligned on one template, with U·G base pairs at each junction (*c*). The third complex contains five different oligonucleotides aligned on a template, with Watson-Crick base pairs at each junction (part *d*). The expected product of the ligation of these five oligonucleotides is an RNA fully complementary to the template RNA. *b–d.* Time courses of multiple-oligonucleotide ligation reactions. Reaction conditions were as in Fig. 5, except that both enzyme and substrate were at 5 µM. Aliquots of each reaction were electrophoresed on 20% polyacrylamide gels prepared in 90% formamide to promote complete denaturation of the samples. Lanes marked 'm' contain end-labelled RNA size markers. *b*, Ligation of four oligonucleotides shown in Fig. 6*a*, top. Lanes 1–5, complete reactions incubated for 0, 5, 30, 60 and 90 min, respectively; lane 6, no template.

90 min; lane 7 no enzyme, 90 min. *c*, Ligation of four oligonucleotides shown in Fig. 6*a*, middle. Lanes 1–7, as in part *b*; lane 8, no second oligonucleotide. *d*, Ligation of five oligonucleotides shown in Fig. 6*a*, bottom. Lanes 1–4, 0, 30, 60 and 120 min, respectively; lane 5, no enzyme, 120 min; lane 6, no template, 120 min; lane 7, no second oligonucleotide, 120 min.

different Watson–Crick base pair. We observed ~5% ligation to full-length product, with larger amounts of shorter products also accumulating (Fig. 6*d*). The reaction was completely dependent on template and enzyme.

Discussion

The ligation reaction that we have characterized differs from all previously described reactions catalysed by group I introns in that base pairing of substrate RNAs to the intron itself is not required. For example, in the polymerase experiments reported by Cech *et al.*[15], the primer to be extended is bound to the internal guide sequence, which is itself covalently joined to the catalytic domain of the intron. In the experiments described here, the template that aligns the oligonucleotides to be ligated is a separate RNA molecule that does not interact with the enzyme by base pairing. One consequence of this difference is much weaker interaction of our substrates with the enzyme. We think that this weakened interaction is the simplest explanation for the rapid turnover we see in the ligation reaction; after

ligation, the double-stranded product and the single guanosine residue are rapidly released from the enzyme, allowing another cycle of catalysis to begin.

In principle, a ligation reaction similar to that described here could have been used by a crude primordial replicase, and could be used now in the design of an RNA replicase. The rapid turnover that we see is important in this respect, because many catalytic cycles will be required for the replication of a long template. Our experiments show that a wide range of template sequences can be used, and the addition of spermidine to the reaction allows the formation of a fully complementary product. But several refinements will be necessary to achieve autocatalytic replication in the laboratory. The efficiency of full-length strand synthesis must be increased, the possible deleterious effects of template secondary structure must be explored and overcome, and a way to separate product and template strands must be found. These problems can now be approached experimentally; their solution should allow the design of a self-replicating RNA molecule.   □

Received 14 April; accepted 8 May 1989.

1. Cech, T. R., Zaug, A. J. & Grobowski, P. J. *Cell* **27**, 487–496 (1981).
2. Kruger, K. *et al. Cell* **31**, 147–157 (1982).
3. Guerrier-Takada, C., Gardiner, K., Marsh, T., Pace, N. & Altman, S. *Cell* **35**, 849–857 (1983).
4. Van der Veen, R. *et al. Cell* **44**, 225–234 (1986).
5. Garriga, G. & Lambowitz, A. M. *Cell* **39**, 631–641 (1984).
6. Cech, T. R. *Science* **236**, 1532–1539 (1987).
7. Sharp, P. A. *Cell* **42**, 397–400 (1985).
8. Darnell, J. E. & Doolittle, W. F. *Proc. natn. Acad. Sci. U.S.A.* **83**, 1271–1275 (1986).
9. Gilbert, W. *Nature* **319**, 618 (1986).
10. Cech, T. R. *Proc. natn. Acad. Sci. U.S.A.* **83**, 4360–4363 (1986).
11. Joyce, G. F. *Nature* **338**, 217–224 (1989).
12. Zaug, A. J. & Cech, T. R. *Science* **231**, 470–475 (1986).
13. Zaug, A. J., Been, M. D. & Cech, T. R. *Nature* **324**, 429–433 (1986).
14. Zaug, A. J. & Cech, T. R. *Biochemistry* **25**, 4478–4482 (1986).
15. Been, M. D. & Cech, T. R. *Science* **239**, 1412–1416 (1988).
16. Waring, R. B., Towner, P., Minter, S. J. & Davies, R. W. *Nature* **321**, 133–139 (1986).
17. Been, M. D. *et al. Cold Springer Harb. Symp. quant. Biol.* **LII**, 147–157 (1987).
18. Doudna, J. A., Cormack, B. P. & Szostak, J. W. *Proc. natn. Acad. Sci. U.S.A.* (in the press).
19. Milligan, J. F., Groebe, D. R., Witherell, G. W. & Uhlenbeck, O. C. *Nucleic Acids Res.* **15**, 8783 (1987).

■ RESEARCH ARTICLES

Unusual Resistance of Peptidyl Transferase to Protein Extraction Procedures

Harry F. Noller, Vernita Hoffarth, Ludwika Zimniak

Peptidyl transferase, the ribosomal activity responsible for catalysis of peptide bond formation, is resistant to vigorous procedures that are conventionally employed to remove proteins from protein–nucleic acid complexes. When the "fragment reaction" was used as a model assay for peptide bond formation, *Escherichia coli* ribosomes or 50S subunits retained 20 to 40 percent activity after extensive treatment with proteinase K and SDS, but lost activity after extraction with phenol or exposure to EDTA. Ribosomes from the thermophilic eubacterium *Thermus aquaticus* remained more than 80 percent active after treatment with proteinase K and SDS, which was followed by vigorous extraction with phenol. This activity is attributable to peptidyl transferase, as judged by specific inhibition by the peptidyl transferase–specific antibiotics chloramphenicol and carbomycin. In contrast, activity is abolished by treatment with ribonuclease T1. These findings support the possibility that 23S ribosomal RNA participates in the peptidyl transferase function.

There is much evidence to support the view that ribosomal RNA (rRNA) participates directly in protein synthesis (*1, 2*), and it has even been argued that the fundamental mechanism underlying translation may be RNA-based (*3, 4*). Indeed, demonstration of the ability of RNA to perform enzymatic catalysis in other biological contexts (*5, 6*) has drawn increased attention to the functional potential of rRNA. However, apart from the well-established role of the 3′ terminus of 16S rRNA in mRNA selection, direct proof of this has been elusive. For example, efforts to carry out steps of protein synthesis with protein-free preparations of rRNA have not been successful [but see (*7*)], possibly because billions of years of co-evolution of ribosomal proteins and rRNA have led to a require-

ment for ribosomal proteins to achieve proper folding and function of the rRNA (*8–10*).

Localization of peptidyl transferase to the large ribosomal subunit. In our efforts to study the biological activity of rRNA, we have chosen as a model system the peptidyl transferase reaction, which is the source of the catalysis of peptide bond formation, and is also the single catalytic activity that has unambiguously been shown to be an integral part of the ribosome structure (*11*). In spite of many attempts by several laboratories, peptidyl transferase activity has never been detected in RNA-free preparations of ribosomal proteins. An important attraction of peptidyl transferase is that it can be monitored with a simplified assay known as the "fragment reaction" (Fig. 1), which measures the transfer of *N*-formyl-methionine from a short fragment of tRNA to the amino group of puromycin to form a model peptide bond (*12*). The fragment reaction requires only the large ribosomal subunit,

appropriate ionic conditions, and 33 percent methanol or ethanol, in addition to the f-Met-oligonucleotide and puromycin substrates. Thus, there is no requirement for the small ribosomal subunit, mRNA, protein factors, guanosine triphosphate (GTP), or even complete tRNA molecules. The authenticity of the model reaction is supported by the stereochemical specificity of the substrates and highly specific inhibition of the reaction by antibiotics that are known peptidyl transferase inhibitors (*13*).

Earlier studies showed that this system can be simplified even further by stepwise removal of ribosomal proteins from the 50S subunit with high concentrations of salt (*14–16*). In one study, removal of approximately half of the proteins from the 50S subunit resulted in loss of peptidyl transferase functions; full activity was restored by reconstitution of the resulting core particles with purified protein L16 (*16*). These same preparations of purified L16 showed no detectable peptidyl transferase activity, however. Another study provided evidence for an L16-dependent, conformational change in similar protein-deficient 50S core particles (*17*). The temperature dependence of the kinetics of this process corresponds to an activation energy of about 30 kcal/mol, suggesting the occurrence of a fairly substantial structural rearrangement. These experiments indicate that protein L16 plays an important role in proper assembly of the core particle. Reconstitution experiments, in which individual components were omitted, showed that proteins L2, L3, L4, L15, L16, and L18, as well as 23S rRNA were essential for reconstitution of peptidyl transferase activity (*18*); of this group, L18 could also be excluded on the basis of other studies (*19*). This list most likely represents an overestimate of the number of proteins actually needed for ca-

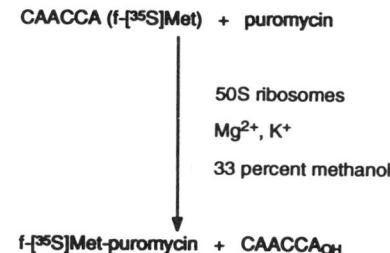

Fig. 1. The "fragment reaction." Peptidyl transferase activity is measured by formation of f-[35S]Met-puromycin from reaction of the CAACCA(f-[35S]Met) oligonucleotide fragment, derived from the 3′ end of f-[35S]Met-tRNA by RNase T1, with puromycin, in the presence of 33 percent methanol (*12*). The oligonucleotide fragment and puromycin serve as peptidyl-tRNA and aminoacyl-tRNA analogues, respectively.

The authors are at the Sinsheimer Laboratories, University of California at Santa Cruz, Santa Cruz, CA 95064. The present address of L. Zimniak is Department of Nephrology, University of Arkansas for Medical Sciences, Little Rock, AR 72205.

Reprinted with permission from *Science*,
Volume 256, pp. 1416-1419. © 1992 by the American Association for the Advancement of Science.

talysis, since some of them may be important only for assembly, and others may be needed for binding the full-length tRNA used in these experiments.

In vitro assembly experiments have shown that correct folding of rRNA depends on sequential, cooperative interactions with ribosomal proteins (8, 9); the above-mentioned L16-dependent event is one such example. Many features of the higher order structure of rRNA are lost during conventional RNA extraction procedures (20). Thus, the requirement for certain ribosomal proteins could be explained by their role in stabilizing the active conformation of 23S rRNA. Our strategy, therefore, was to attempt to remove all of the 50S proteins without disturbing crucial elements of the RNA conformation. This includes avoidance of denaturing conditions, chelating agents, or extremes of pH.

Resistance of peptidyl transferase to protein extraction procedures. Experiments with *Escherichia coli* 70S ribosomes or 50S subunits showed that peptidyl transferase activity survives treatment with 0.5 percent SDS and proteinase K at 1 mg/ml for 1 hour at 37°C, as shown by the formation of N-formyl-[^{35}S]methionyl-puromycin (Fig. 2). However, phenol extraction (Fig. 2) or exposure to 5 mM EDTA abolished the activity.

One possible explanation for the loss of activity on phenol extraction is that, in the mesophilic bacterium *E. coli*, some crucial higher order RNA structural feature might be unstable in the absence of r-proteins. We therefore subjected ribosomes from the thermophilic eubacterium *Thermus aquaticus* to a similar extraction because its rRNA structure might be inherently more robust than that of *E. coli*. Indeed, we found that the peptidyl transferase activity of *T. aquaticus* 50S subunits was resistant not only to treatment with proteinase K and SDS at 37°C (Fig. 2), but also at 40°, 50°, and 60°C (Fig. 3). Surprisingly, loss of activity was seen only when subunits were incubated at 60°C in the absence of SDS and proteinase K; this result may be due to low levels of endogenous ribonuclease in the subunit preparations. As in the case of *E. coli* ribosomes, activity was abolished by treatment with EDTA. In contrast to the behavior of *E. coli* ribosomes, *T. aquaticus* peptidyl transferase was resistant to vigorous extraction with phenol, even in combination with SDS–proteinase K digestion (Fig. 2). Peptidyl transferase activity also withstood extraction with two other detergents known for their ability to disrupt protein-RNA interactions, triisopropylnaphthalene sulfonate (TNS) and *p*-aminosalicylate (PAS) (21). When *T. aquaticus* 50S subunits were extracted with TNS, either alone or in combination with PAS and phenol, they retained significant activity (Table 1). The lower activity observed after PAS treatment may be caused by sequestration of magnesium ions.

An important question is whether the activity that survives the extraction procedures is actually due to peptidyl transferase.

Table 1. Peptidyl transferase activity of *T. aquaticus* 50S subunits following treatments with various detergents (21) and in combination with phenol. *Thermus aquaticus* 50S subunits (300 μg) were treated, as indicated, with 1 percent triisopropyl sulfonic acid, sodium salt (TNS), 5 percent *p*-aminosalicylic acid (PAS), or 0.5 percent SDS plus proteinase K at 1 mg/ml for 2 hours at 37°C in 50 μl of buffer A. After addition of 200 μl of buffer A and 250 μl of neutralized, water-saturated phenol, the mixture was vortexed for 40 minutes at 4°C. A second phenol extraction was carried out for 10 minutes, the aqueous phase was extracted three times with ether, and the RNA-containing material was recovered by precipitation with three volumes of ethanol. The precipitate was redissolved in 50 μl of buffer A and heated for 10 minutes at 42°C, and 1.5 A_{260} units assayed for peptidyl transferase activity (32). The results shown are the net ethyl acetate extractable ^{35}S (counts per minute) after subtracting 182 cpm background from a control reaction lacking ribosomes; the background is the result of an unidentified side product that is resolved from the f-Met-puromycin product as a fast-moving spot on paper electrophoresis (Figs. 2 and 4).

Treatment	Peptidyl transferase activity, ^{35}S	
	cpm	percent
Phenol only	1515	96
1 percent TNS	1469	93
1 percent TNS, 5 percent PAS	896	57
1 percent TNS, phenol	1212	77
1 percent TNS, 5 percent PAS, phenol	359	23
0.5 percent SDS, proteinase K, phenol	1375	87

Fig. 2. Peptidyl transferase activity of *E. coli* and *T. aquaticus* ribosomes after protein extraction. Ribosomes and subunits, prepared as described (20, 31), were suspended in buffer A (5 mM MgCl$_2$, 150 mM NH$_4$Cl, 25 mM tris-HCl, pH 7.5), with 375 μg of *E. coli* 70S ribosomes, 250 μg of *E. coli* 50S subunits, or 150 μg of *T. aquaticus* 50S subunits in a volume of 500 μl for the following treatments. SDS (with or without proteinase K digestion), 0.5 percent SDS, 1 hour at 37°C; PK, proteinase K at 1 mg/ml;

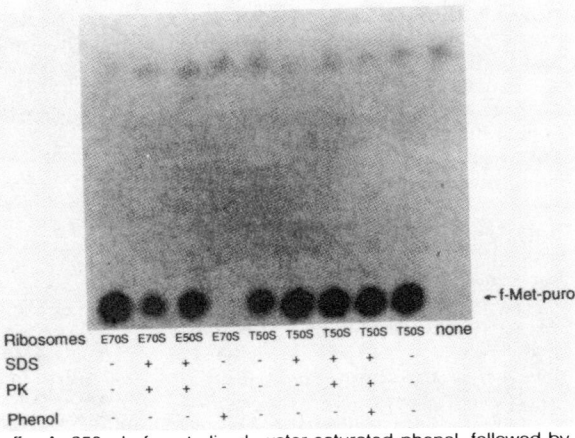

phenol, addition of 200 μl of buffer A, 250 μl of neutralized, water-saturated phenol, followed by vortexing for 45 minutes at 4°C. The phenol phase was removed, the aqueous phase was extracted four times with ether, precipitated with three volumes of ethanol, and resuspended in 50 μl of buffer A. Samples were heated for 10 minutes at 42°C and assayed for peptidyl transferase activity (32). An autoradiograph of the products after high-voltage paper electrophoresis is shown. The position of the f-[^{35}S]Met-puromycin product is shown. The cathode is at the top. E70S, E50S, *E. coli* 70S ribosomes, or 50S subunits, respectively; T50S, *T. aquaticus* 50S subunits.

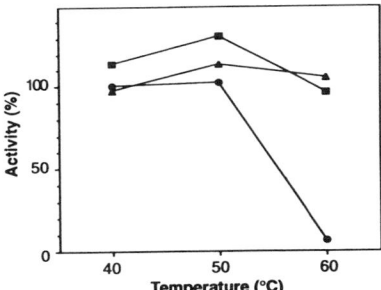

Fig. 3. Peptidyl transferase activity after treatment of *T. aquaticus* 50S subunits with SDS and proteinase K at increasing temperatures. *T. aquaticus* subunits (120 μg) were treated with proteinase K at 1 mg/ml with (■) or without (▲) 0.5 percent SDS, for 1 hour at the indicated temperature. Extracted material was recovered by precipitation with ethanol, redissolved in buffer A, and assayed (32). Control subunits (●) were treated identically except that proteinase K and SDS were omitted. Full (100 percent) activity is defined as that of control subunits incubated for 1 hour at 40°C and corresponds to 1295 cpm.

Authentic eubacterial peptidyl transferase is known to be inhibited specifically by the antibiotics chloramphenicol and carbomycin (13). Formation of f-Met-puromycin by T. aquaticus 50S subunits is indeed inhibited by these two antibiotics (Fig. 4), with carbomycin showing significantly stronger inhibition than chloramphenicol. No inhibition was observed with erythromycin, a 50S subunit-specific antibiotic that is known not to affect peptidyl transferase, or with anisomycin, which inhibits peptidyl transferase in eukaryotic and archaebacterial ribosomes, but not in those of eubacteria. Similar inhibition by these drugs was observed in Thermus 50S subunits treated with SDS, proteinase K, and phenol (Fig. 4), evidence that the extraction-resistant activity is indeed peptidyl transferase.

In contrast to its resistance to extensive digestion with proteinase K, peptidyl transferase activity is highly sensitive to treatment with ribonuclease (RNase). Digestion was effected by RNase T1, which cleaves specifically at guanine residues, so that cleavage of the oligonucleotide substrate (which lacks guanines) could be excluded. Brief digestion of T. aquaticus 50S subunits with RNase T1 caused significant loss of activity, and digestion of the extracted material abolished peptidyl transferase activity (Fig. 4). This result is in agreement with earlier studies on E. coli ribosomes, in which peptidyl transferase was inactivated

by treatment with RNase T1 (22) or with the guanine-specific reagent kethoxal (23).

Protein composition of extracted particles. Treatment of T. aquaticus 50S subunits with SDS and proteinase K, followed by two successive phenol extractions, typically yielded material that was more than 80 percent active, compared to untreated subunits. When ribosomes were labeled with [35S]methionine, the amount of protein remaining in the extracted material was typically about 5 percent of that of untreated subunits. The extraction-resistant polypeptides were removed by acetic acid treatment (24) for analysis by SDS gel electrophoresis (peptidyl transferase activity is irreversibly lost when exposed to acetic acid). A typical result is shown in Fig. 5, which shows the polypeptide material that survives one, two, or three successive phenol extractions after SDS and proteinase K treatment of T. aquaticus ribosomes. Coomassie blue staining (Fig. 5A) or 35S autoradiography (Fig. 5B) show that most of the remaining polypeptide material runs with the SDS front (<10 kD). In addition, three or four slower migrating polypeptides are seen, and their intensity diminishes with successive extraction steps. The decreasing intensity of these bands was not correlated with loss of activity in this experiment; indeed, full peptidyl transferase activity persisted after three successive extractions, when 95 percent of the [35S]me-

thionine-labeled material had been removed (Table 2). It has so far proved difficult to remove the remaining extraction-resistant polypeptide material reproducibly, without resorting to treatments that are disruptive to RNA structure. With certain subunit preparations, more than 99 percent of the 35S-labeled protein was removed under conditions where high activity is normally preserved although this is not typically the case.

The unusual resistance of T. aquaticus peptidyl transferase to protease digestion, ionic detergents, and phenol extraction, in contrast to its sensitivity to ribonuclease, points to the importance of 23S rRNA for this ribosome-catalyzed reaction. However,

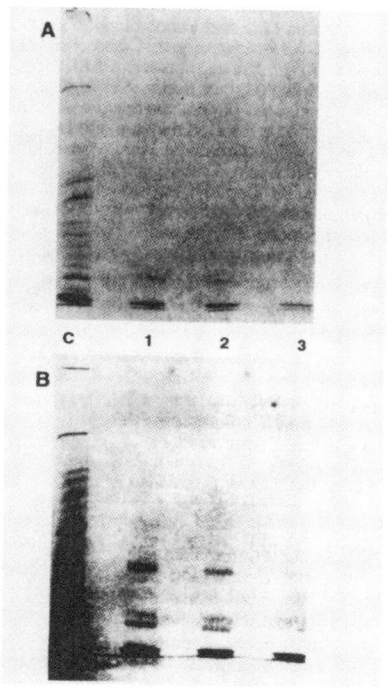

Fig. 5. SDS gel electrophoresis of ribosomal proteins remaining after extraction of ribosomes. 35S-labeled T. aquaticus 70S ribosomes (300 µg; 230,000 cpm) were incubated for 1 hour at 37°C in 50 µl of buffer A containing 0.5 percent SDS and proteinase K at 1 mg/ml. After addition of 200 µl of buffer A, the mixtures were then extracted by vortexing for 40 minutes at 4°C, one to three times with an equal volume of phenol. After four ether extractions, the aqueous phases were precipitated with three volumes of ethanol, redissolved in 100 µl of buffer A, of which 60 µl was extracted with 66 percent acetic acid as described (24), and analyzed by SDS gel electrophoresis (33). The gel was stained with Coomassie blue (**A**) and autoradiographed (**B**). The lanes show proteins associated with (C) untreated ribosomes, and those remaining after one, two, or three successive phenol extractions.

Table 2. Extraction of protein from [35S]methionine-labeled ribosomes. Thermus aquaticus ribosomes labeled in vivo with [35S]methionine (31) were treated as described in Fig. 5. Samples were removed for quantitation of 35S and for the peptidyl transferase assay (32). The peptidyl transferase activity represents ethyl acetate extractable 35S (cpm) after subtraction of 183 cpm background (see legend to Table 1).

Treatment	[35S]Protein remaining		Peptidyl transferase activity	
	cpm	percent	cpm	percent
Control; no treatment	16392	100	1488	100
SDS, proteinase K, 1× phenol	1894	12	1308	88
SDS, proteinase K, 2× phenol	1156	7	1522	102
SDS, proteinase K, 3× phenol	860	5	1638	110

Fig. 4. Sensitivity of T. aquaticus subunits or subunits treated with SDS and proteinase K and then extracted with phenol to peptidyl transferase inhibitors or ribonuclease. Thermus aquaticus 50S subunits, either untreated or extracted as in Fig. 2, were assayed in the presence of 0.2 mM chloramphenicol, 0.2 mM carbomycin, or after treatment for 10 minutes at 37°C with 5 µg of ribonuclease T1 per 225 µg of 50S subunits or the molar equivalent of extracted subunits.

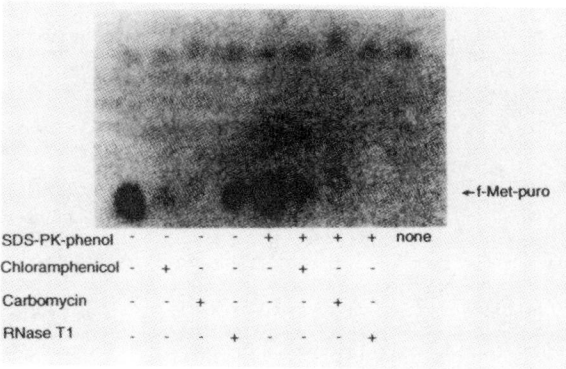

SDS-PK-phenol	-	-	-	-	+	+	+	+	none
Chloramphenicol	-	+	-	-	+	-	-	
Carbomycin	-	-	+	-	-	+	-	
RNase T1	-	-	-	+	-	-	+	

←f-Met-puro

the presence of significant amounts of polypeptide material, even after vigorous extraction, prevents us from concluding that peptide bond formation is catalyzed solely by RNA. The approximately 5 percent of the ribosomal protein remaining could correspond to stoichiometric amounts of as many as two or three intact ribosomal proteins (depending upon molecular masses). If any proteins are, in fact, required for peptidyl transferase function, they must be few in number. Preliminary indications from gel electrophoresis (Fig. 5) suggest that much of the remaining protein is in the form of peptide fragments. Further characterization is required to assess whether any specific protein or its fragments might be present in stoichiometric amounts following extraction. Another important issue is the fraction of active particles in the starting ribosome population; if only a small fraction were active, the remaining protein could belong to an extraction-resistant subpopulation of active particles. While we have no direct measure of the fraction of ribosomes that is active in the peptidyl transferase reaction, there is good evidence that a high proportion of the ribosomes in these preparations is active in other ribosomal functions. The preparations of *E. coli* ribosomes used are virtually fully active in binding tRNA, as judged by complete or nearly complete protection by tRNA of bases in 16S and 23S rRNA from chemical probes (25, 26). Similarly complete protection of bases in 23S rRNA was observed when the CAACCA(f-Met) peptidyl transferase substrate was bound to *E. coli* 50S subunits in the presence of sparsomycin (27). Since activity of *T. aquaticus* ribosomes and subunits in the fragment reaction is similar to that of their *E. coli* counterparts, similar arguments are applicable to these ribosomes as well.

Implication of specific sites in 23S rRNA in the peptidyl transferase function. Our findings are precedented by an abundance of circumstantial evidence implicating 23S rRNA in the peptidyl transferase function (1). Affinity labeling studies, in which tRNAs with various chemically reactive groups attached to their aminoacyl ends have been reacted with ribosomes, show that highly conserved regions of 23S rRNA, most notably the central loop of domain V, are in close proximity to the site of peptide bond formation (28). Chloramphenicol and carbomycin, two antibiotics that are known specifically to inhibit the peptidyl transferase reaction, protect certain bases in the central loop of domain V from chemical probes (29). Point mutations conferring resistance to chloramphenicol and anisomycin, another peptidyl transfer-

ase inhibitor, have also been found in this part of 23S rRNA, and some of them occur at positions that are identical to those that are protected by chloramphenicol and carbomycin (30). Chemical footprinting studies have shown that tRNA interacts, either directly or indirectly, with bases in this same region of 23S rRNA, several of which are again identical with the sites of protection by these antibiotics or of mutations conferring resistance to them (26). Furthermore, it has been shown by stepwise deletion experiments that the part of the tRNA structure responsible for these interactions is the aminoacyl-CCA 3′ end (26). Indeed, N-blocked aminoacylated oligonucleotides, such as the CAACCA-(f-Met) fragment used in our experiments, bound in the presence of sparsomycin and ethanol, protect virtually all of the same bases in 23S rRNA that are protected by normal binding of the intact tRNA (27). Taken together, these results provide convincing evidence for interaction between the CCA end of tRNA and a region of 23S rRNA in and around the central loop of domain V. Finally, the absolute conservation of many bases in this region of 23S rRNA in all known large subunit rRNA sequences, which include eubacterial, archaebacterial, eukaryotic, plastid, and mitochondrial sequences, is in keeping with important functional, as opposed to structural, constraints. Direct proof for the hypothesis that peptide bond formation is catalyzed solely by rRNA will require demonstration of activity with completely protein-free preparations, such as in vitro transcripts of 23S rRNA.

REFERENCES AND NOTES

1. H. F. Noller, *Ann. Rev. Biochem.* **60**, 191 (1991).
2. A. E. Dahlberg, *Cell* **57**, 525 (1989).
3. C. R. Woese, in *Ribosomes: Structure, Function, and Genetics*, G. Chambliss, G. R. Craven, J. Davies, L. Kahan, M. Nomura, Eds. (University Park Press, Baltimore, 1980), pp. 357–376.
4. H. F. Noller and C. R. Woese, *Science* **212**, 403 (1981).
5. T. R. Cech, A. J. Zaug, P. J. Grabowski, *Cell* **27**, 487 (1981).
6. C. Guerrier-Takada, K. Gardiner, T. Marsh, N. Pace, S. Altman, *ibid.* **35**, 848 (1983).
7. However, a complex of 16S, 23S, and 5S rRNA with a limited number of ribosomal proteins has been reported to have low activity in several steps of protein synthesis [D. P. Burma, D. S. Tewari, A. K. Srivastava, *Arch. Biochem. Biophys.* **239**, 427 (1985)].
8. W. A. Held, S. Mizushima, M. Nomura, *J. Biol. Chem.* **248**, 5720 (1973).
9. R. Röhl and K. H. Nierhaus, *Proc. Natl. Acad. Sci. U.S.A.* **79**, 729 (1982).
10. S. Stern, T. Powers, L.-M. Changchien, H. F. Noller, *Science* **244**, 783 (1989).
11. B. E. H. Maden, R. R. Traut, R. E. Monro, *J. Mol. Biol.* **35**, 333 (1968).
12. R. E. Monro and K. A. Marcker, *ibid.* **25**, 347 (1967).
13. R. E. Monro and D. Vazquez, *ibid.* **28**, 161 (1967).
14. T. Staehelin, D. Maglott, R. E. Monro, *Cold Spring Harbor Symp. Quant. Biol.* **34**, 39 (1969).
15. K. H. Nierhaus and V. Montejo, *Proc. Natl. Acad. Sci. U.S.A.* **70**, 1931 (1973).
16. V. G. Moore, R. E. Atchison, G. Thomas, M. Moran, H. F. Noller, *ibid.* **72**, 844 (1975).
17. H. Teraoka and K. H. Nierhaus, *FEBS Lett.* **88**, 223 (1978).
18. H. Hampl, H. Schulze, K. H. Nierhaus, *J. Biol. Chem.* **256**, 2284 (1981).
19. K. H. Nierhaus, *Curr. Topics Microbiol. Immunol.* **97**, 81 (1982).
20. D. Moazed, S. Stern, H. F. Noller, *J. Mol. Biol.* **187**, 399 (1986).
21. J. H. Parish and K. S. Kirby, *Biochim. Biophys. Acta* **129**, 554 (1966).
22. J. Černá, I. Rychlík, J. Jonák, *Eur. J. Biochem.* **34**, 551 (1973).
23. R. E. Atchison and H. F. Noller, unpublished results.
24. M. Siegmann and G. Thomas, *Methods Enzymol.* **146**, 362 (1987).
25. D. Moazed and H. F. Noller, *Cell* **47**, 985 (1986).
26. ———, *ibid.* **57**, 585 (1989).
27. ———, *Proc. Natl. Acad. Sci. U.S.A.* **88**, 3725 (1991).
28. A. Barta, G. Steiner, J. Brosius, H. F. Noller, E. Kuechler, *Proc. Natl. Acad. Sci. U.S.A.* **81**, 3607 (1984); G. Steiner, E. Kuechler, A. Barta, *EMBO J.* **7**, 3949 (1988).
29. D. Moazed and H. F. Noller, *Biochimie* **69**, 879 (1987).
30. B. Vester and R. A. Garrett, *EMBO J.* **7**, 3577 (1988).
31. *Thermus aquaticus* (strain provided by D. Gelfand, Cetus Corp.) was grown to saturation at 70°C overnight in a shaking water bath in the following medium: 1× Castenholz salts (34), 5× Nitsch's trace elements (34), yeast extract at 4 g/liter, bactopeptone at 8 g/liter, and NaCl at 2 g/liter. Ribosomes were prepared from frozen *T. aquaticus* cells as described for *E. coli* ribosomes (20), except that buffers contained 25 mM MgCl₂ for preparation of 70S ribosomes and 2 mM MgCl₂ for 50S subunits. Ribosomes and subunits were stored at −70°C in 25 mM MgCl₂, 100 mM NH₄Cl, 6 mM 2-mercaptoethanol, 0.5 mM EDTA, 20 mM tris-HCl, pH 7.5. For preparation of [³⁵S]methionine-labeled ribosomes, the same procedure was followed, except that cells were grown in media containing 1× Castenholz salts, 10× Nitsch's trace elements, bactopeptone at 0.2 g/liter, and [³⁵S]methionine (10 μCi/ml; 1000 mCi/mmol specific activity).
32. Peptidyl transferase activity was assayed by the fragment reaction (12), typically in a total volume of 50 μl, containing 25 μl of ribosomes in buffer A, CAACCA(f-[³⁵S]Met) at 20,000 cpm, prepared as described (12), 2 mM neutralized puromycin, 10 μl of buffer B (2 M potassium acetate, 100 mM MgCl₂, 250 mM tris-HCl, pH 7.5). The reaction was initiated by addition of 25 μl of methanol, held at 4°C for 10 minutes, and terminated by addition of 50 μl of 0.3 M sodium acetate, pH 5.5, saturated with MgSO₄. The mixture was extracted with 1 ml of ethyl acetate, and 0.5 ml of the extract was dried down, subjected to high-voltage paper electrophoresis in pyridine acetate, pH 3.5 at 3000 V for 2 hours, and autoradiographed. Under these conditions, the initial rate of peptide bond formation was measured, and the assay was linear with respect to ribosome concentration within the amounts used in these experiments.
33. U. K. Laemmli, *Nature* **227**, 680 (1970).
34. Castenholz and Nitsch's salts are described in *Catalogue of Bacteria and Bacteriophages* (American Type Culture Collection, Rockville, MD, ed. 17, 1989).
35. We thank D. Gelfand (Cetus Corp.) for supplying the *T. aquaticus* strain; J. Silverthorne for suggesting the use of triisopropylnaphthalene sulfonic acid (TNS) and *p*-aminosalicylate (PAS); and M. Ares, C. Guthrie, A. Mankin, T. Powers, and J. Silverthorne for helpful discussions. Supported by NIH grant no. GM17129.

24 February 1992; accepted 27 April 1992

Aminoacyl Esterase Activity of the *Tetrahymena* Ribozyme

Joseph A. Piccirilli, Timothy S. McConnell, Arthur J. Zaug, Harry F. Noller, Thomas R. Cech*

Several classes of ribozymes (catalytic RNA's) catalyze reactions at phosphorus centers, but apparently no reaction at a carbon center has been demonstrated. The active site of the *Tetrahymena* ribozyme was engineered to bind an oligonucleotide derived from the 3' end of *N*-formyl-methionyl-tRNA^{fMet}. This ribozyme catalyzes the hydrolysis of the aminoacyl ester bond to a modest extent, 5 to 15 times greater than the uncatalyzed rate. Catalysis involves binding of the oligonucleotide to the internal guide sequence of the ribozyme and requires Mg^{2+} and sequence elements of the catalytic core. The ability of RNA to catalyze reactions with aminoacyl esters expands the catalytic versatility of RNA and suggests that the first aminoacyl tRNA synthetase could have been an RNA molecule.

Several structurally and mechanistically distinct classes of catalytic RNA's have been discovered (1). In every case, the substrate for these ribozymes is RNA or DNA (2), and the reactions they catalyze are limited to transesterification or hydrolysis of phosphate diesters or phosphate monoesters (3). Observations that RNA can form a specific binding site for an amino acid (4) and various organic dyes (5) have suggested that RNA might catalyze chemical reactions in which the substrates are not nucleic acids.

Theories of the origin and evolution of life postulate that RNA played a significant role as both information carrier and catalyst (6, 7). It has even been proposed that a metabolic system composed of RNA catalysts could have existed before the advent of ribosomal protein synthesis (8). More information about the catalytic versatility and substrate repertoire of RNA is needed before the plausibility of such scenarios can be evaluated.

The nuclear ribosomal RNA precursor of *Tetrahymena thermophila* contains a 413-nucleotide intervening sequence (IVS) that can splice itself from the larger RNA. Shortened forms of this IVS catalyze sequence-specific cleavage of exogenous oligonucleotide substrates with either guanosine (G) or water as the nucleophile (9–11). The sequence specificity of the reaction is determined by base pairing between an internal guide sequence (IGS) on the ribozyme and a complementary sequence on the substrate (9, 12). A saturable binding site for guanosine (G site) is located in the

catalytic core of the ribozyme (13). The rate constants for the individual steps in the reaction pathway for RNA cleavage have been described (11).

Because the ribozyme from *Tetrahymena* contains a catalytic core for reactions of phosphate diesters, we hypothesized that this core might contain a favorable arrangement of catalytic groups for analogous reactions at carbon centers. Our approach was to target a carboxylate ester to the catalytic core of the ribozyme, by covalent attachment of an amino acid to an oligonucleotide for which the ribozyme contains a binding site.

Carboxylate ester hydrolysis. We initiated study with an aminoacyl derivative of the hexanucleotide CAACCA, esterified with *N*-formyl-L-methionine at the 2' or 3' position of the terminal ribose (CAAC-CA$_{fMet}$). Use of this particular aminoacyl ester offers several advantages in that (i) it is easily prepared by ribonuclease T1 digestion of *N*-formyl-L-methionyl-tRNA^{fMet} (14), (ii) it can be ^{35}S-labeled in the amino acid portion in vivo, and (iii) *N*-formylation makes it more stable toward spontaneous hydrolysis than an aminoacyl oligonucleotide with a free α-amino group (15). A ribozyme was constructed in which the IGS was modified from GGAGGG to GGG-UUG (16–19) to allow base pairing with the aminoacylated oligonucleotide (Fig. 1, A and B). The guanosine at the 5' end (G22) was left unaltered even though it was not complementary to the A in the substrate, because a G in this position is conserved and contributes to reactivity (20).

This newly engineered ribozyme was tested in the endoribonuclease reaction (50°C, 10 mM MgCl$_2$, pH 7.0) with RNA substrate CAACCUAAAAA, which forms a "matched" duplex with the IGS. Cleavage occurred at the expected site (the UA bond) (21) with the second-order rate constant for

reaction of E · G (the ribozyme · guanosine complex) and free substrate $(k_{cat}/K_m)^S = 2 \times 10^7$ M^{-1}min^{-1} (22). This value is comparable to $(k_{cat}/K_m)^S = 9 \times 10^7$ M^{-1} min^{-1} for the endonuclease reaction studied previously (11). Thus, these base changes within the IGS do not severely impair ribozyme folding or catalytic activity. A second substrate, CAACCAAAAAA was also examined because, like CAACCA$_{fMet}$, it would form a terminal G · A mismatch when bound to the IGS (Fig. 1A). This substrate was also cleaved at the expected site after CAACCA (21); the rate of the actual cleavage step was reduced by at least 30-fold for the guanosine-dependent reaction and 10-fold for the hydrolysis reaction, compared to those for the "matched" substrate (23).

CAACCA$_{fMet}$ was incubated with ribozyme at 30°C in the absence of guanosine. When the ^{32}P-labeled oligonucleotide moiety was monitored, the *pCAACCA$_{fMet}$ (24) was converted to a product of greater electrophoretic mobility (Fig. 2A) (25) that comigrated with authentic *pCAACCA (21). When the ^{35}S-labeled aminoacyl moiety was followed, [^{35}S]fMet was released from CAACCA$_{[35S]fMet}$ (Fig. 2B) (26). In both assays, the amino acid was hydrolyzed from the oligonucleotide with first-order kinetics about five times faster in the presence of ribozyme than in its absence (Fig. 2C) (27). In the presence of guanosine, guanosine analogues, L-arginine, and arginine analogues (28), hydrolysis of the amino acid persisted with no evidence of transfer of *N*-formyl-methionine to a nucleoside or of dipeptide formation (29).

The ribozyme is a metalloenzyme, requiring magnesium ion both for tertiary structure formation and for RNA cleavage activity (30, 31). When CAACCA$_{fMet}$ was incubated with ribozyme in the absence of Mg^{2+}, no enhancement of hydrolysis over the uncatalyzed reaction was observed. Neither Zn^{2+} nor Ca^{2+} substituted for magnesium ion in the esterase reaction. Although Ca^{2+} promotes ribozyme tertiary structure formation, this structure has no measurable catalytic activity in the endonuclease reaction (30). Because the Ca^{2+}-stabilized ri-

Table 1. Deletions that destroy ribozyme activity with RNA also eliminate aminoacyl-esterase activity.

RNA*	Concentration (μM)	k_{obs} (hour^{-1})†
Ribozyme	1	1.00
L-42 Sca I	1	0.21
L-21 Fok I	10–50	0.21
L-21 Mbo II	10–50	0.21
No RNA	—	0.21

*RNA secondary structures are shown in Fig. 3. †Determined by TLC as described in Fig. 2B; the variation was ±10%.

J. A. Piccirilli, T. S. McConnell, A. J. Zaug and T. R. Cech are at the Howard Hughes Medical Institute, Department of Chemistry and Biochemistry, University of Colorado, Boulder, CO 80309. H. F. Noller is at the Sinsheimer Laboratories, University of California, Santa Cruz, CA 95064.

*To whom correspondence should be addressed.

Reprinted with permission from *Science*,
Volume 256, pp. 1420–1424. © 1992 by the American Association for the Advancement of Science.

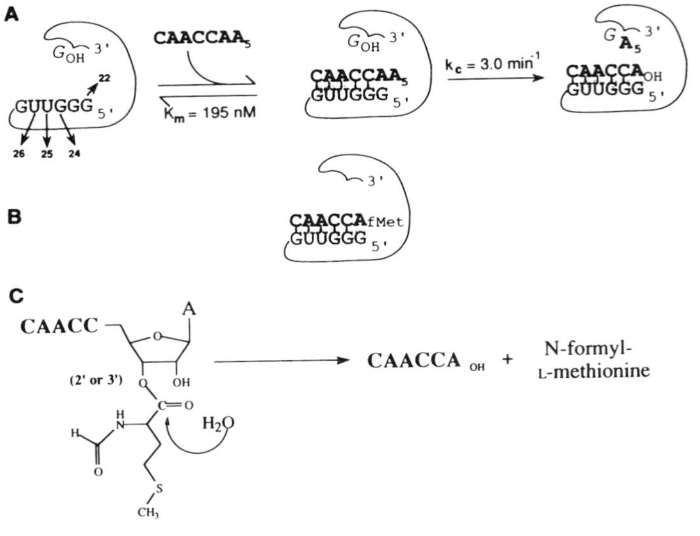

Fig. 1. Endoribonuclease and aminoacyl esterase activities of a ribozyme designed to bind the 3' end of tRNA[fMet]. (**A**) Endoribonuclease reaction. The ribozyme is a shortened version of the *Tetrahymena* IVS RNA missing 21 nucleotides from the 5' end and one nucleotide from the 3' end. This ribozyme was engineered to bind the 3'-terminal sequence of tRNA[fMet] CAACCA by alteration of the IGS at positions 24, 25, and 26 (*16*). The conserved G at position 22 was left unaltered. The ribozyme binds the oligonucleotide substrate S (CAAC-CAAAAAA, bold uppercase letters) by base pairing to the IGS (shaded). The Michaelis constant (K_m) for the reaction was determined under single turnover conditions from a plot of k_{obs} as a function of ribozyme concentration (30°C, 50 mM $MgCl_2$, pH 7.0). The rate constant for the central step (k_c) is the observed rate constant in the presence of saturating enzyme and guanosine. (**B**) Model for ribozyme-aminoacyl oligonucleotide complex. The substrate, derived from ribonuclease T1 digestion of aminoacylated tRNA[fMet], is complementary to the IGS (shaded). The usual phosphoryl group at the cleavage site is replaced with an aminoacyl group derived from *N*-formyl-L-methionine (fMet). Binding of the aminoacyl oligonucleotide to the IGS is expected to target the carboxylate ester to the ribozyme active site. (**C**) Aminoacyl ester hydrolysis, showing nucleophilic attack by H_2O (or OH^-) to release the *N*-formyl-methionyl group from the oligonucleotide. The oligonucleotide product of this reaction is identical to that of the endonuclease reaction shown in (A). The substrate is a mixture of two isomeric forms, the

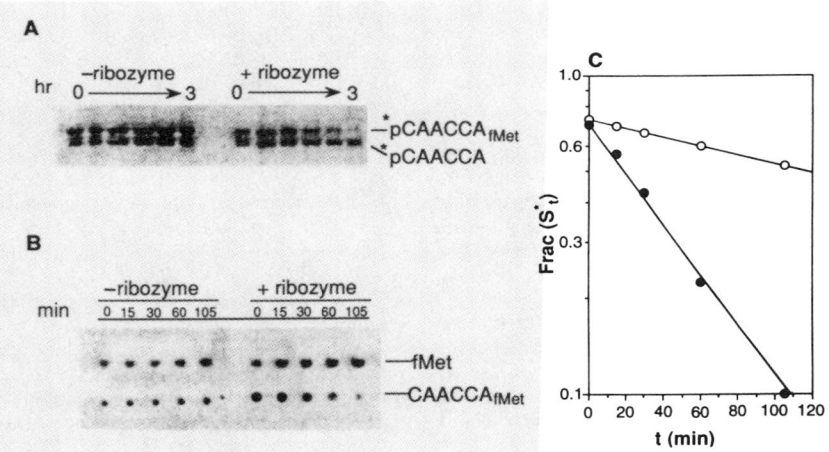

2'-*O*-aminoacylated and 3'-*O*-aminoacylated isomers, that interconvert in aqueous solution (*47*).

bozyme is also inactive in the esterase reaction, Mg^{2+} appears to be directly involved in the chemical step in this case as well.

Requirement for catalytic core and substrate binding site. To ascertain what sequences and structural elements of the ribozyme contribute to its esterase activity, we tested IVS transcripts with different deletions that render the enzyme inactive in the endonuclease reaction (*32*) for their ability to catalyze ester hydrolysis (Fig. 3). IVS transcripts with 3' deletions were made by cutting the DNA template at two restric-

tion sites, producing RNA truncated within the catalytic core. Neither the L-21 Fok I nor the L-21 Mbo II versions of the ribozyme promoted esterase activity, although these RNA constructs still contain the guide sequence. With 50 μM L-21 Mbo II RNA, a concentration that should assure binding of the oligonucleotide based on thermodynamic parameters and that was shown to give greater than 80 percent binding of CAACCAAAAAA by native gel electrophoresis (*33*), no catalysis of aminoacyl bond cleavage was observed (Ta-

ble 1). These results suggest that catalysis was not simply a consequence of Watson-Crick base pairing of CAACCA[fMet] to the IGS, but required additional elements of the ribozyme's tertiary structure.

An L-42 Sca I version of the ribozyme, which lacks part of the 5' end including the IGS, was also incapable of catalyzing ester hydrolysis (Table 1). The catalytic core of this RNA should be correctly folded (*34*). Thus, some portion of the 5' region missing in L-42, presumably the IGS, was essential for hydrolysis of the ester linkage, which is

Fig. 2. Hydrolysis of aminoacyl oligonucleotide linkage catalyzed by the ribozyme. (**A**) Reaction of labeled oligonucleotide moiety analyzed by gel electrophoresis and scanned with a PhosphorImager (Molecular Dynamics). The ribozyme was first incubated in reaction buffer (20 minutes; 50°C). The temperature was reduced to 30°C and the ribozyme was added to p*CAACCA[fMet]. Reaction conditions were 1 μM ribozyme, ~0.1 nM substrate, 50 mM MES buffer [2-(*N*-morpholino)ethanesulfonic acid], pH 7.0, and 50 mM $MgCl_2$. As a control, ribozyme was omitted (left set of lanes). Reactions were quenched with EDTA and urea (buffered to pH 6.0 with potassium phosphate), placed immediately on solid CO_2, and subjected to electrophoresis on a 20 percent denaturing polyacrylamide gel (*25*). (**B**) Labeled amino acid moiety analyzed by thin-layer chromatography (TLC) and scanned with a PhosphorImager (Molecular Dynamics). CAACCA[[35S]fMet] (~50 nM) was incubated with 1 μM ribozyme (or without ribozyme as indicated). The reaction was carried out as described above. Portions were removed at intervals and placed immediately on solid CO_2. After the last sample was removed (105 minutes), the portions were thawed and immediately applied to a cellulose TLC plate. The plate was dried at reduced pressure for about 30 seconds and eluted with *n*-butanol:acetic acid:water (5:3:2). Regardless of the assay used,

there is some product visible at zero time. This is presumably the result of hydrolysis that takes place during substrate preparation. (**C**) Semilogarithmic plot of the data from (B). k_{obs} = 1.0 hour^{-1} (±10 percent) for hydrolysis in the presence of ribozyme (closed circles) and 0.2 hour^{-1} in the absence of ribozyme (open circles) (*48*). Frac S*$_t$ is the fraction of aminoacyl oligonucleotide remaining at time t.

consistent with a requirement for base pairing of the substrate to the IGS to fix the aminoacyl ester at the active site.

When CAACCAAAAAA was assayed as an inhibitor, the rate of ester hydrolysis decreased with increasing concentration of the RNA substrate (Fig. 4A). The concentration dependence was consistent with the measured K_i of 200 nM for CAACCAAAAAA in the endonuclease reaction. The "matched" oligo CAACCUAAAAA inhibited completely at 1 μM concentration, as was expected because of its dissociation constant (~5 nM) (21). This inhibition provided evidence that CAACCA$_{fMet}$ and CAACCAAAAAA bind to the same site, that is, the IGS, as implied by the diagrams in Fig. 1.

The rate of ester hydrolysis shows a hyperbolic dependence on the concentration of ribozyme (Fig. 4B). At low concentration of ribozyme the rate of hydrolysis was first order with respect to ribozyme concentration, while at high ribozyme concentration the reaction rate appeared to become independent of ribozyme concentration. Such Michaelis-Menten reaction kinetics are typical of enzyme-catalyzed reactions in which substrate and catalyst form a complex. The K_m is ~450 nM. For this reaction, we would expect K_m to be equal to the dissociation constant, K_d (35). This value of 450 nM is ~10 to 15 times higher than the K_d of non-aminoacylated CAACCAAAAAA (35), suggesting an effect of the amino acid moiety on binding.

As mentioned above, the reaction was examined in the presence of guanosine and some guanosine analogues. Measurement of ester hydrolysis in the presence of saturating concentrations of guanosine or guanosine-5'-monophosphate (GMP) showed ~25 percent inhibition compared to reactions in the absence of guanosine, while addition of uridine 5'-monophosphate had no effect. 3'-Deoxy-3'-amino-GMP inhibited the reaction, like guanosine. 2'-Deoxy-2'-aminoguanosine stimulated the rate about threefold, increasing the overall rate enhancement to 15 (Fig. 4C). Both of these guanosine analogues occupy the G site (36). The sensitivity of the hydrolysis reaction to G-site occupation is further evidence that the reaction was occurring at the active site described previously. Changes in solvation or conformation that occur when guanosine binds, or direct interaction of the nucleotide with the aminoacyl group are possible ways in which the nucleotides could affect hydrolysis.

Transfer of the aminoacyl group to guanosine or its analogues was not observed, in contrast to nucleotidyl group transfer observed in the endonuclease reaction. This may seem surprising in view of the fact that in the RNA endonuclease reaction, the rate of the chemical step is 1000 times faster with G as the nucleophile than with water as the nucleophile (11). However, closer analysis suggests that, although the aminoacyl group and the guanosine may be proximal, they may not be appropriately oriented for reaction, as described below.

Comparison of esterase and endonuclease mechanisms. We found that the hydrolysis of CAACCA$_{fMet}$ to N-formyl-methionine and CAACCA was accelerated 5-fold by the ribozyme and 15-fold in the presence of one particular guanosine analogue. This modest rate acceleration required magnesium ion and specific sequence elements of the RNA; both of these properties correspond to the endonuclease reaction, suggesting that a similarly folded tertiary structure is necessary for catalysis. Several observations indicate that the reaction occurs at the active site with substrate bound to the IGS: (i) a deleted ribozyme lacking the IGS but having the core sequence elements is inactive, (ii) the endonuclease substrate CAACCAAAAAA is an inhibitor, (iii) the reaction kinetics show saturation behavior, and (iv) the reaction rate is sensitive to bound guanosine or guanosine analogues. Although simple buffers and small molecules are known to catalyze the hydrolysis of carboxylate esters (37), these reactions are not likely to have the sequence specificity inherent in the ribozyme-catalyzed reaction.

By protein enzyme standards and by comparison to RNA-catalyzed reactions with natural substrates, the ribozyme is a modest aminoacyl esterase. The rate advantage obtained in the endonuclease reaction is six to seven orders of magnitude larger

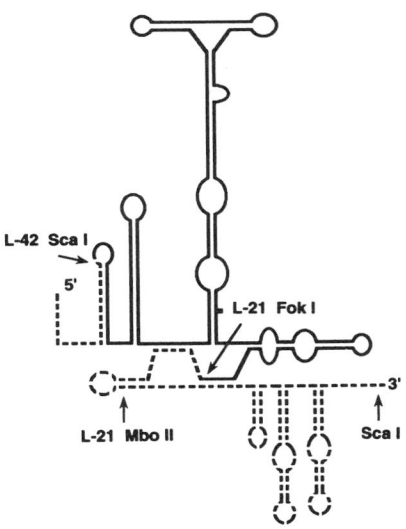

Fig. 3. Outline of secondary structure of the *Tetrahymena* ribozyme showing deleted sequence elements (dashed lines). The standard ribozyme includes the entire outline. L-42 lacks the 5' dashed region and extends to the Sca I site. L-21 Fok I includes the 5' dashed region and the solid region only. L-21 Mbo II is like L-21 Fok I but extends to the Mbo II site.

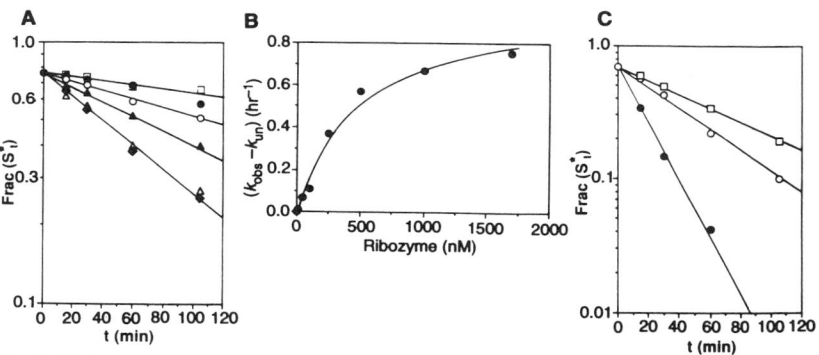

Fig. 4. (**A**) Esterase activity requires pairing of substrate to IGS. Semilogarithmic plots for hydrolysis of *pCAACCA$_{fMet}$ catalyzed by the ribozyme with CAACCAAAAAA at 23 μM (closed circles), 2.7 μM (open circles), 1.3 μM (closed triangles), and 0.7 μM (open triangles). Closed diamonds represent the ribozyme-catalyzed reaction in the absence of CAACCAAAAAA, and open squares represent the uncatalyzed reaction. Reaction protocol and data analysis were carried out with the gel electrophoresis assay (Fig. 2A). (**B**) Ester hydrolysis follows Michaelis-Menten kinetics. $k_{obs} - k_{un}$ plotted as a function of ribozyme concentration; k_{obs} is calculated from the slope of graphs such as those of Figs. 2C and 4, and k_{un} is the uncatalyzed rate of ester hydrolysis. The reactions were done with ~0.1 nM *pCAACCA$_{fMet}$ as indicated in Fig. 2A. The solid line represents the fit to a theoretical curve with K_m = 450 nM and k_c = 1.2 hour⁻¹. (**C**) Sensitivity of ester hydrolysis rate to G-site occupation. Semilogarithmic plots for the hydrolysis of CAACCA$_{[35S]fMet}$ catalyzed by the ribozyme in the absence of guanosine (open circles) or in the presence of 2 mM guanosine (open squares) or 2 mM 2'-deoxy-2'-aminoguanosine (closed circles) (36). Other reaction conditions and quantitation were as in Fig. 2B. Substrates and products were analyzed by TLC and quantitated with the PhosphorImager (Molecular Dynamics).

Fig. 5. Alignment of substrate and guanosine on the ribozyme in (**A**) the RNA endonuclease reaction and (**B**) the esterase reaction. The three-dimensional ribozyme surface is represented by the hatched pattern outside the dark outline. The unhatched interior represents the ribozyme active site. The substrate is bound to the IGS (represented by uppercase italic letters). Guanosine is bound by at least three separate interactions (*13*). The closed thin arrow in (A) represents a favorable pathway for the nucleophile, in-line with the 3'-oxygen-phosphorus bond (*40*). For nucleophilic attack on a carbonyl carbon, shown in (B), this direction of attack is not favorable as indicated by the x through the arrow. For carboxylate

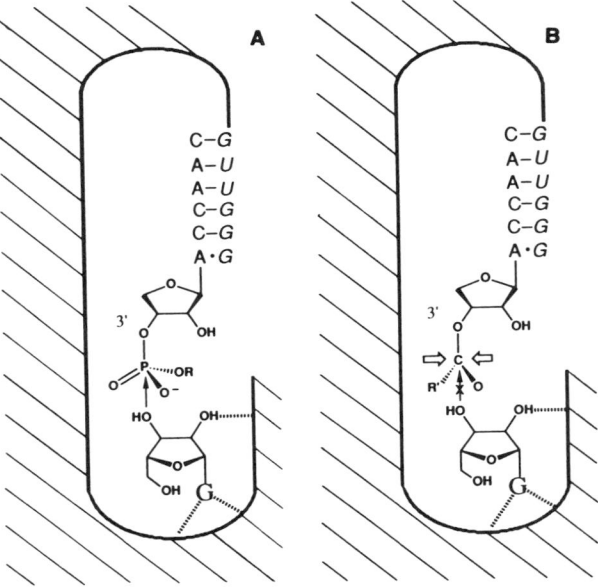

esters, a more favorable direction of nucleophilic attack is orthogonal to the pi face (*42*), shown by the thick open arrows. Assuming free rotation about the 3' oxygen–carbonyl carbon bond, the open arrows define a plane perpendicular to the surface of the page in this model.

than that obtained in the esterase reaction (*38*). However, in the esterase reaction the natural G·U at the cleavage site is replaced with a G·A mispair because the 3' terminal nucleotide of CAACCA$_{fMet}$ is an A. The RNA substrate containing an A at the cleavage site reacts 30 times more slowly than the corresponding U-containing substrate in the guanosine-dependent reaction (*23*). Thus, the ribozyme-substrate combination tested here may not be optimal for the esterase reaction (*39*).

In addition, the intrinsic differences in mechanism between phosphate diester and carboxylate ester reactions may limit the efficiency of aminoesterolysis with this ribozyme. In reactions of phosphate diesters in solution, the nucleophile attacks the phosphorus atom in-line with the labile phosphorus-oxygen bond (*40*). On the ribozyme, nucleophilic attack by the 3'-hydroxyl of guanosine is an in-line, S_N2 (P) reaction as judged by its stereochemistry, that is, inversion of configuration at phosphorus (*41*). Carboxylate esters undergo nucleophilic attack by a different mechanism (*42*). In the first step the nucleophile is added to the carbon-oxygen double bond by attack orthogonal to the pi face, giving a tetrahedral intermediate. In the second step the leaving group is expelled. If the relative orientations of CAACCA$_{fMet}$ and guanosine in the active site are the same as in the endoribonuclease reaction, then the reactants will not be aligned appropriately for acyl group transfer (Fig. 5).

The *Tetrahymena* ribozyme has been selected to splice itself out of pre-ribosomal rRNA, not to catalyze the hydrolysis of aminoacyl ester linkages. In fact, the ribozyme is highly specific for the phosphate at the cleavage site in the RNA substrate. For example, stereospecific substitution of a sulfur for one of the nonbridging phosphate oxygen atoms at the cleavage site reduces activity 1000 times or more (*43*). Thus, it may be possible to create mutant versions of the *Tetrahymena* ribozyme or to find other RNA's that are considerably better at catalyzing this esterase reaction.

Nevertheless, some rate acceleration might be expected because a phosphate diester is thought to mimic the transition state for ester hydrolysis. Antibodies that bind tetrahedral phosphate analogues catalyze reactions at carbon centers that proceed through tetrahedral transition states such as ester or amide hydrolysis (*44*). Since the ribozyme has an active site for a phosphate diester, the ground state or transition state interactions that occur with the phosphate might also stabilize a tetrahedral carboxylate ester transition state.

In particular, a candidate for such an interaction is the ribose 3' oxygen of the substrate (Fig. 5), the leaving group in both the esterase and endonuclease reactions. There is evidence that this oxygen atom is directly coordinated to Mg^{2+} in the transition state of the endonuclease reaction (*31*). Thus, the 3'-oxygen of the aminoacyl ester could be coordinated to a magnesium

ion serving to activate the ester toward nucleophilic attack.

The discovery of catalytic RNA supported earlier speculation regarding a self-replicating system based solely on RNA. The demonstration of catalysis of a reaction at a carbon center by an RNA enzyme suggests that the RNA world could have expanded to include reactions of amino acid and other non-nucleic acid substrates prior to the involvement of proteins. More specifically, RNA might be able to act as an aminoacyl tRNA synthetase. Since RNA can break acyl bonds to oligonucleotides, then, by the principle of microscopic reversibility, RNA must also be able to charge oligonucleotides with amino acids. The substrate for the esterase activity is derived from the acceptor stem of an aminoacyl tRNA. These observations, coupled with the observations of specific amino acid binding by RNA, provide evidence for the chemical plausibility of earlier speculation (*4, 45*) that the first aminoacyl RNA synthetase was an RNA molecule.

Extensively deproteinized large ribosomal subunits retain peptidyl transferase activity (*46*), suggesting that the catalytic activity resides in the RNA component. With tRNA's, mRNA, and ribozymes as tRNA synthetases and peptidyl transferases, it is possible to conceive of information-driven protein synthesis being carried out exclusively by RNA.

REFERENCES AND NOTES

1. S. Altman, *J. Biol. Chem.* **265**, 20053 (1990); T. R. Cech, *Annu. Rev. Biochem.* **59**, 543 (1990); F. Michel, K. Umesono, H. Ozeki, *Gene* **82**, 5 (1989); N. R. Pace and D. Smith, *J. Biol. Chem.* **265**, 3587 (1990); R. H. Symons, *Crit. Rev. Plant Sci.* **10**, 189 (1991).
2. D. Herschlag and T. R. Cech, *Nature* **344**, 405 (1990); D. L. Robertson and G. F. Joyce, *ibid.*, p. 467; A. C. Forster and S. Altman, *Science* **249**, 784 (1990).
3. T. R. Cech, *Science* **236**, 1532 (1987).
4. M. Yarus, *ibid.* **240**, 1751 (1988).
5. A. D. Ellington and J. W. Szostak, *Nature* **346**, 818 (1990).
6. C. Woese, *The Genetic Code: The Molecular Basis for Genetic Expression* (Harper and Row, New York, 1967), p. 186; F. Crick, *J. Mol. Biol.* **38**, 367 (1968); L. E. Orgel, *ibid.*, p. 381.
7. N. R. Pace and T. L. Marsh, *Origins Life* **16**, 97 (1985); L. E. Orgel, *J. Theor. Biol.* **123**, 127 (1986); T. R. Cech, *Proc. Natl. Acad. Sci. U.S.A.* **83**, 4360 (1986); W. Gilbert, *Nature* **319**, 618 (1986); F. H. Westheimer, *ibid.*, p. 534; G. F. Joyce, *ibid.* **338**, 217 (1989).
8. S. A. Benner, A. D. Ellington, A. Tauer, *Proc. Natl. Acad. Sci. U.S.A.* **86**, 7054 (1989); C. M. Visser, *Origins Life* **14**, 291 (1984); _____ and R. M. Kellogg, *J. Mol. Evol.* **11**, 171 (1978); H. B. White III, *ibid.* **7**, 101 (1976); _____, in *The Pyridine Nucleotide Coenzymes*, J. Everse, B. Anderson, K.-S. Yu, Eds. (Academic Press, New York, 1982).
9. A. J. Zaug, M. D. Been, T. R. Cech, *Nature* **324**, 429 (1986).
10. A. J. Zaug, C. A. Grosshans, T. R. Cech, *Biochemistry* **27**, 8924 (1988).
11. D. Herschlag and T. R. Cech, *ibid.* **29**, 10159 (1990); *ibid.*, p. 10172.
12. R. W. Davies, R. B. Waring, J. A. Ray, T. A. Brown,

C. Scazzocchio, Nature 300, 719 (1982); R. B. Waring, P. Towner, S. J. Minter, R. W. Davies, ibid. 321, 133 (1986); M. D. Been and T. R. Cech, Cell 47, 207 (1986).

13. B. L. Bass and T. R. Cech, Nature 308, 820 (1984); F. Michel et al., ibid. 342, 391 (1989).

14. R. E. Monro, Methods Enzymol. 20, 472 (1971).

15. R. Wolfenden, Biochemistry 2, 1090 (1963).

16. The Tetrahymena ribozyme containing an altered guide sequence GGGUUG was prepared by phagemid mutagenesis (17) of plasmid pTZ-21H2 (18) and characterized by dideoxy sequencing (19). Plasmid was cut with Hind III and transcribed with phage T7 RNA polymerase at 30°C. The MgCl$_2$ concentration was then raised from 15 mM to 50 mM, the temperature was raised from 37° to 50°C, and incubation was continued for 2 hours to promote processing of the 3'-terminal hammerhead ribozyme and thereby generate a precise 3' end (18).

17. T. A. Kunkel, J. D. Roberts, R. A. Zakour, Methods Enzymol. 154, 367 (1987).

18. C. G. Grosshans and T. R. Cech, Nucleic Acids Res. 19, 3875 (1991).

19. F. Sanger, S. Nicklen, A. R. Coulson, Proc. Natl. Acad. Sci. U.S.A. 74, 5463 (1977).

20. E. T. Barfod and T. R. Cech, Mol. Cell. Biol. 9, 3657 (1989); J. A. Doudna, B. P. Cormack, J. W. Szostak, Proc. Natl. Acad. Sci. U.S.A. 86, 7402 (1989).

21. J. A. Piccirilli, T. S. McConnell, T. R. Cech, unpublished data.

22. The second-order rate constant for reaction of E · G and S, $(k_{cat}/K_m)^S$, was determined in single turnover experiments with saturating guanosine and 0.5 nM 5'-[^{32}P]-CAACCUAAAAA at 50°C, pH 7.0, 10 mM Mg^{2+}. The value of k_{obs} increased linearly with the concentration of E (1 to 5 nM, 1 mM G).

23. The rate constants (k_c) for the central step representing the conversion of E · G · S ternary complex to E · G · P are 200 min^{-1} and 6 min^{-1} for CAACCUAAAAA and CAACCAAAAAA, respectively. These values might represent the actual chemical conversion, as in the case of the ribozyme with the natural IGS, or could represent an accompanying conformational change or binding of G. The value of k_c = 200 min^{-1} for CAACCUAAAAA was calculated as described in (11) from $(k_{cat}/K_m)^G$ = 2.0 ± 0.5 × 10^5 M^{-1} min^{-1} with the assumption that K_d for guanosine = 1 mM (50°C, 10 mM MgCl$_2$, pH 7.0). The value of k_c = 6 min^{-1} for CAACCAAAAAA was the observed rate of single turnover cleavage in the presence of excess ribozyme and 1 mM G. The values of k_c (-G), the rate of the central step in the guanosine-independent hydrolysis reaction, were 0.2 min^{-1} and 0.02 min^{-1} for CAACCUAAAAA and CAACCAAAAAA, respectively. The second-order rate constant for reaction of E · S and G, $(k_{cat}/K_m)^G$, was determined in single turnover experiments with 200 nM ribozyme and 1 nM 5'-[^{32}P]-CAACCUAAAAA at 50°C, pH 7.0, 10 mM Mg^{2+}. Guanosine concentration was varied from 1 to 5 μM.

24. Because of the hydrolytic instability of the ester linkage in CAACCA$_{fMet}$, the following procedure was used for 5' end-labeling and subsequent isolation of 5'-[^{32}P]-CAACCA$_{fMet}$ (*pCAAC-CA$_{fMet}$). The kinase reaction was performed (37°C, 5 minutes) in 5 μl of buffer [50 mM MES (pH 7.0), 10 mM MgCl$_2$, 0.2 mM spermidine, 0.1 mM EDTA], and contained CAACCA$_{fMet}$ (1.2 pmol), [γ-^{32}P]ATP (adenosine triphosphate) (3 pmol), and polynucleotide kinase (50 units). The reaction was quenched by the addition of an equal volume of sample buffer [90 percent formamide, 20 mM sodium phosphate, 16 mM borate, 20 mM EDTA, 0.02 percent bromophenol blue, and 0.02 percent xylene cyanol (pH 6.0)] and stored on solid CO$_2$ prior to gel electrophoresis. The 5'-labeled oligonucleotide was purified by electrophoresis on a 20 percent polyacrylamide–3.2 M urea gel that

had been equilibrated at 20 W for 2 hours. Electrophoresis was then continued for 4 to 7 hours at 4°C. The buffer was 100 mM sodium phosphate, 83 mM boric acid, and 1 mM EDTA (pH 6.0). *pCAACCA$_{fMet}$ was visualized by autoradiography, excised, and eluted for 1 hour in buffer containing 10 mM ammonium acetate, pH 4.5, and 1 mM EDTA. The eluted aminoacyl oligonucleotide was stored at −70°C.

25. Polyacrylamide gel electrophoresis was carried out on slab gels of 20 percent polyacrylamide and 7 M urea. The buffer was 100 mM tris(hydroxymethyl)aminomethane, 100 mM boric acid, 2 mM EDTA; pH 8.3. Gels were first run at 4°C (cold room) at 15 W before use and then run for ~3 hours after application of the samples. The gels were dried, and radioactivity was quantitated (PhosphorImager).

26. The [^{35}S]-aminoacyl oligonucleotide was isolated by the procedure of Monro (14). Treatment of the oligonucleotide with CHES buffer (pH 9.0) hydrolyzed the acyl linkage. The product did not comigrate with authentic N-formyl-methionine (Sigma) on TLC (silica gel 60; butanol:acetic acid:H$_2$O, 4:1:1). Treatment of this product with 10% HCl for 10 min at 95°C (conditions of deformylation) gave a new product that comigrated with authentic methionine sulfoxide (Sigma). These results suggest that sulfur was oxidized during preparation of CAACCA$_{fMet}$. Nevertheless, the carboxylate ester linkage remained intact, and the same material was active in the puromycin reaction with large ribosomal subunits. Treatment of 5'-[^{32}P]-CAAC-CA$_{fMet}$ with CHES buffer (pH 9.0) converted the aminoacyl oligonucleotide to a product of greater electrophoretic mobility that comigrated with authentic 5'-[^{32}P]-CAACCA.

27. The absolute values for hydrolysis rates obtained using the gel assay were slightly lower than those obtained by TLC, but the relative rate enhancement by ribozyme was the same.

28. L-Arginine binds at the G site with its guanidino group interacting with G-264 in the ribozyme, the same nucleotide that interacts with the guanidino group of G (4). The α-amino group might lie in the portion of the site normally occupied by the 2' or 3' OH of the ribose on G. Thus, the α-amino group on arginine might attack the carbonyl carbon of the ester to give peptide bond formation. Arginine analogues tested were argininamide, homoarginine, agmatine, glycylarginine (Research Plus), and glycylglycylarginine (Sigma).

29. If transesterification with guanosine or its analogues were occurring, an aminoacylated nucleoside or dipeptide should appear as a new radioactive spot on TLC or paper electrophoresis.

30. C. A. Grosshans and T. R. Cech, Biochemistry 28, 6888 (1989); D. W. Celander and T. R. Cech, Science 251, 401 (1991).

31. J. A. Piccirilli, J. S. Vyle, M. H. Caruthers, T. R. Cech, in preparation.

32. W. D. Downs and T. R. Cech, Biochemistry 29, 5605 (1990); J. W. Szostak, Nature 322, 83 (1986).

33. The calculated free energy for the duplex CAACCA/AGUUGGG at 30°C was determined as ΔG° = ΔH° − TΔS°, with ΔH° and ΔS° values (S. M. Freier et al., Proc. Natl. Acad. Sci. U.S.A. 83, 9373 (1986)). The terminal A · G mismatch and the first unpaired A of the ribozyme were included in the calculation. Gel binding experiments were performed at 30°C and 15 mM MgCl$_2$ with 5'-[^{32}P]CAAC-CAAAAAA as described previously [A. M. Pyle, J. A. McSwiggen, T. R. Cech, Proc. Natl. Acad. Sci. U.S.A. 87, 8187 (1990)].

34. A mutant RNA missing the entire sequence upstream of nucleotide 46 can promote specific hydrolysis at the 3' splice site; B. Young, thesis, University of Colorado, Boulder (1990).

35. Assuming a two-step model E + S ↔ E·S → E·P, $K_m = (k_{-1} + k_c)/k_1$ where k_{-1} is the rate constant for E · S → E + S, k_1 is the rate constant for E + S

→ E · S, and $K_d = k_{-1}/k_1$. If $k_c << k_{-1}$, then $K_m = K_d$. In the esterase reaction k_c = 0.02 min^{-1} is expected to be slow compared to k_{-1} (the rate constant for the dissociation of CAACCA$_{fMet}$ from the ribozyme · substrate complex). The rate constant for dissociation of the non-aminoacylated product CAACCA is 0.8 min^{-1}. For the RNA substrate CAACCAAAAA, k_c = 3.0 min^{-1}, K_m = 195 nM, and if k_{-1}(CAACCA) = k_{-1}(CAACCAAAAA) = 0.8 min^{-1}, k_1 can be calculated from the relationship for K_m. This gives K_d = 45 nM. The assumption that k_{-1}(P) = k_{-1}(S) is true for the reaction studied previously (11). K_d = 45 nM is consistent with preliminary data that give K_m = 30 nM for CAACCAAAAA in the hydrolysis reaction where $k_c << k_{-1}$, and $K_m = K_d$.

36. The L-21 Sca I ribozyme is more than 90 percent saturated at the concentrations (2 mM) of 2'-deoxy-2'-amino-guanosine and 3'-deoxy-3'-amino-GMP (D. Herschlag and T. R. Cech, unpublished results). This should also be true for the ribozyme used in our study because guanosine binding is not expected to be affected by alteration of the IGS.

37. W. P. Jencks, Catalysis in Chemistry and Enzymology (Dover, New York, ed. 2, 1987).

38. $k_c/k_u \cong 10^8$ for hydrolysis of CAACCUAAAAA or $\cong 10^7$ for hydrolysis of CAACCAAAAAA, compared to 5 to 15 for the hydrolysis of CAACCA$_{fMet}$; k_c represents the observed pseudo first-order rate constant for reaction of substrate bound to ribozyme with water, and k_u represents the pseudo first-order rate constant for reaction of free substrate with water (estimated for phosphate diesters as described) (11).

39. An L-18 ribozyme, containing a U at position 22 instead of G, converts the terminal G · A mismatch to an U · A base pair in the ribozyme substrate complex. However, both ribozymes catalyzed the deacylation of CAACCA$_{fMet}$ at the same rate. This does not provide the G · U combination that is important for the RNA cleavage reaction.

40. For reviews see: S. J. Benkovic and K. J. Schray, in Transition States of Biochemical Processes, R. D. Gandour Ed. (Plenum, New York, 1978), pp. 493–527; in The Enzymes, P. D. Boyer Ed. (Academic Press, New York, 1973), vol. 8, pp. 201–238.

41. J. A. McSwiggen and T. R. Cech, Science 244, 679 (1989); J. Rajagopal, J. A. Doudna, J. W. Szostak, ibid., p. 692.

42. P. Deslongchamps, P. Allani, D. Fre'hel, A. Malaval, Can. J. Chem. 50, 3405 (1972); H. B. Buergi, J. D. Dunitz, E. Shefter, J. Am. Chem. Soc. 95, 5605 (1973).

43. The pro-Sp oxygen atom; J. A. Piccirilli and T. R. Cech, in preparation.

44. P. G. Schultz, R. A. Lerner, S. J. Benkovic, Chem. Eng. News, 26 (28 May 1990); R. A. Lerner and S. J. Benkovic, BioEssays 9, 107 (1988).

45. A. Weiner and N. Maizels, Proc. Natl. Acad. Sci. U.S.A. 84, 7383 (1987).

46. H. F. Noller, V. Hoffarth, L. Zimniak, Science 256, 1416 (1992).

47. M. Taiji, S. Yokoyama, T. Miyazawa, Biochemistry 24, 5776 (1985); the transacylation rates of 2'-(3')-O-(N-acetyl-L-phenylalanyl-L-phenylalanyl)-adenosine were obtained as k(2' → 3') = 0.41 min^{-1} and k(3' → 2') = 0.18 min^{-1} at pH 7.0, 0°C.

48. The activity was also independent of buffer concentration in the range of 10 to 50 mM.

49. We thank F. Eckstein, L. Orgel, and M. Yarus for gifts of 2'-deoxy-2'-amino-guanosine, 3'-deoxy-3'-amino-GMP, and arginine derivatives, respectively; C. Rusconi for preparation of L-42 Sca I RNA; D. Usher for useful discussion; and D. Herschlag and J. Doudna for critical reading of the manuscript. We also thank the W. M. Keck Foundation for support of RNA science on the Boulder campus.

24 February 1992; accepted 27 April 1992

The Rise and Fall of the RNA World

Gerald F. Joyce

It is generally believed that there was a time when life on earth was based on RNA rather than on DNA and protein. Considering the relevant evidence from geophysics, geology, paleobiology, and molecular biology, it is possible to set the time frame for the existence of RNA-based life to a 400 million year interval beginning 4.0 to 4.2 billion years ago and ending 3.6 to 3.8 billion years ago. The minimum level of biochemical complexity that existed during this time consists of those functions necessary for the establishment and maintenance of an RNA-based evolving system, namely, an RNA unwinding activity, an RNA replicase activity, and a primitive biosynthetic apparatus leading to enrichment of the local environment with activated D mononucleotides.

Received August 16, 1990; revised January 28, 1991

It is generally believed that there was a time in the early history of life on earth when RNA served as both the genetic material and the agent of catalytic function (Woese, 1967; Crick, 1968; Orgel, 1968; Sharp, 1985; Pace and Marsh, 1985; Lewin, 1986; Gilbert, 1986; Cech, 1986). The DNA/protein–based life form that is common to all known terrestrial biology is thought to have descended from an RNA-based life form approximately 4 billion years ago. Very little is known about this postulated RNA-based life form except what can be inferred by examining the role of RNA in contemporary organisms and by studying the behavior of RNA in the laboratory. The lack of first-hand knowledge of the RNA world has not prevented speculation concerning the chronology of events that led to its appearance and ultimate extinction (Darnell and Doolittle, 1986; Weiner and Maizels, 1987; Joyce, 1989; Weiner and Maizels, in press). Nor has it prevented speculation concerning the degree of metabolic complexity that existed during this early period of biological evolution (Alberts, 1986; Orgel, 1986; Benner et al., 1989; Gibson and Lamond, 1990).

Here I consider two issues that may help to set boundary conditions to aid in speculation about the RNA world. First, I will attempt to fix a time frame for the existence of the RNA world. The dawn of the RNA world required the establishment of environmental conditions that are compatible with the solution chemistry of RNA; the end of the RNA world is demarcated by the invention of DNA genomes and the development of a translation apparatus. Considering the relevant evidence from geophysics, geology, paleobiology, and molecular biology, it is possible to fix the time frame for the RNA world to a 400 million year interval centered around the beginning of the early Archean (3.9 billion years ago). Second, I will examine the range of catalytic functions that are essential to the establishment and maintenance of an RNA-based evolving system. These functions constitute the minimum level of biochemical complexity that existed in the RNA world.

Making the World Safe for RNA

The solar system was formed by gravitational collapse of interstellar matter into a rotating disk-shaped cloud of gas and dust (Cameron and Truran, 1977; Cameron, 1978). At the center of this "accretion disk" a protosun appeared. At varying distances from the center, matter became condensed into small planetesimals (≈ 1 to 10 km in diameter), which in turn aggregated to form meteorites and planetary bodies (Goldreich and Ward, 1973).

The age of the solar system is approximately 4.6 billion years, based on isotopic evidence from meteorites and lunar soil samples (Patterson, 1956; Papanastassiou and Wasserburg, 1971). Details of the process of planetary formation remain controversial, but it is generally agreed that the earth was formed by 4.5 billion years ago (Stevenson, 1983). Over the subse-

Based on a paper presented at a symposium on "Molecular Biology and the Origin of Life", Berkeley, California, July 16-18, 1990.

Departments of Chemistry and Molecular Biology, Scripps Research Institute, 10666 N. Torrey Pines Road, La Jolla, CA 92037.

KEY WORDS: origin of life/paleobiology/prebiotic chemistry/RNA enzymes/RNA evolution/RNA world/self-replication

quent 0.3 billion years the earth continued to sweep up material at a brisk rate. The kinetic energy of impacting planetesimals resulted in planetary heating; some of this heat was radiated back to space, but much was retained as buried heat. The buried heat would have been dissipated to a large extent by convection, but, as the planet grew, its internal temperature began to rise faster than could be offset by convective cooling. Eventually a temperature was reached at which at least partial melting of the planet occurred, resulting in downward displacement of relatively dense liquid iron to form the earth's core (Stevenson, 1980).

The process of core formation would have liberated large amounts of heat, sufficient to raise the temperature of the planet by about 1,500°C (Shaw, 1978). The actual increase may have been far more modest, provided that the heat was radiated to space. However, if liquid water was present at the surface, and therefore water vapor was present in the atmosphere, much of the heat would have been reflected back to the surface. This would have caused more water vapor to enter the atmosphere, leading to even greater heat retention and culminating in a runaway greenhouse effect (Zahnle et al., 1988; Kasting, 1988). A runaway greenhouse would result in extremely high surface temperatures, enough to melt the surface rocks and form a global magma ocean. Obviously, a young earth blanketed in magma and steam would have been an inhospitable place for RNA. So too, if liquid water was absent from the surface so that a runaway greenhouse did not occur, there would have been no way to support the solution chemistry of RNA. Existence of the RNA world prior to 4.2 billion years ago seems very unlikely.

The earth's atmosphere and hydrosphere is thought to have been derived from two principal sources. First and foremost, outgassing of volatile components of the primitive mantle led to formation of a nonreducing atmosphere containing mostly CO_2, N_2, SO_2, and H_2O (Walker, 1977; Pollack and Yung, 1980). Second, arrival of material on comets and other impacting bodies added to the inventory, possibly providing a substantial portion of the hydrosphere (Chyba, 1987). By 4.2 billion years ago, the earth's crust had formed. The discovery of 4.2 billion year old detrital zircon in Western Australia is consistent with the presence of a differentiated silicic crust that solidified in the presence of liquid water (Froude et al., 1983). The crustal material was covered by a global ocean, except for volcanic island arcs rising up from the ocean floor.

It is conceivable that prebiotic chemistry leading to the emergence of the RNA world began in the global ocean, in shallow marine environments at the edge of volcanic islands, or in freshwater basins on island surfaces. It has been argued, however, that the global environment would have continued to be in turmoil

until about 3.8 to 4.0 billion years ago because of intense meteoric bombardment that would have "frustrated" any attempt to synthesize complex organic materials (Maher and Stevenson, 1988). If one considers the lunar impact record and extrapolates to estimate the level of meteoric bombardment on the primitive earth, it appears that large impacts (producing a crater ≥ 40 km in diameter) would have occurred roughly every 500 years on the 4.2 billion year old earth. Very large impacts, producing a crater ≥ 265 km in diameter and resulting in a transient 100°C rise in the temperature of the earth's surface, would have occurred about every 50,000 years. Assuming an exponential decline in the level of meteoric bombardment, the interval between very large impacts would have increased to 1 million years by 4.0 billion years ago and to 10 million years by 3.8 billion years ago (Maher and Stevenson, 1988).

There are a number of uncertainties concerning the relation between meteoric impacts and the origins of life. First, the lunar impact record is spotty prior to 4.0 billion years ago, making it difficult to estimate the terrestrial impact flux at very early times. Second, meteoric impacts are stochastic events, so that while very large impacts may have occurred on average once every 50,000 years around 4.2 billion years ago, there may have been longer intervals during which life could have gained a foothold. Third, it is difficult to know whether a "catastrophic" impact would, in fact, have destroyed all life on the planet. Extreme thermophiles, for example, may have survived the transient rise in global temperatures that followed a very large impact event. Organisms located in the deep ocean or some other protected microenvironment may have escaped the brunt of the impact's effects.

Biological diversity is the key to survival under extreme or fluctuating environmental conditions. The question posed by the terrestrial impact data is: At what point was there sufficient time between sizeable impacts for life to have originated so that biological evolution could begin to explore adaptive countermeasures to ensure survival despite repeated environmental upheaval? This is a very difficult question to answer because it is impossible to say how much time would have been required for life to originate and develop appreciable biological diversity. It seems that 50,000 years would have been uncomfortably short, which suggests that life at 4.2 billion years before present would have been doomed to extinction. One million years, on the other hand, may have been enough time, particularly if generation times were short and evolutionary clock rates were rapid. Thus life could have become firmly established by 4.0 billion years ago.

Recognizing the uncertainty that accompanies any discussion of the earth's early history, it seems that the existence of the RNA world prior to 4.2 billion years ago is implausible on several counts. Between 4.2 and 4.0 billion years ago environmental conditions became increasingly favorable, chiefly because of the fall-off in the level of meteoric impacts. One can imagine the RNA world in existence 4.0 billion years ago without contradicting any of the prevailing scientific beliefs concerning the Hadean earth. If one adopts a literal interpretation of the meteoric impact data and believes that life would have required 10 million years or more to become established, then the world would not have been safe for RNA until about 3.8 billion years ago.

Life Before RNA

While there is good evidence to suggest that an RNA-based life form existed at an early stage in evolutionary history, there are several reasons to believe that Darwinian evolution did not begin with RNA (Shapiro, 1984; Joyce et al., 1987; Orgel, 1989; Joyce, 1989). It is difficult to see how activated mononucleotides could have been synthesized under plausible prebiotic conditions. The problem is not so much the availability of purines, pyrimidines, phosphate, and simple sugars, but the difficulty in assembling these components without the aid of biological catalysts.

If one starts with a mixture of D- and L-ribose and a purine base, it is possible to obtain αβ-DL-purine nucleosides in reasonable yield, although the comparable reaction involving a pyrimidine base does not occur (Fuller et al., 1972). If one starts with a mixture of a nucleoside and a source of inorganic phosphate, one can obtain the nucleoside 5'-phosphate, although one also obtains the nucleoside 2'-phosphate, 3'-phosphate, 2',3'-cyclic phosphate, 2',5'-bisphosphate, and 3',5'-bisphosphate (Lohrmann and Orgel, 1968; Lohrmann and Orgel, 1971). Nature, of course, does not have the luxury of starting each reaction with pure materials. This greatly complicates the situation because side-products formed at each step in nucleotide synthesis would be carried over to subsequent reactions. Thus the prebiotic synthesis of DL-ribose would be accompanied by the synthesis of other pentoses as well as the 4-, 6-, and 7-carbon sugars. The synthesis of adenine and guanine would be accompanied by the synthesis of diaminopurine, hypoxanthine, and other related compounds. Rather than condensation between ribose and either adenine or guanine, one should consider condensation between a complex mixture of sugars and a complex mixture of purine-related compounds. For each pairwise combination one would obtain not only the αβ-DL-nucleoside analogue, but also the two anomeric products involving condensation between the base and each of the sugar-hydroxyl groups. Phosphorylation would add another layer of complexity because each sugar-hydroxyl could serve as a phosphorylation site.

The situation on the primitive earth may not have

been quite so difficult. There may have been natural biases in the prebiotic synthesis reactions that favored ribose over other sugars, favored addition of the base at the 1' position, and favored phosphorylation at the 5' position. It has been shown, for example, that condensation of glycoaldehyde-phosphate and formaldehyde yields ribose 2,4-diphosphate as the major pentameric product (Müller et al., 1990). Whether glycoaldehyde-phosphate was more abundant than glycoaldehyde on the primitive earth is not known, but, at least in principle, this reaction demonstrates the possibility of a favored route toward DL-ribose and by analogy toward the DL nucleotides.

Let us consider, then, a very special local environment that had become enriched in DL nucleotides and supported condensation reactions to yield short oligonucleotides. Could self-replication begin in such an environment? The possibility of chemical (i.e., uncatalyzed) self-replication seems remote. Two of the major difficulties are first, how to overcome RNA self-structure so that an existing RNA can act as a template to direct the synthesis of its complement, and second, how to minimize incorporation of closely related nucleotide analogues that would disrupt information transfer between template and complement. The answer to the self-structure problem might lie in an extrinsic unwinding agent, such as a mineral surface or single-strand binding factor, that would disfavor long regions of stable duplex structure (Joyce and Orgel, 1986). The nucleotide analogue problem would be diminished to the extent that the local environment had become enriched in DL nucleotides. However, the fact that the local environment necessarily contains a racemic mixture of DL nucleotides means that a potent inhibitor of template-directed RNA synthesis, namely the opposite enantiomer, would be present in equimolar concentration relative to the enantiomer that is being incorporated (Joyce et al., 1984).

Perhaps RNA self-replication was catalyzed from the start by an RNA enzyme that itself was the substrate for replication. A closed system of this type fulfills the standard notion of the RNA world (Sharp, 1985; Pace and Marsh, 1985; Gilbert, 1986; Cech, 1986). An RNA enzyme might discriminate against nucleotide analogues present in the local environment, including the "wrong" enantiomer, so that template-directed synthesis could proceed in an efficient manner. According to this line of thinking, an RNA enzyme with RNA replicase activity was produced abiotically in the primitive environment, allowing RNA evolution to begin spontaneously. The difficulty with this hypothesis lies in the relation between genome length and replication fidelity.

Very short RNAs, of decamer length or less, are not expected to have appreciable catalytic activity, much less discriminate between closely related nucleotide analogues while functioning as an RNA replicase. Long RNAs, of length 100 residues or more, might be capable of such activity, but are not expected to form spontaneously in the prebiotic environment. (It could be argued that any RNA longer than about a trimer is very unlikely). Is there some intermediate-length RNA that could have formed spontaneously and catalyzed its own replication with sufficiently high fidelity to maintain its base sequence through successive rounds of replication? For an RNA of length ν (the number of positions in the sequence) the minimum required replication fidelity (per position) is approximately $\nu-1/\nu$ (Eigen, 1971). Thus a polymer of 30 residues (30mer) must catalyze its own replication with a fidelity of about 97% per position. Leaving aside the question of what fraction of all 4^{30} possible 30mers would have replicase activity, it seems unlikely that *any* 30mer could catalyze replication with 97% fidelity, considering the complex mixture of nucleotide analogues that it would have had to contend with.

Neither chemical self-replication nor RNA-catalyzed replication offers a plausible explanation for the spontaneous appearance of the RNA world in the prebiotic environment. The conclusion is that life did not start with RNA; RNA followed in the evolutionary footsteps of some other replicating molecule, just as DNA followed in the footsteps of RNA. It has been suggested that RNA was preceded by a simple RNA-like molecule that avoided the problems of template self-structure and competition from closely related analogues (Joyce et al., 1987). It is not unreasonable to believe that there was a succession of several distinct self-replicating entities to bridge the gap between prebiotic chemistry and the RNA world.

Returning to the issue of the time frame for the existence of the RNA world, it is impossible to say what portion of the interval available to the RNA world was, in fact, occupied by a genetic molecule other than RNA. Arguments concerning the suitability of physical conditions on the primitive earth apply to RNA as well as to its likely predecessors. The development of a genetic system based on DNA and protein may have followed a long period of RNA evolution or may have been derived from RNA's predecessor with only a brief intervening reign by RNA. Thus one can only attempt to fix the time interval during which the RNA world could have existed.

The Dawn of Modern Biology

The geologic record extends to the beginning of the early Archean, 3.8 billion years ago (Allaart, 1976). There is abundant fossil evidence to show that life existed on earth 3.5 to 3.6 billion years ago (Lowe, 1980; Walter et al., 1980; Awramik et al., 1983; Byerly et al., 1986; Schopf and Packer, 1987). Two types of fossil evidence are available: microbial stromatolites and cellularly preserved microfossils. Stromatolites are accretionary organosedimentary structures produced by communities of marine-dwelling microorganisms, principally cyanobacteria. They typically have a laminated appearance, resulting from sediment trapping or precipitation of inorganic material by the growing microorganisms (Walter, 1976). Stromatolites, therefore, are complex structures, reflecting the biochemistry and behaviorial responses (e.g., phototaxis, filament orientation, mucus production) of an entire microbial community.

The oldest known stromatolites are found in the lower strata of the Warrawoona Group of Western Australia, dated at 3.556 ± 0.032 billion years before present (Lowe, 1980; Walter et al., 1980; Blake and McNaughton, 1984; Glickson, 1986). The fine structure of the microbial mat suggests that it was laid down by organisms that were phototactic, photoautotrophic, and produced mucus sheaths (Walter, 1983). Comparison of the Warrawoona stromatolites with numerous more recent examples, including modern stromatolites, demonstrates very close structural homology. If one considers the biochemical and behavioral complexity associated with stromatolite formation, this strongly suggests that the Warrawoona stromatolites, like their more recent counterparts, were produced by DNA/protein organisms.

The Warrawoona Group also contains cellularly preserved fossil microorganisms, ranging in age from 3.3 to 3.5 billion years (Awramik et al., 1983; Schopf and Walter, 1983). The authenticity of these fossils is firmly established (Schopf and Packer, 1987). Several distinct taxa have been recognized, including three well-defined species of filamentous microorganisms (genus *Primaevifilum*) and an as-yet-unnamed unicellular organism that forms sheath-enclosed colonies (Schopf, in press). Each of these morphotypes has close analogues among the modern cyanobacteria. The analogy is supported by detailed morphometric analysis and, in the case of the filamentous microorganisms, observation of the way in which septations form during cell division (Schopf, in press). Unless one believes that the Warrawoona microfossils are "cyanobacteria-like" RNA organisms that later found an essentially perfect mimic among the true cyanobacteria, it seems reasonable to conclude that life based on DNA and protein existed on earth at least 3.5 billion years ago.

The only well-characterized rock formation that is older than the Warrawoona Group is the Isua Supracrustal Belt of southwestern Greenland, dated at 3.770 ± 0.042 billion years before present (Allaart, 1976). These are high-grade metamorphic rocks, having undergone reformation at high pressure (3 to 8 kbar) and temperature (450° to 700°C). Not surprisingly, Isua rocks do not contain any fossil evidence of life. However, on the basis of examination of the carbon isotope composition of the contained organic material it has been claimed that they provide indirect evidence of life (Schidlowski et al., 1979, 1983; Schidlowski, 1988).

Living organisms are known to discriminate against ^{13}C relative to ^{12}C during fixation of atmospheric CO_2. This is due primarily to kinetic isotope fractionation by ribulose 1,5-bisphosphate carboxylase-oxygenase (RuBisCO), the enzyme responsible for carbon fixation in C_3 plants (plants in which the first product of photosynthesis is a three-carbon acid). RuBisCO results in isotope compositions ($\delta^{13}C_{PDB}$ values) in the range of -20 to $-40‰$ ($\delta^{13}C_{PDB}$ refers to the permil deviation of the $^{13}C/^{12}C$ ratio of a sample relative to that of the "Peedee belemnite" carbonate standard [Craig, 1957]). Thus organic sediments are depleted in ^{13}C ($\delta^{13}C_{PDB} \approx -15$ to $-40‰$) whereas carbonate sediments are somewhat enriched ($\delta^{13}C_{PDB} \approx -5$ to $+5‰$) compared to atmospheric CO_2 ($\delta^{13}C_{PDB} \approx -7‰$). With very few exceptions, organic and carbonate sediments throughout the geologic record fall within these established ranges (Schidlowski et al., 1983; Hayes et al., 1983). This includes samples taken from the 3.3 to 3.6 billion year old Warrawoona Group, suggesting that biological carbon fixation involving an enzyme with kinetic properties similar to RuBisCO was operating at that time.

The most notable exceptions to the isotope composition data involve samples taken from rocks that have undergone high-grade metamorphism, including those of the Isua Supracrustal Belt. At high temperatures partial reequilibration of carbon isotopes can occur, resulting in elevation of the $\delta^{13}C_{PDB}$ value for the organic fraction and lowering of the $\delta^{13}C_{PDB}$ value for the carbonate fraction (Hayes et al., 1983). It has been claimed that this is precisely what happened to the Isua material, accounting for skewed $\delta^{13}C_{PDB}$ values of $-15‰$ for organic carbon and $-5‰$ for sedimentary carbonate (Schidlowski et al., 1983; Schidlowski, 1988). The implication is that life in Isua time was not very different from life represented by the Warrawoona stromatolites and microfossils, only that the Isua evidence now appears to be marginal because of the intervening metamorphism.

Clearly, the case for biological carbon fixation operating 3.8 billion years ago rests on shaky ground. There is no doubt that the Isua rocks underwent high-grade metamorphism. The H/C ratio of Isua samples is below 0.1, and in some cases as low as 0.01, indicating that most of the hydrocarbons have been converted to graphite (Hayes et al., 1983). A reasonable correlation exists between the extent of dehydrogenation and the degree of upward shift in the $\delta^{13}C_{PDB}$ value for organic sediments. However, this correlation has not been established for H/C ratios below 0.1. Thus, it is difficult to say what the upward shift in

$\delta^{13}C_{PDB}$ value for Isua samples should be. At present there is no reliable evidence to indicate that life, let alone DNA/protein life, existed prior to 3.6 billion years ago.

A very different approach, based on examination of tRNA sequences, has been used to estimate the time of the origin of the genetic code (Eigen et al., 1989). Two types of phylogenies can be constructed from the existing tRNA sequence data base. One involves alignment of tRNAs from different species, all tRNAs in the alignment having the same anticodon. The other involves alignment of tRNAs having different anticodons, all tRNAs in the alignment belonging to the same species. The species alignment has its root in the last common ancestor of the major kingdoms; the anticodon alignment has its root at the time of fixation of the genetic code. By comparing the amount of evolutionary divergence along these two lines, it is possible to determine an upper bound for the time since the fixation of the genetic code relative to the time since the divergence of the major kingdoms (Eigen et al., 1989).

Extracting deep-seated phylogenetic information from tRNA sequences requires a statistical method that goes beyond simple distance measurements and considers the relative positions of entire tRNA sequences within ν-dimensional sequence space (ν = the number of positions in the sequence). One compares all possible quartets that can be formed from a set of sequences and obtains an ensemble average of the tendency of members of the quartet to cluster as two pairs versus three together with one outlier versus all four together. If the amount of evolutionary divergence is low, then all members of the quartet will tend to cluster together in sequence space. If the amount of evolutionary divergence is high, then members of the quartet will tend to cluster randomly. The relation between evolutionary divergence and quartet clustering tendency can be expressed quantitatively, providing a measure of relative mutational distance (Eigen et al., 1988).

This method of "statistical geometry in sequence space" has been applied to the tRNA sequence data base in two ways. First, extant sequences were used to determine the total amount of randomization that has occurred since the origin of the genetic code. Second, a consensus sequence was constructed for each anticodon by aligning tRNAs from various species (weighing the major kingdoms equally). These consensus sequences were then used to determine the amount of randomization that occurred between the time of the origin of the genetic code and the time of divergence of the major kingdoms. If one compares the two results, it appears that 35% of randomization occurred prior to the divergence of the major kingdoms. Assuming that the kingdoms diverged 2.5 ± 0.5 billion years ago, the genetic code must be no older than 3.8 ± 0.6 billion years (Eigen et al., 1989).

The weakness in the above argument stems not from the method of sequence analysis but from the uncertainty in trying to fix a time for the divergence of the major kingdoms. Several authors have attempted to calibrate evolutionary distances by tying specific branch points in the phylogenetic tree to known events in the geological record (Kimua and Ohta, 1973; Hori, 1975; Ochman and Wilson, 1987). For example, the presence of substantial levels of molecular oxygen beginning about 1.7 to 2.0 billion years ago (as indicated by the redox state of iron-mineral deposits) must have predated the appearance and rapid diversification of obligate aerobes. However, there is insufficient evidence to date the time of the divergence of the kingdoms, and even the very rough estimate of 2.5 ± 0.5 billion years ago should be viewed with skepticism.

If one summarizes the information presented here, one finds that a current best estimate for the time of existence of the RNA world is a 400 million year interval beginning 4.0 to 4.2 billion years ago and ending 3.6 to 3.8 billion years ago (see Fig. 1). The upper bound is set by the time required for the earth's crust to stabilize and for liquid water to be available. The range from 4.0 to 4.2 billion years ago reflects the possible significance of meteoric impacts that may have frustrated early attempts to initiate biological evolution. The lower bound is set by the earliest fossil evidence of life that closely resembles known DNA/protein–based organisms. The range from 3.6 to 3.8 billion years ago reflects uncertainty about the carbon isotope record and the possible significance of the tRNA sequence data. It is important to note that the aforementioned time interval reflects the time when the RNA world could have existed, but that a substantial portion of this interval may have been taken up by a genetic molecule other than RNA.

Catalytic Function in the RNA World

There are two general categories of catalytic function relevant to the existence of the RNA world. The first includes those activities necessary to establish and maintain an RNA-based evolving system. The second includes those activities that led to the development of a genetic system based on DNA and protein. Here I consider only the first category in an attempt to define the minimum level of biochemical complexity that existed in the RNA world. The second category requires a more extensive treatment that is beyond the scope of this paper. Several excellent reviews have appeared recently that delve into the problem of the origin of the genetic code (Fitch and Upper, 1987; Weiner and Maizels, 1987; Orgel, 1989; Lacey and Staves, 1990).

The primary catalytic task that an RNA-based organism would have faced is the replication of its own genetic material. Specifically, an RNA-dependent RNA polymerase activity would have been required, together with whatever accessory functions were necessary for the polymerase to operate accurately and efficiently on the genomic RNA. In addition, an adequate supply of the four mononucleotides must have been maintained and these mononucleotides must have been converted to a form that could be utilized by the polymerase. Beyond that, a number of evolutionary innovations could be imagined, particularly if those innovations enhanced the rate or fidelity of replication or helped to ensure a stable supply of activated starting materials.

Several authors have pointed to the catalytic function of a group I ribozyme as a model for a primordial replicase (Sharp, 1985; Lewin, 1986; Gilbert, 1986; Cech, 1986; Szostak, 1986). Group I ribozymes, of which the *Tetrahymena* ribozyme is the best-studied example, catalyze sequence-specific phosphoester trans-

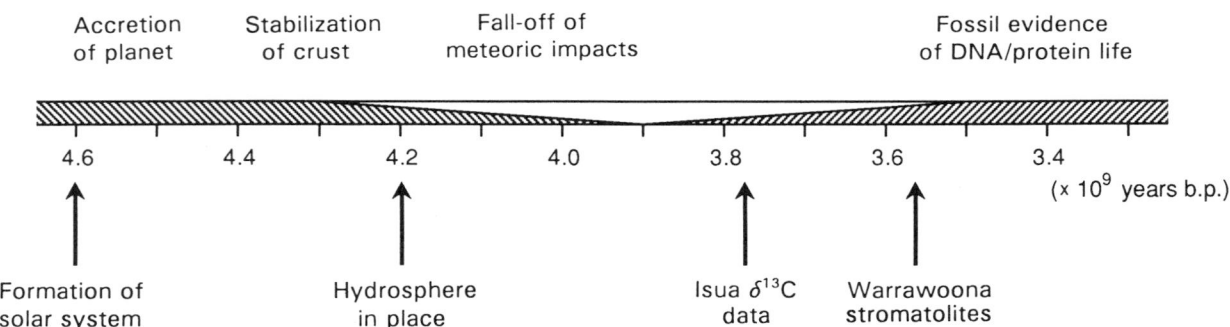

Figure 1.   Time frame for the existence of the RNA world.

The hatched line at the left indicates the time during which geophysical conditions were incompatible with the solution chemistry of RNA. The hatched line at the right indicates the time for which fossil evidence of DNA/protein life is available. The time window for the existence of the RNA world (shown in white) lies between approximately 4.0–4.2 and 3.6–3.8 billion years ago.

fer reactions involving nucleic acid substrates (Zaug and Cech, 1985; Kay and Inoue, 1987). The ribozyme contains an internal template region that confers specificity for a complementary substrate (Been and Cech, 1986; Waring et al., 1986). On one hand, the template can be externalized and generalized to contain virtually any sequence (Doudna and Szostak, 1989). Thus the molecule has many of the characteristics of an RNA-dependent RNA polymerase including, most important, the ability to catalyze nucleotidyl-transfer reactions in a template-directed manner. On the other hand, the ribozyme is unable to move processively along the template, does not operate on template regions that adopt stable self-structure, and is not expected to achieve >99% fidelity per residue as would be required to produce additional copies of itself. One should not take the comparison between a group I ribozyme and an RNA replicase too literally.

Another useful model for a primordial RNA replicase is T7 RNA polymerase. This enzyme, although evolved to function as a DNA-dependent RNA polymerase, can operate as an RNA replicase under special circumstances (Konarska and Sharp, 1989). It is the only "RNA replicase" for which a crystal structure is available (Sousa et al., 1989). T7 RNA polymerase has solved the problem of processivity, apparently by providing a channel for the sliding template and by relying on the energy of hydrolysis of incoming nucleotide triphosphates to drive translocation in the forward direction. The template strand and product strand are thought to exit the polymerase at two distinct sites (Chung et al., 1990). This allows the product strand to adopt its own self-structure rather than forming a complete duplex with the template strand. The error rate of T7 RNA polymerase is not known, but by analogy to other polymerases that lack an exonuclease function it is likely to be about 10^{-4} per position. As our knowledge about the crystal structure of T7 RNA polymerase becomes refined and as the structures of various small RNAs become available, it will be interesting to consider how nucleotides could substitute for amino acids within the active site of the polymerase.

With regard to the characteristics of an RNA replicase in the RNA world, the level of catalytic sophistication depends on the amount of evolutionary progress that was made prior to the emergence of RNA. At a minimum, one could imagine that life before RNA was based on a genetic molecule that replicated spontaneously and that the primary catalytic requirement was for an unwinding activity that helped to overcome template self-structure. Thus the first ribozyme may have been an unwinding enzyme rather than an RNA replicase. At the other extreme, one could imagine that life before RNA utilized a sophisticated replicase enzyme to copy an RNA-like genetic

molecule and that this replicase function was carried over intact to the RNA world. If the RNA world began with an unwinding enzyme rather than a full-fledged replicase, then there must have been a way to bias the synthesis of activated nucleotides to favor utilization of D-ribose over all other sugars. As an RNA replicase developed, the burden of maintaining sugar specificity would have been shared between the primitive biosynthetic apparatus and an increasingly specific replicase enzyme.

Life in the RNA world would have been under intense selection pressure to improve the fidelity of replication. Beginning with a saturated solution of pure, activated D mononucleotides, nonenzymatic template-directed polymerization proceeds with an error rate of about 10^{-2} per position (Inoue and Orgel, 1983). Under plausible prebiotic conditions the fidelity of replication would probably be considerably lower. Thus genome size would have been limited to considerably less than 100 nucleotides. If the fidelity of replication could be improved, either by enriching the local environment with the proper starting materials or by making template-directed synthesis less tolerant of base mismatch, then genome size could increase. Increased genome size would allow development of more sophisticated catalysts, which in turn would lead to further enhancement of replication fidelity and a further increase in genome size. In this way an RNA-based evolving system could bootstrap its way toward greater metabolic complexity.

Enriching the local environment with the four activated mononucleotides need not have required a complex biosynthetic apparatus. As mentioned previously, phosphorylation of glycoaldehyde strongly biases the condensation reaction between glycoaldehyde and formaldehyde to favor synthesis of ribose compared to other sugars (Müller et al., 1990). Thus a glycoaldehyde phosphorylase activity would have been selectively advantageous in the RNA world. Similarly, there may have been other ways to bias the spontaneous chemistry of the primitive environment to favor the production of materials required for nucleotide synthesis. As genome size increased and metabolism became more sophisticated, RNA organisms would have come to depend less on the chemistry of the primitive environment and more on their own catalytic abilities.

If an RNA unwinding activity existed and an adequate supply of the four mononucleotides was available, then the limiting factor on replication fidelity would have been the template-directed polymerization reaction itself. Modern RNA replicases operate with an error rate that is about 10^2-fold lower than occurs during nonenzymatic template-directed polymerization. Thus protein replicases can support an RNA genome consisting of 10^3 to 10^4 residues (Eigen, 1971). It is not known how modern RNA replicases achieve this enhanced fidelity, let alone whether a ribozyme could operate in a similar manner. In principle, ribozymes could enhance the fidelity of template-directed polymerization in several ways, for example by constraining the replicating fork to a well-defined duplex structure that is intolerant of base mismatch or by specifically destabilizing G · U wobble pairs. Fidelity could be enhanced further by the addition of an exonuclease domain that provides a proofreading function. No such proofreading mechanism exists among the known RNA replicases, but 3'-exoribonuclease activity is within the demonstrated abilities of catalytic RNA (Zaug et al., 1986). It is conceivable that error rates as low as 10^{-5} to 10^{-6} per position were attained in the RNA world, permitting genomes the size of the smallest bacteria.

Life in the RNA world would have faced a number of challenges. Two top priorities would have been to protect the genetic material from spontaneous hydrolysis and to ensure that all of the components of the genetic system were localized within a common microenvironment. An answer to both of these challenges, possibly arrived at prior to the onset of the RNA world, would have been to compartmentalize the system within a lipid bilayer (Morowitz et al., 1988) or some sort of organizing center (Gibson and Lamond, 1990). Another important challenge would have been to make use of metal ions, amino acids, and other components of the primitive environment to broaden the range of catalytic function. At some point RNA organisms began to dabble in the use of short peptides, leading eventually to the development of protein synthesis. Other "experiments" led to the discovery of DNA, which provided a more stable repository for genetic information. By 3.6 to 3.8 billion years ago all of these events had come to pass; the RNA world had fallen and the DNA/protein world had risen in its place.

REFERENCES

Alberts BM (1986): The function of the hereditary materials: biological catalysts reflect the cell's evolutionary history. Am Zoologist 26:781-796

Allaart JH (1976): The pre-3760 Myr old supracrustal rocks of the Isua area, central West Greenland and the associated occurrence of quartz banded ironstone, in Windley BF (ed) The Early History of the Earth, London, Wiley, pp 177-189

Awramik SM, Schopf JW, Walter MR (1983): Filamentous fossil bacteria 3.5 × 10⁹ years old from the Archaen of Western Australia. Precambrian Res 20:357-374

Been MD, Cech TR (1986): One binding site determines sequence specificity of Tetrahymena pre-rRNA self-splicing, trans-splicing, and RNA enzyme activity. Cell 47:207-216

Benner SA, Ellington AD, Tauer A (1989): Modern metabolism as a palimpsest of the RNA world. Proc Natl Acad Sci USA 86:7054-7058

Blake TS, McNaughton NJ (1984): A geochronological framework for the Pilbara region, in Muhling JR, Groves DI, Blake TS (eds)

Archaen and Proterozoic Basins. Univ. Western Australia, Geology Dept. and Univ. Extension, p 1-22

Byerly GR, Lowe DR, Walsh MM (1986): Stromatolites from the 3,300-3,500 Myr Swaziland Supergroup, Barberton Mountain Land, South Africa. Nature 319:489-491

Cameron AGW (1978): Physics of the primitive solar accretion disk. Moon and Planets 18:5-40

Cameron AGW, Truran JW (1977): The supernova trigger for the formation of the solar system. Icarus 30:447-461

Cech TR (1986): A model for the RNA-catalyzed replication of RNA. Proc Natl Acad Sci USA 83:4360-4363

Chung YJ, Sousa R, Rose JP, Lafer E, Wang BC (1990): Crystallographic structure of phage T7 RNA polymerase at a resolution of 4.0 A, in Wu FY-H, Wu C-W (eds) Structure and Function of Nucleic Acids and Proteins, New York, Raven Press, pp 55-59

Chyba CF (1987): The cometary contribution to the oceans of the primitive earth. Nature 330:632-635

Craig H (1957): Isotopic standards for carbon and oxygen and correction factors for mass spectrophotometric analysis of carbon dioxide. Geochim Cosmochim Acta 12:133-149

Crick FHC (1968): The origin of the genetic code. J Mol Biol 38:367-379

Darnell JE, Doolittle WF (1986): Speculations on the early course of evolution. Proc Natl Acad Sci USA 83:1271-1275

Doudna JA, Szostak JW (1989): RNA-catalysed synthesis of complementary-strand RNA. Nature 339:519-522

Eigen M (1971): Selforganization of matter and the evolution of biological macromolecules. Naturwissenschaften 58:465-523

Eigen M, Lindemann BF, Tietze M, Winkler-Oswatitsch R, Dress A, von Haeseler A (1989): How old is the genetic code? Statistical geometry of tRNA provides an answer. Science 244:673-679

Eigen M, Winkler-Oswatitsch R, Dress A (1988): Statistical geometry in sequence space: a method of quantitative comparative sequence analysis. Proc Natl Acad Sci USA 85:5913-5917

Fitch WM, Upper K (1987): The phylogeny of tRNA sequences provides evidence for ambiguity reduction in the origin of the genetic code. Cold Spring Harbor Symp Quant Biol 52:759-767

Froude DO, Ireland TR, Kinny PD, Williams IS, Compston W (1983): Ion microprobe identification of 4,100-4,200 Myr-old terrestrial zircons. Nature 304:616-618

Fuller WD, Sanchez RA, Orgel LE (1972): Studies in prebiotic synthesis. VI. Synthesis of purine nucleosides. J Mol Biol 67:25-33

Gibson TJ, Lamond AI (1990): Metabolic complexity in the RNA world and implications for the origin of protein synthesis. J Mol Evol 30:7-15

Gilbert W (1986): The RNA world. Nature 319:618

Glickson A (1986): The oldest age determined for Pilbara greenstones. Bur Mineral Resources Res Newslett 4:7

Goldreich P, Ward WR (1973): The formation of planetesimals. Astrophys J 183:1051-1061

Hayes JM, Kaplan IR, Wedeking KW (1983): Precambrian organic geochemistry, preservation of the record, in Schopf JW (ed) Earth's Earliest Biosphere, Princeton University Press, p 93-134

Hori H (1975): Evolution of 5S RNA. J Mol Evol 7:75-86

Inoue T, Orgel LE (1983): A nonenzymatic RNA polymerase model. Science 219:859-862

Joyce GF (1989): RNA evolution and the origins of life. Nature 338:217-224

Joyce GF, Orgel LE (1986): Non-enzymic template-directed synthesis on RNA random copolymers: poly(C,G) templates. J Mol Biol 188:433-441

Joyce GF, Schwartz AW, Miller SL, Orgel LE (1987): The case for an ancestral genetic system involving simple analogues of the nucleotides. Proc Natl Acad Sci USA 84:4398-4402

Joyce GF, Visser GM, van Boeckel CAA, van Boom JH, Orgel LE, van Westrenen J (1984): Chiral selection in poly(C)-directed synthesis of oligo(G). Nature 310:602-604

Kasting JF (1988): Runaway and moist greenhouse atmospheres and the evolution of Earth and Venus. Icarus 74:472-494

Kay PS, Inoue T (1987): Catalysis of splicing-related reactions between dinucleotides by a ribozyme. Nature 327:343-346

Kimua M, Ohta T (1973): Eukaryotes-prokaryotes divergence estimated by 5S ribosomal RNA sequences. Nature New Biol 243:199-200

Konarska MM, Sharp PA (1989): Replication of RNA by the DNA-dependent RNA polymerase of phage T7. Cell 57:423-431

Lacey JC, Staves MP (1990): Was there a universal tRNA before specialized tRNAs came into existence? Orig Life 20:303-308

Lewin R (1986): RNA catalysis gives fresh perspective on the origin of life. Science 231:545-546

Lohrmann R, Orgel LE (1968): Prebiotic synthesis: Phosphorylation in aqueous solution. Science 161:64-66

Lohrmann R, Orgel LE (1971): Urea-inorganic phosphate mixtures as prebiotic phosphorylating agents. Science 171:490-494

Lowe DR (1980): Stromatolites 3,400-Myr old from the Archean of Western Australia. Nature 284:441-443

Maher K, Stevenson DJ (1988): Impact frustration of the origin of life. Nature 331:612-614

Morowitz HJ, Heinz B, Deamer D (1988): The chemical logic of a minimum protocell. Orig Life 18:281-287

Müller D, Pitsch S, Kittaka A, Wagner E, Wintner CE, Eschenmoser A (1990): Chemie von α-aminonitrilen. Aldomerisierung von Glykolaldehydphosphat zu racemischen hexose-2,4,6-triphosphaten und (in gegenwart von formaldehyd) racemischen pentose-2,4-diphosphaten: und rac.-ribose-2,4-diphosphat sind die reaktionshauptprodukte. Helv Chim Acta 73:1410-1468

Ochman H, Wilson AC (1987): Evolution in bacteria: evidence for a universal substitution rate in cellular genomes. J Mol Evol 26:74-86

Orgel LE (1968): Evolution of the genetic apparatus. J Mol Biol 38:381-393

Orgel LE (1986): RNA catalysis and the origins of life. J Theor Biol 123:127-149

Orgel LE (1989a): The origin of polynucleotide-directed protein synthesis. J Mol Evol 29:465-474

Orgel LE (1989b): Was RNA the first genetic polymer?, in Grunberg-Manago M, Clark BFC, Zachau HG (eds) Evolutionary Tinkering in Gene Expression. London, Plenum, pp 215-224

Pace NR, Marsh TL (1985): RNA catalysis and the origin of life. Orig Life 16:97-116

Papanastassiou DA, Wasserburg GJ (1971): Lunar chronology and evolution for Rb-Sr studies of Apollo 11 and 12 samples. Earth Planet Sci Lett 11:37-62

Patterson CC (1956): Age of meteorites and the earth. Geochim Cosmochim Acta 10:230-237

Pollack JB, Yung YL (1980): Origin and evolution of planetary atmospheres. Annu Rev Earth Planet Sci 8:425-487

Schidlowski M (1988): A 3,800-million-year isotopic record of life from carbon in sedimentary rocks. Nature 333:313-318

Schidlowski M, Appel PWU, Eichmann R, Junge CE (1979): Carbon isotope geochemistry of the 3.7×10^9 yr old Isua sediments, West

Greenland: implications for the Archean carbon and oxygen cycles. Geochim Cosmochim Acta 43:189-199

Schidlowski M, Hayes JM, Kaplan IR (1983): Isotopic inferences of ancient biochemistries: carbon, sulfur, hydrogen, and nitrogen, in Schopf JW (ed) Earth's Earliest Biosphere. Princeton University Press, pp 149-186

Schopf JW (in press): Paleobiology of the Archean, in Schopf JW, Klein C (eds) The Proterozoic Biosphere. Cambridge University Press

Schopf JW, Packer BM (1987): Early Archean (3.3-billion to 3.5 billion-year old) microfossils from Warrawoona Group, Australia. Science 237:70-73

Schopf JW, Walter MR (1983): Archaen microfossils: new evidence of ancient microbes, in Schopf JW (ed) Earth's Earliest Biosphere. Princeton University Press, p 214-239

Shapiro R (1984): The improbability of prebiotic nucleic acid synthesis. Orig Life 14:565-570

Sharp PA (1985): On the origin of RNA splicing and introns. Cell 42:397-400

Shaw GH (1978): Effects of core formation. Phys Earth Planet Interiors 16:361-369

Sousa R, Rose JP, Chung YJ, Lafer EM, Wang B-C (1989): Single crystals of bacteriophage T7 RNA polymerase. Proteins 5:266-270

Stevenson DJ (1980): Core formation dynamics and primordial planetary dynamos. Proc 11th Lunar Sci Conf, pp 1091-1093

Stevenson DJ (1983): The nature of the earth prior to the oldest known rock record: the Hadean earth, in Schopf JW (ed) Earth's Earliest Biosphere. Princeton University Press, pp 32-40

Szostak JW (1986): Enzymatic activity of the conserved core of a group I self-splicing intron. Nature 322:83-86

Walker JCG (1977): Evolution of the Atmosphere. New York, Macmillan, Ch 5.

Walter MR (1976): Introduction, in Walter MR (ed) Stromatolites. Amsterdam, Elsevier, p 1-3

Walter MR (1983): Archaen stromatolites: evidence of the earth's earliest benthos, in Schopf JW (ed) Earth's Earliest Biosphere. Princeton University Press, p 187-213

Walter MR, Buick R, Dunlop JSR (1980): Stromatolites 3,400-3,500 Myr old from the North Pole area, Western Australia. Nature 284:443-445

Waring RB, Towner P, Minter SJ, Davies RW (1986): Splice-site selection by a self-splicing RNA of Tetrahymena. Nature 321:133-139

Weiner AM, Maizels N (1987): 3' terminal tRNA-like structures tag genomic RNA molecules for replication: implications for the origin of protein synthesis. Proc Natl Acad Sci USA 84:7383-7387

Weiner AM, Maizels N (in press): The genomic tag model for the origin of protein synthesis: further evidence from the molecular fossil record. Proceedings of the International Symposium on Evolution of Life, Tokyo, 1990. Berlin, Springer-Verlag, pp 51-66

Woese C (1967): The Genetic Code. New York, Harper & Row, Ch 7

Zahnle KJ, Kasting JF, Pollack JB (1988): Evolution of a steam atmosphere during Earth's accretion. Icarus 74:62-97

Zaug AJ, Been MD, Cech TR (1986): The Tetrahymena ribozyme acts like an RNA restriction endonuclease. Nature 324:429-433

Zaug AJ, Cech TR (1985): Oligomerization of intervening sequence RNA molecules in the absence of proteins. Science 229:1060-1064

REFERENCES
ON THE
ORIGINS OF LIFE

REFERENCES ON THE ORIGIN OF LIFE

The following references deal with major concepts of biogenesis. They are listed alphabetically by author and are grouped according to the six sections of this book. An annual supplemental bibliography to the origins-of-life literature was published from 1970 through 1984 in the pages of *Origins of Life and Evolution of the Biosphere*, the journal of the International Society for the Study of the Origin of Life (ISSOL). (See Roy and Powers, 1990, below for a cumulative keyword subject index to this bibliography.)

INTRODUCTION

Bernal, J. D. (1949) The physical basis of life. *Proc. Phys. Soc. A* **62**:537-558.

Bernal, J. D. (1951) *The Physical Basis of Life*. London: Routledge and Kegan Paul.

Bernal, J. D. (1965) Molecular matrices for living systems. In: *The Origins of Prebiological Systems and of Their Molecular Matrices*, Proceedings of the 2nd International Symposium on the Origin of Life on the Earth, Wakulla Springs, Florida, 1963, S. W. Fox, ed., New York: Academic Press, pp. 65-88.

Bernal, J. D. (1967) *The Origin of Life*. London: Weidenfeld and Nicolson.

Bertalanffy, L. von (1952) *Problems of Life*. London: Watts.

Bertalanffy, L. von (1969) Chance or law. In: *Beyond Reductionism – New Perspectives in the Life Sciences: The Alpbach Symposium 1968*, A. Koestler and J. R. Smythies, eds., London: Hutchinson, pp. 56-84.

Bertalanffy, L. von (1973) *General System Theory – Foundations, Development, Applications*, revised ed. New York: G. Braziller.

Brack, A. (1990) The origin of life on Earth. *Grana* **30**:505-509.

Breder, C. M. (1942) A consideration of evolutionary hypotheses in reference to the origin of life. *Zoologica: New York Zoological Society* **27**:131-143.

Buvet, R. and Ponnamperuma, C. (eds.)(1971) *Chemical Evolution and the Origin of Life*, Proceedings of the 3rd International Symposium on the Origin of Life on the Earth, Pont-à-Mousson, France, 1970. Amsterdam: North-Holland.

Cairns-Smith, A. G. (1985) *Seven Clues to the Origin of Life: A Scientific Detective Story*. Cambridge, U.K.: The Cambridge University Press.

Clark, F. and Synge, A. (eds./trans.) (1959) *The Origin of Life on the Earth*, Proceedings of the 1st International Symposium on the Origin of Life on the Earth, Moscow, 1957. New York: Pergamon Press.

Crick, F. H. C. (1981) *Life Itself: Its Origin and Nature*. New York: Simon and Schuster.

Day, W. (1979) *Genesis on Planet Earth: The Search For Life's Beginning*, 2nd ed. New Haven: Yale University Press.

de Duve, C. (1991) *Blueprint For a Cell: The Nature and Origin of Life*. Burlington, NC: N. Patterson.

Dose, K., Fox, S. W., Deborin, G. A. and Pavlovskaya, T. E. (1974) *The Origin of Life and Evolutionary Biochemistry*. New York: Plenum Press.

Dyson, F. J. (1985) *Origins of Life*. Cambridge, U. K.: The Cambridge University Press.

Farley, J. (1977) *The Spontaneous Generation Controversy From Descartes to Oparin*. Baltimore: Johns Hopkins University Press.

Ferris, J. P. and Usher, D. (1988) Origins of life. In: *Biochemistry*, G. Zubay, ed., New York: Macmillan, pp. 1191-1241.

Fleischaker, G. R. (1990) Origins of life: An operational definition. *Orig. Life Evol. Biosphere* **20**:127-137.

Fleischaker, G. R. (1991) The myth of the putative "organism." *Uroboros* **1** (2):23-43.

Folsome, C. E. (1979) *The Origin of Life: A Warm Little Pond*. San Francisco: W. H. Freeman.

Fox, S. W. (ed.) (1965) *The Origins of Prebiological Systems and of Their Molecular Matrices*, Proceedings of the 2nd International Symposium on the Origin of Life on the Earth, Wakulla Springs, Florida, 1963. New York: Academic Press.

Fox, S. W. and Dose, K. (1972) *Molecular Evolution and the Origin of Life*. San Francisco: W. H. Freeman.

Haldane, J. B. S. (1929) The origin of life. Reprinted in: J. D. Bernal, *The Origin of Life* (1967). London: Weidenfeld and Nicolson, pp. 242-249.

Haldane, J. B. S. (1954) Origins of life. *New Biology* 16:12-27.

Hartmann, H., Lawless, J. G. and Morrison, P. (1987) *Search for the Universal Ancestors: The Origin of Life*. Palo Alto, CA: Blackwell Scientific Publications.

Horgan, J. (1991) In the beginning. *Sci. Am.* 261 (February):116-125.

Kamminga, H. (1988) The problem of the origin of life in the context of developments in biology. *Orig. Life Evol. Biosphere* 18:1-11.

Lazcano, A. (in press) The transition from non-living to living. In: *Early Life on Earth: A Nobel Symposium*, S. Bengtson, ed., New York: Columbia University Press.

Miller, S. L. (1974) The first laboratory synthesis of organic compounds under primitive Earth conditions. In: *The Heritage of Copernicus: Theories "Pleasing to the Mind,"* J. Neyman, ed., Cambridge, MA: The MIT Press, pp. 228-242.

Miller, S. L. and Orgel, L. E. (1974) *The Origins of Life on the Earth*. Englewood Cliffs, NJ: Prentice -Hall.

Morowitz, H. J. (1992) *Beginnings of Cellular Life: Metabolism Recapitulates Biogenesis*. New Haven, CT: Yale University Press.

Muller, H. J. (1929) The gene as the basis of life. *Proc. Int. Congr. of Plant Sci.* 1:897-921.

Oparin, A. I. (1924) The Origin of Life [*Proiskhozhdenie zhizny*. Moscow: Izd. Moskovshii Rabochii]. English translation in: J. D. Bernal, *The Origin of Life* (1967). London: Weidenfeld and Nicolson, pp. 199-234.

Oparin, A. I. (1938) *Origin of Life*, S. Morgulis, trans., New York: Macmillan.

Oparin, A. I. (1957) *The Origin of Life on the Earth*, 3rd ed., A. Synge, trans., New York: Academic Press.

Oparin, A. I. (1968) *Genesis and Evolutionary Development of Life*. New York: Academic Press.

Oparin, A. I. (1972) The appearance of life in the universe. In: *Exobiology*, C. Ponnamperuma, ed., Amsterdam: North-Holland, pp. 1-15.

Oró, J. (1976) Prebiological chemistry and the origin of life: A personal account. In: *Reflections on Biochemistry*, A. Kornberg, L. Cornudella, B. L. Horecker and J. Oró, eds., Oxford: Pergamon Press.

Oró, J., Miller, S. L. and Lazcano, A. (1990) The origin and early evolution of life on Earth. *Annu. Rev. Earth Planet. Sci.* 18:317-356.

Pasteur, L. (1864) (lecture to the Sorbonne on spontaneous generation). In: *Oeuvres de Pasteur*, Volume 2 (1922), S. Pasteur Valley-Padot, ed., Paris: Masson & Cie.

Perret, J. (1952) Biochemistry and bacteria. *New Biology* 12:68-96.

Pirie, N. W. (1953) Ideas and assumptions about the origin of life. *Discovery* 14:238-242.

Roy, A. C. and Powers, J. V. (1990) Chemical evolution and the origin of life 1970-1986: A cumulative keyword subject index. *Orig. Life Evol. Biosphere* 20:425-456.

Shapiro, R. (1985) *Origins: The Possibilities of Science for the Genesis of Life on Earth*. New York: Summit Books.

Smith, T. and Morowitz, H. J. (1982) Between history and physics. *J. Mol. Evol.* 18:265-282.

Swenson, R. (1989) The Earth as an incommensurate field at the geo-cosmic interface: Fundamentals to a theory of general evolution. In: *Geo-Cosmic Relations: The Earth and Its Macro-Environment*, G. J. M. Tomassen, W. de Graaf, A. A. Knoop and R. Hengeveld, eds., Wageningen, The Netherlands: Pudoc Science Publishers, pp. 299-306.

Swenson, R. and Turvey, M. T. (1991) Thermodynamic reasons for perception-action cycles. *Ecological Psychology* 3 (4):317-348.

Troland, L. T. (1917) Biological enigmas and the theory of enzyme action. *The American Naturalist* 51:321-350.

Urey, H. C. (1952) On the early chemical history of the earth and the origin of life. *Proc. Natl. Acad. Sci. USA* 38:351-363.

Vernadsky, V. I. (1986) *The Biosphere* [abridged, based on the original (1929) French edition]. London: Synergetic Press.

Weiss, P. A. (1969) The living system: Determinism stratified. In: *Beyond Reductionism – New Perspectives in the Life Sciences: The Alpbach Symposium 1968*, A. Koestler and J. R. Smythies, eds., London: Hutchinson, pp. 3-55.

Weiss, P. A. (1971) *The Science of Life: The Living System*. New York: Futura.

Weiss, P. A. (1978) Causality: Linear or systemic? In: *Psychology and Biology of Language and Thought – Essays in Honor of Eric Lenneberg*, G. A. Miller and E. Lenneberg, eds., New York: Academic Press, pp. 13-26.

THE EARLY-EARTH ENVIRONMENT

Abelson, P. H. (1966) Chemical events on the primitive earth. *Proc. Natl. Acad. Sci. USA* **55**:1365-1372.

Arrhenius, G. (1987) The first 800 million years: Environmental models for early Earth. *Earth, Moon and Planets* **37**:187-199.

Bernal, J. D. (1951) *The Physical Basis of Life*. London: Routledge and Kegan Paul.

Bernal, J. D. (1967) *The Origin of Life*. London: Weidenfeld and Nicolson.

Boss, A. P. (1986) The origin of the moon. *Science* **231**:341-345.

Cairns-Smith, A. G. (1978) Precambrian solution photochemistry, inverse segregation and banded iron formations. *Nature* **276**:807-808.

Calvin, M. (1969) *Chemical Evolution: Molecular Evolution Towards the Origin of Living Systems on the Earth and Elsewhere*. New York: Oxford University Press.

Canuto, V. M., Levine, J. S., Augustsson, T. R. and Imhoff, C. L. (1982) UV radiation from the young sun and oxygen and ozone levels in the prebiological palaeoatmosphere. *Nature* **296**:816-820.

Chamberlin, T. C. and Chamberlin, R. T. (1908) Early terrestrial conditions that may have favored organic synthesis. *Science* **28**:897-911.

Chang, S. (1982) Prebiotic organic matter: Possible pathways for synthesis in a geological context. *Phys. Earth Planet. Interiors* **29**:261-280.

Chang, S. (1988) Planetary environments and the conditions of life. *Phil. Trans. R. Soc. Lond. A* **325**:601-610.

Chyba, C. F. (1987) Cometary contribution to the oceans of the primitive Earth. *Nature* **330**:632-635.

Chyba, C. F. (1990) Impact delivery and erosion of planetary oceans in the early inner Solar System. *Nature* **343**:129-133.

Chyba, C. F. and Sagan, C. (1992) Endogenous production, exogenous delivery, and impact-shock synthesis of organic molecules: An inventory for the origins of life. *Nature* **355**:125-132.

Cloud, P. (1972) A working model of the primitive Earth. *Amer. J. Sci.* **272**:537-548.

Cloud, P. (1974) Evolution of ecosystems. *Amer. Scientist* **62**:54-56.

Cloud, P. (1976) Beginnings of biospheric evolution and their biogeochemical consequences. *Paleobiology* **2**:351-387.

Cloud, P. (1988) *Oasis in Space: Earth History From the Beginning*. New York: Norton.

Corliss, J. B., Baross, J. A. and Hoffmann, S. E. (1981) An hypothesis concerning the relationship between submarine hot springs and the origin of life on Earth. *Oceanol. Acta* **4** (suppl.):59-69.

Cronin, J. R., Pizzarello, S. and Cruikshank, D. P. (1988) Organic matter in carbonaceous chondrites, planetary satellites, asteroids and comets. In: *Meteorites and the Early Solar System*, J. F. Kerridge and M.S. Matthews, eds., Tucson: University of Arizona Press.

Delsemme, A.H. (1984) The cometary connection with prebiotic chemistry. *Origins of Life* **14**:51-60.

Haldane, J. B. S. (1929) The origin of life. Reprinted in: J. D. Bernal, *The Origin of Life* (1967). London: Weidenfeld and Nicolson, pp. 242-249.

Hargraves, R. B. (1976) Precambrian geologic history. *Science* **193**:363-371.

Hayatsu, R. and Anders, E. (1981) Organic compounds in meteorites and their origins. *Top. Curr. Chem.* **99**:1-37.

Hayes, J. M. (1967) Organic constituents of meteorites – a review. *Geochim. Cosmochim. Acta* **31**:1395-1440.

Holland, H. D. (1978) *The Chemistry of the Atmosphere and Oceans.* New York: Wiley.

Holland, H. D. (1984) *The Chemical Evolution of the Atmosphere and Oceans.* Princeton, NJ: Princeton University Press.

Kahru, J. and Epstein, S. (1986) The implication of the oxygen isotope records in coexisting cherts and phosphates. *Geochim. Cosmochim. Acta* **50**:1745-1756.

Kasting, J. F. (1987) Theoretical constraints on oxygen and carbon dioxide concentrations in the precambrian atmosphere. *Precambr. Res.* **34**:205-229.

Kasting, J. F. (1990) Bolide impacts and the oxidation state of carbon in the Earth's early atmosphere. *Orig. Life Evol. Biosphere* **20**:199-231.

Kasting, J. F. and Ackerman, T. F. (1986) Climatic consequences of very high carbon dioxide levels in the Earth's early atmosphere. *Science* **234**:1383-1385.

Kasting, J. F., Toon, O. B. and Pollack, J. B. (1988) How climate evolved on the terrestrial planets. *Sci. Am.* **258** (February):90-97.

Kuma, K., Paplawsky, B., Gedulin, B. and Arrhenius, G. (1989) Mixed-valence hydroxides as bioorganic host minerals. *Orig. Life Evol. Biosphere* **19**:573-602.

Lazcano, A. and Oró, J. (1981) Cometary material and the origin of life on Earth. In: *Comets and the Origin of Life*, C. Ponnamperuma, ed., Dordrecht, Holland: D. Reidel, pp. 191-225.

Levine, J. S. (1982) The photochemistry of the paleoatmosphere. *J. Mol. Evol.* **18**:161-172.

Levine, J. S. (1985) The photochemistry of the early atmosphere. In: *The Photochemistry of Atmospheres: Earth, The Other Planets, and Comets*, J. S. Levine, ed., New York: Academic Pres, pp. 3-38.

Levine, J. S., Augustsson, T. R. and Natarajan, M. (1982) The prebiological paleoatmosphere: Stability and composition. *Origins of Life* **12**:245-259.

Maher, K. A. and Stevenson, D. J. (1988) Impact frustration of the origin of life. *Nature* **331**:612-614.

Miller, S. L., Urey, H. C. and Oró, J. (1976) Origin of organic compounds on the primitive Earth and in meteorites. *J. Mol. Evol.* **9**:59-72.

Mueller, G. (1953) The properties and theories of genesis of the carbonaceous complex within the Cold Bokkeveld meteorite. *Geochim. Geophys. Acta* **4**:1-10.

Nisbet, E. G. (1985) The geological setting of the earliest life forms. *J. Mol. Evol.* **21**:289-298.

Nisbet, E. G. (1987) *The Young Earth: An Introduction to Archaean Geology.* Winchester, MA: Allen and Unwin.

Oberbeck, V. R. and Fogleman, G. (1988) Estimates of the maximum time required to originate life. *Orig. Life Evol. Biosphere* **19**:549-560.

Oparin, A. I. (1924) The Origin of Life [*Proiskhozhdenie zhizny.* Moscow: Izd. Moskovshii Rabochii]. English translation in: J. D. Bernal, *The Origin of Life* (1967). London: Weidenfeld and Nicolson, pp. 199-234.

Oró, J. (1961) Comets and the formation of biochemical compounds on the primitive Earth. *Nature* **190**:389-390.

Oró, J. (1965) Stages and mechanisms of prebiological organic synthesis. In: *The Origins of Prebiological Systems and of Their Molecular Matrices*, S. W. Fox, ed., New York: Academic Press, pp. 137-171.

Pollack, J. B. and Yung, Y. L. (1980) Origin and evolution of planetary atmospheres. *Annu. Rev. Earth Planet. Sci.* **8**:425-487.

Rubey, W. W. (1951) Geological history of sea water – an attempt to state the problem. *Geol. Soc. Am. Bull.* **62**:1111-1148.

Sagan, C. and Mullen, G. (1972) Earth and Mars: Evolution of atmospheres and surface temperatures. *Science* **177**:52-56.

Schopf, J. W. (ed.) (1983) *Earth's Earliest Biosphere: Its Origin and Evolution.* Princeton, NJ: Princeton University Press.

Sleep, N. H., Zahnle, K., Kasting, J. F. and Morowitz, H. J. (1989) Annihilation of ecosystems by large asteroid impacts on the early Earth. *Nature* **342**:139-142.

Swenson, R. (1991) End-directed physics and evolutionary ordering: Obviating the problem of the population of one. In: *Cybernetics of Complex Systems: Self-Organization, Evolution, and Social Change,* F. Geyer, ed, Salinas, CA: Intersystems Publications, pp. 41-59.

Swenson, R. and Turvey, M. T. (1991) Thermodynamic reasons for perception-action cycles. *Ecological Psychology* **3** (4):317-348.

Towe, K. M. (1981) Environmental conditions surrounding the origin and early archean evolution of life: A hypothesis. *Precambr. Res.* **16**:1-10.

Urey, H. C. (1952) On the early chemical history of the Earth and the origin of life. *Proc. Natl. Acad. Sci. USA* **38**:351-363.

Urey, H. C. (1959) Primitive planetary atmospheres and the origin of life. In: *The Origin of Life on the Earth,* F. Clark and A. Synge, eds./trans., New York: Pergamon Press, pp. 16-22.

Waldrop, M. M. (1990) Goodbye to the warm little pond? *Science* **250**:1078-1080.

Walker, J. C. G. (1983) Possible limits on the composition of the Archaean ocean. *Nature* **302**:518-520.

Walker, J. C. G. (1986) *Earth History.* Boston: Jones and Bartlett Publishers.

Wetherill, G. W. (1985) Occurrence of giant impacts during the growth of the terrestrial planets. *Science* **228**:877-879.

Wood, J. A. and Chang, S. (1985) *The Cosmic History of the Biogenic Elements and Compounds.* Washington, DC: NASA Publication 476, Government Printing Office.

PREBIOTIC CHEMISTRY

Allamandola, L. J., Sandford, S. A. and Wopenka, B. (1987) Interstellar polycyclic aromatic hydrocarbons and carbon in interplanetary dust particles and meteorites. *Science* **237**:56-59.

Anders, E. (1989) Pre-biotic organic matter from comets and asteroids. *Nature* **342**:255-257.

Anders, E., Hayatsu, R. and Studier, M. H. (1973) Organic compounds in meteorites. *Science* **182**:781-790.

Bada, J. L. and Miller, S. L. (1968) Ammonium ion concentration in the primitive ocean. *Science* **159**:423-425.

Baly, E. Ch. C., Heilbron, I. M. and Barker, W. F. (1921) Photocatalysis. Part I: The synthesis of formaldehyde and carbohydrates from carbon dioxide and water. *J. Chem. Soc.* **119**:1025-1035.

Bar-Nun, A., Bar-Nun, N., Bauer, S. H. and Sagan, C. (1970) Shock synthesis of amino acids in simulated prebiotic environments. *Science* **168**:470-473.

Bar-Nun, A. and Hartman, H. (1978) Synthesis of organic compounds from carbon monoxide and water by UV photolysis. *Origins of Life* **9**:93-101.

Bernal, J. D. (1951) *The Physical Basis of Life.* London: Routledge and Kegan Paul.

Briggs, M. H. (1963) Evidence of an extraterrestrial origin for some organic constituents of meteorites. *Nature* **197**:1290.

Broda, E. (1971) The origins of bacterial respiration. In: *Chemical Evolution and the Origin of Life,* R. Buvet and C. Ponnamperuma, eds., Amsterdam: North-Holland, pp. 446-452.

Cairns-Smith, A. G. (1982) *Genetic Takeover and the Mineral Origins of Life.* Cambridge, U.K.: The Cambridge University Press.

Calvin, M. (1956) Chemical evolution and the origin of life. *American Scientist* **44**:248-263.

Calvin, M. (1969) *Chemical Evolution: Molecular Evolution towards the Origin of Living Systems on the Earth and Elsewhere.* Oxford: Oxford University Press.

Chittenden, G. J. F. and Schwartz, A. W. (1981) Prebiotic photosynthetic reactions. *BioSystems* **14**:15-32.

Chyba, C. F. and Sagan, C. (1991) Electrical energy sources for organic synthesis on the early Earth. *Orig. Life Evol. Biosphere* **21**:3-17.

Chyba, C. F., Thomas, P. J., Brookshaw, L. and Sagan, C. (1990) Cometary delivery of organic molecules to the early Earth. *Science* **249**:366-373.

Clark, B. C. (1988) Primeval procreative comet pond. *Orig. Life Evol. Biosphere* **18**:209-238.

Corliss, J. B., Baross, J. A. and Hoffmann, S. E. (1981) An hypothesis concerning the relationship between submarine hot springs and the origin of life on Earth. *Oceanol. Acta* **4** (suppl.):59-69.

Cronin, J. R. and Moore, C. B. (1971) Amino acid analysis of the Murchison, Murray and Allende carbonaceous chondrites. *Science* **172**:1327-1329.

Cronin, J. R., Pizzarello, S. and Moore, C. B. (1979) Amino acids in an Antarctic carbonaceous chondrite. *Science* **206**:335-337.

Eichberg, J., Sherwood, E., Epps, D. E. and Oró, J. (1977) Cyanamide mediated syntheses under plausible primitive Earth conditions. IV: The synthesis of acylglycerols. *J. Mol. Evol.* **10**:221-230.

Epps, D. E., Nooner, D. W., Eichberg, J., Sherwood, E. and Oró, J. (1979) Cyanamide mediated syntheses under plausible primitive Earth conditions. VI: The synthesis of glycerol and glycerophosphate. *J. Mol. Evol.* **14**:235-241.

Epstein, S., Krishnamurthy, R. V., Cronin, J. R., Pizzarello, S. and Yuen, G. U. (1987) Unusual stable isotope ratios in amino acid and carboxylic acid extracts from the Murchison meteorite. *Nature* **326**:477-479.

Ferris, J. P. (1984) The chemistry of life's origin. *Chem. Eng. News* **62**:22-35.

Ferris, J. P. (1987) Prebiotic synthesis: Problems and challenges. *Cold Spring Harbor Symp. Quant. Biol.* **52**:29-35.

Ferris, J. P. and Hagan, W. J. (1984) HCN and chemical evolution: The possible role of cyano compounds in prebiotic synthesis. *Tetrahedron* **40**:1093-1120.

Ferris, J. P. and Hagan, W. J. (1986) The adsorption and reaction of adenine nucleotides on montmorillonite. *Origins of Life* **17**:69-84.

Ferris, J. P., Huang, C. H. and Hagan, W. J. (1988) Montmorillonite: A multifunctional mineral catalyst for the prebiological formation of phosphate esters. *Orig. Life Evol. Biosphere* **18**:121-133.

Ferris, J. P., Joshi, P. C., Edelson, E. H. and Lawless, J. G. (1978) HCN: A plausible source of purines, pyrimidines and amino acids on the primitive Earth. *J. Mol. Evol.* **11**:293-311.

Fox, S. W. and Harada, K. (1958) Thermal copolymerization of amino acids to a product resembling protein. *Science* **128**:1214.

Garrison, W. M., Morrison, D. C., Hamilton, J. G., Benson, A. A. and Calvin, M. (1951) Reduction of carbon dioxide in aqueous solutions by ionizing radiation. *Science* **114**:416-418.

Getoff, N. (1962) Reduction of carbon dioxide in aqueous solution by UV light. *Z. Naturforsch.* **17B**:87-90.

Greenberg, J. M. (1981) Chemical evolution of interstellar dust – a source of prebiotic material? In: *Comets and the Origin of Life.* C. Ponnamperuma, ed., Dordrecht, Holland: D. Reidel, pp. 111-127.

Groth, W. and Suess, H. (1938) Bemerkungen zür Photochemie der Erdatmosphäre. *Naturwissenschaften* **26**:77.

Hahn, J. H., Zenobi, R., Bada, J. L. and Zare, R. N. (1988) Application of two-step laser mass spectrometry to cosmogeochemistry: Direct analysis of meteorites. *Science* **239**:1523-1525.

Hargreaves, W. R., Mulvihill, S. and Deamer, D. W. (1977) Synthesis of phospholipids and membranes in prebiotic conditions. *Nature* 266:78-80.

Hayatsu, R. and Anders, E. (1981) Organic compounds in meteorites and their origins. *Top. Curr. Chem.* 99:1-37.

Henderson-Sellers, A. and Schwartz, A. W. (1980) Chemical evolution and ammonia in the early Earth's atmosphere. *Nature* 287:526-528.

Hulsof, J. and Ponnamperuma, C. (1976) Prebiotic condensing agents in an aqueous medium: A review of condensing agents. *Origins of Life* 7:197-224.

Kenyon, D. H. and Steinman, G. (1969) *Biochemical Predestination.* New York: McGraw-Hill.

Kerridge, J. F. (1982) Isotopic composition of C, H, N in carbonaceous chondrite polymer using stepwise combustion. *Proc. Lunar Planet. Sci. Conf.* 13:381-382.

Kerridge, J. F. (1983) Isotopic composition of carbonaceous-chondrite kerogen: Evidence for an interstellar origin of organic matter in meteorites. *Earth Planet. Sci. Lett.* 64:186-200.

Kolodny, Y., Kerridge, J. F. and Kaplan, I. R. (1980) Deuterium in carbonaceous chondrites. *Earth Planet. Sci. Lett.* 46:149-158.

Krumbein, W. E. and Schellnhuber, H.-J. (1990) Geophysiology of carbonates as a function of bioplanets. In: *Facets of Modern Biogeochemistry – Festschrift for F. T. Degens,* V. Ittekkot, S. Kempe, W. Michaelif and A. Spitzy, eds., Berlin: Springer-Verlag, pp. 5-22.

Kvenvolden, K. A. (ed.) (1974) *Geochemistry and the Origin of Life.* Stroudsburg, PA: Dowden, Hutchinson and Ross.

Kvenvolden, K. A., Lawless, J., Pering, K., Peterson, E., Flores, J., Ponnamperuma, C., Kaplan, I. R. and Moore, C. (1970) Evidence for extraterrestrial aminoacids and hydrocarbons in the Murchison meteorite. *Nature* 228:923-926.

Levine, J. S. (1982) The photochemistry of the paleoatmosphere. *J. Mol. Evol.* 18:161-172.

Levine, J. S. and Augustsson, T. R. (1985) The photochemistry of biogenic gases in the early and present atmosphere. *Origins of Life* 15:299-318.

Lohrmann, R. and Orgel, L. E. (1968) Prebiotic synthesis: Phosphorylation in aqueous solution. *Science* 161:64-66.

Lotka, A. J. (1924) *Elements of Mathematical Biology.* New York: Dover.

Marcus, J. N. and Olsen, M. A. (1991) Biological implications of organic compounds in comets. In: *Comets in the Post-Halley Era,* R. L. Newburn, M. Neugebauer and J. Rahe, eds., Boston: Kluwer Academic Publishers, pp. 439-462.

Matthews, C. N. (1990) Origins of life: Polymers before monomers? In: *Environmental Evolution: Effects of the Origin and Evolution of Life on Planet Earth,* L. Margulis and L. Olenzenski, ed., Cambridge, MA: The MIT Press, pp. 29-38.

Matthews, C. N. (1992) Dark matter and the solar system: Hydrogen cyanide polymers. *Orig. Life Evol. Biosphere* (in press).

Miller, S. L. (1953) Production of amino acids under possible primitive Earth conditions. *Science* 117:528-529.

Miller, S. L. (1955) Production of some organic compounds under possible primitive Earth conditions. *J. Am. Chem. Soc.* 77:2351-2361.

Miller, S. L. (1984) The prebiotic synthesis of organic molecules and polymers. *Adv. Chem. Phys.* 55:85-107.

Miller, S. L. (1987) Which organic compounds could have occurred on the prebiotic Earth? *Cold Spring Harbor Symp. Quant. Biol.* 52:17-27.

Miller, S. L. and Bada, J. L. (1988) Submarine hot springs and the origin of life. *Nature* 334:609-611.

Miller, S. L. and Orgel, L. E. (1974) *The Origins of Life on the Earth.* Englewood Cliffs, NJ: Prentice-Hall.

Miller, S. L. and Urey, H. C. (1959) Organic compound synthesis on the primitive Earth. *Science* 130:245-251.

Miller, S. L., Urey, H. C. and Oró, J. (1976) Origin of organic compounds on the primitive Earth and in meteorites. *J. Mol. Evol.* 9:59-72.

Oberbeck, V. R. and Fogleman, G. (1990) Impact constraints on the requirement for chemical evolution. *Orig. Life Evol. Biosphere* 20:181-195.

Oberbeck, V. R., McKay, C. P., Scattergood, T. W., Carle, G. C. and Valentin, J. R. (1989) The role of cometary particle coalescence in chemical evolution. *Orig. Life Evol. Biosphere* 19:39-56.

Oró, J. (1961) Mechanism of synthesis of adenine from hydrogen cyanide under possible primitive Earth conditions. *Nature* 191:1193-1194.

Oró, J. (1965) Stages and mechanisms of prebiological organic synthesis. In: *The Origins of Prebiological Systems and of Their Molecular Matrices*, S. W. Fox, ed., New York: Academic Press, pp. 131-171.

Oró, J. and Kimball, A. P. (1961) Synthesis of purines under possible primitive Earth conditions. I: Adenine from hydrogen cyanide. *Arch. Biochem. Biophys.* 94:217-227.

Oró, J., Mills, T. and Lazcano, A. (1992) The cometary contributions to prebiotic chemistry. *Adv. Space Res.* 12 (4):33-41.

Oró, J., Sherwood, E., Eichberg, J. and Epps, D. E. (1978) Formation of phospholipids under primitive Earth conditions and the role of membranes in prebiological evolution. In: *Light-Transducing Membranes: Structure, Function and Evolution*, D. W. Deamer, ed., New York: Academic Press, pp. 1-19.

Pinto, J. P., Gladstone, G. R. and Yung, Y. L. (1980) Photochemical production of formaldehyde in Earth's primitive atmosphere. *Science* 210:183-185.

Ponnamperuma, C. and Chang, S. (1971) The role of phosphates in chemical evolution. In: *Chemical Evolution and the Origin of Life*, R. Buvet and C. Ponnamperuma, eds., Amsterdam: North-Holland, pp. 216-223.

Ponnamperuma, C., Shimoyama, H. and Friebele, E. (1982) Clay and the origin of life. *Origins of Life* 12:9-40.

Rao, M., Eichberg, J. and Oró, J. (1982) Synthesis of phosphatidylcholine under possible primitive Earth conditions. *J. Mol. Evol.* 18:196-202.

Shock, E. L. (1990) Do amino acids equilibrate in hydrothermal fluids? *Geochim. Cosmochim. Acta* 54:1185-1189.

Shock, E. L. and Schulte, M. D. (1990) Amino-acid synthesis in carbonaceous meteorites by aqueous alteration of polycyclic aromatic hydrocarbons. *Nature* 343:728-731.

Stribling, R. and Miller, S. L. (1987) Energy yields for hydrogen cyanide and formaldehyde syntheses: The HCN and amino acid concentration in the primitive ocean. *Origins of Life* 17:261-273.

Swenson, R. (1989) Emergent attractors and the law of maximum entropy production: Foundations to a theory of general evolution. *Systems Research* 6:187-197.

Swenson, R. (1991) Order, evolution, and natural law: Fundamental relations in complex system theory. In: *Cybernetics and Applied Systems*, C. Negoita, ed., New York: Marcel Dekker, pp. 125-148.

Swenson, R. and Turvey, M. T. (1991) Thermodynamic reasons for perception-action cycles. *Ecological Psychology* 3 (4):317-348.

Usher, D. A. (1977) Early chemical evolution of nucleic acids: A theoretical model. *Science* 196:311-313.

Vernadsky, V. I. (1944) Problems in biogeochemistry II. *Trans. Conn. Acad. Arts Sciences* 35:483-517.

Wächtershäuser, G. (1988) An all-purine precursor of nucleic acids. *Proc. Natl. Acad. Sci. USA* 85:1134-1135.

Wächtershäuser, G. (1990) Evolution of the first metabolic cycles. *Proc. Natl. Acad. Sci. USA* 87:200-204.

Weber, A. L. (1991) Origin of fatty acid synthesis: Thermodynamics and kinetics of reaction pathways. *J. Mol. Evol.* 32:93-100.

Westheimer, F. H. (1987) Why nature chose phosphates. *Science* 235:1173-1178.

SELF-ASSEMBLY OF SUPRAMOLECULAR SYSTEMS

Alexander, J. and Bridges, C. B. (1928) Some physicochemical aspects of life, mutation and evolution. In: *Colloid Chemistry Theoretical and Applied*, J. Alexander, ed., New York: The Chemical Catalog Co., pp. 9-58.

Baeza, I., Ibanez, M., Lazcano, A., Santiago, C., Arguello, C., Wong, C. and Oró, J. (1987) Liposomes with polyribonucleotides as model of precellular systems. *Origins of Life* **17**:321-331.

Bangham, A. D. (1968) Membrane models with phospholipids. In: *Progress in Biophysics and Molecular Biology*, Volume 18, J. A. V. Butler and D. Noble, eds., Oxford: Pergamon Press.

Bangham, A. D. (1970) The liposome as a membrane model. In: *Permeability and Function of Biological Membranes*, L. Bolis, A. Katchalsky, R. D. Keynes, W. R. Lowenstein and B. A. Pethica, eds., Amsterdam: North-Holland, pp. 195-206.

Bangham, A. D., Standish, M. M. and Watkins, J. B. (1965) Diffusion of univalent ions across the lamellae of swollen phospholipids. *J. Mol. Biol.* **13**:238-252.

Bernal, J. D. (1967) *The Origin of Life.* London: Weidenfeld and Nicolson.

Bonner, W. A., Blair, N. E. and Dirbas, F. M. (1981) Experiments on the abiotic amplification of optical activity. *Origins of Life* **11**:119-134.

Brooke, S. and Fox, S. W. (1977) Compartmentalization in proteinoid microspheres. *BioSystems* **9**:1-22.

Cavalier-Smith, T. (1987) The origin of cells: A symbiosis between genes, catalysts and membranes. *Cold Spring Harbor Symp. Quant. Biol.* **52**:805-824.

Deamer, D. W. (ed.) (1978) *Light-Transducing Membranes: Structure, Function and Evolution.* New York: Academic Press.

Deamer, D. W. (1985) Boundary structures are formed by organic components of the Murchison carbonaceous chondrite. *Nature* **317**:92-794.

Deamer, D. W. (1986) Role of amphiphilic compounds in the evolution of membrane structure on the early Earth. *Origins of Life* **17**:3-25.

Deamer, D. W. and Barchfeld, G. L. (1982) Encapsulation of macromolecules by lipid vesicles under simulated prebiotic conditions. *J. Mol. Evol.* **18**:203-206.

Deamer, D. W. and Oró, J. (1980) Role of lipids in prebiotic structures. *BioSystems* **12**:167-175.

Deamer, D. W. and Pashley, R. (1989) Amphiphilic molecules in organic extracts of the Murchison carbonaceous chondrite: Surface activity and membrane formation. *Orig. Life Evol. Biosphere* **19**:21-38.

Eisenberg, M. and McLaughlin, S. (1976) Lipid bilayers as models of biological membranes. *BioScience* **26**:436-443.

Fajszi, C. and Czege. J. (1981) Critical evaluation of mathematical models for the amplification of chirality. *Origins of Life* **11**:143-162.

Fleischaker, G. R. (1990) Origins of life: An operational definition. *Orig. Life Evol. Biosphere* **20**:127-137.

Folsome, C. E. (1976) Synthetic organic microstructures and the origins of cellular life. *Naturwissenschaften* **63**:303-306.

Folsome, C. E., Allen, R. D. and Ichinose, N. K. (1975) Organic microstructures as products of Miller-Urey electrical discharges. *Precambrian Research* **2**:263-275.

Folsome, C. E. and Morowitz, H. J. (1969) Prebiological membranes: Synthesis and properties. *Space Life Sciences* **1**:538-544.

Fox, S. W. (ed.) (1965) *The Origins of Prebiological Systems and of Their Molecular Matrices.* New York: Academic Press.

Fox, S. W. and Dose, K. (1972) *Molecular Evolution and the Origin of Life.* San Francisco: W. H. Freeman.

Fox, S. W. and Harada, K. (1958) Thermal copolymerization of amino acids to a product resembling protein. *Science* **128**:1214.

Fraser, C. L. and Folsome, C. E. (1975) Exponential kinetics of formation of organic microstructures. *Origins of Life* **6**:429-433.

Goldacre, R. J. (1958) Surface films: Their collapse on compression, the shapes and sizes of cells, and the origin of life. In: *Surface Phenomena in Biology and Chemistry*, J. F. Danielli, K. G. A. Pankhurst and A. C. Riddiford, eds., New York: Pergamon Press, pp. 12-27.

Hammes, G. and Schullery, S. E. (1970) Structure of macromolecular aggregates. II: Construction of model membranes from phospholipids and polypeptides. *Biochemistry* **9**:2555-2563.

Joyce, G. F., Visser, G. M., van Boeckel, C. A. A., van Boom, J. H., Orgel, L. E. and van Westrenen, J. (1984) Chiral selection in poly(C)-directed synthesis of oligo(G). *Nature* 310:602-604.

Kenyon, D. H. and Steinman, G. (1969) *Biochemical Predestination*. New York: McGraw-Hill, pp. 214-217.

Koch, A. L. (1985) Primeval cells: Possible energy-generating and cell-division mechanisms. *J. Mol. Evol.* 21:270-277.

Maturana, H. R. and Varela, F. J. (1973) Autopoiesis – The organization of the living. In: *Autopoiesis and Cognition – The Realization of the Living* (1980), H. R. Maturana and F. J. Varela, eds., Dordrecht, Holland: D. Reidel, pp. 73-141.

Minchin, E. A. (1915) The evolution of the cell. [Presidential address, Section D.] *Ann. Report of the British Assoc. for the Advancement of Science*, pp. 437-464.

Morowitz, H. J. (1981) Phase separation, charge separation and biogenesis. *BioSystems* 14:41-47.

Morowitz, H. J., Deamer, D. W. and Smith, T. (1991) Biogenesis as an evolutionary process. *J. Mol. Evol.* 33:207-208.

Morowitz, H. J., Heinz, B. and Deamer, D. W. (1988) The chemical logic of a minimal protocell. *Orig. Life Evol. Biosphere* 18:281-287.

Oparin, A. I. (1924) The Origin of Life [*Proiskhozhdenie zhizny*. Moscow: Izd. Moskovshii Rabochii]. English translation in: J. D. Bernal, *The Origin of Life* (1967). London: Weidenfeld and Nicolson, pp. 199-234.

Oró, J., Holzer, G., Rao, M. and Tornabene, T (1980) Membrane lipids and the origin of life. In: *Origin of Life*, Y. Wolman, ed., Dordrecht, Holland: D. Reidel, pp. 313-322.

Oró, J., Sherwood, E., Eichberg, J. and Epps, D. E. (1978) Formation of phospholipids under primitive Earth conditions and the role of membranes in prebiological evolution. In: *Light-Transducing Membranes: Structure, Function and Evolution*, D. W. Deamer, ed., New York: Academic Press, pp. 1-19.

Raven, J. A. and Smith, F. A. (1982) Solute transport at the plasmalemma and early evolution of cells. *BioSystems* 15:13-26.

Schwegler, H. and Tarumi, K. (1986) The "protocell": A mathematical model of self-maintenance. *BioSystems* 19:307-315.

Shah, D. O. (1972) The origin of membranes and related surface phenomena. In: *Exobiology*, C. Ponnamperuma, ed., Amsterdam: North-Holland, pp. 235-265.

Singer, S. J. and Nicolson, G. L. (1972) The fluid mosaic model of the structure of cell membranes. *Science* 175:720-731.

Stillwell, W. (1980) Facilitated diffusion as a method for selective accumulation of materials from the primordial oceans by a lipid-vesicle protocell. *Origins of Life* 10:277-292.

Tabushi, I. (1986) Minimum requirements for single cell activity. *Adv. Space Res.* 6:45-52.

Tanford, C. (1978) The hydrophobic effect and the organization of living matter. *Science* 200:1012-1018.

Varela, F. J., Maturana, H. R. and Uribe, R. (1974) Autopoiesis: The organization of living systems, its characterization and a model. *BioSystems* 5:187-196.

Wald, G. (1957) Origin of optical activity. *Ann. York Acad. Sci.* 69:352-368.

Wilson, A. T. (1960) Synthesis of macromolecules under possible primeval Earth conditions. *Nature* 188:1007-1009.

Yanagawa, H., Kobayashi, Y. and Egami, F. (1980) Characterization of marigranules and marisomes, organized particles with elastin-like structures. *J. Biochem.* 87:855-869.

Yanagawa, H., Ogawa, Y., Kojima, K. and Ito, M. (1988) Construction of protocellular structures under simulated primitive Earth conditions. *Orig. Life Evol. Biosphere* 18:179-207.

Young, R. S. (1965) Morphology and chemistry of microspheres from proteinoid. In: *The Origins of Prebiological Systems and of Their Molecular Matrices*, S. W. Fox, ed., New York: Academic Press, pp. 347-357.

ENERGETICS OF LIFE'S ORIGIN

Baltscheffsky, H. (1971) Inorganic pyro-phosphate and the origin and evolution of biological energy transformation. In: *Chemical Evolution and the Origin of Life*, R. Buvet and C. Ponnamperuma, eds., Amsterdam: North-Holland, pp. 466-474.

Baltscheffsky, M. (1977) Conversion of solar energy into energy-rich phosphate compounds. In: *Living Systems as Energy Converters*, R. Buvet, ed., Amsterdam: North-Holland Biomedical Press, pp. 199-207.

Baltscheffsky, M. and Nyrén, P. (1984) The synthesis and utilization of inorganic pyrophosphate. In: *Bioenergetics*, L. Ernster, ed., Amsterdam: Elsevier, pp. 187-206.

Broda, E. (1975) *The Evolution of the Bioenergetic Processes*. New York: Pergamon Press.

Calvin, M. (1959) Evolution of enzymes and the photosynthetic apparatus. *Science* 130:1170-1174.

Chittenden, G. J. F. and Schwartz, A. W. (1981) Prebiotic photosynthetic reactions. *BioSystems* 14:15-32.

Chyba, C. F. and Sagan, C. (1991) Electrical energy sources for organic synthesis on the early Earth. *Orig. Life Evol. Biosphere* 21:3-17.

Corliss, J. B., Baross, J. A. and Hoffmann, S. E. (1981) An hypothesis concerning the relationship between submarine hot springs and the origin of life on Earth. *Oceanol. Acta* 4 (suppl.):59-69.

Deamer, D. W. (1992) Polycyclic aromatic hydrocarbons: Primitive pigment systems in the prebiotic environment. *Adv. Space Res.* 12:183-189.

Deamer, D. W. and Harang, E. (1990) Light-dependent pH gradients are generated in liposomes containing ferrocyanide. *BioSystems* 24:1-4.

Deamer, D. W. and Oró, J. (1980) Role of lipids in prebiotic structures. *BioSystems* 12:167-175.

Ferris, J. P. (1968) Cyanovinyl phosphate: A prebiological phosphorylating agent? *Science* 161:53-54.

Gest, H. (1980) Hypothesis: The evolution of biological energy-transducing systems. *Microbiol. Lett.* 7:73-77.

Gust, D. and Moore, T. A. (1989) Mimicking photosynthesis. *Science* 244:35-41.

Harold, F. M. (1986) *The Vital Force: A Study in Bioenergetics*. New York: W. H. Freeman.

Inoue, T. and Orgel, L. E. (1983) A nonenzymatic RNA polymerase model. *Science* 219:859-862.

King, C. C. (1990) Did membrane electrochemistry precede translation? *Orig. Life Evol. Biosphere* 20:15-25.

Krasnovsky, A. A. (1976) Chemical evolution of photosynthesis. *Origins of Life* 7:133-143.

Lipmann, F. (1941) Metabolic generation and utilization of phosphate bond energy. *Adv. Enzymol.* 1:99-162.

Lipmann, F. (1965) Projecting backward from the present stage of evolution of biosynthesis. In: *The Origins of Prebiological Systems and of Their Molecular Matrices*, S. W. Fox, ed., New York: Academic press, pp. 259-280.

Mercer, J. A. and Mauzerall, D. C. (1984) Photochemistry of porphyrins: A model for the origin of photosynthesis. *Photochem. Photobiol.* 39:397-405.

Mitchell, P. (1961) Coupling of phosphorylation to electron and hydrogen transfer by a chemiosmotic type of mechanism. *Nature* 191:144-148.

Mitchell, P. (1981) Davy's electrochemistry: Nature's protochemistry. *Chemistry in Britain* 17:14-23.

Morowitz, H. J. (1981) Phase separation, charge separation and biogenesis. *BioSystems* 14:41-47.

Nicholls, D. G. (1982) *Bioenergetics: An Introduction to the Chemiosmotic Theory*. London: Academic Press.

Oesterhelt, D. and Stoeckenius, W. (1973) Functions of a new photoreceptor membrane. *Proc. Natl. Acad. Sci. USA* 70:2853-2857.

Olson, J. M. and Pierson, B. K. (1987) Origin and evolution of photosynthetic reaction centers. *Origins of Life* **17**:419-130.

Raven, J. A. and Smith, F. A. (1976) The evolution of chemiosmotic energy coupling. *J. Theor. Biol.* **57**:301-312.

Robertson, R. M. (1967) The separation of protons and electrons as a fundamental biological process. *Endeavour* **26**:134-139.

Robertson, R. M. (1983) *The Lively Membranes.* Cambridge, U.K.: The Cambridge University Press.

Saygin, Ö. (1981) Nonenzymatic photo-phosphorylation with visible light: A possible mode of prebiotic ATP formation. *Naturwissenschaften* **68**:617-619.

Schwartz, A. W. and Orgel, L. E. (1985) Template-directed synthesis of novel, nucleic acid-like structures. *Science* **228**:585-587.

Seta, P., Bienvenue, E., Moore, A. L., Mathis, P., Bensasson, R. V., Liddell, P., Pessiki, P. J., Joy, A., Moore, T. A. and Gust, D. (1985) Photodriven transmembrane charge separation and electron transfer by a carotenoporphyrin-quinone triad. *Nature* **316**:653-655.

Smith, T. F. and Morowitz, H. J. (1982) Between history and physics. *J. Mol. Evol.* **18**:265-282.

Stillwell, W. (1977) On the origin of photophosphorylation. *J. Theor. Biol.* **65**:479-497.

Stoeckenius, W. (1978) Speculations about the evolution of halobacteria and of chemiosmotic mechanisms. In: *Light-Transducing Membranes: Structure, Function and Evolution.* D. W. Deamer, ed., New York: Academic Press, pp. 127-139.

Stribling, R. and Miller, S. L. (1987) Energy yields for hydrogen cyanide and formaldehyde syntheses: The HCN and amino acid concentrations in the primitive ocean. *Origins of Life* **17**:261-273.

Usher, D. A. and McHale, A. H. (1976) Nonenzymic joining of oligoadenylates on a polyuridylic acid template. *Science* **192**:53-54.

Wächtershäuser, G. (1988) Before enzymes and templates: A theory of surface metabolism. *Microbiol. Rev.* **52**:452-484.

Wächtershäuser, G. (1988) Pyrite formation, the first energy source for life: A hypothesis. *Syst. Appl. Microbiol.* **10**:207-210.

Wächtershäuser, G. (1990) Evolution of the first metabolic cycles. *Proc. Natl. Acad. Sci. USA* **87**:200-204.

Weber, A. L. (1987) The triose model: Glyceraldehyde as a source of energy and monomers for prebiotic condensation reactions. *Origins of Life* **17**:107-119.

Weber, A. L. and Hsu, V. (1990) Energy-rich glyceric acid oxygen esters: Implications for the origin of glycolysis. *Orig. Life Evol. Biosphere* **20**:145-150.

Westheimer, F. H. (1987) Why nature chose phosphates. *Science* **235**:1173-1178.

Wilson, T. H. and Lin, E. C. C. (1980) Evolution of membrane bioenergetics. *J. Supramol. Struct.* **13**:421-446.

Wilson, T. H. and Maloney, P. C. (1976) Speculations on the evolution of ion transport mechanisms. *Fed. Amer. Soc. Exp. Biol. Proc.* **35**:2174-2179.

Woese, C. R. (1979) A proposal concerning the origin of life on the planet Earth. *J. Mol. Evol.* **13**:95-101.

BIOINFORMATIONAL MOLECULES

Alberts, B. M. (1986) The function of the hereditary materials: Biological catalyses reflect the cell's evolutionary history. *American Zoologist* **26**:781-796.

Bass, B. L. and Cech, T. R. (1984) Specific interaction between the self-splicing RNA of *Tetrahymena* and its guanosine substrate: Implications for biological catalysis by RNA. *Nature* **308**:820-826.

Been, M. D. and Cech, T. R. (1988) RNA as an RNA polymerase: Net elongation of an RNA primer catalysed by the *Tetrahymena* ribozyme. *Science* **239**:1412-1415.

Brack, A. (1987) Selective emergence and survival of early polypeptides in water. *Origins of Life* **17**:367-379.

Brack, A. and Barbier, B. (1990) Chemical activity of simple basic peptides. *Orig. Life Evol. Biosphere* **20**:139-144.

Brack, A. and Orgel, L. E. (1975) Beta structures of alternating polypeptides and their possible prebiotic significance. *Nature* **256**:383-387.

Cairns-Smith, A. G. (1966) The origin of life and the nature of the primitive gene. *J. Theoret. Biol.* **10**:53-88.

Cairns-Smith, A. G. (1982) *Genetic Takeover and the Mineral Origins of Life*. Cambridge, U.K.: The Cambridge University Press.

Cech, T. R. (1986) A model for the RNA-catalyzed replication of RNA. *Proc. Natl. Acad. Sci. USA* **83**:4360-4363.

Cech, T. R. (1989) Ribozyme self-replication? *Nature* **339**:507-508.

Cech, T. R., Zaug, A. J. and Grabowski, P. J. (1981) *In vitro* splicing of the ribosomal RNA precursor of Tetrahymena: Involvement of a guanosine nucleotide in the excision of the intervening sequence. *Cell* **27**:487-496.

Crick, F. H. C. (1968) The origin of the genetic code. *J. Mol. Biol.* **38**:367-379.

Darnell, J. E. and Doolittle, W. F. (1986) Speculations on the early course of evolution. *Proc. Natl. Acad. Sci. USA* **83**:1271-1275.

Doudna, J. A., Couture, S. and Szostak, J. W. (1991) A multisubunit enzyme that is a catalyst of and template for complementary strand RNA synthesis. *Science* **251**:1605-1608.

Doudna, J. A. and Szostak, J. W. (1989) RNA-catalysed synthesis of complementary-strand RNA. *Nature* **339**:519-522.

Eigen, M. and Schuster, P. (1979) *The Hypercycle: A Principle of Natural Self-Organization*. Berlin: Springer-Verlag.

Eigen, M. and Schuster, P. (1982) Stages of emerging life — Five principles of early organization. *J. Mol. Evol.* **19**:47-61.

Ferris, J. P., Huang, C.-H. and Hagan, W. J. (1988) Montmorillonite: A multifunctional mineral catalyst for the prebiological formation of phosphate esters. *Orig. Life Evol. Biosphere* **18**:121-133.

Fox, S. W. and Harada, K. (1958) Thermal copolymerization of amino acids to a product resembling protein. *Science* **128**:1214.

Gilbert, W. (1986) The RNA world. *Nature* **319**:618.

Guerrier-Takada, C., Gardiner, K., Marsh, T., Pace, N. and Altman, S. (1983) The RNA moiety of ribonuclease P is the catalytic subunit of the enzyme. *Cell* **35**:849-857.

Horowitz, N. H. (1945) On the evolution of biochemical syntheses. *Proc. Natl. Acad. Sci. USA* **31**:153-157.

Inoue, T. and Orgel, L. E. (1983) A nonenzymatic RNA polymerase model. *Science* **219**:859-862.

Joyce, G. F. (1989) RNA evolution and the origins of life. *Nature* **338**:217-224.

Joyce, G. F. (1991) The rise and fall of the RNA world. *The New Biologist* **3**:399-407.

Joyce, G. F., Inoue, T. and Orgel, L. E. (1984) Nonenzymatic template-directed synthesis on RNA random copolymers – Poly (C,U) templates. *J. Mol. Biol.* **176**:279-306.

Joyce, G. F. and Orgel, L. E. (1986) Non-enzymic template-directed synthesis on RNA random copolymers – Poly (C,G) templates. *J. Mol. Biol.* **188**:433-441.

Joyce, G. F., Schwartz, A. W., Miller, S. L. and Orgel, L. E. (1987) The case for an ancestral genetic system involving simple analogues of the nucleotides. *Proc. Natl. Acad. Sci. USA* **84**:4398-4402.

Joyce, G. F., Visser, G. M., van Boeckel, C. A. A., van Boom, J. H., Orgel, L. E. and van Westrenen, J. (1984) Chiral selection in poly(C)-directed synthesis of oligo(G). *Nature* **310**:602-604.

Kauffman, S. A. (1986) Autocatalytic sets of proteins. *J. Theor. Biol.* **119**:1-24.

Kiedrowski, G. von (1986) A self-replicating hexadeoxynucleotide. *Angew. Chem. Int. Ed. Engl.* **25** (10):932-935.

Kruger, K., Grabowski, P. J., Zaug, A. J., Sands, J., Gottschling, D. E. and Cech, T. R. (1982) Self-splicing RNA: Autoexcision and autocyclization of the ribosomal RNA intervening sequence of *Tetrahymena*. *Cell* 31:147-157.

Lacey, J. C. (1990) Ribonucleic acids may be catalysts for the preferential synthesis of L-amino acid peptides: A minireview. *J. Mol. Evol.* 31:244-248.

Lacey, J. C., Hawkins, A. F., Thomas, R. D. and Watkins, C. L. (1988) Differential distribution of D- and L-amino acids between the 2′ and 3′ positions of the AMP residue at the 3′ terminus of transfer ribonucleic acid. *Proc. Natl. Acad. Sci. USA* 85:4996-5000.

Lacey, J. C. and Mullins, D. W. (1983) Experimental studies related to the origin of the genetic code and the process of protein synthesis – A review. *Origins of Life* 13:3-42.

Lazcano, A. (1986) Prebiotic evolution and the origin of cells. *Treb. Soc. Cat. Biol.* 39:73-103.

Lazcano, A. (in press) The transition from non-living to living. In: *Early Life on Earth: A Nobel Symposium*, S. Bengtson, ed., New York: Columbia University Press.

Lazcano, A., Fastag, J., Gariglio, P., Ramirez, C. and Oró, J. (1988) On the early evolution of RNA polymerase. *J. Mol. Evol.* 27:365-376.

Lazcano, A., Fox, G. and Oró, J. (in press) Life before DNA: The origin and evolution of early Archaen cells. In: *The Evolution of Metabolic Function*, R. P. Mortlock, ed., Boca Raton, FL: CRC Press.

Lazcano, A., Guerrero, R., Margulis, L. and Oró, J. (1988) The evolutionary transition from RNA to DNA in early cells. *J. Mol. Evol.* 27:283-290.

Lazcano, A., Llaca, V., Cappello, R., Calverde, V. and Oró J. (1992) The origin and early evolution of nucleic acid polymerases. *Adv. Space Res.* 12 (4):207-216.

Muller, H. J. (1929) The gene as the basis of life. *Proc. Int. Congr. of Plant Sci.* 1:897-921.

Nisbet, E. G. (1986) RNA and hot-water springs. *Nature* 322:206.

Noller, H. F., Hoffarth, V. and Zimniak, L. (1992) Unusual resistance to protein extraction procedures. *Science* 256:1416-1419.

North, G. (1987) Back to the RNA world – and beyond. *Nature* 328:18-19.

Orgel, L. E. (1968) Evolution of the genetic apparatus. *J. Mol. Biol.* 38:381-393.

Orgel, L. E. (1986) Did template-directed nucleation precede molecular replication? *Origins of Life* 17:27-34.

Orgel, L. E. (1986) Molecular replication and the origins of life. In: *The Lesson of Quantum Theory*, J. de Baer, E. Dal and O. Ulfbeck, eds., Amsterdam: Elsevier Science Publishers, pp. 283-294.

Orgel, L. E. (1986) RNA catalysis and the origins of life. *J. Theor. Biol.* 123:127-149.

Orgel, L. E. (1987) Evolution of the genetic apparatus: A review. *Cold Spring Harbor Symp. Quant. Biol.* 52:9-16.

Orgel, L. E. and Lohrmann, R. (1974) Prebiotic chemistry and nucleic acid replication. *Acc. Chem. Res.* 7:368-377.

Pace, N. R. (1992) New horizons for RNA catalysis. *Science* 256:1402-1403.

Pace, N. R. and Marsh, T. L. (1985) RNA catalysis and the origin of life. *Origins of Life* 16:97-116.

Perello, M., Barbier, B. and Brack, A. (1991) Hydrolysis of oligoribonucleotides by alpha-helical basic peptides. *Int. J. Peptide Protein Res.* 38:154-160.

Piccirilli, J. A., McConnell, T. S., Zaug, A. J., Noller, H. F. and Cech, T. R. (1992) Aminoacyl esterase activity of the *Tetrahymena* ribozyme. *Science* 256:1420-1424.

Sarkar, S. (1991) *What is Life?* revisited. *BioScience* 41:631-634.

Schrödinger, E. (1944) *What is Life?* (Chapter VI: Order, disorder and entropy). Cambridge, U.K.: The Cambridge University Press (pp. 68-75).

Schwartz, A. W. and Orgel, L. E. (1985) Template-directed synthesis of novel, nucleic acid-like structures. *Science* 228:585-587.

Schwartz, A. W., Visscher, J., van der Woerd, R. and Bakker, C. G. (1987) In search of RNA ancestors. *Cold Spring Harbor Symp. Quant. Biol.* **52**:37-39.

Sharp, P. A. (1985) On the origin of RNA splicing and introns. *Cell* **42**:399-400.

Sharp, P. A. (1989) RNA evolution and the origins of life. *Nature* **338**:217-224.

Sulston, J., Lohrmann, R., Orgel, L. E. and Miles, H. T. (1968) Nonenzymatic synthesis of oligoadenylates on a polyadenylic acid template. *Proc. Natl. Acad. Sci. USA* **59**:726-733.

Tjivikua, T., Ballester, P. and Rebek, J. (1990) A self-replicating system. *J. Am. Chem. Soc.* **112**:1249-1250.

Troland, L. T. (1917) Biological enigmas and the theory of enzyme action. *The American Naturalist* **51**:321-350.

Usher, D. A. (1977) Early chemical evolution of nucleic acids: A theoretical model. *Science* **196**:311-313.

Usher, D. A. and McHale, A. H. (1976) Hydrolytic stability of helical RNA: A selective advantage for the natural 3',5'- bond. *Proc. Natl. Acad. Sci. USA* **73**:1149-1153.

Usher, D. A. and McHale, A. H. (1976) Nonenzymic joining of oligoadenylates on a polyuridylic acid template. *Science* **192**:53-54.

Wächtershäuser, G. (1988) An all-purine precursor of nucleic acids. *Proc. Natl. Acad. Sci. USA* **85**:1134-1135.

Waldrop, M. M. (1989) Did life really start out in an RNA world? *Science* **246**:1248-1249.

Waldrop, M. M. (1992) Finding RNA makes protein gives 'RNA world' a big boost. *Science* **256**:1396-1397.

Watson, J. D. and Crick, F. H. C. (1953) Genetical implications of the structure of deoxyribose nucleic acid. *Nature* **171**:964-967.

Watson, J. D. and Crick, F. H. C. (1953) Structure of deoxyribose nucleic acid. *Nature* **171**:737-738.

Weber, A. L. (1989) Model of early self-replication based on covalent complementarity for a copolymer of glycerate-3-phosphate and glycerol-3-phosphate. *Orig. Life Evol. Biosphere* **19**:179-186.

Weiner, A. M. and Maizels, N. (1987) tRNA-like structures tag the 3' ends of genomic RNA molecules for replication: Implications for the origin of protein synthesis. *Proc. Natl. Acad. Sci. USA* **84**:7383-7387.

Westheimer, F. H. (1986) Polyribonucleic acids as enzymes. *Nature* **319**:534-536.

White, D. H. (1980) A theory for the origin of a self-replicating chemical system. I: Natural selection of the autogen from short, random oligomers. *J. Mol. Evol.* **16**:121-147.

Woese, C. R. (1967) *The Genetic Code – The Molecular Basis for Genetic Expression.* New York: Harper and Row.

Woese, C. R. (1972) The emergence of genetic organization. In: *Exobiology*, C. Ponnamperuma, ed., Amsterdam: North-Holland, pp. 301-341.

Wong, J. T. F. (1981) Coevolution of genetic code and amino acid biosynthesis. *Trends Biochem. Sci.* **6**:33-35.

Yarus, M. (1988) A specific amino acid binding site composed of RNA. *Science* **240**:1751-1758.

Zaug, A. J. and Cech, T. R. (1986) The intervening sequence RNA of *Tetrahymena* is an enzyme. *Science* **231**:470-475.

INDEX OF AUTHORS CITED

Entries in italic are papers reprinted in this volume.

Berkner, LV and Marshall, LC (1965), 181

Bernal, JD (1949), 5, 147, 148, 155, 200, 201, 202, 403

Bernal, JD (1951), 95, 136, 138, 147, 148, 155, 403, 405, 407

Bernal, JD (1965), 403

Bernal, JD (1967), 12, 227, 231, 403, 405, 411

Bertalanffy, L von (1952), 403

Bertalanffy, L von (1969), 403

Bertalanffy, L von (1973), 403

Berthelot, M (1868), 209, 212

Berzelius, JJ (1834), 209, 212

Biegel, CM and Gould, JM (1981), 267, 269

Bienvenue, E, Seta, P, Hofmanova, A, Gavach, C and Momenteau, M (1984), 327

Bishop, MJ, Lohrmann, R and Orgel, LE (1972), 354

Bjerrum, N (1923), 185, 188

Blake, RD and Fresco, JR (1973), 349, 351

Blake, TS and McNaughton, NJ (1984), 394, 398

Bligh, EG and Dyer, WJ (1959), 207

Bloch, K (1959), 187

Bloch, K (1981), 187

Bloomfield, VA, Crothers, DM and Tinoco, I, Jr., (1974), 351

Blum, HF (1951), 84, 95

Bock, RM (1967), 351

Bockris, J O'M and Reddy, AKN (1970), 315

Bonner, WA, Blair, NE and Dirbas, FM (1981), 228, 231, 411

Borer, PN, Dengler, B, Tinoco, I, Jr. and Uhlenbeck, OC (1974), 354

Borer, PN, Kan, LS and Ts'o, POP (1975), 351

Borisov, AA, Galochkin, VT and Mulenko, SA, Oraevskii, AN, Starodubtsev, EF and Suchkov, AF (1978), 181

Boss, AP (1986), 405

Bossard, A, Raulin, F, Mourey, D and Toupance, G (1981), 306

Bossard, A and Toupance, G (1981), 103, 110, 301, 306

Boyer, PD, Chance, B, Ernster, L, Mitchell, P, Racker, E and Slater, EC (1977), 317, 322

Brack, A (1987), 414

Brack, A (1990), 403

Brack, A and Barbier, B (1990), 414

Brack, A and Orgel, LE (1975), 415

Brack, AL (1976), 200, 202

Braterman, PS, Cairns-Smith, AG and Sloper, R (1983), 122, 200, 203

Bratt, SR, Solomon, SC and Head, JW (1985), 122

Bratt, SR, Solomon, SC, Head, JW and Thuber, CH (1985), 122

Braun, V (1975), 239

Breder, CM (1942), 403

Bredereck, H, Schmötzer, G and Becher, HS (1956), 158

Bridson, PK, Fakhrai, H, Lohrmann, R, Orgel, LE and van Roode, M (1981), 361

Bridson, PK and Orgel, LE (1980), 358, 373, 374

Briggs, MH (1963), 407

Brimblecombe, P and Walker, JCG (1982), 114

Briner, E and Baerfuss, A (1919), 301, 306

Briner, E, Desbaillets, J and Paillard, H (1938), 301, 306

Briner, E and Hoefer, H (1940), 301, 306

Broda, E (1971), 407

Broda, E (1975), 413

Broecker, WS (1974), 113, 114

Bronshten, VA (1983), 130, 220

Brooke, S and Fox, SW (1977), 411

Brown, DM and Todd, AR (1952), 354

Brown, WT (1981), 220

Brownlee, DE (1985), 129, 223

Bruice, TC and Benkovic, SJ (1966), 351

Bruna, PJ, Buenker, RJ and Peyerimhoff, SD (1976), 181

Brunner, J, Skrabal, P and Hauser, H (1976), 239

Buchner, 74

Buchwald, SL, Friedman, JM and Knowles, JR (1984), 188

Buchwald, SL, Hansen, DE, Hassett, A and Knowles, JR (1982), 188

Buergi, HB, Dunitz, JD and Shefter, E (1973), 389

Bullard, E (1954), 155

Bunton, CA and Farber, SJ (1969), 370

Burke, JM, Irvine, KD, Kaneko, KJ, Kerker, BJ, Oettgen, AB, Tierney, WM, Williamson, CL, Zaug, AJ and Cech, TR (1986), 374

Burma, DP, Tewari, DS and Srivastava, AK (1985), 384

Burnham, AK, Braun, RL, Gregg, HR and Samoun, AM (1987), 220

Burton, M (1947), 145

Buskirk, R (1973), 353, 354

Butcher, W and Westheimer, FH (1955), 188

Butler, ET and Chamberlin, MJ (1982), 370

Button, A et al (1982), 114

Buvet, R and Ponnamperuma, C (1971), 212, 403

Byerly, GR, Lowe, DR and Walsh, MM (1986), 394, 398

C

Cafiso, JD and Hubbell, WL (1983), 267, 269

Cairns-Smith, AG (1966), 415

Cairns-Smith, AG (1978), 405

Cairns-Smith, AG (1982), 136, 138, 200, 203, 338, 339, 407, 415

Cairns-Smith, AG (1985), 403

Calvert, JG, Kerr, JA, Demerjian, KL and McQuigg, RD (1972), 181

Calvin, M (1956), 408

Calvin, M (1959), 413

Calvin, M (1969), 334, 405, 408

Calvin, M (1981), 327

Calvo, KC (1985), 188

Calvo, KC and Berg, JM (1984), 188

Calvo, KC and Westheimer, FH (1983), 188

Calvo, KC and Westheimer, FH (1984), 188

Cameron, AGW (1978), 391, 398

Cameron, AGW and Truran, JW (1977), 391, 398

Cameron, EM (1982), 114

Canuto, VM, Levine, JS, Augustsson, TR and Imhoff, CL (1982), 405

Canuto, VM and Levine, JS (1982), 130

Capezzuto, P, Cramarossa, F, Ferraro, G, Maione, P and Molinari, E (1973), 301, 306

Carlson, RW and Lugmair, GW (1979), 122

Carlson, RW and Lugmair, GW (1988), 122

Carslaw, HS and Jaeger, JC (1947), 250, 256

Carver, JH (1981), 117

Cavalier-Smith, T (1987), 230, 231, 411

Cech, TR (1983), 374

Cech, TR (1985), 374, 375

Cech, TR (1986), 188

Cech, TR (1986), 337, 338, 339, 380, 388, 391, 393, 396, 398, 415

Cech, TR (1987), 380, 388

Cech, TR (1989), 415

Cech, TR (1990), 388

Cech, TR, Zaug, AJ and Grabowski, PJ (1981), 370, 374, 380, 384, 415

Cech, TR et al (1983), 370

Celander, DW and Cech, TR (1991), 389

Ceplecha, Z (1977), 129, 220